Introduction to Coupled Theories in Solid Mechanics

Introduction to Coupled Theories in Solid Mechanics

Lallit Anand
Department of Mechanical Engineering,
Massachusetts Institute of Technology,
Cambridge, MA 02139, USA

Eric M. Stewart
Mechanical and Materials Engineering
Department, University of Cincinnati,
Cincinnati, OH 45221, USA

Shawn A. Chester
Mechanical and Industrial Engineering
Department, New Jersey Institute of Technology,
Newark, NJ 07102, USA

OXFORD
UNIVERSITY PRESS

OXFORD
UNIVERSITY PRESS

Great Clarendon Street, Oxford, OX2 6DP,
United Kingdom

Oxford University Press is a department of the University of Oxford.
It furthers the University's objective of excellence in research, scholarship,
and education by publishing worldwide. Oxford is a registered trade mark of
Oxford University Press in the UK and in certain other countries.

Published in the United States of America by Oxford University Press
198 Madison Avenue, New York, NY 10016, United States of America
British Library Cataloguing in Publication Data
Data available

Library of Congress Control Number: 2025940553

ISBN 9780198986218

DOI: 10.1093/oso/9780198986218.001.0001

Printed and bound by
CPI Group (UK) Ltd, Croydon, CR0 4YY

Introduction to Coupled Theories in Solid Mechanics

Introduction to coupled theories in solid mechanics is concerned with the formulation of some coupled theories of solid mechanics which form the basis for a wide range of engineering applications. These theories require a knowledge of the foundational balance laws of mechanics, thermodynamics, and electrodynamics of continua, together with a knowledge of the structure and properties of the materials which are provided by particular constitutive models for the so-called "smart-materials", and the materials used in the many applications that involve thermo-, poro-, electro- and/or magneto-mechanical coupling.

This book reviews the basic kinematics, balance laws, thermodynamics and constitutive equations for the large deformation theories of elasticity, viscoelasticity, thermoelasticity, poroelasticity, electroelasticity, and magnetoelasticity. The governing partial differential equations are formulated, and suitable boundary conditions are discussed. For specificity, attention is directed towards formulating coupled theories for elastomeric materials.

The large deformation coupled theories discussed in this book have been numerically implemented in the open-source finite element program FEniCS, and representative simulations which illustrate key coupled behaviors are presented for each theory. The FEniCS codes for the representative simulations shown in this book are posted online at `https://solidmechanicscoupledtheories.github.io`.

Lallit Anand, Warren and Towneley Rohsenow Professor of Mechanical Engineering, Massachusetts Institute of Technology.

Eric M. Stewart, Assistant Professor, Mechanical and Materials Engineering, University of Cincinnati.

Shawn A. Chester, Professor, Mechanical and Industrial Engineering, New Jersey Institute of Technology.

Preface

The purpose of this book

There is a growing need for Ph.D. students and other researchers in Solid Mechanics to have a knowledge of the coupled-physics response of materials in modern technologies which employ so-called "smart materials" for sensors and actuators, as well as devices used in the broad areas of soft-robotics, energy, and the environment. The purpose of the book, which is divided into eleven parts, is to provide an introduction to the modern large-deformation theories of elasticity, viscoelasticity, thermoelasticity, poroelasticity, electro-elasticity and electro-viscoelasticity, and magneto-viscoelasticity, of soft elastomeric materials. Numerical implementation of the theories in a finite element program is also discussed, and representative simulations which illustrate key coupled behaviors are presented for each theory.

The eleven parts of the book are:

- Part I: Finite elasticity of elastomeric materials
- Part II: Viscoelasticity of elastomeric materials
- Part III: Thermoelasticity of elastomeric materials
- Part IV: Poroelasticity of elastomeric gels
- Part V: Thermally-responsive elastomeric gels
- Part VI: Cahn-Hilliard theory for species diffusion coupled with elastic deformations
- Part VII: Electro-elasticity of dielectric elastomers
- Part VIII: Electro-viscoelasticity of dielectric elastomers
- Part IX: Electro-chemo-elasticity of ionic polymers
- Part X: Magneto-viscoelasticity of hard-magnetic soft-elastomers
- Part XI: Magneto-viscoelasticity of soft-magnetic soft-elastomers

For whom is this book meant?

This book is aimed at individuals who have a reasonable background in continuum mechanics and are embarking on research in the challenging and exciting frontier areas of solid mechanics involving coupled theories and their engineering applications.

Historical issues and attributions

The subject matter covered has been developed by many leading figures in the renaissance of Continuum Mechanics since the second half of the twentieth century ($\sim$1950s) and it is difficult to provide correct historical references and attributions to all the ideas discussed in this book. Our emphasis is on basic concepts and central results, not on the history of our subject. We have attempted to cite the contributions most central to our presentation, and we apologize in advance if we have not done so faultlessly.

Contents

XII APPENDICES

Introduction

Coupled Theories in Solid Mechanics refers to mathematical theories that model the interaction between multiple physical fields within a solid material. Such theories address how the response of a solid material in the mechanical domain (e.g., displacement field) is influenced by and influences the response of the material in another domain (e.g., thermal, electrical, or magnetic fields). Examples of **coupled phenomena** in solid mechanics include:

- *Thermo-mechanical coupling*: Interactions between mechanical deformation and temperature changes.
- *Poro-mechanical coupling*: Interactions between the deformation of a solid skeleton and the movement of a fluid within its nano/microscale pores.
- *Thermo-poro-mechanical coupling*: Interactions between the deformation of a solid skeleton, movement of a fluid within nano/microscale pores, and temperature changes.
- *Chemo-mechanical coupling*: Interactions between diffusion of a chemical species (e.g., hydrogen, oxygen, Li-ions) and chemical reactions with mechanical behavior.
- *Electro-mechanical coupling*: Interactions between mechanical deformation and electric fields.
- *Magneto-mechanical coupling*: Interactions between mechanical deformation and magnetic fields.
- . . . and many more couplings and cross-couplings.

A complete **coupled theory** consists of:

- a system of coupled partial differential equations (PDEs) and boundary conditions for the different fields,
- with the coupling achieved through suitable thermodynamically consistent constitutive relations which define how one field affects another.

Properly formulated and numerically implemented coupled theories can provide a deep understanding of the response of materials and structures under conditions where multiple physical processes interact simultaneously. Such theories are important and practically useful for the design of many multifunctional materials, sensors, actuators, and structures.

This book, **Introduction to Coupled Theories in Solid Mechanics**, is concerned with the formulation of *some* (and not all) coupled theories of solid mechanics which form the basis for a wide range of applications. These theories require a knowledge of the foundational balance laws of mechanics, thermodynamics, and electrodynamics of continua, together with a knowledge of the structure and properties of the materials which are provided by particular constitutive models for the so-called smart-materials, and the materials used in the many applications that involve thermo-, poro-, electro- and/or magneto-mechanical coupling.

Introduction to Coupled Theories in Solid Mechanics. Lallit Anand, Eric M. Stewart, Shawn A. Chester, Oxford University Press.
 DOI: 10.1093/oso/9780198986218.003.0001

This book reviews the basic kinematics, balance laws, and constitutive equations for the large deformation theories of elasticity, thermoelasticity, poroelasticity, electroelasticity, and magnetoelasticity. The governing partial differential equations are formulated and suitable boundary conditions are discussed. For specificity,[1]

- *attention will be directed towards formulating such theories for large deformations of elastomeric materials.*

The large-deformation coupled theories discussed in this book have been numerically implemented in the FEniCSx version of the open-source finite element program FEniCS.[2] Applications of the numerically implemented theories in the analysis of sensors and actuators based on such coupled phenomena, and problems related to soft-robotics, energy, and the environment are discussed.

This book, which covers some topics that are in the mainstream of current research in coupled theories in solid mechanics, is divided into eleven parts. The first two parts,

- Part I: Finite elasticity of elastomeric materials

and

- Part II: Viscoelasticity of elastomeric materials

provide the main principles and equations of nonlinear continuum mechanics in the purely mechanical context of finite deformation elasticity and viscoelasticity, which form the foundations for the coupled theories elaborated upon in the later parts of the book:

- Part III: Thermoelasticity of elastomeric materials
- Part IV: Poroelasticity of elastomeric gels
- Part V: Thermally responsive elastomeric gels
- Part VI: Cahn–Hilliard theory for species diffusion coupled with elastic deformations
- Part VII: Electro-elasticity of dielectric elastomers
- Part VIII: Electro-viscoelasticity of dielectric elastomers
- Part IX: Electro-chemo-elasticity of ionic polymers
- Part X: Magneto-viscoelasticity of hard-magnetic soft-elastomers
- Part XI: Magneto-viscoelasticity of soft-magnetic soft-elastomers

In each of these parts of the book,

- we first describe the basic theory,
- we then describe the numerical implementation of the theory in the open-source finite element program FEniCS, and

[1] For a modern treatment of the *small-deformation theories* of thermoelasticity, poroelasticity, and piezoelectricity—which are applicable to a much wider class of materials than elastomers—see the recent book by Anand and Govindjee (2020).

[2] The numerical implementation of these large-deformation coupled theories is challenging—and also a moving target because computer architectures and computational science and engineering are evolving rapidly.

- we finally show several example simulations which demonstrate the capability of the theory and its numerical implementation to model key coupled behaviors.

To help disseminate the coupled theories discussed in this book, the FEniCSx codes used for all the representative simulations shown in this book are posted online (Stewart et al., 2025):

`https://solidmechanicscoupledtheories.github.io.`

Our hope is that the theory and the numerical implementations presented in this book will help in the formulation of other rigorous coupled theories of importance in science and engineering.

Part I

Finite elasticity of elastomeric materials

1 Finite elasticity of elastomeric materials

This chapter discusses the classical theory of finite deformations of elastic solids under isothermal conditions. This theory is of central importance in continuum mechanics. In addition to being important in its own right, it is also important because it helps delineate the central steps in the construction of sound constitutive theories for finite deformation of solids.[1]

1.1 Notation

We follow the notation of Gurtin et al. (2010). Specifically:

- The operators ∇ and Div denote the gradient and divergence with respect to the material point $\mathbf{X}$ in the reference configuration, and $\Delta = \mathrm{Div}\,\nabla$ denotes the referential Laplace operator.
- The operators grad, div, and div grad denote the gradient, divergence, and Laplace operators with respect to the point $\mathbf{x} = \boldsymbol{\chi}(\mathbf{X}, t)$ in the deformed body.
- A superposed dot denotes the material time-derivative.
- Throughout, we write $\mathbf{F}^{-1} = (\mathbf{F})^{-1}$, $\mathbf{F}^{-\top} = (\mathbf{F})^{-\top}$, etc.
- We write $\mathrm{tr}\,\mathbf{A}$, $\mathrm{sym}\,\mathbf{A}$, $\mathrm{skw}\,\mathbf{A}$, $\mathbf{A}_0$, and $\mathrm{sym}_0\mathbf{A}$ respectively, for the trace, symmetric, skew, deviatoric, and symmetric–deviatoric parts of a tensor $\mathbf{A}$. The deviatoric part of a tensor $\mathbf{A}$ is also occasionally written as $\mathrm{dev}\mathbf{A}$, which is of course the same as $\mathbf{A}_0$.
- Also, the inner product of tensors $\mathbf{A}$ and $\mathbf{B}$ is denoted by $\mathbf{A}:\mathbf{B}$, and the magnitude of $\mathbf{A}$ by $|\mathbf{A}| = \sqrt{\mathbf{A}:\mathbf{A}}$.

1.2 Kinematics

We consider a homogeneous body $\mathcal{B}$ identified with the region of space it occupies in a fixed *reference configuration*, and denote by $\mathbf{X}$ an arbitrary material point of $\mathcal{B}$. A *motion* of $\mathcal{B}$ is then a smooth one-to-one mapping

$$\mathbf{x} = \boldsymbol{\chi}(\mathbf{X}, t) \qquad (1.2.1)$$

which gives the place $\mathbf{x}$ occupied by the material point $\mathbf{X}$ at time t in the *deformed body* $\mathcal{B}_t$.

[1] This chapter is excepted from Gurtin et al. (2010).

Introduction to Coupled Theories in Solid Mechanics. Lallit Anand, Eric M. Stewart, Shawn A. Chester, Oxford University Press.
 DOI: 10.1093/oso/9780198986218.003.0002

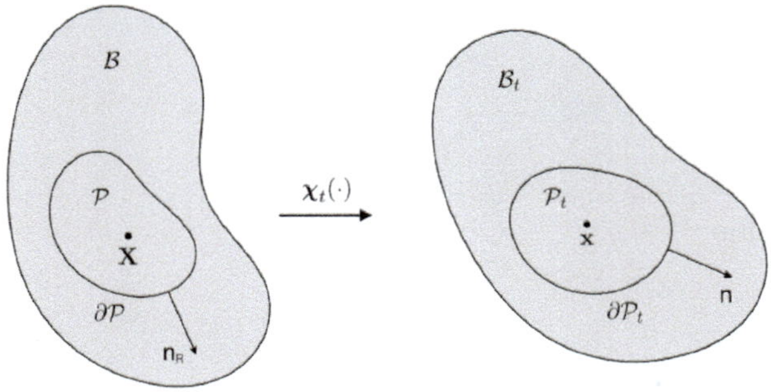

Fig. 1.1 Reference undeformed configuration $\mathcal{B}$ and a current deformed configuration $\mathcal{B}_t$ of a body.

Whenever convenient, we will write

$$\chi_t(\cdot) = \chi(\cdot, t)$$

to denote the **deformation** of the body at a fixed time *t*.

Throughout we shall denote an arbitrary part of the reference body by $\mathcal{P}$, with $\mathbf{n}_R$ the outward unit normal on the boundary $\partial\mathcal{P}$ of $\mathcal{P}$. Also, we denote by $\mathcal{P}_t = \chi(\mathcal{P}, t)$ the image of $\mathcal{P}$ in the deformed body, with $\mathbf{n}$ the outward unit normal on the boundary $\partial\mathcal{P}_t$ of $\mathcal{P}_t$, cf., Fig. 1.1.

Corresponding to the motion (1.2.1), the *displacement*

$$\mathbf{u}(\mathbf{X}, t) = \chi(\mathbf{X}, t) - \mathbf{X}, \tag{1.2.2}$$

deformation gradient, *velocity*, and *acceleration* are given by,

$$\mathbf{F} = \nabla\chi \equiv \frac{\partial\chi(\mathbf{X}, t)}{\partial\mathbf{X}} = \mathbf{1} + \nabla\mathbf{u}(\mathbf{X}, t), \qquad \dot{\mathbf{u}} \equiv \dot{\chi} = \frac{\partial\chi(\mathbf{X}, t)}{\partial t}, \qquad \text{and}$$
$$\ddot{\mathbf{u}} \equiv \ddot{\chi} = \frac{\partial^2\chi(\mathbf{X}, t)}{\partial t^2}, \tag{1.2.3}$$

respectively.

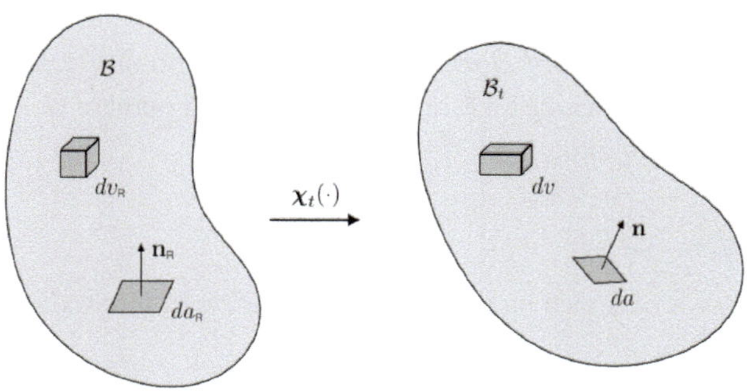

Fig. 1.2 Transformation of infinitesimal volume and area elements.

The relation between a volume element $dv_{\rm R}$ in the reference body $\mathcal{B}$ and the corresponding volume element dv in the deformed body $\mathcal{B}_t$, cf. Fig. 1.2, is

$$dv = J dv_{\rm R}, \qquad \text{with} \qquad J = \det \mathbf{F}. \tag{1.2.4}$$

We restrict our attention to the physically relevant case,

$$J = \det \mathbf{F} > 0. \tag{1.2.5}$$

Further, the relation between an elemental oriented area element $\mathbf{n}_{\rm R} da_{\rm R}$ in the reference body $\mathcal{B}$ and its image $\mathbf{n}\, da$ in the deformed body $\mathcal{B}_t$, cf. Fig. 1.2, is given by Nanson's relation

$$\mathbf{n}\, da = \mathbf{F}^{\rm C}\, \mathbf{n}_{\rm R}\, da_{\rm R}, \qquad \text{where} \qquad \mathbf{F}^{\rm C} \stackrel{\text{def}}{=} J\mathbf{F}^{-\top} \tag{1.2.6}$$

is the cofactor of $\mathbf{F}$, which in turn gives the mapping relation

$$da = \jmath\, da_{\rm R} \qquad \text{with} \qquad \jmath \stackrel{\text{def}}{=} |\mathbf{F}^{\rm C}\, \mathbf{n}_{\rm R}|, \tag{1.2.7}$$

between the area elements $da_{\rm R}$ in the reference body and da in the deformed body, where $\jmath$ represents an *areal Jacobian.*

Basic to our discussion of constitutive equations for elastic solids is the *polar decomposition*

$$\mathbf{F} = \mathbf{R}\mathbf{U} = \mathbf{V}\mathbf{R}, \tag{1.2.8}$$

of the *deformation gradient* into a *proper orthogonal tensor* or *rotation* $\mathbf{R}$, together with a *symmetric and positive-definite right stretch tensor* $\mathbf{U}$, and another *symmetric and positive-definite left stretch tensor* $\mathbf{V}$, satisfying

$$\mathbf{U} = \sqrt{\mathbf{F}^\top\mathbf{F}} \qquad \text{and} \qquad \mathbf{V} = \sqrt{\mathbf{F}\mathbf{F}^\top}. \tag{1.2.9}$$

The tensors

$$\mathbf{C} = \mathbf{U}^2 = \mathbf{F}^\top\mathbf{F} \qquad \text{and} \qquad \mathbf{B} = \mathbf{V}^2 = \mathbf{F}\mathbf{F}^\top, \tag{1.2.10}$$

respectively, are referred to as the *right* and *left Cauchy–Green tensors.* Important consequences of (1.2.10) and the polar decomposition (1.2.8) are the relations

$$\mathbf{V} = \mathbf{R}\mathbf{U}\mathbf{R}^\top \qquad \text{and} \qquad \mathbf{B} = \mathbf{R}\mathbf{C}\mathbf{R}^\top. \tag{1.2.11}$$

1.3 Balance of linear and angular momentum expressed spatially

The balance of linear and angular momentum implies that there exists a tensor field $\mathbf{T}$ in the *deformed body* $\mathcal{B}_t$, called the **Cauchy stress**, which has the following properties (cf., Gurtin et al., 2010):

(i) For each unit vector $\mathbf{n}$ the traction $\mathbf{t}(\mathbf{n})$ is given by

$$\mathbf{t}(\mathbf{n}) = \mathbf{T}\mathbf{n}; \tag{1.3.1}$$

(ii) $\mathbf{T}$ is symmetric,

$$\mathbf{T} = \mathbf{T}^{\top}; \tag{1.3.2}$$

and

(iii) $\mathbf{T}$ satisfies the local equation of motion

$$\operatorname{div}\mathbf{T} + \mathbf{b}_0 = \rho\dot{\mathbf{v}}, \tag{1.3.3}$$

where $\dot{\mathbf{v}}(\mathbf{x}, t) = \ddot{\chi}(\underbrace{\chi^{-1}(\mathbf{x}, t)}_{\mathbf{X}}, t)$ is the spatial description of the acceleration $\ddot{\chi}(\mathbf{X}, t)$, and $\mathbf{b}_0$ is a non-inertial body force.

When working with solids the use of a purely spatial description can be problematic. On the other hand, solids typically possess stress-free reference configurations with respect to which one may measure the displacement and deformation gradient and develop constitutive equations. For that reason we now reformulate the basic balance laws for balance of linear and angular momentum in a referential setting.

1.4 Linear and angular momentum balances expressed referentially

With $\mathbf{T}$ the symmetric Cauchy stress, the contact force on a part $\mathcal{P}_t$ of $\mathcal{B}_t$ in the deformed body is given by

$$\int_{\partial\mathcal{P}_t} \mathbf{T}\mathbf{n}\, da.$$

Using Nanson's relation (1.2.6) for the transformation of oriented area elements, this force can be expressed as an integral over the surface $\partial\mathcal{P}$, defined implicitly by $\partial\mathcal{P}_t = \chi_t(\partial\mathcal{P})$:

$$\int_{\partial\mathcal{P}_t} \mathbf{T}\mathbf{n}\, da = \int_{\partial\mathcal{P}} J\,\mathbf{T}\,\mathbf{F}^{-\top}\mathbf{n}_{\mathrm{R}}\, da_{\mathrm{R}}. \tag{1.4.1}$$

If we define

$$\mathbf{T}_{\mathrm{R}} \stackrel{\text{def}}{=} J\mathbf{T}\mathbf{F}^{-\top}, \tag{1.4.2}$$

equation (1.4.1) may be written as

$$\int_{\partial\mathcal{P}} \mathbf{T}_{\mathrm{R}}\mathbf{n}_{\mathrm{R}}\, da_{\mathrm{R}} = \int_{\partial\mathcal{P}_t} \mathbf{T}\mathbf{n}\, da. \tag{1.4.3}$$

- Thus, $\mathbf{T}_{\mathrm{R}}$ represents the stress measured per unit area in the reference body; $\mathbf{T}_{\mathrm{R}}$ is referred to as the **Piola stress.**

Next, we define

$$\mathbf{b}_{0\mathrm{R}} = J\mathbf{b}_0, \tag{1.4.4}$$

so that $\mathbf{b}_{0\mathrm{R}}$ represents the non-inertial body force measured per unit volume in the reference body. Thus, by (1.2.4),

$$\int_{\mathcal{P}_t} \mathbf{b}_0\, dv = \int_{\mathcal{P}} \mathbf{b}_{0\mathrm{R}}\, dv_{\mathrm{R}}. \tag{1.4.5}$$

Since $\ddot{\mathbf{u}}$ represents the referential description of $\dot{\mathbf{v}}$, we conclude using the balance of mass, $\rho_{\mathrm{R}} = \rho J$, where ρ_{R} is the mass density in the reference body, that

$$\overline{\int_{\mathcal{P}_t} \rho\mathbf{v}\, dv}^{\boldsymbol{\cdot}} = \int_{\mathcal{P}_t} \rho\dot{\mathbf{v}}\, dv = \int_{\mathcal{P}} \rho_{\mathrm{R}}\ddot{\mathbf{u}}\, dv_{\mathrm{R}} = \overline{\int_{\mathcal{P}} \rho_{\mathrm{R}}\dot{\mathbf{u}}\, dv_{\mathrm{R}}}^{\boldsymbol{\cdot}}. \tag{1.4.6}$$

Using (1.4.3), (1.4.5), and (1.4.6), the balance of linear momentum in the deformed body, viz.

$$\int_{\partial\mathcal{P}_t} \mathbf{T}\mathbf{n}\, da + \int_{\mathcal{P}_t} \mathbf{b}_0\, dv = \overline{\int_{\mathcal{P}_t} \rho\mathbf{v}\, dv}^{\boldsymbol{\cdot}}, \tag{1.4.7}$$

when expressed referentially, yields the following **balance of linear momentum**:

$$\int_{\partial\mathcal{P}} \mathbf{T}_{\mathrm{R}}\mathbf{n}_{\mathrm{R}}\, da_{\mathrm{R}} + \int_{\mathcal{P}} \mathbf{b}_{0\mathrm{R}}\, dv_{\mathrm{R}} = \overline{\int_{\mathcal{P}} \rho_{\mathrm{R}}\dot{\mathbf{u}}\, dv_{\mathrm{R}}}^{\boldsymbol{\cdot}}. \tag{1.4.8}$$

Next, since $\mathcal{P}$ is a part of the reference body $\mathcal{B}$, and using Div to represent the referential divergence operator, the divergence theorem yields

$$\int_{\partial\mathcal{P}} \mathbf{T}_{\mathrm{R}}\mathbf{n}_{\mathrm{R}}\, da_{\mathrm{R}} = \int_{\mathcal{P}} \operatorname{Div}\mathbf{T}_{\mathrm{R}}\, dv_{\mathrm{R}},$$

and use of (1.4.6) and (1.4.8) implies that

$$\int_{\mathcal{P}} (\operatorname{Div}\mathbf{T}_{\mathrm{R}} + \mathbf{b}_{0\mathrm{R}} - \rho_{\mathrm{R}}\ddot{\mathbf{u}})\, dv_{\mathrm{R}} = \mathbf{0}.$$

Since the material region $\mathcal{P}$ is arbitrary, we arrive at the referential relation expressing the local form of **balance of linear momentum**:

$$\mathrm{Div}\,\mathbf{T}_{\mathrm{R}} + \mathbf{b}_{0\mathrm{R}} = \rho_{\mathrm{R}}\ddot{\mathbf{u}}. \tag{1.4.9}$$

For later use it is convenient to introduce a *generalized body force*

$$\mathbf{b}_{\mathrm{R}} \overset{\text{def}}{=} \mathbf{b}_{0\mathrm{R}} + \mathbf{i}_{\mathrm{R}}, \qquad \mathbf{i}_{\mathrm{R}} \overset{\text{def}}{=} -\rho_{\mathrm{R}}\ddot{\mathbf{u}}, \tag{1.4.10}$$

in which $\mathbf{i}_{\mathrm{R}}$ may be referred to as the *inertial body force*. Using (1.4.10), balance of linear momentum (1.4.9) may be written as a *force balance*,

$$\mathrm{Div}\,\mathbf{T}_{\mathrm{R}} + \mathbf{b}_{\mathrm{R}} = \mathbf{0}. \tag{1.4.11}$$

Further, since balance of angular momentum requires that the Cauchy stress $\mathbf{T}$ be symmetric, the first Piola stress $\mathbf{T}_{\mathrm{R}}$ as defined in (1.4.2) must satisfy

$$\mathbf{T}_{\mathrm{R}}\mathbf{F}^{\top} = \mathbf{F}\mathbf{T}_{\mathrm{R}}^{\top}. \tag{1.4.12}$$

Equation (1.4.12) represents the local **balance of moments**, expressed referentially.

Summarizing, balance of linear and angular momentum when expressed referentially, implies that there exists a tensor field $\mathbf{T}_{\mathrm{R}}$ in the **reference body** $\mathcal{B}$, called the **Piola stress**, such that:

(i) For each unit vector $\mathbf{n}_{\mathrm{R}}$ the traction $\mathbf{t}_{\mathrm{R}}(\mathbf{n}_{\mathrm{R}})$ is given by

$$\mathbf{t}_{\mathrm{R}}(\mathbf{n}_{\mathrm{R}}) = \mathbf{T}_{\mathrm{R}}\mathbf{n}_{\mathrm{R}}; \tag{1.4.13}$$

(ii) $\mathbf{T}_{\mathrm{R}}$ satisfies the local equation of motion

$$\mathrm{Div}\,\mathbf{T}_{\mathrm{R}} + \mathbf{b}_{0\mathrm{R}} = \rho_{\mathrm{R}}\ddot{\mathbf{u}}, \tag{1.4.14}$$

where ρ_{R} is the mass density in the reference body, and $\mathbf{b}_{0\mathrm{R}} = J\mathbf{b}_0$, is the body force per unit volume of the reference body; and

(iii) $\mathbf{T}_{\mathrm{R}}$ satisfies,

$$\mathbf{T}_{\mathrm{R}}\mathbf{F}^{\top} = \mathbf{F}\mathbf{T}_{\mathrm{R}}^{\top}, \tag{1.4.15}$$

which represents a balance of moments.

1.5 Balance of energy. Entropy imbalance. Free-energy imbalance

Our discussion of thermodynamics involves the following fields:

ε_{R} the internal energy density per unit reference volume,

η_{R} the entropy density per unit reference volume,

$\mathbf{q}_{\mathrm{R}}$ the heat flux per unit reference area,

q_{R} the external heat supply per unit reference volume,

ϑ the absolute temperature ($\vartheta > 0$),

and follows the discussion in Gurtin et al. (2010). Consider a material region $\mathcal{P}$. Then, we take the balance law for energy as

$$\overline{\int_{\mathcal{P}} \varepsilon_{\mathrm{R}}\, dv_{\mathrm{R}}}^{\boldsymbol{\cdot}} = -\int_{\partial\mathcal{P}} \mathbf{q}_{\mathrm{R}}\cdot\mathbf{n}_{\mathrm{R}}\, da_{\mathrm{R}} + \int_{\mathcal{P}} q_{\mathrm{R}}\, dv_{\mathrm{R}} + \int_{\partial\mathcal{P}} (\mathbf{T}_{\mathrm{R}}\mathbf{n}_{\mathrm{R}})\cdot\dot{\mathbf{u}}\, da_{\mathrm{R}} + \int_{\mathcal{P}} \mathbf{b}_{\mathrm{R}}\cdot\dot{\mathbf{u}}\, dv_{\mathrm{R}}, \tag{1.5.1}$$

where recall that $\mathbf{b}_{\mathrm{R}}$ is the generalized body force defined in (1.4.10).

Applying the divergence theorem to the terms in (1.5.1) involving integrals over the boundary $\partial\mathcal{P}$ of $\mathcal{P}$, we obtain

$$\int_{\mathcal{P}} \Big(\dot{\varepsilon}_{\mathrm{R}} - (\mathrm{Div}\,\mathbf{T}_{\mathrm{R}} + \mathbf{b}_{\mathrm{R}})\cdot\dot{\mathbf{u}} - \mathbf{T}_{\mathrm{R}}\!:\!\dot{\mathbf{F}} + \mathrm{Div}\,\mathbf{q}_{\mathrm{R}} - q_{\mathrm{R}}\Big)\, dv_{\mathrm{R}} = 0, \tag{1.5.2}$$

which upon use of the force balance (1.4.11) and using the fact that (1.5.2) must hold for all parts $\mathcal{P}$, gives the local form of the energy balance as

$$\dot{\varepsilon}_{\mathrm{R}} = \mathbf{T}_{\mathrm{R}}\!:\!\dot{\mathbf{F}} - \mathrm{Div}\,\mathbf{q}_{\mathrm{R}} + q_{\mathrm{R}}. \tag{1.5.3}$$

Also, the second law takes the form of an entropy imbalance

$$\overline{\int_{\mathcal{P}} \eta_{\mathrm{R}}\, dv_{\mathrm{R}}}^{\boldsymbol{\cdot}} \geq -\int_{\partial\mathcal{P}} \frac{\mathbf{q}_{\mathrm{R}}\cdot\mathbf{n}_{\mathrm{R}}}{\vartheta}\, da_{\mathrm{R}} + \int_{\mathcal{P}} \frac{q_{\mathrm{R}}}{\vartheta}\, dv_{\mathrm{R}}, \tag{1.5.4}$$

in which case the local entropy imbalance has the form

$$\dot{\eta}_{\mathrm{R}} \geq -\mathrm{Div}\left(\frac{\mathbf{q}_{\mathrm{R}}}{\vartheta}\right) + \frac{q_{\mathrm{R}}}{\vartheta}. \tag{1.5.5}$$

Then, in view of the local energy balance (1.5.3),

$$\begin{aligned}-\mathrm{Div}\left(\frac{\mathbf{q}_{\mathrm{R}}}{\vartheta}\right) + \frac{q_{\mathrm{R}}}{\vartheta} &= \frac{1}{\vartheta}(-\mathrm{Div}\,\mathbf{q}_{\mathrm{R}} + q_{\mathrm{R}}) + \frac{1}{\vartheta^2}\,\mathbf{q}_{\mathrm{R}}\cdot\nabla\vartheta,\\ &= \frac{1}{\vartheta}\Big(\dot{\varepsilon}_{\mathrm{R}} - \mathbf{T}_{\mathrm{R}}\!:\!\dot{\mathbf{F}} + \frac{1}{\vartheta}\,\mathbf{q}_{\mathrm{R}}\cdot\nabla\vartheta\Big),\end{aligned}$$

and this with the local entropy imbalance (1.5.5) implies that

$$(\dot{\varepsilon}_{\mathrm{R}} - \vartheta\dot{\eta}_{\mathrm{R}}) - \mathbf{T}_{\mathrm{R}}:\dot{\mathbf{F}} + \frac{1}{\vartheta}\mathbf{q}_{\mathrm{R}}\cdot\nabla\vartheta \le 0. \tag{1.5.6}$$

Introducing the Helmholtz free energy per unit reference volume

$$\psi_{\mathrm{R}} = \varepsilon_{\mathrm{R}} - \vartheta\eta_{\mathrm{R}}, \tag{1.5.7}$$

(1.5.6) yields the following local free-energy imbalance

$$\dot{\psi}_{\mathrm{R}} + \eta_{\mathrm{R}}\dot{\vartheta} - \mathbf{T}_{\mathrm{R}}:\dot{\mathbf{F}} + \frac{1}{\vartheta}\mathbf{q}_{\mathrm{R}}\cdot\nabla\vartheta \le 0. \tag{1.5.8}$$

Note that the dissipation per unit reference volume per unit time is given by

$$\mathcal{D}_{\mathrm{R}} = \mathbf{T}_{\mathrm{R}}:\dot{\mathbf{F}} - \dot{\psi}_{\mathrm{R}} - \eta_{\mathrm{R}}\dot{\vartheta} - \frac{1}{\vartheta}\mathbf{q}_{\mathrm{R}}\cdot\nabla\vartheta \ge 0. \tag{1.5.9}$$

1.5.1 Free-energy imbalance under isothermal conditions

Under *isothermal conditions*, where $\dot{\vartheta}$ and $\nabla\vartheta$ are zero, the local free-energy imbalance (1.5.8) reduces to

$$\dot{\psi}_{\mathrm{R}} - \mathbf{T}_{\mathrm{R}}:\dot{\mathbf{F}} \le 0, \tag{1.5.10}$$

and the dissipation per unit reference volume per unit time is given by

$$\mathcal{D}_{\mathrm{R}} = \mathbf{T}_{\mathrm{R}}:\dot{\mathbf{F}} - \dot{\psi}_{\mathrm{R}} \ge 0. \tag{1.5.11}$$

1.5.2 The second Piola stress

Next, we introduce a new stress measure which allows us to express the free-energy imbalance (1.5.10) in a form more amenable to applications. Recall from (1.4.2) that the first Piola stress is defined by

$$\mathbf{T}_{\mathrm{R}} = J\mathbf{T}\mathbf{F}^{-\top}. \tag{1.5.12}$$

We now introduce a new stress measure

$$\mathbf{T}_{\mathrm{RR}} \stackrel{\mathrm{def}}{=} \mathbf{F}^{-1}\mathbf{T}_{\mathrm{R}} = J\,\mathbf{F}^{-1}\mathbf{T}\mathbf{F}^{-\top}, \tag{1.5.13}$$

called the **second Piola stress**. Note that since $\mathbf{T}$ is symmetric, the second Piola stress is also **symmetric**.

Next, we write the local free-energy imbalance (1.5.10) in terms of the second Piola stress. Specifically the stress power $\mathbf{T}_{\text{R}}:\dot{\mathbf{F}}$ may be written as:

$$\mathbf{T}_{\text{R}}:\dot{\mathbf{F}} = \mathbf{F}\mathbf{T}_{\text{RR}}:\dot{\mathbf{F}} \tag{1.5.14}$$

$$= \mathbf{T}_{\text{RR}}:\mathbf{F}^{\top}\dot{\mathbf{F}} \tag{1.5.15}$$

$$= \frac{1}{2}\mathbf{T}_{\text{RR}}:\left(\mathbf{F}^{\top}\dot{\mathbf{F}} + \dot{\mathbf{F}}^{\top}\mathbf{F}\right) \qquad \text{(since } \mathbf{T}_{\text{RR}} \text{ is symmetric)} \tag{1.5.16}$$

$$= \frac{1}{2}\mathbf{T}_{\text{RR}}:\dot{\mathbf{C}}, \tag{1.5.17}$$

where in the last step the identity $\dot{\mathbf{C}} = \overline{\mathbf{F}^{\top}\mathbf{F}}^{\boldsymbol{\cdot}} = \mathbf{F}^{\top}\dot{\mathbf{F}} + \dot{\mathbf{F}}^{\top}\mathbf{F}$ has been used.

Then, use of (1.5.17) in the free-energy imbalance (1.5.10) yields

$$\dot{\psi}_{\text{R}} - \frac{1}{2}\mathbf{T}_{\text{RR}}:\dot{\mathbf{C}} \leq 0. \tag{1.5.18}$$

For later use we note from (1.5.18) that the dissipation rate per unit reference volume is

$$\mathcal{D}_{\text{R}} = \frac{1}{2}\mathbf{T}_{\text{RR}}:\dot{\mathbf{C}} - \dot{\psi}_{\text{R}} \geq 0. \tag{1.5.19}$$

1.6 Change in frame

A *change in frame*, at each *fixed time t*, is a transformation—defined by a rotation $\mathbf{Q}(t)$ and a spatial point $\mathbf{y}(t)$—which transforms *spatial points* $\mathbf{x}$ to spatial points

$$\mathbf{x}^* = \mathcal{F}(\mathbf{x}), \tag{1.6.1}$$

$$= \mathbf{y}(t) + \mathbf{Q}(t)(\mathbf{x} - \mathbf{o}), \tag{1.6.2}$$

the function $\mathcal{F}$ represents a rigid mapping of the observed space into itself, with $\mathbf{o}$ a fixed spatial origin. By (1.6.2) the transformation law for the motion $\mathbf{x} = \boldsymbol{\chi}(\mathbf{X}, t)$ has the form

$$\boldsymbol{\chi}^*(\mathbf{X}, t) = \mathbf{y}(t) + \mathbf{Q}(t)(\boldsymbol{\chi}(\mathbf{X}, t) - \mathbf{o}). \tag{1.6.3}$$

Then the basic kinematical fields transform as follows under a change in frame (cf., e.g. Gurtin et al. 2010),

$$\mathbf{F}^* = \mathbf{Q}\mathbf{F}, \quad \mathbf{R}^* = \mathbf{Q}\mathbf{R}, \quad \mathbf{U}^* = \mathbf{U} \quad \mathbf{V}^* = \mathbf{Q}\mathbf{V}\mathbf{Q}^{\top}, \quad \mathbf{C}^* = \mathbf{C}, \quad \mathbf{B}^* = \mathbf{Q}\mathbf{B}\mathbf{Q}^{\top}, \tag{1.6.4}$$

and the Cauchy, Piola, and second Piola stresses transform according to

$$\mathbf{T}^* = \mathbf{Q}\mathbf{T}\mathbf{Q}^\top, \quad \mathbf{T}^*_{\mathrm{R}} = \mathbf{Q}\mathbf{T}_{\mathrm{R}}, \quad \mathbf{T}^*_{\mathrm{RR}} = \mathbf{T}_{\mathrm{RR}}. \tag{1.6.5}$$

Note that:

- the Cauchy stress is frame-indifferent;
- the Piola stress is neither frame-indifferent nor invariant;
- the second Piola stress is invariant.

Note that since $\psi^*_{\mathrm{R}} = \psi_{\mathrm{R}}$, all quantities appearing in the free-energy imbalance (1.5.18) are **invariant** under a change in frame.

1.7 Constitutive theory for elastic materials

Since elastic materials are assumed to have a constitutive response that only depends on the current kinematic state of the material, we assume the following special set of constitutive equations for ψ_{R} and $\mathbf{T}_{\mathrm{RR}}$:[2]

$$\begin{aligned} \psi_{\mathrm{R}} &= \bar{\psi}_{\mathrm{R}}(\mathbf{C}), \\ \mathbf{T}_{\mathrm{RR}} &= \bar{\mathbf{T}}_{\mathrm{RR}}(\mathbf{C}), \end{aligned} \tag{1.7.1}$$

where in writing $(1.7.1)_1$ we wish to make clear that dependence of the free energy on the deformation gradient $\mathbf{F}$ is taken to appear only through the right Cauchy–Green tensor $\mathbf{C} = \mathbf{F}^\top\mathbf{F}$.

As noted in the previous section, ψ, $\mathbf{C}$ and $\mathbf{T}_{\mathrm{RR}}$ are invariant under a change in frame. Hence the constitutive equations are invariant under a change in frame, and the principle of material frame indifference is automatically satisfied.

1.7.1 Thermodynamic restrictions

Recalling also that elastic materials do not dissipate energy and since

$$\overline{\bar{\psi}_{\mathrm{R}}(\mathbf{C})}^{\boldsymbol{\cdot}} = \frac{\partial \bar{\psi}_{\mathrm{R}}(\mathbf{C})}{\partial \mathbf{C}} : \dot{\mathbf{C}}, \tag{1.7.2}$$

satisfaction of (1.5.19) with $\mathcal{D}_{\mathrm{R}} = 0$ requires that the constitutive equations (1.7.1) satisfy

$$\left(\frac{1}{2}\bar{\mathbf{T}}_{\mathrm{RR}}(\mathbf{C}) - \frac{\partial \bar{\psi}_{\mathrm{R}}(\mathbf{C})}{\partial \mathbf{C}}\right) : \dot{\mathbf{C}} = 0. \tag{1.7.3}$$

Since this expression must be satisfied at all $\mathbf{C}$ and for all $\dot{\mathbf{C}}$, we see that[3]

[2] We distinguish functions from their values by using overset $\hat{\bullet}$, $\bar{\bullet}$, $\tilde{\bullet}$, etc., as convenient, for example $\psi = \tilde{\psi}$.
[3] This logical argument is generally known as the *Coleman–Noll procedure* (Coleman and Noll, 1963).

- *the free energy determines the second Piola stress via the* **stress relation**:

$$\mathbf{T}_{\text{RR}} = \bar{\mathbf{T}}_{\text{RR}}(\mathbf{C}) = 2\,\frac{\partial \bar{\psi}_{\text{R}}(\mathbf{C})}{\partial \mathbf{C}}. \tag{1.7.4}$$

Next, $(1.5.13)_1$ and (1.7.4) imply that the first Piola stress is given by a constitutive equation of the form

$$\mathbf{T}_{\text{R}} = 2\mathbf{F}\,\frac{\partial \bar{\psi}_{\text{R}}(\mathbf{C})}{\partial \mathbf{C}}. \tag{1.7.5}$$

Similarly, $(1.5.13)_2$ and (1.7.4) imply that the Cauchy stress is given by a constitutive equation of the form

$$\mathbf{T} = 2J^{-1}\mathbf{F}\,\frac{\partial \bar{\psi}_{\text{R}}(\mathbf{C})}{\partial \mathbf{C}}\mathbf{F}^{\top}. \tag{1.7.6}$$

Materials consistent with (1.7.4), (1.7.5), and (1.7.6) are commonly termed **hyperelastic**.

1.8 Summary of basic equations. Initial/boundary-value problems

The basic field equations describing the motion of an elastic body consist of the kinematical relations defining the deformation gradient and right Cauchy–Green tensor, the relation determining the Piola stress, and the local balance of linear and angular momentum:

$$\begin{gathered} \mathbf{F} = \mathbf{1} + \nabla\mathbf{u}, \qquad \mathbf{C} = \mathbf{F}^{\top}\mathbf{F}, \\ \mathbf{T}_{\text{R}} = 2\mathbf{F}\,\frac{\partial \bar{\psi}_{\text{R}}(\mathbf{C})}{\partial \mathbf{C}}, \\ \operatorname{Div}\mathbf{T}_{\text{R}} + \mathbf{b}_{0\text{R}} = \rho_{\text{R}}\ddot{\mathbf{u}}, \\ \mathbf{T}_{\text{R}}\mathbf{F}^{\top} = \mathbf{F}\mathbf{T}_{\text{R}}^{\top}. \end{gathered} \tag{1.8.1}$$

These equations hold at all material points $\mathbf{X} \in \mathcal{B}$.

In the quasi-static setting $(1.8.1)_4$ is replaced by the *equation of equilibrium*,

$$\operatorname{Div}\mathbf{T}_{\text{R}} + \mathbf{b}_{0\text{R}} = \mathbf{0}. \tag{1.8.2}$$

1.8.1 A typical initial/boundary-value problem

Let $\partial\mathcal{B}_{\mathrm{u}}$ and $\partial\mathcal{B}_{\mathrm{t}_{\text{R}}}$ be mutually exclusive complementary subsurfaces of the boundary $\partial\mathcal{B}$, so that $\partial\mathcal{B}_{\mathrm{u}} \cup \partial\mathcal{B}_{\mathrm{t}_{\text{R}}} = \partial\mathcal{B}$. As possible boundary conditions one might specify the displacement on $\partial\mathcal{B}_{\mathrm{u}}$ and the surface traction on $\partial\mathcal{B}_{\mathrm{t}_{\text{R}}}$:

$$\left.\begin{aligned} \mathbf{u} &= \bar{\mathbf{u}} \quad \text{a prescribed function on } \partial\mathcal{B}_{\mathrm{u}} \text{ for all } t \geq 0, \\ \mathbf{T}_{\mathrm{R}}\mathbf{n}_{\mathrm{R}} &= \bar{\mathbf{t}}_{\mathrm{R}} \quad \text{a prescribed function on } \partial\mathcal{B}_{\mathrm{t_R}} \text{ for all } t \geq 0. \end{aligned}\right\} \tag{1.8.3}$$

Standard *initial conditions* involve a specification of the initial deformation and the velocity:

$$\mathbf{u}(\mathbf{X},0) = \mathbf{u}_0(\mathbf{X}), \qquad \dot{\mathbf{u}}(\mathbf{X},0) = \dot{\mathbf{u}}_0(\mathbf{X}), \tag{1.8.4}$$

with $\mathbf{u}_0$ and $\dot{\mathbf{u}}_0$ prescribed functions on $\mathcal{B}$.

The *initial/boundary-value problem* corresponding to the prescribed data

$$\{\mathbf{b}_{0\mathrm{R}}, \rho_{\mathrm{R}}, \bar{\mathbf{u}}, \bar{\mathbf{t}}_{\mathrm{R}}, \mathbf{u}_0, \dot{\mathbf{u}}_0\}$$

then consists in finding a displacement $\mathbf{u}(\mathbf{X},t)$—defined for $\mathbf{X}$ in $\mathcal{B}$ and $t \geq 0$—that satisfies the field equations (1.8.1) for $\mathbf{X}$ in $\mathcal{B}$ and $t \geq 0$, the boundary conditions (1.8.3) on $\partial\mathcal{B}$ for $t \geq 0$, and the initial conditions (1.8.4) for $\mathbf{X}$ in $\mathcal{B}$.

Also of importance are quasi-static boundary-value problems involving solutions of the basic field equations and boundary conditions. Since initial conditions are irrelevant to such problems, the data consists of $\{\mathbf{b}_{\mathrm{R}}, \bar{\mathbf{u}}, \bar{\mathbf{t}}_{\mathrm{R}}\}$ and the corresponding *boundary-value problem* consists in finding a displacement $\mathbf{u}$ satisfying the field equations (1.8.1)—with the linear momentum balance replaced by the equilibrium balance (1.8.2)—and the natural time-independent analogs of the boundary conditions (1.8.3).

REMARK

Another form of boundary condition arises when one specifies the surface traction $\mathbf{Tn}$ on the deformed surface

$$\partial\mathcal{B}_{\mathrm{t}}(t) = \boldsymbol{\chi}(\partial\mathcal{B}_{\mathrm{t_R}}, t).$$

A simple example of this type of condition arises when one considers the effects of a uniform pressure p_0:

$$\mathbf{Tn} = -p_0\mathbf{n} \quad \text{on } \partial\mathcal{B}_{\mathrm{t}}(t). \tag{1.8.5}$$

Using Nanson's relation (1.2.6),

$$\mathbf{n}\,da = J\mathbf{F}^{-\top}\mathbf{n}_{\mathrm{R}}\,da_{\mathrm{R}} = \mathbf{F}^{\mathrm{C}}\,\mathbf{n}_{\mathrm{R}}\,da_{\mathrm{R}},$$

the boundary condition (1.8.5) may be expressed as

$$\mathbf{T}_{\mathrm{R}}\mathbf{n}_{\mathrm{R}} = -p_0\mathbf{F}^{\mathrm{C}}\,\mathbf{n}_{\mathrm{R}} \quad \text{on } \partial\mathcal{B}_{\mathrm{t_R}} \tag{1.8.6}$$

and furnishes an example of a **configuration-dependent boundary condition.**

1.9 Constitutive theory for slightly compressible elastic materials

In considering special constitutive equations we shall limit our considerations to models for elastomeric materials. Elastomeric materials are usually modeled as *isotropic and incompressible* materials. However, *no material is completely incompressible.* Thus, for reasons of more accurate modeling, as well as for computational reasons, it is desirable to account for the slight compressibility of materials. A theory for slightly compressible elastic materials can be developed based on a multiplicative decomposition

$$\mathbf{F} = \mathbf{F}^{\text{vol}}\bar{\mathbf{F}} \tag{1.9.1}$$

of the deformation gradient $\mathbf{F}$ into volumetric and isochoric factors $\mathbf{F}^{\text{vol}}$ and $\bar{\mathbf{F}}$—as proposed by Flory (1961), where

$$\mathbf{F}^{\text{vol}} = J^{1/3}\mathbf{1} \qquad \text{and} \qquad \bar{\mathbf{F}} = J^{-1/3}\mathbf{F}. \tag{1.9.2}$$

Decomposition (1.9.1) and the representations (1.9.2) lead to a multiplicative decomposition of the right Cauchy–Green tensor in the form:

$$\begin{aligned} \mathbf{C} &= \mathbf{C}^{\text{vol}}\bar{\mathbf{C}}, \\ \mathbf{C}^{\text{vol}} &= J^{2/3}\mathbf{1}, \\ \bar{\mathbf{C}} &= J^{-2/3}\mathbf{C}. \end{aligned} \tag{1.9.3}$$

Similarly, we also have a multiplicative decomposition of the left Cauchy–Green tensor in the form:

$$\begin{aligned} \mathbf{B} &= \mathbf{B}^{\text{vol}}\bar{\mathbf{B}}, \\ \mathbf{B}^{\text{vol}} &= J^{2/3}\mathbf{1}, \\ \bar{\mathbf{B}} &= J^{-2/3}\mathbf{B}. \end{aligned} \tag{1.9.4}$$

We now develop a theory for slightly compressible materials in an alternative form

$$\begin{aligned} \psi_{\text{R}} &= \bar{\psi}_{\text{R}}(\mathbf{C}), \\ &= \tilde{\psi}_{\text{R}}(\bar{\mathbf{C}}, J), \end{aligned} \tag{1.9.5}$$

where $\bar{\mathbf{C}}$ is the isochoric part of $\mathbf{C}$, and J represents the volumetric part of the deformation.

Accordingly, we may use the relations (1.9.5) with (1.7.4), (1.7.5), and (1.7.6) to determine expressions for the Piola and Cauchy stresses in an elastic solid with free energy

$\psi_{\mathrm{R}} = \tilde{\psi}_{\mathrm{R}}(\bar{\mathbf{C}}, J)$. To achieve this, we first note that, by $J = \sqrt{\det \mathbf{C}}$, the chain rule, and the symmetry of $\mathbf{C}$,

$$\begin{aligned}
\frac{\partial J}{\partial \mathbf{C}} &= \frac{\partial \sqrt{\det \mathbf{C}}}{\partial \mathbf{C}} \\
&= \frac{1}{2\sqrt{\det \mathbf{C}}} \frac{\partial(\det \mathbf{C})}{\partial \mathbf{C}} \\
&= \frac{\det \mathbf{C}}{2\sqrt{\det \mathbf{C}}} \mathbf{C}^{-1} \\
&= \tfrac{1}{2} J \mathbf{C}^{-1}.
\end{aligned} \tag{1.9.6}$$

Similarly, by $\bar{\mathbf{C}} = J^{-2/3}\mathbf{C}$,

$$\begin{aligned}
\frac{\partial \bar{\mathbf{C}}}{\partial \mathbf{C}} &= \frac{\partial (J^{-2/3}\mathbf{C})}{\partial \mathbf{C}} \\
&= J^{-2/3} \frac{\partial \mathbf{C}}{\partial \mathbf{C}} + \mathbf{C} \otimes \frac{\partial (J^{-2/3})}{\partial \mathbf{C}} \\
&= J^{-2/3} \mathbb{I}^{\mathrm{sym}} - \tfrac{1}{3} J^{-5/3} \mathbf{C} \otimes \frac{\partial(\det \mathbf{C})}{\partial \mathbf{C}} \\
&= J^{-2/3} (\mathbb{I}^{\mathrm{sym}} - \tfrac{1}{3} \mathbf{C} \otimes \mathbf{C}^{-1}) \\
&= J^{-2/3} (\mathbb{I}^{\mathrm{sym}} - \tfrac{1}{3} J^{2/3} \bar{\mathbf{C}} \otimes \mathbf{C}^{-1}) \\
&= J^{-2/3} (\mathbb{I}^{\mathrm{sym}} - \tfrac{1}{3} \bar{\mathbf{C}} \otimes (J^{-2/3}\mathbf{C})^{-1}) \\
&= J^{-2/3} (\mathbb{I}^{\mathrm{sym}} - \tfrac{1}{3} \bar{\mathbf{C}} \otimes \bar{\mathbf{C}}^{-1}),
\end{aligned} \tag{1.9.7}$$

where $\mathbb{I}^{\mathrm{sym}}$ denotes the symmetric fourth-order identity tensor $(\mathbb{I}^{\mathrm{sym}})_{ijkl} = \frac{1}{2}(\delta_{ik}\delta_{jl} + \delta_{il}\delta_{jk})$, and the tensor product $\mathbf{A} \otimes \mathbf{G}$ of two second order tensors $\mathbf{A}$ and $\mathbf{G}$ is the fourth-order tensor defined by

$$(\mathbf{A} \otimes \mathbf{G})\mathbf{M} = (\mathbf{G} : \mathbf{M})\mathbf{A}, \qquad (A_{ij}G_{kl})M_{kl} = (G_{kl}M_{kl})A_{ij},$$

for any second-order tensor $\mathbf{M}$.

Thus, by (1.7.4), (1.9.5), (1.9.6), and (1.9.7), the second Piola stress has the form

$$\begin{aligned}\mathbf{T}_{\text{RR}} &= 2\frac{\partial\tilde{\psi}_{\text{R}}(\bar{\mathbf{C}},J)}{\partial\mathbf{C}}\\ &= 2J^{-2/3}(\mathbb{I}-\tfrac{1}{3}\bar{\mathbf{C}}^{-1}\otimes\bar{\mathbf{C}})\frac{\partial\tilde{\psi}_{\text{R}}(\bar{\mathbf{C}},J)}{\partial\bar{\mathbf{C}}}+J\frac{\partial\tilde{\psi}_{\text{R}}(\bar{\mathbf{C}},J)}{\partial J}\mathbf{C}^{-1}\\ &= \frac{2}{J^{2/3}}\left[\frac{\partial\tilde{\psi}_{\text{R}}(\bar{\mathbf{C}},J)}{\partial\bar{\mathbf{C}}}-\frac{1}{3}\left(\bar{\mathbf{C}}:\frac{\partial\tilde{\psi}_{\text{R}}(\bar{\mathbf{C}},J)}{\partial\bar{\mathbf{C}}}\right)\bar{\mathbf{C}}^{-1}\right]+J\frac{\partial\tilde{\psi}_{\text{R}}(\bar{\mathbf{C}},J)}{\partial J}\mathbf{C}^{-1}.\end{aligned}\tag{1.9.8}$$

Next, using (1.9.8) in the expression $\mathbf{T}_{\text{R}}=\mathbf{F}\mathbf{T}_{\text{RR}}$ relating the Piola stresses, we obtain

$$\mathbf{T}_{\text{R}} = 2\,J^{-1/3}\left[\bar{\mathbf{F}}\frac{\partial\tilde{\psi}_{\text{R}}(\bar{\mathbf{C}},J)}{\partial\bar{\mathbf{C}}}-\frac{1}{3}\left(\bar{\mathbf{C}}:\frac{\partial\tilde{\psi}_{\text{R}}(\bar{\mathbf{C}},J)}{\partial\bar{\mathbf{C}}}\right)\bar{\mathbf{F}}^{-\top}\right]+J\frac{\partial\tilde{\psi}_{\text{R}}(\bar{\mathbf{C}},J)}{\partial J}\mathbf{F}^{-\top}.\tag{1.9.9}$$

Further, by (1.9.9), $\bar{\mathbf{F}}=J^{-1/3}\mathbf{F}$, and the relation $\mathbf{T}=J^{-1}\mathbf{T}_{\text{R}}\mathbf{F}^{\top}$, the Cauchy stress is given by

$$\mathbf{T} = 2\,J^{-1}\left[\bar{\mathbf{F}}\frac{\partial\tilde{\psi}_{\text{R}}(\bar{\mathbf{C}},J)}{\partial\bar{\mathbf{C}}}\bar{\mathbf{F}}^{\top}-\frac{1}{3}\left(\bar{\mathbf{C}}:\frac{\partial\tilde{\psi}_{\text{R}}(\bar{\mathbf{C}},J)}{\partial\bar{\mathbf{C}}}\right)\mathbf{1}\right]+\frac{\partial\tilde{\psi}_{\text{R}}(\bar{\mathbf{C}},J)}{\partial J}\mathbf{1}.\tag{1.9.10}$$

Computing the trace of $\mathbf{T}$ as given by (1.9.10), we find that

$$\begin{aligned}\operatorname{tr}\mathbf{T} &= 2\,J^{-1}\operatorname{tr}\left[\bar{\mathbf{F}}\frac{\partial\tilde{\psi}_{\text{R}}(\bar{\mathbf{C}},J)}{\partial\bar{\mathbf{C}}}\bar{\mathbf{F}}^{\top}-\frac{1}{3}\left(\frac{\partial\tilde{\psi}_{\text{R}}(\bar{\mathbf{C}},J)}{\partial\bar{\mathbf{C}}}:\bar{\mathbf{C}}\right)\mathbf{1}\right]+3\frac{\partial\tilde{\psi}_{\text{R}}(\bar{\mathbf{C}},J)}{\partial J}\\ &= 2\,J^{-1}\left[\bar{\mathbf{F}}^{\top}:\frac{\partial\tilde{\psi}_{\text{R}}(\bar{\mathbf{C}},J)}{\partial\bar{\mathbf{C}}}\bar{\mathbf{F}}^{\top}-\bar{\mathbf{C}}:\frac{\partial\tilde{\psi}_{\text{R}}(\bar{\mathbf{C}},J)}{\partial\bar{\mathbf{C}}}\right]+3\frac{\partial\tilde{\psi}_{\text{R}}(\bar{\mathbf{C}},J)}{\partial J}\\ &= 2\,J^{-1}\left[\bar{\mathbf{F}}^{\top}\bar{\mathbf{F}}:\frac{\partial\tilde{\psi}_{\text{R}}(\bar{\mathbf{C}},J)}{\partial\bar{\mathbf{C}}}-\bar{\mathbf{C}}:\frac{\partial\tilde{\psi}_{\text{R}}(\bar{\mathbf{C}},J)}{\partial\bar{\mathbf{C}}}\right]+3\frac{\partial\tilde{\psi}_{\text{R}}(\bar{\mathbf{C}},J)}{\partial J}\\ &= 2\,J^{-1}\left[\bar{\mathbf{C}}:\frac{\partial\tilde{\psi}_{\text{R}}(\bar{\mathbf{C}},J)}{\partial\bar{\mathbf{C}}}-\bar{\mathbf{C}}:\frac{\partial\tilde{\psi}_{\text{R}}(\bar{\mathbf{C}},J)}{\partial\bar{\mathbf{C}}}\right]+3\frac{\partial\tilde{\psi}_{\text{R}}(\bar{\mathbf{C}},J)}{\partial J}\\ &= 3\frac{\partial\tilde{\psi}_{\text{R}}(\bar{\mathbf{C}},J)}{\partial J}.\end{aligned}$$

Thus, the alternative constitutive equation (1.9.5) determining the free energy ψ_{R} as a function $\tilde{\psi}_{\text{R}}$ of the isochoric component $\bar{\mathbf{C}}$ of the right Cauchy–Green tensor $\mathbf{C}$, the volumetric determinant J, leads naturally to a decomposition of the Cauchy stress tensor $\mathbf{T}$ into a sum

$$\mathbf{T} = \mathbf{T}_0 - p\mathbf{1},\tag{1.9.11}$$

of a deviatoric component

$$\mathbf{T}_0 = 2\,J^{-1}\left[\bar{\mathbf{F}}\frac{\partial\tilde{\psi}_{\text{R}}(\bar{\mathbf{C}},J)}{\partial\bar{\mathbf{C}}}\bar{\mathbf{F}}^{\top}-\frac{1}{3}\left(\frac{\partial\tilde{\psi}_{\text{R}}(\bar{\mathbf{C}},J)}{\partial\bar{\mathbf{C}}}:\bar{\mathbf{C}}\right)\mathbf{1}\right],\tag{1.9.12}$$

and a spherical component with pressure

$$p = -\frac{\partial \tilde{\psi}_{\mathrm{R}}(\bar{\mathbf{C}}, J)}{\partial J}. \tag{1.9.13}$$

1.9.1 **Separable free energy into isochoric and volumetric parts**

For a slightly compressible material it seems reasonable to expect that the deviatoric component $\mathbf{T}_0$ of the Cauchy stress be inversely proportional to the volumetric Jacobian J (thereby accounting for the change in volume between the reference and spatial configurations) and that the pressure p be independent of the isochoric factor $\bar{\mathbf{C}}$ of the right Cauchy–Green tensor $\mathbf{C}$. Necessary and sufficient for the satisfaction of these requirements is the assumption that $\tilde{\psi}$ has the separable form

$$\tilde{\psi}_{\mathrm{R}}(\bar{\mathbf{C}}, J) = \underbrace{\breve{\psi}_{\mathrm{R}}(\bar{\mathbf{C}})}_{\text{isochoric}} + \underbrace{\psi_{\mathrm{R}}^{\mathrm{vol}}(J)}_{\text{volumetric}}. \tag{1.9.14}$$

This being the case, (1.9.9) and (1.9.10) simplify somewhat to

$$\mathbf{T}_{\mathrm{R}} = 2\,J^{-1/3}\left[\bar{\mathbf{F}}\frac{\partial \breve{\psi}_{\mathrm{R}}(\bar{\mathbf{C}})}{\partial \bar{\mathbf{C}}} - \frac{1}{3}\left(\bar{\mathbf{C}} : \frac{\partial \breve{\psi}_{\mathrm{R}}(\bar{\mathbf{C}})}{\partial \bar{\mathbf{C}}}\right)\bar{\mathbf{F}}^{-\top}\right] + J\frac{\partial \psi_{\mathrm{R}}^{\mathrm{vol}}(J)}{\partial J}\mathbf{F}^{-\top}, \tag{1.9.15}$$

and

$$\mathbf{T} = 2\,J^{-1}\left[\bar{\mathbf{F}}\frac{\partial \breve{\psi}_{\mathrm{R}}(\bar{\mathbf{C}})}{\partial \bar{\mathbf{C}}}\bar{\mathbf{F}}^{\top} - \frac{1}{3}\left(\bar{\mathbf{C}} : \frac{\partial \breve{\psi}_{\mathrm{R}}(\bar{\mathbf{C}})}{\partial \bar{\mathbf{C}}}\right)\mathbf{1}\right] + \frac{\partial \psi_{\mathrm{R}}^{\mathrm{vol}}(J)}{\partial J}\mathbf{1}. \tag{1.9.16}$$

1.9.2 **Slightly compressible isotropic materials**

Specializing the separable energy (1.9.14) to the case of an *isotropic material*, the free energy (1.9.14) may alternatively be expressed, using a classical representation theorem (cf., e.g. Gurtin et al., 2010), as

$$\psi_{\mathrm{R}} = \underbrace{\bar{\psi}_{\mathrm{R}}(\bar{I}_1, \bar{I}_2)}_{\text{isochoric}} + \underbrace{\psi_{\mathrm{R}}^{\mathrm{vol}}(J)}_{\text{volumetric}}, \tag{1.9.17}$$

where

$$\bar{I}_1 = \operatorname{tr}\bar{\mathbf{C}} \qquad \text{and} \qquad \bar{I}_2 = \frac{1}{2}[(\operatorname{tr}\bar{\mathbf{C}})^2 - \operatorname{tr}(\bar{\mathbf{C}})] \tag{1.9.18}$$

are the principal invariants of $\bar{\mathbf{C}}$. Using (1.9.15) and (1.9.16) the Piola stress and the Cauchy stress may be obtained as

$$\mathbf{T}_{\mathrm{R}} = 2\,J^{-1/3}\left[\bar{\mathbf{F}}\frac{\partial \bar{\psi}_{\mathrm{R}}(\bar{I}_1, \bar{I}_2)}{\partial \bar{\mathbf{C}}} - \frac{1}{3}\left(\bar{\mathbf{C}} : \frac{\partial \bar{\psi}_{\mathrm{R}}(\bar{I}_1, \bar{I}_2)}{\partial \bar{\mathbf{C}}}\right)\bar{\mathbf{F}}^{-\top}\right] + J\,\frac{\partial \psi_{\mathrm{R}}^{\mathrm{vol}}(J)}{\partial J}\,\mathbf{F}^{-\top}, \tag{1.9.19}$$

and

$$\mathbf{T} = 2\,J^{-1}\left[\bar{\mathbf{F}}\frac{\partial\bar{\psi}_{\mathrm{R}}(\bar{I}_1,\bar{I}_2)}{\partial\bar{\mathbf{C}}}\bar{\mathbf{F}}^{\top} - \frac{1}{3}\left(\bar{\mathbf{C}} : \frac{\partial\bar{\psi}_{\mathrm{R}}(\bar{I}_1,\bar{I}_2)}{\partial\bar{\mathbf{C}}}\right)\mathbf{1}\right] + \frac{\partial\psi_{\mathrm{R}}^{\mathrm{vol}}(J)}{\partial J}\,\mathbf{1}. \tag{1.9.20}$$

Thus, since

$$\frac{\partial\bar{I}_1}{\partial\bar{\mathbf{C}}} = \mathbf{1} \qquad \text{and} \qquad \frac{\partial\bar{I}_2}{\partial\bar{\mathbf{C}}} = \bar{I}_1\mathbf{1} - \bar{\mathbf{C}},$$

we find that the Cauchy stress is given by

$$\mathbf{T} = 2\,J^{-1}\,\mathrm{dev}\left(\frac{\partial\bar{\psi}_{\mathrm{R}}(\bar{I}_1,\bar{I}_2)}{\partial\bar{I}_1}\,\bar{\mathbf{B}} + \frac{\partial\bar{\psi}_{\mathrm{R}}(\bar{I}_1,\bar{I}_2)}{\partial\bar{I}_2}\,(\bar{I}_1\mathbf{1} - \bar{\mathbf{B}})\,\bar{\mathbf{B}}\right) + \frac{\partial\psi_{\mathrm{R}}^{\mathrm{vol}}(J)}{\partial J}\,\mathbf{1}, \tag{1.9.21}$$

where $\mathrm{dev}\mathbf{A} \overset{\mathrm{def}}{=} \mathbf{A} - (1/3)(\mathrm{tr}\mathbf{A})\mathbf{1}$ is the deviatoric operator. Finally, using

$$\mathbf{T}_{\mathrm{R}} = J\,\mathbf{T}\mathbf{F}^{-\top} = J^{2/3}\mathbf{T}\bar{\mathbf{F}}^{-\top},$$

the Piola stress is given by

$$\mathbf{T}_{\mathrm{R}} = 2\,J^{-1/3}\,\mathrm{dev}\left(\frac{\partial\bar{\psi}_{\mathrm{R}}(\bar{I}_1,\bar{I}_2)}{\partial\bar{I}_1}\,\bar{\mathbf{F}} + \frac{\partial\bar{\psi}_{\mathrm{R}}(\bar{I}_1,\bar{I}_2)}{\partial\bar{I}_2}\,(\bar{I}_1\mathbf{1} - \bar{\mathbf{B}})\,\bar{\mathbf{F}}\right) + J\frac{\partial\psi_{\mathrm{R}}^{\mathrm{vol}}(J)}{\partial J}\,\mathbf{F}^{-\top}. \tag{1.9.22}$$

1.10 Neo-Hookean and Arruda–Boyce free energies for slightly compressible isotropic materials

The literature on special free-energy functions which have been proposed to model elastomeric materials is *vast*; for reviews see Boyce and Arruda (2000), Marckmann and Verron (2006), Beda (2014), and Anand and Govindjee (2020, Chapter 31). In this section we summarize two prominent free-energy functions which have found significant use in practice.

Specifically, for the remainder of this part we limit our attention to the classical Neo-Hookean and Arruda–Boyce free energies which neglect any dependence on the $\bar{I}_2$ invariant. We summarize these free energies and the resulting expressions for the Cauchy stress and the Piola stress below:[4]

[4] These models are the building blocks for the constitutive relations for "smart soft" materials such as electroactive and magnetoactive polymers which we will consider in later chapters of this book.

1. In elastomeric materials the major part of the isochoric free energy change $\bar{\psi}_{\mathrm{R}}(\bar{I}_1)$ arises from an entropic contribution. Let

$$\bar{\lambda} \stackrel{\text{def}}{=} \frac{1}{\sqrt{3}}\sqrt{\bar{I}_1} \tag{1.10.1}$$

define an *effective stretch*. Then, classical statistical mechanics models of rubber elasticity, cf., Treloar (1975), Arruda and Boyce (1993), and Anand et al. (2022, Chapter 22.A) provide the following *estimates* for the isochoric part of the free-energy change due to mechanical stretching:

(i) For small to moderate values of $\bar{\lambda}$, based on Gaussian statistics,

$$\bar{\psi}_{\mathrm{R}}(\bar{I}_1) = \frac{3}{2}G_0\,(\bar{\lambda}^2 - 1), \tag{1.10.2}$$

where

$$G_0 = Nk_B\vartheta \tag{1.10.3}$$

is the shear modulus, where k_B is Boltzmann's constant, N represents the number of polymer chains per unit reference volume, and ϑ the absolute temperature.

(ii) For larger values of $\bar{\lambda}$ it is necessary to use non-Gaussian statistics to account for the limited extensibility of the polymer chains, and for these circumstances the free-energy change is given by

$$\bar{\psi}_{\mathrm{R}}(\bar{I}_1) = G_0\lambda_L^2\left[\left(\frac{\bar{\lambda}}{\lambda_L}\right)\beta + \ln\left(\frac{\beta}{\sinh\beta}\right) - \left(\frac{1}{\lambda_L}\right)\beta_0 - \ln\left(\frac{\beta_0}{\sinh\beta_0}\right)\right], \tag{1.10.4}$$

with

$$\beta \stackrel{\text{def}}{=} \mathcal{L}^{-1}\left(\frac{\bar{\lambda}}{\lambda_L}\right), \qquad \text{and} \qquad \beta_0 \stackrel{\text{def}}{=} \mathcal{L}^{-1}\left(\frac{1}{\lambda_L}\right), \tag{1.10.5}$$

where $\mathcal{L}^{-1}$ is the inverse of the Langevin function

$$\mathcal{L}(\bullet) = \coth(\bullet) - (\bullet)^{-1}.$$

This functional form for the change in free energy involves two material parameters:

- $G_0 = Nk_B\vartheta$, the ground-state shear modulus, and
- λ_L, the network locking stretch.

2. As with the incompressible free energy, the literature contains a wide array of options for the volumetric energy $\psi_{\mathrm{R}}^{v}(J)$. Here we follow Anand (1996) and assume that[5]

$$\psi_{\mathrm{R}}^{\mathrm{vol}}(J) = \frac{1}{2}K(\ln J)^2 \tag{1.10.6}$$

where K is a bulk modulus.

REMARK

The following Padé approximation to $\mathcal{L}^{-1}(z)$ considerably simplifies numerical calculations (Cohen, 1991),

$$\mathcal{L}^{-1}(z) \approx z\,\frac{3 - z^2}{1 - z^2} \qquad \text{with} \qquad z \overset{\text{def}}{=} \frac{\bar{\lambda}}{\lambda_L}. \tag{1.10.7}$$

1.10.1 Cauchy stress

Using (1.9.21) we obtain the following expressions for the Cauchy stress from the Neo-Hookean and Arruda–Boyce energies:

- Neo-Hookean:

$$\mathbf{T} = J^{-1}G_0\,\bar{\mathbf{B}}_0 + K\left(\frac{\ln J}{J}\right)\mathbf{1}. \tag{1.10.8}$$

For later convenience we rewrite this as

$$\mathbf{T} = J^{-1}G_0\,\bar{\mathbf{B}}_0 - p\mathbf{1}, \tag{1.10.9}$$

where

$$p \overset{\text{def}}{=} -\frac{1}{3}\operatorname{tr}\mathbf{T} = -K\left(\frac{\ln J}{J}\right), \tag{1.10.10}$$

represents a pressure.

- Arruda–Boyce:

$$\mathbf{T} = J^{-1}\,G\,\bar{\mathbf{B}}_0 - p\mathbf{1}, \qquad \text{where} \qquad G \overset{\text{def}}{=} G_0\left(\frac{\lambda_L}{3\bar{\lambda}}\right)\mathcal{L}^{-1}\left(\frac{\bar{\lambda}}{\lambda_L}\right) \tag{1.10.11}$$

is a *generalized shear modulus*. Note that since $\mathcal{L}^{-1}(\bar{\lambda}/\lambda_L) \to \infty$ as $\bar{\lambda} \to \lambda_L$, the generalized modulus $G \to \infty$ as $\bar{\lambda} \to \lambda_L$, *a situation which reflects the limited chain-extensibility of elastomeric materials*. Also, it can be shown that $G \to G_0$ as $\lambda_L \to \infty$, and we recover the Neo-Hookean result.

[5] Some other common choices are discussed in Appendix 1.B.

1.10.2 Piola stress

Next, using $\mathbf{T}_{\mathrm{R}} = J\mathbf{T}\mathbf{F}^{-\top}$, the corresponding expressions for the Piola stress are:

- Neo-Hookean:

$$\mathbf{T}_{\mathrm{R}} = J^{-2/3} G_0 \left(\mathbf{F} - \frac{1}{3}(\operatorname{tr}\mathbf{C})\mathbf{F}^{-\top} \right) - Jp\,\mathbf{F}^{-\top}. \tag{1.10.12}$$

- Arruda–Boyce:

$$\mathbf{T}_{\mathrm{R}} = J^{-2/3} G \left(\mathbf{F} - \frac{1}{3}(\operatorname{tr}\mathbf{C})\mathbf{F}^{-\top} \right) - Jp\,\mathbf{F}^{-\top}. \tag{1.10.13}$$

1.11 Governing partial differential equation. Boundary conditions

Neglecting inertia, the local force balance requires that

$$\left.\begin{aligned} &\operatorname{Div}\mathbf{T}_{\mathrm{R}} + \mathbf{b}_{0\mathrm{R}} = \mathbf{0}, \\ &\frac{p}{K} + \left(\frac{\ln J}{J}\right) = 0, \end{aligned}\right\} \quad \text{in } \mathcal{B} \tag{1.11.1}$$

with $\mathbf{T}_{\mathrm{R}}$ given by

$$\begin{aligned} \mathbf{T}_{\mathrm{R}} &= J\,\mathbf{T}\mathbf{F}^{-\top}, \\ \mathbf{T} &= J^{-1}\,G\,\bar{\mathbf{B}}_0 - p\mathbf{1}, \\ G &= G_0 \left(\frac{\lambda_L}{3\bar{\lambda}}\right) \mathcal{L}^{-1}\left(\frac{\bar{\lambda}}{\lambda_L}\right). \end{aligned} \tag{1.11.2}$$

In writing (1.11.1) we have *split* the standard equation of equilibrium into two equations because this helps in numerically dealing with the near-incompressibility of the material. In our numerical finite element implementation of the theory, rather than just using the displacement $\mathbf{u}$, as in standard displacement-based finite element procedures, we will use $(\mathbf{u}, p)$ as separate degrees of freedom.

With $\partial\mathcal{B}_{\mathrm{u}}$ and $\partial\mathcal{B}_{\mathrm{t_R}}$ denoting complementary subsurfaces of the boundary $\partial\mathcal{B}$ of the body $\mathcal{B}$, we consider a pair of possible boundary conditions:

$$\begin{aligned} \mathbf{u} &= \bar{\mathbf{u}} \quad \text{on } \partial\mathcal{B}_{\mathrm{u}}, \\ \mathbf{T}_{\mathrm{R}}\mathbf{n}_{\mathrm{R}} &= \bar{\mathbf{t}}_{\mathrm{R}} \quad \text{on } \partial\mathcal{B}_{\mathrm{t_R}}. \end{aligned} \tag{1.11.3}$$

In the boundary conditions above $\bar{\mathbf{u}}$ and $\bar{\mathbf{t}}_{\mathrm{R}}$ are prescribed displacement and tractions, both functions of $(\mathbf{X})$. Standard initial conditions involve the specifications

$$\mathbf{u}(\mathbf{X}, 0) = \mathbf{u}_0(\mathbf{X}) \quad \text{in } \mathcal{B}, \tag{1.11.4}$$

with $\mathbf{u}_0$ a prescribed function of $\mathbf{X} \in \mathcal{B}$.

REMARK

As discussed earlier, another form of boundary condition arises when one specifies the surface traction $\mathbf{T}\mathbf{n}$ on the deformed surface

$$\partial\mathcal{B}_{\mathrm{t}}(t) = \chi(\partial\mathcal{B}_{\mathrm{t}_{\mathrm{R}}}, t).$$

A simple example of this type of condition arises when one considers the effects of a uniform pressure, $p_0\mathbf{n}$:

$$\mathbf{T}\mathbf{n} = -p_0\mathbf{n} \quad \text{on } \partial\mathcal{B}_{\mathrm{t}}(t). \tag{1.11.5}$$

Using Nanson's relation (1.2.6), viz.

$$\mathbf{n}\, da = J\mathbf{F}^{-\top}\mathbf{n}_{\mathrm{R}}\, da_{\mathrm{R}} = \mathbf{F}^{\mathrm{C}}\,\mathbf{n}_{\mathrm{R}}\, da_{\mathrm{R}}, \tag{1.11.6}$$

the boundary condition (1.11.5) may be expressed as

$$\mathbf{T}_{\mathrm{R}}\mathbf{n}_{\mathrm{R}} = \underbrace{-p_0\mathbf{F}^{\mathrm{C}}\,\mathbf{n}_{\mathrm{R}}}_{\bar{\mathbf{t}}_{\mathrm{R}}} \quad \text{on } \partial\mathcal{B}_{\mathrm{t}_{\mathrm{R}}}. \tag{1.11.7}$$

1.A Derivatives of the invariants of a tensor

In this appendix we verify the computation of the derivatives of the principal invariants of a tensor $\mathbf{A}$:

$$I_1(\mathbf{A}) = \mathrm{tr}\,\mathbf{A}, \qquad I_2(\mathbf{A}) = \frac{1}{2}[(\mathrm{tr}\,\mathbf{A})^2 - \mathrm{tr}\,(\mathbf{A}^2)], \qquad I_3(\mathbf{A}) = \det\mathbf{A}.$$

1. Using the definition of the derivative of a scalar, vector, or tensor field, the derivative of $I_1(\mathbf{A}) = \mathrm{tr}\,\mathbf{A}$ is

$$\left(\frac{\partial I_1(\mathbf{A})}{\partial \mathbf{A}}\right) : \mathbf{H} = \frac{d}{d\alpha}\,[I_1(\mathbf{A} + \alpha\mathbf{H})]\,|_{\alpha=0} = \mathrm{tr}\,\mathbf{H} = \mathbf{1} : \mathbf{H}.$$

So

$$\frac{\partial I_1(\mathbf{A})}{\partial \mathbf{A}} = \mathbf{1}.$$

2. Next, we calculate the derivative of $I_2(\mathbf{A}) = \frac{1}{2}\left[(\text{tr}\,\mathbf{A})^2 - \text{tr}(\mathbf{A}^2)\right]$. Since

$$
\begin{aligned}
I_2(\mathbf{A}+\alpha\mathbf{H}) &= \frac{1}{2}\left[(\text{tr}(\mathbf{A}+\alpha\mathbf{H}))^2 - \text{tr}((\mathbf{A}+\alpha\mathbf{H})^2)\right] \\
&= \frac{1}{2}\left[(\text{tr}\mathbf{A}+\alpha\text{tr}\,\mathbf{H})^2 - \text{tr}((\mathbf{A}+\alpha\mathbf{H})(\mathbf{A}+\alpha\mathbf{H}))\right] \\
&= \frac{1}{2}\big[(\text{tr}\mathbf{A})^2 + 2\alpha\text{tr}\,\mathbf{A}\text{tr}\,\mathbf{H} + \alpha^2(\text{tr}\,\mathbf{H})^2 - \text{tr}(\mathbf{A}^2) + \alpha\text{tr}(\mathbf{A}\mathbf{H}+\mathbf{H}\mathbf{A}) \\
&\quad + \alpha^2\text{tr}(\mathbf{H}^2)\big],
\end{aligned}
$$

we have

$$
\begin{aligned}
\left(\frac{\partial I_2(\mathbf{A})}{\partial \mathbf{A}}\right) : \mathbf{H} = \frac{d}{d\alpha}\left[I_2(\mathbf{A}+\alpha\mathbf{H})\right]\big|_{\alpha=0} &= (\text{tr}\,\mathbf{A})(\text{tr}\,\mathbf{H}) - \text{tr}(\mathbf{A}\mathbf{H}), \\
&= ((\text{tr}\,\mathbf{A})\mathbf{1} - \mathbf{A}^\top) : \mathbf{H}.
\end{aligned}
$$

So

$$
\frac{\partial I_2(\mathbf{A})}{\partial \mathbf{A}} = (\text{tr}\,\mathbf{A})\mathbf{1} - \mathbf{A}^\top.
$$

3. Recall that the determinant of the tensor $(\mathbf{A} - \omega\mathbf{1})$ admits the representation

$$
\det(\mathbf{A} - \omega\mathbf{1}) = -\omega^3 + I_1(\mathbf{A})\omega^2 - I_2(\mathbf{A})\omega + I_3(\mathbf{A}).
$$

Then, with $\omega = -1$

$$
\det(\mathbf{1} + \mathbf{A}) = 1 + \text{tr}(\mathbf{A}) + o(|\mathbf{A}|),
$$

Thus, for $\mathbf{A}$ invertible and $\mathbf{H}$ a tensor,

$$
\begin{aligned}
\det(\mathbf{A}+\alpha\mathbf{H}) &= \det\left[(\mathbf{1}+\alpha\mathbf{H}\mathbf{A}^{-1})\mathbf{A}\right], \\
&= (\det\mathbf{A})\det(\mathbf{1}+\alpha\mathbf{H}\mathbf{A}^{-1}), \\
&= (\det\mathbf{A})[1+\alpha\text{tr}(\mathbf{H}\mathbf{A}^{-1}) + o(|\alpha\mathbf{H})|], \\
&= \det(\mathbf{A}) + \alpha(\det\mathbf{A})\text{tr}(\mathbf{H}\mathbf{A}^{-1}) + o(|\alpha\mathbf{H}|).
\end{aligned}
$$

Therefore,

$$
\begin{aligned}
\left(\frac{\partial\det(\mathbf{A})}{\partial \mathbf{A}}\right) : \mathbf{H} = \frac{d}{d\alpha}\left[I_3(\mathbf{A}+\alpha\mathbf{H})\right]\big|_{\alpha=0} &= (\det\mathbf{A})\text{tr}(\mathbf{H}\mathbf{A}^{-1}) \\
&= \left((\det\mathbf{A})\mathbf{A}^{-\top}\right) : \mathbf{H}.
\end{aligned}
$$

Hence

$$
\frac{\partial\det(\mathbf{A})}{\partial \mathbf{A}} = (\det\mathbf{A})\mathbf{A}^{-\top}.
$$

1.B Models for the volumetric energy

As with the incompressible free energy, the literature contains a wide array of options for the volumetric energy term $\psi^v_{\mathrm{R}}(J)$. A few common models are given in Table 1.1, in which K is the ground-state bulk modulus in all the models and α is an additional dimensionless parameter (cf., e.g., Anand and Govindjee, 2020, Section 31.7).

Table 1.1 Sample volumetric energy expressions and their derivatives.

Model	$\psi^v_{\mathrm{R}}(J)$	$\dfrac{\partial \psi^v_{\mathrm{R}}}{\partial J}$
1	$\frac{1}{2}K(J-1)^2$	$K(J-1)$
2	$\frac{1}{2}K\,(\ln J)^2$	$K\frac{\ln J}{J}$
3	$\frac{1}{\alpha^2}K\,(\cosh(\alpha(J-1))-1)$	$K\frac{1}{\alpha}\sinh(\alpha(J-1))$
4	$\frac{1}{4}K\,(J^2-1-2\ln J)$	$K\frac{1}{2}\left(J-\frac{1}{J}\right)$
5	$\frac{1}{2\alpha}K\exp\left(\alpha\,(\ln J)^2\right)$	$K\frac{\ln J}{J}\exp\left(\alpha\,(\ln J)^2\right)$

REMARKS

1. Model 1 is very common and easy to work with. It provides a linear pressure–volume relation. However, the material may be compressed to zero volume ($J=0$) with finite energy, which is not physically realistic, and this can cause difficulties in numerical finite element computations. Nonetheless, for materials with large bulk modulus at modest pressures, it works well.
2. Model 2 repairs the problem with Model 1 in that it requires infinite energy to compress the material to zero volume. However, at high dilations ($J \to \infty$) the accompanying pressure goes to zero ($p \to 0$) instead of ($p \to -\infty$). Thus caution should be exercised when using this model in the presence of large positive dilation.
3. Model 3 was proposed in Bischoff et al. (2001) to better model observed pressure–volume nonlinearities, which it does nicely. However, like Model 1, it allows the material to be compressed to zero volume with finite energy.
4. Model 4 was proposed in Simo and Taylor (1991) and is a particularization of a more general model due to Ogden (1972). Like Model 2, it requires infinite energy to compress material to zero volume. However, unlike Model 2, here $J \to \infty$ implies $p \to -\infty$. These two features make it quite suitable for numerical finite element computations.
5. Lastly Model 5 (see e.g. Montella et al. (2016)) has the same limiting behaviors as Model 4, but can model higher degrees of nonlinearity in the pressure–volume response.

1.C Linearization of the Piola stress relation from the Neo-Hookean model for small deformations

Let $\mathbf{H} \overset{\text{def}}{=} \nabla\mathbf{u}$ denote the displacement gradient. Deformations are considered to be *small* when the displacement gradient is small everywhere in the body, $|\mathbf{H}| \ll 1$. We are interested here in the linearized "small deformation" form of the constitutive equation for the Piola stress, eq. (1.10.12), for the compressible Neo-Hookean model, viz.

$$\mathbf{T}_{\mathrm{R}} = J^{-2/3}G\left(\mathbf{F} - \frac{1}{3}(\operatorname{tr}\mathbf{C})\mathbf{F}^{-\top}\right) - J\,p\,\mathbf{F}^{-\top} \qquad \text{with} \qquad p = -K\left(\frac{\ln J}{J}\right), \tag{1.C.1}$$

where we have written the ground-state shear modulus $G_0 = N k_B \vartheta$ simply as G. We show below that the linearized version of (1.C.1) for small deformations is

$$\mathbf{T} = 2G\mathbf{E}_0 + K(\operatorname{tr}\mathbf{E})\,\mathbf{1}, \tag{1.C.2}$$

which is precisely the classical stress–strain relation for isotropic linear elasticity.[6] Thus, operationally,

- the compressible Neo-Hookean model (1.C.1) may be considered a generalization of the classic model (1.C.2) of isotropic linear elasticity to moderately large deformations using the ground-state values of the shear and bulk moduli G and K.[7]

The implication of this result is that any numerical implementation of a theory based on (1.C.1) for compressible elastomeric materials, may also be applied to the simulation of the mechanical response of other materials such as metals and ceramics—where the elasticity is energetic and not entropic in nature—for moderately large elastic deformations, within of course the limits of onset of plasticity, fracture, or other inelastic phenomena in these materials.

Verification of the result (1.C.2):
First, given any function $f(\mathbf{Y})$, where f can take on scalar, vector, or tensor values and $\mathbf{Y}$ can be a scalar, vector, or tensor argument, the linear approximation of f at a point $\mathbf{Y}_o$ is given by

$$\operatorname{Lin}_{\mathbf{Y}_o} f(\mathbf{Y}) = f(\mathbf{Y}_o) + \left.\frac{d}{d\alpha} f\left(\mathbf{Y}_o + \alpha(\mathbf{Y} - \mathbf{Y}_o)\right)\right|_{\alpha=0}. \tag{1.C.3}$$

This expression is nothing more than the first two terms of the Taylor series expansion for f at $\mathbf{Y}_o$. When linearizing kinematic expressions to obtain their small displacement gradient

[6] Recall that in (1.C.2) G is a phenomenological parameter representing the shear modulus of an isotropic linear elastic material which is not necessarily associated with a statistical mechanical interpretation of $G = N k_B \vartheta$ related to changes in entropy.

[7] The Neo-Hookean generalization (1.C.1) of the linear elasticity relation (1.C.2) is very different with the commonly used generalization of the small strain relation (1.C.2) to large strains using the second Piola stress $\mathbf{T}_{\mathrm{RR}}$ and the Green strain $\mathbf{E}^G \overset{\text{def}}{=} (1/2)(\mathbf{C} - \mathbf{1})$, viz.

$$\mathbf{T}_{\mathrm{RR}} = 2G\mathbf{E}_0^G + K(\operatorname{tr}\mathbf{E}^G)\,\mathbf{1}.$$

counterparts, we apply this definition with $\mathbf{Y} = \mathbf{H}$ and $\mathbf{Y}_o = \mathbf{0}$, so that

$$\mathrm{Lin}_{\mathbf{0}} f(\mathbf{H}) = f(\mathbf{0}) + \frac{d}{d\alpha} f(\alpha\mathbf{H})\bigg|_{\alpha=0}. \tag{1.C.4}$$

Thus, the linearization of the right Cauchy–Green tensor $\mathbf{C}$ for small displacement gradient is given by

$$\begin{aligned}
\mathrm{Lin}_{\mathbf{0}}\mathbf{C}(\mathbf{H}) &= \mathbf{C}(\mathbf{0}) + \frac{d}{d\alpha}\mathbf{C}(\alpha\mathbf{H})\bigg|_{\alpha=0}, \\
&= \mathbf{1} + \frac{d}{d\alpha}\left[(\mathbf{1}+\alpha\mathbf{H})^{\top}(\mathbf{1}+\alpha\mathbf{H})\right]\bigg|_{\alpha=0}, \\
&= \mathbf{1} + [\mathbf{H}+\mathbf{H}^{\top}].
\end{aligned} \tag{1.C.5}$$

Additional linearizations which we shall use below are:

Deformation gradient	$\mathrm{Lin}_{\mathbf{0}}\mathbf{F}(\mathbf{H}) = \mathbf{1}+\mathbf{H}$	(1.C.6)
Inverse of deformation gradient	$\mathrm{Lin}_{\mathbf{0}}\mathbf{F}^{-1}(\mathbf{H}) = \mathbf{1}-\mathbf{H}$	(1.C.7)
Inverse transpose of deformation gradient	$\mathrm{Lin}_{\mathbf{0}}\mathbf{F}^{-\top}(\mathbf{H}) = \mathbf{1}-\mathbf{H}^{\top}$	(1.C.8)
Jacobian	$\mathrm{Lin}_{\mathbf{0}}J(\mathbf{H}) = 1+\mathrm{tr}\,\mathbf{H}.$	(1.C.9)

We now determine the "linear" form of the constitutive equation (1.C.1):

1. First,

$$\begin{aligned}
-\frac{1}{3}(\mathrm{tr}\,\mathbf{C})\mathbf{F}^{-\top} &\approx -\frac{1}{3}\mathrm{tr}(\mathbf{1}+\mathbf{H}+\mathbf{H}^{\top})(\mathbf{1}-\mathbf{H}^{\top}) \\
&\approx -\frac{1}{3}(3+\mathrm{tr}(\mathbf{H}+\mathbf{H}^{\top}))(\mathbf{1}-\mathbf{H}^{\top}) \\
&\approx -(1+\frac{1}{3}\mathrm{tr}(\mathbf{H}+\mathbf{H}^{\top}))(\mathbf{1}-\mathbf{H}^{\top}) \\
&\approx -\mathbf{1}+\mathbf{H}^{\top}-\frac{1}{3}\mathrm{tr}(\mathbf{H}+\mathbf{H}^{\top})\mathbf{1},
\end{aligned}$$

and hence

$$\begin{aligned}
\mathbf{F}-\frac{1}{3}(\mathrm{tr}\,\mathbf{C})\mathbf{F}^{-\top} &\approx (\mathbf{1}+\mathbf{H})-\mathbf{1}+\mathbf{H}^{\top}-\frac{1}{3}\mathrm{tr}(\mathbf{H}+\mathbf{H}^{\top})\mathbf{1} \\
&\approx \mathbf{H}+\mathbf{H}^{\top}-\frac{1}{3}\mathrm{tr}(\mathbf{H}+\mathbf{H}^{\top})\mathbf{1} \\
&\approx 2\left(\frac{1}{2}(\mathbf{H}+\mathbf{H}^{\top})-\frac{1}{3}\mathrm{tr}(\mathbf{H})\mathbf{1}\right),
\end{aligned}$$

which upon introducing the small strain tensor $\mathbf{E}$, its trace $\mathrm{tr}\,\mathbf{E}$, and deviator $\mathbf{E}_0$ defined by

$$\mathbf{E} \stackrel{\text{def}}{=} \frac{1}{2}(\mathbf{H}+\mathbf{H}^{\top}) \quad \text{with} \quad \mathrm{tr}\,\mathbf{E} = \mathrm{tr}\,\mathbf{H} \quad \text{and} \quad \mathbf{E}_0 \stackrel{\text{def}}{=} \mathbf{E}-\frac{1}{3}(\mathrm{tr}\,\mathbf{E})\mathbf{1}, \tag{1.C.10}$$

respectively, gives

$$\mathbf{F} - \frac{1}{3}(\operatorname{tr}\mathbf{C})\mathbf{F}^{-\top} \approx 2\mathbf{E}_0. \tag{1.C.11}$$

Thus, together with the approximation

$$J^{-2/3} \approx 1 - \frac{2}{3}\operatorname{tr}\mathbf{H},$$

we obtain

$$J^{-2/3}G\left(\mathbf{F} - \frac{1}{3}(\operatorname{tr}\mathbf{C})\mathbf{F}^{-\top}\right) \approx 2\,G\mathbf{E}_0. \tag{1.C.12}$$

2. Next, consider the term

$$\begin{aligned}\frac{\ln J}{J} &= \ln(1+\operatorname{tr}\mathbf{H})(1+\operatorname{tr}\mathbf{H})^{-1}\\ &\approx \operatorname{tr}\mathbf{H}(1-\operatorname{tr}\mathbf{H}) \approx \operatorname{tr}\mathbf{H}\\ &\approx \operatorname{tr}\mathbf{E},\end{aligned} \tag{1.C.13}$$

so

$$p \approx -K\operatorname{tr}\mathbf{E},$$

and hence

$$Jp\mathbf{F}^{-\top} \approx (1+\operatorname{tr}\mathbf{H})(-K(\operatorname{tr}\mathbf{E}))(1-\mathbf{H}^{\top}) \approx -K(\operatorname{tr}\mathbf{E})\,\mathbf{1}. \tag{1.C.14}$$

3. Thus, using (1.C.12) and (1.C.14) we obtain

$$J^{-2/3}G\left(\mathbf{F} - \frac{1}{3}(\operatorname{tr}\mathbf{C})\mathbf{F}^{-\top}\right) - Jp\mathbf{F}^{-\top} \approx 2\,G\mathbf{E}_0 + K(\operatorname{tr}\mathbf{E})\,\mathbf{1}. \tag{1.C.15}$$

4. Similarly,

$$\mathbf{T}_{\mathrm{R}} = J\mathbf{T}\mathbf{F}^{-\top} = (1+\operatorname{tr}\mathbf{H})\mathbf{T}(1-\mathbf{H}^{\top}) \approx \mathbf{T}. \tag{1.C.16}$$

5. Thus the linearized form of the Piola stress relation (1.C.1) for the Neo-Hookean material is

$$\mathbf{T} = 2G\mathbf{E}_0 + K(\operatorname{tr}\mathbf{E})\,\mathbf{1}, \tag{1.C.17}$$

which is the desired result (1.C.2).

2 Numerical implementation of finite elasticity

2.1 Weak form of the governing partial differential equation

Consider a test field $\mathbf{u}_{\text{test}}(\mathbf{X})$ in $\mathcal{B}$ which satisfies $\mathbf{u}_{\text{test}}(\mathbf{X}) = \mathbf{0}$ on $\partial\mathcal{B}_{\mathbf{u}}$. If the balance equation $\operatorname{Div}\mathbf{T}_{\text{R}} + \mathbf{b}_{0\text{R}} = \mathbf{0}$ on $\mathcal{B}$ and the traction condition $(\bar{\mathbf{t}}_{\text{R}} - \mathbf{T}_{\text{R}}\mathbf{n}_{\text{R}}) = \mathbf{0}$ on $\partial\mathcal{B}_{\mathbf{t}_{\text{R}}}$ are satisfied, then clearly

$$\int_{\mathcal{B}} (\operatorname{Div}\mathbf{T}_{\text{R}}) \cdot \mathbf{u}_{\text{test}}\, dv_{\text{R}} + \int_{\mathcal{B}} \mathbf{b}_{0\text{R}} \cdot \mathbf{u}_{\text{test}}\, dv_{\text{R}} + \int_{\partial\mathcal{B}_{\mathbf{t}_{\text{R}}}} (\bar{\mathbf{t}}_{\text{R}} - \mathbf{T}_{\text{R}}\mathbf{n}_{\text{R}}) \cdot \mathbf{u}_{\text{test}}\, da_{\text{R}} = 0, \tag{2.1.1}$$

or equivalently

$$\int_{\partial\mathcal{B}_{\mathbf{t}_{\text{R}}}} \mathbf{T}_{\text{R}}\mathbf{n}_{\text{R}} \cdot \mathbf{u}_{\text{test}}\, da_{\text{R}} - \int_{\mathcal{B}} \operatorname{Div}\mathbf{T}_{\text{R}} \cdot \mathbf{u}_{\text{test}}\, dv_{\text{R}} - \int_{\mathcal{B}} \mathbf{b}_{0\text{R}} \cdot \mathbf{u}_{\text{test}}\, dv_{\text{R}} - \int_{\partial\mathcal{B}_{\mathbf{t}_{\text{R}}}} \bar{\mathbf{t}}_{\text{R}} \cdot \mathbf{u}_{\text{test}}\, da_{\text{R}} = 0 \tag{2.1.2}$$

holds for all admissible test functions $\mathbf{u}_{\text{test}}$. Thus, using $\mathbf{u}_{\text{test}} = \mathbf{0}$ on $\partial\mathcal{B}_{\mathbf{u}}$ we may extend the first boundary integral in the equation above to the whole boundary $\partial\mathcal{B}$, and we may rewrite this equation as

$$\int_{\partial\mathcal{B}} \mathbf{T}_{\text{R}}\mathbf{n}_{\text{R}} \cdot \mathbf{u}_{\text{test}}\, da_{\text{R}} - \int_{\mathcal{B}} \operatorname{Div}\mathbf{T}_{\text{R}} \cdot \mathbf{u}_{\text{test}}\, dv_{\text{R}} - \int_{\mathcal{B}} \mathbf{b}_{0\text{R}} \cdot \mathbf{u}_{\text{test}}\, dv_{\text{R}} - \int_{\partial\mathcal{B}_{\mathbf{t}_{\text{R}}}} \bar{\mathbf{t}}_{\text{R}} \cdot \mathbf{u}_{\text{test}}\, da_{\text{R}} = 0. \tag{2.1.3}$$

Next, by the divergence theorem

$$\int_{\partial\mathcal{B}} \mathbf{T}_{\text{R}}\mathbf{n}_{\text{R}} \cdot \mathbf{u}_{\text{test}}\, da_{\text{R}} = \int_{\partial\mathcal{B}} (\mathbf{T}_{\text{R}}^{\top}\mathbf{u}_{\text{test}}) \cdot \mathbf{n}\, da_{\text{R}} = \int_{\mathcal{B}} \operatorname{Div}\mathbf{T}_{\text{R}} \cdot \mathbf{u}_{\text{test}}\, dv_{\text{R}} + \int_{\mathcal{B}} \mathbf{T}_{\text{R}} \colon \nabla\mathbf{u}_{\text{test}}\, dv_{\text{R}}, \tag{2.1.4}$$

use of which in (2.1.3) gives

$$\int_{\mathcal{B}} \mathbf{T}_{\text{R}} \colon \nabla\mathbf{u}_{\text{test}}\, dv_{\text{R}} - \int_{\mathcal{B}} \mathbf{b}_{0\text{R}} \cdot \mathbf{u}_{\text{test}}\, dv_{\text{R}} - \int_{\partial\mathcal{B}_{\mathbf{t}_{\text{R}}}} \bar{\mathbf{t}}_{\text{R}} \cdot \mathbf{u}_{\text{test}}\, da_{\text{R}} = 0. \tag{2.1.5}$$

Introduction to Coupled Theories in Solid Mechanics. Lallit Anand, Eric M. Stewart, Shawn A. Chester, Oxford University Press.
 DOI: 10.1093/oso/9780198986218.003.0003

Hence, if the balance equation $\operatorname{Div} \mathbf{T}_{\mathrm{R}} + \mathbf{b}_{0\mathrm{R}} = \mathbf{0}$ in $\mathcal{B}$ and the traction condition $(\bar{\mathbf{t}}_{\mathrm{R}} - \mathbf{T}_{\mathrm{R}}\mathbf{n}_{\mathrm{R}}) = \mathbf{0}$ on $\partial\mathcal{B}_{\mathbf{t}_{\mathrm{R}}}$ are satisfied, then the weak form (2.1.5) holds for all admissible test fields $\mathbf{u}_{\text{test}}$ which vanish on $\partial\mathcal{B}_{\mathbf{u}}$.

In addition, we consider an arbitrary test field $p_{\text{test}}(\mathbf{X})$ in $\mathcal{B}$, and take the weak form of the pressure equation

$$\frac{p}{K} + \left(\frac{\ln J}{J}\right) = 0,$$

as

$$\int_{\mathcal{B}} \left[\frac{p}{K} + \frac{\ln J}{J}\right] p_{\text{test}}\, dv_{\mathrm{R}} = 0. \tag{2.1.6}$$

Summarizing, the weak forms of the governing equations are

$$\begin{aligned} \int_{\mathcal{B}} \mathbf{T}_{\mathrm{R}} : \nabla \mathbf{u}_{\text{test}}\, dv_{\mathrm{R}} - \int_{\mathcal{B}} \mathbf{b}_{0\mathrm{R}} \cdot \mathbf{u}_{\text{test}}\, dv_{\mathrm{R}} - \int_{\partial\mathcal{B}_{\mathbf{t}_{\mathrm{R}}}} \bar{\mathbf{t}}_{\mathrm{R}} \cdot \mathbf{u}_{\text{test}}\, da_{\mathrm{R}} &= 0, \\ \int_{\mathcal{B}} \left[\frac{p}{K} + \frac{\ln J}{J}\right] p_{\text{test}}\, dv_{\mathrm{R}} &= 0. \end{aligned} \tag{2.1.7}$$

2.2 **Solution procedure in the FEniCS code**

Briefly, in FEniCS we use the following degrees of freedom

$$\{\mathbf{u},\, p\} \tag{2.2.1}$$

to calculate the *residuals* which correspond to the weak forms of the governing equations, viz.

$$\begin{aligned} R_{\mathbf{u}} &= \int_{\mathcal{B}} \mathbf{T}_{\mathrm{R}} : \nabla \mathbf{u}_{\text{test}}\, dv_{\mathrm{R}} - \int_{\mathcal{B}} \mathbf{b}_{0\mathrm{R}} \cdot \mathbf{u}_{\text{test}}\, dv_{\mathrm{R}} - \int_{\partial\mathcal{B}_{\mathbf{t}_{\mathrm{R}}}} \bar{\mathbf{t}}_{\mathrm{R}} \cdot \mathbf{u}_{\text{test}}\, da_{\mathrm{R}}, \\ R_{p} &= \int_{\mathcal{B}} \left[\frac{p}{K} + \frac{\ln J}{J}\right] p_{\text{test}}\, dv_{\mathrm{R}}. \end{aligned} \tag{2.2.2}$$

FEniCS then differentiates these residuals with respect to the separate degrees of freedom to obtain *tangents*. At each time step in an incremental time-stepping procedure, FEniCS uses the residuals and tangents in a global Newton–Raphson solver to determine the values of the degrees of freedom $\{\mathbf{u}, p\}$ for which the residuals are zero to within some small numerical tolerance, and which therefore satisfy the governing equations as well as the initial and boundary conditions in a weak sense.

REMARK

A mixed finite element method is utilized to solve for the displacement $\mathbf{u}$ and pressure p, concurrently. To avoid numerical instabilities we use an element known as the

- "Taylor–Hood" element which *combines a continuous piecewise quadratic interpolation for the displacement and a continuous piecewise linear interpolation for the pressure.*

To numerically enforce the near-incompressibility, a typical value of $K \approx 10^3 G_0$ is used. This approach for dealing with the "near-incompressibility" is a *pragmatic* one. It is known to work well in the context of solving the Navier–Stokes equations for incompressible fluids (Langtangen and Logg, 2017, p. 62). We adopt it here as well for solid mechanics.

The literature on numerically dealing with the near-incompressibility of elastomeric materials is vast. For references to the literature, and detailed discussions see, for example, Häggblad and Sundberg (1983), Sussman and Bathe (1987), Bathe (1996), Bonet and Wood (1997), Wriggers (2016), Schröder et al. (2017), and Kadapa and Hossain (2022).

2.3 Algorithm for calculation of the residuals

The algorithm used in our FEniCS code to calculate the residuals in (2.2.2) is summarized below.

1. Begin the step with known values of

$$\{\mathbf{u}_{\text{old}},\, p_{\text{old}}\}, \tag{2.3.1}$$

   where for the first step these quantities are given by the initial conditions for $\mathbf{u} = \mathbf{0}$ and $p = 0$.

2. Calculate necessary kinematical quantities for the current step as,

$$\begin{aligned} &\mathbf{F} = \mathbf{1} + \nabla\,\mathbf{u}, \quad J = \det \mathbf{F}, \quad \mathbf{C} = \mathbf{F}^{\top}\mathbf{F}, \quad \mathbf{B} = \mathbf{F}\mathbf{F}^{\top}, \\ &\bar{\mathbf{C}} = J^{-2/3}\,\mathbf{C}, \quad \bar{\mathbf{B}} = J^{-2/3}\,\mathbf{B}, \\ &\bar{\lambda} = \sqrt{(\operatorname{tr}\bar{\mathbf{C}})/3} = \sqrt{(\operatorname{tr}\bar{\mathbf{B}})/3}. \end{aligned} \tag{2.3.2}$$

3. Calculate an intermediate factor ζ for the generalized shear modulus G in the Arruda–Boyce model using the Pade approximation (1.10.7),

$$\begin{aligned} &\beta \overset{\text{def}}{=} \mathcal{L}^{-1}\left(\frac{\bar{\lambda}}{\lambda_L}\right) \approx x\,\frac{3 - x^2}{1 - x^2} \qquad \text{where} \qquad x \overset{\text{def}}{=} \frac{\bar{\lambda}}{\lambda_L}, \\ &\zeta \overset{\text{def}}{=} \left(\frac{\lambda_L}{3\bar{\lambda}}\right)\beta. \end{aligned} \tag{2.3.3}$$

4. Calculate the Cauchy stress

$$\mathbf{T} = J^{-1} G \bar{\mathbf{B}}_0 - p\mathbf{1}, \tag{2.3.4}$$

with

$$G = G_0 \, \zeta.$$

5. Calculate the Piola stress

$$\mathbf{T}_{\mathrm{R}} = J\mathbf{T}\mathbf{F}^{-\top}. \tag{2.3.5}$$

6. Use the quantities from steps 1–5 to compute the residuals as

$$\begin{aligned} R_{\mathbf{u}} &= \int_{\mathcal{B}} \mathbf{T}_{\mathrm{R}} : \nabla \mathbf{u}_{\text{test}} \, dv_{\mathrm{R}} - \int_{\mathcal{B}} \mathbf{b}_{0\mathrm{R}} \cdot \mathbf{u}_{\text{test}} \, dv_{\mathrm{R}} - \int_{\partial \mathcal{B}_{\mathrm{t_R}}} \bar{\mathbf{t}}_{\mathrm{R}} \cdot \mathbf{u}_{\text{test}} \, da_{\mathrm{R}}, \\ R_p &= \int_{\mathcal{B}} \left[\frac{p}{K} + \frac{\ln J}{J} \right] p_{\text{test}} \, dv_{\mathrm{R}}. \end{aligned} \tag{2.3.6}$$

7. FEniCS solves for the current values of $\{\mathbf{u},\, p\}$.
8. Set these current values as the "old" values for the next step in time,

$$\{\mathbf{u},\, p\} \implies \{\mathbf{u}_{\text{old}},\, p_{\text{old}}\}. \tag{2.3.7}$$

9. Repeat steps 1 through 8 for desired number of time steps.

3 Representative simulations of finite elasticity

A discussion of the numerical implementation is most readily digested through examples. Accordingly, in this section we present representative simulations to cement the concepts and show their practical utility though the solution of several boundary-value problems.

- The codes used for all the representative simulations shown in this book are posted online (Stewart et al., 2025):

  `https://solidmechanicscoupledtheories.github.io`.

In our numerical simulations in FEniCS we will use the basic units listed as SI (mm) in Table 3.1. Particular values of the material parameters for a vulcanized natural rubber are listed in Table 3.2.

Table 3.1 Basic units.

Quantity	SI (m)	SI (mm)
Length	m	mm
Mass	kg	kg
Time	s	s
Mass density	kg/m^3	kg /mm^3
Force	N	mN
Stress	Pa	kPa

Table 3.2 Material parameters for a vulcanized natural rubber (Anand and Govindjee, 2020).

Quantity	Value SI (m)	Value SI (mm)
$\rho_{\rm R}$, mass density	900 kg/m^3	0.9×10^{-6} kg/mm^3
G_0, ground-state shear modulus	280×10^3 Pa	280 kPa
λ_L, locking stretch	5.12	5.12
K, bulk modulus $(1000 \times G_0)$	280×10^6 Pa	280×10^3 kPa

Introduction to Coupled Theories in Solid Mechanics. Lallit Anand, Eric M. Stewart, Shawn A. Chester, Oxford University Press.
 DOI: 10.1093/oso/9780198986218.003.0004

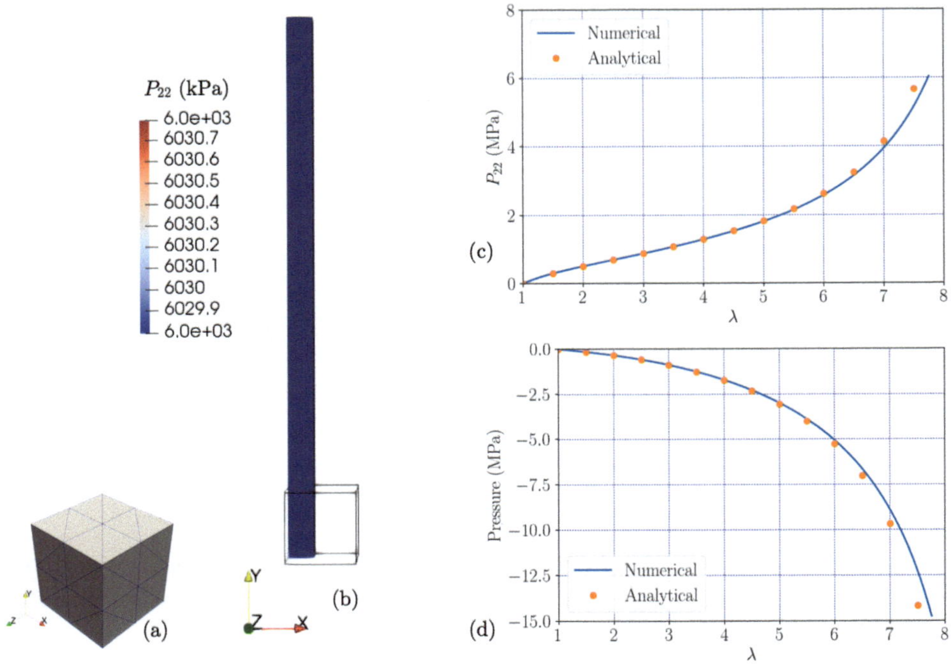

Fig. 3.1 (a) Finite element mesh used in this example. (b) The reference and deformed configurations at maximum stretch showing the axial component of the Piola stress. (c) A comparison of the numerical simulation against the analytical solution for the axial stress as a function of stretch. (d) A comparison of the numerical simulation against the analytical solution for the pressure as a function of stretch. The difference between the numerical solution and the analytical solution arises because the analytical solution assumes complete incompressibility, while the numerical solution allows for a slight compressibility.

3.1 Simple extension

This first numerical example is the most familiar and tractable boundary-value problem used in all of mechanics—homogeneous simple extension. There is value to compare numerical solutions against analytical solutions. Thus, consider a homogeneous *incompressible* deformation with deformation gradient

$$[\mathbf{F}] = \begin{bmatrix} \lambda^{-1/2} & 0 & 0 \\ 0 & \lambda & 0 \\ 0 & 0 & \lambda^{-1/2} \end{bmatrix}, \tag{3.1.1}$$

where λ is the stretch in the $\mathbf{e}_2$-direction of a rectangular Cartesian coordinate system with orthonormal basis $\{\mathbf{e}_i | i = 1, 2, 3\}$. For an incompressible isotropic elastic material the Piola stress is given by

$$\mathbf{T}_{\text{R}} = G\mathbf{F} - p\mathbf{F}^{-\top}, \tag{3.1.2}$$

and for the homogeneous deformation under consideration, after determining the scalar p by using the condition that the lateral faces of the body perpendicular to the $\mathbf{e}_1$- and $\mathbf{e}_2$-directions are traction-free, the analytical result for the stress component $P_{22} = \mathbf{e}_2 \cdot \mathbf{T}_{\rm R}\mathbf{e}_2$ is

$$P_{22} = G\left(\lambda - \lambda^{-2}\right). \tag{3.1.3}$$

We will compare the result from our corresponding numerical simulation against this analytical result.

For the numerical simulation the initial geometry of the body is a simple cube. We use symmetry conditions and only mesh 1/4 of the body. Figure 3.1(a) shows the modeled domain which has an edge length of 10 mm, and is meshed with 2 tetrahedral elements along each edge. In order to model homogeneous simple extension, referring to Fig. 3.1(a), the top face of the cubical domain is prescribed a displacement ramp, while the exterior lateral faces are traction-free, and the others faces are prescribed zero displacement. Figure 3.1(b) shows the reference body and deformed configuration at the maximum applied stretch $\lambda = 7.8$. Figure 3.2(c) shows a plot of the resulting Piola stress component P_{22} as a function of stretch λ. The numerical result is in very good agreement with the analytical result. There are some minor differences between the numerical and analytical results at the largest stretches, but these we attribute to the use of a finite bulk modulus in the numerical solution, while the analytical result assumes complete incompressibility. One can see that difference more clearly in Fig. 3.2(d) where we compare the numerically calculated mean-normal pressure, $-(1/3)\,\mathrm{tr}\,\mathbf{T}$, for a slightly compressible material against the analytical result for a completely incompressible material.

3.2 Shear of a cube

The second numerical example we consider is the large deformation shear of a cube. As shown in Fig. 3.2(a) we consider a cube of edge length of $L_0 = 1$ mm, meshed with eight tetrahedral elements along each edge. As boundary conditions: (i) the bottom face of the cube is fixed; (ii) the top face is subjected to a displacement history in which the u_2 and u_3 components are held fixed while the u_1 component is prescribed a sinusoidal history

$$u_1 = \gamma L_0 \qquad \text{with} \qquad \gamma(t) = \sin(2\pi t/T) \qquad t \in [0, T],$$

for a total period of $T = 10$s; and (iii) the four lateral faces of the cube are traction-free. The prescribed boundary conditions result in sinusoidal shearing of the cube starting from $\gamma = 0$ and then cycling between $\gamma = 1$ and $\gamma = -1$.

Figure 3.2(b) shows the reference body and deformed configuration at $\gamma = 1$, together with contours of the P_{12} component of the Piola stress. Figure 3.2(c) shows the resulting nonlinear curve of the reaction shear stress on the top face of the cube as a function of the amount of shear γ.

REMARK

The boundary conditions used in our numerical simulation are not the same as those considered in the classical problem of homogeneous "simple shear" which is widely studied in analytical continuum mechanics (cf., e.g., Anand and Govindjee, 2020, Section 30.14.2). In our simulation the side faces of the cube were not prescribed appropriate boundary conditions in order to match the mathematical description of "simple shear." As a result, the deformation is inhomogenous—which is abundantly clear from the deformed geometry in Fig. 3.2(b) and also the contours of P_{12} in this figure which show that shear stress is not uniform in the body.

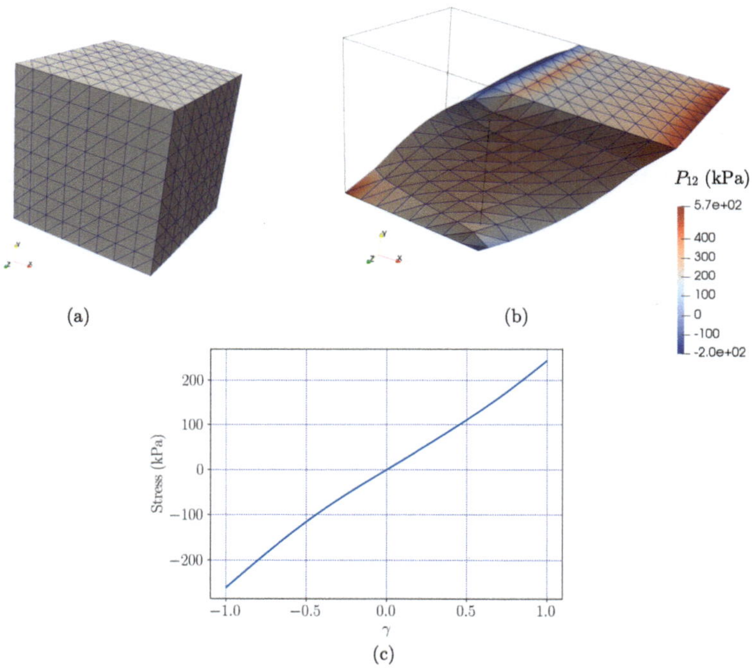

Fig. 3.2 (a) The finite element mesh used in this example. (b) The reference and deformed configuration at maximum deformation $\gamma = 1$ showing the P_{12} component of the Piola stress. (c) The shear stress P_{12} on the top face as a function of the amount of shear γ.

3.3 **Fixed-end torsion**

Next, we consider a classical example in finite elasticity of fixed-end torsion of a right circular cylinder. In the numerical simulation the initial radius of the cylinder is taken as $R_0 = 12.7\,\text{mm}$ (1/2 inch) and the initial length of the cylinder is taken as $L_0 = 25.4\,\text{mm}$ (1 inch); this geometry is meshed with tetrahedral elements, as shown in Fig. 3.3(a). The boundary conditions used to affect simple torsion are that the bottom face of the cylinder at

$x_1 = 0$ is fixed, the top face at $x_1 = 25.4$ mm is prescribed a total twist of $\theta = 2.5$ rad without allowing for any axial displacement, while the lateral face of the cylinder is traction-free.

Figure 3.3(b) shows the deformed configuration after the full twist of 2.5 rad has been applied, highlighting the inhomogeneous nature of the axial stress P_{11} field that develops during simple torsion. Figure 3.3(c) shows the twisting moment as a function of the twist per unit length, θ/L_0, and Fig. 3.3(d) shows the corresponding compressive axial force which is generated due to the nonlinear material behavior. This axial stress response—called the Poynting (1909) effect—is well-known in the finite elasticity literature (Rivlin, 1948; Rivlin and Saunders, 1951).[1] For a recent discussion and an analytical solution for simple torsion in finite elasticity, using a particular analytically tractable form of the free-energy function, see Anand and Govindjee (2020, Section 31.6).

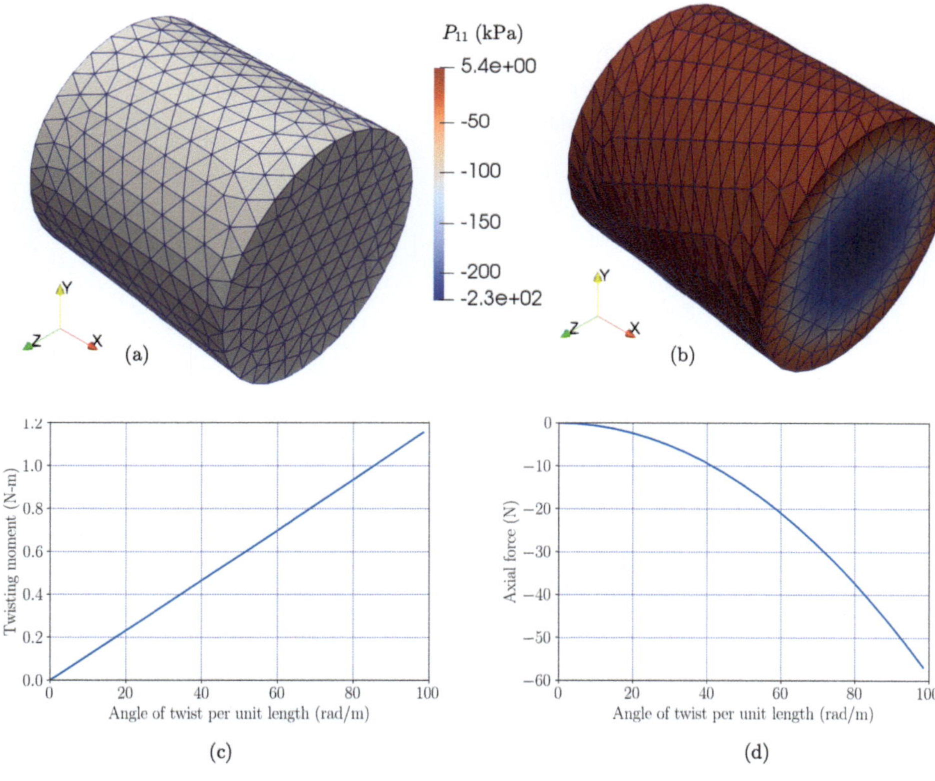

Fig. 3.3 (a) The initial configuration with the finite element mesh. (b) The deformed configuration at maximum twist showing contours of the compressive axial stress P_{11}. (c) The twisting moment versus twist per unit length curve. (d) The compressive axial force versus twist per unit length curve.

[1] The terminology "Poynting effect" actually refers to the phenomenon of the extension of a cylinder when it is twisted under axial free-end conditions, rather than the development of a compressive normal stress when the cylinder is twisted under fixed-end conditions. Both phenomena are manifestations of the same underlying non-nonlinear elasticity of the material, just different boundary conditions.

3.4 Extension of a plate with a hole

The linear elastic analysis of tension of a plate with a hole is a very common example used to demonstrate the phenomenon of stress concentration. Here we perform a numerical analysis of the problem of a plate with a hole undergoing *very large* elastic extension.

In the numerical simulation the full geometry is a plate with an initial length $L_0 = 30\,\mathrm{mm}$, height $H_0 = 20\,\mathrm{mm}$, thickness of $T_0 = 1\,\mathrm{mm}$, and a centered hole of initial diameter $D_0 = 6\,\mathrm{mm}$, as shown in Fig. 3.4(a). Using the symmetry of the problem, we mesh only one quarter of the full geometry with one element through the thickness, as shown in Fig. 3.4(a). On the quarter plate geometry the two faces of symmetry of the plate are prescribed appropriate symmetry boundary conditions, the face of the hole and the lateral top face of the specimen are traction-free, while the right face is prescribed a displacement in the $\mathbf{e}_1$-direction to impose a stretch of $\lambda = 3$.

Figure 3.4(b) shows the deformed configuration of the plate at a stretch of $\lambda = 3$, together with contours of the effective stretch $\bar{\lambda}$. Note the *stretch concentration* of $\bar{\lambda} = 4.4$ on the boundaries of the deformed hole. Figure 3.4(c) shows a plot of the far-field Piola stress P_{11} versus axial stretch λ during the extension of the plate.

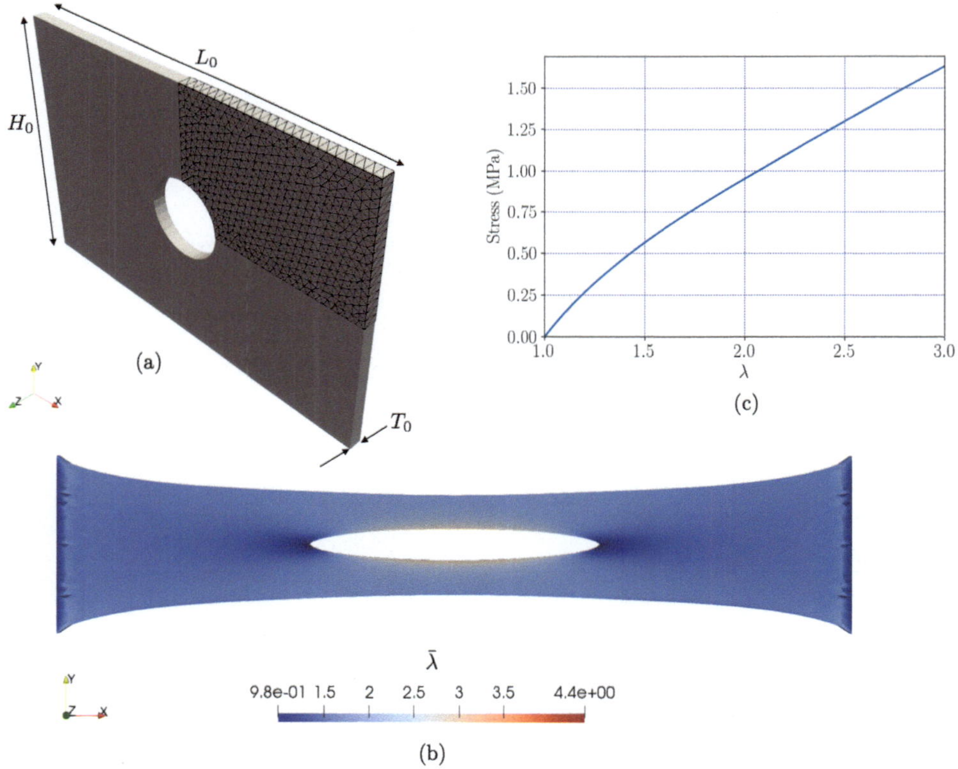

Fig. 3.4 (a) The reference configuration and the finite element mesh used in the simulation. (b) The deformed configuration at a stretch of $\lambda = 3$ together with contours of the effective stretch $\bar{\lambda}$. (c) The corresponding far-field axial Piola stress versus prescribed stretch λ curve.

3.5 Inflation of a cylindrical tube

In this example we consider the inflation of a cylindrical tube subjected to an internal pressure under axial plane strain conditions. The full geometry is a cylinder with initial inner radius $R_0 = 10\,\text{mm}$, wall thickness $T_0 = 1\,\text{mm}$, and an initial axial length $L_0 = 5\,\text{mm}$. Using the symmetry of the problem, we mesh only one-quarter of the full geometry, as shown in Fig. 3.5(a), with two elements through the thickness of the wall of the cylinder. The two faces of symmetry on the meshed domain are prescribed appropriate symmetry boundary conditions, the front and back faces of the cylinder in the $\mathbf{e}_3$-direction are prescribed zero displacements, the outer surface of the cylinder is traction-free, and the inner surface of the cylinder is prescribed an inflation pressure which is linearly ramped from 0 kPa to 45 kPa at a rate of 1 kPa/s.

Figure 3.5(a) also shows the deformed configuration of the cylinder at a maximum effective stretch of $\bar{\lambda} = 4.0$ on the inner wall. Figure 3.4(b) shows the corresponding hoop stretch R/R_0 of the internal wall of the cylinder as a function of the applied internal pressure.

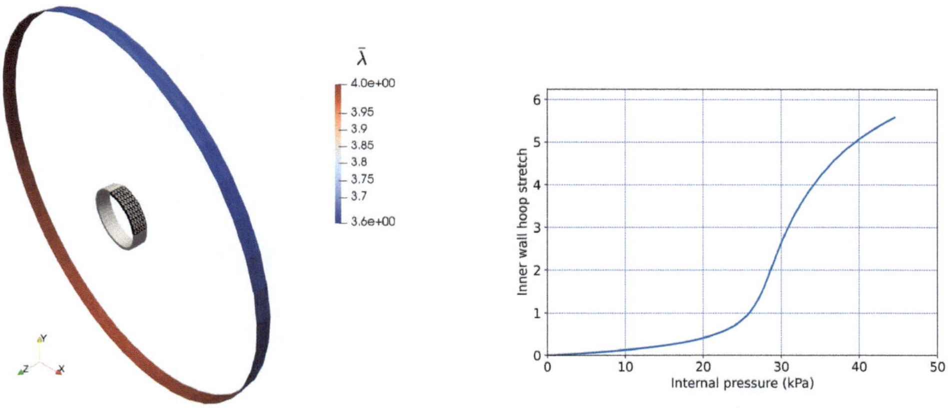

(a) (b)

Fig. 3.5 (a) The initial geometry and mesh used in this example, together with the deformed configuration showing contours of the effective stretch when the maximum $\bar{\lambda} = 4.0$. The simulation results have been reflected about both symmetry planes to show the full geometry. (b) The corresponding hoop stretch R/R_0 of the internal wall of the cylinder as a function of the applied internal pressure.

3.6 Inflation of a hollow sphere

Next we study the inflation of a spherical shell. The full geometry is a sphere with initial inner radius $R_0 = 10\,\text{mm}$ and wall thickness $T_0 = 1\,\text{mm}$. Using the symmetry of the problem we mesh only one-eighth of the full geometry, as shown in Fig. 3.6(a), with two elements through the thickness of the wall of the sphere. The three faces of symmetry of this domain are prescribed appropriate symmetry boundary conditions, the outer surface of the sphere is traction-free, while the inner surface is prescribed an inflation pressure which is linearly ramped from 0 kPa to 55 kPa at a rate of 1 kPa/s. Figure 3.6(a) also shows the deformed configuration of the sphere when the effective stretch is $\bar{\lambda} = 1.2$ on the inner surface. The numerical simulation failed to converge at this stage of deformation, because of the well-known *snap-through* instability of hyperelastic thin spherical shells when inflated under pressure control.[2] Figure 3.6(b) shows the corresponding hoop stretch R/R_0 of the internal surface of the sphere as a function of the applied internal pressure.

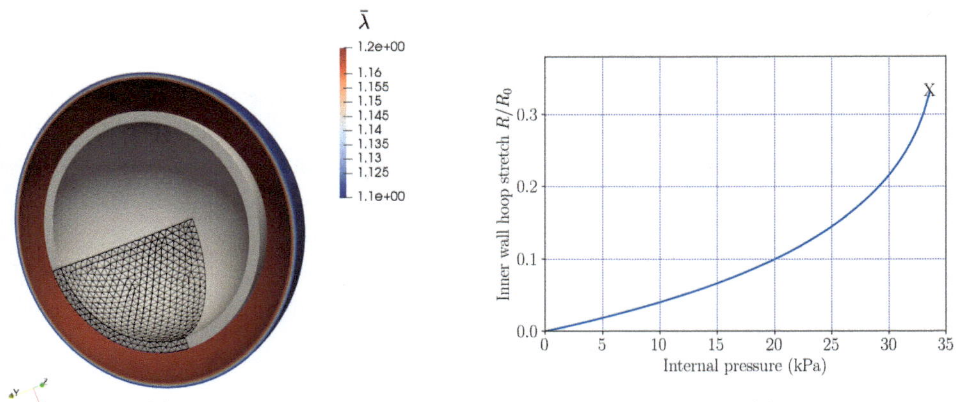

(a) (b)

Fig. 3.6 (a) The initial geometry and mesh used in this example, together with the deformed configuration when the maximum effective stretch is $\bar{\lambda} = 1.2$. The simulation results have been reflected about two symmetry planes to show one-half of the full geometry. (b) The corresponding hoop stretch R/R_0 of the internal wall of the sphere as a function of the applied internal pressure; "X" marks the point where the simulation failed to converge.

3.7 A block under pressure on a part of its face

This is a classical bench-mark problem in computational solid mechanics, known as the "footing problem" (Schröder et al., 2021). It consists of a rectangular block subjected to a constant pressure loading on a part of its upper surface. The full geometry of the body is a rectangular block with initial length and width of 100 mm and height of 50 mm. A uniform pressure is applied on a central square region of edge-length 25 mm of the top face, as shown in Fig. 3.7(a).

Using the symmetry of the problem, computations were carried out on only one quarter of the block marked by the thick black outline in Fig. 3.7(a). The mesh consists of tetrahedral elements, with 10×10 elements on the top face where the pressure is applied, and 6 elements in the depth direction, Fig. 3.7(b). As boundary conditions: (i) the bottom of the computational domain is fixed in the $\mathbf{e}_3$-direction; (ii) the two faces of symmetry of the domain are prescribed appropriate symmetry conditions; (iii) the square region of the top face outlined in blue—the "load face"—is subjected to a uniform pressure in the $\mathbf{e}_3$-direction, and is restricted from moving in the $\mathbf{e}_1$- and $\mathbf{e}_2$-directions; (iv) all other external surfaces of the domain are traction-free. The prescribed pressure is linearly ramped from 0 MPa to 1.5 MPa over 30 s.

[2] Such instabilities are difficult to capture in quasi-static numerical simulations using standard finite element techniques. See Liu et al. (2022) for a recent sophisticated approach using an "arc-length" iteration method to numerically simulate large inflation of thin nonlinear elastic shells.

Figure 3.6(c) shows the final deformed configuration together with contours of the effective stretch $\bar{\lambda}$. Figure 3.6(d) shows the vertical displacement of the center point of the top surface—shown in red in Fig. 3.7(a)—as a function of time.

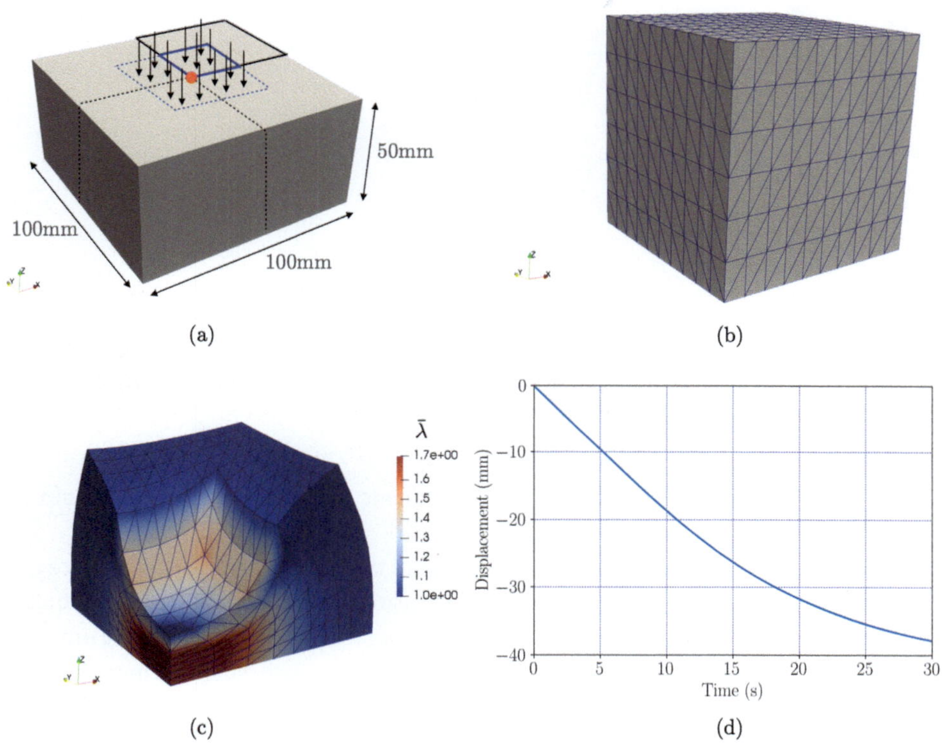

Fig. 3.7 (a) The initial geometry, symmetry planes, and pressure loading used for this example. (b) The corresponding mesh of one quarter of the full geometry. (c) The deformed configuration (only one quarter shown) at maximum deformation showing the inhomogeneous effective stretch $\bar{\lambda}$. (d) The displacement of the top center point as a function of time.

3.8 Column buckling

As a final example we study the buckling of a column with a square cross-section under fixed-end boundary conditions. We consider a column of initial height $H_0 = 20$ mm, and a square cross-section of edge length $W_0 = 1$ mm, Fig. 3.8(a). Since buckling phenomena are known to be highly imperfection-sensitive we consider two cases: (i) a geometrically "perfect" column, and (ii) a column with a "geometric imperfection" along its length.

(i) *Geometrically "perfect" column*: Even though the column is geometrically "perfect," the manner in which one discretizes the geometry is also known to effect the buckling characteristics of the column. To demonstrate this we employ two meshes: both meshes have 4 elements along the edges of the cross-section, but the first mesh uses 20 elements along the height of the column, and the second mesh uses 50 elements.

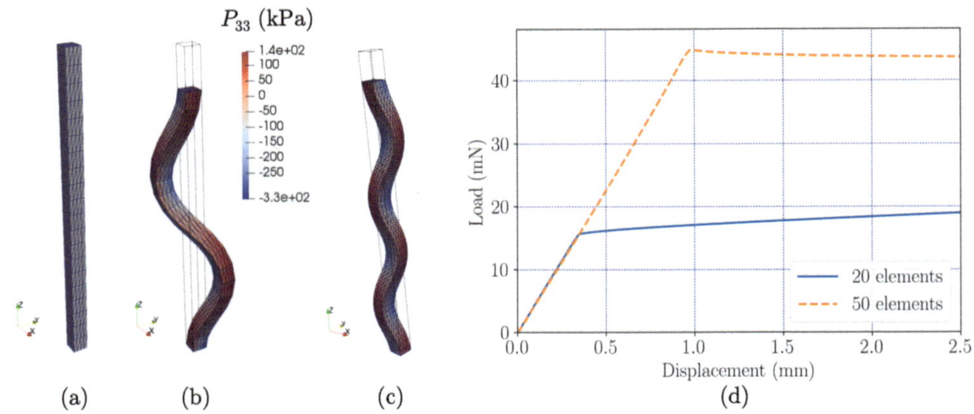

Fig. 3.8 Buckling of a "perfect" column. (a) The initial geometry and mesh for the case of 20 elements along the height of the column. (b) Deformed configuration with contours of the P_{33} stress component at maximum deformation for the mesh using 20 elements along the height. (c) Deformed configuration with contours of the P_{33} stress component at maximum deformation for the mesh using 50 elements along the height. (d) Plot of the compressive load on the top face as a function of the applied displacement for meshes with 20 and 50 elements along the height of the column.

Figure 3.8(a) shows the mesh with 20 elements along its height; the column with 50 elements is not shown.

As boundary conditions: the bottom face of the column is fixed, the top face of the column is prescribed a displacement in the negative $\mathbf{e}_3$-direction direction to compress the column while the transverse displacements of this face are fixed, and all other faces of the column are traction-free. The prescribed compressive displacement is ramped from 0 mm to -2.5 mm over a period of 1 s.

Figures 3.8(b) and 3.8(c) show the deformed configurations of the columns with the two different meshes at the maximum applied displacement, together with contours of the P_{33} component of the Piola stress. Figure 3.8(d) shows a plot of the compressive load on the top face as a function of the applied displacement for the two simulations. Instantly noticeable from these figures is that buckled configurations are *very different from each other*, and the magnitude of the buckling load for the mesh with 50 elements along its height is *more than two times higher* than the buckling load for the mesh with 20 elements along its height.

These large differences in buckling loads and buckled shapes can only be attributable to the numerical differences caused by the different spatial finite element discretizations. In order to avoid spurious results due to different mesh discretizations in numerical studies of buckling, it is common practice to introduce either geometric or material imperfections. We consider next the buckling of a column with a small geometric imperfection.

(ii) *Column with a "geometric imperfection"*: Thus, consider a column with nominally the same geometry and material properties as in the previous case, but with a small geometric imperfection which is introduced by modifying the initial coordinates of $\mathbf{X}^0$ of a "perfect column" to the coordinates $\mathbf{X}$ of an "imperfect column" using the mapping relation

$$X_1 = X_1^0 - \left(\frac{\delta}{2}\right)\left[1 - \cos\left(\frac{2\pi X_3}{H_0}\right)\right], \quad X_2 = X_2^0 - \left(\frac{\delta}{2}\right)\left[1 - \cos\left(\frac{2\pi X_3}{H_0}\right)\right],$$

$$X_3 = X_3^0, \quad \text{with} \quad \delta = 0.005 W_0,$$

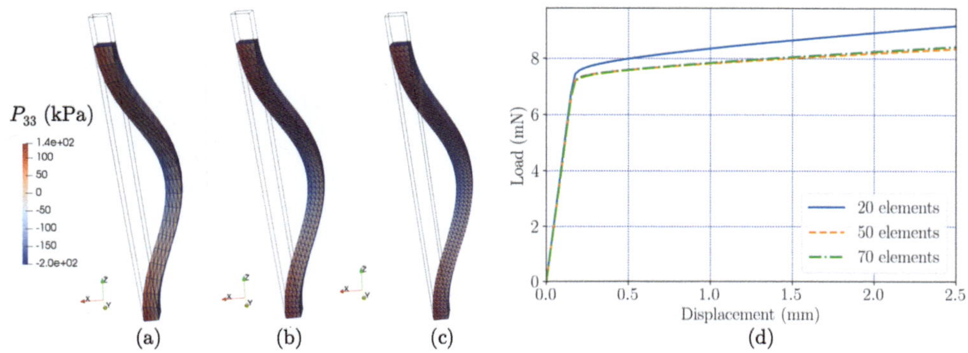

Fig. 3.9 Buckling of an "imperfect" column. Deformed configuration with contours of the P_{33} stress component at maximum deformation for the mesh using (a) 20 elements, (b) 50 elements, and (c) 70 elements along the height. (d) Plot of the compressive load on the top face as a function of the applied displacement for meshes with 20, 50, and 70 elements along the height of the column.

where $W_0 = 1\,\text{mm}$ and $H_0 = 20\,\text{mm}$ are, respectively, the width of the cross-section and the height of the perfect column. This functional form of the geometric imperfection results in a maximum imperfection of magnitude δ at the mid-height of the column, and zero imperfection at the top and bottom of the column. In addition, in order to study mesh convergence, we ran simulations with 20, 50, and 70 elements along the height of the column.

The simulation results shown in Fig. 3.9 clearly show that the buckled configurations are all similar, and that the plots of the load on the top face as a function of the applied displacement, as well as the critical buckling load, converge as the mesh is refined—the simulation results using 50 and 70 elements along the height of the column are indistinguishable from each other.

Part II

Viscoelasticity of Elastomeric Materials

4 Viscoelasticity of elastomeric materials

4.1 Introduction

The constitutive response of most solid materials deviates from an ideal elastic response in various ways. A major deviation occurs when a material exhibits both elastic as well as dissipative viscous characteristics, and such a response is called *viscoelastic*. Some phenomena exhibited by viscoelastic solids are:

- if the stretch is held constant then the stress decreases with time—*stress–relaxation*;
- if the stress is held constant then the stretch increases with time—*creep*;
- the stress–stretch curve depends on the rate of application of the stretch—*stretch-rate sensitivity*;
- there is a strong tendency for near-full *recovery* of the stretch upon unloading to zero stress; and
- if a cyclic stretch is applied then there is a *phase shift* in the stress response which results in *hysteresis*.

Despite their name, elastomeric materials typically exhibit a *pronounced* viscoelastic response. At the microscopic level such materials may be considered to be made from several interpenetrating polymer networks: (i) A *ground network* formed by *strongly cross-linked* macromolecules. (ii) Several *transient networks* based on *temporary entanglements* of macromolecules and their side-groups. The ground network is associated with the elastic (or time-independent) response of the material, and the secondary transient networks are responsible for the viscous (or time-dependent) response. On application of macroscopic deformation the ground network stretches and drags the mobile transient sub-networks along with it. After a sufficient amount of time at a constant deformation, the sub-networks relax towards a state at which it produces no viscous overstress.

In this chapter we formulate a large-deformation theory of viscoelasticity for isotropic elastomeric materials. The theory considered here is in the spirit of the existing theories of viscoelasticity of elastomeric materials in the literature which use internal variables with attendant evolution equations to capture time-dependence (cf., e.g., Sidoroff, 1974; Lubliner, 1985; Bergstrom and Boyce, 1998; Reese and Govindjee, 1998; Wijaya et al., 2023, and the numerous references therein), but it differs in several details of the continuum-mechanical formulation.[1]

[1] We do not consider hereditary stress–relaxation-integral (or creep-integral) forms of viscoelasticity theories (cf., e.g., Green and Rivlin, 1957; Pipkin and Rogers, 1968), because in contrast to such theories the theories using internal variables have been found in recent years to be numerically more tractable, and therefore preferred in applications.

Introduction to Coupled Theories in Solid Mechanics. Lallit Anand, Eric M. Stewart, Shawn A. Chester, Oxford University Press.
 DOI: 10.1093/oso/9780198986218.003.0005

4.2 A single Maxwell element-based viscoelasticity theory

We begin in this section by formulating a theory which is conceptually based on a classical Maxwell-type rheological model of a spring in series with dashpot element, cf. Fig. 4.1. This spring–dashpot model motivates a multiplicative decomposition $\mathbf{F} = \mathbf{F}^e\mathbf{F}^v$ of the deformation gradient $\mathbf{F}$, in which $\mathbf{F}^e$ represents the distortion due to the deformation of the spring, and $\mathbf{F}^v$ represents the distortion of the dashpot—more on this below.

In order to represent the viscoelastic response of actual materials, in the next section we will generalize this single Maxwell element-based theory to a theory based on *multiple* spring–dashpot Maxwell elements acting in parallel.[2]

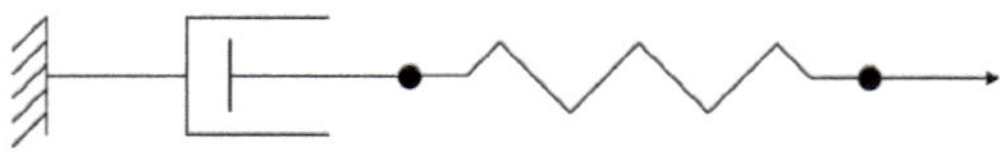

Fig. 4.1 Maxwell-type spring–dashpot rheological element for finite deformation viscoelasticity which motivates a multiplicative decomposition, $\mathbf{F} = \mathbf{F}^e\mathbf{F}^v$ of the deformation gradient $\mathbf{F}$.

4.2.1 Kinematics

Consider a macroscopically homogeneous body $\mathcal{B}$ with the region of space it occupies in a fixed reference configuration, and denote by $\mathbf{X}$ an arbitrary material point of $\mathcal{B}$. A motion of $\mathcal{B}$ is then a smooth one-to-one mapping

$$\mathbf{x} = \boldsymbol{\chi}(\mathbf{X}, t), \tag{4.2.1}$$

with displacement

$$\mathbf{u}(\mathbf{X}, t) = \boldsymbol{\chi}(\mathbf{X}, t) - \mathbf{X}, \tag{4.2.2}$$

and deformation gradient, velocity, acceleration, and velocity gradient are respectively given by

$$\mathbf{F} = \nabla\boldsymbol{\chi} = \mathbf{1} + \nabla\mathbf{u}, \qquad \dot{\mathbf{u}} = \frac{\partial\boldsymbol{\chi}}{\partial t}, \qquad \ddot{\mathbf{u}} = \frac{\partial^2\boldsymbol{\chi}}{\partial t^2}, \qquad \mathbf{L} = \operatorname{grad}\dot{\mathbf{u}} = \dot{\mathbf{F}}\mathbf{F}^{-1}. \tag{4.2.3}$$

We base our theory on a multiplicative decomposition of the deformation gradient $\mathbf{F}$ (Kröner, 1960; Lee, 1969),

$$\mathbf{F} = \mathbf{F}^e\mathbf{F}^v. \tag{4.2.4}$$

[2] Rheological models based on springs and dashpots serve as useful visual/conceptual aids—no actual material is made of springs and dashpots.

The central idea underlying this assumption is that although the specific details of microstructure and its changes are not visible at the macroscopic level, we can account for microscopic elastic and viscous distortions of the material by introducing *elastic and viscous distortion tensors* $\mathbf{F}^e$ and $\mathbf{F}^v$.[3]

It is important to understand the differences between the tensor fields $\mathbf{F}$, $\mathbf{F}^e$, and $\mathbf{F}^v$ in the multiplicative decomposition $\mathbf{F} = \mathbf{F}^e\mathbf{F}^v$ which is shown schematically in Fig. 4.2. First of all, while $\mathbf{F} = \nabla\chi$ is the gradient of a point field χ, in general there is no point field χ^v such that $\mathbf{F}^v = \nabla\chi^v$, nor is there a point field χ^e such that $\mathbf{F}^e = \nabla\chi^e$. Thus we can at most describe the physical nature of the tensor fields $\mathbf{F}^e$ and $\mathbf{F}^v$ through their pointwise mapping properties as *linear transformations*. With this in mind, recall that $\mathrm{d}\mathbf{x} = \mathbf{F}\mathrm{d}\mathbf{X}$. Hence,

$$\mathrm{d}\mathbf{x} = \mathbf{F}^e\mathbf{F}^v\mathrm{d}\mathbf{X}.$$

For want of a better notation, let

$$\mathrm{d}\mathbf{l} = \mathbf{F}^v\mathrm{d}\mathbf{X},$$

so that,

$$\mathrm{d}\mathbf{x} = \mathbf{F}^e\mathrm{d}\mathbf{l}.$$

The output of the linear transformation $\mathbf{F}^v$ must therefore coincide with the input of the linear transformation $\mathbf{F}^e$; that is,

$$\text{the range of } \mathbf{F}^v = \text{the domain of } \mathbf{F}^e.$$

We refer to this common vector space as the *intermediate space*.

As is standard, we assume that

$$J \stackrel{\text{def}}{=} \det \mathbf{F} > 0, \tag{4.2.5}$$

and hence, using (4.2.4),

$$J = J^e J^v \qquad \text{where} \qquad J^e \stackrel{\text{def}}{=} \det\ \mathbf{F}^e > 0 \qquad \text{and} \qquad J^v \stackrel{\text{def}}{=} \det \mathbf{F}^v > 0, \tag{4.2.6}$$

so that both $\mathbf{F}^e$ and $\mathbf{F}^v$ are invertible.

Polar decompositions

The right and left polar decompositions of $\mathbf{F}$ are

$$\mathbf{F} = \mathbf{R}\mathbf{U} = \mathbf{V}\mathbf{R}, \tag{4.2.7}$$

[3] Such a multiplicative decomposition of $\mathbf{F}$ is widely used in formulating large-deformation theories of *inelasticity* of materials. For example, in elasto-(visco)plasticity as $\mathbf{F} = \mathbf{F}^e\mathbf{F}^p$ (cf., e.g. Kröner, 1960; Lee, 1969; Anand, 1985; Weber and Anand, 1990; Gurtin et al., 2010) and in viscoelasticity as $\mathbf{F} = \mathbf{F}^e\mathbf{F}^v$ (cf., e.g. Sidoroff, 1974; Lubliner, 1985; Reese and Govindjee, 1998).

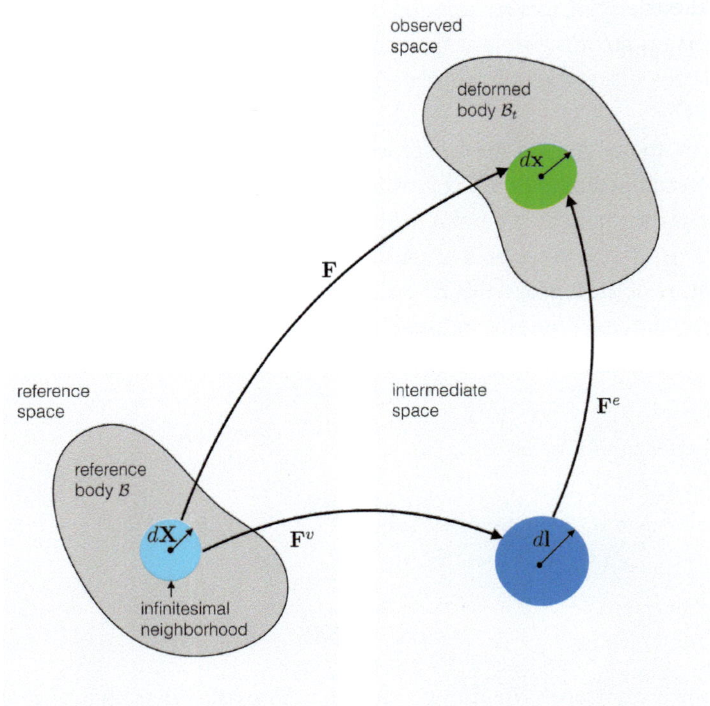

Fig. 4.2 Schematic of the decomposition $\mathbf{F} = \mathbf{F}^e\mathbf{F}^v$.

where $\mathbf{R}$ is a rotation (proper orthogonal tensor), and $\mathbf{U}$ and $\mathbf{V}$ are symmetric and positive-definite tensors with

$$\mathbf{U} = \sqrt{\mathbf{F}^\top \mathbf{F}} \qquad \text{and} \qquad \mathbf{V} = \sqrt{\mathbf{F}\mathbf{F}^\top}. \tag{4.2.8}$$

Also, the right and left Cauchy–Green tensors are given by

$$\mathbf{C} = \mathbf{U}^2 = \mathbf{F}^\top \mathbf{F} \qquad \text{and} \qquad \mathbf{B} = \mathbf{V}^2 = \mathbf{F}\,\mathbf{F}^\top. \tag{4.2.9}$$

Similarly, the right and left polar decompositions of $\mathbf{F}^e$ are

$$\mathbf{F}^e = \mathbf{R}^e\mathbf{U}^e = \mathbf{V}^e\mathbf{R}^e, \tag{4.2.10}$$

where $\mathbf{R}^e$ is a rotation (proper orthogonal tensor), and $\mathbf{U}^e$ and $\mathbf{V}^e$ are symmetric and positive-definite tensors with

$$\mathbf{U}^e = \sqrt{\mathbf{F}^{e\top} \mathbf{F}^e} \qquad \text{and} \qquad \mathbf{V}^e = \sqrt{\mathbf{F}^e\mathbf{F}^{e\top}}. \tag{4.2.11}$$

Also, the right and left elastic Cauchy–Green tensors are given by

$$\mathbf{C}^e = \mathbf{U}^{e2} = \mathbf{F}^{e\top} \mathbf{F}^e \qquad \text{and} \qquad \mathbf{B}^e = \mathbf{V}^{e2} = \mathbf{F}^e\mathbf{F}^{e\top}. \tag{4.2.12}$$

Velocity gradient

Next, by $(4.2.3)_4$ and (4.2.4),

$$\mathbf{L} = \mathbf{L}^e + \mathbf{F}^e\mathbf{L}^v\mathbf{F}^{e\,-1}, \tag{4.2.13}$$

with

$$\mathbf{L}^e = \dot{\mathbf{F}}^e\mathbf{F}^{e-1} \qquad \text{and} \qquad \mathbf{L}^v = \dot{\mathbf{F}}^v\mathbf{F}^{v-1}. \tag{4.2.14}$$

As is standard, we define the total, elastic, and viscous stretching and spin tensors through

$$\begin{aligned} \mathbf{D} &= \operatorname{sym}\mathbf{L}, & \mathbf{W} &= \operatorname{skw}\mathbf{L}, \\ \mathbf{D}^e &= \operatorname{sym}\mathbf{L}^e, & \mathbf{W}^e &= \operatorname{skw}\mathbf{L}^e, \\ \mathbf{D}^v &= \operatorname{sym}\mathbf{L}^v, & \mathbf{W}^v &= \operatorname{skw}\mathbf{L}^v, \end{aligned} \tag{4.2.15}$$

so that $\mathbf{L} = \mathbf{D} + \mathbf{W}$, $\mathbf{L}^e = \mathbf{D}^e + \mathbf{W}^e$, and $\mathbf{L}^v = \mathbf{D}^v + \mathbf{W}^v$.

We now make two basic kinematical assumptions concerning viscous flow:

(i) First, from the outset we constrain the theory by limiting our discussion to circumstances under which the material may be idealized as *isotropic*. For isotropic theories utilizing a multiplicative decomposition of $\mathbf{F}$, it is widely assumed that the viscous flow is *irrotational* (cf., e.g., Gurtin and Anand, 2005; Gurtin et al., 2010, Section 97), in the sense that

$$\mathbf{W}^v = \mathbf{0}. \tag{4.2.16}$$

Then, $\mathbf{L}^v \equiv \mathbf{D}^v$, and

$$\dot{\mathbf{F}}^v = \mathbf{D}^v\mathbf{F}^v. \tag{4.2.17}$$

(ii) Next, we assume that viscous flow does not induce changes in volume. Consistent with this we assume that

$$J^v = 1, \tag{4.2.18}$$

and hence that

$$\operatorname{tr}\mathbf{D}^v = 0. \tag{4.2.19}$$

Finally, using $(4.2.3)_4$, (4.2.4), and (4.2.16) we may write (4.2.13) for future use as

$$(\nabla\dot{\mathbf{u}})\mathbf{F}^{-1} = \dot{\mathbf{F}}^e\mathbf{F}^{e\,-1} + \mathbf{F}^e\,\mathbf{D}^v\,\mathbf{F}^{e\,-1}. \tag{4.2.20}$$

4.2.2 **Principle of virtual power**

We derive macroscopic and microscopic force balances via the principle of virtual power (cf., e.g., Germain, 1973; Gurtin, 2002; Gurtin et al., 2010). In developing our theory we take the "rate-like" kinematical descriptors to be $\dot{\mathbf{u}}$, $\dot{\mathbf{F}}^e$, and $\mathbf{D}^v$. In exploiting the principle of virtual power we note that the rates $\dot{\mathbf{u}}$, $\dot{\mathbf{F}}^e$, and $\mathbf{D}^v$ are not independent—they are constrained by (4.2.20).

We denote by $\mathcal{P}$ an arbitrary part of the reference body $\mathcal{B}$, and with $\mathbf{n}_{\mathrm{R}}$ the outward unit normal on the boundary $\partial\mathcal{P}$ of $\mathcal{P}$. With each motion of the body we associate macroscopic and microscopic force systems. The macroscopic system is defined by:

(i) A traction $\mathbf{t}_{\mathrm{R}}(\mathbf{n}_{\mathrm{R}})$, for each unit vector $\mathbf{n}_{\mathrm{R}}$, that expends power over the velocity $\dot{\mathbf{u}}$.

(ii) An external body force $\mathbf{b}_{\mathrm{R}}$ that also expends power over $\dot{\mathbf{u}}$. The body force $\mathbf{b}_{\mathrm{R}}$ presumed to account for inertia; that is, granted the underlying frame is inertial,

$$\mathbf{b}_{\mathrm{R}} = \mathbf{b}_{0\mathrm{R}} - \rho_{\mathrm{R}}\ddot{\mathbf{u}}, \tag{4.2.21}$$

with $\mathbf{b}_{0\mathrm{R}}$ the non-inertial body force and ρ_{R} the mass density in the reference configuration.

(iii) An elastic stress $\mathbf{S}^e$ that expends power over the elastic distortion rate $\dot{\mathbf{F}}^e$.

The microscopic system is defined by:

(i) A viscous stress $\mathbf{T}^v$ that expends power over the viscous stretching $\mathbf{D}^v$.

We characterize the force systems through the manner in which these forces expend power. That is, given any part $\mathcal{P}$, through the specification of $\mathcal{W}_{\mathrm{ext}}(\mathcal{P})$ the power expended on $\mathcal{P}$ by material *external* to $\mathcal{P}$, and $\mathcal{W}_{\mathrm{int}}(\mathcal{P})$ a concomitant expenditure of power within $\mathcal{P}$. Specifically,

$$\left.\begin{aligned}\mathcal{W}_{\mathrm{ext}}(\mathcal{P}) &= \int_{\partial\mathcal{P}} \mathbf{t}_{\mathrm{R}}(\mathbf{n}_{\mathrm{R}})\cdot\dot{\mathbf{u}}\,da_{\mathrm{R}} + \int_{\mathcal{P}} \mathbf{b}_{\mathrm{R}}\cdot\dot{\mathbf{u}}\,dv_{\mathrm{R}},\\ \mathcal{W}_{\mathrm{int}}(\mathcal{P}) &= \int_{\mathcal{P}} \left(\mathbf{S}^e : \dot{\mathbf{F}}^e + \mathbf{T}^v : \mathbf{D}^v\right) dv_{\mathrm{R}}.\end{aligned}\right\} \tag{4.2.22}$$

Note that in (4.2.22) the stress powers $\mathbf{S}^e : \dot{\mathbf{F}}^e$ and $\mathbf{T}^v : \mathbf{D}^v$ are per unit volume of the intermediate space. However, since $J^v = 1$, these stress powers are also the stress powers per unit volume of the reference space.

Assume that, at some arbitrarily chosen but *fixed time*, the fields $\mathbf{u}$, $\mathbf{F}^e$ (and hence $\mathbf{F}$ and $\mathbf{F}^v$) are known, and consider the fields $\dot{\mathbf{u}}$, $\dot{\mathbf{F}}^e$, and $\mathbf{D}^v$ as virtual velocities to be specified independently in a manner consistent with (4.2.20). That is, denoting the virtual fields by $\tilde{\mathbf{u}}$, $\tilde{\mathbf{F}}^e$, $\tilde{\mathbf{D}}^v$ to differentiate them from fields associated with the actual motion of the body, we require that

$$(\nabla\tilde{\mathbf{u}})\mathbf{F}^{-1} = \tilde{\mathbf{F}}^e\mathbf{F}^{e\,-1} + \mathbf{F}^e\,\tilde{\mathbf{D}}^v\,\mathbf{F}^{e\,-1}. \tag{4.2.23}$$

Further, we define a generalized virtual velocity to be a list

$$\mathcal{V} = (\tilde{\mathbf{u}}, \tilde{\mathbf{F}}^e, \tilde{\mathbf{D}}^v),$$

consistent with (4.2.23).

We refer to a macroscopic virtual field $\mathcal{V}$ as *rigid* if it satisfies

$$\nabla\tilde{\mathbf{u}} \equiv \tilde{\mathbf{F}} = \boldsymbol{\Omega}\mathbf{F}, \tag{4.2.24}$$

with $\boldsymbol{\Omega}$ a spatially constant skew tensor, together with

$$\tilde{\mathbf{F}}^e = \boldsymbol{\Omega}\mathbf{F}^e, \quad \tilde{\mathbf{D}}^v = 0. \tag{4.2.25}$$

Next, writing

$$\begin{aligned}
\mathcal{W}_{\text{ext}}(\mathcal{P}, \mathcal{V}) &= \int_{\partial\mathcal{P}} \mathbf{t}_{\mathrm{R}}(\mathbf{n}_{\mathrm{R}}) \cdot \tilde{\mathbf{u}}\, da_{\mathrm{R}} + \int_{\mathcal{P}} \mathbf{b}_{\mathrm{R}} \cdot \tilde{\mathbf{u}}\, dv_{\mathrm{R}}, \\
\mathcal{W}_{\text{int}}(\mathcal{P}, \mathcal{V}) &= \int_{\mathcal{P}} \left(\mathbf{S}^e : \tilde{\mathbf{F}}^e + \mathbf{T}^v : \tilde{\mathbf{D}}^v\right) dv_{\mathrm{R}},
\end{aligned} \tag{4.2.26}$$

respectively, for the external and internal expenditures of virtual power, the *principle of virtual power* consists of two basic requirements:

(V1) Given any part P,

$$\mathcal{W}_{\text{ext}}(\mathcal{P}, \mathcal{V}) = \mathcal{W}_{\text{int}}(\mathcal{P}, \mathcal{V}) \quad \text{for all generalized virtual velocities } \mathcal{V}. \tag{4.2.27}$$

(V2) Given any part $\mathcal{P}$,

$$\mathcal{W}_{\text{int}}(\mathcal{P}, \mathcal{V}) = 0 \quad \text{whenever } \mathcal{V} \text{ is a rigid macroscopic virtual velocity.} \tag{4.2.28}$$

To deduce the consequences of the principle of virtual power, assume that (4.2.27) and (4.2.28) are satisfied. Note that in applying the virtual balance we are at liberty to choose any $\mathcal{V}$ consistent with the constraint (4.2.23).

Macroscopic force and moment balances

Consider a generalized virtual velocity with $\tilde{\mathbf{D}}^v \equiv 0$, so that $(\nabla\tilde{\mathbf{u}})\mathbf{F}^{v-1} = \tilde{\mathbf{F}}^e$. For this choice of $\mathcal{V}$, (4.2.27) yields

$$\int_{\partial\mathcal{P}} \mathbf{t}_{\mathrm{R}}(\mathbf{n}_{\mathrm{R}}) \cdot \tilde{\mathbf{u}}\, da_{\mathrm{R}} + \int_{\mathcal{P}} \mathbf{b}_{\mathrm{R}} \cdot \tilde{\mathbf{u}}\, dv_{\mathrm{R}} = \int_{\mathcal{P}} \mathbf{S}^e : \tilde{\mathbf{F}}^e\, dv_{\mathrm{R}} = \int_{\mathcal{P}} (\mathbf{S}^e\mathbf{F}^{v\,-\top}) : \nabla\,\tilde{\mathbf{u}}\, dv_{\mathrm{R}}, \tag{4.2.29}$$

which, upon introducing a stress-measure

$$\mathbf{T}_{\mathrm{R}} \stackrel{\text{def}}{=} \mathbf{S}^e\mathbf{F}^{v\,-\top}, \tag{4.2.30}$$

may be rewritten as

$$\int_{\partial\mathcal{P}} \mathbf{t}_{\mathrm{R}}(\mathbf{n}_{\mathrm{R}}) \cdot \tilde{\mathbf{u}}\, da_{\mathrm{R}} = \int_{\mathcal{P}} \Big(\mathbf{T}_{\mathrm{R}} : \nabla\, \tilde{\mathbf{u}} - \mathbf{b}_{\mathrm{R}} \cdot \tilde{\mathbf{u}}\Big)\, dv_{\mathrm{R}}, \tag{4.2.31}$$

and using the divergence theorem we may conclude that

$$\int_{\partial\mathcal{P}} (\mathbf{t}_{\mathrm{R}}(\mathbf{n}_{\mathrm{R}}) - \mathbf{T}_{\mathrm{R}}\mathbf{n}_{\mathrm{R}}) \cdot \tilde{\mathbf{u}}\, da_{\mathrm{R}} + \int_{\mathcal{P}} (\mathrm{Div}\ \mathbf{T}_{\mathrm{R}} + \mathbf{b}_{\mathrm{R}}) \cdot \tilde{\mathbf{u}}\, dv_{\mathrm{R}} = 0.$$

Since this relation must hold for all parts $\mathcal{P}$ and all $\tilde{\mathbf{u}}$, standard variational arguments yield the traction condition

$$\mathbf{t}_{\mathrm{R}}(\mathbf{n}_{\mathrm{R}}) = \mathbf{T}_{\mathrm{R}}\mathbf{n}_{\mathrm{R}}, \tag{4.2.32}$$

and the local macroscopic force balance

$$\mathrm{Div}\ \mathbf{T}_{\mathrm{R}} + \mathbf{b}_{0\,\mathrm{R}} = \rho_{\mathrm{R}}\ddot{\mathbf{u}}, \tag{4.2.33}$$

respectively, where in writing (4.2.33) we have used (4.2.21).

Next, we deduce the consequences of requirement (V2) of the principle of virtual power. Using (4.2.24), (4.2.25), $(4.2.26)_2$, and (4.2.30), requirement (V2) of the principle of virtual power leads to the requirement that

$$\begin{aligned} 0 &= \int_{\mathcal{P}} \mathbf{S}^e : \tilde{\mathbf{F}}^e\, dv_{\mathrm{R}} \\ &= \int_{\mathcal{P}} (\mathbf{S}^e\mathbf{F}^{e\top}) : \boldsymbol{\Omega}\, dv_{\mathrm{R}} \\ &= \int_{\mathcal{P}} \big[(\mathbf{S}^e\mathbf{F}^{v\,-\top})\,\mathbf{F}^{\top}\big] : \boldsymbol{\Omega}\, dv_{\mathrm{R}} \\ &= \int_{\mathcal{P}} (\mathbf{T}_{\mathrm{R}}\mathbf{F}^{\top}) : \boldsymbol{\Omega}\, dv_{\mathrm{R}}. \end{aligned} \tag{4.2.34}$$

Since the part $\mathcal{P}$ is arbitrary, we obtain that $(\mathbf{T}_{\mathrm{R}}\mathbf{F}^{\top}) : \boldsymbol{\Omega} = 0$ for all skew tensors $\boldsymbol{\Omega}$, which implies that $(\mathbf{T}_{\mathrm{R}}\mathbf{F}^{\top})$ is *symmetric*:

$$\mathbf{T}_{\mathrm{R}}\mathbf{F}^{\top} = \mathbf{F}\mathbf{T}_{\mathrm{R}}^{\top}. \tag{4.2.35}$$

In view of (4.2.32), (4.2.33), and (4.2.35),

- *the stress* $\mathbf{T}_{\mathrm{R}}$ *may be viewed as the classical Piola stress, with* (4.2.32) *the classical traction condition, and* (4.2.33) *and* (4.2.35) *representing the local macroscopic force and moment balances in the reference body.*

Next, as is standard, the Piola stress $\mathbf{T}_{\mathrm{R}}$ is related to the symmetric Cauchy stress $\mathbf{T}$ in the deformed body by

$$\mathbf{T}_{\mathrm{R}} = J\,\mathbf{T}\,\mathbf{F}^{-\top}, \tag{4.2.36}$$

so that

$$\mathbf{T} = J^{-1}\mathbf{T}_{\mathrm{R}}\mathbf{F}^{\top}. \tag{4.2.37}$$

Using (4.2.30) in (4.2.37), together with $\mathbf{F}^{e\top} = \mathbf{F}^{v\,-\top}\,\mathbf{F}^{\top}$, yields that

$$\mathbf{T} \stackrel{\text{def}}{=} J^{e-1}\,\mathbf{S}^{e}\mathbf{F}^{e\top}. \tag{4.2.38}$$

It is convenient to introduce two new stress measures:

- The *elastic second Piola stress*,

$$\mathbf{T}^{e} \stackrel{\text{def}}{=} J^{e}\,\mathbf{F}^{e\,-1}\mathbf{T}\mathbf{F}^{e\,-\top}, \tag{4.2.39}$$

  which is *symmetric* on account of the symmetry of the Cauchy stress $\mathbf{T}$.

- The *Mandel stress*,

$$\mathbf{M}^{e} \stackrel{\text{def}}{=} \mathbf{C}^{e}\mathbf{T}^{e} = J^{e}\mathbf{F}^{e\top}\mathbf{T}\mathbf{F}^{e\,-\top}, \tag{4.2.40}$$

  which in general is *not symmetric*.

Thus, using the definitions (4.2.38)–(4.2.40) we find that

$$\mathbf{F}^{e\,-1}\mathbf{S}^{e} = \mathbf{T}^{e} \quad \text{and} \quad \mathbf{F}^{e\top}\mathbf{S}^{e} = \mathbf{M}^{e}. \tag{4.2.41}$$

Microscopic force balance

To discuss the microscopic counterpart of the macroscopic force balance, we consider a generalized virtual velocity with $\tilde{\mathbf{u}} \equiv \mathbf{0}$. Choose the virtual field $\tilde{\mathbf{D}}^{v}$ arbitrarily, and let

$$\tilde{\mathbf{F}}^{e} = -\mathbf{F}^{e}\tilde{\mathbf{D}}^{v},$$

so that

$$\mathbf{S}^{e} : \tilde{\mathbf{F}}^{e} = -(\mathbf{F}^{e\top}\mathbf{S}^{e}) : \tilde{\mathbf{D}}^{v} = -(\mathrm{sym}_0\mathbf{M}^{e}) : \tilde{\mathbf{D}}^{v} \tag{4.2.42}$$

where we have used $(4.2.41)_2$ and the fact that $\tilde{\mathbf{D}}^{v}$ is symmetric and deviatoric. The power balance (4.2.27) and (4.2.42) yield the *microscopic virtual-power relation*

$$0 = \int_{\mathcal{P}} \Big(\mathbf{T}^{v} - \mathrm{sym}_0\mathbf{M}^{e}\Big) : \tilde{\mathbf{D}}^{v} dv_{\mathrm{R}} \tag{4.2.43}$$

to be satisfied for all $\tilde{\mathbf{D}}^v$ and all $\mathcal{P}$. This yields the *microscopic force balance*

$$\text{sym}_0\mathbf{M}^e = \mathbf{T}^v, \tag{4.2.44}$$

which characterizes the interaction between internal forces associated with the elastic and viscous responses of the material.

Actual external and internal powers

Using the traction condition (4.2.32) the actual external expenditure of power $(4.2.22)_1$ may be written as

$$\mathcal{W}_{\text{ext}}(\mathcal{P}) = \int_{\partial\mathcal{P}} (\mathbf{T}_{\text{R}}\mathbf{n}_{\text{R}}) \cdot \dot{\mathbf{u}}\, da_{\text{R}} + \int_{\mathcal{P}} \mathbf{b}_{\text{R}} \cdot \dot{\mathbf{u}}\, dv_{\text{R}}. \tag{4.2.45}$$

Also, using $(4.2.12)_1$, the stress power $\mathbf{S}^e : \dot{\mathbf{F}}^e$ may be alternatively written as

$$\mathbf{S}^e : \dot{\mathbf{F}}^e = \tfrac{1}{2}\,\mathbf{T}^e : \dot{\mathbf{C}}^e, \tag{4.2.46}$$

using which the actual internal expenditure of power $(4.2.22)_2$ may be written as

$$\mathcal{W}_{\text{int}}(\mathcal{P}) = \int_{\mathcal{P}} \left(\tfrac{1}{2}\,\mathbf{T}^e : \dot{\mathbf{C}}^e + \mathbf{T}^v : \mathbf{D}^v \right) dv_{\text{R}}. \tag{4.2.47}$$

Finally, power balance requires that

$$\mathcal{W}_{\text{ext}}(\mathcal{P}) = \mathcal{W}_{\text{int}}(\mathcal{P}). \tag{4.2.48}$$

4.2.3 **Free-energy imbalance**

Under isothermal conditions the two laws of thermodynamics reduce to the statement that *the temporal increase in free energy of any part $\mathcal{P}$ is less than or equal to the power expended on $\mathcal{P}$.* Precisely, letting ψ_{R} denote the free energy per unit reference volume, this requirement takes the form of a *free-energy imbalance* (Gurtin et al., 2010)

$$\overline{\int_{\mathcal{P}} \psi_{\text{R}}\, dv_{\text{R}}}^{\,\boldsymbol{\cdot}} \leq \mathcal{W}_{\text{ext}}(\mathcal{P}). \tag{4.2.49}$$

Since $\mathcal{W}_{\text{ext}}(\mathcal{P}) = \mathcal{W}_{\text{int}}(\mathcal{P})$, using (4.2.47) we obtain

$$\int_{\mathcal{P}} \left(\dot{\psi}_{\text{R}} - \tfrac{1}{2}\,\mathbf{T}^e : \dot{\mathbf{C}}^e - \mathbf{T}^v : \mathbf{D}^v \right) dv_{\text{R}} \leq 0, \tag{4.2.50}$$

which upon using the fact that (4.2.50) must hold for all parts $\mathcal{P}$, yields the following local *free-energy imbalance under isothermal conditions*

$$\dot{\psi}_{\text{R}} - \tfrac{1}{2}\,\mathbf{T}^e : \dot{\mathbf{C}}^e - \mathbf{T}^v : \mathbf{D}^v \leq 0. \tag{4.2.51}$$

REMARK

For brevity we have not discussed invariance properties of the various fields appearing in our theory. However, such considerations are straightforward and extensively elaborated upon in Gurtin et al. (2010) in the context of a elasto-(visco)plasticity theory. Here, we simply note that all quantities in the free-energy imbalance (4.2.51) are invariant under a change in frame.

4.2.4 **Constitutive theory**

Guided by the free-energy imbalance (4.2.51), we consider the following pair of constitutive equations for the free energy ψ_{R} and the elastic second Piola stress $\mathbf{T}^e$,

$$\psi_{\mathrm{R}} = \hat{\psi}_{\mathrm{R}}(\mathbf{C}^e), \qquad \mathbf{T}^e = \hat{\mathbf{T}}^e(\mathbf{C}^e). \tag{4.2.52}$$

From $(4.2.52)_1$,

$$\dot{\psi}_{\mathrm{R}} = \frac{\partial \hat{\psi}_{\mathrm{R}}(\mathbf{C}^e)}{\partial \mathbf{C}^e} : \dot{\mathbf{C}}^e.$$

Hence, for thermodynamic consistency, the constitutive equations (4.2.52) must satisfy

$$\left(\frac{\partial \hat{\psi}_{\mathrm{R}}(\mathbf{C}^e)}{\partial \mathbf{C}^e} - \frac{1}{2}\hat{\mathbf{T}}^e(\mathbf{C}^e) \right) : \dot{\mathbf{C}}^e - \mathbf{T}^v : \mathbf{D}^v \le 0. \tag{4.2.53}$$

We assume that the free-energy function $\hat{\psi}_{\mathrm{R}}(\mathbf{C}^e)$ delivers the stresses $\mathbf{T}^e$ through the state relation

$$\mathbf{T}^e = 2\,\frac{\partial \hat{\psi}_{\mathrm{R}}(\mathbf{C}^e)}{\partial \mathbf{C}^e}. \tag{4.2.54}$$

We are then left with the following reduced dissipation inequality

$$\mathcal{D} = \mathbf{T}^v : \mathbf{D}^v \ge 0. \tag{4.2.55}$$

Isotropic free energy

For isotropic materials the free-energy functions $\hat{\psi}_{\mathrm{R}}(\mathbf{C}^e)$ is an isotropic function of $\mathbf{C}^e$. An immediate consequence is that this free-energy function has the representation

$$\hat{\psi}_{\mathrm{R}}(\mathbf{C}^e) = \tilde{\psi}_{\mathrm{R}}(\mathcal{I}_{\mathbf{C}^e}), \tag{4.2.56}$$

with

$$\mathcal{I}_{\mathbf{C}^e} = \Big(I_1(\mathbf{C}^e), I_2(\mathbf{C}^e), I_3(\mathbf{C}^e) \Big)$$

the list of principal invariants of $\mathbf{C}^e$. Thus, from (4.2.54) it follows that

$$\mathbf{T}^e = 2\,\frac{\partial\tilde{\psi}_{\mathrm{R}}(\mathcal{I}_{\mathbf{C}^e})}{\partial\mathbf{C}^e}, \tag{4.2.57}$$

and that $\mathbf{T}^e$ is an *isotropic function of* $\mathbf{C}^e$. Hence, $\mathbf{T}^e$ and $\mathbf{C}^e$ commute,

$$\mathbf{C}^e\mathbf{T}^e = \mathbf{T}^e\mathbf{C}^e. \tag{4.2.58}$$

Then, since the Mandel stresses are defined by, cf. (4.2.40),

$$\mathbf{M}^e = \mathbf{C}^e\mathbf{T}^e, \tag{4.2.59}$$

we find that for isotropic materials the Mandel stresses $\mathbf{M}^e$ is *symmetric*.

Further, from (4.2.39) the Cauchy stress is given by $\mathbf{T} = J^{e-1}\mathbf{F}^e\mathbf{T}^e\mathbf{F}^{e\top}$. Hence, using (4.2.57)

$$\mathbf{T} = J^{e-1}\left(2\mathbf{F}^e\,\frac{\partial\tilde{\psi}_{\mathrm{R}}(\mathcal{I}_{\mathbf{C}^e})}{\partial\mathbf{C}^e}\,\mathbf{F}^{e\top}\right). \tag{4.2.60}$$

Constitutive equations for the viscous response

Recall the reduced dissipation inequality (4.2.55), viz.

$$\mathcal{D} = \mathbf{T}^v : \mathbf{D}^v \geq 0. \tag{4.2.61}$$

Upon using the microforce balance (4.2.44), together with the symmetry of the Mandel stress discussed in reference to (4.2.59), this dissipation inequality may be re-expressed as

$$\mathcal{D} = \mathbf{M}^e_0 : \mathbf{D}^v \geq 0. \tag{4.2.62}$$

Let

$$\mathbf{N}^v = \frac{\mathbf{D}^v}{|\mathbf{D}^v|} \qquad \text{when} \qquad \mathbf{D}^v \neq \mathbf{0}, \tag{4.2.63}$$

define the *viscous flow direction*, and let

$$\dot{\bar{\gamma}}^v \stackrel{\text{def}}{=} \sqrt{2}|\mathbf{D}^v|, \tag{4.2.64}$$

define an *equivalent viscous shear strain rate*. Then the viscous stretching $\mathbf{D}^v$ may be written as

$$\mathbf{D}^v = \frac{1}{\sqrt{2}}\,\dot{\bar{\gamma}}^v\mathbf{N}^v. \tag{4.2.65}$$

Using eq. (4.2.65) in the dissipation, inequality (4.2.62) may be re-expressed as

$$\mathcal{D} = \left(\frac{1}{\sqrt{2}}\mathbf{M}^e_0 : \mathbf{N}^v\right)\dot{\bar{\gamma}}^v \geq 0. \tag{4.2.66}$$

Guided by (4.2.66),

- we assume that viscous flow direction $\mathbf{N}^v$ and the deviatoric Mandel stress are *co-directional*, that is

$$\mathbf{N}^v = \frac{\mathbf{M}_0^e}{|\mathbf{M}_0^e|}. \tag{4.2.67}$$

Then,

$$\mathcal{D} = \bar{\tau}\dot{\bar{\gamma}}^v \geq 0, \tag{4.2.68}$$

where

$$\bar{\tau} \stackrel{\text{def}}{=} \frac{1}{\sqrt{2}}|\mathbf{M}_0^e|, \tag{4.2.69}$$

defines an *equivalent shear stress*, and the dissipation $\mathcal{D}$ is then trivially non-negative.

Further, in view of (4.2.68) we assume that whenever there is viscous flow, that is whenever $\dot{\bar{\gamma}}^v > 0$, the equivalent shear stress $\bar{\tau}$ is constrained to satisfy

$$\bar{\tau} = \mathcal{S}(\dot{\bar{\gamma}}^v), \tag{4.2.70}$$

where $\mathcal{S}(\dot{\bar{\gamma}}^v)$ represents a *shear flow strength* of the material at a given $\dot{\bar{\gamma}}^v$. We assume that

$$\mathcal{S}(\dot{\bar{\gamma}}^v) \text{ is a positive-valued strictly increasing function of } \dot{\bar{\gamma}}^v, \tag{4.2.71}$$

so that the function $\mathcal{S}(\dot{\bar{\gamma}}^v)$ is *invertible*, with inverse

$$\dot{\bar{\gamma}}^v = \mathcal{F}(\bar{\tau}) \geq 0. \tag{4.2.72}$$

To summarize, using the constitutive assumptions above, the evolution equation for $\mathbf{F}^v$ is

$$\dot{\mathbf{F}}^v = \mathbf{D}^v\,\mathbf{F}^v, \tag{4.2.73}$$

with

$$\mathbf{D}^v = \dot{\bar{\gamma}}^v\left(\frac{\mathbf{M}_0^e}{2\,\bar{\tau}}\right), \tag{4.2.74}$$

and the scalar equivalent shear strain rate $\dot{\bar{\gamma}}^v$ given by (4.2.72).

4.2.5 Specialization of the constitutive equations

Free energy

For slightly compressible materials, with $J^e = \sqrt{\det \mathbf{C}^e}$ denoting the elastic volumetric Jacobian and $\bar{\mathbf{C}}^e \stackrel{\text{def}}{=} (J^e)^{-2/3}\mathbf{C}^e$ denoting the distortional elastic right Cauchy–Green tensor, we take the free energy to be given as a sum of a contribution from a distortional energy and a volumetric energy,

$$\psi_{\mathrm{R}} = \underbrace{\psi_{\mathrm{R}}(\mathcal{I}_{\bar{\mathbf{C}}^e})}_{\text{Distortional energy}} + \underbrace{\psi_{\mathrm{R}}^{\mathrm{vol}}(J^e)}_{\text{Volumetric energy}}, \tag{4.2.75}$$

where $\mathcal{I}_{\bar{\mathbf{C}}^e}$ is a list of the principal invariants of $\bar{\mathbf{C}}^e$. As a simple model for the distortional energy we adopt a Neo-Hookean form,

$$\psi_{\text{R}}(\mathcal{I}_{\bar{\mathbf{C}}^e}) = \tfrac{1}{2}\, G \left(\bar{I}_1^e - 3\right), \tag{4.2.76}$$

where

$$\bar{I}_1^e \overset{\text{def}}{=} \operatorname{tr} \bar{\mathbf{C}}^e \tag{4.2.77}$$

is the first invariant of $\bar{\mathbf{C}}^e$, and $G > 0$ is a shear modulus. Also, the volumetric energy is taken as

$$\psi_{\text{R}}^{\text{vol}}(J^e) = \tfrac{1}{2}\, K \,(J^e - 1)^2, \tag{4.2.78}$$

where $K > 0$ is a bulk modulus.[4] Then, using (4.2.57), this free energy (after a bit of algebra) gives the elastic second Piola stress as

$$\mathbf{T}^e = \frac{2}{(J^e)^{2/3}} \left[\frac{\partial \psi_{\text{R}}(\mathcal{I}_{\bar{\mathbf{C}}^e})}{\partial \bar{\mathbf{C}}^e} - \frac{1}{3}\left(\bar{\mathbf{C}}^e : \frac{\partial \psi_{\text{R}}(\mathcal{I}_{\bar{\mathbf{C}}^e})}{\partial \bar{\mathbf{C}}^e} \right) \bar{\mathbf{C}}^{e-1} \right] + J^e \frac{\partial \psi_{\text{R}}^{\text{vol}}(J^e)}{\partial J^e} \bar{\mathbf{C}}^{e-1}, \tag{4.2.79}$$

which for the slightly compressible Neo-Hookean energy under consideration gives

$$\mathbf{T}^e = (J^e)^{-2/3} G \left[\mathbf{1} - \frac{1}{3} (\operatorname{tr} \bar{\mathbf{C}}^e) \bar{\mathbf{C}}^{e-1} \right] + K\, J^e (J^e - 1) \bar{\mathbf{C}}^{e-1}. \tag{4.2.80}$$

Then, using (4.2.59), the Mandel stress is given by

$$\mathbf{M}^e = G\, \bar{\mathbf{C}}_0^e + K\, (J^e)^{5/3} (J^e - 1) \mathbf{1}, \tag{4.2.81}$$

which shows that it is symmetric.

Finally, using $\mathbf{T} = J^{e-1} \mathbf{F}^e \mathbf{T}^e \mathbf{F}^{e\top}$, the Cauchy stress is given by

$$\mathbf{T} = (J^e)^{-1}\, G\, \bar{\mathbf{B}}_0^e + K\, (J^e)^{2/3}\, (J^e - 1) \mathbf{1}. \tag{4.2.82}$$

Note that in the stress relations (4.2.80), (4.2.81), and (4.2.37),

$$J^e = J,$$

because we have assumed viscous flow to be incompressible.

[4] We note that forms such as $\frac{1}{2}\, K\, (J^e - 1)^2$ or $\frac{1}{2}\, K\, (\ln J^e)^2$ for $\psi_{\text{R}}^{\text{vol}}(J^e)$ as used in other chapters produce essentially identical results since values of J^e are always close to unity for slightly compressible materials.

Viscous flow rule

As a simple special case for the shear flow strength function $\mathcal{S}(\dot{\bar{\gamma}}^v)$ we assume that it is given by

$$\mathcal{S}(\dot{\bar{\gamma}}^v) = \eta\,\dot{\bar{\gamma}}^v, \tag{4.2.83}$$

so that strength relation (4.2.70) has the familiar form, often referred to as linearly viscous,

$$\bar{\tau} = \eta\,\dot{\bar{\gamma}}^v, \tag{4.2.84}$$

with $\eta > 0$ a constant *viscosity* parameter.

In this case the evolution equation for $\mathbf{F}^v$ is

$$\dot{\mathbf{F}}^v = \mathbf{D}^v\,\mathbf{F}^v, \tag{4.2.85}$$

with

$$\mathbf{D}^v = \frac{\mathbf{M}_0^e}{2\,\eta}. \tag{4.2.86}$$

Using (4.2.81) and (4.2.86) we may rewrite the viscous stretching relation (4.2.86) alternatively as

$$\mathbf{D}^v = \frac{1}{2\,\tau_r}\bar{\mathbf{C}}_0^e \qquad \text{with} \qquad \tau_r \overset{\text{def}}{=} \frac{\eta}{G}, \tag{4.2.87}$$

where τ_r represents a *relaxation time.*

The complete set of material parameters appearing in this simple viscoelasticity theory are

$$\{G, K, \eta\}, \tag{4.2.88}$$

or alternatively in terms of the relaxation time $\tau_r = \eta/G$ as the list

$$\{G, K, \tau_r\}. \tag{4.2.89}$$

REMARKS

1. The strength function (4.2.83) with a constant Newtonian viscosity, η, is the *simplest* strength function for viscoelastic materials. Another commonly used form of the strength function is the power-law relation (cf., e.g., Norton, 1929; Dorn, 1955)

$$\mathcal{S}(\dot{\bar{\gamma}}^v) = \mathrm{S}_0\left(\frac{\dot{\bar{\gamma}}^v}{\dot{\gamma}_0}\right)^m, \tag{4.2.90}$$

in which $\mathrm{S}_0 > 0$ is a *reference flow strength,* $\dot{\gamma}_0 > 0$ is a *reference shear rate,* and $m \in (0, 1]$ is *rate-sensitivity parameter.*

The limit $m \to 1$ of (4.2.90) recovers the *linearly viscous response*,

$$\mathcal{S}(\dot{\bar{\gamma}}^v) = \eta \dot{\bar{\gamma}}^v \qquad \text{with} \qquad \eta \equiv \frac{\mathrm{S}_0}{\dot{\gamma}_0}, \tag{4.2.91}$$

as in (4.2.83). On the other hand, the limit $m \to 0$ is

$$\mathcal{S}(\dot{\bar{\gamma}}^v) = \mathrm{S}_0, \tag{4.2.92}$$

which is commonly used to model *rate-independent, perfectly plastic materials*.
The strength relation corresponding to (4.2.90) is

$$\bar{\tau} = \mathrm{S}_0 \left(\frac{\dot{\bar{\gamma}}^v}{\dot{\gamma}_0} \right)^m, \tag{4.2.93}$$

a relation which is easily inverted to obtain

$$\dot{\bar{\gamma}}^v = \dot{\gamma}_0 \left(\frac{\bar{\tau}}{\mathrm{S}_0} \right)^{1/m}. \tag{4.2.94}$$

This *power-law* form of the strength relation with $0 < m < 1$ is widely used to describe the (non-hardening) *viscoplasticity* of solids (cf., e.g., Anand, 1982, 1985; Narayan and Anand, 2018).

2. An extension of the simple power-law form (4.2.93) which is useful over a large range of strain rates is (Garofalo, 1963; Brown et al., 1989)

$$\bar{\tau} = \mathrm{S}_0 \sinh^{-1} \left[\left(\frac{\dot{\bar{\gamma}}^v}{\dot{\gamma}_0} \right)^m \right], \tag{4.2.95}$$

with inverted form

$$\dot{\bar{\gamma}}^v = \dot{\gamma}_0 \left[\sinh \left(\frac{\bar{\tau}}{\mathrm{S}_0} \right) \right]^{1/m}. \tag{4.2.96}$$

See Brown et al. (1989) for additional extensions of this hyperbolic-sine type functional form, which account for thermal effects as well as strain-hardening.

4.3 A theory based on a multi-branch multiplicative decomposition of $\mathbf{F}$

In this section we summarize a straightforward generalization of the single Maxwell element-based theory of viscoelasticity formulated in Section 4.2, to a theory based on multiple spring–dashpot Maxwell elements in parallel. The theory that we consider is based on a *multi-branch multiplicative decomposition* of the deformation gradient $\mathbf{F}$ of the form,

$$\mathbf{F} = \mathbf{F}^{e(\alpha)}\mathbf{F}^{v(\alpha)} \qquad \alpha = 0, 1, \ldots, M, \tag{4.3.1}$$

where $\mathbf{F}^{v(\alpha)}$ and $\mathbf{F}^{e(\alpha)}$ represent *viscous and elastic distortions*, respectively. The theory considered here is in the spirit of the existing theories of viscoelasticity of elastomeric materials in the literature which use internal variables with attendant evolution equations to capture time-dependence (cf., e.g., Sidoroff, 1974; Lubliner, 1985; Bergstrom and Boyce, 1998; Reese and Govindjee, 1998; Wijaya et al., 2023, and the numerous references therein), but it differs in several details of the continuum-mechanical formulation. For a detailed formal derivation of the multi-mechanism theory that we consider here, see Mao et al. (2017). The theory relates the following basic fields:

$\mathbf{x} = \boldsymbol{\chi}(\mathbf{X}, t)$,	motion;
$\mathbf{F} = \nabla\boldsymbol{\chi}, \quad J = \det\mathbf{F} > 0$,	deformation gradient;
$\mathbf{C} = \mathbf{F}^{\top}\mathbf{F}$,	right Cauchy–Green tensor;
$\mathbf{B} = \mathbf{F}\mathbf{F}^{\top}$,	left Cauchy–Green tensor;
$\bar{\mathbf{F}} = J^{-1/3}\mathbf{F}, \quad \det\bar{\mathbf{F}} = 1$,	isochoric part of $\mathbf{F}$;
$\bar{\mathbf{C}} = \bar{\mathbf{F}}^{\top}\bar{\mathbf{F}}, \quad \det\bar{\mathbf{C}} = 1$,	isochoric part of $\mathbf{C}$;
$\bar{\mathbf{B}} = \bar{\mathbf{F}}\bar{\mathbf{F}}^{\top}, \quad \det\bar{\mathbf{B}} = 1$,	isochoric part of $\mathbf{B}$;
$\mathcal{I}_{\bar{\mathbf{C}}}$ and $\mathcal{I}_{\bar{\mathbf{B}}}$,	list of principal invariants of $\bar{\mathbf{C}}$ and $\bar{\mathbf{B}}$;
$\bar{\lambda} = \sqrt{\operatorname{tr}\bar{\mathbf{C}}/3} = \sqrt{\operatorname{tr}\bar{\mathbf{B}}/3}$,	effective distortional stretch;
$\mathbf{F} = \mathbf{F}^{e(\alpha)}\mathbf{F}^{v(\alpha)} \quad \alpha = 1, \ldots, M$,	multi–branch multiplicative decomposition of $\mathbf{F}$;
$\mathbf{F}^{v(\alpha)}, \ J^{v(\alpha)} = \det\mathbf{F}^{v(\alpha)} = 1$,	isochoric viscous distortion for the αth branch;
$\mathbf{C}^{v(\alpha)} = \mathbf{F}^{v(\alpha)\top}\mathbf{F}^{v(\alpha)}$,	viscous right Cauchy–Green tensors;
$\mathbf{F}^{e(\alpha)}, \ J^{e(\alpha)} = \det\mathbf{F}^{e(\alpha)} > 0$,	elastic distortion for the αth branch;
$\mathbf{C}^{e(\alpha)} = \mathbf{F}^{e(\alpha)\top}\mathbf{F}^{e(\alpha)}$,	elastic right Cauchy–Green tensors;
$\mathbf{B}^{e(\alpha)} = \mathbf{F}^{e(\alpha)}\mathbf{F}^{e(\alpha)\top}$,	elastic left Cauchy–Green tensors;
$\bar{\mathbf{F}}^{e(\alpha)} = \bar{\mathbf{F}}(\mathbf{F}^{v(\alpha)})^{-1} \ \det\bar{\mathbf{F}}^{e(\alpha)} = 1$,	isochoric part of $\mathbf{F}^{e(\alpha)}$;
$\bar{\mathbf{C}}^{e(\alpha)} = \bar{\mathbf{F}}^{e(\alpha)\top}\bar{\mathbf{F}}^{e(\alpha)}, \ \det\bar{\mathbf{C}}^{e(\alpha)} = 1$,	isochoric elastic right Cauchy–Green tensors;
$\bar{\mathbf{B}}^{e(\alpha)} = \bar{\mathbf{F}}^{e(\alpha)}\bar{\mathbf{F}}^{e(\alpha)\top}, \ \det\bar{\mathbf{B}}^{e(\alpha)} = 1$,	isochoric elastic left Cauchy–Green tensors;
$\mathcal{I}_{\bar{\mathbf{C}}^{e(\alpha)}}$,	list of principal invariants of $\bar{\mathbf{C}}^{e(\alpha)}$;
$\mathbf{T}^{(\alpha)}$,	Cauchy stress for the αth branch;
$\mathbf{T}^{e(\alpha)} = J^{e(\alpha)}\,\mathbf{F}^{e(\alpha)-1}\mathbf{T}^{(\alpha)}\mathbf{F}^{e(\alpha)-\top}$,	elastic second Piola stress for the αth branch;
$\mathbf{M}^{e(\alpha)} = \mathbf{C}^{e(\alpha)}\mathbf{T}^{e(\alpha)}$,	Mandel stress for the αth branch;

$\mathbf{T} = \sum_\alpha \mathbf{T}^{(\alpha)}$, Cauchy stress;

$\mathbf{T}_{\rm R} = J\mathbf{T}\mathbf{F}^{-\top}$, Piola stress;

$\psi_{\rm R}$, free energy density per unit reference volume.

- We limit our attention to situations under which the material may be idealized to be *isotropic*. Accordingly, all constitutive functions considered below are presumed to be isotropic in character.
- Also, when convenient, we shall use the shorthand

$$\sum_\alpha = \sum_{\alpha=0}^{M}, \tag{4.3.2}$$

1. **Free energy:**
We assume an additive decomposition of the free energy,

$$\psi_{\rm R} = \sum_\alpha \tilde{\psi}_{\rm R}^{(\alpha)}(\mathcal{I}_{\mathbf{C}^{e(\alpha)}}), \tag{4.3.3}$$

where $\tilde{\psi}_{\rm R}^{(\alpha)}$ represents the contribution to the free energy from the αth branch, with $\mathcal{I}_{\mathbf{C}^{e(\alpha)}}$ a list of the principal invariants of $\mathbf{C}^{e(\alpha)}$.

2. **Free-energy imbalance**: The multi-branch extension of (4.2.51) may be shown to be given by

$$\dot{\psi}_{\rm R} - \sum_\alpha \tfrac{1}{2}\mathbf{T}^{e(\alpha)} : \dot{\mathbf{C}}^{e(\alpha)} - \textstyle\sum_\alpha \mathbf{M}_0^{e(\alpha)} : \mathbf{D}^{v(\alpha)} \le 0. \tag{4.3.4}$$

3. **Elastic second Piola stress. Mandel Stress. Piola stress. Cauchy stress:**
For each branch the elastic second Piola stress is given by

$$\mathbf{T}^{e(\alpha)} = 2\,\frac{\partial \tilde{\psi}_{\rm R}^{(\alpha)}(\mathcal{I}_{\mathbf{C}^{e(\alpha)}})}{\partial \mathbf{C}^{e(\alpha)}}, \tag{4.3.5}$$

the symmetric Mandel stress by

$$\mathbf{M}^{e(\alpha)} = \mathbf{C}^{e(\alpha)}\mathbf{T}^{e(\alpha)}, \tag{4.3.6}$$

and the contribution $\mathbf{T}^{(\alpha)}$ to the Cauchy stress $\mathbf{T}$ by

$$\mathbf{T}^{(\alpha)} = (J^{e(\alpha)})^{-1}\Big(2\mathbf{F}^{e(\alpha)}\frac{\partial \tilde{\psi}_{\rm R}^{(\alpha)}(\mathcal{I}_{\mathbf{C}^{e(\alpha)}})}{\partial \mathbf{C}^{e(\alpha)}}\mathbf{F}^{e(\alpha)\top}\Big). \tag{4.3.7}$$

The total Cauchy stress $\mathbf{T} = \sum_\alpha \mathbf{T}^{(\alpha)}$ is then given by

$$\mathbf{T} = \sum_\alpha (J^{e(\alpha)})^{-1}\Big(2\mathbf{F}^{e(\alpha)}\frac{\partial \tilde{\psi}_{\rm R}^{(\alpha)}(\mathcal{I}_{\mathbf{C}^{e(\alpha)}})}{\partial \mathbf{C}^{e(\alpha)}}\mathbf{F}^{e(\alpha)\top}\Big), \tag{4.3.8}$$

and the total Piola stress $\mathbf{T}_{\mathrm{R}}$ is

$$\mathbf{T}_{\mathrm{R}} = J\mathbf{T}\mathbf{F}^{-\top}. \tag{4.3.9}$$

4. **Viscous flow rules:**
The evolution equation for each $\mathbf{F}^{v(\alpha)}$ is[5]

$$\dot{\mathbf{F}}^{v\,(\alpha)} = \mathbf{D}^{v(\alpha)}\,\mathbf{F}^{v(\alpha)}, \tag{4.3.10}$$

with the viscous stretching given by

$$\mathbf{D}^{v(\alpha)} = \dot{\bar{\gamma}}^{v(\alpha)}\left(\frac{\mathbf{M}_0^{e(\alpha)}}{2\,\bar{\tau}^{(\alpha)}}\right), \tag{4.3.11}$$

where

$$\bar{\tau}^{(\alpha)} \stackrel{\text{def}}{=} \frac{1}{\sqrt{2}}\,|\mathbf{M}_0^{e(\alpha)}| \tag{4.3.12}$$

defines an *equivalent shear stress* for each α, and

$$\dot{\bar{\gamma}}^{v(\alpha)} \stackrel{\text{def}}{=} \sqrt{2}\,|\mathbf{D}^{v(\alpha)}| \tag{4.3.13}$$

is an *equivalent viscous shear strain rate.*

We assume that $\bar{\tau}^{(\alpha)}$ satisfies the strength relation

$$\bar{\tau}^{(\alpha)} = \mathcal{S}^{(\alpha)}(\dot{\bar{\gamma}}^{v(\alpha)}), \tag{4.3.14}$$

with $\mathcal{S}^{(\alpha)}(\dot{\bar{\gamma}}^{v(\alpha)})$ a positive-valued strictly increasing function of $\dot{\bar{\gamma}}^{v(\alpha)}$, so that the function $\mathcal{S}^{(\alpha)}(\dot{\bar{\gamma}}^{v(\alpha)})$ is *invertible*, with

$$\dot{\bar{\gamma}}^{v(\alpha)} = \mathcal{F}^{(\alpha)}(\bar{\tau}^{(\alpha)}) \geq 0. \tag{4.3.15}$$

5. **Total dissipation rate:**
The total rate of viscous dissipation per unit reference volume is

$$\mathcal{D} = \sum_{\alpha} \mathcal{S}^{(\alpha)}(\dot{\bar{\gamma}}^{v(\alpha)})\,\dot{\bar{\gamma}}^{v(\alpha)} \geq 0. \tag{4.3.16}$$

[5] For isotropic theories utilizing a multiplicative decomposition of $\mathbf{F}$, it is widely assumed that the viscous flow is *irrotational*, in the sense that $\mathbf{W}^{v(\alpha)} = \mathbf{0}$ (cf., e.g., Gurtin et al., 2010, Section 97 in the context of plasticity theory).

4.3.1 Specialization of the constitutive equations

In order to model the response of *solid* viscoelastic elastomers we introduce the following specializations:

(i) We take the first branch, $\alpha = 0$, to represent the *equilibrium elastic response* of the material. We do not associate any viscous relaxation with this network. Accordingly we set $\mathbf{F}^{v(0)} = \mathbf{1}$, and hence

$$\mathbf{F}^{e(0)} = \mathbf{F}. \tag{4.3.17}$$

(ii) The remaining branches, $\alpha = 1, \dots, M$, represent the *non-equilibrium viscoelastic response* of the material.

Henceforth, when it is important to distinguish between the zeroth branch and the other branches of the rheological model, we will use the descriptive terms "equilibrium" for branch $\alpha = 0$ and "non-equilibrium" for branches $\alpha = 1, \dots, M$.

Note that an important feature of the multi-branch decomposition $\mathbf{F} = \mathbf{F}^{e(\alpha)}\mathbf{F}^{v(\alpha)}$, with $\det \mathbf{F}^{v(\alpha)} = 1$, is that

$$J^{e(\alpha)} = J \quad \text{for all} \quad \alpha = 1, \dots, M. \tag{4.3.18}$$

4.3.2 Free energy

We take the free energy to be given as a sum of a contribution from a volumetric energy and a distortional energy, where the distortional energy is the sum of contributions from the equilibrium branch and contributions from non-equilibrium branches,

$$\psi_{\mathrm{R}} = \underbrace{\psi_{\mathrm{R}}^{\mathrm{eq\,vol}}(J)}_{\text{Volumetric energy}} \underbrace{+\,\psi_{\mathrm{R}}^{\mathrm{eq}}(\mathcal{I}_{\bar{\mathbf{C}}}) + \sum_{\alpha} \psi_{\mathrm{R}}^{\mathrm{neq}(\alpha)}(\mathcal{I}_{\bar{\mathbf{C}}^{e(\alpha)}})}_{\text{Distortional energy}}, \tag{4.3.19}$$

where now instead of (4.3.2) the shorthand summation notation stands for

$$\sum_{\alpha} = \sum_{\alpha=1}^{M}. \tag{4.3.20}$$

Figure 4.3 shows a spring–dashpot representation of the constitutive assumption (4.3.19).

Special forms for the equilibrium and non-equilibrium free energies in equation (4.3.19) are taken as follows:

1. *Equilibrium free energy*: For a nearly incompressible material, with $J = \sqrt{\det \mathbf{C}}$ and $\bar{\mathbf{C}} \stackrel{\text{def}}{=} J^{-2/3}\mathbf{C}$, the volumetric equilibrium free energy is given by

$$\psi_{\mathrm{R}}^{\mathrm{eq\,vol}}(J) = \tfrac{1}{2} K (J-1)^2. \tag{4.3.21}$$

Also, with

$$\bar{\lambda} \stackrel{\text{def}}{=} \frac{1}{\sqrt{3}} \sqrt{\operatorname{tr} \bar{\mathbf{C}}} \tag{4.3.22}$$

defining an *effective distortional stretch* we assume that $\bar{\psi}_{\mathrm{R}}^{\mathrm{eq}}(\bar{\mathbf{C}})$ is given by (Arruda and Boyce, 1993)

$$\bar{\psi}_{\mathrm{R}}^{\mathrm{eq}}(\bar{\mathbf{C}}) = G_{0_{\mathrm{eq}}}\lambda_L^2\left[\left(\frac{\bar{\lambda}}{\lambda_L}\right)\beta + \ln\left(\frac{\beta}{\sinh\beta}\right) - \left(\frac{1}{\lambda_L}\right)\beta_0 - \ln\left(\frac{\beta_0}{\sinh\beta_0}\right)\right], \quad (4.3.23)$$

with

$$\beta = \mathcal{L}^{-1}\left(\bar{\lambda}/\lambda_L\right), \qquad \beta_0 = \mathcal{L}^{-1}\left(1/\lambda_L\right), \quad (4.3.24)$$

where $\mathcal{L}^{-1}$ is the function inverse of the Langevin function, which is defined as

$$\mathcal{L}(\bullet) = \coth(\bullet) - (\bullet)^{-1}. \quad (4.3.25)$$

The equilibrium free energy involves three material parameters:

- K, a bulk modulus,
- $G_{0_{\mathrm{eq}}}$, a ground-state equilibrium shear modulus, and
- λ_L, a network locking stretch.

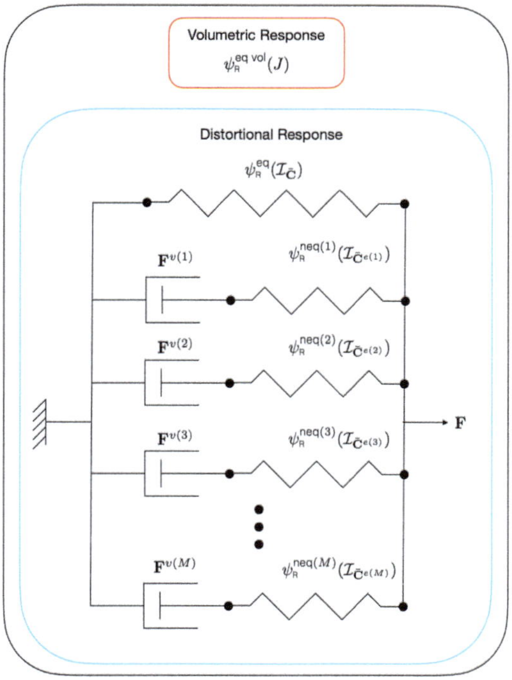

Fig. 4.3 A spring-dashpot representation of the free-energy ψ_{R}. The free energy is given as a sum of a contribution from a volumetric energy and a distortional energy, where the distortional energy is the sum of contributions from an equilibrium branch and contributions from non-equilibrium branches.

Using (4.3.7) the free energy (4.3.23) gives the equilibrium Cauchy stress as

$$\mathbf{T}^{\text{eq}} = J^{-1} G_{\text{eq}} \bar{\mathbf{B}}_0 - p\mathbf{1},$$
$$G_{\text{eq}} \stackrel{\text{def}}{=} G_{0\text{eq}} \left(\frac{\lambda_L}{3\bar{\lambda}}\right) \mathcal{L}^{-1}\left(\frac{\bar{\lambda}}{\lambda_L}\right), \tag{4.3.26}$$
$$p \stackrel{\text{def}}{=} -K(J-1).$$

2. *Non-equilibrium free energy*: To model the non-equilibrium contributions to the free energy we adopt a simple Neo-Hookean form

$$\psi_{\text{R}}^{\text{neq}(\alpha)}(\mathcal{I}_{\bar{\mathbf{C}}^{e(\alpha)}}) = \tfrac{1}{2} G_{\text{neq}}^{(\alpha)} \left(\bar{I}_1^{e(\alpha)} - 3\right), \tag{4.3.27}$$

where $G_{\text{neq}}^{(\alpha)}$ represent shear moduli, and

$$\bar{I}_1^{e(\alpha)} = \operatorname{tr} \bar{\mathbf{C}}^{e(\alpha)}. \tag{4.3.28}$$

Then using (4.3.5) the elastic second Piola stress from each non-equilibrium branch is given by

$$\mathbf{T}^{e(\alpha)} = (J^{e(\alpha)})^{-2/3} G_{\text{neq}} \left[\mathbf{1} - \frac{1}{3}(\operatorname{tr}\mathbf{C}^e)\mathbf{C}^{e-1}\right], \tag{4.3.29}$$

so that the Mandel stress associated with each of the non-equilibrium branches $\mathbf{M}^{e(\alpha)} = \mathbf{C}^{e(\alpha)}\mathbf{T}^{e(\alpha)}$ is

$$\mathbf{M}_0^{e(\alpha)} = G_{\text{neq}}^{(\alpha)} \bar{\mathbf{C}}_0^{e(\alpha)}. \tag{4.3.30}$$

From (4.3.7) and (4.3.29) the contribution to the Cauchy stress from each non-equilibrium branch is

$$\mathbf{T}^{(\alpha)} = (J^{e(\alpha)})^{-1} G_{\text{neq}}^{(\alpha)} \bar{\mathbf{B}}_0^{e(\alpha)}. \tag{4.3.31}$$

The total Cauchy stress

Summarizing the calculations above gives the total Cauchy stress as

$$\mathbf{T} = \mathbf{T}^{\text{eq}} + \sum_{\alpha} \mathbf{T}^{(\alpha)}, \qquad \text{where}$$
$$\mathbf{T}^{\text{eq}} = J^{-1} G_{\text{eq}} \bar{\mathbf{B}}_0 - p\mathbf{1} \quad \text{with} \quad G_{\text{eq}} = G_{0\text{eq}} \left(\frac{\lambda_L}{3\bar{\lambda}}\right) \mathcal{L}^{-1}\left(\frac{\bar{\lambda}}{\lambda_L}\right), \quad p = -K(J-1),$$
$$\mathbf{T}^{(\alpha)} = J^{-1} G_{\text{neq}}^{(\alpha)} \bar{\mathbf{B}}_0^{e(\alpha)}, \tag{4.3.32}$$

where we have used (4.3.18), that is $J^{e(\alpha)} = J$, in writing (4.3.32).

4.3.3 Viscous flow rules

The evolution equation for $\mathbf{F}^{v(\alpha)}$ in the non-equilibrium branches is taken as

$$\dot{\mathbf{F}}^{v(\alpha)} = \mathbf{D}^{v(\alpha)}\,\mathbf{F}^{v(\alpha)}, \tag{4.3.33}$$

with the viscous stretching given by

$$\mathbf{D}^{v(\alpha)} = \dot{\bar{\gamma}}^{v(\alpha)}\left(\frac{\mathbf{M}_0^{e(\alpha)}}{2\,\bar{\tau}^{(\alpha)}}\right), \tag{4.3.34}$$

where

$$\dot{\bar{\gamma}}^{v(\alpha)} \stackrel{\text{def}}{=} \sqrt{2}\,|\mathbf{D}^{v(\alpha)}| \tag{4.3.35}$$

is an *equivalent shear strain rate* for each branch and

$$\bar{\tau}^{(\alpha)} \stackrel{\text{def}}{=} \frac{1}{\sqrt{2}}\,|\mathbf{M}_0^{e(\alpha)}| \tag{4.3.36}$$

defines an *equivalent shear stress* for each branch. We assume that each $\bar{\tau}^{(\alpha)}$ satisfies a strength relation of the form

$$\bar{\tau}^{(\alpha)} = \mathcal{S}^{(\alpha)}(\dot{\bar{\gamma}}^{v(\alpha)}), \tag{4.3.37}$$

and as a special case we assume that

$$\mathcal{S}^{(\alpha)}(\dot{\bar{\gamma}}^{v(\alpha)}) = \eta^{(\alpha)}\dot{\bar{\gamma}}^{v(\alpha)}, \tag{4.3.38}$$

with $\eta^{(\alpha)} > 0$ a constant *viscosity* parameter for each α, so that

$$\bar{\tau}^{(\alpha)} = \eta^{(\alpha)}\dot{\bar{\gamma}}^{v(\alpha)} \qquad \Longrightarrow \qquad \dot{\bar{\gamma}}^{v(\alpha)} = \frac{\bar{\tau}^{(\alpha)}}{\eta^{(\alpha)}}. \tag{4.3.39}$$

With this choice of the strength relation the dissipation inequality is

$$\mathcal{D} = \sum_{\alpha} \eta^{(\alpha)}\,(\dot{\bar{\gamma}}^{v(\alpha)})^2 \geq 0. \tag{4.3.40}$$

which is always satisfied since we have chosen positive viscosities, $\eta^{(\alpha)} > 0$.

Note that using (4.3.30) and (4.3.39) we may rewrite the viscous stretching relation (4.3.34) alternatively as

$$\mathbf{D}^{v(\alpha)} = \frac{1}{2\,\tau_r^{(\alpha)}}\bar{\mathbf{C}}_0^{e(\alpha)} \qquad \text{with} \qquad \tau_r^{(\alpha)} \stackrel{\text{def}}{=} \frac{\eta^{(\alpha)}}{G_{\text{neq}}^{(\alpha)}}, \tag{4.3.41}$$

where $\tau_r^{(\alpha)}$ represents a *relaxation time* for the αth non-equilibrium branch.

The complete set of material parameters appearing in our theory is

$$\{G_{0_{\text{eq}}}, \lambda_L, K\} \qquad \text{and} \qquad \{G_{\text{neq}}^{(\alpha)}, \eta^{(\alpha)}\} \quad \text{for} \quad \alpha = 1, \ldots, M, \tag{4.3.42}$$

or alternatively in terms of the relaxation times $\tau_r^{(\alpha)} = \eta^{(\alpha)}/G_{\text{neq}}^{(\alpha)}$,

$$\{G_{0_{\text{eq}}}, \lambda_L, K\} \qquad \text{and} \qquad \{G_{\text{neq}}^{(\alpha)}, \tau_r^{(\alpha)}\} \quad \text{for} \quad \alpha = 1, \ldots, M. \tag{4.3.43}$$

4.3.4 Summary of the constitutive equations

The Piola stress $\mathbf{T}_{\text{R}}$ is given by

$$\mathbf{T}_{\text{R}} = J\,\mathbf{T}\mathbf{F}^{-\top} \qquad \text{with} \qquad \mathbf{T} = \mathbf{T}^{\text{eq}} + \sum_{\alpha} \mathbf{T}^{(\alpha)}, \tag{4.3.44}$$

in which the equilibrium and non-equilibrium contributions to the Cauchy stress are given by

$$\begin{aligned} \mathbf{T}^{\text{eq}} &= J^{-1} G_{\text{eq}}\, \bar{\mathbf{B}}_0 - p\mathbf{1} \quad \text{with} \quad G_{\text{eq}} = G_{0_{\text{eq}}} \left(\frac{\lambda_L}{3\bar{\lambda}}\right) \mathcal{L}^{-1}\left(\frac{\bar{\lambda}}{\lambda_L}\right), \quad p = -K\,(J-1), \\ \mathbf{T}^{(\alpha)} &= J^{-1} G_{\text{neq}}^{(\alpha)} \bar{\mathbf{B}}_0^{e(\alpha)}. \end{aligned} \tag{4.3.45}$$

Also, the evolution of the internal variables $\mathbf{F}^{v(\alpha)}$ is specified by

$$\dot{\mathbf{F}}^{v(\alpha)} = \mathbf{D}^{v(\alpha)}\,\mathbf{F}^{v(\alpha)}, \qquad \mathbf{D}^{v(\alpha)} = \frac{1}{2\,\tau_r^{(\alpha)}} \bar{\mathbf{C}}_0^{e(\alpha)} \qquad \text{with} \qquad \tau_r^{(\alpha)} \stackrel{\text{def}}{=} \frac{\eta^{(\alpha)}}{G_{\text{neq}}^{(\alpha)}}. \tag{4.3.46}$$

4.3.5 Constitutive equations re-expressed

Consider eq. $(4.3.45)_2$ for the Cauchy stress in the non-equilibrium branches. Fixing attention on one of these branches and suppressing the index α, we can write:

$$\mathbf{T} = J^{-5/3} G_{\text{neq}} \left(\mathbf{B}^e - \frac{1}{3}(\operatorname{tr}\mathbf{B}^e)\mathbf{1} \right). \tag{4.3.47}$$

Thus, noting that $\mathbf{B}^e = \mathbf{F}\mathbf{C}^{v-1}\mathbf{F}^\top$ and $\operatorname{tr}\mathbf{B}^e = \mathbf{C}:\mathbf{C}^{v-1}$, we obtain

$$\mathbf{T} = J^{-1} G_{\text{neq}} \left(\bar{\mathbf{F}}\mathbf{C}^{v-1}\bar{\mathbf{F}}^\top - \frac{1}{3}(\bar{\mathbf{C}}:\mathbf{C}^{v-1})\mathbf{1} \right), \tag{4.3.48}$$

which shows that the Cauchy stress $\mathbf{T}$ may be calculated from a knowledge of $\mathbf{F}$ and $\mathbf{C}^v$. This observation is important in developing a time-integration procedure for viscoelasticity, in which evolution of viscoelastic flow phrased in terms of an evolution equation for $\mathbf{C}^v$, rather than an evolution equation for $\mathbf{F}^v$. We consider this below.

From (4.3.46), for a single branch, the evolution of $\mathbf{F}^v$ is

$$\dot{\mathbf{F}}^v = \mathbf{D}^v\,\mathbf{F}^v, \qquad \mathbf{D}^v = \frac{\bar{\mathbf{C}}_0^e}{2\tau_r}, \qquad \tau_r \stackrel{\text{def}}{=} \frac{\eta}{G_{\text{neq}}}. \tag{4.3.49}$$

Next, from $\mathbf{C}^v = \mathbf{F}^{v\top}\mathbf{F}^v$ we obtain that $\dot{\mathbf{C}}^v = \mathbf{F}^{v\top}\,(2\mathbf{D}^v)\,\mathbf{F}^v$. Hence, using $(4.3.49)_2$,

$$\dot{\mathbf{C}}^v = \mathbf{F}^{v\top}\left(\frac{\bar{\mathbf{C}}_0^e}{\tau_r}\right)\mathbf{F}^v. \tag{4.3.50}$$

Next, since $\bar{\mathbf{C}}^e = J^{-2/3}\left(\mathbf{F}^{v-\top}\mathbf{C}\mathbf{F}^{v-1}\right) = \left(\mathbf{F}^{v-\top}\bar{\mathbf{C}}\mathbf{F}^{v-1}\right)$ and $\operatorname{tr}\bar{\mathbf{C}}^e = J^{-2/3}\,(\mathbf{C}\!:\!\mathbf{C}^{v-1}) = \bar{\mathbf{C}}\!:\!\mathbf{C}^{v-1}$ we can write

$$\bar{\mathbf{C}}_0^e = \left[\left(\mathbf{F}^{v-\top}\bar{\mathbf{C}}\mathbf{F}^{v-1}\right) - \frac{1}{3}\left(\bar{\mathbf{C}}\!:\!\mathbf{C}^{v-1}\right)\mathbf{1}\right], \tag{4.3.51}$$

and use of this in (4.3.50) gives the following evolution equation

$$\dot{\mathbf{C}}^v = \frac{1}{\tau_r}\left(\bar{\mathbf{C}} - \frac{1}{3}\left(\bar{\mathbf{C}}\!:\!\mathbf{C}^{v-1}\right)\mathbf{C}^v\right). \tag{4.3.52}$$

Finally, using (4.3.47) and reinstating the index α, equation $(4.3.45)_2$ for the non-equilibrium Cauchy stresses may be replaced by

$$\mathbf{T}^{(\alpha)} = J^{-1}\,G_{\text{neq}}^{(\alpha)}\left(\bar{\mathbf{F}}(\mathbf{C}^{v(\alpha)})^{-1}\bar{\mathbf{F}}^{\top} - \frac{1}{3}(\bar{\mathbf{C}}\!:(\mathbf{C}^{v(\alpha)})^{-1})\mathbf{1}\right), \tag{4.3.53}$$

and using (4.3.52) the evolution equations (4.3.46) for $\mathbf{F}^{v(\alpha)}$ may be replaced by the following evolution equations for $\mathbf{C}^{v(\alpha)}$

$$\dot{\mathbf{C}}^{v(\alpha)} = \frac{1}{\tau_r^{(\alpha)}}\left(\bar{\mathbf{C}} - \frac{1}{3}\left(\bar{\mathbf{C}}\!:(\mathbf{C}^{v(\alpha)})^{-1}\right)\mathbf{C}^{v(\alpha)}\right). \tag{4.3.54}$$

Note that in (4.3.53) and (4.3.54) the dependence on the internal variables $\mathbf{F}^{v(\alpha)}$ enters only through the *symmetric tensors* $\mathbf{C}^{v(\alpha)} = (\mathbf{F}^{v(\alpha)})^{\top}\mathbf{F}^{v(\alpha)}$ (cf., e.g., Wijaya et al., 2023; Lefévre et al., 2024).

4.3.6 **Governing equations**

We neglect body forces. In the absence of body forces the governing linear momentum balance equation is taken as

$$\left.\begin{aligned} \operatorname{Div}\mathbf{T}_{\mathrm{R}} &= \rho_{\mathrm{R}}\ddot{\mathbf{u}}, \\ \frac{p}{K} + (J-1) &= 0, \end{aligned}\right\} \quad \text{in } \mathcal{B}. \tag{4.3.55}$$

In writing (4.3.55) we have *split* the standard equation of motion into two equations because this helps in numerically dealing with the near-incompressibility of the material.

The Piola stress $\mathbf{T}_{\mathrm{R}}$ is given by

$$\mathbf{T}_{\mathrm{R}} = J\,\mathbf{T}\mathbf{F}^{-\top} \qquad \text{with} \qquad \mathbf{T} = \mathbf{T}^{\mathrm{eq}} + \sum_{\alpha} \mathbf{T}^{(\alpha)}, \tag{4.3.56}$$

in which the equilibrium and non-equilibrium contributions to the Cauchy stress are given by

$$\begin{aligned} \mathbf{T}^{\mathrm{eq}} &= J^{-1} G_{\mathrm{eq}} \bar{\mathbf{B}}_0 - p\mathbf{1} \quad \text{with} \quad G_{\mathrm{eq}} = G_{0\mathrm{eq}} \left(\frac{\lambda_L}{3\bar{\lambda}}\right) \mathcal{L}^{-1}\left(\frac{\bar{\lambda}}{\lambda_L}\right), \\ \mathbf{T}^{(\alpha)} &= J^{-1} G^{(\alpha)}_{\mathrm{neq}} \left(\bar{\mathbf{F}} (\mathbf{C}^{v(\alpha)})^{-1} \bar{\mathbf{F}}^{\top} - \frac{1}{3} (\bar{\mathbf{C}} : (\mathbf{C}^{v(\alpha)})^{-1}) \mathbf{1} \right), \end{aligned} \tag{4.3.57}$$

and the evolution of the internal variables $\mathbf{C}^{v(\alpha)}$ is specified by

$$\dot{\mathbf{C}}^{v(\alpha)} = \frac{1}{\tau_r^{(\alpha)}} \left(\bar{\mathbf{C}} - \frac{1}{3} \left(\bar{\mathbf{C}} : (\mathbf{C}^{v(\alpha)})^{-1} \right) \mathbf{C}^{v(\alpha)} \right). \tag{4.3.58}$$

Next, for a time interval $t \in [0, T]$, with $\partial\mathcal{B}_{\mathrm{u}}$ and $\partial\mathcal{B}_{\mathrm{t_R}}$ denoting complementary subsurfaces of the boundary $\partial\mathcal{B}$ of the body $\mathcal{B}$, we consider a pair of possible boundary conditions:

$$\begin{aligned} \mathbf{u} &= \bar{\mathbf{u}} \quad \text{on } \partial\mathcal{B}_{\mathrm{u}} \times [0, T], \\ \mathbf{T}_{\mathrm{R}}\mathbf{n}_{\mathrm{R}} &= \bar{\mathbf{t}}_{\mathrm{R}} \quad \text{on } \partial\mathcal{B}_{\mathrm{t_R}} \times [0, T]. \end{aligned} \tag{4.3.59}$$

In the boundary conditions above $\bar{\mathbf{u}}$ and $\bar{\mathbf{t}}_{\mathrm{R}}$ are prescribed displacement and tractions, both functions of $(\mathbf{X}, t)$.

Standard initial conditions involve the specifications

$$\mathbf{u}(\mathbf{X}, 0) = \mathbf{u}_0(\mathbf{X}), \qquad \dot{\mathbf{u}}(\mathbf{X}, 0) = \dot{\mathbf{u}}_0(\mathbf{X}), \qquad \mathbf{C}^{v(\alpha)}(\mathbf{X}, 0) = \mathbf{1} \qquad \text{in } \mathcal{B}, \tag{4.3.60}$$

with $\mathbf{u}_0$ and $\dot{\mathbf{u}}_0$ prescribed functions on $\mathcal{B}$.

5 Numerical implementation of finite viscoelasticity

5.1 Weak form of the governing equations

Consider a test field $\mathbf{u}_{\text{test}}(\mathbf{X})$ in $\mathcal{B}$ which satisfies $\mathbf{u}_{\text{test}}(\mathbf{X}) = \mathbf{0}$ on $\partial\mathcal{B}_{\mathrm{u}}$. If the balance equation $\operatorname{Div}\mathbf{T}_{\mathrm{R}} - \rho_{\mathrm{R}}\,\ddot{\mathbf{u}} = \mathbf{0}$ in $\mathcal{B}$ and the traction condition $(\bar{\mathbf{t}}_{\mathrm{R}} - \mathbf{T}_{\mathrm{R}}\mathbf{n}_{\mathrm{R}}) = \mathbf{0}$ on $\partial\mathcal{B}_{\mathbf{t}_{\mathrm{R}}}$ are satisfied, then it is straightforward to show that the weak form

$$\int_{\mathcal{B}} \mathbf{T}_{\mathrm{R}} \colon \nabla\mathbf{u}_{\text{test}}\, dv_{\mathrm{R}} + \int_{\mathcal{B}} \rho_{\mathrm{R}}\ddot{\mathbf{u}} \cdot \mathbf{u}_{\text{test}}\, dv_{\mathrm{R}} - \int_{\partial\mathcal{B}_{\mathbf{t}_{\mathrm{R}}}} \bar{\mathbf{t}}_{\mathrm{R}} \cdot \mathbf{u}_{\text{test}}\, da_{\mathrm{R}} = 0 \tag{5.1.1}$$

holds for all admissible test fields $\mathbf{u}_{\text{test}}$ which vanish on $\partial\mathcal{B}_{\mathrm{u}}$. Also, the weak form of the pressure equation $(4.3.55)_2$ is

$$\int_{\mathcal{B}} \left[\frac{p}{K} + (J-1)\right] p_{\text{test}}\, dv_{\mathrm{R}} = 0, \tag{5.1.2}$$

where $p_{\text{test}}(\mathbf{X})$ is a test field in $\mathcal{B}$.

5.2 Time-integration procedure

The time domain $t \in [0, T]$ is discretized in a series of N_t time steps,

$$t_{\mathrm{n}+1} = t_{\mathrm{n}} + \Delta t, \qquad \mathrm{n} = 0, 1, \ldots, N_t, \tag{5.2.1}$$

such that $t_{N_t} = T$. Then, in order to solve the residuals based on the weak forms (5.1.1) and (5.1.2) in a $(\mathbf{u}, p)$ based finite element solution procedure, given

$$(\mathbf{u}_{\mathrm{n}}, \dot{\mathbf{u}}_{\mathrm{n}}, \ddot{\mathbf{u}}_{\mathrm{n}}, \mathbf{C}^{v(\alpha)}_{\mathrm{n}}, \mathbf{T}_{\mathrm{Rn}})$$

at time t_{n}, we need to calculate

$$(\dot{\mathbf{u}}_{\mathrm{n}+1}, \ddot{\mathbf{u}}_{\mathrm{n}+1}, \mathbf{C}^{v(\alpha)}_{\mathrm{n}+1}, \mathbf{T}_{\mathrm{Rn}+1})$$

in terms of and $(\mathbf{u}_{\mathrm{n}+1}, p_{\mathrm{n}+1})$ at time $t_{\mathrm{n}+1} = t_{\mathrm{n}} + \Delta t$.

Introduction to Coupled Theories in Solid Mechanics. Lallit Anand, Eric M. Stewart, Shawn A. Chester, Oxford University Press.
 DOI: 10.1093/oso/9780198986218.003.0006

5.3 Calculation of $\mathbf{C}_{\mathrm{n+1}}^{v(\alpha)}$ and $\mathbf{T}_{\mathrm{Rn+1}}$

Recall the evolution equation (4.3.58) for the internal variable $\mathbf{C}^{v(\alpha)}$, viz.

$$\dot{\mathbf{C}}^{v(\alpha)} = \frac{1}{\tau_r^{(\alpha)}}\left(\bar{\mathbf{C}} - \frac{1}{3}\left(\bar{\mathbf{C}}\,{:}\,(\mathbf{C}^{v(\alpha)})^{-1}\right)\mathbf{C}^{v(\alpha)}\right). \tag{5.3.1}$$

A fully-implicit time integration of this evolution equation (5.3.1) using a backward Euler rule gives,

$$\begin{aligned}\mathbf{C}_{\mathrm{n+1}}^{v(\alpha)} &= \mathbf{C}_{\mathrm{n}}^{v(\alpha)} + \Delta t\,\dot{\mathbf{C}}{v}_{\mathrm{n+1}}^{(\alpha)},\\ &= \mathbf{C}_{\mathrm{n}}^{v(\alpha)} + \Delta t\,\frac{1}{\tau_r^{(\alpha)}}\left(\bar{\mathbf{C}}_{\mathrm{n+1}} - \frac{1}{3}\left(\bar{\mathbf{C}}_{\mathrm{n+1}}\,{:}\,(\mathbf{C}_{\mathrm{n+1}}^{v(\alpha)})^{-1}\right)\mathbf{C}_{\mathrm{n+1}}^{v(\alpha)}\right),\end{aligned} \tag{5.3.2}$$

which may be rearranged as

$$\left[1 + \frac{\Delta t}{3\tau_r^{(\alpha)}}\left(\bar{\mathbf{C}}_{\mathrm{n+1}}\,{:}\,(\mathbf{C}_{\mathrm{n+1}}^{v(\alpha)})^{-1}\right)\right]\mathbf{C}_{\mathrm{n+1}}^{v(\alpha)} = \mathbf{C}_{\mathrm{n}}^{v(\alpha)} + \frac{\Delta t}{\tau_r^{(\alpha)}}\bar{\mathbf{C}}_{\mathrm{n+1}}, \tag{5.3.3}$$

and then written as

$$\mathbf{C}_{\mathrm{n+1}}^{v(\alpha)} = \beta\left(\mathbf{C}_{\mathrm{n}}^{v(\alpha)} + \frac{\Delta t}{\tau_r^{(\alpha)}}\bar{\mathbf{C}}_{\mathrm{n+1}}\right) \qquad \text{with} \qquad \beta \stackrel{\text{def}}{=} \left(1 + \frac{\Delta t}{3\tau_r^{(\alpha)}}\left(\bar{\mathbf{C}}_{\mathrm{n+1}}\,{:}\,(\mathbf{C}_{\mathrm{n+1}}^{v(\alpha)})^{-1}\right)\right)^{-1} \tag{5.3.4}$$

Now, consider a non-singular tensor $\mathbf{A}$ and define its distortional part $\bar{\mathbf{A}}$ by

$$\bar{\mathbf{A}} \stackrel{\text{def}}{=} (\det \mathbf{A})^{-1/3}\mathbf{A} \qquad \text{so that} \qquad \det(\bar{\mathbf{A}}) = 1. \tag{5.3.5}$$

Then for a real scalar a, the tensor $a\mathbf{A}$ has the property

$$\overline{a\mathbf{A}} = \bar{\mathbf{A}} \qquad (\text{since } \det(a\mathbf{A}) = a^3\det(\mathbf{A})). \tag{5.3.6}$$

Now, we desire that $\mathbf{C}_{\mathrm{n+1}}^{v(\alpha)}$ satisfies the volume-preserving constraint $\det(\mathbf{C}_{\mathrm{n+1}}^{v(\alpha)}) = 1$, that is

$$\mathbf{C}_{\mathrm{n+1}}^{v(\alpha)} = \overline{\mathbf{C}_{\mathrm{n+1}}^{v(\alpha)}}.$$

Thus, taking the distortional part of the right-hand side of eq. (5.3.4) and using the property (5.3.6) we obtain

$$\overline{\beta\left(\mathbf{C}_{\mathrm{n}}^{v(\alpha)} + \frac{\Delta t}{\tau_r^{(\alpha)}}\bar{\mathbf{C}}_{\mathrm{n+1}}\right)} = \overline{\left(\mathbf{C}_{\mathrm{n}}^{v(\alpha)} + \frac{\Delta t}{\tau_r^{(\alpha)}}\bar{\mathbf{C}}_{\mathrm{n+1}}\right)}, \tag{5.3.7}$$

which gives the update formula

$$\mathbf{C}^{v(\alpha)}_{n+1} = \left(\det\left[\mathbf{C}^{v(\alpha)}_{n} + \frac{\Delta t}{\tau_r^{(\alpha)}}\bar{\mathbf{C}}_{n+1}\right]\right)^{-1/3}\left(\mathbf{C}^{v(\alpha)}_{n} + \frac{\Delta t}{\tau_r^{(\alpha)}}\bar{\mathbf{C}}_{n+1}\right). \tag{5.3.8}$$

This elegant update formula for $\mathbf{C}^{v(\alpha)}$—which is *fully-implicit, volume-preserving, and iteration-free*—is due to Shutov et al. (2013).

Finally, recalling (4.3.56) and (4.3.57), the total Piola stress at the end of the step is given by

$$\begin{aligned}
\mathbf{T}_{\mathrm{R}n+1} &= J_{n+1}\,\mathbf{T}_{n+1}\mathbf{F}^{-\top}_{n+1}, \qquad \text{with}\\
\mathbf{T}_{n+1} &= \mathbf{T}^{\mathrm{eq}}_{n+1} + \sum_{\alpha}\mathbf{T}^{(\alpha)}_{n+1}, \qquad \text{where}\\
\mathbf{T}^{\mathrm{eq}}_{n+1} &= J^{-1}_{n+1}\,(G_{\mathrm{eq}})_{n+1}\,(\bar{\mathbf{B}}_0)_{n+1} - p_{n+1}\,\mathbf{1} \qquad \text{with}\\
&\qquad (G_{\mathrm{eq}})_{n+1} = G_{0_{\mathrm{eq}}}\left(\frac{\lambda_L}{3\bar{\lambda}_{n+1}}\right)\mathcal{L}^{-1}\left(\frac{\bar{\lambda}_{n+1}}{\lambda_L}\right),\\
\mathbf{T}^{(\alpha)}_{n+1} &= J^{-1}_{n+1}\,G^{(\alpha)}_{\mathrm{neq}}\left(\bar{\mathbf{F}}_{n+1}(\mathbf{C}^{v(\alpha)}_{n+1})^{-1}\bar{\mathbf{F}}^{\top}_{n+1} - \frac{1}{3}(\bar{\mathbf{C}}_{n+1} : (\mathbf{C}^{v(\alpha)}_{n+1})^{-1})\mathbf{1}\right).
\end{aligned} \tag{5.3.9}$$

5.4 Calculation of $\ddot{\mathbf{u}}_{n+1}$ and $\dot{\mathbf{u}}_{n+1}$

Next, to solve the weak form (5.1.1), a kinematic relationship is required between $(\ddot{\mathbf{u}}_{n+1}, \dot{\mathbf{u}}_{n+1})$ and $\mathbf{u}_{n+1}$ so that the inertia term in the residual (5.1.1) becomes a function solely of $\mathbf{u}_{n+1}$. Using the notation

$$\mathbf{a} \equiv \ddot{\mathbf{u}} \qquad \text{and} \qquad \mathbf{v} \equiv \dot{\mathbf{u}}$$

for the acceleration and velocity, respectively, we approximate these terms using the Newmark time-integration scheme (Newmark, 1959):[1]

$$\begin{aligned}
\mathbf{a}_{n+1} &= \frac{\mathbf{u}_{n+1} - \mathbf{u}_n - \mathbf{v}_n\,\Delta t}{\beta\Delta t^2} - \frac{(1-2\beta)}{2\beta}\,\mathbf{a}_n,\\
\mathbf{v}_{n+1} &= \mathbf{v}_n + \Delta t\,[(1-\gamma)\mathbf{a}_n + \gamma\,\mathbf{a}_{n+1}],
\end{aligned} \tag{5.4.1}$$

where γ and β are parameters which can be chosen in the ranges $\gamma \in (0, 1]$ and $\beta \in (0, 0.5]$ to yield different time-integration schemes.[2]

[1] See Appendix 6.A for a slightly more detailed discussion of the Newmark method.
[2] The parameter β here is not the same as the parameter β in eq. (5.3.34).

The numerical solution of dynamical problems in solid mechanics often leads to time-discrete solutions that contain high frequency oscillations of small amplitude. Although it is possible that these oscillations are also present in the real motion, it is more likely that they are the result of the approximations inherently present in the discrete time-integration procedure. Such behavior can be damped out and the solution stabilized by introducing a material viscosity into the simulation with characteristic relaxation time on the same order as the characteristic time of the high-frequency oscillations. An alternative approach is to introduce *numerical dissipation* or artificial viscosity without sacrificing the overall accuracy of the time integration. One can use an enhanced version of the Newmark method called the *α-method* (Hilber et al., 1977) to dampen such high-frequency oscillations.

In the α-method the Newmark kinematic approximations (5.4.1) are retained for evaluating $\mathbf{a}_{\mathrm{n}+1}$ and $\mathbf{v}_{\mathrm{n}+1}$. However, one introduces a positive-valued parameter $\alpha \in [0, 1)$, and replaces $\mathbf{u}_{\mathrm{n}+1}$ and $\mathbf{v}_{\mathrm{n}+1}$ with $\mathbf{u}$ and $\mathbf{v}$ linearly interpolated to an intermediate time

$$t_\alpha = (1-\alpha)t_{\mathrm{n}+1} + \alpha t_{\mathrm{n}} = t_{\mathrm{n}+1} - \alpha\Delta t$$

between time steps $\mathrm{n}+1$ and n, that is

$$\begin{aligned} \mathbf{u}_\alpha &= (1-\alpha)\mathbf{u}_{\mathrm{n}+1} + \alpha\mathbf{u}_{\mathrm{n}}, \\ \mathbf{v}_\alpha &= (1-\alpha)\mathbf{v}_{\mathrm{n}+1} + \alpha\mathbf{v}_{\mathrm{n}}. \end{aligned} \tag{5.4.2}$$

In principle, the algorithmic parameters α, β, and γ of the α-method may be independent. However, care must be taken to achieve consistency, stability, and favorable dissipation properties of the time-integration procedure; cf. Erlicher et al. (2002). Note that:

1. If $\alpha = 0$, the method reverts back to the Newmark method.
2. If

$$\alpha = 0, \qquad \gamma = 1/2, \qquad \beta = 1/4, \tag{5.4.3}$$

   then one obtains the trapezoidal time-integration scheme which is known to be unconditionally stable, and hence is preferred in many numerical simulations.
3. If the parameters α, γ, and β are selected such that,

$$\alpha \leq \tfrac{1}{2}, \qquad \gamma = \tfrac{1}{2} + \alpha, \qquad \beta = \frac{1}{4}(\gamma + \tfrac{1}{2})^2, \tag{5.4.4}$$

   then this results in an unconditionally stable and second-order accurate scheme, provided the time step remains constant.

For further detailed discussions regarding the choices for (α, γ, β)—and also other time-integration schemes—see Erlicher et al. (2002) and Bonet et al. (2021).

REMARK

In the representative simulations for visco-elastomers shown in Chapter 6, the presence of viscoelasticity provides sufficient damping for numerical stability and we therefore use $\alpha = 0$. For completeness, however, the numerical implementation in this chapter and the accompanying FEniCS codes still include the α-method relations.

5.5 Solution procedure in the FEniCS code

For ease of notation we will write

$$(\bullet) \equiv (\bullet)_{\mathrm{n+1}} \qquad \text{and} \qquad (\bullet)_{\mathrm{old}} \equiv (\bullet)_{\mathrm{n}},$$

for all time-varying quantities $(\bullet)$ evaluated at times $\mathrm{n}+1$ and n, respectively.

In FEniCS we will use the following degrees of freedom

$$\{\mathbf{u},\, p\}, \tag{5.5.1}$$

as well as their known values from the previous step

$$\{\mathbf{u}_{\mathrm{old}},\, p_{\mathrm{old}}\}, \tag{5.5.2}$$

to calculate the *residuals* resulting from the weak forms of the governing equations, viz.

$$\begin{aligned} R_{\mathbf{u}} &= \int_{\mathcal{B}} \mathbf{T}_{\mathrm{R}} : \nabla \mathbf{u}_{\mathrm{test}}\, dv_{\mathrm{R}} + \int_{\mathcal{B}} \rho_{\mathrm{R}} \ddot{\mathbf{u}} \cdot \mathbf{u}_{\mathrm{test}}\, dv_{\mathrm{R}} - \int_{\partial \mathcal{B}_{\mathbf{t}_{\mathrm{R}}}} \bar{\mathbf{t}}_{\mathrm{R}} \cdot \mathbf{u}_{\mathrm{test}}\, da_{\mathrm{R}}, \\ R_p &= \int_{\mathcal{B}} \left[\frac{p}{K} + (J-1) \right] p_{\mathrm{test}}\, dv_{\mathrm{R}}, \end{aligned} \tag{5.5.3}$$

along with the time-integration updates for the velocity and acceleration, cf. (5.4.1). FEniCS then automatically differentiates these residuals with respect to the separate degrees of freedom to obtain *tangents*. At each step in time, FEniCS uses the residuals and tangents in a global Newton–Raphson solver to determine current values of the degrees of freedom $\{\mathbf{u},\, p\}$ for which the residuals are zero (to within some small numerical tolerance) and which therefore satisfy the governing equations as well as the initial and boundary conditions (in a weak sense).

REMARK

A mixed finite element method is utilized to solve for the displacement $\mathbf{u}$ and the pressure p, concurrently. To avoid numerical instabilities we use a class of elements known as "Taylor–Hood" elements which combine second-order quadratic interpolation for the displacement and first-order linear interpolation for the scalar pressure. We also use FEniCS quadrature-representation elements for the internal variables $\mathbf{C}^{v(\alpha)}$.

5.6 Algorithm for calculation of the residuals

The algorithm we follow in our code to calculate the residuals in (5.5.3) is summarized below.

1. Begin the step with known values of

$$\left\{\mathbf{u}_{\text{old}},\, \dot{\mathbf{u}}_{\text{old}},\, \ddot{\mathbf{u}}_{\text{old}},\, p_{\text{old}},\, \mathbf{C}^{v(\alpha)}_{\text{old}}\right\}, \tag{5.6.1}$$

   where for the first step these quantities are given by the initial conditions for $\mathbf{u}$, $\dot{\mathbf{u}}$, $\ddot{\mathbf{u}}$, p, $\mathbf{C}^{v(\alpha)}$. We also have temporary values of the degrees of freedom at the end of the step,

$$\left\{\mathbf{u},\, p\right\}. \tag{5.6.2}$$

2. Calculate the acceleration and velocity using the Newmark scheme:

$$\begin{aligned} \ddot{\mathbf{u}} &= \frac{\mathbf{u} - \mathbf{u}_{\text{old}} - \dot{\mathbf{u}}_{\text{old}}\,\Delta t}{\beta \Delta t^2} - \frac{(1-2\beta)}{2\beta}\,\ddot{\mathbf{u}}_{\text{old}}, \\ \dot{\mathbf{u}} &= \dot{\mathbf{u}}_{\text{old}} + \Delta t\left[(1-\gamma)\ddot{\mathbf{u}}_{\text{old}} + \gamma\ddot{\mathbf{u}}\right]. \end{aligned} \tag{5.6.3}$$

3. Calculate the intermediate step values of $\mathbf{u}$ and p, respectively denoted by $\mathbf{u}_{\text{avg}}$ and p_{avg}, for use in the α-method of time integration

$$\begin{aligned} \mathbf{u}_{\text{avg}} &= \alpha\mathbf{u}_{\text{old}} + (1-\alpha)\mathbf{u}, \\ p_{\text{avg}} &= \alpha p_{\text{old}} + (1-\alpha)p. \end{aligned} \tag{5.6.4}$$

4. Calculate necessary kinematical quantities for the current step:

$$\begin{aligned} &\mathbf{F} = \mathbf{1} + \nabla\mathbf{u}_{\text{avg}}, \quad J = \det\mathbf{F}, \quad \mathbf{C} = \mathbf{F}^{\top}\mathbf{F}, \quad \bar{\mathbf{F}} = J^{-1/3}\mathbf{F}, \\ &\bar{\mathbf{B}} = \bar{\mathbf{F}}\bar{\mathbf{F}}^{\top}, \quad \bar{\lambda} = \sqrt{(\text{tr}\bar{\mathbf{C}})/3}. \end{aligned} \tag{5.6.5}$$

5. Update the internal variables $\mathbf{C}^{v(\alpha)}$:

$$\mathbf{C}^{v(\alpha)} = \left(\det\left[\mathbf{C}^{v(\alpha)}_{\text{old}} + \frac{\Delta t}{\tau_r^{(\alpha)}}\bar{\mathbf{C}}\right]\right)^{-1/3}\left(\mathbf{C}^{v(\alpha)}_{\text{old}} + \frac{\Delta t}{\tau_r^{(\alpha)}}\bar{\mathbf{C}}\right). \tag{5.6.6}$$

6. Calculate the Piola stress,

$$\begin{aligned}
\mathbf{T}_{\text{R}} &= J\,\mathbf{T}\mathbf{F}^{-\top}, \qquad \text{with} \\
\mathbf{T} &= \mathbf{T}^{\text{eq}} + \sum_{\alpha} \mathbf{T}^{(\alpha)}, \qquad \text{where} \\
\mathbf{T}^{\text{eq}} &= J^{-1} G_{\text{eq}}\,\bar{\mathbf{B}}_0 - p\,\mathbf{1}, \quad G_{\text{eq}} = G_{0\,\text{eq}}\left(\frac{\lambda_L}{3\bar{\lambda}}\right)\mathcal{L}^{-1}\left(\frac{\bar{\lambda}}{\lambda_L}\right), \\
\mathbf{T}^{(\alpha)} &= J^{-1}\,G^{(\alpha)}_{\text{neq}}\left(\bar{\mathbf{F}}(\mathbf{C}^{v(\alpha)})^{-1}\bar{\mathbf{F}}^{\top} - \frac{1}{3}(\bar{\mathbf{C}}:(\mathbf{C}^{v(\alpha)})^{-1})\mathbf{1}\right).
\end{aligned} \tag{5.6.7}$$

In our numerical implementation we compute the inverse of the Langevin function $\mathcal{L}^{-1}(x)$ in eq. (5.6.7) using a Padé approximation (Cohen, 1991)

$$\mathcal{L}^{-1}(x) \approx x\,\frac{3 - x^2}{1 - x^2}. \tag{5.6.8}$$

7. Compute the residuals:

$$\begin{aligned}
R_{\mathbf{u}} &= \int_{\mathcal{B}} \mathbf{T}_{\text{R}} : \nabla \mathbf{u}_{\text{test}}\, dv_{\text{R}} + \int_{\mathcal{B}} \rho_{\text{R}}\ddot{\mathbf{u}} \cdot \mathbf{u}_{\text{test}}\, dv_{\text{R}} - \int_{\partial\mathcal{B}_{\mathbf{t}_{\text{R}}}} \breve{\mathbf{t}}_{\text{R}} \cdot \mathbf{u}_{\text{test}}\, da_{\text{R}}, \\
R_p &= \int_{\mathcal{B}} \left[\frac{p_{\text{avg}}}{K} + (J-1)\right] p_{\text{test}}\, dv_{\text{R}}.
\end{aligned} \tag{5.6.9}$$

8. FEniCS solves for the current values of $\{\mathbf{u},\, p\}$.
9. Set the current values as the "old" values for the next step in time,

$$\{\mathbf{u},\, \dot{\mathbf{u}},\, \ddot{\mathbf{u}},\, p,\, \mathbf{C}^{v(\alpha)}\} \implies \{\mathbf{u}_{\text{old}},\, \dot{\mathbf{u}}_{\text{old}},\, \ddot{\mathbf{u}}_{\text{old}},\, p_{\text{old}},\, \mathbf{C}^{v(\alpha)}_{\text{old}}\}. \tag{5.6.10}$$

10. Repeat the steps above for desired number of steps in time.

6 Representative simulations of finite viscoelasticity

In this chapter we present several representative finite element simulations in FEniCS which demonstrate key features of the viscoelastic response of elastomers as well as its interaction with inertial effects. We present numerical simulations using two different elastomeric materials:[1]

(i) An acrylic polymer VHB-4910 produced by 3M™. This is a very soft viscoelastic electro-active polymer whose large deformation capacity may be harnessed for use in making actuators and sensors (cf., e.g. Pelrine et al., 2000). Table 6.1 lists material parameters—taken from Wang et al. (2016)—for modeling the large deformation viscoelastic response of this material using a 3-branch generalized Maxwell model.

Table 6.1 Material parameters for a 3-branch generalized Maxwell model for VHB-4910 from Wang et al. (2016). Note that for the non-equilibrium branches the relation times $\tau_r^{(\alpha)}$ may be converted to viscosities $\eta^{(\alpha)}$ using $\eta^{(\alpha)} = \tau_r^{(\alpha)} \times G_{\text{neq}}^{(\alpha)}$.

Quantity	Value SI (m)	Value SI (mm)
ρ_{R}, mass density	1000 kg/m^3	1×10^{-6} kg/mm^3
$G_{0\,\text{eq}}$, ground-state equilibrium shear modulus	15.36×10^3 Pa	15.36 kPa
K, bulk modulus	15.36×10^6 Pa	15.36×10^3 kPa
λ_L, locking stretch	5.85	5.85
$G_{\text{neq}}^{(1)}$, non-equilibrium shear modulus	26.06×10^3 Pa	26.06 kPa
$\tau_r^{(1)}$, relaxation time	0.6074 s	0.6074 s
$G_{\text{neq}}^{(2)}$, non-equilibrium shear modulus	26.53×10^3 Pa	26.53 kPa
$\tau_r^{(2)}$, relaxation time	6.56 s	6.56 s
$G_{\text{neq}}^{(3)}$, non-equilibrium shear modulus	10.83×10^3 Pa	10.83 kPa
$\tau_r^{(3)}$, relaxation time	61.25 s	61.25 s

(ii) An acrylonitrile butadiene rubber, also known as nitrile rubber or simply NBR. This material is widely used in many engineering applications to make transmission belts, O-rings, gaskets, oil seals, V-belts, and fuel and oil handling hoses. Table 6.2 lists

[1] In our FEniCS codes, we have selected the SI (mm) system of units which measures distance in mm, mass in kg, and time in s. Thus, the values of material parameters in the SI (mm) columns of Table 6.1 and Table 6.2 are the numbers which appear directly in the FEniCS codes.

Introduction to Coupled Theories in Solid Mechanics. Lallit Anand, Eric M. Stewart, Shawn A. Chester, Oxford University Press.
 DOI: 10.1093/oso/9780198986218.003.0007

Table 6.2 Material parameters for a 5-branch generalized Maxwell model for NBR. The equilibrium shear modulus is fitted to experimental data from Miehe and Goktepe (2005); the Maxwell branch parameters are from Linder et al. (2011). We have set the locking stretch λ_L to a large value to approximate a Neo-Hookean equilibrium branch.

Quantity	Value SI (m)	Value SI (mm)
ρ_R, mass density	$1300\ \text{kg/m}^3$	$1.3 \times 10^{-6}\ \text{kg/mm}^3$
$G_{0\text{eq}}$, ground-state equilibrium shear modulus	0.4×10^6 Pa	400 kPa
K, bulk modulus	0.4×10^9 Pa	0.4×10^3 MPa
λ_L, locking stretch	10.0	10.0
$G^{(1)}_{\text{neq}}$, non-equilibrium shear modulus	0.5357×10^6 Pa	535.7 kPa
$\tau^{(1)}$, relaxation time	1 s	1 s
$G^{(2)}_{\text{neq}}$, non-equilibrium shear modulus	0.0762×10^6 Pa	76.2 kPa
$\tau^{(2)}$, relaxation time	10 s	10 s
$G^{(3)}_{\text{neq}}$, non-equilibrium shear modulus	0.1205×10^6 Pa	120.5 kPa
$\tau^{(3)}$, relaxation time	100 s	100 s
$G^{(4)}_{\text{neq}}$, non-equilibrium shear modulus	0.0213×10^6 Pa	21.3 kPa
$\tau^{(4)}$, relaxation time	1000 s	1000 s
$G^{(5)}_{\text{neq}}$, non-equilibrium shear modulus	0.0229×10^6 Pa	22.9 kPa
$\tau^{(5)}$, relaxation time	10000 s	10000 s

material parameters for an HNBR50 nitrile rubber—adapted from Miehe and Goktepe (2005) and Linder et al. (2011)—for modeling the large deformation viscoelastic response of this material using a 5-branch generalized Maxwell model.

REMARKS

1. The material parameters which we have listed in Tables 6.1 and 6.2 are *representative* values for two very different elastomeric materials—VHB-4910 and HNBR50—which we have *extracted* from published data in the literature (Wang et al., 2016; Miehe and Goktepe, 2005; Linder et al., 2011). We use these material parameters in this section to *qualitatively* demonstrate the major features of the viscoelastic response of elastomeric materials.
2. Inertial effects are only considered in the final two simulations: damped oscillatory response of a viscoelastic beam and dynamic column buckling. In all the other simulations, inertial effects are neglected by removing the inertia term from the weak form (5.1.1) of the equation of motion.

6.1 **Simulations using material parameters for VHB-4910**

Simple extension at different stretch rates

As a first example we consider homogeneous simple extension of a viscoelastic elastomer at different stretch rates. Some experimental data for the viscoelastic response of an acrylic elastomer VHB-4910 in simple extension are shown in Fig. 6.1:

- Figure 6.1(a) shows the monotonic extension response to a maximum stretch of $\lambda = 9$ at a quasistatic rate of $\dot{\lambda} = 0.9 \times 10^{-4}$ /s. At this slow stretch rate all viscoelastic effects have relaxed, and the response shown in Fig. 6.1(a) essentially represents the "equilibrium" response of VHB-4910.
- Figure 6.1(b) shows the loading–unloading response to a maximum stretch of $\lambda = 2.5$ at several stretch rates $\dot{\lambda} = 0.05,\ 0.03,\ 0.01$ /s. This figure clearly demonstrates the rate-sensitivity of the stress-stretch response of the material.

The experimental data for VHB-4910 shown in Fig. 6.1(a) is from Wang et al. (2016), while the experimental data for nominally the same material in Fig. 6.1(b) is from Hossain et al. (2012).

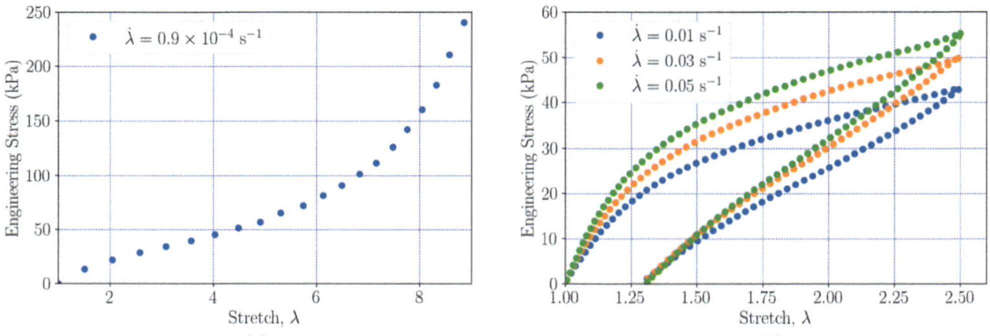

Fig. 6.1 Experimental data for VHB-4910 (filled circles) in simple extension: (a) Monotonic extension response to a maximum stretch of $\lambda = 9$ at a quasistatic rate of $\dot{\lambda} = 0.9 \times 10^{-4}$ /s. (b) Load–unload response to a maximum stretch of $\lambda = 2.5$ at several stretch rates $\dot{\lambda} = 0.05,\ 0.03,\ 0.01$ /s. The experimental data in (a) is from Wang et al. (2016), while the experimental data in (b) is from Hossain et al. (2012).

For the numerical simulation the initial geometry of the body is a simple cube. We use symmetry conditions and only mesh 1/4 of the body. Figure 6.2(a) shows the modeled domain which has an edge length of 1 mm, and is meshed with 1 tetrahedral element along each edge. In order to model homogeneous simple extension, referring to Fig. 6.2(a), the top face of the cubic domain is prescribed a displacement ramp, while the exterior lateral faces are traction-free, and the other faces are prescribed zero displacement. Inertial effects are neglected in this simulation. Figure 6.2(b) shows the reference body and deformed configuration at the maximum applied stretch $\lambda = 9$. Figure 6.2(c) shows a plot of the resulting Piola stress component $T_{\mathrm{R}\,22}$ as a function of the applied stretch λ. The numerical result using the material parameters listed in Table 6.1 is in very good agreement with the experimental result from Wang et al. (2016).

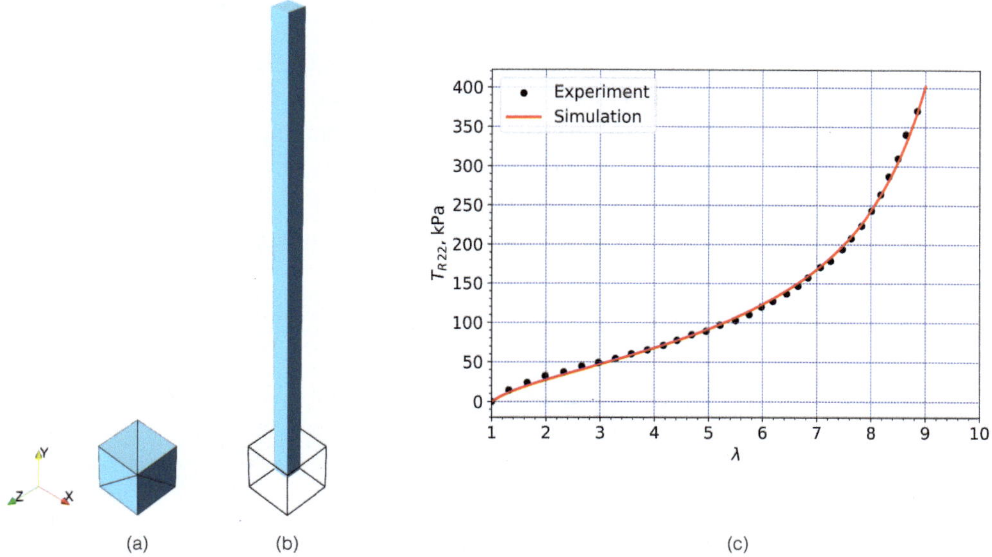

Fig. 6.2 (a) Finite element mesh. (b) The reference configuration and the deformed configurations at maximum stretch. (c) A comparison of the numerical simulation result (solid line) against the experimental data (filled circles) from Wang et al. (2016) for the axial Piola stress $T_{R\,22}$ as a function of axial stretch λ at a quasi-static stretch rate of $\dot{\lambda} = 0.9 \times 10^{-4}$ s.

Figure 6.3 shows a comparison of the numerical simulation results against the experimental data for the load–unload response to a maximum stretch of $\lambda = 2.5$ at several stretch rates $\dot{\lambda} = 0.05,\ 0.03,\ 0.01\,/\mathrm{s}$. The numerical results (solid lines) using the material parameters listed in Table 6.1 are in quite good agreement with the experimental results (filled circles) from Hossain et al. (2012).

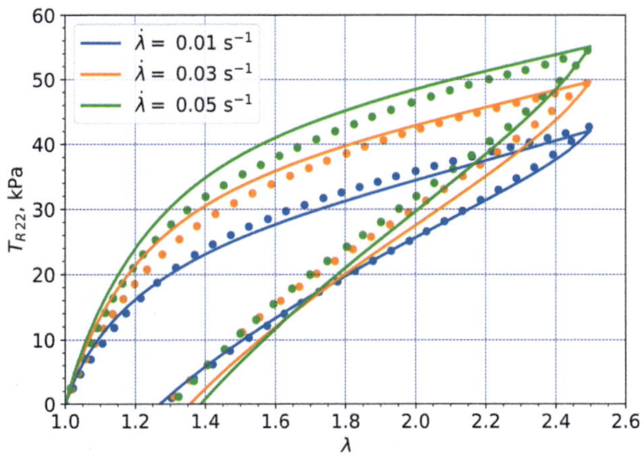

Fig. 6.3 Comparison of the numerical simulation results (solid lines) against the experimental data (filled circles) from Hossain et al. (2012) for the load-unload response to a maximum stretch of $\lambda = 2.5$ at several stretch rates $\dot{\lambda} = 0.05,\ 0.03,\ 0.01\,/\mathrm{s}$.

Stress–relaxation

As a second example we simulate a stress–relaxation test. We use the same geometry as in Fig. 6.2(a) and material properties as in Table 6.1, and in order to simulate stress–relaxation the top face of the cubical domain is prescribed a displacement ramp to an applied stretch of $\lambda = 1.5$ in 0.1 s and then held constant for a total time of $t = 10$ s, cf. Fig. 6.4(a). Inertial effects are neglected in this simulation. Figure 6.4(b) shows a plot of the resulting Piola stress $T_{\mathrm{R}\,22}$ as a function of time. In this material the stress relaxes from a value of ~ 80 kPa to a value of ~ 30 kPa in ~ 10 s.

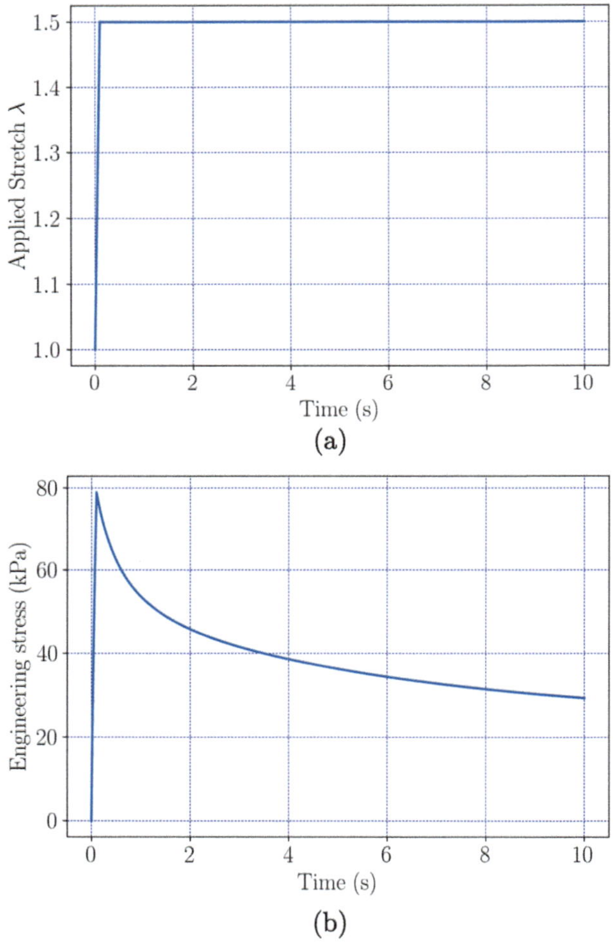

Fig. 6.4 A simulation of stress relaxation in VHB-4910: (a) Applied stretch and (b) stress response versus time.

Creep

Next, we simulate a creep test. We use the same geometry as in Fig. 6.2(a) and material properties as in Table 6.1, and in order to simulate creep, the top face of the cubical domain is prescribed a ramp in the normal traction of $T_{\mathrm{R}\,22} = 50$ kPa in 0.1 s and then held constant for a total time of $t = 10$ s, cf. Fig. 6.5(a). Inertial effects are neglected in this simulation.

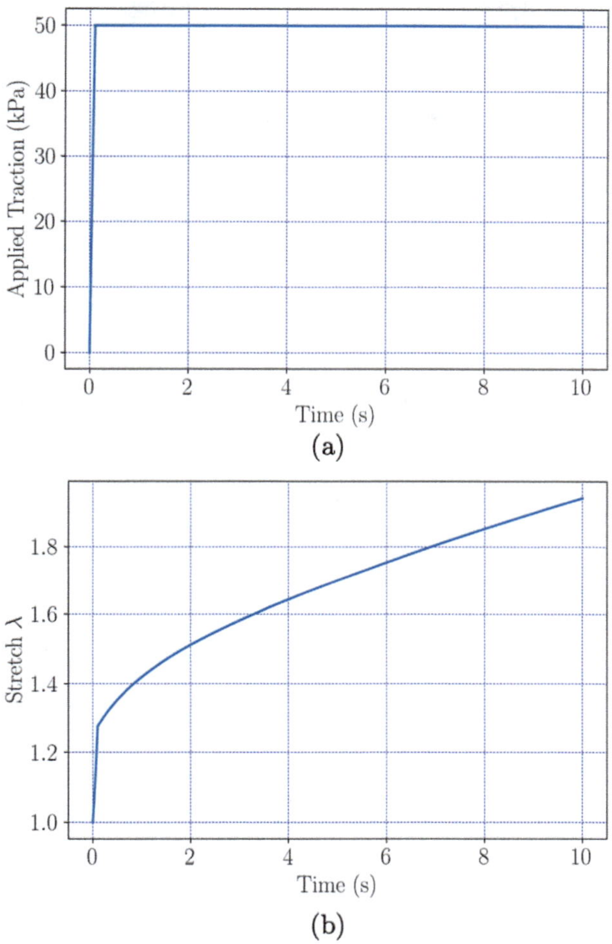

Fig. 6.5 A simulation of creep in VHB-4910: (a) Applied stress and (b) stretch response versus time.

Figure 6.5(b) shows a plot of the resulting axial stretch λ as a function of time. At a stress of $T_{\text{R}\,22} = 50\,\text{kPa}$, this material creeps to a stretch of $\lambda \sim 2$ in $\sim 10\,\text{s}$.

Staircase stretch–hold

Next, we simulate a complex staircase type stretch–hold test. We use the same geometry as in Fig. 6.2(a) and material properties as in Table 6.1, and in order to simulate the stretch–hold experiments the top face of the initial cube of edge length 1 mm is prescribed a displacement history listed in Table 6.3.

This applied stretch history consists of:

- Increments and decrements in stretch of 0.25.
- An absolute value of the stretch rate of 0.2/s when the stretch is increased or decreased.
- Hold-times of 100 s.

Table 6.3 History of applied displacement to the initial unit cube.

Time, s	0	5	105	110	210	215	315	320	420
Disp., mm	0	0.25	0.25	0.5	0.5	0.75	0.75	1.0	1.0
Time, s	425	525	530	630	635	735	740	840	
Disp., mm	0.75	0.75	0.5	0.5	0.25	0.25	0	0	

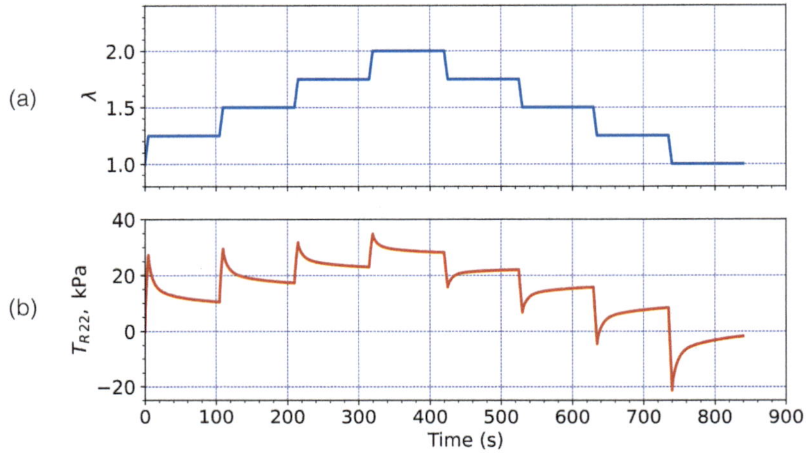

Fig. 6.6 Simulation of a staircase stretch–hold experiment for VHB-4910: (a) Input stretch history. (b) Output stress–relaxation events.

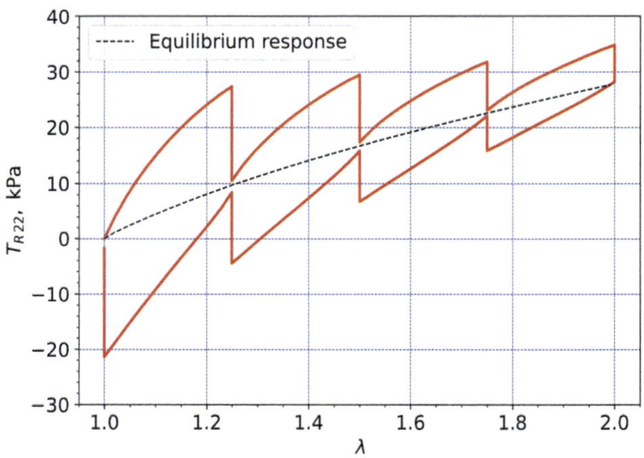

Fig. 6.7 Stress versus stretch curve from the stretch–hold experiment (red) plotted against the quasi-static "equilibrium" curve (dashed black line).

This applied stretch history is shown in Fig. 6.6(a), and Fig. 6.6(b) shows a plot of the resulting stress–relaxation events in the axial stress $T_{R\,22}$ as a function of time.

Figure 6.7 shows a cross-plot of the stress versus stretch curve from the results in Fig. 6.6. Note from this figure that VHB-4910 almost completely relaxes after each 100 s hold time towards an "equilibrium" curve (dashed black line). The stress *decreases* towards the equilibrium curve in the initial stage of the simulation as the stretch is being increased, but the stress *increases* towards the equilibrium curve in the later stage of the simulation as the stretch is being decreased.

Sinusoidal axial stretch

Here, we simulate a sinusoidal axial stretch test. We use the same geometry as in Fig. 6.2(a) and material properties as in Table 6.1, and in order to simulate the sinusoidal axial stretch the top face of the initial cube of edge length 1 mm is prescribed a sinusoidal displacement history

$$u_2 = u_{\max} \sin(2\pi t/T)$$

with $u_{\max} = 0.5\,\text{mm}$ for two cycles with a period of $T = 1$ s. The prescribed displacement boundary condition results in sinusoidal axial extension of the cube starting from $\lambda = 1$ and then cycling between $\lambda = 1.5$ and $\lambda = 0.5$. The Piola stress component $T_{\text{R}\,22}$ versus stretch curve is shown in Fig. 6.8.

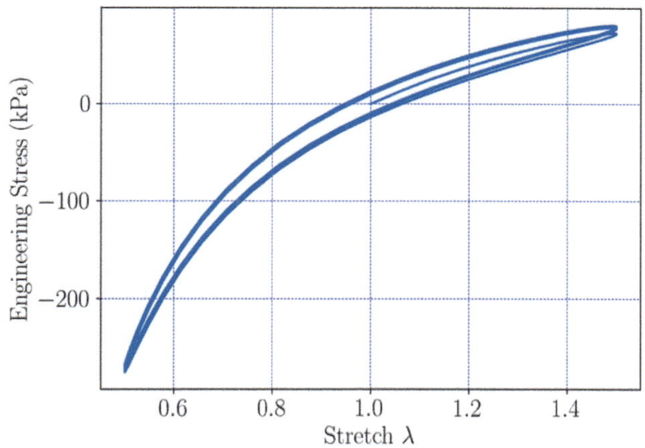

Fig. 6.8 Stress versus stretch curve from the sinusoidal axial extension/compression simulation.

Sinusoidal shear

The next numerical example we consider is the large deformation sinusoidal shear of a cube. As shown in Fig. 6.9(a) we consider a cube of edge length of $L_0 = 1$ mm, meshed with six elements along each edge in the (X,Y)-plane and four elements in the (Y, Z)-plane. We use the material properties listed in Table 6.1. As boundary conditions: (i) the bottom face of the cube is fixed; (ii) the top face is subjected to a displacement history in which the u_2 and u_3 components are held fixed while the u_1 component is prescribed a sinusoidal history

$$u_1 = \gamma L_0 \qquad \text{with} \qquad \gamma(t) = \sin(2\pi t/T),$$

for two cycles with a period of $T = 1$ s; and (iii) the four lateral faces of the cube are traction-free. The prescribed boundary conditions result in sinusoidal shearing of the cube starting

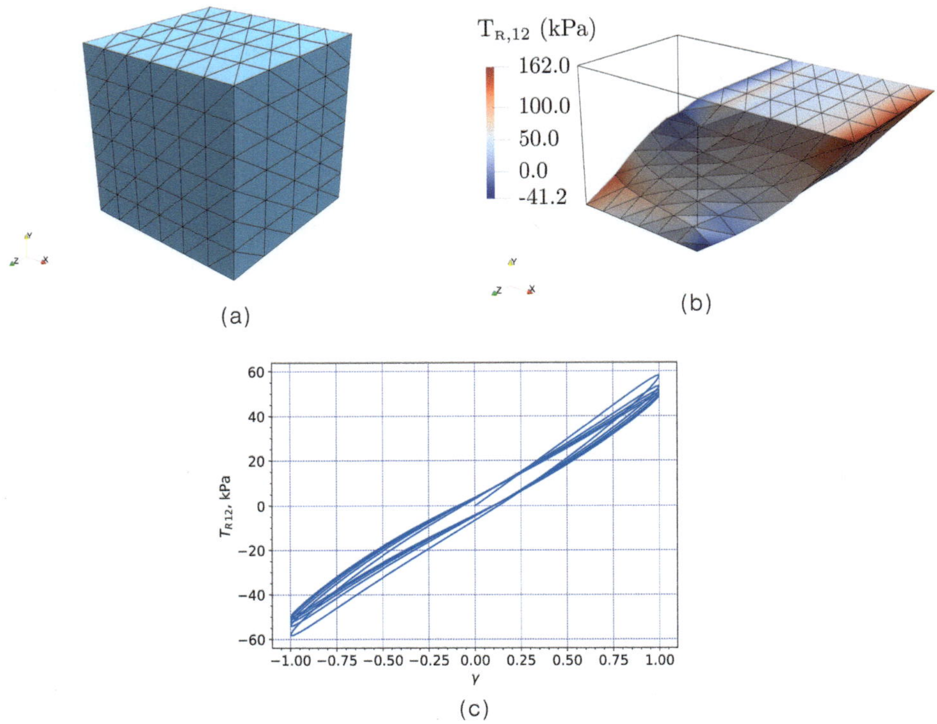

Fig. 6.9 (a) The finite element mesh used in this example. (b) The reference and deformed configuration at maximum deformation $\gamma = 1$ showing the $T_{\mathrm{R}\,12}$ component of the Piola stress in kPa. (c) The shear component of the Piola stress $T_{\mathrm{R}\,12}$ on the top face as a function of the amount of shear γ.

from $\gamma = 0$ and then cycling between $\gamma = 1$ and $\gamma = -1$. Inertial effects are neglected in this simulation.

Figure 6.9(b) shows the deformed configuration at $\gamma = 1$, together with contours of the $T_{\mathrm{R}\,12}$ component of the Piola stress. Figure 6.9(c) shows the resulting hysteresis loop of the shear stress on the top face of the cube as a function of the amount of shear γ.[2]

6.2 Simulations using material parameters for nitrile rubber HNBR50

A block under pressure on a part of its face

Here we present a classical bench-mark problem in computational solid mechanics, known as the "footing problem" (Schröder et al., 2021). We use the material properties for a nitrile rubber listed in Table 6.2.

[2] The boundary conditions used in our numerical simulation are not the same as those considered in the classical problem of homogeneous "simple shear" which is widely studied in analytical continuum mechanics. In our simulation the side faces of the cube were not prescribed appropriate boundary conditions in order to match the mathematical description of "simple shear." As a result, the deformation is inhomogenous—which is abundantly clear from the deformed geometry in Fig. 6.9(b) and also the contours of T_{12} in this figure which show that shear stress is not uniform in the body.

The geometry in this problem consists of a rectangular block subjected to a constant pressure loading on a part of its upper surface. The full geometry of the body is a rectangular block with initial length and width of 100 mm and height of 50 mm. A uniform pressure is applied on a central square region of edge-length 25 mm of the top face, as shown in Fig. 6.10(a).

Using the symmetry of the problem computations were carried out on only one quarter of the block marked by the thick black outline in Fig. 6.10(a). The mesh consists of tetrahedral elements, with 10×10 elements on the top face where the pressure is applied, and 6 elements in the depth direction, Fig. 6.10(b). As boundary conditions: (i) the bottom of the computational domain is fixed in the $\mathbf{e}_3$-direction; (ii) the two faces of symmetry of the domain are prescribed appropriate symmetry conditions; (iii) the square region of the top face outlined in blue—the "load face"—is subjected to a uniform pressure in the $\mathbf{e}_3$-direction, and is restricted from moving in the $\mathbf{e}_1$- and $\mathbf{e}_2$-directions; (iv) all other external surfaces of the domain are traction-free. The prescribed pressure is (i) linearly increased from 0 kPa to 1500 kPa over 30 s; (ii) linearly decreased back to 0 kPa over 30 s; and then (iii) held at zero pressure for another 400 s for a total time of 460 s; cf. Fig. 6.10(c).

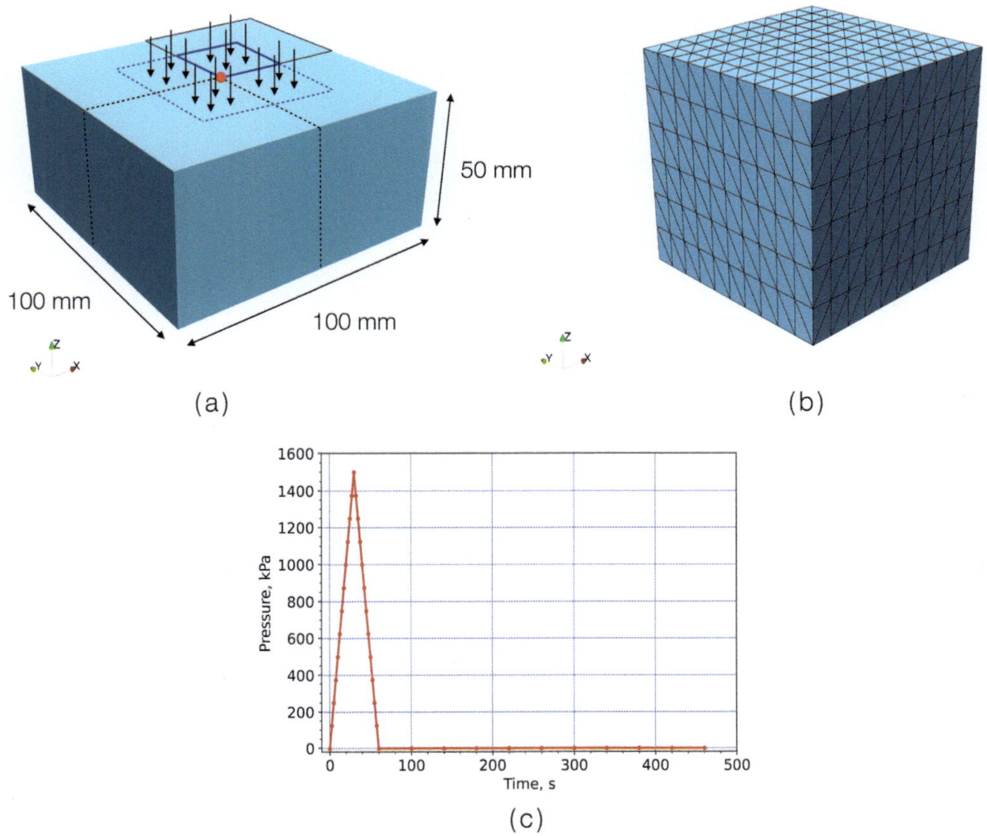

Fig. 6.10 (a) The initial geometry, symmetry planes, and pressure loading used for this example. (b) The corresponding mesh of one quarter of the full geometry. (c) The pressure–time profile.

Figure 6.11(a) shows the vertical displacement of the center point of the top surface—shown in red in Fig. 6.10(a)—as a function of the applied pressure. Figures 6.11(b), (c), and (d) show the deformed configurations with contours of the effective stretch $\bar{\lambda}$ at times $t = 30,\ 60,\ 460$ s. Note that:

- at a time of $t = 30$ s at peak pressure of 1500 kPa, the displacement of the center point of the top surface is ≈ -25 mm;
- at a time of $t = 60$ s when the pressure has been reduced to back to 0 kPa, the displacement of the center point of the top surface is ≈ -3 mm; and
- at a time of $t = 460$ s after being held at zero pressure for 400 s, the displacement of the center point of the top surface is essentially 0 mm, and the specimen has recovered back to its original configuration.

This example nicely demonstrates a major characteristic of viscoelastic materials—a strong tendency for full **recovery** of the stretch upon unloading to zero stress.

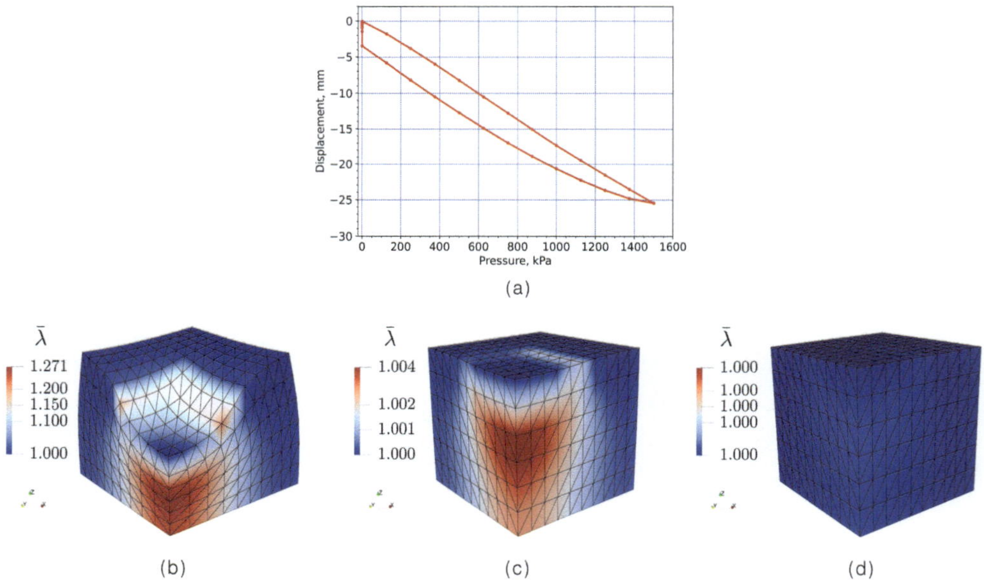

Fig. 6.11 (a) The vertical displacement of the center point of the top surface, shown in red in Fig. 6.10(a), as a function of the applied pressure. (b), (c), and (d) show the deformed configurations with contours of the effective stretch $\bar{\lambda}$ at times $t = 30,\ 60,\ 460$ s, respectively.

Indentation of a block with a spherical rigid indenter

Next, we present a simulation which is similar to the "footing problem" considered in the previous example, except that the loading of the top face is modeled through contact with a spherical rigid indenter.

As in the previous example, the geometry consists of a rectangular block of nitrile rubber with initial length and width of 100 mm, and a height of 50 mm. A spherical rigid indenter with a radius of $r_{\text{ind}} = 10\,\text{mm}$ is pushed into the material at the center of the top face as shown in Fig. 6.12(a).

Using the symmetry of the problem, computations were carried out on only one quarter of the block marked by the thick black outline in Fig. 6.12(a). The mesh consists of 13,075 tetrahedral elements which are finer near where contact with the spherical indenter occurs, Fig. 6.12(b). As boundary conditions: (i) the bottom of the computational domain is fixed in the $\mathbf{e}_3$-direction; (ii) the two faces of symmetry of the domain are prescribed appropriate symmetry conditions; (iii) all other external surfaces of the domain are traction-free. The initial position of the center of the spherical indenter is $\mathbf{x}_{\text{ind}} = (0, 0, 60)$ mm, and the motion of the indenter is such that the z-coordinate of the indenter is (i) linearly decreased from 60 mm to 50 mm over a time of 60 s, corresponding to an indentation depth of 10 mm, and (ii) then held in place for another 400 s for a total time of 460 s; cf. Fig. 6.12(c).

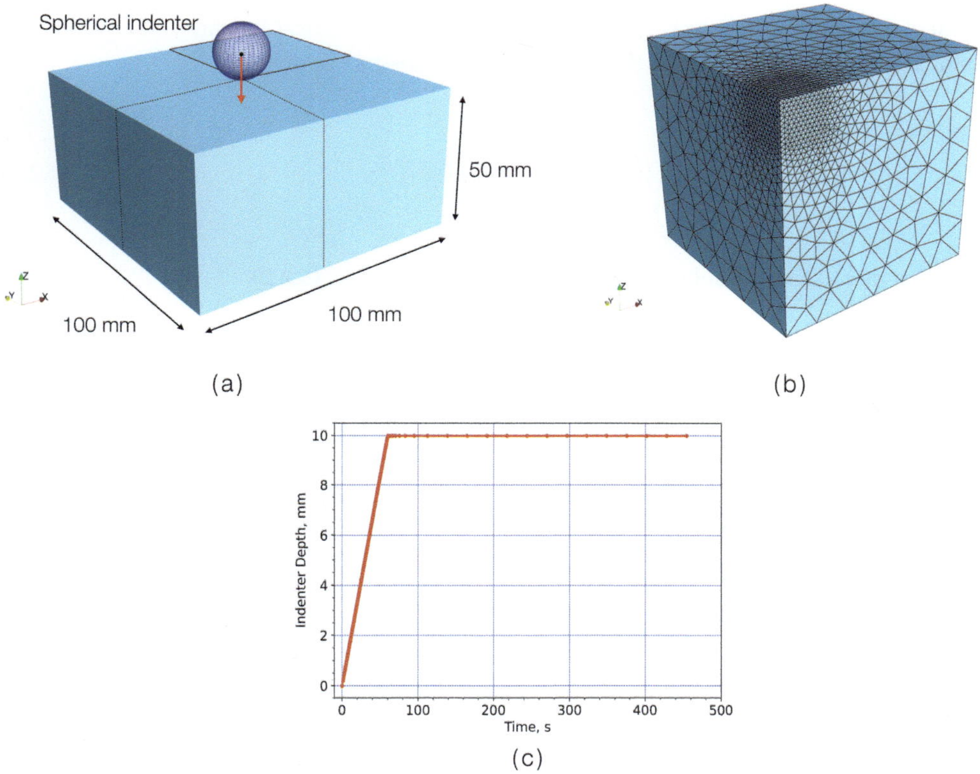

Fig. 6.12 (a) The initial geometry, symmetry planes, and spherical indenter used for this example. (b) The corresponding mesh of one quarter of the full geometry. (c) The indenter depth versus time profile.

The contact between the spherical indenter and the elastomeric block is modeled using a penalty energy approach for modeling frictionless contact of rigid spheres (or circles) with deformable bodies (Plummer et al., 2024). With reference to Fig. 6.13, let

$$\boldsymbol{\delta} \stackrel{\text{def}}{=} \mathbf{x} - \mathbf{x}_{\text{ind}}, \tag{6.2.1}$$

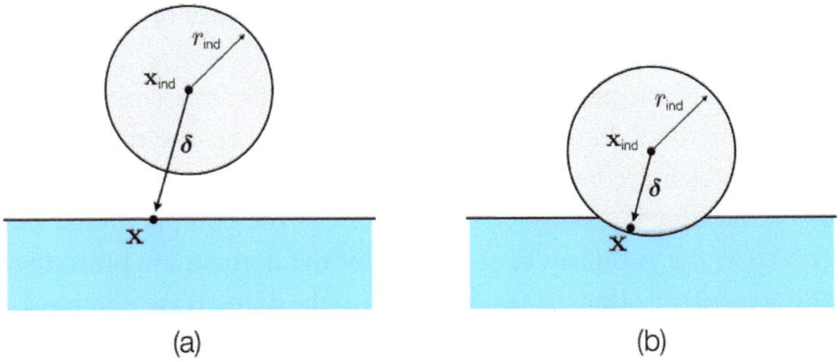

Fig. 6.13 Illustrations of situations where the inter-penetration measure d_{pen} is (a) zero and (b) positive.

denote the vector from the center $\mathbf{x}_{ind}$ of the rigid indenter to a point $\mathbf{x}$ on the surface of the deformable elastomer, let r_{ind} denote the radius of the spherical indenter, and introduce a scalar measure of penetration of the sphere into the substrate as

$$d_{pen} \stackrel{\text{def}}{=} \langle r_{ind}^2 - |\boldsymbol{\delta}|^2 \rangle. \tag{6.2.2}$$

Note that due to the Macaulay bracket in the definition of d_{pen} (6.2.2),

- d_{pen} is *zero* if the point $\mathbf{x}$ is *outside* of the indenter, cf. Fig. 6.13(a), and
- d_{pen} is *positive* if the point $\mathbf{x}$ is *inside* the indenter, and it increases as $\mathbf{x}$ and $\mathbf{x}_{ind}$ get closer together; cf. Fig. 6.13(b).

Using d_{pen}, we define a penetration energy, per unit referential area of the deformable body, by

$$\mathcal{E}_{R,\,pen} \stackrel{\text{def}}{=} \begin{cases} 0 & \text{if} \quad |\boldsymbol{\delta}| > r_{ind}, \\ \frac{1}{2}\, k_{pen}\, d_{pen}^2 & \text{if} \quad |\boldsymbol{\delta}| < r_{ind}, \end{cases} \tag{6.2.3}$$

so that this energy is zero when there is no penetration, and otherwise it increases quadratically with d_{pen}. In writing (6.2.3) we have introduced a numerical contact penalty coefficient, k_{pen}, with units of $\mathrm{J/m^6}$. In this simulation we take $k_{pen} = 100\,\mu\mathrm{J/mm^6} = 10^{14}\,\mathrm{J/m^6}$.

Next, the contact traction corresponding to the penalty energy in (6.2.3) is taken to be given by

$$\bar{\mathbf{t}}_{R,contact} \stackrel{\text{def}}{=} -\frac{\partial \mathcal{E}_{R,\,pen}}{\partial \boldsymbol{\delta}}. \tag{6.2.4}$$

Thus, if $|\boldsymbol{\delta}| < r_{\text{ind}}$, then the penalty energy $\mathcal{E}_{\text{R, pen}}$ is non-zero and the contact traction is given by

$$\begin{aligned}\bar{\mathbf{t}}_{\text{R,contact}} &= -k_{\text{pen}} d_{\text{pen}} \frac{\partial d_{\text{pen}}}{\partial \boldsymbol{\delta}} \\ &= -k_{\text{pen}} d_{\text{pen}} \frac{\partial}{\partial \boldsymbol{\delta}} \left(r_{\text{ind}}^2 - \boldsymbol{\delta} \cdot \boldsymbol{\delta} \right) \\ &= 2\, k_{\text{pen}} d_{\text{pen}}\, \boldsymbol{\delta}, \end{aligned} \tag{6.2.5}$$

so that the contact traction at a surface point $\mathbf{x}$ of the deformable body is always directed along the vector direction $\boldsymbol{\delta}$ when contact occurs.[3]

Summarizing, the contact traction is taken to be given by[4]

$$\bar{\mathbf{t}}_{\text{R,contact}} = \begin{cases} 0 & \text{if} \quad |\boldsymbol{\delta}| > r_{\text{ind}}, \\ 2\, k_{\text{pen}} d_{\text{pen}} \boldsymbol{\delta} & \text{if} \quad |\boldsymbol{\delta}| < r_{\text{ind}}. \end{cases} \tag{6.2.6}$$

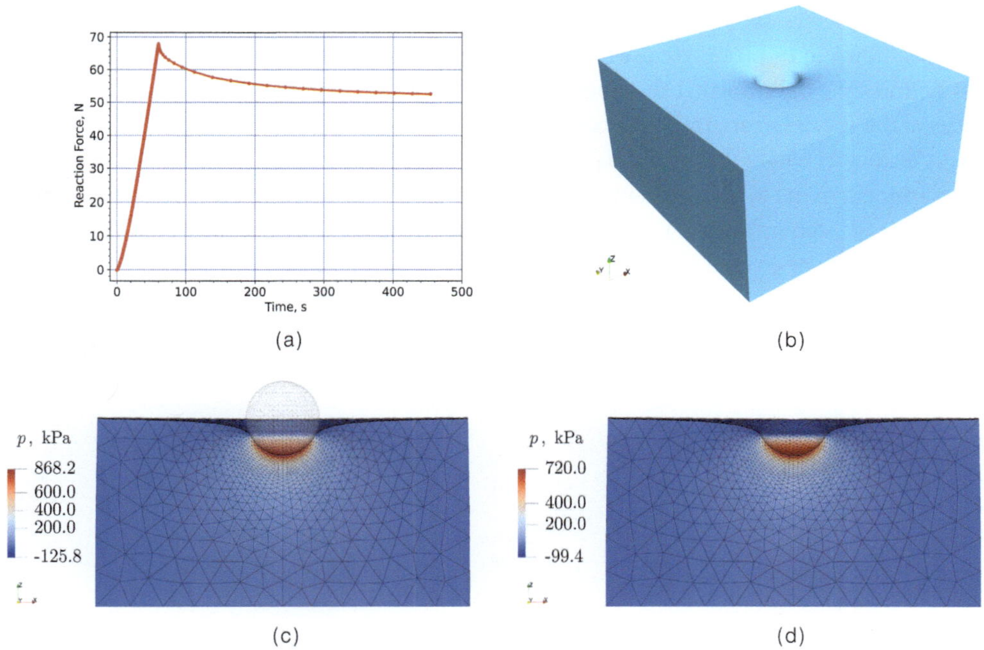

Fig. 6.14 (a) The reaction force exerted on the spherical indenter over time. (b) The deformed geometry of the block at full indenter depth at $t = 60\,\text{s}$. (c) The deformed configuration of half of the block at $t = 60\,\text{s}$ with a visualization of the indenter, and with contours of the pressure, and (d) without the spherical indenter.

[3] The numerical factor of 2 is inconsequential, it can be absorbed in k_{pen}.

[4] In the FEniCS code, the two cases for Eq. 6.2.6 are automatically handled by the Macaulay bracket in the definition of the penetration depth measure $d_{\text{pen}} = \langle r_{\text{ind}}^2 - |\boldsymbol{\delta}|^2 \rangle$.

The corresponding contribution to the weak form of the mechanical balance from this boundary traction due to contact is

$$-\int_{\partial\mathcal{B}_{\mathbf{t}_R}} \bar{\mathbf{t}}_{\mathrm{R,contact}} \cdot \mathbf{u}_{\mathrm{test}}\, da_{\mathrm{R}}. \tag{6.2.7}$$

Figure 6.14(a) shows the reaction force exerted on the spherical indenter during the ramp-and-hold indentation process. Note the relaxation of the reaction force during the holding period. Figure 6.14(b) shows the deformed geometry of the block at full indenter depth, time $t = 60\,\mathrm{s}$. Then, Figs. 6.14(c) and (d) show the deformed configuration of half of the block with contours of the effective stretch $\bar{\lambda}$ at time $t = 60$ s with and without a visualization of the spherical indenter, respectively.

Cyclic shearing of a vibration isolator

The next numerical example we consider is the cyclic shearing of a specimen which might serve as an elastomeric vibration isolator. The geometry of the specimen, which has a varying hyperbolic transversal section and a circular planar cross-section, together with its dimensions is shown in Fig. 6.15. We use the material properties for nitrile rubber listed in Table 6.2. As boundary conditions: (i) the bottom (X, Z)-face of the isolator is fixed; (ii) the lateral faces of the specimen are traction-free; and (iii) the top (X, Z)-face is subjected to a displacement history in which the u_2 and u_3 components are held fixed, and subjected to a saw-tooth displacement in the $\mathbf{e}_1$-direction in the range $u_1 \in [10, 10]\,\mathrm{mm}$ using the displacement amplitude history listed in Table 6.4, with the displacement increased or decreased in the saw-tooth history at a rate of 40 mm/min, or $|\dot{u}_1| = 10/15 = 0.6667\,\mathrm{mm/s}$.

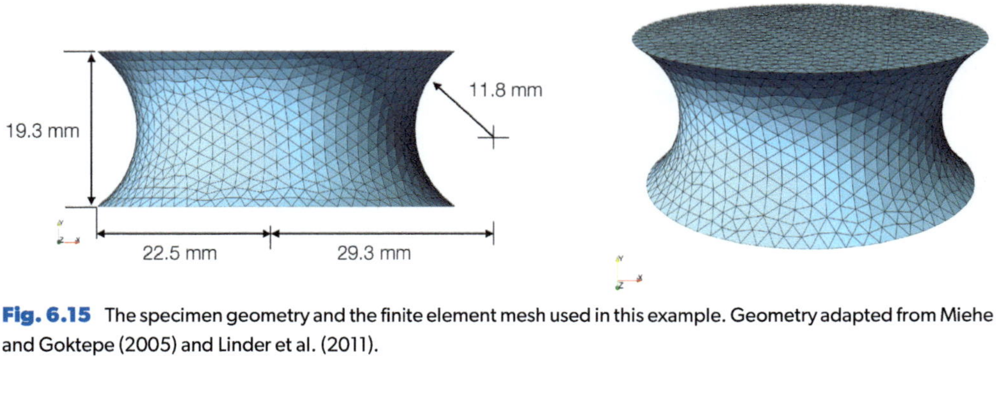

Fig. 6.15 The specimen geometry and the finite element mesh used in this example. Geometry adapted from Miehe and Goktepe (2005) and Linder et al. (2011).

Figures 6.16(a) and (b) show the deformed configurations of the vibration isolator at $u_1 = 10\,\mathrm{mm}$ and $u_1 = -10\,\mathrm{mm}$ together with contours of the T_{12} component of the Cauchy stress in kPa, and Fig. 6.16(c) shows the hysteresis loop generated when the shear force (N) is plotted as a function of the nominal shear strain $\gamma \overset{\mathrm{def}}{=} u_1/19.3\,\mathrm{mm}$.

Table 6.4 History of applied displacement.

Time, s	Displacement amplitude, mm
0	0
15	10
45	-10
75	10
105	-10
120	0

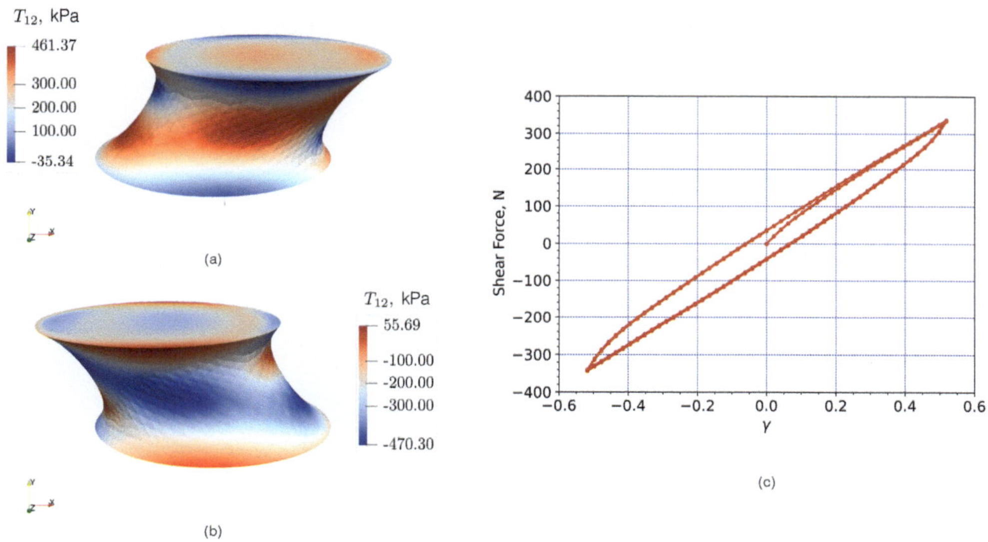

Fig. 6.16 (a) The deformed configuration at maximum positive displacement of $u_1 = 10$ mm, together with contours of the T_{12} component of the Cauchy stress in kPa. (b) The deformed configuration at maximum negative displacement of $u_1 = -10$ mm, together with contours of the T_{12} component of the Cauchy stress. (c) The shear reaction force in N as a function of the applied nominal shear strain $\gamma = u_1/19.3$ mm.

Damped oscillatory response of a viscoelastic beam

In this example we consider the damped oscillatory response of a cantilevered viscoelastic beam when subjected to an impulsive shear force applied to its free face, cf. Fig. 6.17(a). We consider a beam of length of $L = 20$ mm, height $H = 2$ mm, and width $W = 2$ mm. The beam is meshed with 12 tetrahedral elements along its length, 6 elements along its height and 2 elements along its thickness. As boundary conditions: (i) the left face of the beam is fixed; (ii) the four lateral faces of the beam are traction-free; and (iii) with reference to Fig. 6.17(b), the right face is subjected to an impulsive shear traction history with a maximum value of $T_{R12\,max} = 2$ kPa over a time period of 0.008 s, and then the beam is allowed to oscillate freely for an additional 0.392 s, for a total simulation time of 0.4 s. The prescribed boundary conditions result in damped oscillatory bending of the beam in the e_2-direction.

REMARK

In this example and the next, we include inertial effects and therefore use a slightly modified list of material properties for NBR from those listed in Table 6.2. Specifically, in order to observe viscoelastic dissipation effects which occur on a timescale similar to that associated with inertial effects, we modified the value of the relaxation time $\tau_r^{(1)}$ in Table 6.2, from $\tau_r^{(1)} = 1.0$ s to a much *smaller* value of $\tau_r^{(1)} = 1 \times 10^{-3}$ s, while keeping all other values of the material parameters the same as those listed in Table 6.2.[5]

Figure 6.17(c) shows a plot of the tip displacement u_2 as a function of time, and Figs. 6.17(d)–(e) show contours of the axial Cauchy stress component T_{11} on the deformed configuration of the beam at the points of maximum positive and negative displacement, respectively. This simulation demonstrates the effects of both inertia and viscoelasticity in the vibration response of a beam.

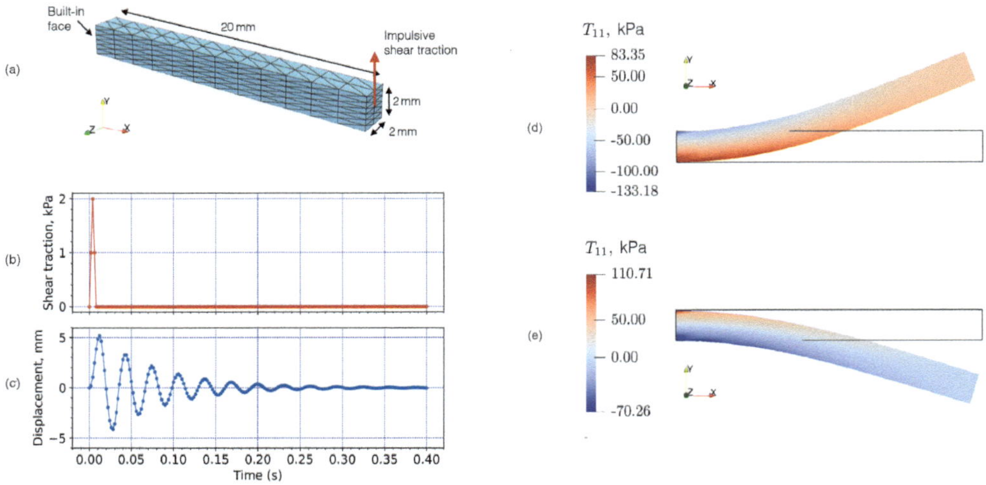

Fig. 6.17 (a) The geometry of the beam and the finite element mesh used in this example. (b) The impulsive shear traction history. (c) The tip displacement of the beam as a function of time. The deformed configuration of the beam is shown at (d) time $t = 0.012$ s and (e) $t = 0.028$ s with contours of the axial Cauchy stress component T_{11}.

Column buckling

As a final example we study the buckling of a column with a square cross-section under fixed-end boundary conditions. We consider a column of initial height $H_0 = 20$ mm, and a square cross-section of edge length $W_0 = 1$ mm; cf. Fig. 6.18(a). Since buckling phenomena are known to be highly imperfection-sensitive we consider a column with a small "geometric

[5] The timescale of inertial effects may be roughly estimated using the elastic shear wave speed c_s and a characteristic dimension ℓ of the part under consideration. With $G_{0\text{eq}}$ the ground-state equilibrium shear modulus and ρ_R the mass density, $c_s = \sqrt{G_{0\text{eq}}/\rho_R}$, and the time it takes a shear wave to traverse the body is then $\tau_{\text{inertia}} \approx \ell/c_s$. For our NBR beam, with $G_{0\text{eq}} = 400$ kPa, $\rho_R = 1300$ kg/m^3, and $\ell = 20$ mm, we obtain the estimate $\tau_{\text{inertia}} \approx 1$ ms.

imperfection" along its length. The geometric imperfection is introduced by modifying the initial coordinates of $\mathbf{X}^0$ of a "perfect column" to the coordinates $\mathbf{X}$ of an "imperfect column" using the mapping relation

$$X_1 = X_1^0 - \left(\frac{\delta}{2}\right)\left[1 - \cos\left(\frac{2\pi X_3}{H_0}\right)\right], \quad X_2 = X_2^0 - \left(\frac{\delta}{2}\right)\left[1 - \cos\left(\frac{2\pi X_3}{H_0}\right)\right],$$
$$X_3 = X_3^0, \quad \text{with} \quad \delta = 0.005 W_0.$$

This functional form of the geometric imperfection results in a maximum imperfection of magnitude δ at the mid-height of the column, and zero imperfection at the top and bottom of the column. We employ a mesh with 4 elements along the edges of the cross-section and use 50 elements along the height of the column. We use the same material properties as those discussed in the Remark on page 100.

As boundary conditions: the bottom face of the column is fixed, the top face of the column is prescribed a displacement in the negative $\mathbf{e}_3$-direction to compress the column, while the transverse displacements of this face are fixed. All other faces of the column are traction-free. The prescribed compressive displacement is ramped from 0 mm to -0.75 mm over a period of 0.02 s and then held constant for another 0.02 s, cf. Fig. 6.18(c). We use a "smoothstep"

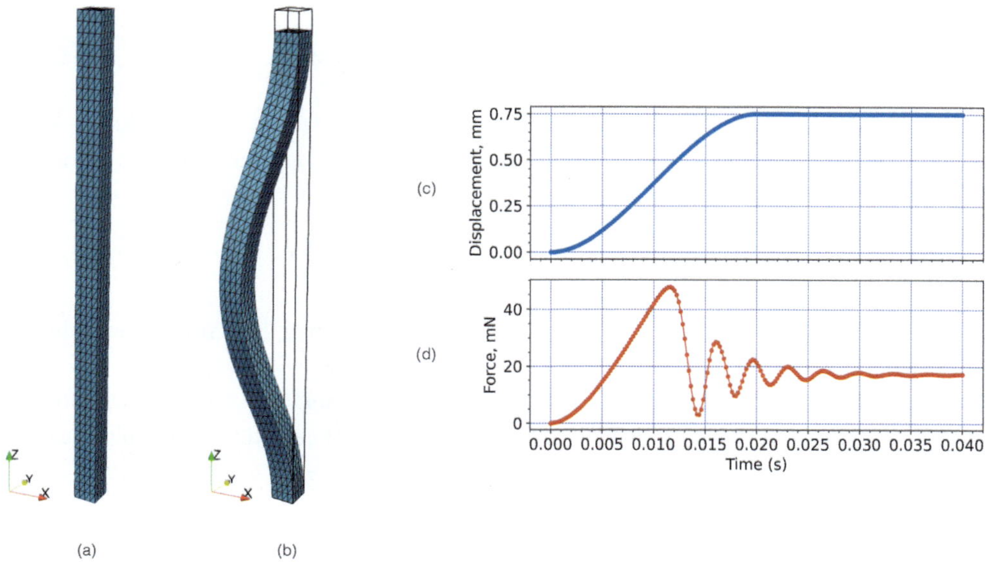

Fig. 6.18 (a) Geometry and mesh for buckling of an "imperfect" column made from a viscoelastic material. (b) Buckled configuration at maximum compressive displacement. (c) The prescribed compressive displacement as a function of time. (d) Plot of the magnitude of the compressive force on the top face as a function of time.

function to apply this displacement history so that there are no sudden changes in velocity.[6] Figure 6.18(b) shows the buckled configuration of the column at the maximum applied displacement.

Figure 6.18(d) shows a plot of the magnitude of the compressive force on the top face of the column as a function of time. An examination of the numerical simulation results shows that the beam starts to deviate from its initial "straight" configuration at a time $t \approx 0.0088\,\text{s}$ while the compressive force is still rising, and upon further compression the load continues to increase until it reaches a peak value of $\approx 47.5\,\text{mN}$ at which time the beam "buckles" substantially from its initially straight configuration and the load drops dramatically to a value of $\approx 2.5\,\text{mN}$. As the beam is further compressed axially it vibrates laterally, and the axial force fluctuates until eventually the vibrations die down by a time of $t \approx 0.04\,\text{s}$ and the compressive force settles at $\approx 17.5\,\text{mN}$; it is this final configuration which is shown in Fig. 6.18(b).

This is another simulation which demonstrates the effects of both inertia and viscoelasticity in the buckling response of a beam.

6.A Appendix. Newmark time-integration scheme for the inertia term

Using the notation

$$\mathbf{v} \equiv \dot{\mathbf{u}} \qquad \text{and} \qquad \mathbf{a} \equiv \ddot{\mathbf{u}}$$

for the velocity and acceleration, respectively, consider the time-discrete solution of (5.1.1) subject to the initial conditions defined at time $t = 0$,

$$\mathbf{u}(0) = \mathbf{u}_0, \qquad \mathbf{v}(0) = \mathbf{v}_0 \qquad \mathbf{a}(0) = \mathbf{a}_0. \tag{6.A.1}$$

Given these initial conditions, a solution for $\mathbf{u}(t)$, $\mathbf{v}(t)$, and $\mathbf{a}(t)$ is required during some interval $t \in [0, T]$. With a numerical time-integration strategy in mind, the time domain is discretized in a series of N_t time steps,

$$t_{n+1} = t_n + \Delta t, \qquad n = 0, 1, \ldots, N_t, \tag{6.A.2}$$

such that $t_{N_t} = T$. For simplicity in what follows, it will be assumed that Δt is *constant* during the chosen time period.

The problem now becomes one in which a time-discrete solution is to be found for the vectors $\mathbf{u}_{n+1}$, $\mathbf{v}_{n+1}$, and $\mathbf{a}_{n+1}$ satisfying the residual $(5.5.3)_1$ at each time step. This equation

[6] The "smoothstep" function that we use is,

$$\text{smoothstep}(x) \overset{\text{def}}{=} \begin{cases} 0, & x \le 0 \\ 3x^2 - 2x^3, & 0 \le x \le 1 \\ 1, & 1 \le x \end{cases}$$

represents a nonlinear algebraic equation which is a function of $\mathbf{u}_{n+1}$, $\mathbf{v}_{n+1}$, and $\mathbf{a}_{n+1}$. To solve this equation a kinematic relationship is required between $(\mathbf{a}_{n+1}, \mathbf{v}_{n+1})$ and $\mathbf{u}_{n+1}$ so that inertia term in the residual $(5.5.3)_1$ becomes a function solely of $\mathbf{u}_{n+1}$. This is furnished by expanding the displacement $\mathbf{u}$ as a third-order derivative of $\mathbf{u}$ with respect to time, and by assuming a linear variation of acceleration over time, as detailed below.

Expanding the position $\mathbf{u}$ and the velocity $\mathbf{v}$ as a Taylor series about time t gives

$$\begin{aligned} \mathbf{u}_{t+\Delta t} &\approx \mathbf{u}_t + \Delta t \left.\frac{d\mathbf{u}}{dt}\right|_t + \frac{\Delta t^2}{2}\left.\frac{d^2\mathbf{u}}{dt^2}\right|_t + \frac{\Delta t^3}{6}\left.\frac{d^3\mathbf{u}}{dt^3}\right|_t + \ldots, \\ \left.\frac{d\mathbf{u}}{dt}\right|_{t+\Delta t} &\approx \left.\frac{d\mathbf{u}}{dt}\right|_t + \Delta t\left.\frac{d^2\mathbf{u}}{dt^2}\right|_t + \frac{\Delta t^2}{2}\left.\frac{d^3\mathbf{u}}{dt^3}\right|_t + \ldots. \end{aligned} \tag{6.A.3}$$

The above equations can be truncated and rewritten in terms of the velocity $\mathbf{v}$ and acceleration $\mathbf{a}$ with the substitution of (as yet unknown) parameters β and γ which compensate for the truncation error to give,

$$\begin{aligned} \mathbf{u}_{t+\Delta t} &\approx \mathbf{u}_t + \mathbf{v}_t\,\Delta t + \frac{1}{2}\,\mathbf{a}_t\Delta t^2 + \beta\left.\frac{d^3\mathbf{u}}{dt^3}\right|_t \Delta t^3, \\ \mathbf{v}_{t+\Delta t} &\approx \mathbf{v}_t + \mathbf{a}_t\Delta t + \gamma\left.\frac{d^3\mathbf{u}}{dt^3}\right|_t \Delta t^2. \end{aligned} \tag{6.A.4}$$

The third-order derivative is now approximated by assuming a linear variation of the acceleration $\mathbf{a}$ over the time interval Δt, to give

$$\begin{aligned} \left.\frac{d^3\mathbf{u}}{dt^3}\right|_t &= \left(\left.\frac{d^2\mathbf{u}}{dt^2}\right|_{t+\Delta t} - \left.\frac{d^2\mathbf{u}}{dt^2}\right|_t\right)\frac{1}{\Delta t}, \\ &= \left(\mathbf{a}_{t+\Delta t} - \mathbf{a}_t\right)\frac{1}{\Delta t}. \end{aligned} \tag{6.A.5}$$

Substitution of (6.A.5) into (6.A.4) gives

$$\begin{aligned} \mathbf{u}_{t+\Delta t} &\approx \mathbf{u}_t + \mathbf{v}_t\,\Delta t + \frac{1}{2}(1-2\beta)\,\mathbf{a}_t\Delta t^2 + \beta\mathbf{a}_{t+\Delta t}\Delta t^2, \\ \mathbf{v}_{t+\Delta t} &\approx \mathbf{v}_t + (1-\gamma)\mathbf{a}_t\Delta t + \gamma\mathbf{a}_{t+\Delta t}\Delta t. \end{aligned} \tag{6.A.6}$$

Thus, reverting back to our time-stepping notation we obtain the widely used Newmark family of time-stepping schemes (Newmark, 1959):

$$\begin{aligned} \mathbf{a}_{n+1} &= \frac{\mathbf{u}_{n+1} - \mathbf{u}_n - \mathbf{v}_n\,\Delta t}{\beta\Delta t^2} - \frac{(1-2\beta)}{2\beta}\,\mathbf{a}_n, \\ \mathbf{v}_{n+1} &= \mathbf{v}_n + \Delta t\left[(1-\gamma)\mathbf{a}_n + \gamma\mathbf{a}_{n+1}\right], \end{aligned} \tag{6.A.7}$$

where γ and β are parameters which can be chosen in the ranges $\gamma \in (0, 1]$ and $\beta \in (0, 0.5]$ to yield different time-integration schemes. Equation (6.A.7) establishes the relationship between $\mathbf{a}_{n+1}$ and $\mathbf{v}_{n+1}$ with $\mathbf{u}_{n+1}$, which can now be used to advance the solution in time from t_n to t_{n+1}. Different choices for β and γ determine the stability and accuracy of the time-integration scheme. For our purposes a specific choice of the Newmark family is of particular interest:

- Parameter values: $\gamma = 1/2$ and $\beta = 1/4$ This choice gives

$$\begin{aligned} \mathbf{a}_{n+1} &= \frac{\mathbf{u}_{n+1} - \mathbf{u}_n - \mathbf{v}_n\,\Delta t}{(\Delta t^2)/4} - \mathbf{a}_n, \\ \mathbf{v}_{n+1} &= \mathbf{v}_n + \frac{\Delta t}{2}\left[\mathbf{a}_n + \mathbf{a}_{n+1}\right]. \end{aligned} \tag{6.A.8}$$

  This choice is known as the *average acceleration scheme*, where the use of the average acceleration is clearly evident in $(6.A.8)_2$ in advancing $\mathbf{v}$ from n to $n+1$. This scheme can also be interpreted as a *trapezoidal integration of the acceleration* to give appropriate increments in velocity and position.

  - The average acceleration scheme is an *implicit scheme* which is known to be *unconditionally stable*, and preferred in many numerical simulation procedures.

Part III

Thermoelasticity of elastomeric materials

7 Thermoelasticity of elastomeric materials

7.1 Introduction

It is well-known that heating or cooling an unconfined solid specimen generally leads to dimensional changes of the specimen. For confined specimens, the deformation produced by heating may generate complex stress distributions and the peak magnitudes of such thermally induced stresses are often quite large. Conversely, temperature changes and distributions generated by the mechanical loading of solids may also be important.

In this chapter we discuss a framework for the coupled thermal and mechanical response of solids, restricting our attention to situations in which the deformation is elastic.[1] The general framework—known as the theory of thermoelasticity—is broad enough to describe both metals and elastomeric materials.[2]

For specificity, we will focus our attention on a large-deformation thermoelasticity theory for elastomeric materials. This class of materials exhibit the classical Gough–Joule effects (Gough, 1805; Joule, 1859) which may be summarized as:

- "A rubber under a constant uniaxial stress contracts on heating."
- "Under adiabatic conditions a rubber increases in temperature when stretched."
- "Under isothermal conditions a rubber gives out heat when stretched."

These classical experimental observation on the response of rubber-like materials go back in time to the beginnings of the discipline of "Thermodynamics." We will show that the thermo-mechanically coupled theory that we will formulate and numerically implement can reproduce these well-established experimental observations.

Note that the theoretical framework for coupled mechanical and thermal effects that we shall discuss in this chapter is a prototypical framework which is useful for other coupled theories such as coupling diffusion of chemical species with elasticity, or electrical fields with elasticity, and so forth.

[1] We neglect viscoelastic effects.

[2] This chapter is based on the theory of Thermoelasticity in chapters 55 through 59 of the book *The Mechanics and Thermodynamics of Continua* by Gurtin et al. (2010).

For earlier classical works on thermoelasticity see, for example, Biot (1956), Boley and Weiner (1960), Green and Adkins (1960), Nowacki (1962), Coleman and Noll (1963), Gurtin (1967), Carlson (1972) and the references therein. In the context of thermomechanical description of rubber-like materials see Flory (1961), Chadwick (1974), Chadwick and Creasy (1984), Oden (1969), Argyris and Doltsinis (1981), Miehe (1995), Holzapfel and Simo (1996).

Introduction to Coupled Theories in Solid Mechanics. Lallit Anand, Eric M. Stewart, Shawn A. Chester, Oxford University Press.
 DOI: 10.1093/oso/9780198986218.003.0008

7.2 Kinematics

We consider a macroscopically homogeneous body $\mathcal{B}$ with the region of space it occupies in a fixed reference configuration, and denote by $\mathbf{X}$ an arbitrary material point of $\mathcal{B}$. A motion of $\mathcal{B}$ is a smooth one-to-one mapping

$$\mathbf{x} = \boldsymbol{\chi}(\mathbf{X}, t), \tag{7.2.1}$$

with displacement

$$\mathbf{u}(\mathbf{X}, t) = \boldsymbol{\chi}(\mathbf{X}, t) - \mathbf{X}, \tag{7.2.2}$$

deformation gradient, velocity, and acceleration given by,

$$\mathbf{F} = \nabla\boldsymbol{\chi} = \mathbf{1} + \nabla\mathbf{u}, \qquad \dot{\mathbf{u}} = \frac{\partial\boldsymbol{\chi}}{\partial t}, \qquad \ddot{\mathbf{u}} = \frac{\partial^2\boldsymbol{\chi}}{\partial t^2}. \tag{7.2.3}$$

We assume that

$$J = \det\mathbf{F} > 0, \tag{7.2.4}$$

so that $\mathbf{F}$ is invertible, and admits the polar decompositions

$$\mathbf{F} = \mathbf{R}\mathbf{U} \qquad \text{and} \qquad \mathbf{F} = \mathbf{V}\mathbf{R}, \tag{7.2.5}$$

into a rotation tensor $\mathbf{R}$, together with a symmetric and positive-definite right and left stretch tensors $\mathbf{U}$ and $\mathbf{V}$, satisfying

$$\mathbf{U} = \sqrt{\mathbf{F}^{\top}\mathbf{F}} \qquad \text{and} \qquad \mathbf{V} = \sqrt{\mathbf{F}\mathbf{F}^{\top}}. \tag{7.2.6}$$

The symmetric and positive-definite right and left Cauchy–Green tensors are defined as

$$\mathbf{C} \stackrel{\text{def}}{=} \mathbf{F}^{\top}\mathbf{F} \qquad \text{and} \qquad \mathbf{B} \stackrel{\text{def}}{=} \mathbf{F}\mathbf{F}^{\top}, \tag{7.2.7}$$

respectively.

For later use we note that the deformation gradient $\mathbf{F}$ may be decomposed multiplicatively into volumetric and isochoric factors $\mathbf{F}^{\text{vol}}$ and $\bar{\mathbf{F}}$ as,

$$\mathbf{F} = \mathbf{F}^{\text{vol}}\bar{\mathbf{F}}, \tag{7.2.8}$$

where

$$\mathbf{F}^{\text{vol}} \stackrel{\text{def}}{=} J^{1/3}\mathbf{1} \qquad \text{and} \qquad \bar{\mathbf{F}} \stackrel{\text{def}}{=} J^{-1/3}\mathbf{F}. \tag{7.2.9}$$

This leads to a multiplicative decomposition of the right Cauchy–Green tensor $\mathbf{C}$ of the form,

$$\mathbf{C} = \mathbf{C}^{\text{vol}}\bar{\mathbf{C}}, \qquad \mathbf{C}^{\text{vol}} = J^{2/3}\mathbf{1}, \qquad \bar{\mathbf{C}} = \bar{\mathbf{F}}^{\top}\bar{\mathbf{F}}. \tag{7.2.10}$$

Similarly, the left Cauchy–Green tensor $\mathbf{B}$ can be multiplicatively decomposed as

$$\mathbf{B} = \mathbf{B}^{\text{vol}}\bar{\mathbf{B}}, \qquad \mathbf{B}^{\text{vol}} = J^{2/3}\mathbf{1}, \qquad \bar{\mathbf{B}} = \bar{\mathbf{F}}\bar{\mathbf{F}}^{\top}. \tag{7.2.11}$$

We also introduce the invariants

$$\bar{I}_1(\bar{\mathbf{C}}) \stackrel{\text{def}}{=} \operatorname{tr}\bar{\mathbf{C}}, \qquad \bar{I}_2(\bar{\mathbf{C}}) \stackrel{\text{def}}{=} \tfrac{1}{2}\left[(\operatorname{tr}\bar{\mathbf{C}})^2 - \operatorname{tr}(\bar{\mathbf{C}}^2)\right], \tag{7.2.12}$$

and

$$\bar{I}_1(\bar{\mathbf{B}}) \stackrel{\text{def}}{=} \operatorname{tr}\bar{\mathbf{B}}, \qquad \bar{I}_2(\bar{\mathbf{B}}) \stackrel{\text{def}}{=} \tfrac{1}{2}\left[(\operatorname{tr}\bar{\mathbf{B}})^2 - \operatorname{tr}(\bar{\mathbf{B}}^2)\right]. \tag{7.2.13}$$

Note that the two sets of invariants $\{\bar{I}_1(\bar{\mathbf{C}}), \bar{I}_2(\bar{\mathbf{C}})\}$ and $\{\bar{I}_1(\bar{\mathbf{B}}), \bar{I}_2(\bar{\mathbf{B}})\}$ are equal to each other.

Throughout we shall denote an arbitrary part of the reference body by $\mathcal{P}$, with $\mathbf{n}_{\text{R}}$ the outward unit normal on the boundary $\partial\mathcal{P}$ of $\mathcal{P}$. Also, we denote by $\mathcal{P}_t = \chi(\mathcal{P}, t)$ the image of $\mathcal{P}$ in the deformed body, with $\mathbf{n}$ the outward unit normal on the boundary $\partial\mathcal{P}_t$ of $\mathcal{P}_t$.

7.3 Frame-indifference

A *change in frame*, at each *fixed time t*, is a transformation—defined by a rotation $\mathbf{Q}(t)$ and a spatial point $\mathbf{y}(t)$—which transforms *spatial points* $\mathbf{x}$ to spatial points

$$\mathbf{x}^* = \mathcal{F}(\mathbf{x}), \tag{7.3.1}$$

$$= \mathbf{y}(t) + \mathbf{Q}(t)(\mathbf{x} - \mathbf{o}); \tag{7.3.2}$$

the function $\mathcal{F}$ represents a rigid mapping of the observed space into itself, with $\mathbf{o}$ a fixed spatial origin. By (7.3.2) the transformation law for the motion $\mathbf{x} = \chi(\mathbf{X}, t)$ has the form

$$\chi^*(\mathbf{X}, t) = \mathbf{y}(t) + \mathbf{Q}(t)(\chi(\mathbf{X}, t) - \mathbf{o}). \tag{7.3.3}$$

Hence the deformation gradient $\mathbf{F}$ transforms according to

$$\mathbf{F}^* = \mathbf{Q}\mathbf{F}. \tag{7.3.4}$$

7.4 Balance of forces and moments

As is standard, we introduce a referential stress measure, the Piola stress $\mathbf{T}_{\text{R}}$, which is related to the spatial Cauchy stress $\mathbf{T}$ by

$$\mathbf{T}_{\text{R}} = J\,\mathbf{T}\mathbf{F}^{-\top}, \tag{7.4.1}$$

and in terms of which:

(a) The surface traction $\mathbf{t}_R(\mathbf{n}_R)$ on an oriented area element in the reference body is given by the traction relation

$$\mathbf{t}_R(\mathbf{n}_R) = \mathbf{T}_R\mathbf{n}_R. \tag{7.4.2}$$

(b) $\mathbf{T}_R$ satisfies the force balance

$$\mathrm{Div}\,\mathbf{T}_R + \mathbf{b}_R = \mathbf{0}, \tag{7.4.3}$$

with

$$\mathbf{b}_R = \mathbf{b}_{0R} - \rho_R\ddot{\mathbf{u}}, \tag{7.4.4}$$

where $\mathbf{b}_{0R}$ represents the conventional body force per unit volume of the reference body, and $(-\rho_R\ddot{\mathbf{u}})$ represents the inertial body force, where ρ_R is the mass density of the referential body and $\ddot{\mathbf{u}}$ the acceleration.

(iii) $\mathbf{T}_R$ satisfies,

$$\mathbf{T}_R\mathbf{F}^\top = \mathbf{F}\mathbf{T}_R^\top, \tag{7.4.5}$$

which represents a balance of moments.

Standard considerations give that the Cauchy stress and the Piola stress transform under a change in frame as (Gurtin et al., 2010)

$$\mathbf{T}^* = \mathbf{Q}\mathbf{T}\mathbf{Q}^\top \qquad \text{and} \qquad \mathbf{T}_R^* = \mathbf{Q}\mathbf{T}_R. \tag{7.4.6}$$

7.5 Balance of energy. Entropy imbalance

Our discussion of thermodynamics involves the following fields:[3]

ε_R	the internal energy density per unit reference volume,
η_R	the entropy density per unit reference volume,
$\mathbf{q}_R$	the heat flux per unit reference area,
q_R	the external heat supply per unit reference volume,
ϑ	the absolute temperature ($\vartheta > 0$),

[3] Much of the discussion of thermodynamics in this section is the same as in Section 1.5, but is repeated here for convenience in order to make this Chapter self-contained.

and follows the discussion in Gurtin et al. (2010). Consider a material region $\mathcal{P}$. Then, we take the balance law for energy as

$$\overline{\int_{\mathcal{P}} \varepsilon_{\rm R}\, dv_{\rm R}}^{\,\cdot} = -\int_{\partial\mathcal{P}} \mathbf{q}_{\rm R}\cdot\mathbf{n}_{\rm R}\, da_{\rm R} + \int_{\mathcal{P}} q_{\rm R}\, dv_{\rm R} + \int_{\partial\mathcal{P}} (\mathbf{T}_{\rm R}\mathbf{n}_{\rm R})\cdot\dot{\mathbf{u}}\, da_{\rm R} + \int_{\mathcal{P}} \mathbf{b}_{\rm R}\cdot\dot{\mathbf{u}}\, dv_{\rm R}. \tag{7.5.1}$$

Applying the divergence theorem to the terms in (7.5.1) involving integrals over the boundary $\partial\mathcal{P}$ of $\mathcal{P}$, we obtain

$$\int_{\mathcal{P}} \Big(\dot{\varepsilon}_{\rm R} - (\mathrm{Div}\,\mathbf{T}_{\rm R} + \mathbf{b}_{\rm R})\cdot\dot{\mathbf{u}} - \mathbf{T}_{\rm R} : \dot{\mathbf{F}} + \mathrm{Div}\,\mathbf{q}_{\rm R} - q_{\rm R}\Big)\, dv_{\rm R} = 0, \tag{7.5.2}$$

which upon use of the force balance (7.4.3), and using the fact that (7.5.2) must hold for all parts $\mathcal{P}$, gives the local form of the energy balance as

$$\dot{\varepsilon}_{\rm R} = \mathbf{T}_{\rm R} : \dot{\mathbf{F}} - \mathrm{Div}\,\mathbf{q}_{\rm R} + q_{\rm R}. \tag{7.5.3}$$

Also, the second law takes the form of an entropy imbalance

$$\overline{\int_{\mathcal{P}} \eta_{\rm R}\, dv_{\rm R}}^{\,\cdot} \geq -\int_{\partial\mathcal{P}} \frac{\mathbf{q}_{\rm R}\cdot\mathbf{n}_{\rm R}}{\vartheta}\, da_{\rm R} + \int_{\mathcal{P}} \frac{q_{\rm R}}{\vartheta}\, dv_{\rm R}, \tag{7.5.4}$$

in which case the local entropy imbalance has the form

$$\dot{\eta}_{\rm R} \geq -\mathrm{Div}\left(\frac{\mathbf{q}_{\rm R}}{\vartheta}\right) + \frac{q_{\rm R}}{\vartheta}. \tag{7.5.5}$$

Then, in view of the local energy balance (7.5.3),

$$\begin{aligned} -\mathrm{Div}\left(\frac{\mathbf{q}_{\rm R}}{\vartheta}\right) + \frac{q_{\rm R}}{\vartheta} &= \frac{1}{\vartheta}(-\mathrm{Div}\,\mathbf{q}_{\rm R} + q_{\rm R}) + \frac{1}{\vartheta^2}\,\mathbf{q}_{\rm R}\cdot\nabla\vartheta, \\ &= \frac{1}{\vartheta}\Big(\dot{\varepsilon}_{\rm R} - \mathbf{T}_{\rm R} : \dot{\mathbf{F}} + \frac{1}{\vartheta}\,\mathbf{q}_{\rm R}\cdot\nabla\vartheta\Big), \end{aligned}$$

and this with the local entropy imbalance (7.5.5) implies that

$$(\dot{\varepsilon}_{\rm R} - \vartheta\dot{\eta}_{\rm R}) - \mathbf{T}_{\rm R} : \dot{\mathbf{F}} + \frac{1}{\vartheta}\,\mathbf{q}_{\rm R}\cdot\nabla\vartheta \leq 0. \tag{7.5.6}$$

Introducing the Helmholtz free energy per unit reference volume

$$\psi_{\rm R} \stackrel{\rm def}{=} \varepsilon_{\rm R} - \vartheta\eta_{\rm R}, \tag{7.5.7}$$

(7.5.6) yields the following local free-energy imbalance

$$\dot{\psi}_{\rm R} + \eta_{\rm R}\dot{\vartheta} - \mathbf{T}_{\rm R} : \dot{\mathbf{F}} + \frac{1}{\vartheta}\,\mathbf{q}_{\rm R}\cdot\nabla\vartheta \leq 0. \tag{7.5.8}$$

Further, introducing a symmetric second Piola stress by

$$\mathbf{T}_{\mathrm{RR}} \stackrel{\text{def}}{=} J\mathbf{F}^{-1}\mathbf{T}\mathbf{F}^{-\top}, \tag{7.5.9}$$

and noting that the rate of change of the right Cauchy–Green tensor $\mathbf{C}$ is

$$\dot{\mathbf{C}} = \mathbf{F}^{\top}\dot{\mathbf{F}} + \dot{\mathbf{F}}^{\top}\mathbf{F}, \tag{7.5.10}$$

we have

$$\begin{aligned}
\mathbf{T}_{\mathrm{RR}} : \dot{\mathbf{C}} &= (J\mathbf{F}^{-1}\mathbf{T}\mathbf{F}^{-\top}) : \left(\mathbf{F}^{\top}\dot{\mathbf{F}} + \dot{\mathbf{F}}^{\top}\mathbf{F}\right) \\
&= 2\,(J\mathbf{F}^{-1}\mathbf{T}\mathbf{F}^{-\top}) : \left(\mathbf{F}^{\top}\dot{\mathbf{F}}\right) \qquad \text{(since } \mathbf{T}_{\mathrm{RR}} \text{ is symmetric)} \\
&= 2\,(J\mathbf{T}\mathbf{F}^{-\top}) : \dot{\mathbf{F}} \\
&= 2\,\mathbf{T}_{\mathrm{R}} : \dot{\mathbf{F}},
\end{aligned}$$

or

$$\mathbf{T}_{\mathrm{R}} : \dot{\mathbf{F}} = \frac{1}{2}\mathbf{T}_{\mathrm{RR}} : \dot{\mathbf{C}}. \tag{7.5.11}$$

Using (7.5.11) in (7.5.8) the local free-energy imbalance may be written as

$$\dot{\psi}_{\mathrm{R}} + \eta_{\mathrm{R}}\dot{\vartheta} - \frac{1}{2}\mathbf{T}_{\mathrm{RR}} : \dot{\mathbf{C}} + \frac{1}{\vartheta}\mathbf{q}_{\mathrm{R}} \cdot \nabla\vartheta \leq 0. \tag{7.5.12}$$

We note that under a change in frame

$$\psi_{\mathrm{R}},\ \eta_{\mathrm{R}},\ \vartheta,\ \mathbf{q}_{\mathrm{R}},\ \text{and } \nabla\vartheta \text{ are invariant,} \tag{7.5.13}$$

ψ_{R}, η_{R}, and ϑ because they are scalars, and $\mathbf{q}_{\mathrm{R}}$ and $\nabla\vartheta$ since they are referential vector fields. Using the standard transformation rule (7.3.4) for the deformation gradient $\mathbf{F}$, the transformation rule (7.4.6) for the Cauchy stress, the definition (7.2.7) of $\mathbf{C}$, and the definition (7.5.9) of $\mathbf{T}_{\mathrm{RR}}$, it is straightforward to show that

$$\mathbf{C} \text{ and } \mathbf{T}_{\mathrm{RR}} \text{ are invariant under a change in frame.} \tag{7.5.14}$$

Thus with the invariance properties discussed above, all quantities in the free-energy imbalance (7.5.12) are invariant under a change in frame.

7.6 Constitutive theory

7.6.1 Basic constitutive equations

Guided by the free-energy imbalance (7.5.12), we assume that the free energy ψ_R, entropy η_R, the second Piola stress $\mathbf{T}_{RR}$, and the heat flux $\mathbf{q}_R$ are given by the following set of constitutive equations:

$$\begin{aligned} \psi_R &= \hat{\psi}_R(\mathbf{C}, \vartheta), \\ \eta_R &= \hat{\eta}_R(\mathbf{C}, \vartheta), \\ \mathbf{T}_{RR} &= \hat{\mathbf{T}}_{RR}(\mathbf{C}, \vartheta), \\ \mathbf{q}_R &= \hat{\mathbf{q}}_R(\mathbf{C}, \vartheta, \nabla\vartheta), \end{aligned} \tag{7.6.1}$$

together with an auxiliary constitutive equation for the internal energy

$$\varepsilon_R = \hat{\varepsilon}_R(\mathbf{C}, \vartheta) \equiv \hat{\psi}_R(\mathbf{C}, \vartheta) + \vartheta\, \hat{\eta}_R(\mathbf{C}, \vartheta). \tag{7.6.2}$$

Here, for simplicity, from the outset we have assumed that the constitutive equations for the free energy, entropy, and stress are independent of $\nabla\vartheta$, and that they depend only on the mechanical stretch $\mathbf{C}$ and the temperature ϑ. Also, using the transformation rules (7.5.13) and (7.5.14) we find that the constitutive equations (7.6.1) are invariant under a change in frame.

7.6.2 Thermodynamic restrictions

With a view toward determining the restrictions imposed by the local free-energy imbalance (7.5.12) on the constitutive equations, note that

$$\dot{\psi}_R = \frac{\partial \hat{\psi}_R}{\partial \mathbf{C}} : \dot{\mathbf{C}} + \frac{\partial \hat{\psi}_R}{\partial \vartheta}\, \dot{\vartheta}. \tag{7.6.3}$$

Substituting (7.6.1) and (7.6.3) into (7.5.12) we find that

$$\Big(\frac{\partial \hat{\psi}_R}{\partial \mathbf{C}} - \frac{1}{2}\hat{\mathbf{T}}_{RR}\Big) : \dot{\mathbf{C}} + \Big(\frac{\partial \hat{\psi}_R}{\partial \vartheta} + \hat{\eta}_R\Big)\, \dot{\vartheta} + \frac{1}{\vartheta}\hat{\mathbf{q}}_R(\mathbf{C}, \vartheta, \nabla\vartheta) \cdot \nabla\vartheta \le 0. \tag{7.6.4}$$

This inequality is to hold for all values of $\mathbf{C}$, $\dot{\mathbf{C}}$, ϑ, $\dot{\vartheta}$, and $\nabla\vartheta$. Since $\dot{\mathbf{C}}$ and $\dot{\vartheta}$ appear linearly, their "coefficients" must vanish, for otherwise $\dot{\mathbf{C}}$ and $\dot{\vartheta}$ may be chosen to violate (7.6.4). We are therefore led to the thermodynamic restriction that the free energy determines the stress $\mathbf{T}_{RR}$ and the entropy η_R through the "state relations"

$$\begin{aligned} \mathbf{T}_{\mathrm{RR}} &= 2\,\frac{\partial \hat{\psi}_{\mathrm{R}}(\mathbf{C},\vartheta)}{\partial \mathbf{C}}, \\ \eta_{\mathrm{R}} &= -\frac{\partial \hat{\psi}_{\mathrm{R}}(\mathbf{C},\vartheta)}{\partial \vartheta}. \end{aligned} \tag{7.6.5}$$

Further, as is standard, we assume that the response function $\hat{\mathbf{q}}_{\mathrm{R}}$ satisfies the *heat-conduction inequality*

$$\hat{\mathbf{q}}_{\mathrm{R}}(\mathbf{C},\vartheta,\nabla\vartheta)\cdot\nabla\vartheta < 0 \quad \text{when} \quad \nabla\vartheta \neq 0. \tag{7.6.6}$$

Next, recalling the stress relations

$$\mathbf{T}_{\mathrm{RR}} = J\,\mathbf{F}^{-1}\mathbf{T}\mathbf{F}^{-\top} \quad \text{and} \quad \mathbf{T}_{\mathrm{R}} = J\,\mathbf{T}\mathbf{F}^{-\top}, \tag{7.6.7}$$

$(7.6.5)_1$ gives the Cauchy stress as

$$\mathbf{T} = J^{-1}\left[2\mathbf{F}\frac{\partial \hat{\psi}_{\mathrm{R}}(\mathbf{C},\vartheta)}{\partial \mathbf{C}}\mathbf{F}^{\top}\right], \tag{7.6.8}$$

and the Piola stress as

$$\mathbf{T}_{\mathrm{R}} = 2\mathbf{F}\frac{\partial \hat{\psi}_{\mathrm{R}}(\mathbf{C},\vartheta)}{\partial \mathbf{C}}. \tag{7.6.9}$$

7.6.3 Further consequences of thermodynamics

In view of $(7.6.1)_1$ and (7.6.5) we have the *first Gibbs relation*

$$\dot{\psi}_{\mathrm{R}} = \frac{1}{2}\mathbf{T}_{\mathrm{RR}}:\dot{\mathbf{C}} - \eta_{\mathrm{R}}\dot{\vartheta}, \tag{7.6.10}$$

which with (7.5.7), yields the *second Gibbs relation*

$$\dot{\varepsilon}_{\mathrm{R}} = \vartheta\dot{\eta}_{\mathrm{R}} + \frac{1}{2}\mathbf{T}_{\mathrm{RR}}:\dot{\mathbf{C}}. \tag{7.6.11}$$

Using the balance of energy (7.5.3), the stress–power relation (7.5.11) and the second Gibbs relation (7.6.11), we arrive at the entropy balance

$$\vartheta\dot{\eta}_{\mathrm{R}} = -\mathrm{Div}\,\mathbf{q}_{\mathrm{R}} + q_{\mathrm{R}}. \tag{7.6.12}$$

Granted the thermodynamically restricted constitutive relations (7.6.5), this balance is equivalent to the balance of energy.

Next, as is standard, the *specific heat* or *heat capacity* per unit reference volume is defined by

$$c_v \stackrel{\text{def}}{=} \frac{\partial \varepsilon_{\text{R}}}{\partial \vartheta}. \tag{7.6.13}$$

Use of (7.5.7) and (7.6.5) gives

$$c_v = \vartheta \frac{\partial \eta_{\text{R}}}{\partial \vartheta} = -\vartheta \frac{\partial^2 \psi_{\text{R}}}{\partial \vartheta^2}. \tag{7.6.14}$$

Further, from (7.6.5) and (7.6.14),

$$\vartheta \dot{\eta}_{\text{R}} = -\vartheta \frac{\partial^2 \hat{\psi}_{\text{R}}}{\partial \vartheta \partial \mathbf{C}} : \dot{\mathbf{C}} + c_v \dot{\vartheta}. \tag{7.6.15}$$

Then using (7.6.14) and (7.6.15) in (7.6.12) gives the following partial differential equation for the temperature

$$c_v \dot{\vartheta} = -\operatorname{Div} \mathbf{q}_{\text{R}} + q_{\text{R}} + \vartheta \frac{\partial^2 \hat{\psi}_{\text{R}}}{\partial \vartheta \partial \mathbf{C}} : \dot{\mathbf{C}}. \tag{7.6.16}$$

which using the state relation $(7.6.5)_1$ for $\mathbf{T}_{\text{RR}}$ may be written as the *heat equation*

$$c_v \dot{\vartheta} = -\operatorname{Div} \mathbf{q}_{\text{R}} + q_{\text{R}} + \frac{1}{2} \vartheta \, \mathbf{M} : \dot{\mathbf{C}}, \tag{7.6.17}$$

where

$$\mathbf{M}(\mathbf{C}, \vartheta) \stackrel{\text{def}}{=} \frac{\partial \mathbf{T}_{\text{RR}}}{\partial \vartheta} \tag{7.6.18}$$

is a *stress–temperature modulus*. The last term on the right-hand side of (7.6.17) characterizes the *thermoelastic coupling effect* where the power

$$\mathcal{H} \stackrel{\text{def}}{=} \frac{1}{2} \vartheta \, \mathbf{M} : \dot{\mathbf{C}} \tag{7.6.19}$$

is called the *thermoelastic heating*.

Note that the partial differential equation (7.6.17) is generic, its derivation requires no assumptions other than frame-indifference and consistency with thermodynamics.

7.7 Specialization of the free-energy function

7.7.1 Nearly incompressible material

The constitutive equations considered thus far are fairly general, applicable to both metals and elastomers. With a view towards modeling the thermoelastic response of elastomeric materials, which are *approximately incompressible*, we develop the theory in an alternative form

$$\begin{aligned}\psi_R &= \hat{\psi}_R(\mathbf{C}, \vartheta),\\ &= \tilde{\psi}_R(\bar{\mathbf{C}}, J, \vartheta),\end{aligned} \tag{7.7.1}$$

where $\bar{\mathbf{C}}$ is the isochoric part of $\mathbf{C}$, and J represents the volumetric part of the deformation.

We may now use the relations $(7.6.5)_1$, (7.6.9), and (7.6.8) to determine expressions for the Piola stress and the Cauchy stress in a thermoelastic solid with free energy $\psi = \tilde{\psi}_R(\bar{\mathbf{C}}, J, \vartheta)$. Following the derivation in Section 1.9 we obtain:

$$\mathbf{T}_R = 2\,J^{-1/3}\left[\bar{\mathbf{F}}\frac{\partial\tilde{\psi}_R(\bar{\mathbf{C}}, J, \vartheta)}{\partial\bar{\mathbf{C}}} - \frac{1}{3}\left(\bar{\mathbf{C}} : \frac{\partial\tilde{\psi}_R(\bar{\mathbf{C}}, J, \vartheta)}{\partial\bar{\mathbf{C}}}\right)\bar{\mathbf{F}}^{-\top}\right] + J\frac{\partial\tilde{\psi}_R(\bar{\mathbf{C}}, J, \vartheta)}{\partial J}\mathbf{F}^{-\top}, \tag{7.7.2}$$

and

$$\mathbf{T} = 2\,J^{-1}\left[\bar{\mathbf{F}}\frac{\partial\tilde{\psi}_R(\bar{\mathbf{C}}, J, \vartheta)}{\partial\bar{\mathbf{C}}}\bar{\mathbf{F}}^{\top} - \frac{1}{3}\left(\bar{\mathbf{C}} : \frac{\partial\tilde{\psi}_R(\bar{\mathbf{C}}, J, \vartheta)}{\partial\bar{\mathbf{C}}}\right)\mathbf{1}\right] + \frac{\partial\tilde{\psi}_R(\bar{\mathbf{C}}, J, \vartheta)}{\partial J}\mathbf{1}. \tag{7.7.3}$$

7.7.2 Separable free energy

For an approximately incompressible material, we assume that $\tilde{\psi}$ may be expressed in the *separable* form

$$\tilde{\psi}_R(\bar{\mathbf{C}}, J, \vartheta) = \breve{\psi}_R(\bar{\mathbf{C}}, \vartheta) + \psi_R^{\mathrm{vol}}(J, \vartheta). \tag{7.7.4}$$

This being the case, (7.7.2) and (7.7.3) simplify somewhat to

$$\mathbf{T}_R = 2\,J^{-1/3}\left[\bar{\mathbf{F}}\frac{\partial\breve{\psi}_R(\bar{\mathbf{C}}, \vartheta)}{\partial\bar{\mathbf{C}}} - \frac{1}{3}\left(\bar{\mathbf{C}} : \frac{\partial\breve{\psi}_R(\bar{\mathbf{C}}, \vartheta)}{\partial\bar{\mathbf{C}}}\right)\bar{\mathbf{F}}^{-\top}\right] + J\frac{\partial\psi_R^{\mathrm{vol}}(J, \vartheta)}{\partial J}\mathbf{F}^{-\top}, \tag{7.7.5}$$

and

$$\mathbf{T} = 2\,J^{-1}\left[\bar{\mathbf{F}}\frac{\partial\breve{\psi}_R(\bar{\mathbf{C}}, \vartheta)}{\partial\bar{\mathbf{C}}}\bar{\mathbf{F}}^{\top} - \frac{1}{3}\left(\bar{\mathbf{C}} : \frac{\partial\breve{\psi}_R(\bar{\mathbf{C}}, \vartheta)}{\partial\bar{\mathbf{C}}}\right)\mathbf{1}\right] + \frac{\partial\psi_R^{\mathrm{vol}}(J, \vartheta)}{\partial J}\mathbf{1}. \tag{7.7.6}$$

7.7.3 Isotropic nearly incompressible material

Specializing the separable energy (7.7.4) to the case of an *isotropic material*, the free energy (7.7.4) may alternatively be expressed, using a classical representation theorem (Gurtin et al., 2010), as

$$\psi_{\rm R} = \underbrace{\bar{\psi}_{\rm R}(\bar{I}_1, \bar{I}_2, \vartheta)}_{\text{isochoric}} + \underbrace{\psi_{\rm R}^{\rm vol}(J, \vartheta)}_{\text{volumetric}}, \tag{7.7.7}$$

where

$$\bar{I}_1 = \operatorname{tr} \bar{\mathbf{C}} \qquad \text{and} \qquad \bar{I}_2 = \frac{1}{2}[(\operatorname{tr} \bar{\mathbf{C}})^2 - \operatorname{tr}(\bar{\mathbf{C}})] \tag{7.7.8}$$

are the principal invariants of $\bar{\mathbf{C}}$. Using (7.7.5) and (7.7.6) the Piola stress and the Cauchy stress may be obtained as

$$\mathbf{T}_{\rm R} = 2\,J^{-1/3}\left[\bar{\mathbf{F}}\frac{\partial \bar{\psi}_{\rm R}(\bar{I}_1, \bar{I}_2, \vartheta)}{\partial \bar{\mathbf{C}}} - \frac{1}{3}\left(\bar{\mathbf{C}} \colon \frac{\partial \bar{\psi}_{\rm R}(\bar{I}_1, \bar{I}_2, \vartheta)}{\partial \bar{\mathbf{C}}}\right)\bar{\mathbf{F}}^{-\top}\right] + J\,\frac{\partial \psi_{\rm R}^{\rm vol}(J, \vartheta)}{\partial J}\,\mathbf{F}^{-\top}, \tag{7.7.9}$$

and

$$\mathbf{T} = 2\,J^{-1}\left[\bar{\mathbf{F}}\frac{\partial \bar{\psi}_{\rm R}(\bar{I}_1, \bar{I}_2, \vartheta)}{\partial \bar{\mathbf{C}}}\bar{\mathbf{F}}^{\top} - \frac{1}{3}\left(\bar{\mathbf{C}} \colon \frac{\partial \bar{\psi}_{\rm R}(\bar{I}_1, \bar{I}_2, \vartheta)}{\partial \bar{\mathbf{C}}}\right)\mathbf{1}\right] + \frac{\partial \psi_{\rm R}^{\rm vol}(J, \vartheta)}{\partial J}\,\mathbf{1}. \tag{7.7.10}$$

Thus, since

$$\frac{\partial \bar{I}_1}{\partial \bar{\mathbf{C}}} = \mathbf{1} \qquad \text{and} \qquad \frac{\partial \bar{I}_2}{\partial \bar{\mathbf{C}}} = \bar{I}_1\mathbf{1} - \bar{\mathbf{C}},$$

we find that the Cauchy stress is given by

$$\mathbf{T} = 2\,J^{-1}\operatorname{dev}\left(\frac{\partial \bar{\psi}_{\rm R}(\bar{I}_1, \bar{I}_2, \vartheta)}{\partial \bar{I}_1}\,\bar{\mathbf{B}} + \frac{\partial \bar{\psi}_{\rm R}(\bar{I}_1, \bar{I}_2, \vartheta)}{\partial \bar{I}_2}\,(\bar{I}_1\mathbf{1} - \bar{\mathbf{B}})\,\bar{\mathbf{B}}\right) + \frac{\partial \psi_{\rm R}^{\rm vol}(J, \vartheta)}{\partial J}\,\mathbf{1}, \tag{7.7.11}$$

where $\operatorname{dev}\mathbf{A} \stackrel{\text{def}}{=} \mathbf{A} - (1/3)(\operatorname{tr}\mathbf{A})\mathbf{1}$ is the deviatoric operator. Finally, using

$$\mathbf{T}_{\rm R} = J\,\mathbf{T}\mathbf{F}^{-\top} = J^{2/3}\mathbf{T}\bar{\mathbf{F}}^{-\top},$$

the Piola stress is given by

$$\mathbf{T}_{\rm R} = 2\,J^{-1/3}\operatorname{dev}\left(\frac{\partial \bar{\psi}_{\rm R}(\bar{I}_1, \bar{I}_2, \vartheta)}{\partial \bar{I}_1}\,\bar{\mathbf{F}} + \frac{\partial \bar{\psi}_{\rm R}(\bar{I}_1, \bar{I}_2, \vartheta)}{\partial \bar{I}_2}\,(\bar{I}_1\mathbf{1} - \bar{\mathbf{B}})\,\bar{\mathbf{F}}\right) + J\,\frac{\partial \psi_{\rm R}^{\rm vol}(J, \vartheta)}{\partial J}\,\mathbf{F}^{-\top}. \tag{7.7.12}$$

7.7.4 Neo-Hookean and Arruda–Boyce type free energies

Many reasonably successful classical models for the mechanical response of elastomeric materials are based on neglecting a dependence of the free energy on the second invariant $\bar{I}_2$ in (7.7.7), so that it may be taken in the simpler form

$$\psi_{\mathrm{R}} = \underbrace{\bar{\psi}_{\mathrm{R}}(\bar{I}_1, \vartheta)}_{\text{isochoric}} + \underbrace{\psi_{\mathrm{R}}^{\mathrm{vol}}(J, \vartheta)}_{\text{volumetric}}. \tag{7.7.13}$$

We now consider special forms for such a free energy which generalize the classical isothermal Neo-Hookean and Arruda–Boyce energies considered in Section 1.10 to account for thermal effects (Anand, 1996):

1. In elastomeric materials the major part of the isochoric free energy change $\bar{\psi}_{\mathrm{R}}(\bar{I}_1, \vartheta)$ arises from an entropic contribution. Let

$$\bar{\lambda} \stackrel{\text{def}}{=} \frac{1}{\sqrt{3}}\sqrt{\operatorname{tr}\bar{\mathbf{C}}} \tag{7.7.14}$$

define an *effective stretch*. Then, classical statistical mechanics models of rubber elasticity (Treloar (cf., 1975); Arruda and Boyce (cf., 1993)) provide the following estimates for the isochoric part of the free-energy change due to mechanical stretching:

(i) For small to moderate values of $\bar{\lambda}$, based on Gaussian statistics,

$$\bar{\psi}_{\mathrm{R}}(\bar{I}_1, \vartheta) = \frac{3}{2}(N\, k_B \vartheta)\,(\bar{\lambda}^2 - 1), \tag{7.7.15}$$

where k_B is Boltzmann's constant and N represents the number of polymer chains per unit reference volume.

(ii) For larger values of $\bar{\lambda}$ it is necessary to use non-Gaussian statistics to account for the limited extensibility of the polymer chains, and for these circumstances the free-energy change is given by

$$\bar{\psi}_{\mathrm{R}}(\bar{I}_1, \vartheta) = (N k_B \vartheta)\lambda_L^2 \left[\left(\frac{\bar{\lambda}}{\lambda_L}\right)\beta + \ln\left(\frac{\beta}{\sinh\beta}\right) - \left(\frac{1}{\lambda_L}\right)\beta_0 - \ln\left(\frac{\beta_0}{\sinh\beta_0}\right)\right], \tag{7.7.16}$$

with

$$\beta \stackrel{\text{def}}{=} \mathcal{L}^{-1}\left(\frac{\bar{\lambda}}{\lambda_L}\right), \qquad \text{and} \qquad \beta_0 \stackrel{\text{def}}{=} \mathcal{L}^{-1}\left(\frac{1}{\lambda_L}\right), \tag{7.7.17}$$

where $\mathcal{L}^{-1}$ is the inverse of the Langevin function

$$\mathcal{L}(\bullet) = \coth(\bullet) - (\bullet)^{-1}.$$

This functional form for the change in free energy involves two material parameters:
- N, the number of polymer chains per unit reference volume, and
- λ_L, the network locking stretch.[4]

[4] In the non-Gaussian model, the network locking stretch λ_L is related to the number of links n in a freely jointed chain by $\lambda_L = \sqrt{n}$.

2. For the volumetric energy $\psi_{\text{R}}^{\text{vol}}(J,\vartheta)$ we follow Anand (1996) and assume that,

$$\begin{aligned}\psi_{\text{R}}^{\text{vol}}(J,\vartheta) = &\frac{1}{2}K(\ln J)^2\\ &-3K\alpha(\ln J)(\vartheta-\vartheta_0)\\ &+c_v(\vartheta-\vartheta_0)-c_v\vartheta\ln\left(\frac{\vartheta}{\vartheta_0}\right),\end{aligned} \tag{7.7.18}$$

where ϑ_0 is a reference temperature, and the quantities c_v, K, and α are the specific heat, bulk modulus, and coefficient of thermal expansion, respectively. On the right-hand side of (7.7.18): (i) the first term is a contribution due to purely mechanical volume change; (ii) the second term is a contribution due to combined thermo-mechanical expansion; and (iii) the terms on the third line collectively represent a purely thermal contribution related to a temperature-dependent specific heat of the material. For simplicity, the coefficients c_v, K, and α are presumed to be positive constants. However, in general these coefficients are functions of both the temperature and the deformation.

Then, using (7.7.15) through (7.7.18), and writing

$$G_0 \stackrel{\text{def}}{=} Nk_B\vartheta, \tag{7.7.19}$$

for a temperature-dependent *ground-state shear modulus*, we obtain the estimates:

(i) For small to moderate values of $\bar{\lambda}$, we consider the simple *Neo-Hookean* form (see e.g., Treloar, 1975),

$$\begin{aligned}\psi_{\text{R}} = &\frac{3}{2}G_0\left(\bar{\lambda}^2-1\right)\\ &+\frac{1}{2}K(\ln J)^2\\ &-3K\alpha(\ln J)(\vartheta-\vartheta_0)\\ &+c_v(\vartheta-\vartheta_0)-c_v\vartheta\ln\left(\frac{\vartheta}{\vartheta_0}\right).\end{aligned} \tag{7.7.20}$$

(ii) For larger values of $\bar{\lambda}$, we consider the *Arruda–Boyce* form (Arruda and Boyce, 1993; Anand, 1996),

$$\begin{aligned}\psi_{\text{R}} = &G_0\,\lambda_L^2\left[\left(\frac{\bar{\lambda}}{\lambda_L}\right)\beta+\ln\left(\frac{\beta}{\sinh\beta}\right)-\left(\frac{1}{\lambda_L}\right)\beta_0-\ln\left(\frac{\beta_0}{\sinh\beta_0}\right)\right]\\ &+\frac{1}{2}K(\ln J)^2\\ &-3K\alpha(\ln J)(\vartheta-\vartheta_0)\\ &+c_v(\vartheta-\vartheta_0)-c_v\vartheta\ln\left(\frac{\vartheta}{\vartheta_0}\right).\end{aligned} \tag{7.7.21}$$

This functional form for ψ_{R} involves the following material parameters:

- N, the number of polymer chains per unit reference volume;
- $G_0 = N k_B \vartheta$, the ground-state shear modulus;[5]
- λ_L, the network locking stretch;
- K, the bulk modulus;
- α, the coefficient of thermal expansion; and
- c_v the specific heat.

Cauchy stress

Using (7.7.11) we obtain the following expressions for the Cauchy stress from the Neo-Hookean and Arruda–Boyce energies:

- Neo-Hookean:

$$\mathbf{T} = J^{-1} G_0 \,\bar{\mathbf{B}}_0 + K \,\frac{(\ln J - 3\alpha(\vartheta - \vartheta_0))}{J}\,\mathbf{1}. \tag{7.7.22}$$

For later convenience we rewrite this as

$$\mathbf{T} = J^{-1} G_0 \,\bar{\mathbf{B}}_0 - p\mathbf{1}, \tag{7.7.23}$$

where

$$p \overset{\text{def}}{=} -\frac{1}{3}\operatorname{tr}\mathbf{T} = -K\left[\frac{\ln J - 3\alpha(\vartheta - \vartheta_0)}{J}\right], \tag{7.7.24}$$

represents a pressure.

- Arruda–Boyce:

$$\mathbf{T} = J^{-1} G \,\bar{\mathbf{B}}_0 - p\,\mathbf{1}, \qquad \text{where} \qquad G \overset{\text{def}}{=} G_0 \left(\frac{\lambda_L}{3\bar{\lambda}}\right) \mathcal{L}^{-1}\left(\frac{\bar{\lambda}}{\lambda_L}\right), \tag{7.7.25}$$

is a *generalized shear modulus*. Note that since $\mathcal{L}^{-1}(\bar{\lambda}/\lambda_L) \to \infty$ as $\bar{\lambda} \to \lambda_L$, the generalized modulus $G \to \infty$ as $\bar{\lambda} \to \lambda_L$, *a situation which reflects the limited chain-extensibility of elastomeric materials*. Note also that as $\lambda_L \to \infty$, it can be shown that $G \to G_0$, and we recover the Neo-Hookean result.

Piola stress

Next, using $\mathbf{T}_R = J\mathbf{T}\mathbf{F}^{-\top}$, the corresponding expressions for the Piola stress are:

- Neo-Hookean:

$$\mathbf{T}_R = J^{-2/3} G_0 \left(\mathbf{F} - \frac{1}{3}(\operatorname{tr}\mathbf{C})\mathbf{F}^{-\top}\right) - Jp\,\mathbf{F}^{-\top}. \tag{7.7.26}$$

[5] This is actually a "derived" material parameter in terms of N, k_B, and ϑ.

- Arruda–Boyce:

$$\mathbf{T}_{\mathrm{R}} = J^{-2/3} G \left(\mathbf{F} - \frac{1}{3} (\operatorname{tr} \mathbf{C}) \mathbf{F}^{-\top} \right) - Jp\, \mathbf{F}^{-\top}. \tag{7.7.27}$$

REMARK

The following Padé approximation to $\mathcal{L}^{-1}(z)$ considerably simplifies numerical calculations (Cohen, 1991),

$$\mathcal{L}^{-1}(z) \approx z \, \frac{3 - z^2}{1 - z^2} \qquad \text{with} \qquad z \stackrel{\text{def}}{=} \frac{\bar{\lambda}}{\lambda_L}. \tag{7.7.28}$$

Entropy

Using $(7.6.5)_2$ we obtain the following expressions for the entropy:

- Neo-Hookean:

$$\eta_{\mathrm{R}} = 3K\alpha \ln J + c_v \ln \left(\frac{\vartheta}{\vartheta_0} \right) - \frac{3}{2} N k_B \left(\bar{\lambda}^2 - 1 \right). \tag{7.7.29}$$

- Arruda–Boyce:

$$\begin{aligned} \eta_{\mathrm{R}} = {} & 3K\alpha \ln J + c_v \ln \left(\frac{\vartheta}{\vartheta_0} \right) \\ & - N\, k_B\, \lambda_L^2 \left[\left(\frac{\bar{\lambda}}{\lambda_L} \right) \beta + \ln \left(\frac{\beta}{\sinh \beta} \right) - \left(\frac{1}{\lambda_L} \right) \beta_0 - \ln \left(\frac{\beta_0}{\sinh \beta_0} \right) \right]. \end{aligned} \tag{7.7.30}$$

REMARK

Henceforth we will only consider the Arruda–Boyce model. The Neo-Hookean model is a special case of this model when the locking stretch λ_L takes a very large value, say $\lambda_L \approx 100$.

7.7.5 **Specialization of the equation for heat flux**

For isotropic materials the heat flux in the *spatial description* is taken to obey a Fourier-type relation,

$$\mathbf{q} = -\kappa \operatorname{grad} \vartheta, \tag{7.7.31}$$

with $k > 0$ a positive-valued scalar thermal conductivity. Using the standard mapping results

$$\mathbf{F}^{-\top} \nabla \vartheta = \operatorname{grad} \vartheta \qquad \text{and} \qquad \mathbf{q}_{\mathrm{R}} = J \mathbf{F}^{-1} \mathbf{q}, \tag{7.7.32}$$

we obtain the referential heat flux as

$$\mathbf{q}_{\mathrm{R}} = -\mathbf{K}\nabla\vartheta, \qquad \text{where} \qquad \mathbf{K} = J\,\kappa\,\mathbf{C}^{-1} \tag{7.7.33}$$

is a positive-definite thermal conductivity tensor.

7.7.6 The heat equation

The energy balance equation (7.6.17) is

$$c_v\dot{\vartheta} = -\mathrm{Div}\,\mathbf{q} + q_{\mathrm{R}} + \frac{1}{2}\,\vartheta\,\mathbf{M}:\dot{\mathbf{C}}, \tag{7.7.34}$$

where

$$\mathbf{M} \stackrel{\text{def}}{=} \frac{\partial \mathbf{T}_{\mathrm{RR}}}{\partial\vartheta} \tag{7.7.35}$$

is the stress–temperature modulus. From the stress relation $\mathbf{T}_{\mathrm{RR}} = \mathbf{F}^{-1}\mathbf{T}_{\mathrm{R}}$, and the relation (7.7.27) for $\mathbf{T}_{\mathrm{R}}$ using the Arruda–Boyce model we obtain

$$\mathbf{T}_{\mathrm{RR}} = J^{-2/3}G\left(\mathbf{1} - \frac{1}{3}(\mathrm{tr}\,\mathbf{C})\mathbf{C}^{-1}\right) - Jp\,\mathbf{C}^{-1}. \tag{7.7.36}$$

Thus, the stress–temperature modulus $\mathbf{M}$ in (7.7.35) is

$$\begin{aligned} \mathbf{M} &= J^{-2/3}\frac{\partial G}{\partial\vartheta}\left(\mathbf{1} - \frac{1}{3}(\mathrm{tr}\,\mathbf{C})\mathbf{C}^{-1}\right) - J\,\frac{\partial p}{\partial\vartheta}\,\mathbf{C}^{-1}, \qquad \text{with} \\ \frac{\partial G}{\partial\vartheta} &= Nk_B\left(\frac{\lambda_L}{3\bar{\lambda}}\right)\mathcal{L}^{-1}\left(\frac{\bar{\lambda}}{\lambda_L}\right), \quad \text{and} \\ \frac{\partial p}{\partial\vartheta} &= \frac{3K\alpha}{J}, \end{aligned} \tag{7.7.37}$$

or

$$\mathbf{M} = J^{-2/3}\left[Nk_B\left(\frac{\lambda_L}{3\bar{\lambda}}\right)\mathcal{L}^{-1}\left(\frac{\bar{\lambda}}{\lambda_L}\right)\right]\left(\mathbf{1} - \frac{1}{3}(\mathrm{tr}\,\mathbf{C})\mathbf{C}^{-1}\right) - (3K\alpha)\,\mathbf{C}^{-1}. \tag{7.7.38}$$

The last term on the right-hand side of (7.7.34) characterizes the *thermoelastic coupling effect* where the power

$$\begin{aligned} \mathcal{H} &\stackrel{\text{def}}{=} \frac{1}{2}\vartheta\,\mathbf{M}:\dot{\mathbf{C}}, \\ &= \underbrace{J^{-2/3}\left[Nk_B\left(\frac{\lambda_L}{3\bar{\lambda}}\right)\mathcal{L}^{-1}\left(\frac{\bar{\lambda}}{\lambda_L}\right)\right]\left(\mathbf{1} - \frac{1}{3}(\mathrm{tr}\,\mathbf{C})\mathbf{C}^{-1}\right):\dot{\mathbf{C}}}_{\text{isochoric chain stretching contribution}} \\ &\quad + \underbrace{(-3K\alpha)\,\mathbf{C}^{-1}:\dot{\mathbf{C}}}_{\text{volumetric thermal expansion contribution}} \end{aligned} \tag{7.7.39}$$

is called the *thermoelastic heating*. This heating term has two contributions:

- The first term on the right-hand side is due to entropy changes from isochoric chain stretching.
- The second term on the right-hand side is due to entropy changes from volumetric thermal expansion.

It is the first term on the right of (7.7.39) which accounts for the classical Gough–Joule effect: "Under adiabatic conditions a rubber increases in temperature when stretched."

REMARK

Note that unlike in metals for which the coefficient of thermal expansion α is quite small, for elastomeric materials the term $(-3K\alpha)$ in the second term in the thermoelastic heating $\mathcal{H}$ can be quite large because of the large value of the coefficient of thermal expansion α of such materials.

7.8 Governing partial differential equations. Boundary conditions

1. Neglecting conventional and inertial body forces, and assuming the Arruda–Boyce model, the local force balance is represented by the pair of equations

$$\left.\begin{aligned} \operatorname{Div}\mathbf{T}_{\mathrm{R}} = \mathbf{0}, \\ \frac{p}{K} + \left[\frac{\ln J - 3\alpha(\vartheta - \vartheta_0)}{J}\right] = 0, \end{aligned}\right\} \quad \text{in } \mathcal{B} \tag{7.8.1}$$

with $\mathbf{T}_{\mathrm{R}}$ given by

$$\begin{aligned} \mathbf{T}_{\mathrm{R}} &= J^{-2/3}G\left(\mathbf{F} - \frac{1}{3}(\operatorname{tr}\mathbf{C})\mathbf{F}^{-\top}\right) - J\,p\,\mathbf{F}^{-\top}, \\ G &= G_0\left(\frac{\lambda_L}{3\bar{\lambda}}\right)\mathcal{L}^{-1}\left(\frac{\bar{\lambda}}{\lambda_L}\right), \\ G_0 &= Nk_B\vartheta. \end{aligned} \tag{7.8.2}$$

In writing (7.8.1) we have split the standard force balance into two equations because this helps in dealing with the near-incompressibility of the material.

With $\partial\mathcal{B}_{\mathrm{u}}$ and $\partial\mathcal{B}_{\mathrm{t_R}}$ denoting complementary subsurfaces of the boundary $\partial\mathcal{B}$ of the body $\mathcal{B}$, we consider a pair of possible boundary conditions:

$$\begin{aligned} \mathbf{u} &= \bar{\mathbf{u}} \quad \text{on } \partial\mathcal{B}_{\mathrm{u}}, \\ \mathbf{T}_{\mathrm{R}}\mathbf{n}_{\mathrm{R}} &= \bar{\mathbf{t}}_{\mathrm{R}} \quad \text{on } \partial\mathcal{B}_{\mathrm{t_R}}. \end{aligned} \tag{7.8.3}$$

In the boundary conditions above $\bar{\mathbf{u}}$ and $\bar{\mathbf{t}}_R$ are prescribed displacement and tractions, both functions of $(\mathbf{X}, t)$.

2. The energy balance equation is (7.7.34), viz.

$$c_v \dot{\vartheta} = -\mathrm{Div}\,\mathbf{q}_R + q_R + \frac{1}{2}\vartheta\,\mathbf{M}:\dot{\mathbf{C}}, \tag{7.8.4}$$

with the heat flux given by

$$\mathbf{q}_R = -\mathbf{K}\nabla\vartheta, \qquad \text{where} \qquad \mathbf{K} = J\,\kappa\,\mathbf{C}^{-1} \tag{7.8.5}$$

and the stress-temperature modulus $\mathbf{M}$ given by

$$\mathbf{M} = J^{-2/3}\left[N k_B \left(\frac{\lambda_L}{3\bar{\lambda}}\right)\mathcal{L}^{-1}\left(\frac{\bar{\lambda}}{\lambda_L}\right)\right]\left(\mathbf{1} - \frac{1}{3}(\mathrm{tr}\,\mathbf{C})\mathbf{C}^{-1}\right) - (3K\alpha)\,\mathbf{C}^{-1}. \tag{7.8.6}$$

With $\partial\mathcal{B}_\vartheta$ and $\partial\mathcal{B}_{\mathbf{q}_R}$ denoting complementary subsurfaces of $\partial\mathcal{B}$, we consider a pair of possible boundary conditions:

$$\left.\begin{aligned} \vartheta &= \bar{\vartheta}_R \quad \text{on } \partial\mathcal{B}_\vartheta, \\ \mathbf{q}_R\cdot\mathbf{n}_R &= \bar{q}_R \quad \text{on } \partial\mathcal{B}_{\mathbf{q}_R}, \end{aligned}\right. \tag{7.8.7}$$

with $\bar{\vartheta}$ and $\bar{q}_R$ prescribed functions of $\mathbf{X}$ and t.

Standard initial conditions involve the specifications

$$\mathbf{u}(\mathbf{X}, 0) = \mathbf{u}_0(\mathbf{X}) \quad \text{and} \qquad \vartheta(\mathbf{X}, 0) = \vartheta_0(\mathbf{X}) \quad \text{on } \mathcal{B}, \tag{7.8.8}$$

with $\mathbf{u}_0$ and ϑ_0 prescribed functions of $\mathbf{X} \in \mathcal{B}$.

REMARKS

1. Another form of boundary condition arises when one specifies the surface traction $\mathbf{Tn}$ on the deformed surface

$$\partial\mathcal{B}_{\mathrm{t}}(t) = \boldsymbol{\chi}(\partial\mathcal{B}_{\mathrm{t}_R}, t).$$

A simple example of this type of condition arises when one considers the effects of a uniform pressure $-p_0\mathbf{n}$:

$$\mathbf{Tn} = -p_0\mathbf{n} \quad \text{on } \partial\mathcal{B}_{\mathrm{t}}(t). \tag{7.8.9}$$

Using Nanson's relation (1.2.6), viz.

$$\mathbf{n}\,da = J\mathbf{F}^{-\top}\mathbf{n}_R\,da_R = \mathbf{F}^{\mathrm{C}}\,\mathbf{n}_R\,da_R, \tag{7.8.10}$$

the boundary condition (7.8.9) may be expressed as

$$\mathbf{T}_{\rm R}\mathbf{n}_{\rm R} = \underbrace{-p_0\mathbf{F}^{\rm C}\,\mathbf{n}_{\rm R}}_{\bar{\mathbf{t}}_{\rm R}} \quad \text{on } \partial\mathcal{B}_{\mathbf{t}_{\rm R}}. \tag{7.8.11}$$

2. Analogously, another form of boundary condition arises when one specifies the surface heat flux $\mathbf{q}\cdot\mathbf{n}$ on the deformed surface

$$\partial\mathcal{B}_{\mathbf{q}}(t) = \boldsymbol{\chi}(\partial\mathcal{B}_{\mathbf{q}_{\rm R}}, t),$$

that is,

$$\mathbf{q}\cdot\mathbf{n} = \bar{q} \quad \text{on } \partial\mathcal{B}_{\mathbf{q}}(t). \tag{7.8.12}$$

Using the areal mapping relation (1.2.7), viz

$$da = \jmath\, da_{\rm R} \qquad \text{with} \qquad \jmath \stackrel{\text{def}}{=} |\mathbf{F}^{\rm C}\,\mathbf{n}_{\rm R}|, \tag{7.8.13}$$

the boundary condition (7.8.12) may be expressed as

$$\mathbf{q}_{\rm R}\cdot\mathbf{n}_{\rm R} = \underbrace{\jmath\,\bar{q}}_{\bar{q}_{\rm R}} \quad \text{on } \partial\mathcal{B}_{\mathbf{q}_{\rm R}}. \tag{7.8.14}$$

Some common (idealized) boundary conditions for the spatial normal heat flux in (7.8.12) are as follows:

(a) **Perfectly insulated surface**:

$$\bar{q}(\mathbf{x}, t) = 0. \tag{7.8.15}$$

(b) **Uniform heat flux**:

$$\bar{q}(\mathbf{x}, t) = q_0. \tag{7.8.16}$$

(c) **Convection boundary condition**:
At a solid–fluid interface the heat flux is related linearly to the difference between the temperature of the surface $\vartheta(\mathbf{x}, t)$ and the far field ambient fluid (sink) temperature ϑ^{∞},

$$\bar{q}(\mathbf{x}, t) = h(\mathbf{x}, t)\,(\vartheta(\mathbf{x}, t) - \vartheta^{\infty})\,, \tag{7.8.17}$$

where $h(\mathbf{x}, t)$ is called the *film coefficient*. This relation is known as *Newton's law of cooling*.

(d) **Radiation boundary condition**:
If the surface of the body is exposed to a high-temperature source, it will receive heat by radiation in accordance with a law of the form

$$\bar{q}(\mathbf{x}, t) = A(\mathbf{x}, t)\left((\vartheta(\mathbf{x}, t))^4 - (\vartheta^{\infty})^4\right), \tag{7.8.18}$$

where $\vartheta(\mathbf{x}, t)$ is the temperature of the surface, ϑ^{∞} is the temperature of the source, and $A(\mathbf{x}, t)$ is the *effective* radiation parameter (that is, the emissivity of the source, including the view factor, times the Stefan–Boltzmann constant).

Items 1 and 2 represent examples of *configuration-dependent boundary conditions.*

8 Numerical implementation of thermoelasticity of elastomeric materials

8.1 Weak forms of the governing partial differential equations

1. **Mechanical problem:**
Consider a test field $\mathbf{u}_{\text{test}}(\mathbf{X})$ in $\mathcal{B}$ which satisfies $\mathbf{u}_{\text{test}}(\mathbf{X}) = \mathbf{0}$ on $\partial\mathcal{B}_{\mathbf{u}}$. If the balance equation $\operatorname{Div}\mathbf{T}_{\mathrm{R}} = \mathbf{0}$ on $\mathcal{B}$ and the traction condition $(\bar{\mathbf{t}}_{\mathrm{R}} - \mathbf{T}_{\mathrm{R}}\mathbf{n}_{\mathrm{R}}) = \mathbf{0}$ on $\partial\mathcal{B}_{\mathbf{t}_{\mathrm{R}}}$ are satisfied, then clearly

$$\int_{\mathcal{B}} (\operatorname{Div}\mathbf{T}_{\mathrm{R}}) \cdot \mathbf{u}_{\text{test}}\, dv_{\mathrm{R}} + \int_{\partial\mathcal{B}_{\mathbf{t}_{\mathrm{R}}}} (\bar{\mathbf{t}}_{\mathrm{R}} - \mathbf{T}_{\mathrm{R}}\mathbf{n}_{\mathrm{R}}) \cdot \mathbf{u}_{\text{test}}\, da_{\mathrm{R}} = 0, \tag{8.1.1}$$

or equivalently

$$\int_{\partial\mathcal{B}_{\mathbf{t}_{\mathrm{R}}}} \mathbf{T}_{\mathrm{R}}\mathbf{n}_{\mathrm{R}} \cdot \mathbf{u}_{\text{test}}\, da_{\mathrm{R}} - \int_{\mathcal{B}} (\operatorname{Div}\mathbf{T}_{\mathrm{R}}) \cdot \mathbf{u}_{\text{test}}\, dv_{\mathrm{R}} - \int_{\partial\mathcal{B}_{\mathbf{t}_{\mathrm{R}}}} \bar{\mathbf{t}}_{\mathrm{R}} \cdot \mathbf{u}_{\text{test}}\, da_{\mathrm{R}} = 0 \tag{8.1.2}$$

holds for all admissible test functions $\mathbf{u}_{\text{test}}$. Thus, using $\mathbf{u}_{\text{test}} = \mathbf{0}$ on $\partial\mathcal{B}_{\mathbf{u}}$ we may extend the first boundary integral in the equation above to the whole boundary $\partial\mathcal{B}$ and we may rewrite this equation as

$$\int_{\partial\mathcal{B}} \mathbf{T}_{\mathrm{R}}\mathbf{n}_{\mathrm{R}} \cdot \mathbf{u}_{\text{test}}\, da_{\mathrm{R}} - \int_{\mathcal{B}} (\operatorname{Div}\mathbf{T}_{\mathrm{R}}) \cdot \mathbf{u}_{\text{test}}\, dv_{\mathrm{R}} - \int_{\partial\mathcal{B}_{\mathbf{t}_{\mathrm{R}}}} \bar{\mathbf{t}}_{\mathrm{R}} \cdot \mathbf{u}_{\text{test}}\, da_{\mathrm{R}} = 0. \tag{8.1.3}$$

Next, by the divergence theorem

$$\int_{\partial\mathcal{B}} \mathbf{T}_{\mathrm{R}}\mathbf{n}_{\mathrm{R}} \cdot \mathbf{u}_{\text{test}}\, da_{\mathrm{R}} = \int_{\partial\mathcal{B}} (\mathbf{T}_{\mathrm{R}}^{\top}\mathbf{u}_{\text{test}}) \cdot \mathbf{n}\, da_{\mathrm{R}} = \int_{\mathcal{B}} \operatorname{Div}\mathbf{T}_{\mathrm{R}} \cdot \mathbf{u}_{\text{test}}\, dv_{\mathrm{R}} + \int_{\mathcal{B}} \mathbf{T}_{\mathrm{R}} : \nabla\mathbf{u}_{\text{test}}\, dv_{\mathrm{R}}, \tag{8.1.4}$$

use of which in (8.1.3) gives

$$\int_{\mathcal{B}} \mathbf{T}_{\mathrm{R}} : \nabla\mathbf{u}_{\text{test}}\, dv_{\mathrm{R}} - \int_{\partial\mathcal{B}_{\mathbf{t}_{\mathrm{R}}}} \bar{\mathbf{t}}_{\mathrm{R}} \cdot \mathbf{u}_{\text{test}}\, da_{\mathrm{R}} = 0. \tag{8.1.5}$$

Introduction to Coupled Theories in Solid Mechanics. Lallit Anand, Eric M. Stewart, Shawn A. Chester, Oxford University Press.
 DOI: 10.1093/oso/9780198986218.003.0009

Hence, if the balance equation $\operatorname{Div}\mathbf{T}_{\mathrm{R}} = \mathbf{0}$ in $\mathcal{B}$ and the traction condition $(\bar{\mathbf{t}}_{\mathrm{R}} - \mathbf{T}_{\mathrm{R}}\mathbf{n}_{\mathrm{R}}) = \mathbf{0}$ on $\partial\mathcal{B}_{\mathbf{t}_{\mathrm{R}}}$ are satisfied, then the weak form (8.1.5) holds for all admissible test fields $\mathbf{u}_{\mathrm{test}}$ which vanish on $\partial\mathcal{B}_{\mathbf{u}}$.

In addition, with $p_{\mathrm{test}}(\mathbf{X})$ a test field in $\mathcal{B}$ the weak form of the pressure equation

$$\frac{p}{K} + \left(\frac{\ln J - 3\alpha(\vartheta - \vartheta_0)}{J}\right) = 0 \qquad \text{in } \mathcal{B},$$

is

$$\int_{\mathcal{B}} \left[\frac{p}{K} + \frac{\ln J - 3\alpha(\vartheta - \vartheta_0)}{J}\right] p_{\mathrm{test}}\, dv_{\mathrm{R}} = 0. \tag{8.1.6}$$

2. **Energy balance problem**:
Next, consider a scalar test field $\vartheta_{\mathrm{test}}(\mathbf{X})$ in $\mathcal{B}$ which satisfies $\vartheta_{\mathrm{test}}(\mathbf{X}) = \mathbf{0}$ on $\partial\mathcal{B}_{\vartheta}$. In the absence of external heat sources q_{R}, if the balance equation

$$c_v\, \dot{\vartheta} + \operatorname{Div} \mathbf{q}_{\mathrm{R}} - \frac{1}{2}\vartheta\, \mathbf{M} : \dot{\mathbf{C}} = 0 \qquad \text{in } \mathcal{B},$$

and the boundary condition $\mathbf{q}_{\mathrm{R}} \cdot \mathbf{n}_{\mathrm{R}} = \bar{q}_{\mathrm{R}}$ on $\partial\mathcal{B}_{\mathbf{q}_{\mathrm{R}}}$ are satisfied, then

$$\int_{\mathcal{B}} c_v \dot{\vartheta}\, \vartheta_{\mathrm{test}}\, dv_{\mathrm{R}} + \int_{\mathcal{B}} (\operatorname{Div} \mathbf{q}_{\mathrm{R}})\, \vartheta_{\mathrm{test}}\, dv_{\mathrm{R}} - \int_{\mathcal{B}} \left(\frac{1}{2}\vartheta\, \mathbf{M} : \dot{\mathbf{C}}\right) \vartheta_{\mathrm{test}}\, dv_{\mathrm{R}}$$
$$+ \int_{\partial\mathcal{B}_{\mathbf{q}_{\mathrm{R}}}} (\bar{q}_{\mathrm{R}} - \mathbf{q}_{\mathrm{R}} \cdot \mathbf{n}_{\mathrm{R}})\, \vartheta_{\mathrm{test}}\, da_{\mathrm{R}} = 0, \tag{8.1.7}$$

or equivalently

$$\int_{\mathcal{B}} c_v\, \dot{\vartheta}\, \vartheta_{\mathrm{test}}\, dv_{\mathrm{R}} + \int_{\mathcal{B}} (\operatorname{Div} \mathbf{q}_{\mathrm{R}})\, \vartheta_{\mathrm{test}}\, dv_{\mathrm{R}} - \int_{\mathcal{B}} \left(\frac{1}{2}\vartheta\, \mathbf{M} : \dot{\mathbf{C}}\right) \vartheta_{\mathrm{test}}\, dv_{\mathrm{R}}$$
$$- \int_{\partial\mathcal{B}_{\mathbf{q}_{\mathrm{R}}}} \mathbf{q}_{\mathrm{R}} \cdot \mathbf{n}_{\mathrm{R}}\, \vartheta_{\mathrm{test}}\, da_{\mathrm{R}} + \int_{\partial\mathcal{B}_{\mathbf{q}_{\mathrm{R}}}} \bar{q}_{\mathrm{R}}\, \vartheta_{\mathrm{test}}\, da_{\mathrm{R}} = 0 \tag{8.1.8}$$

holds for all test fields $\vartheta_{\mathrm{test}}$. Thus, since $\vartheta_{\mathrm{test}} = 0$ on $\partial\mathcal{B}_{\vartheta}$ the first boundary integral in expression above may be extended to all of $\partial\mathcal{B}$,

$$\int_{\mathcal{B}} c_v\, \dot{\vartheta}\, \vartheta_{\mathrm{test}}\, dv_{\mathrm{R}} + \int_{\mathcal{B}} (\operatorname{Div} \mathbf{q}_{\mathrm{R}})\, \vartheta_{\mathrm{test}}\, dv_{\mathrm{R}} - \int_{\mathcal{B}} \left(\frac{1}{2}\vartheta\, \mathbf{M} : \dot{\mathbf{C}}\right) \vartheta_{\mathrm{test}}\, dv_{\mathrm{R}}$$
$$- \int_{\partial\mathcal{B}} \mathbf{q}_{\mathrm{R}} \cdot \mathbf{n}_{\mathrm{R}}\, \vartheta_{\mathrm{test}}\, da_{\mathrm{R}} + \int_{\partial\mathcal{B}_{\mathbf{q}_{\mathrm{R}}}} \bar{q}_{\mathrm{R}}\, \vartheta_{\mathrm{test}}\, da_{\mathrm{R}} = 0. \tag{8.1.9}$$

Next, by the divergence theorem

$$\int_{\partial\mathcal{B}} \mathbf{q}_{\mathrm{R}} \cdot \mathbf{n}_{\mathrm{R}}\, \vartheta_{\mathrm{test}}\, da_{\mathrm{R}} = \int_{\partial\mathcal{B}} (\vartheta_{\mathrm{test}}\, \mathbf{q}_{\mathrm{R}}) \cdot \mathbf{n}_{\mathrm{R}}\, da_{\mathrm{R}}$$

$$= \int_{\mathcal{B}} \vartheta_{\mathrm{test}}\, \mathrm{Div}\, \mathbf{q}_{\mathrm{R}}\, dv_{\mathrm{R}} + \int_{\mathcal{B}} \mathbf{q}_{\mathrm{R}} \cdot \nabla\vartheta_{\mathrm{test}}\, dv_{\mathrm{R}}. \tag{8.1.10}$$

Use of (8.1.10) in (8.1.9) gives the weak form

$$\int_{\mathcal{B}} c_v\, \dot{\vartheta}\, \vartheta_{\mathrm{test}}\, dv_{\mathrm{R}} - \int_{\mathcal{B}} \left(\frac{1}{2}\vartheta\, \mathbf{M}\colon \dot{\mathbf{C}}\right) \vartheta_{\mathrm{test}}\, dv_{\mathrm{R}} - \int_{\mathcal{B}} \mathbf{q}_{\mathrm{R}} \cdot \nabla\vartheta_{\mathrm{test}}\, dv_{\mathrm{R}} + \int_{\partial\mathcal{B}_{q_{\mathrm{R}}}} \bar{q}_{\mathrm{R}}\, \vartheta_{\mathrm{test}}\, da_{\mathrm{R}} = 0. \tag{8.1.11}$$

We use first-order finite difference approximations to write

$$\dot{\vartheta} = \frac{\vartheta - \vartheta_{\mathrm{old}}}{\Delta t} \qquad \text{and} \qquad \dot{\mathbf{C}} = \frac{\mathbf{C} - \mathbf{C}_{\mathrm{old}}}{\Delta t}, \tag{8.1.12}$$

where ϑ is the current temperature, so that the weak form (8.1.11) may be expressed as

$$\int_{\mathcal{B}} c_v \left(\frac{\vartheta - \vartheta_{\mathrm{old}}}{\Delta t}\right) \vartheta_{\mathrm{test}}\, dv_{\mathrm{R}} - \int_{\mathcal{B}} \frac{1}{2}\vartheta\, \mathbf{M}\colon \left(\frac{\mathbf{C} - \mathbf{C}_{\mathrm{old}}}{\Delta t}\right) \vartheta_{\mathrm{test}}\, dv_{\mathrm{R}}$$

$$- \int_{\mathcal{B}} \mathbf{q}_{\mathrm{R}} \cdot \nabla\vartheta_{\mathrm{test}}\, dv_{\mathrm{R}} + \int_{\partial\mathcal{B}_{q_{\mathrm{R}}}} \bar{q}_{\mathrm{R}}\, \vartheta_{\mathrm{test}}\, da_{\mathrm{R}} = 0. \tag{8.1.13}$$

Summarizing, the weak forms of the governing equations for the coupled problem are

$$\begin{gathered}
\int_{\mathcal{B}} \mathbf{T}_{\mathrm{R}}\colon \nabla\mathbf{u}_{\mathrm{test}}\, dv_{\mathrm{R}} - \int_{\partial\mathcal{B}_{t_{\mathrm{R}}}} \bar{\mathbf{t}}_{\mathrm{R}} \cdot \mathbf{u}_{\mathrm{test}}\, da_{\mathrm{R}} = 0, \\
\int_{\mathcal{B}} \left[\frac{p}{K} + \frac{\ln J - 3\alpha(\vartheta - \vartheta_0)}{J}\right] p_{\mathrm{test}}\, dv_{\mathrm{R}} = 0, \\
\int_{\mathcal{B}} c_v \left(\frac{\vartheta - \vartheta_{\mathrm{old}}}{\Delta t}\right) \vartheta_{\mathrm{test}}\, dv_{\mathrm{R}} - \int_{\mathcal{B}} \frac{1}{2}\vartheta\, \mathbf{M}\colon \left(\frac{\mathbf{C} - \mathbf{C}_{\mathrm{old}}}{\Delta t}\right) \vartheta_{\mathrm{test}}\, dv_{\mathrm{R}} \\
- \int_{\mathcal{B}} \mathbf{q}_{\mathrm{R}} \cdot \nabla\vartheta_{\mathrm{test}}\, dv_{\mathrm{R}} + \int_{\partial\mathcal{B}_{q_{\mathrm{R}}}} \bar{q}_{\mathrm{R}}\, \vartheta_{\mathrm{test}}\, da_{\mathrm{R}} = 0.
\end{gathered} \tag{8.1.14}$$

8.2 Solution procedure in the FEniCS code

Briefly, in FEniCS we use the following degrees of freedom

$$\{\mathbf{u},\, p,\, \vartheta\} \tag{8.2.1}$$

and their known values from the previous step

$$\{\mathbf{u}_{\text{old}},\, p_{\text{old}},\, \vartheta_{\text{old}}\}, \tag{8.2.2}$$

to calculate the *residuals* which correspond to the weak forms of the governing equations, viz.

$$\begin{aligned}
R_{\mathbf{u}} &= \int_{\mathcal{B}} \mathbf{T}_{\text{R}} \colon \nabla \mathbf{u}_{\text{test}}\, dv_{\text{R}} - \int_{\partial\mathcal{B}_{\mathbf{t}_{\text{R}}}} \bar{\mathbf{t}}_{\text{R}} \cdot \mathbf{u}_{\text{test}}\, da_{\text{R}}, \\
R_{p} &= \int_{\mathcal{B}} \left[\frac{p}{K} + \frac{\ln J - 3\alpha(\vartheta - \vartheta_0)}{J} \right] p_{\text{test}}\, dv_{\text{R}}, \\
R_{\vartheta} &= \int_{\mathcal{B}} c_v\, (\vartheta - \vartheta_{\text{old}})\, \vartheta_{\text{test}}\, dv_{\text{R}} - \int_{\mathcal{B}} \frac{1}{2} \vartheta\, \mathbf{M} \colon (\mathbf{C} - \mathbf{C}_{\text{old}})\, \vartheta_{\text{test}}\, dv_{\text{R}} \\
&\quad - \int_{\mathcal{B}} \Delta t\, \mathbf{q}_{\text{R}} \cdot \nabla \vartheta_{\text{test}}\, dv_{\text{R}} + \int_{\partial\mathcal{B}_{\mathbf{q}_{\text{R}}}} \Delta t\, \bar{q}_{\text{R}}\, \vartheta_{\text{test}}\, da_{\text{R}}.
\end{aligned} \tag{8.2.3}$$

FEniCS then differentiates these residuals with respect to the degrees of freedom to obtain *tangents*. At each time step in an incremental time-stepping procedure, FEniCS uses the residuals and tangents in a global Newton–Raphson solver to determine the values of the degrees of freedom $\{\mathbf{u}, p, \vartheta\}$ for which the residuals are zero to within some small numerical tolerance, and which therefore satisfy the governing equations as well as the initial and boundary conditions in a weak sense.

REMARK

A mixed finite element method is utilized to solve for the displacement $\mathbf{u}$, pressure p, and the temperature ϑ, concurrently. To avoid the numerical instabilities we use a class of elements known as “Taylor–Hood” elements which combine a continuous piecewise quadratic interpolation for the displacement and continuous piecewise linear interpolation for the pressure and the temperature.

8.3 Algorithm for calculation of the residuals

The algorithm used in our FEniCS code to calculate the residuals in (8.2.3) is summarized below:

1. Begin the step with known values of

$$\{\mathbf{u}_{\rm old},\, p_{\rm old},\, \vartheta_{\rm old}\}, \tag{8.3.1}$$

where for the first step these quantities are given by the initial conditions for $\mathbf{u}$ and ϑ_0.

2. Calculate necessary kinematical quantities for the current step as,

$$\begin{aligned} &\mathbf{F} = \mathbf{1} + \nabla\mathbf{u}, \quad J = \det\mathbf{F}, \quad \mathbf{C} = \mathbf{F}^{\top}\mathbf{F}, \quad \mathbf{B} = \mathbf{F}\mathbf{F}^{\top}, \\ &\bar{\mathbf{C}} = J^{-2/3}\,\mathbf{C}, \quad \bar{\mathbf{B}} = J^{-2/3}\,\mathbf{B}, \\ &\bar{\lambda} = \sqrt{(\operatorname{tr}\bar{\mathbf{C}})/3} = \sqrt{(\operatorname{tr}\bar{\mathbf{B}})/3}. \end{aligned} \tag{8.3.2}$$

3. Calculate an intermediate factor ζ for the generalized shear modulus using the Padé approximation (1.10.7),

$$\begin{aligned} &\beta \overset{\text{def}}{=} \mathcal{L}^{-1}\left(\frac{\bar{\lambda}}{\lambda_L}\right) \approx x\,\frac{3 - x^2}{1 - x^2} \qquad \text{where} \qquad x \overset{\text{def}}{=} \frac{\bar{\lambda}}{\lambda_L} \\ &\zeta \overset{\text{def}}{=} \left(\frac{\lambda_L}{3\bar{\lambda}}\right)\beta. \end{aligned} \tag{8.3.3}$$

4. Calculate the Piola stress

$$\mathbf{T}_{\rm R} = J^{-2/3} G \left(\mathbf{F} - \frac{1}{3}(\operatorname{tr}\mathbf{C})\mathbf{F}^{-\top}\right) - J\,p\,\mathbf{F}^{-\top}, \tag{8.3.4}$$

with

$$G = (N k_B \vartheta)\,\zeta.$$

5. Calculate the stress–temperature tensor

$$\mathbf{M} = J^{-2/3}\left[N k_B \left(\frac{\lambda_L}{3\bar{\lambda}}\right)\mathcal{L}^{-1}\left(\frac{\bar{\lambda}}{\lambda_L}\right)\right]\left(\mathbf{1} - \frac{1}{3}(\operatorname{tr}\mathbf{C})\mathbf{C}^{-1}\right) - (3K\alpha)\,\mathbf{C}^{-1}. \tag{8.3.5}$$

6. Calculate the heat flux,

$$\mathbf{q}_{\mathrm{R}} = -\mathbf{K}\nabla\vartheta \qquad \text{with} \qquad \mathbf{K} = J\kappa\,\mathbf{C}^{-1}. \tag{8.3.6}$$

7. Use the quantities from steps 1–6 to compute the residuals as

$$\begin{aligned}
R_{\mathbf{u}} &= \int_{\mathcal{B}} \mathbf{T}_{\mathrm{R}} : \nabla \mathbf{u}_{\text{test}}\, dv_{\mathrm{R}} - \int_{\partial\mathcal{B}_{\mathbf{t}_{\mathrm{R}}}} \bar{\mathbf{t}}_{\mathrm{R}} \cdot \mathbf{u}_{\text{test}}\, da_{\mathrm{R}},\\
R_p &= \int_{\mathcal{B}} \left[\frac{p}{K} + \frac{\ln J - 3\alpha(\vartheta - \vartheta_0)}{J} \right] p_{\text{test}}\, dv_{\mathrm{R}},\\
R_\vartheta &= \int_{\mathcal{B}} c_v\, (\vartheta - \vartheta_{\text{old}})\, \vartheta_{\text{test}}\, dv_{\mathrm{R}} - \int_{\mathcal{B}} \frac{1}{2} \vartheta\, \mathbf{M} : (\mathbf{C} - \mathbf{C}_{\text{old}})\, \vartheta_{\text{test}}\, dv_{\mathrm{R}}\\
&\quad - \int_{\mathcal{B}} \Delta t\, \mathbf{q}_{\mathrm{R}} \cdot \nabla \vartheta_{\text{test}}\, dv_{\mathrm{R}} + \int_{\partial\mathcal{B}_{q_{\mathrm{R}}}} \Delta t\, \bar{q}_{\mathrm{R}}\, \vartheta_{\text{test}}\, da_{\mathrm{R}}.
\end{aligned} \tag{8.3.7}$$

8. FEniCS solves for the current values of $\{\mathbf{u},\, p,\, \vartheta\}$.
9. Set these current values as the "old" values for the next step in time,

$$\{\mathbf{u},\, p,\, \vartheta\} \implies \{\mathbf{u}_{\text{old}},\, p_{\text{old}},\, \vartheta_{\text{old}}\}. \tag{8.3.8}$$

10. Repeat steps 1–9 for desired number of steps in time.

9 Representative simulations of thermoelasticity of elastomeric materials

In our numerical simulations in FEniCS we will use the basic units listed as SI (mm) in Table 9.1. Representative values of the material parameters for a vulcanized natural rubber are listed in Table 9.2.

Table 9.1 Basic units.

Quantity	SI (m)	SI (mm)
Length	m	mm
Mass	kg	kg
Time	s	s
Mass Density	kg/m^3	kg /mm^3
Force	N	mN
Stress	Pa	kPa
Energy	J	μJ
Temperature	K	K
Specific heat per unit mass	J/(kg K)	μJ/(kg K)
Thermal conductivity	W/(m K)	μW/(mm K)

Table 9.2 Representative material parameters for a vulcanized natural rubber (Anand, 1996).

Quantity	Value SI (m)	Value SI (mm)
ρ_R, mass density	915 kg/m^3	0.915×10^{-6} kg/mm^3
G_0, ground-state shear modulus	280×10^3 Pa	280 kPa
ϑ_0, reference temperature	298 K	298 K
k_B, Boltzmann's constant	1.38×10^{-23} J/K	1.38×10^{-17} μJ/K
$N = G_0/(k_B\vartheta_0)$, number of chains per unit ref. vol.	#/m^3	#/mm^3

continued

Introduction to Coupled Theories in Solid Mechanics. Lallit Anand, Eric M. Stewart, Shawn A. Chester, Oxford University Press.
 DOI: 10.1093/oso/9780198986218.003.0010

Table 9.2 *continued*

Quantity	Value SI (m)	Value SI (mm)
λ_L, locking stretch	5.12	5.12
K, bulk modulus ($1000 \times G_0$)	280×10^6 Pa	280×10^3 kPa
α, coefficient of thermal expansion	180×10^{-6}/K	180×10^{-6}/K
c, specific heat per unit mass	2010 J/(kg K)	2010×10^6 μJ/(kg K)
$c_v = \rho_R c$, specific heat per unit volume	1839×10^3 J/(m^3 K)	1839 μJ/(mm^3 K)
κ, thermal conductivity	0.16 W/(m K)	$0.16 \times 10^3 \mu$W/(mm K)

9.1 Constrained heating

As a first example consider a very basic thermo-mechanically coupled problem of "constrained heating" in which thermal expansion leads to generation of stress in a geometrically constrained body. As a simple case we consider a square domain under plane strain conditions. Figure 9.1(a) shows the initial domain which has an initial edge length of 10 mm, and is meshed with 6 triangular elements along each edge. The initial conditions are $\mathbf{u}_0 = \mathbf{0}$ and $\vartheta_0 = 298$ K, and the boundary conditions are as follows:

- Mechanical boundary conditions: (i) the left and right faces are prescribed $\bar{u}_1 = 0$; (ii) the bottom face is prescribed $\bar{u}_2 = 0$; and (iii) the top face of the domain is traction-free.
- Thermal boundary conditions: (i) the left, right, and bottom faces of the domain are prescribed zero heat flux (perfectly insulating) boundary conditions; and (ii) the top face is prescribed a temperature history in the form

$$\bar{\vartheta}(t) = \vartheta_0 + 50[1 - \exp(-t/20)],$$

  so that the temperature on the top face increases smoothly and saturates at 50 K above the initial temperature of $\vartheta_0 = 298$ K.

Figure 9.1(b) shows the temperature field on the deformed configuration at a simulation time of 400 s, which is an ample amount of time for heat to have conducted from the top surface into the depth of the domain. Figure 9.1(c) shows the resulting inhomogeneous distribution of Piola stress component P_{11} which is generated in the body due to the constrained heating—note that the stress is *compressive*.

Figure 9.1(d) shows the temperature history at three locations shown as red dots in Fig. 9.1(a)—specifically, the top, center, and the bottom of the domain. The effects of heat conduction from the top surface down into the domain are clearly visible in Fig. 9.1(d), and it is this inhomogeneous temperature field which is the cause for the resulting inhomogeneous stress field observed in Fig. 9.1(c). Finally, Fig. 9.1(e) shows the pressure p generated at the bottom center of the domain as a function of time.

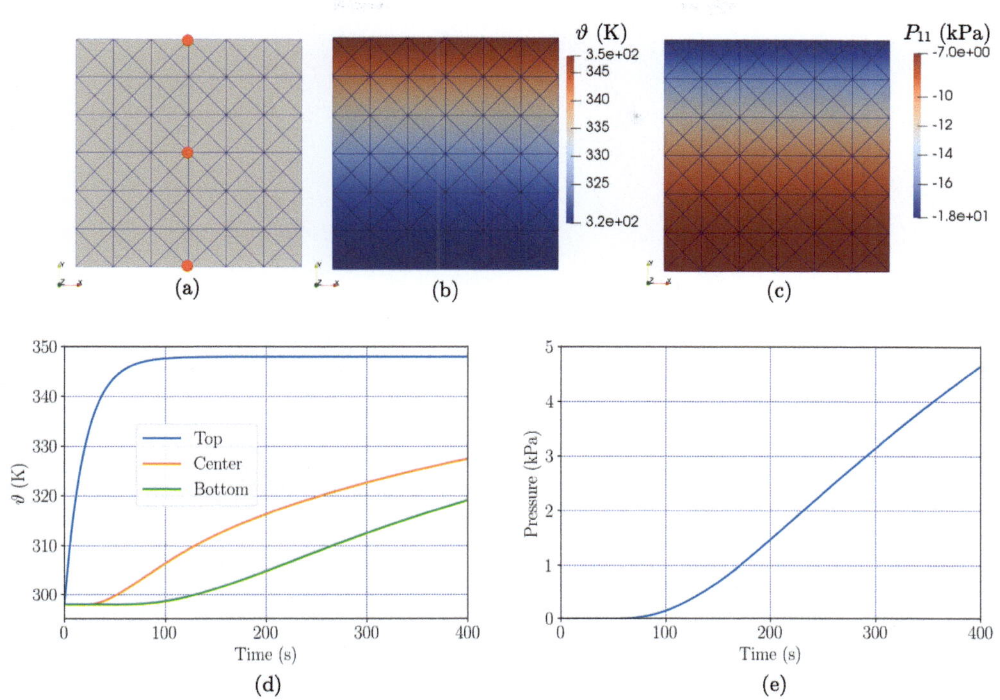

Fig. 9.1 (a) The initial geometry and the plane strain mesh used in the simulation. The three points where data is extracted for plotting are shown as red dots. (b) The inhomogeneous temperature field after 400 s. (c) The inhomogeneous stress field P_{11} after 400 s. (d) Evolution of the temperature at the three locations indicated by red dots in (a). (e) Evolution of the pressure at the bottom of the domain.

9.2 Gough–Joule effect: Under adiabatic conditions a rubber increases in temperature when stretched

This example highlights one of the Gough–Joule effects—"under adiabatic conditions a rubber increases in temperature when stretched" (Gough, 1805; Joule, 1859). Recall the heat equation (7.8.4), viz.

$$c_v\dot{\vartheta} = -\mathrm{Div}\,\mathbf{q}_\mathrm{R} + q_\mathrm{R} + \mathcal{H}, \qquad \text{with}$$

$$\mathcal{H} = \underbrace{J^{-2/3}\left[Nk_B\left(\frac{\lambda_L}{3\bar{\lambda}}\right)\mathcal{L}^{-1}\left(\frac{\bar{\lambda}}{\lambda_L}\right)\right]\left(\mathbf{1} - \frac{1}{3}(\mathrm{tr}\,\mathbf{C})\mathbf{C}^{-1}\right):\dot{\mathbf{C}}}_{\text{isochoric chain stretching contribution}}$$

$$+ \underbrace{(-3K\alpha)\,\mathbf{C}^{-1}:\dot{\mathbf{C}}}_{\text{volumetric termal expansion contribution}} \quad . \tag{9.2.1}$$

With $\mathbf{q}_\mathrm{R}$ and q_R both zero under adiabatic conditions, it is the first term of the thermoelastic heating $\mathcal{H}$ in (9.2.1) which accounts for this Gough–Joule effect.

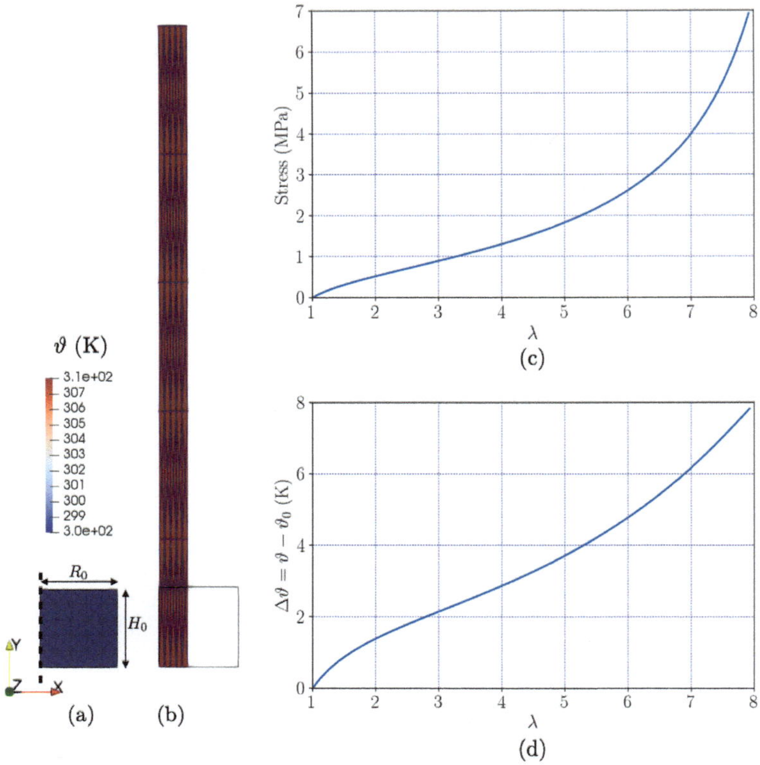

Fig. 9.2 (a) The axisymmetric mesh showing the temperature field on the initial body, and (b) the deformed body. (c) The corresponding stress–stretch and (d) the temperature rise due to adiabatic heating.

We consider homogeneous extension of a body in the form of a right circular cylinder of initial radius $R_0 = 10$ mm and height $H_0 = 10$ mm, and conduct an axisymmetric simulation. Figure 9.2(a) shows the finite element mesh which consists of 5 triangular elements along each edge. The initial conditions are $\mathbf{u}_0 = \mathbf{0}$ and $\vartheta_0 = 298$ K, and the boundary conditions are as follows:

- Mechanical boundary conditions: (i) the left face is prescribed $\bar{u}_1 = 0$ due to symmetry; (ii) the bottom face is prescribed $\bar{u}_2 = 0$; (iii) the right face of the domain is traction-free; and (iv) the top face is prescribed a linear displacement ramp to reach $\bar{u}_2 = 70$ mm over period of 100 s.
- Thermal boundary conditions: the left, right, bottom, and top faces of the domain are prescribed zero heat flux boundary conditions.

Figure 9.2(b) shows the deformed configuration at the maximum deformation, together with contours of temperature field. Due to the adiabatic conditions there is a temperature rise, which because the deformation is homogeneous is uniform throughout the body. Figure 9.2(c) shows the corresponding axial Piola stress as a function of the applied stretch, and Fig. 9.2(d) shows that the there is a resulting measurable temperature rise of ≈ 8 K.

9.3 **Gough–Joule effect: A rubber under a constant uniaxial stretch contracts on heating**

This numerical example highlights another one of the Gough–Joule effects—"a rubber under a constant uniaxial stress will contract upon heating" (Gough, 1805; Joule, 1859). The primary contributing factor that gives rise to this Gough–Joule effect is that the shear modulus $G_0 = Nk_B\vartheta$ increases as the temperature increases, and hence the material stiffens.

We consider a homogeneous body in the form of a right circular cylinder of initial radius $R_0 = 10\,\text{mm}$ and height $H_0 = 10\,\text{mm}$, and conduct an axisymmetric simulation. Figure 9.3 shows the finite element mesh which consists of 20 triangular elements along each edge. The initial conditions for displacement are $\mathbf{u}_0 = \mathbf{0}$, and for temperature $\vartheta_0 = 298\,\text{K}$. To clearly observe this phenomenon, the simulation is carried out in *two steps*: in the first step a traction is applied to place the body in a state of uniaxial stress causing elongation, and in the second step the surface temperature is increased leading to an observable contraction. Referring to Fig. 9.3, the boundary conditions are as follows:

- Step 1:
  - Mechanical boundary conditions: (i) the left face is prescribed $\bar{u}_1 = 0$ due to symmetry; (ii) the bottom face is prescribed $\bar{u}_2 = 0$; (iii) the right face is traction-free; and (iv) the top face is prescribed a linear traction ramp to reach a traction of $\bar{t}_2 = 2\,\text{MPa}$ over 50 s.
  - Thermal boundary conditions: (i) the left face is prescribed zero flux due to symmetry, and (ii) all other faces are prescribed a constant temperature of $\bar{\vartheta} = 298\,\text{K}$.
- Step 2:
  - Mechanical boundary conditions: The mechanical boundary conditions are maintained from the previous step, that is (i) the left face is prescribed $\bar{u}_1 = 0$ due to symmetry; (ii) the bottom face is prescribed $\bar{u}_2 = 0$; (iii) the right face is traction-free; and (iv) the top face is maintained at a constant traction of $\bar{t}_2 = 2\,\text{MPa}$.
  - Thermal boundary conditions: (i) the left face is prescribed zero flux due to symmetry, and (ii) all other faces are prescribed a linear temperature ramp from $\vartheta_0 = 298\,\text{K}$ to $\vartheta = 348\,\text{K}$ over a period of 50 s, and then held constant for another 200 s.

Figures 9.4(a)–(d) show a snapshots of the temperature field on the initial body at 0 s, as well as the deformed body at times 50 s, 100 s, and 350 s, respectively. Figure 9.4(b) at 50 s shows the deformation due to the applied traction. Figure 9.4(c) at 100 s shows the heat conduction into the body due to surface heating, accompanied by a contraction in length due to an increase of the shear modulus of the material as the temperature increases. Figure 9.4(d) at 350 s shows that the temperature in the body is now essentially uniform, and that the body has contracted further in length from its length in Fig. 9.4(c).

Using time-history data extracted from the location on the top face indicated by a red dot on Fig. 9.3, Fig. 9.5(a) shows the axial stretch as a function of the prescribed stress, Fig. 9.5(b) shows the stretch as a function of time, and Fig. 9.5(c) shows the stretch as a function of the prescribed surface temperature. As expected, these results show that as the temperature increases the body contracts upon heating.

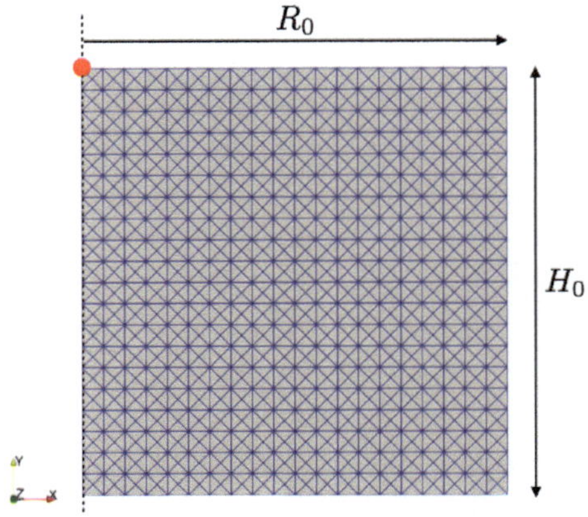

Fig. 9.3 Axisymmetric geometry and mesh used for two-step simulation of the Gough–Joule effect that a rubber under a constant uniaxial stretch contracts on heating. Data used for plotting is extracted from the location on the top surface indicated in red.

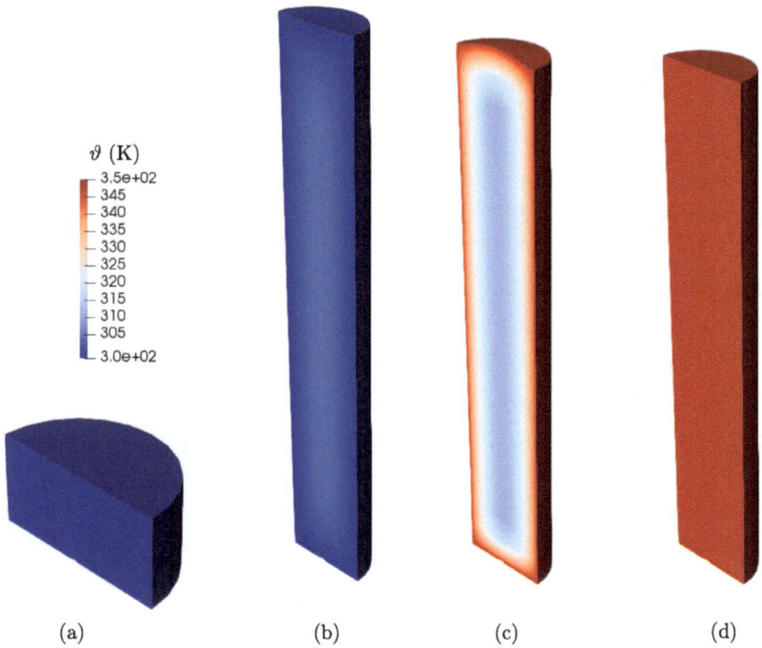

Fig. 9.4 (a)–(d) The temperature field on the deformed body at 0 s, 50 s, 100 s, and 350 s, respectively. The axisymmetric body has been rotationally extruded about the axis of symmetry to show one-half of the full geometry.

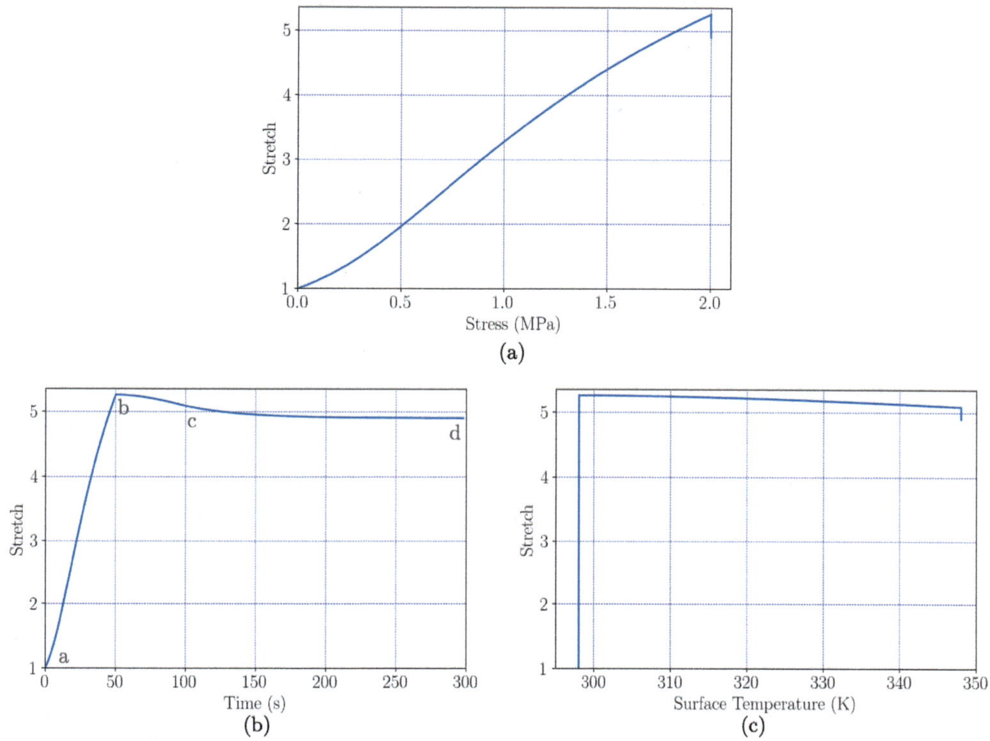

Fig. 9.5 (a) The axial stretch as a function of the prescribed stress; (b) the axial stretch as a function of time, with points labelled a–d corresponding to the deformed shapes in Fig. 9.4; and (c) the axial stretch as a function of the prescribed surface temperature.

9.4 Thermoelastic bilayer beam

This numerical example considers the classic problem of bending of a thermoelastic bilayer beam in which strips of two materials with different coefficients of thermal expansion are bonded together, and upon heating or cooling the bimaterial strip bends due to the differential thermal expansion of the strips. The resulting deflection is used in many applications for actuation or sensing purposes.

Figure 9.6(a) shows a schematic of the initial configuration of a cantilevered bilayer beam. It is a rectangular domain of length $L_0 = 100\,\text{mm}$ and height $H_0 = 1\,\text{mm}$, with a built-in left face. The beam is split into two along its height, with the top and bottom halves having the same material properties except for the coefficient of thermal expansion: the top half has a value $\alpha = 0\,/\text{K}$, while the bottom half has a value $\alpha = 180 \times 10^{-6}\,/\text{K}$.

The simulation is carried out under plane strain conditions using a finite element mesh which has 200 triangular elements along the length of the beam, and 6 elements along its height. Figure 9.6(b) shows the full mesh, but due to the large aspect ratio of the beam, a zoom-in is provided for a clearer picture of the mesh. The initial conditions are $\mathbf{u}_0 = \mathbf{0}$ and $\vartheta_0 = 298\,\mathrm{K}$, and the boundary conditions are as follows:

- Mechanical boundary conditions: (i) the left face is fixed; and (ii) all other faces are traction-free.
- Thermal boundary conditions: (i) the left face is prescribed a zero heat flux; and (ii) all other faces are prescribed a linear temperature increase from $\vartheta_0 = 298\,\mathrm{K}$ to $\vartheta = 348\,\mathrm{K}$ over a period of 1 hr.

Figure 9.6(c) shows the deformed body at the end of the simulation, together with contours of the temperature field; due to the very thin geometry the temperature field is essentially uniform. This figure clearly shows the large bending deformation due to the mismatch in the thermal expansion coefficients. Finally, Fig. 9.6(d) shows a plot of the vertical displacement of the tip of the beam normalized by the beam length, u_2/L_0, as a function of the prescribed surface temperature.

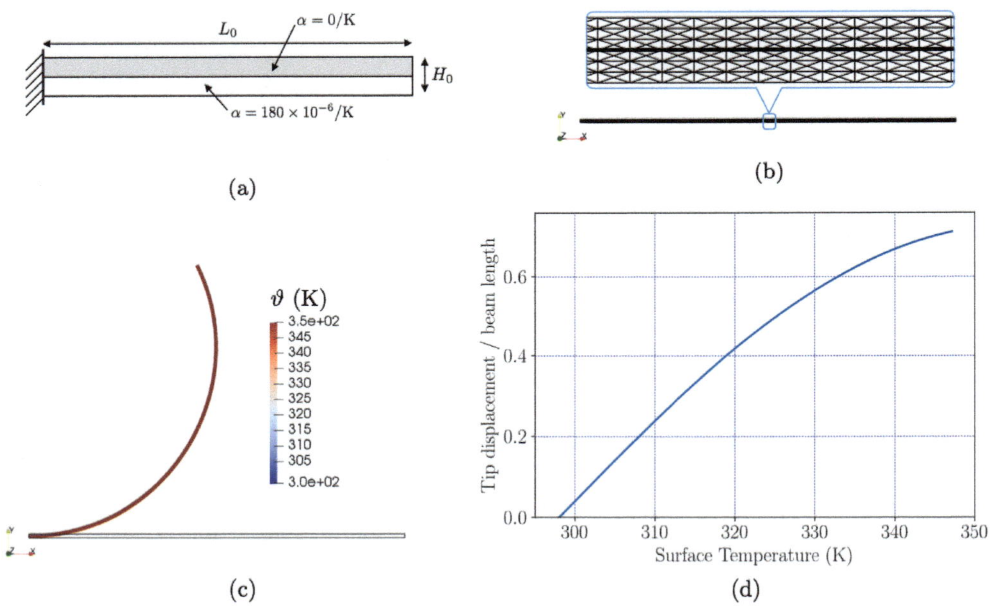

Fig. 9.6 (a) Schematic (not to scale) of a bilayer beam of length $L_0 = 100\,\mathrm{mm}$ and height $H_0 = 1\,\mathrm{mm}$ with mechanical constraints at its left end and different coefficients of thermal expansion. (b) Full mesh and magnified image of a portion of the mesh. (c) Deformed configuration of the bilayer beam after heating. (d) Vertical displacement of the tip of the beam normalized by the beam length, u_2/L_0, as a function of the prescribed surface temperature.

9.5 Circular plate subject to a surface heat flux

In this example we simulate the bending of a circular plate due to non-uniform heating when it is subjected to a heat flux on one of its faces. This example highlights the effects of a configuration-dependent heat flux boundary condition. Figure 9.7 shows an axisymmetric

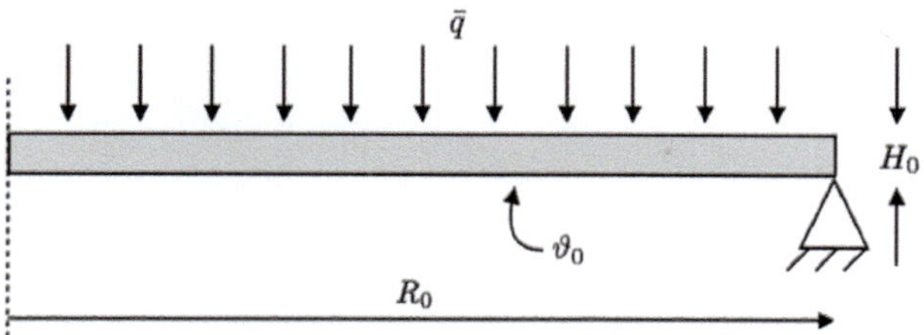

Fig. 9.7 Schematic of the axisymmetric simulation of a plate subjected to a heat flux of $\bar{q} = -10^4$ W/m^2 on its top surface while the bottom surface is maintained at a constant temperature of $\vartheta_0 = 298$ K.

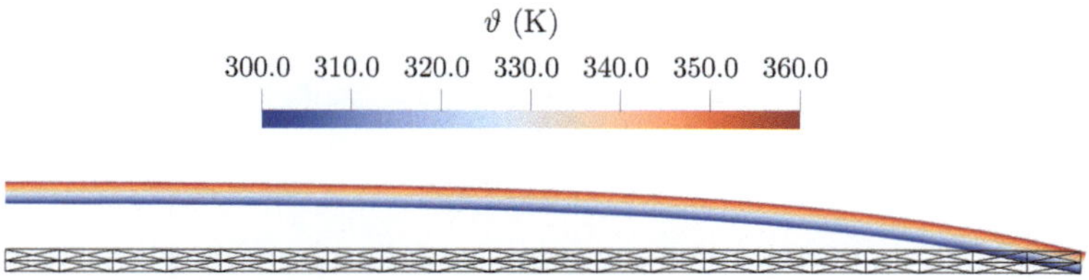

(a)

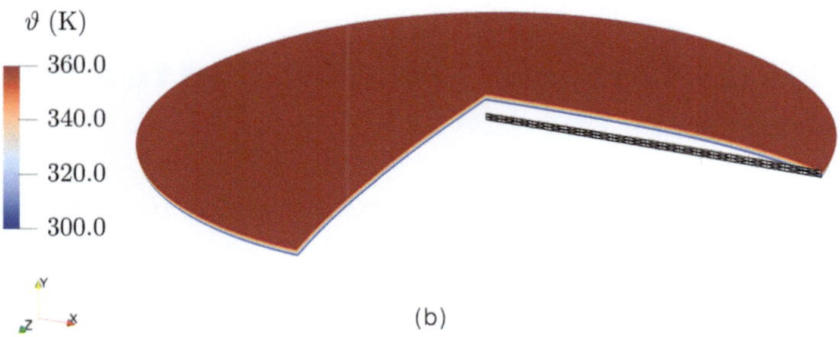

(b)

Fig. 9.8 (a) Undeformed body with finite-element mesh for the axisymmetric simulation, as well as the deformed body with contours of temperature field. (b) 270° revolution of the deformed body with contours of temperature field.

schematic of a circular plate with an initial radius $R_0 = 50$ mm and height $H_0 = 1$ mm. The initial conditions are $\mathbf{u}_0 = \mathbf{0}$ and $\vartheta_0 = 298$ K, and the boundary conditions are:

- Mechanical boundary conditions: (i) the left face is prescribed $\bar{u}_1 = 0$ due to symmetry; (ii) the bottom right point of the plate is fixed; and (iii) all other faces are traction-free.
- Thermal boundary conditions: (i) the bottom face is subjected to a prescribed temperature $\bar{\vartheta} = \vartheta_0$; (ii) the left and right faces are prescribed zero heat flux: and (iii) the top face is prescribed a heat flux of $\bar{q} = -10^4$ W/m^2 in the $\mathbf{e}_2$-direction.

Figure 9.8(a) shows the axisymmetric mesh used in the simulation. The mesh has 20 triangular elements along the radius and 2 along the total height, leading to a total of 160 elements. This figure also shows the deformed geometry at the end of the simulation, together with the contours of the temperature field. The out-of-plane bending deformation due to the constrained non-uniform thermal expansion of the plate is clearly visible in this figure. For a better sense of the bulged shape of the plate, Fig 9.8(b) shows a 270° revolution of the deformed body.

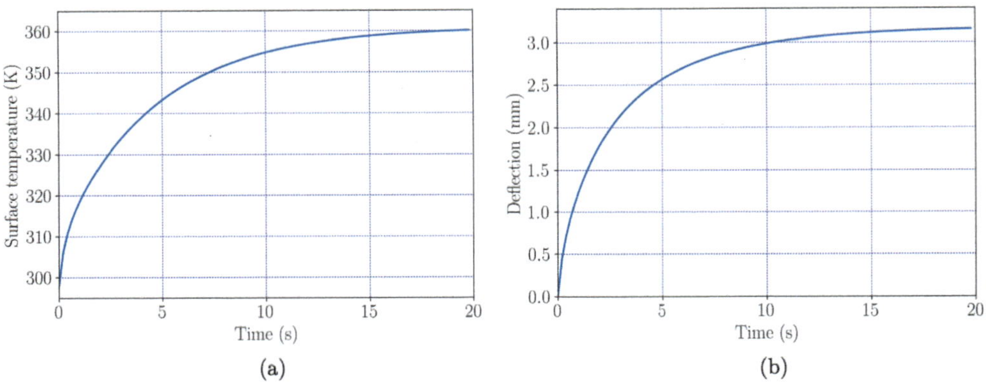

Fig. 9.9 (a) Temperature rise of the mid-point of the top surface of the plate. (b) Corresponding vertical deflection of that point of the plate.

For a more quantitative indication of the out-of-plane deflection of the plate, Fig. 9.9(a) shows the temperature rise of the mid-point of the top surface of the plate, and Fig. 9.9(b) shows the corresponding vertical deflection of that point of the plate.

9.6 Wrinkling of a membrane subjected to a pressure and a temperature change

In this example we study the wrinkling of a membrane subjected to a pressure and a temperature change. This example is motivated by the idea of a "solar sail" which is essentially a kite driven by solar radiation pressure.[1] We consider a kite which has been deployed in a location in which one face of the kite catches the wind, while the ambient temperature changes over time. To be specific, we consider that the kite is subjected to[2]

- pressures on the order of 100 Pa, and
- temperatures in the range of 200 K to 323 K.

Figure 9.10 shows a schematic of the initial geometry which is modeled: a planar square plate with an edge length $L_0 = 200$ mm and thickness $T_0 = 1$ mm. The points labeled in blue in this figure denote where the kite is tethered.

[1] In 1873 Maxwell, on the basis of his electromagnetic theory, predicted that, "... in a medium in which waves are propagated there is a pressure in the direction normal to the waves and numerically equal to the energy in unit volume." In 1960 NASA picked up on this idea of "radiation pressure" and considered the possibility of a constructing a spacecraft with a large reflecting "solar sail" which might use the Sun's radiation pressure for propulsion. The solar sail would experience a constant but small force from solar radiation, accelerate slowly, but by being accelerated continuously, it would eventually reach great speeds. A spacecraft with small total mass and a sail with a large area would be necessary to obtain a usable acceleration. The first such **LightSail** spacecraft was launched in 2015. For a history of the LightSail idea see the Wikipedia article `https://en.wikipedia.org/wiki/LightSail`, and the website of the Planetary Society `https://www.planetary.org`.

[2] This pressure and temperature range is **not** representative of the conditions encountered by a "solar sail."

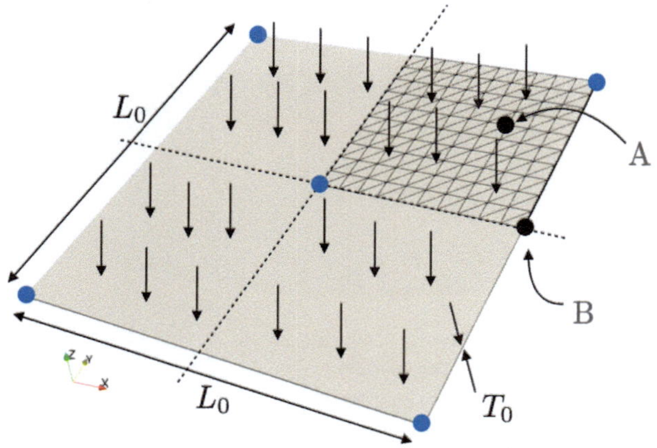

Fig. 9.10 The initial geometry, symmetry planes, and pressure loading used in this example. The points labeled in blue denote where the kite is tethered and therefore constrained. The points labeled in black and denoted by "A" and "B" are used for extracting and plotting time-history data.

We make use of symmetry and mesh only one quarter of the geometry. The mesh consists of 10 tetrahedral elements along each edge, with 2 elements through the thickness. The initial conditions are $\mathbf{u}_0 = \mathbf{0}$ and $\vartheta_0 = 273\,\text{K}$, and the boundary conditions are as follows:

- Mechanical boundary conditions: (i) The two faces of symmetry of the modeled domain are prescribed appropriate symmetry boundary conditions. (ii) Two points, indicated in blue, are pinned to model tethers. (iii) A pressure is applied to the top (visible) face. It is linearly ramped from 0 Pa to 10 Pa over a period of 100 s. (iv) All other faces are traction-free.
- Thermal boundary conditions: (i) The two faces of symmetry are prescribed zero heat flux. (ii) All other faces are prescribed a linear change in temperature from $\vartheta_0 = 273\,\text{K}$ to either $\vartheta = 323\,\text{K}$ or $\vartheta = 200\,\text{K}$ over 100 s.

Figure 9.11 shows the deformed configuration at 323 K together with the contours of the displacement component u_3 at the end of the simulation at 100 s. The initially flat membrane has substantially *wrinkled*, with a maximum deflection of $u_3 \approx -16\,\text{mm}$, which is more than an order of magnitude larger than the initial thickness $T_0 = 1\,\text{mm}$ of the membrane.

Figures 9.12(a) and (b) show the variation of the displacement component u_3 as a function of the applied pressure for the two locations marked A and B in Fig. 9.10 for final temperatures of 323 K and 200 K, respectively. These plots emphasize the evolution of the non-uniform deformation of the membrane, and highlight the impact that thermal expansion/contraction may have. Specifically, for a final temperature of 323 K the maximum displacement is $\approx -16\,\text{mm}$, while for a final temperature of 200 K, the maximum displacement is $\approx -10\,\text{mm}$.

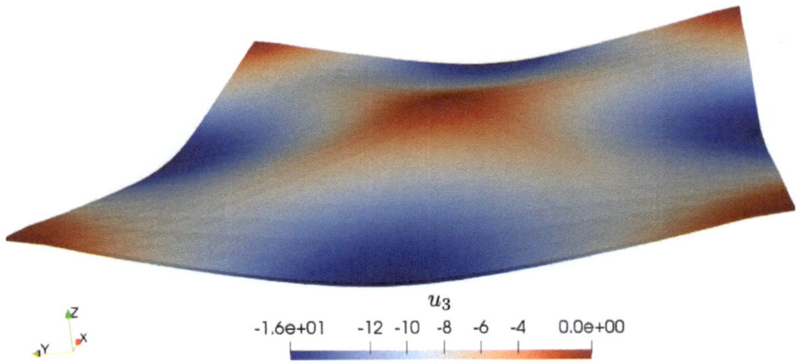

Fig. 9.11 Deformed configuration of the body after heating to 323 K together with contours of the displacement component u_3 at the end of the simulation at 100 s. The quarter-symmetry computational domain has been reflected about both symmetry planes to show the full geometry.

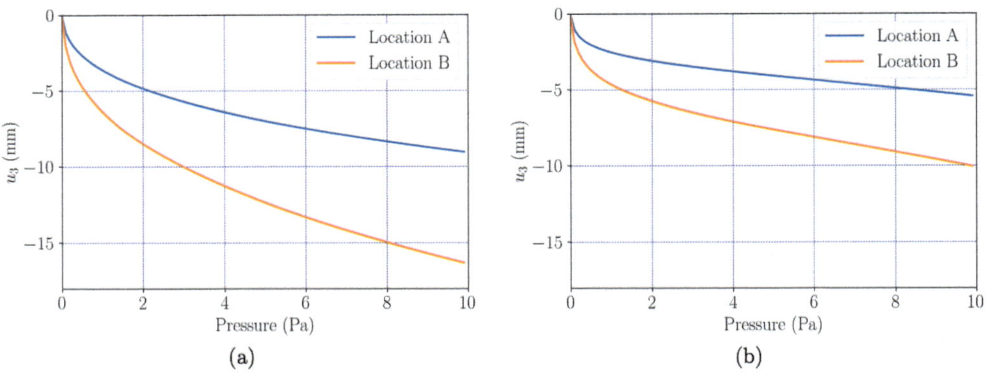

Fig. 9.12 Variation of the displacement component u_3 as a function of the applied pressure for the two locations marked A and B in Fig. 9.10 for a final temperature of (a) 323 K and (b) 200 K.

Part IV

Poroelasticity of elastomeric gels

10 Poroelasticity of elastomeric gels

10.1 Introduction

There are numerous elastomeric materials which can absorb large quantities of suitable fluids without the essential skeletal network structure of the elastomer being disrupted by the action of the fluid. Such a polymer network, together with the fluid molecules, forms a swollen aggregate called an elastomeric *gel*. Elastomeric gels are ubiquitous; they are found in foods and medicines, and they find use in several important and diverse applications, including valves for microfluidic devices and tissue engineering. Indeed, many body parts in humans and other animals are gel-like in constitution.[1]

Early studies on the swelling of gels are due to Tanaka and co-workers (c.f., e.g., Tanaka and Fillmore, 1979), and over the past two decades there have been several notable attempts to formulate a coupled deformation–diffusion theory for describing more complete aspects of the response of gels, including swelling and drying, squeezing of fluid by applied mechanical deformation, and forced permeation;[2] (cf., e.g., Durning and Morman, 1993; Baek and Srinivasa, 2004; Dolbow et al., 2005; Hong et al., 2008; Doi, 2009; Duda et al., 2010; Chester and Anand, 2010, 2011, and references to the vast literature therein).

In this chapter we review the basic laws for the balance of forces, the balance of fluid content, the balance of energy, and the entropy imbalance, and we formulate a frame-indifferent and thermodynamically consistent coupled deformation–diffusion theory for elastomeric gels.[3] In discussing special constitutive equations, we limit our attention to isotropic materials under isothermal conditions, and consider a model based on the following major ingredients:

(i) First, an essential kinematical ingredient of our theory is a multiplicative decomposition

$$\mathbf{F} = \mathbf{F}^e \mathbf{F}^s, \quad \text{with} \quad \mathbf{F}^s = \lambda^s \mathbf{1}, \quad \lambda^s > 0, \tag{10.1.1}$$

of the deformation gradient $\mathbf{F}$ into elastic and swelling parts $\mathbf{F}^e$ and $\mathbf{F}^s$, respectively, with the swelling taken to be isotropic, where λ^s is the swelling stretch. Correspondingly, with $J = \det \mathbf{F} > 0$ denoting the determinant of the deformation gradient, we have

[1] Gels can also be made from colloidal solutions, and such gels are called *colloidal gels*. Here we restrict our attention to elastomeric gels, and within this class of materials we further restrict our attention to *non-ionic* gels.

[2] We use the words permeation and diffusion interchangeably.

[3] This chapter is based on the papers by Chester and Anand (2010) and Chester and Anand (2011).

Introduction to Coupled Theories in Solid Mechanics. Lallit Anand, Eric M. Stewart, Shawn A. Chester, Oxford University Press.
 DOI: 10.1093/oso/9780198986218.003.0011

$$J = J^e J^s, \quad \text{where} \quad J^e = \det \mathbf{F}^e > 0 \quad \text{and} \quad J^s = \det \mathbf{F}^s = (\lambda^s)^3 > 0. \tag{10.1.2}$$

The quantity

$$J^s - 1 \tag{10.1.3}$$

represents the change in volume per unit reference volume due to swelling. We assume that this change arises entirely due to the change in the fluid content. Thus, with Ω denoting the volume of a mole of fluid molecules, and c_{R} denoting the number of moles of fluid molecules absorbed by the elastomer per unit reference volume, we have the important constitutive *swelling constraint*,

$$J^s = 1 + \Omega c_{\mathrm{R}}.$$

Further, since many elastomeric materials are somewhat compressible, a proper accounting for such limited compressibility is important for modeling the response of elastomeric materials. Accordingly, we formulate our theory to include some mechanical compressibility of the gel, and *do not* assume that $J^e = 1$. This relaxation of complete mechanical incompressibility also substantially aids in the numerical implementation of our theory.

(ii) Second, we use the Flory (1942)–Huggins (1942) theory for the free energy change due to mixing of the fluid with the polymer network (cf., e.g., Doi, 1996, 2009). This theory contains a parameter, called the χ-parameter (or interaction parameter), which characterizes the interaction between the solvent and the underlying polymer network (Flory, 1953).

(iii) Finally, in order to model the change in configurational entropy of the polymer chains we use a non-Gaussian statistical-mechanical model which accounts for the limited extensibility of polymer chains.

10.2 Kinematics

Consider a fluid-free (dry) macroscopically homogeneous elastomeric body. In what follows, the spatially continuous fields that define our continuum theory represent averages meant to apply at length scales which are large compared to the length scales associated with the molecular network and its microscopic-scale pore structure. We identify such a macroscopically homogeneous body $\mathcal{B}$ with the region of space it occupies in a fixed *reference configuration*, and denote by $\mathbf{X}$ an arbitrary material point of $\mathcal{B}$. A motion of $\mathcal{B}$ is a smooth one-to-one mapping

$$\mathbf{x} = \boldsymbol{\chi}(\mathbf{X}, t), \tag{10.2.1}$$

with displacement

$$\mathbf{u}(\mathbf{X},t) = \boldsymbol{\chi}(\mathbf{X},t) - \mathbf{X}, \tag{10.2.2}$$

deformation gradient, *velocity*, and *velocity gradient*, given by

$$\mathbf{F} = \nabla\boldsymbol{\chi} = \mathbf{1} + \nabla\mathbf{u}, \quad \mathbf{v} = \dot{\mathbf{u}}, \quad \mathbf{L} = \operatorname{grad}\mathbf{v} = \dot{\mathbf{F}}\mathbf{F}^{-1}. \tag{10.2.3}$$

We base the theory on a multiplicative decomposition of the deformation gradient

$$\mathbf{F} = \mathbf{F}^e\mathbf{F}^s, \quad \text{with} \quad \mathbf{F}^s = \lambda^s\mathbf{1}, \quad \lambda^s > 0. \tag{10.2.4}$$

Here, suppressing the argument t:

- $\mathbf{F}^s(\mathbf{X})$ represents the local distortion of the material at $\mathbf{X}$ due to swelling, and λ^s is the *swelling stretch*. This local deformation accounts for the swelling of the material due to absorbed fluid molecules which are pinned to the *coherent polymer network structure* that resides in the *intermediate space* at $\mathbf{X}$ (as represented by the range of $\mathbf{F}^s(\mathbf{X})$);
- $\mathbf{F}^e(\mathbf{X})$ represents the subsequent stretching and rotation of this coherent swollen network structure, and thereby represents a corresponding mechanical elastic distortion.

We refer to $\mathbf{F}^s$ and $\mathbf{F}^e$ as the *swelling and elastic distortions*.

Consistent with our requirement that

$$J \stackrel{\text{def}}{=} \det\mathbf{F} > 0, \tag{10.2.5}$$

we assume that

$$J^e \stackrel{\text{def}}{=} \det\mathbf{F}^e > 0 \quad \text{and} \quad J^s \stackrel{\text{def}}{=} \det\mathbf{F}^s > 0, \tag{10.2.6}$$

so that both $\mathbf{F}^e$ and $\mathbf{F}^s$ are invertible.

It is important to fully understand the differences between the tensor fields $\mathbf{F}$, $\mathbf{F}^e$, and $\mathbf{F}^s$ in the multiplicative decomposition $\mathbf{F} = \mathbf{F}^e\mathbf{F}^s$ which is shown schematically in Fig. 10.1. First of all, while $\mathbf{F} = \nabla\boldsymbol{\chi}$ is the gradient of a point field, in general there is no point field $\boldsymbol{\chi}^s$ such that $\mathbf{F}^s = \nabla\boldsymbol{\chi}^s$, nor is there a point field $\boldsymbol{\chi}^e$ such that $\mathbf{F}^e = \nabla\boldsymbol{\chi}^e$. Thus we can at most describe the physical nature of the tensor fields $\mathbf{F}^e$ and $\mathbf{F}^s$ through their pointwise mapping properties as **linear transformations**. With this in mind, recall that $\mathbf{dx} = \mathbf{F}\mathbf{dX}$. Hence,

$$\mathbf{dx} = \mathbf{F}^e\mathbf{F}^s\mathbf{dX}.$$

For want of a better notation, let $\mathbf{dl}$ denote $\mathbf{F}^s\mathbf{dX}$,

$$\mathbf{dl} = \mathbf{F}^s\mathbf{dX},$$

so that,

$$\mathbf{dx} = \mathbf{F}^e\mathbf{dl}.$$

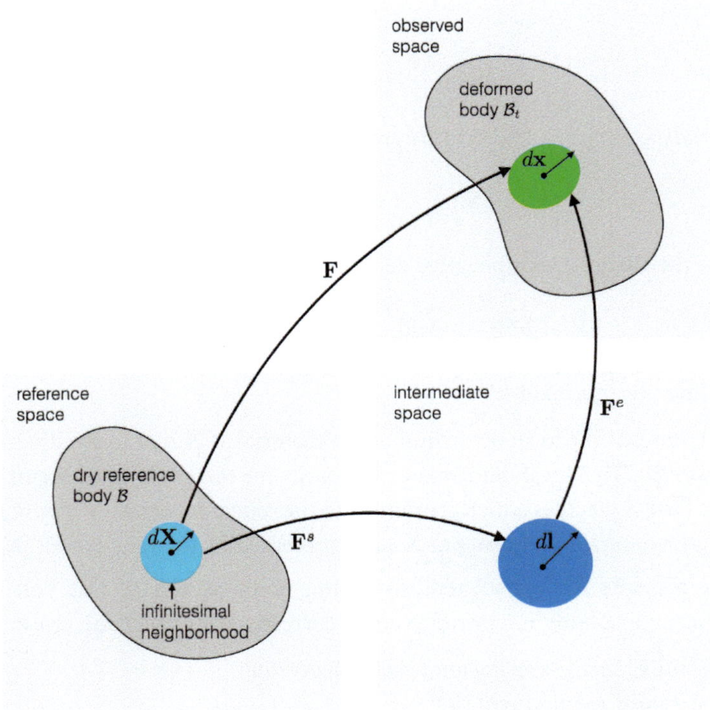

Fig. 10.1 Schematic of the $\mathbf{F} = \mathbf{F}^e\mathbf{F}^s$ decomposition.

The output of the linear transformation $\mathbf{F}^s$ must therefore coincide with the input of the linear transformation $\mathbf{F}^e$; that is,

$$\text{the range of } \mathbf{F}^s = \text{the domain of } \mathbf{F}^e.$$

We refer to this common vector space as the **intermediate space**.

Next, as is standard,

$$\mathbf{F} = \mathbf{R}\mathbf{U} = \mathbf{V}\mathbf{R} \quad \text{and} \quad \mathbf{F}^e = \mathbf{R}^e\mathbf{U}^e = \mathbf{V}^e\mathbf{R}^e \tag{10.2.7}$$

denote the right and left polar decompositions of $\mathbf{F}$ and $\mathbf{F}^e$, respectively, with $\mathbf{U}$, $\mathbf{V}$, $\mathbf{U}^e$, and $\mathbf{V}^e$ symmetric and positive definite tensors, and $\mathbf{R}$ and $\mathbf{R}^e$ rotations. Also, the tensors

$$\begin{aligned} \mathbf{C} = \mathbf{U}^2 = \mathbf{F}^{\top}\mathbf{F}, \quad \mathbf{B} = \mathbf{V}^2 = \mathbf{F}\mathbf{F}^{\top}, \quad \text{and} \\ \mathbf{C}^e = \mathbf{U}^{e2} = \mathbf{F}^{e\top}\mathbf{F}^e, \quad \mathbf{B}^e = \mathbf{V}^{e2} = \mathbf{F}^e\mathbf{F}^{e\top}, \end{aligned} \tag{10.2.8}$$

denote the total and elastic, right and left Cauchy–Green tensors.

Next, by $(10.2.3)_3$ and (10.2.4),

$$\mathbf{L} = \mathbf{L}^e + \mathbf{F}^e\mathbf{L}^s\mathbf{F}^{e-1}, \tag{10.2.9}$$

with

$$\mathbf{L}^e = \dot{\mathbf{F}}^e \mathbf{F}^{e-1}, \quad \mathbf{L}^s = \dot{\mathbf{F}}^s \mathbf{F}^{s-1}. \tag{10.2.10}$$

As is standard, we define the elastic and swelling stretching and spin tensors through

$$\begin{aligned} \mathbf{D}^e &= \operatorname{sym} \mathbf{L}^e, \quad \mathbf{W}^e = \operatorname{skw} \mathbf{L}^e, \\ \mathbf{D}^s &= \operatorname{sym} \mathbf{L}^s, \quad \mathbf{W}^s = \operatorname{skw} \mathbf{L}^s, \end{aligned} \tag{10.2.11}$$

so that $\mathbf{L}^e = \mathbf{D}^e + \mathbf{W}^e$ and $\mathbf{L}^s = \mathbf{D}^s + \mathbf{W}^s$.

Further, from (10.2.4), $(10.2.10)_2$, and (10.2.11)

$$\mathbf{D}^s = \big(\dot{\lambda}^s \lambda^{s-1}\big)\mathbf{1} \quad \text{and} \quad \mathbf{W}^s = \mathbf{0}; \tag{10.2.12}$$

and since

$$\dot{J}^s = J^s \operatorname{tr} \mathbf{D}^s, \tag{10.2.13}$$

we also have

$$\mathbf{D}^s = \frac{1}{3}\big(\dot{J}^s J^{s-1}\big)\mathbf{1}. \tag{10.2.14}$$

10.3 Frame-indifference

A *change in frame*, at each *fixed time t*, is a transformation—defined by a rotation $\mathbf{Q}(t)$ and a spatial point $\mathbf{y}(t)$—which transforms *spatial points* $\mathbf{x}$ to spatial points

$$\begin{aligned} \mathbf{x}^* &= \mathcal{F}(\mathbf{x}), && (10.3.1) \\ &= \mathbf{y}(t) + \mathbf{Q}(t)(\mathbf{x} - \mathbf{o}); && (10.3.2) \end{aligned}$$

the function $\mathcal{F}$ represents a rigid mapping of the observed space into itself, with $\mathbf{o}$ a fixed spatial origin. By (10.3.2) the transformation law for the motion $\mathbf{x} = \boldsymbol{\chi}(\mathbf{X}, t)$ has the form

$$\boldsymbol{\chi}^*(\mathbf{X}, t) = \mathbf{y}(t) + \mathbf{Q}(t)(\boldsymbol{\chi}(\mathbf{X}, t) - \mathbf{o}). \tag{10.3.3}$$

Hence the deformation gradient $\mathbf{F}$ transforms according to

$$\mathbf{F}^* = \mathbf{Q}\mathbf{F}. \tag{10.3.4}$$

The reference configuration and the intermediate structural space are independent of the choice of such changes in frame; thus the fields

$$\mathbf{F}^s = \lambda^s \mathbf{1}, \; J^s, \text{ and } \mathbf{L}^s \equiv \mathbf{D}^s = \frac{1}{3}\big(\dot{J}^s J^{s-1}\big)\mathbf{1} \quad \text{are invariant under a change in frame,} \tag{10.3.5}$$

which also of course follows from the fact that $\mathbf{F}^s$ is spherical. This observation, (10.2.4), and (10.3.4) yield the transformation law

$$(\mathbf{F}^e)^* = \mathbf{Q}\mathbf{F}^e. \tag{10.3.6}$$

10.4 Balance of forces and moments

Throughout, we denote by $\mathcal{P}$ an arbitrary *part* (subregion) of the reference body $\mathcal{B}$, with $\mathbf{n}_R$ the outward unit normal on the boundary $\partial\mathcal{P}$ of $\mathcal{P}$.

Since timescales associated with fluid diffusion are usually considerably longer than those associated with wave propagation, *we neglect all inertial effects*. Then, standard considerations of balance of forces and moments, when expressed referentially, give:

(a) There exists a stress tensor $\mathbf{T}_R$, called the Piola stress, such that the surface traction on an element of the surface $\partial\mathcal{P}$ of $\mathcal{P}$, is given by

$$\mathbf{t}_R(\mathbf{n}_R) = \mathbf{T}_R\mathbf{n}_R. \tag{10.4.1}$$

(b) $\mathbf{T}_R$ satisfies the macroscopic force balance

$$\mathrm{Div}\,\mathbf{T}_R + \mathbf{b}_{0R} = \mathbf{0}, \tag{10.4.2}$$

where $\mathbf{b}_R$ is an external body force per unit reference volume, which, consistent with neglect of inertial effects, is taken to be time-independent.

(c) $\mathbf{T}_R$ obeys the symmetry condition

$$\mathbf{T}_R\mathbf{F}^\top = \mathbf{F}\mathbf{T}_R^\top, \tag{10.4.3}$$

which represents a balance of moments.

Further, under a change in frame, $\mathbf{T}_R$ transforms as

$$\mathbf{T}_R^* = \mathbf{Q}\mathbf{T}_R. \tag{10.4.4}$$

Finally, as is standard, the Piola stress $\mathbf{T}_R$ is related to the standard symmetric Cauchy stress $\mathbf{T}$ in the deformed body by

$$\mathbf{T}_R = J\,\mathbf{T}\,\mathbf{F}^{-\top}, \tag{10.4.5}$$

so that

$$\mathbf{T} = J^{-1}\mathbf{T}_R\mathbf{F}^\top. \tag{10.4.6}$$

10.5 **Fluid content. Balance law for the fluid content**

Let

$$c_{\mathrm{R}}(\mathbf{X}, t) \tag{10.5.1}$$

denote the *number of moles of fluid molecules absorbed by the polymer network, reckoned per unit volume of the dry reference configuration*. We call c_{R} the *fluid content.*[4]

Define a *fluid flux* $\mathbf{j}_{\mathrm{R}}$, measured in moles per unit area per unit time, so that $-\int_{\partial\mathcal{P}} \mathbf{j}_{\mathrm{R}} \cdot \mathbf{n}_{\mathrm{R}} da_{\mathrm{R}}$ represents the number of moles of fluid molecules entering $\mathcal{P}$ across $\partial\mathcal{P}$, per unit time. In this case the *balance law for fluid content* takes the form

$$\overline{\int_{\mathcal{P}} c_{\mathrm{R}}\, dv_{\mathrm{R}}}^{\,\boldsymbol{\cdot}} = -\int_{\partial\mathcal{P}} \mathbf{j}_{\mathrm{R}} \cdot \mathbf{n}_{\mathrm{R}}\, da_{\mathrm{R}}, \tag{10.5.2}$$

for every part $\mathcal{P}$. Bringing the time derivative in (10.5.2) inside the integral and using the divergence theorem on the integral over $\partial\mathcal{P}$, we find that

$$\int_{\mathcal{P}} (\dot{c}_{\mathrm{R}} + \mathrm{Div}\,\mathbf{j}_{\mathrm{R}})\, dv_{\mathrm{R}} = 0. \tag{10.5.3}$$

Since $\mathcal{P}$ is arbitrary, this leads to a (local) *balance law for fluid content*

$$\dot{c}_{\mathrm{R}} = -\mathrm{Div}\,\mathbf{j}_{\mathrm{R}}. \tag{10.5.4}$$

10.5.1 **Constitutive equation for J^s**

Now,

$$J^s - 1 \tag{10.5.5}$$

represents the change in volume per unit reference volume due to swelling. *We assume constitutively that this change arises entirely due to the change in the fluid content*, so that with

- Ω denoting the volume of a mole of fluid molecules,

we have the important constitutive assumption that,

$$J^s = 1 + \Omega c_{\mathrm{R}}, \tag{10.5.6}$$

or equivalently that

$$\dot{J}^s = \Omega \dot{c}_{\mathrm{R}}. \tag{10.5.7}$$

[4] The fluid–solid mixture is treated as a single homogenized continuum body which allows for a mass flux of the fluid.

Using $(10.2.4)_2$, $J^s = \det \mathbf{F}^s$ is also expressible as

$$J^s = (\lambda^s)^3, \tag{10.5.8}$$

and hence the constraint (10.5.6) may also be stated as

$$\lambda^s = (1 + \Omega c_{\mathrm{R}})^{1/3}. \tag{10.5.9}$$

10.6 Balance of energy. Entropy imbalance

Our discussion of thermodynamics involves the following fields:

- ε_{R} the internal energy density per unit reference volume,
- η_{R} the entropy density per unit reference volume,
- $\mathbf{q}_{\mathrm{R}}$ the heat flux per unit reference area,
- q_{R} the external heat supply per unit reference volume,
- ϑ the absolute temperature ($\vartheta > 0$),
- $\mathbf{j}_{\mathrm{R}}$ the fluid flux per unit reference area,
- μ_ε internal energy per mole of diffusing fluid,
- μ_η entropy per mole of diffusing fluid.

Consider a material region $\mathcal{P}$. Then, consistent with our omission of inertial effects, we neglect kinetic energy, and take the balance law for energy as

$$\overline{\int_{\mathcal{P}} \varepsilon_{\mathrm{R}}\, dv_{\mathrm{R}}}^{\boldsymbol{\cdot}} = -\int_{\partial\mathcal{P}} \mathbf{q}_{\mathrm{R}} \cdot \mathbf{n}_{\mathrm{R}}\, da_{\mathrm{R}} + \int_{\mathcal{P}} q_{\mathrm{R}}\, dv_{\mathrm{R}} + \int_{\partial\mathcal{P}} (\mathbf{T}_{\mathrm{R}} \mathbf{n}_{\mathrm{R}}) \cdot \dot{\boldsymbol{\chi}}\, da_{\mathrm{R}} + \int_{\mathcal{P}} \mathbf{b}_{0\mathrm{R}} \cdot \dot{\boldsymbol{\chi}}\, dv_{\mathrm{R}} - \int_{\partial\mathcal{P}} \mu_\varepsilon \mathbf{j}_{\mathrm{R}} \cdot \mathbf{n}_{\mathrm{R}} da_{\mathrm{R}}, \tag{10.6.1}$$

where the term involving $\mu_\varepsilon \mathbf{j}_{\mathrm{R}}$ in (10.6.1) represents the flux of energy carried into $\mathcal{P}$ by the flux of fluid molecules.

Applying the divergence theorem to the terms in (10.6.1) involving integrals over the boundary $\partial\mathcal{P}$ of $\mathcal{P}$, we obtain

$$\int_{\mathcal{P}} \Big(\dot{\varepsilon}_{\mathrm{R}} - (\mathrm{Div}\, \mathbf{T}_{\mathrm{R}} + \mathbf{b}_{0\mathrm{R}}) \cdot \dot{\boldsymbol{\chi}} - \mathbf{T}_{\mathrm{R}} : \dot{\mathbf{F}} + \mathrm{Div}\, \mathbf{q}_{\mathrm{R}} - q_{\mathrm{R}} + \mu_\varepsilon\, \mathrm{Div}\, \mathbf{j}_{\mathrm{R}} + \mathbf{j}_{\mathrm{R}} \cdot \nabla \mu_\varepsilon \Big)\, dv_{\mathrm{R}} = 0, \tag{10.6.2}$$

which upon use of the balance laws (10.4.2) and (10.5.4), and using the fact that (10.6.2) must hold for all parts $\mathcal{P}$, gives the local form of the energy balance as

$$\dot{\varepsilon}_{\mathrm{R}} = \mathbf{T}_{\mathrm{R}} : \dot{\mathbf{F}} + \mu_\varepsilon \dot{c}_{\mathrm{R}} - \mathrm{Div}\, \mathbf{q}_{\mathrm{R}} + q_{\mathrm{R}} - \mathbf{j}_{\mathrm{R}} \cdot \nabla \mu_\varepsilon. \tag{10.6.3}$$

Next, the second law takes the form of an entropy imbalance

$$\overline{\int_{\mathcal{P}} \eta_{\mathrm{R}}\, dv_{\mathrm{R}}}^{\boldsymbol{\cdot}} \geq -\int_{\partial\mathcal{P}} \frac{\mathbf{q}_{\mathrm{R}}}{\vartheta} \cdot \mathbf{n}_{\mathrm{R}}\, da_{\mathrm{R}} + \int_{\mathcal{P}} \frac{q_{\mathrm{R}}}{\vartheta}\, dv_{\mathrm{R}} - \int_{\partial\mathcal{P}} \mu_\eta \mathbf{j}_{\mathrm{R}} \cdot \mathbf{n}_{\mathrm{R}} da_{\mathrm{R}}, \tag{10.6.4}$$

where the term involving $\mu_\eta \mathbf{j}_{\mathrm{R}}$ in (10.6.4) represents the flux of entropy carried into $\mathcal{P}$ by the flux of fluid molecules. Correspondingly, the local entropy imbalance has the form

$$\dot{\eta}_{\mathrm{R}} \geq -\mathrm{Div}\left(\frac{\mathbf{q}_{\mathrm{R}}}{\vartheta}\right) + \frac{q_{\mathrm{R}}}{\vartheta} - \mathrm{Div}\,(\mu_\eta \mathbf{j}_{\mathrm{R}})\,. \tag{10.6.5}$$

Next consider the term on the right-hand side of (10.6.5),

$$\begin{aligned}
&-\mathrm{Div}\left(\frac{\mathbf{q}_{\mathrm{R}}}{\vartheta}\right) + \frac{q_{\mathrm{R}}}{\vartheta} - \mathrm{Div}\,(\mu_\eta \mathbf{j}_{\mathrm{R}})\\
&\quad = \frac{1}{\vartheta}(-\mathrm{Div}\,\mathbf{q}_{\mathrm{R}} + q_{\mathrm{R}}) + \frac{1}{\vartheta^2}\mathbf{q}_{\mathrm{R}} \cdot \nabla\vartheta - (\mu_\eta \mathrm{Div}\,\mathbf{j}_{\mathrm{R}} + \mathbf{j}_{\mathrm{R}} \cdot \nabla\mu_\eta)\,,\\
&\quad = \frac{1}{\vartheta}\left(-\mathrm{Div}\,\mathbf{q}_{\mathrm{R}} + q_{\mathrm{R}} - \vartheta\,\mu_\eta \mathrm{Div}\,\mathbf{j}_{\mathrm{R}} - \vartheta\,\mathbf{j}_{\mathrm{R}} \cdot \nabla\mu_\eta\right) + \frac{1}{\vartheta^2}\mathbf{q}_{\mathrm{R}} \cdot \nabla\vartheta,\\
&\quad = \frac{1}{\vartheta}\left(-\mathrm{Div}\,\mathbf{q}_{\mathrm{R}} + q_{\mathrm{R}} + \vartheta\mu_\eta \dot{c}_{\mathrm{R}} - \vartheta\mathbf{j}_{\mathrm{R}} \cdot \nabla\mu_\eta\right) + \frac{1}{\vartheta^2}\mathbf{q}_{\mathrm{R}} \cdot \nabla\vartheta,
\end{aligned} \tag{10.6.6}$$

where in writing the last equation we have used the balance of fluid content (10.5.4). Further, from the local energy balance (10.6.3) we have that

$$-\mathrm{Div}\,\mathbf{q}_{\mathrm{R}} + q_{\mathrm{R}} = \dot{\varepsilon}_{\mathrm{R}} - \mathbf{T}_{\mathrm{R}} : \dot{\mathbf{F}} - \mu_\varepsilon \dot{c}_{\mathrm{R}} + \mathbf{j}_{\mathrm{R}} \cdot \nabla\mu_\varepsilon, \tag{10.6.7}$$

use of which in (10.6.6) gives

$$\begin{aligned}
&-\mathrm{Div}\left(\frac{\mathbf{q}_{\mathrm{R}}}{\vartheta}\right) + \frac{q_{\mathrm{R}}}{\vartheta} - \mathrm{Div}\,(\mu_\eta \mathbf{j}_{\mathrm{R}})\\
&= \frac{1}{\vartheta}\Big(\dot{\varepsilon}_{\mathrm{R}} - \mathbf{T}_{\mathrm{R}} : \dot{\mathbf{F}} - (\mu_\varepsilon - \vartheta\mu_\eta)\dot{c}_{\mathrm{R}} + \mathbf{j}_{\mathrm{R}} \cdot \nabla\mu_\varepsilon - \vartheta\mathbf{j}_{\mathrm{R}} \cdot \nabla\mu_\eta\Big) + \frac{1}{\vartheta^2}\mathbf{q}_{\mathrm{R}} \cdot \nabla\vartheta\\
&= \frac{1}{\vartheta}\Big(\dot{\varepsilon}_{\mathrm{R}} - \mathbf{T}_{\mathrm{R}} : \dot{\mathbf{F}} - (\mu_\varepsilon - \vartheta\mu_\eta)\dot{c}_{\mathrm{R}} + \mathbf{j}_{\mathrm{R}} \cdot \nabla(\mu_\varepsilon - \vartheta\mu_\eta) + \mu_\eta \mathbf{j}_{\mathrm{R}} \cdot \nabla\vartheta\Big) + \frac{1}{\vartheta^2}\mathbf{q}_{\mathrm{R}} \cdot \nabla\vartheta\,,
\end{aligned} \tag{10.6.8}$$

which upon introducing the notation

$$\begin{aligned}
\mu &\stackrel{\text{def}}{=} (\mu_\varepsilon - \vartheta\mu_\eta),\\
\mathbf{q}_{\mathrm{R}}^{\mathrm{net}} &\stackrel{\text{def}}{=} \mathbf{q}_{\mathrm{R}} + \vartheta\mu_\eta \mathbf{j}_{\mathrm{R}},
\end{aligned} \tag{10.6.9}$$

for a *chemical potential* and a *net heat flux*, respectively, may be written as

$$-\mathrm{Div}\left(\frac{\mathbf{q}_{\mathrm{R}}}{\vartheta}\right) + \frac{q_{\mathrm{R}}}{\vartheta} - \mathrm{Div}\,(\mu_\eta \mathbf{j}_{\mathrm{R}}) = \frac{1}{\vartheta}\Big(\dot{\varepsilon}_{\mathrm{R}} - \mathbf{T}_{\mathrm{R}} : \dot{\mathbf{F}} - \mu\dot{c}_{\mathrm{R}} + \mathbf{j}_{\mathrm{R}} \cdot \nabla\mu\Big) + \frac{1}{\vartheta^2}\mathbf{q}_{\mathrm{R}}^{\mathrm{net}} \cdot \nabla\vartheta. \tag{10.6.10}$$

In the expression $(10.6.9)_2$ for the effective heat flux the term $\vartheta\mu_\eta \mathbf{j}_R$ represents the heat flux due to the diffusing fluid (cf., e.g., Salvadori et al., 2018). Equation (10.6.10), together with the local entropy imbalance (10.6.5), implies that

$$\dot{\varepsilon}_R - \vartheta\dot{\eta}_R - \mathbf{T}_R : \dot{\mathbf{F}} - \mu\dot{c}_R + \frac{1}{\vartheta}\,\mathbf{q}_R^{\text{net}} \cdot \nabla\vartheta + \mathbf{j}_R \cdot \nabla\mu \leq 0. \tag{10.6.11}$$

Introducing the Helmholtz free energy

$$\psi_R = \varepsilon_R - \vartheta\eta_R, \tag{10.6.12}$$

(10.6.11) yields the following local free-energy imbalance

$$\dot{\psi}_R + \eta_R\dot{\vartheta} - \mathbf{T}_R : \dot{\mathbf{F}} - \mu\dot{c}_R + \frac{1}{\vartheta}\,\mathbf{q}_R^{\text{net}} \cdot \nabla\vartheta + \mathbf{j}_R \cdot \nabla\mu \leq 0. \tag{10.6.13}$$

10.7 Free-energy imbalance under isothermal conditions

Henceforth, in this chapter, we restrict our attention to *isothermal conditions*, $\vartheta =$ constant, in which case the free-energy imbalance (10.6.13) reduces to

$$\dot{\psi}_R - \mathbf{T}_R : \dot{\mathbf{F}} - \mu\dot{c}_R + \mathbf{j}_R \cdot \nabla\mu \leq 0. \tag{10.7.1}$$

10.7.1 Two new stress measures and an alternate expression for the stress power

Using (10.2.4), $(10.2.10)_2$, and (10.4.5), the stress power $\mathbf{T}_R : \dot{\mathbf{F}}$ admits the decomposition

$$\begin{aligned}
\mathbf{T}_R : \dot{\mathbf{F}} &= \mathbf{T}_R : (\dot{\mathbf{F}}^e\mathbf{F}^s + \mathbf{F}^e\dot{\mathbf{F}}^s) \\
&= (\mathbf{T}_R\mathbf{F}^{s\top}) : \dot{\mathbf{F}}^e + (\mathbf{F}^{e\top}\mathbf{T}_R) : \dot{\mathbf{F}}^s, \\
&= (\mathbf{T}_R\mathbf{F}^{s\top}) : \dot{\mathbf{F}}^e + (\mathbf{F}^{e\top}\mathbf{T}_R\mathbf{F}^{s\top}) : \mathbf{L}^s, \\
&= (J\mathbf{T}\mathbf{F}^{e-\top}) : \dot{\mathbf{F}}^e + (J\mathbf{F}^{e\top}\mathbf{T}\mathbf{F}^{e-\top}) : \mathbf{L}^s, \\
&= J^s \left[(J^e\mathbf{T}\mathbf{F}^{e-\top}) : \dot{\mathbf{F}}^e + (J^e\mathbf{F}^{e\top}\mathbf{T}\mathbf{F}^{e-\top}) : \mathbf{L}^s \right], \\
&= J^s \left[(J^e\mathbf{F}^{e-1}\mathbf{T}\mathbf{F}^{e-\top}) : \mathbf{F}^{e\top}\dot{\mathbf{F}}^e + (J^e\mathbf{F}^{e\top}\mathbf{T}\mathbf{F}^{e-\top}) : \mathbf{L}^s \right].
\end{aligned} \tag{10.7.2}$$

In view of (10.7.2) we introduce two new stress measures:

- The *elastic second Piola stress*,

$$\mathbf{T}^e \stackrel{\text{def}}{=} J^e \mathbf{F}^{e-1} \mathbf{T} \mathbf{F}^{e-\top}, \tag{10.7.3}$$

  which is *symmetric* on account of the symmetry of $\mathbf{T}$.
- The *Mandel stress*,

$$\mathbf{M}^e \stackrel{\text{def}}{=} J^e \mathbf{F}^{e\top} \mathbf{T} \mathbf{F}^{e-\top} \equiv \mathbf{C}^e \mathbf{T}^e, \tag{10.7.4}$$

  which in general is not symmetric.

Then, the stress power (10.7.2) may be written as

$$\mathbf{T}_{\text{R}} : \dot{\mathbf{F}} = J^s \mathbf{T}^e : \mathbf{F}^{e\top} \dot{\mathbf{F}}^e + J^s \mathbf{M}^e : \mathbf{L}^s. \tag{10.7.5}$$

Noting that the material derivative of the elastic right Cauchy–Green tensor $\mathbf{C}^e$ is

$$\dot{\mathbf{C}}^e = \mathbf{F}^{e\top} \dot{\mathbf{F}}^e + \dot{\mathbf{F}}^{e\top} \mathbf{F}^e, \tag{10.7.6}$$

which, recalling the symmetry of $\mathbf{T}^e$, gives the stress-power decomposition

$$\mathbf{T}_{\text{R}} : \dot{\mathbf{F}} = J^s \left[\frac{1}{2} \mathbf{T}^e : \dot{\mathbf{C}}^e + \mathbf{M}^e : \mathbf{L}^s \right]. \tag{10.7.7}$$

Next, recalling $(10.2.12)_2$ and (10.2.14), viz.

$$\mathbf{L}^s = \frac{1}{3} (\dot{J}^s J^{s-1}) \mathbf{1}, \tag{10.7.8}$$

we may write the stress power (10.7.7) as

$$\mathbf{T}_{\text{R}} : \dot{\mathbf{F}} = J^s \left(\frac{1}{2} \mathbf{T}^e : \dot{\mathbf{C}}^e + \frac{1}{3} (\dot{J}^s J^{s-1}) \mathbf{1} : \mathbf{M}^e \right) \tag{10.7.9}$$

or

$$\mathbf{T}_{\text{R}} : \dot{\mathbf{F}} = J^s \left(\frac{1}{2} \mathbf{T}^e : \dot{\mathbf{C}}^e \right) + \frac{1}{3} (\operatorname{tr} \mathbf{M}^e) \dot{J}^s. \tag{10.7.10}$$

Hence, using (10.7.10) in (10.7.1), the local free-energy imbalance may be written as

$$\dot{\psi}_{\text{R}} - \frac{1}{2} (J^s \mathbf{T}^e) : \dot{\mathbf{C}}^e - \frac{1}{3} (\operatorname{tr} \mathbf{M}^e) \dot{J}^s - \mu \dot{c}_{\text{R}} + \mathbf{j}_{\text{R}} \cdot \nabla \mu \leq 0. \tag{10.7.11}$$

Further, upon enforcing the constraint (10.5.7), the free-energy imbalance (10.7.11) becomes

$$\dot{\psi}_{\mathrm{R}} - \frac{1}{2}(J^s\mathbf{T}^e):\dot{\mathbf{C}}^e - \mu^{\mathrm{net}}\,\dot{c}_{\mathrm{R}} + \mathbf{j}_{\mathrm{R}}\cdot\nabla\mu \le 0, \tag{10.7.12}$$

where we have written

$$\mu^{\mathrm{net}} \stackrel{\mathrm{def}}{=} \mu + \Omega\left(\frac{1}{3}\mathrm{tr}\,\mathbf{M}^e\right). \tag{10.7.13}$$

for a *net* chemical potential.[5]

We note that under a change in frame

$$\psi_{\mathrm{R}},\ c_{\mathrm{R}},\ \mu,\ \mathbf{j}_{\mathrm{R}},\ \text{ and } \nabla\mu \text{ are invariant,} \tag{10.7.14}$$

ψ_{R}, c_{R}, and μ because they are scalars, and $\mathbf{j}_{\mathrm{R}}$ and $\nabla\mu$ since they are referential vector fields. Using (10.3.6), (10.4.4), the definition $(10.2.8)_2$ of $\mathbf{C}^e$ and the definitions (10.7.3) and (10.7.4) of $\mathbf{T}^e$ and $\mathbf{M}^e$, it is straightforward to show that

$$\mathbf{C}^e,\ \mathbf{T}^e,\ \text{and } \mathbf{M}^e \text{ are invariant under a change in frame.} \tag{10.7.15}$$

Thus with the invariance properties discussed above, all quantities in the free-energy imbalance (10.7.12) are invariant under a change in frame.

10.8 Constitutive theory

10.8.1 Basic constitutive equations

Guided by the free-energy imbalance (10.7.12), we assume that the free energy ψ_{R}, the stress $\mathbf{T}^e$, the net chemical potential μ^{net}, and the fluid flux $\mathbf{j}_{\mathrm{R}}$ are given by the following set of constitutive equations:

$$\begin{aligned}
\psi_{\mathrm{R}} &= \bar{\psi}_{\mathrm{R}}(\mathbf{C}^e, c_{\mathrm{R}}),\\
\eta_{\mathrm{R}} &= \bar{\eta}_{\mathrm{R}}(\mathbf{C}^e, c_{\mathrm{R}}),\\
\mathbf{T}^e &= \bar{\mathbf{T}}^e(\mathbf{C}^e, c_{\mathrm{R}}),\\
\mu^{\mathrm{net}} &= \bar{\mu}^{\mathrm{net}}(\mathbf{C}^e, c_{\mathrm{R}}),\\
\mathbf{j}_{\mathrm{R}} &= \bar{\mathbf{j}}_{\mathrm{R}}(\mathbf{C}^e, c_{\mathrm{R}}, \nabla\mu).
\end{aligned} \tag{10.8.1}$$

[5] Note that $\mathrm{tr}\,\mathbf{M}^e = J^e\mathrm{tr}\,\mathbf{T}$.

For simplicity, from the outset we have assumed that the constitutive equations for the free energy, stress, and chemical potential are independent of $\nabla\mu$, and depend only on the mechanical stretch $\mathbf{C}^e$ and the fluid content c_{R}. Using the transformation rules (10.7.14) and (10.7.15) we find that the constitutive equations (10.8.1) are frame-indifferent.

10.8.2 Thermodynamic restrictions

With a view toward determining the restrictions imposed by the local free-energy imbalance (10.7.12), note that

$$\dot{\psi}_{\mathrm{R}} = \frac{\partial\bar{\psi}_{\mathrm{R}}}{\partial\mathbf{C}^e} : \dot{\mathbf{C}}^e + \frac{\partial\bar{\psi}_{\mathrm{R}}}{\partial c_{\mathrm{R}}}\,\dot{c}_{\mathrm{R}}. \tag{10.8.2}$$

Substituting (10.8.1) and (10.8.2) into (10.7.12) we find that

$$\left(\frac{\partial\bar{\psi}_{\mathrm{R}}}{\partial\mathbf{C}^e} - \frac{1}{2}J^s\,\bar{\mathbf{T}}^e\right) : \dot{\mathbf{C}}^e + \left(\frac{\partial\bar{\psi}_{\mathrm{R}}}{\partial c_{\mathrm{R}}} - \bar{\mu}^{\mathrm{net}}\right)\dot{c}_{\mathrm{R}} + \bar{\mathbf{j}}_{\mathrm{R}}(\mathbf{C}^e, c_{\mathrm{R}}, \nabla\mu)\cdot\nabla\mu \le 0. \tag{10.8.3}$$

This inequality is to hold for all values of $\mathbf{C}^e$, $\dot{\mathbf{C}}^e$, c_{R}, $\dot{c}_{\mathrm{R}}$, and $\nabla\mu$. Since $\dot{\mathbf{C}}^e$ and $\dot{c}_{\mathrm{R}}$ appear linearly, their "coefficients" must vanish, for otherwise $\dot{\mathbf{C}}^e$ and $\dot{c}_{\mathrm{R}}$ may be chosen to violate (10.8.3). We are therefore led to the thermodynamic restriction that the free energy determines the stress $\mathbf{T}^e$ and the chemical potential μ through the "state relations"

$$\begin{aligned} \mathbf{T}^e &= 2\,J^{s-1}\,\frac{\partial\bar{\psi}_{\mathrm{R}}(\mathbf{C}^e, c_{\mathrm{R}})}{\partial\mathbf{C}^e}, \\ \mu &= \frac{\partial\bar{\psi}_{\mathrm{R}}(\mathbf{C}^e, c_{\mathrm{R}})}{\partial c_{\mathrm{R}}} - \Omega\left(\frac{1}{3}\mathrm{tr}\,\mathbf{M}^e\right), \end{aligned} \tag{10.8.4}$$

where in writing $(10.8.4)_2$ we have used the definition (10.7.13) for the net chemical potential. Further, as is standard, we assume that the response function $\bar{\mathbf{j}}_{\mathrm{R}}$ satisfies the *fluid-transport inequality*

$$\bar{\mathbf{j}}_{\mathrm{R}}(\mathbf{C}^e, c_{\mathrm{R}}, \nabla\mu)\cdot\nabla\mu < 0 \quad \text{when} \quad \nabla\mu \neq 0. \tag{10.8.5}$$

Next, recalling from (10.7.3), (10.7.4), and (10.4.5) that

$$\mathbf{T}^e = J^e\mathbf{F}^{e-1}\mathbf{T}\mathbf{F}^{e-\top}, \quad \mathbf{M}^e \stackrel{\text{def}}{=} J^e\mathbf{F}^{e\top}\mathbf{T}\mathbf{F}^{e-\top} = \mathbf{C}^e\mathbf{T}^e, \quad \text{and} \quad \mathbf{T}_{\mathrm{R}} = J\,\mathbf{T}\mathbf{F}^{-\top}, \tag{10.8.6}$$

$(10.8.4)_1$ gives the Cauchy stress as

$$\mathbf{T} = J^{-1}\left[2\mathbf{F}^e\frac{\partial\bar{\psi}_{\mathrm{R}}(\mathbf{C}^e, c_{\mathrm{R}})}{\partial\mathbf{C}^e}\mathbf{F}^{e\top}\right], \tag{10.8.7}$$

and the Piola stress as

$$\mathbf{T}_{\mathrm{R}} = 2\mathbf{F}^e \frac{\partial \bar{\psi}_{\mathrm{R}}(\mathbf{C}^e, c_{\mathrm{R}})}{\partial \mathbf{C}^e} \mathbf{F}^{s-\top}. \tag{10.8.8}$$

Also, the mean normal stress $(1/3)\mathrm{tr}\,\mathbf{M}^e$ appearing in $(10.8.4)_2$ is then given by

$$\frac{1}{3}\,\mathrm{tr}\,\mathbf{M}^e = \frac{1}{3}\mathbf{C}^e : \left(2J^{s-1} \frac{\partial \bar{\psi}_{\mathrm{R}}(\mathbf{C}^e, c_{\mathrm{R}})}{\partial \mathbf{C}^e} \right). \tag{10.8.9}$$

10.9 Specialization of the free-energy function

The constitutive equations considered thus far are fairly general. With a view towards applications to elastomeric gels we now specialize the theory by imposing additional constitutive assumptions. We confine our attention to materials for which $\bar{\psi}(\mathbf{C}^e, c_{\mathrm{R}})$ is an isotropic function of $\mathbf{C}^e$, and has the representation

$$\bar{\psi}_{\mathrm{R}}(\mathbf{C}^e, c_{\mathrm{R}}) = \tilde{\psi}_{\mathrm{R}}(\mathcal{I}_{\mathbf{C}^e}, c_{\mathrm{R}}), \tag{10.9.1}$$

where

$$\mathcal{I}_{\mathbf{C}^e} = \big(I_1(\mathbf{C}^e), I_2(\mathbf{C}^e), I_3(\mathbf{C}^e)\big)$$

is the list of principal invariants of $\mathbf{C}^e$. Further, we assume that the free energy ψ_{R} may be written in a separable form as (cf., e.g., Flory, 1953)

$$\tilde{\psi}_{\mathrm{R}}(\mathcal{I}_{\mathbf{C}^e}, c_{\mathrm{R}}) = \mu^{\mathrm{ref}} c_{\mathrm{R}} + \psi^{\mathrm{mix}}(c_{\mathrm{R}}) + \psi^{\mathrm{mech}}(\mathcal{I}_{\mathbf{C}^e}, c_{\mathrm{R}}), \tag{10.9.2}$$

where:

- μ^{ref} is a reference chemical potential of the unmixed pure solvent;
- $\psi_{\mathrm{R}}^{\mathrm{mix}}(c_{\mathrm{R}})$ is the change in free energy due to mixing of the solvent with the polymer network; and
- $\psi^{\mathrm{mech}}(\mathcal{I}_{\mathbf{C}^e}, c_{\mathrm{R}})$ is the contribution to the change in the free energy due to the deformation of the polymer network.

Estimate for ψ^{mix}

In the literature on swelling of elastomers, the quantity

$$\phi \stackrel{\mathrm{def}}{=} \frac{1}{1 + \Omega c_{\mathrm{R}}} = \frac{1}{J^s} = (\lambda^s)^{-3}, \quad 0 < \phi \le 1, \tag{10.9.3}$$

is called the *polymer volume fraction*. Note that

- *the dry state corresponds to $\phi = 1$, and $\phi < 1$ represents a swollen state.*

Following Doi (1996, 2009), we adopt the following form of the Flory (1942)–Huggins (1942) theory for the contribution to the free energy due to mixing

$$\psi^{\text{mix}} = \frac{R\vartheta}{\Omega}\frac{1}{\phi}\Big((1-\phi)\ln(1-\phi) + \chi\phi(1-\phi)\Big), \tag{10.9.4}$$

where R is the universal gas constant, ϑ is the constant temperature under consideration, and χ is a dimensionless parameter (called the χ-parameter, or interaction parameter), which represents the disaffinity between the polymer and the fluid:

- if χ is increased, the fluid molecules are expelled from the gel and the gel shrinks, while if χ is decreased, the gel swells.[6]

The expression (10.9.4) for the mixing energy may be expressed in terms of c_{R} as

$$\psi^{\text{mix}}(c_{\text{R}}) = \frac{R\vartheta}{\Omega}\left(\Omega c_{\text{R}}\ln\left(\frac{\Omega c_{\text{R}}}{1+\Omega c_{\text{R}}}\right) + \chi\left(\frac{\Omega c_{\text{R}}}{1+\Omega c_{\text{R}}}\right)\right). \tag{10.9.5}$$

Estimate for ψ^{mech}

In elastomeric materials, the major part of ψ^{mech} arises from an entropic contribution. Let

$$\bar{\lambda} \overset{\text{def}}{=} \frac{1}{\sqrt{3}}\sqrt{\operatorname{tr}\mathbf{C}} = \frac{1}{\sqrt{3}}\sqrt{(\operatorname{tr}\mathbf{C}^e)\,(1+\Omega c_{\text{R}})^{2/3}} \tag{10.9.6}$$

define an *effective total stretch*. Then, classical statistical mechanics models of rubber elasticity (cf., Treloar (1975); Arruda and Boyce (1993)) provide the following estimates for the entropy change due to mechanical stretching:

1. For small to moderate values of $\bar{\lambda}$, based on Gaussian statistics,

$$\eta_{\text{R}}^{\text{mech}} = -\frac{3}{2}N\,k_B\,(\bar{\lambda}^2 - 1) + N\,k_B\ln J, \tag{10.9.7}$$

where k_B is Boltzmann's constant and N represents the number of polymer chains per unit reference volume.

2. For larger values of $\bar{\lambda}$ it is necessary to use non-Gaussian statistics to account for the limited extensibility of the polymer chains, and for these circumstances the entropy change is given by

$$\begin{aligned}\eta_{\text{R}}^{\text{mech}} = {} & -Nk_B\lambda_L^2\left[\left(\frac{\bar{\lambda}}{\lambda_L}\right)\beta + \ln\left(\frac{\beta}{\sinh\beta}\right) - \left(\frac{1}{\lambda_L}\right)\beta_0 - \ln\left(\frac{\beta_0}{\sinh\beta_0}\right)\right] \\ & + Nk_B\left(\frac{\lambda_L}{3}\beta_0\right)\ln J,\end{aligned} \tag{10.9.8}$$

[6] For reference, a typical value of χ which promotes pronounced swelling is $\chi \approx 0.1$, while $\chi \approx 0.5$ is a borderline value for the swelling of a dry polymer.

with

$$\beta \overset{\text{def}}{=} \mathcal{L}^{-1}\left(\frac{\bar{\lambda}}{\lambda_L}\right), \quad \text{and} \quad \beta_0 \overset{\text{def}}{=} \mathcal{L}^{-1}\left(\frac{1}{\lambda_L}\right), \tag{10.9.9}$$

where $\mathcal{L}^{-1}$ is the inverse of the Langevin function

$$\mathcal{L}(\bullet) = \coth(\bullet) - (\bullet)^{-1}.$$

This functional form for the change in entropy involves two material parameters: N, the number of polymer chains per unit reference volume, and λ_L, the *network locking stretch.*[7]

Next, we take ψ^{mech} to also have an *energetic component*,

$$\varepsilon_{\text{R}} = J^s\left[\frac{1}{2}K(\ln J^e)^2\right], \tag{10.9.10}$$

where $K > 0$ is the bulk modulus of the gel. This term is intended to reflect changes in the internal energy associated with volumetric mechanical deformation of the swollen elastomer. The term within square brackets in (10.9.10) represents a contribution to the free energy measured per unit volume of the intermediate space, and multiplication by J^s converts this to a contribution per unit volume of the reference space.

Then, using (10.9.7) through (10.9.10), and writing

$$G_0 \overset{\text{def}}{=} N\, k_B\, \vartheta, \tag{10.9.11}$$

for a temperature-dependent *ground-state shear modulus,* we obtain the estimate

$$\begin{aligned}\psi^{\text{mech}}(\mathcal{I}_{\mathbf{C}^e}, c_{\text{R}}) = G_0\lambda_L^2&\left[\left(\frac{\bar{\lambda}}{\lambda_L}\right)\beta + \ln\left(\frac{\beta}{\sinh\beta}\right) - \left(\frac{1}{\lambda_L}\right)\beta_0 - \ln\left(\frac{\beta_0}{\sinh\beta_0}\right)\right] \\ &- G_0\left(\frac{\lambda_L}{3}\beta_0\right)\ln J + J^s\left[\frac{1}{2}K(\ln J^e)^2\right],\end{aligned} \tag{10.9.12}$$

where, recalling (10.9.6),

$$\bar{\lambda} = \frac{1}{\sqrt{3}}\sqrt{(\text{tr}\,\mathbf{C}^e)(1+\Omega c_{\text{R}})^{2/3}}, \tag{10.9.13}$$

is an *effective stretch.*

[7] In the non-Gaussian model, the network locking stretch λ_L is related to the number of links n in a freely jointed chain by $\lambda_L = \sqrt{n}$.

REMARK

It is important to note that the definition of effective stretch (10.9.13) is *different* from the definition (1.10.1) of effective stretch in Chapter 1 where the effective stretch for the nearly incompressible elastomers was based on the isochoric part $\bar{\mathbf{C}}$ of $\mathbf{C}$, whereas here—in a theory for gels which swell significantly by imbibing fluids—it is based on the total $\mathbf{C}$. The entropy change expressions (10.9.7) and (10.9.8) for gels are also different from their counterparts in Chapter 1 for dry elastomers, by factors involving $\ln J$.

Total free energy. Stress. Chemical potential

Thus, using (10.9.5) and (10.9.12) in (10.9.2), a simple form of the free-energy function which accounts for the combined effects of mixing, swelling, and elastic stretching is

$$\begin{aligned}\psi_{\mathrm{R}} = \mu^{\mathrm{ref}} c_{\mathrm{R}} &+ \frac{R\vartheta}{\Omega}\left(\Omega c_{\mathrm{R}} \ln\left(\frac{\Omega c_{\mathrm{R}}}{1+\Omega c_{\mathrm{R}}}\right) + \chi\left(\frac{\Omega c_{\mathrm{R}}}{1+\Omega c_{\mathrm{R}}}\right)\right)\\ &+ G_0\lambda_L^2\left[\left(\frac{\bar{\lambda}}{\lambda_L}\right)\beta + \ln\left(\frac{\beta}{\sinh\beta}\right) - \left(\frac{1}{\lambda_L}\right)\beta_0 - \ln\left(\frac{\beta_0}{\sinh\beta_0}\right)\right]\\ &- G_0\left(\frac{\lambda_L}{3}\beta_0\right)\ln J + J^s\left[\frac{1}{2}K(\ln J^e)^2\right].\end{aligned} \tag{10.9.14}$$

Hence, using (10.8.7), we find that the Cauchy stress tensor is given by

$$\mathbf{T} = J^{-1}\left[G\,\phi^{-2/3}\mathbf{B}^e - G_0\zeta_0\mathbf{1}\right] + \left(\frac{K\ln J^e}{J^e}\right)\mathbf{1}, \tag{10.9.15}$$

where

$$G \stackrel{\text{def}}{=} G_0\,\zeta, \quad \text{with} \quad \zeta \stackrel{\text{def}}{=} \left(\frac{\lambda_L}{3\bar{\lambda}}\right)\mathcal{L}^{-1}\left(\frac{\bar{\lambda}}{\lambda_L}\right), \quad \text{and} \quad \zeta_0 \stackrel{\text{def}}{=} \left(\frac{\lambda_L}{3}\right)\mathcal{L}^{-1}\left(\frac{1}{\lambda_L}\right), \tag{10.9.16}$$

is a *generalized shear modulus*, and $\mathbf{B}^e$ is the elastic left Cauchy–Green tensor. Note that since $\mathcal{L}^{-1}(z) \to \infty$ as $z \to 1$, the stretch-dependent shear modulus $G \to \infty$ as $(\bar{\lambda}/\lambda_L) \to 1$.

Next, since

$$\mathbf{B} = (\lambda^s)^2\mathbf{B}^e = \phi^{-2/3}\mathbf{B}^e, \tag{10.9.17}$$

(10.9.15) reduces to

$$\mathbf{T} = J^{-1}\left[G_0\left(\zeta\mathbf{B} - \zeta_0\mathbf{1}\right)\right] - p\,\mathbf{1}, \tag{10.9.18}$$

where we have introduced the notation

$$p \stackrel{\text{def}}{=} -\frac{K \ln J^e}{J^e}, \tag{10.9.19}$$

as a "pressure" associated with the mechanical deformation $\mathbf{F}^e$ of the swollen gel. Hence the Piola stress, $\mathbf{T}_{\mathrm{R}} = J\mathbf{T}\mathbf{F}^{-\top}$, is given by[8]

$$\mathbf{T}_{\mathrm{R}} = G_0 \left[\zeta \mathbf{F} - \zeta_0 \mathbf{F}^{-\top}\right] - Jp\,\mathbf{F}^{-\top}. \tag{10.9.20}$$

Also, using $(10.8.4)_2$ and (10.9.14), the chemical potential μ is given by

$$\begin{aligned}
\mu = \mu^{\text{ref}} &+ R\vartheta\left(\ln\left(\frac{\Omega c_{\mathrm{R}}}{1+\Omega c_{\mathrm{R}}}\right) + \frac{1}{(1+\Omega c_{\mathrm{R}})} + \chi \frac{1}{(1+\Omega c_{\mathrm{R}})^2} \right) \\
&+ \frac{1}{3}\Omega\, G\,(1+\Omega c_{\mathrm{R}})^{-1/3}\,\mathrm{tr}\,\mathbf{C}^e - \Omega\, G_0 \zeta_0\,(1+\Omega c_{\mathrm{R}})^{-1} \\
&+ \Omega\left[\frac{1}{2}K(\ln J^e)^2)\right] - \Omega\left(\frac{1}{3}\mathrm{tr}\,\mathbf{M}^e\right).
\end{aligned} \tag{10.9.21}$$

ASIDE

Further, from (10.8.9) and (10.9.18) and using

$$\phi = J^{s-1} = (1+\Omega c_{\mathrm{R}})^{-1},$$

we obtain

$$\frac{1}{3}\mathrm{tr}\,\mathbf{M}^e = \frac{1}{3}J^e \mathrm{tr}\,\mathbf{T} = \frac{1}{3}\phi G \mathrm{tr}\,\mathbf{B} - G_0\zeta_0\phi + K(\ln J^e), \tag{10.9.22}$$

so that

$$-\Omega\left(\frac{1}{3}\mathrm{tr}\,\mathbf{M}^e\right) = -\Omega\frac{1}{3}G\phi \mathrm{tr}\,\mathbf{B} + \Omega G_0\zeta_0\phi - \Omega K(\ln J^e). \tag{10.9.23}$$

Also, recalling from (10.9.13) that

$$\bar{\lambda}^2 = \frac{1}{3}(\mathrm{tr}\,\mathbf{C}^e)(1+\Omega c_{\mathrm{R}})^{2/3} = \frac{1}{3}\mathrm{tr}\,\mathbf{C} = \frac{1}{3}\mathrm{tr}\,\mathbf{B},$$

we have

$$\begin{aligned}
-\Omega\left(\frac{1}{3}\mathrm{tr}\,\mathbf{M}^e\right) &= -\Omega\frac{1}{3}\phi G \mathrm{tr}\,\mathbf{C} + \Omega G_0\zeta_0\phi - \Omega K(\ln J^e), \\
&= -\Omega\frac{1}{3}G(1+\Omega c_{\mathrm{R}})^{-1/3}(\mathrm{tr}\,\mathbf{C}^e) + \Omega G_0\zeta_0(1+\Omega c_{\mathrm{R}})^{-1} - \Omega K(\ln J^e), \\
&= -\Omega\frac{1}{3}G(1+\Omega c_{\mathrm{R}})^{-1/3}(\mathrm{tr}\,\mathbf{C}^e) + \Omega G_0\zeta_0(1+\Omega c_{\mathrm{R}})^{-1} + J^e\,p\,\Omega.
\end{aligned} \tag{10.9.24}$$

---

[8] It is useful to compare this expression for the Piola stress with the expression (1.10.13) in Chapter 1.

Use of (10.9.24) in (10.9.21) gives

$$\mu = \mu^{\text{ref}} + R\vartheta\Big(\ln(1-\phi) + \phi + \chi\phi^2\Big) + J^e\,p\,\Omega + \Omega\left[\frac{1}{2}K(\ln J^e)^2\right], \tag{10.9.25}$$

which may be written equivalently in terms of c_{R} instead of ϕ as

$$\begin{aligned}\mu = \mu^{\text{ref}} + R\vartheta&\left[\ln\left(1 - \frac{1}{1+\Omega c_{\text{R}}}\right) + \frac{1}{1+\Omega c_{\text{R}}} + \chi\left(\frac{1}{1+\Omega c_{\text{R}}}\right)^2\right]\\ &+ J^e\,p\,\Omega + \Omega\left[\frac{1}{2}K(\ln J^e)^2\right].\end{aligned} \tag{10.9.26}$$

REMARK

For a theory based on Gaussian statistics for configurational changes in entropy, cf. (10.9.7), the generalized shear modulus is no longer stretch-dependent, so that $G = G_0$, and the expressions (10.9.18) and (10.9.20) for the Cauchy stress and the Piola stress reduce to

$$\begin{aligned}\mathbf{T} &= J^{-1}\left[G_0\,(\mathbf{B} - \mathbf{1})\right] - p\mathbf{1},\\ \mathbf{T}_{\text{R}} &= G_0\left[\mathbf{F} - \mathbf{F}^{-\top}\right] - Jp\,\mathbf{F}^{-\top},\end{aligned} \tag{10.9.27}$$

where

$$p \stackrel{\text{def}}{=} -\frac{K\ln J^e}{J^e} \tag{10.9.28}$$

represents a pressure due to the mechanical deformation $\mathbf{F}^e$, while the expression (10.9.25) for the chemical potential remains unchanged,

$$\mu = \mu^{\text{ref}} + R\vartheta\Big(\ln(1-\phi) + \phi + \chi\phi^2\Big) + J^e\,p\,\Omega + \Omega\left[\frac{1}{2}K(\ln J^e)^2\right]. \tag{10.9.29}$$

10.9.1 Specialization of the equations for the fluid flux

For isotropic materials, the fluid flux in the *spatial description* is taken as

$$\mathbf{j} = -m\,\text{grad}\,\mu, \tag{10.9.30}$$

in which we assume that the mobility m at a given temperature ϑ is given by

$$m = \frac{Dc}{R\vartheta}, \tag{10.9.31}$$

where $D > 0$, a constant, represents a diffusion coefficient, and $c = c_{\text{R}}/J$ is the fluid concentration measured in moles of fluid per unit deformed volume.

Next, using the standard mapping results

$$\mathbf{F}^{-\top}\nabla\mu = \operatorname{grad}\mu \quad \text{and} \quad \mathbf{j}_{\text{R}} = J\mathbf{F}^{-1}\mathbf{j}, \tag{10.9.32}$$

we obtain the referential fluid flux as

$$\mathbf{j}_{\text{R}} = -\mathbf{M}^{\text{mob}}\nabla\mu \quad \text{with} \quad \mathbf{M}^{\text{mob}} = \frac{Dc_{\text{R}}}{R\vartheta}\mathbf{C}^{-1}. \tag{10.9.33}$$

10.10 Governing partial differential equations. Boundary conditions

The governing partial differential equations consist of:

1. Neglecting body forces and inertia, the local force balance is given by the pair of equations

$$\begin{aligned} \operatorname{Div}\mathbf{T}_{\text{R}} &= \mathbf{0}, \\ \frac{p}{K} + \frac{\ln J^e}{J^e} &= 0, \end{aligned} \tag{10.10.1}$$

in $\mathcal{B}$, with $\mathbf{T}_{\text{R}}$ given by

$$\begin{aligned} \mathbf{T}_{\text{R}} &= G_0\left[\zeta\mathbf{F} - \zeta_0\mathbf{F}^{-\top}\right] - Jp\,\mathbf{F}^{-\top}, \\ G_0 &= Nk_B\vartheta, \\ \zeta &= \left(\frac{\lambda_L}{3\bar{\lambda}}\right)\mathcal{L}^{-1}\left(\frac{\bar{\lambda}}{\lambda_L}\right), \\ \zeta_0 &= \left(\frac{\lambda_L}{3}\right)\mathcal{L}^{-1}\left(\frac{1}{\lambda_L}\right). \end{aligned} \tag{10.10.2}$$

In writing (10.10.1) we have split the standard force balance equation into two equations because this helps in dealing with the near "elastic" incompressibility of the material. In our numerical finite element implementation, in addition to using the displacement $\mathbf{u}$, we will also use p as a separate degree of freedom.

With $\partial\mathcal{B}_{\mathbf{u}}$ and $\partial\mathcal{B}_{\mathbf{t}_{\text{R}}}$ denoting complementary subsurfaces of the boundary $\partial\mathcal{B}$ of the body $\mathcal{B}$, we consider a pair of possible boundary conditions:

$$\begin{aligned} \mathbf{u} &= \bar{\mathbf{u}} \quad \text{on } \partial\mathcal{B}_{\mathbf{u}}, \\ \mathbf{T}_{\text{R}}\mathbf{n}_{\text{R}} &= \bar{\mathbf{t}}_{\text{R}} \quad \text{on } \partial\mathcal{B}_{\mathbf{t}_{\text{R}}}. \end{aligned} \tag{10.10.3}$$

In the boundary conditions above $\bar{\mathbf{u}}$ and $\bar{\mathbf{t}}_{\text{R}}$ are prescribed displacement and tractions, both functions of $(\mathbf{X}, t)$.

2. The balance equation (10.5.4) for the fluid content is

$$\dot{c}_{\mathrm{R}} + \mathrm{Div}\, \mathbf{j}_{\mathrm{R}} = 0, \tag{10.10.4}$$

with the fluid flux given by

$$\mathbf{j}_{\mathrm{R}} = -\mathbf{M}^{\mathrm{mob}} \nabla \mu, \tag{10.10.5}$$

where

$$\mathbf{M}^{\mathrm{mob}} = \mathbf{C}^{-1} \left(\frac{D c_{\mathrm{R}}}{R\vartheta} \right) \tag{10.10.6}$$

is the mobility tensor, and the chemical potential μ given by

$$\begin{aligned} \mu = R\vartheta \left[\ln\left(1 - \frac{1}{1+\Omega c_{\mathrm{R}}}\right) + \frac{1}{1+\Omega c_{\mathrm{R}}} + \frac{\chi}{(1+\Omega c_{\mathrm{R}})^2} \right] \\ + J^e\, p\Omega + \Omega \left[\frac{1}{2} K (\ln J^e)^2 \right], \end{aligned} \tag{10.10.7}$$

where we have set the reference chemical potential of the fluid to zero, $\mu^{\mathrm{ref}} = 0$.

The free energy supplied to the body $\mathcal{B}$ by the flux of the fluid is given by

$$-\int_{\partial\mathcal{B}} \mu \mathbf{j}_{\mathrm{R}} \cdot \mathbf{n}_{\mathrm{R}} da_{\mathrm{R}}. \tag{10.10.8}$$

Thus, with $\partial\mathcal{B}_\mu$ and $\partial\mathcal{B}_{\mathrm{j_R}}$ denoting complementary subsurfaces of $\partial\mathcal{B}$, we consider a pair of possible boundary conditions:

$$\begin{aligned} \mu &= \bar{\mu} \quad \text{on } \partial\mathcal{B}_\mu, \\ \mathbf{j}_{\mathrm{R}} \cdot \mathbf{n}_{\mathrm{R}} &= \bar{j}_{\mathrm{R}} \quad \text{on } \partial\mathcal{B}_{\mathrm{j_R}}, \end{aligned} \tag{10.10.9}$$

with $\bar{\mu}$ and $\bar{j}_{\mathrm{R}}$ prescribed functions of $\mathbf{X}$ and t on $\partial\mathcal{B}$.

Standard initial conditions involve the specifications

$$\mathbf{u}(\mathbf{X},0) = \mathbf{u}_0(\mathbf{X}) \quad \text{and} \quad \mu(\mathbf{X},0) = \mu_0(\mathbf{X}) \quad \text{on } \mathcal{B}, \tag{10.10.10}$$

with $\mathbf{u}_0$ and μ_0 prescribed functions of $\mathbf{X} \in \mathcal{B}$.

REMARK

As phrased above, the basic degrees of freedom in the boundary value problem are the displacement field, the pressure field, and the chemical potential,

$$(\mathbf{u}, p, \mu).$$

It turns out, however, that we also need knowledge of the fluid content c_R to obtain a solution. Given $(\mathbf{u}, p, \mu)$ the concentration c_R may be calculated from the expression (10.10.7) for the chemical potential which may be written as,

$$\mu - R\vartheta\left(\ln\left(1 - \frac{1}{1+\Omega c_R}\right) + \frac{1}{1+\Omega c_R} + \frac{\chi}{(1+\Omega c_R)^2}\right) - J^e p\Omega - \frac{1}{2}\frac{\Omega}{K}(J^e p)^2 = 0, \tag{10.10.11}$$

which serves as an implicit equation for c_R.

REMARKS

1. Care must be taken if the body is initially dry, that is initial conditions for which $c_R = 0$ or equivalently $\phi = 1$. In this case to prevent an initial value of the chemical potential of $\mu_0 = -\infty$, in numerical calculations an initially "dry" body is idealized as having a value of $c_R = 0.001$ or equivalently $\phi = 0.999$.
2. In applications for the case in which the environment consists of a pure and incompressible liquid, the chemical potential boundary condition is given by

$$\bar{\mu} = \mu^{\text{ref}} + p^{\text{ext}}\Omega, \tag{10.10.12}$$

   where p^{ext} is the hydrostatic pressure of the liquid in the external environment, and Ω is the molar volume of the fluid. Also, if a portion of the boundary is impermeable to the liquid, then on that portion the prescribed normal flux $\bar{j}$ vanishes.

10.A Case of an initially swollen body

In this appendix we consider a situation in which the reference configuration of a gel is taken as a configuration which has been homogeneously and isotropically *pre-swollen* under traction-free conditions. Such a situation may arise due to a particular process for making the gel, or very simply the need to perform numerical simulations where the starting point is a previously freely swollen gel.

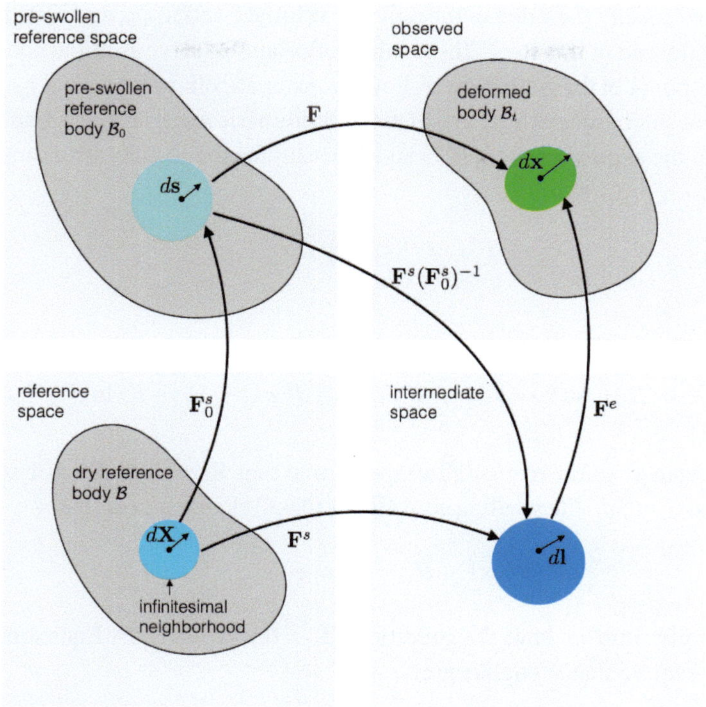

Fig. 10.2 Schematic of the $\mathbf{F} = \mathbf{F}^e\mathbf{F}^s(\mathbf{F}_0^s)^{-1}$ decomposition for the case when the reference configuration is taken as a homogeneously pre-swollen body.

The major change from what was described earlier in Section 10.2 for an initially dry reference configuration is an accounting for the pre-swollen reference configuration which we identify by the region of space $\mathcal{B}_0$, as shown schematically in the top left panel of Fig. 10.2. Thus, with $\mathbf{F}$ now denoting the deformation gradient from the pre-swollen reference configuration $\mathcal{B}_0$ to the deformed body $\mathcal{B}_t$, we adopt the following kinematical decomposition,

$$\mathbf{F}(\mathbf{X},t) = \mathbf{F}^e(\mathbf{X},t)\mathbf{F}^s(\mathbf{X},t)\mathbf{F}_0^{s-1}\,, \tag{10.A.1}$$

where

$$\mathbf{F}_0^s = \lambda_0 \mathbf{1} \quad \text{with} \quad \lambda_0 \equiv \phi_0^{-1/3} \tag{10.A.2}$$

represents the spherical swelling from the dry reference state to the pre-swollen state. Here, $\lambda_0 > 1$ denotes the isotropic stretch in the pre-swollen state and $\phi_0 \in (0,1)$ denotes the corresponding spatially constant polymer volume fraction. Use of (10.A.2) in (10.A.1) gives

$$\mathbf{F} = \phi_0^{1/3}\mathbf{F}^e\mathbf{F}^s. \tag{10.A.3}$$

Consider an initially dry body which has been homogeneously swollen to equilibrium under traction-free conditions. To characterize the state of the body at the end of this pre-swelling step, we need to specify the value of the polymer volume fraction ϕ_0, and also the value of the pressure p_0 at the end of this step.[9] These values of ϕ_0 and p_0 serve as initial conditions for the subsequent response of the gel. In what follows we estimate values for ϕ_0 and p_0. For simplicity we limit our considerations to a gel in which the elastomeric material obeys Gaussian statistics.

Thus, recall the expressions (10.9.27) and (10.9.29) for the Cauchy stress and the chemical potential, viz.

$$\mathbf{T} = J^{-1} G_0 (\mathbf{B} - \mathbf{1}) - p\mathbf{1}, \tag{10.A.4}$$

and

$$\mu = \mu^{\text{ref}} + R\vartheta\Big(\ln(1-\phi) + \phi + \chi\phi^2 \Big) + J^e\, p\,\Omega + \Omega \left[\frac{1}{2} K (\ln J^e)^2 \right]. \tag{10.A.5}$$

For homogeneous traction-free swelling we assume that $\mathbf{F}_0^s = \lambda_0 \mathbf{1}$, $\mathbf{F}^e \approx \mathbf{1}$ so that $J^e \approx 1$, and hence that $J \approx J_0^s$. The swelling constraint (10.5.6) then requires that

$$J_0^s = \lambda_0^3 = \phi_0^{-1}. \tag{10.A.6}$$

Further, at equilibrium we have the conditions $\mathbf{T} = \mathbf{0}$ and $\mu = \mu^{\text{ref}}$. Equations (10.A.4) and (10.A.5) then require that at equilibrium

$$\mathbf{0} = J^{-1} G_0 (\mathbf{B} - \mathbf{1}) - p_0 \mathbf{1}, \tag{10.A.7}$$

and

$$0 = \ln(1-\phi_0) + \phi_0 + \chi\phi_0^2 + p_0 \frac{\Omega}{R\vartheta}. \tag{10.A.8}$$

Since we consider isotropic swelling, (10.A.7) is equivalent to

$$0 = \lambda_0^{-3} G_0 (\lambda_0^2 - 1) - p_0. \tag{10.A.9}$$

This gives

$$p_0 = G_0 (\lambda_0^{-1} - \lambda_0^{-3}). \tag{10.A.10}$$

Thus, note that even though the total stress is zero, $\mathbf{T} = \mathbf{0}$, the initial pressure p_0 is non-zero; it represents an internal residual pressure which exists to balance the stress caused by the isotropic pre-stretch λ_0 of the polymer network.

Finally, use of (10.A.10) in (10.A.8), and using (10.A.6), gives the following *implicit equation* for the polymer volume fraction ϕ_0,

$$0 = \ln(1-\phi_0) + \phi_0 + \chi\phi_0^2 + \frac{G_0\Omega}{R\vartheta} \phi_0 \left(\phi_0^{2/3} - 1 \right). \tag{10.A.11}$$

[9] The pressure p_0 arises because of the pre-stretch of the polymer network.

11 Numerical implementation of poroelasticity of elastomeric gels

11.1 Weak forms of the governing partial differential equations

1. **Mechanical problem**:
Consider a test field $\mathbf{u}_{\text{test}}(\mathbf{X})$ in $\mathcal{B}$ which satisfies $\mathbf{u}_{\text{test}}(\mathbf{X}) = \mathbf{0}$ on $\partial\mathcal{B}_{\mathbf{u}}$. If the balance equation $\text{Div}\,\mathbf{T}_{\text{R}} = \mathbf{0}$ on $\mathcal{B}$ and the traction condition $(\bar{\mathbf{t}}_{\text{R}} - \mathbf{T}_{\text{R}}\mathbf{n}_{\text{R}}) = \mathbf{0}$ on $\partial\mathcal{B}_{\mathbf{t}_{\text{R}}}$ are satisfied, then it is straightforward to show that the weak form

$$\int_{\mathcal{B}} \mathbf{T}_{\text{R}} : \nabla\mathbf{u}_{\text{test}}\, dv_{\text{R}} - \int_{\partial\mathcal{B}_{\mathbf{t}_{\text{R}}}} \bar{\mathbf{t}}_{\text{R}} \cdot \mathbf{u}_{\text{test}}\, da_{\text{R}} = 0 \tag{11.1.1}$$

holds for all admissible test fields $\mathbf{u}_{\text{test}}$ which vanish on $\partial\mathcal{B}_{\mathbf{u}}$. In addition, considering an arbitrary test field $p_{\text{test}}(\mathbf{X})$ in $\mathcal{B}$ the weak form of the constitutive equation

$$\frac{p}{K} + \frac{\ln J^e}{J^e} = 0,$$

is

$$\int_{\mathcal{B}} \left[\frac{p}{K} + \frac{\ln J^e}{J^e}\right] p_{\text{test}}\, dv_{\text{R}} = 0. \tag{11.1.2}$$

2. **Mass balance problem**:
Next, consider a scalar test field $\mu_{\text{test}}(\mathbf{X})$ in $\mathcal{B}$ which satisfies $\mu_{\text{test}}(\mathbf{X}) = \mathbf{0}$ on $\partial\mathcal{B}_{\mu}$. If the balance equation

$$\dot{c}_{\text{R}} + \text{Div}\,\mathbf{j}_{\text{R}} = 0 \qquad \text{in } \mathcal{B},$$

and the boundary condition $\mathbf{j}_{\text{R}} \cdot \mathbf{n} = \bar{\jmath}_{\text{R}}$ on $\partial\mathcal{B}_{\mathbf{j}_{\text{R}}}$ are satisfied, then

$$\int_{\mathcal{B}} \dot{c}_{\text{R}}\, \mu_{\text{test}}\, dv_{\text{R}} + \int_{\mathcal{B}} (\text{Div}\,\mathbf{j}_{\text{R}})\, \mu_{\text{test}}\, dv_{\text{R}} + \int_{\partial\mathcal{B}_{\mathbf{j}_{\text{R}}}} (\bar{\jmath}_{\text{R}} - \mathbf{j}_{\text{R}} \cdot \mathbf{n})\, \mu_{\text{test}}\, da_{\text{R}} = 0, \tag{11.1.3}$$

Introduction to Coupled Theories in Solid Mechanics. Lallit Anand, Eric M. Stewart, Shawn A. Chester, Oxford University Press.
 DOI: 10.1093/oso/9780198986218.003.0012

or equivalently

$$\int_{\mathcal{B}} \dot{c}_{\mathrm{R}}\, \mu_{\text{test}}\, dv_{\mathrm{R}} + \int_{\mathcal{B}} (\operatorname{Div} \mathbf{j}_{\mathrm{R}})\, \mu_{\text{test}}\, dv_{\mathrm{R}} - \int_{\partial\mathcal{B}_{\mathbf{j}_{\mathrm{R}}}} \mathbf{j}_{\mathrm{R}} \cdot \mathbf{n}_{\mathrm{R}}\, \mu_{\text{test}}\, da_{\mathrm{R}} + \int_{\partial\mathcal{B}_{\mathbf{j}_{\mathrm{R}}}} \bar{\jmath}_{\mathrm{R}}\, \mu_{\text{test}}\, da_{\mathrm{R}} = 0 \tag{11.1.4}$$

holds for all test fields μ_{test}. Thus, since $\mu_{\text{test}} = 0$ on $\partial\mathcal{B}_{\mu}$ the first boundary integral in expression above may be extended to all of $\partial\mathcal{B}$,

$$\int_{\mathcal{B}} \dot{c}_{\mathrm{R}}\, \mu_{\text{test}}\, dv_{\mathrm{R}} + \int_{\mathcal{B}} (\operatorname{Div} \mathbf{j}_{\mathrm{R}})\, \mu_{\text{test}}\, dv_{\mathrm{R}} - \int_{\partial\mathcal{B}} \mathbf{j}_{\mathrm{R}} \cdot \mathbf{n}_{\mathrm{R}}\, \mu_{\text{test}}\, da_{\mathrm{R}} + \int_{\partial\mathcal{B}_{\mathbf{j}_{\mathrm{R}}}} \bar{\jmath}_{\mathrm{R}}\, \mu_{\text{test}}\, da_{\mathrm{R}} = 0. \tag{11.1.5}$$

Next, by the divergence theorem

$$\int_{\partial\mathcal{B}} \mathbf{j}_{\mathrm{R}} \cdot \mathbf{n}_{\mathrm{R}}\, \mu_{\text{test}}\, da_{\mathrm{R}} = \int_{\partial\mathcal{B}} (\mu_{\text{test}}\, \mathbf{j}_{\mathrm{R}}) \cdot \mathbf{n}\, da_{\mathrm{R}} = \int_{\mathcal{B}} \mu_{\text{test}} \operatorname{Div} \mathbf{j}_{\mathrm{R}}\, dv_{\mathrm{R}} + \int_{\mathcal{B}} \mathbf{j}_{\mathrm{R}} \cdot \nabla \mu_{\text{test}}\, dv_{\mathrm{R}}. \tag{11.1.6}$$

Use of (11.1.6) in (11.1.5) gives the weak form

$$\int_{\mathcal{B}} \dot{c}_{\mathrm{R}}\, \mu_{\text{test}}\, dv_{\mathrm{R}} - \int_{\mathcal{B}} \mathbf{j}_{\mathrm{R}} \cdot \nabla \mu_{\text{test}}\, dv_{\mathrm{R}} + \int_{\partial\mathcal{B}_{\mathbf{j}_{\mathrm{R}}}} \bar{\jmath}_{\mathrm{R}}\, \mu_{\text{test}}\, da_{\mathrm{R}} = 0. \tag{11.1.7}$$

We use a first-order finite difference approximation to write

$$\dot{c}_{\mathrm{R}} = \frac{c_{\mathrm{R}} - c_{\mathrm{R\,old}}}{\Delta t}, \tag{11.1.8}$$

where c_{R} is the value of the fluid content at the end of the step, so that the weak form (11.1.7) may be expressed as

$$\int_{\mathcal{B}} \frac{c_{\mathrm{R}} - c_{\mathrm{R\,old}}}{\Delta t}\, \mu_{\text{test}}\, dv_{\mathrm{R}} - \int_{\mathcal{B}} \mathbf{j}_{\mathrm{R}} \cdot \nabla \mu_{\text{test}}\, dv_{\mathrm{R}} + \int_{\partial\mathcal{B}_{\mathbf{j}_{\mathrm{R}}}} \bar{\jmath}_{\mathrm{R}}\, \mu_{\text{test}}\, da_{\mathrm{R}} = 0. \tag{11.1.9}$$

3. **Weak form of the implicit equation for c_{R}:**
As discussed in the Remark on page 168, we may calculate the species concentration c_{R} using the implicit equation

$$\frac{\mu}{R\vartheta} - \left(\ln\left(1 - \frac{1}{1+\Omega c_{\mathrm{R}}}\right) + \frac{1}{1+\Omega c_{\mathrm{R}}} + \frac{\chi}{(1+\Omega c_{\mathrm{R}})^2} \right) - \frac{\Omega J^e}{R\vartheta}\, p - \frac{1}{2} \frac{\Omega}{K R\vartheta} (J^e p)^2 = 0. \tag{11.1.10}$$

However, on account of the inability of the FEniCS framework to solve nonlinear equations at the Gauss-point level, we solve equation (11.1.10) in a *weak sense* by introducing a test function c_{test},

$$\int_{\mathcal{B}} \left[\frac{\mu}{R\vartheta} - \left(\ln\left(1 - \frac{1}{1+\Omega c_{\text{R}}}\right) + \frac{1}{1+\Omega c_{\text{R}}} + \frac{\chi}{(1+\Omega c_{\text{R}})^2} \right) - \frac{\Omega J^e}{R\vartheta} p - \frac{1}{2}\frac{\Omega}{KR\vartheta}(J^e p)^2 \right] c_{\text{test}}\, dv_{\text{R}} = 0. \tag{11.1.11}$$

As expected, similar to equation (11.1.2) for the pressure, there are no boundary conditions associated with equation (11.1.11) since it is not a governing partial differential equation.

Summarizing, the complete set of weak forms of the coupled problem are:

$$\begin{aligned}
&\int_{\mathcal{B}} \mathbf{T}_{\text{R}} : \nabla \mathbf{u}_{\text{test}}\, dv_{\text{R}} - \int_{\partial\mathcal{B}_{\mathbf{t}_{\text{R}}}} \bar{\mathbf{t}}_{\text{R}} \cdot \mathbf{u}_{\text{test}}\, da_{\text{R}} = 0,\\
&\int_{\mathcal{B}} \left[\frac{p}{K} + \frac{\ln J^e}{J^e} \right] p_{\text{test}}\, dv_{\text{R}} = 0,\\
&\int_{\mathcal{B}} \frac{c_{\text{R}} - c_{\text{R}\,\text{old}}}{\Delta t} \mu_{\text{test}}\, dv_{\text{R}} - \int_{\mathcal{B}} \mathbf{j}_{\text{R}} \cdot \nabla \mu_{\text{test}}\, dv_{\text{R}} + \int_{\partial\mathcal{B}_{j_{\text{R}}}} \bar{j}_{\text{R}}\, \mu_{\text{test}}\, da_{\text{R}} = 0,\\
&\int_{\mathcal{B}} \left[\frac{\mu}{R\vartheta} - \left(\ln\left(1 - \frac{1}{1+\Omega c_{\text{R}}}\right) + \frac{1}{1+\Omega c_{\text{R}}} + \frac{\chi}{(1+\Omega c_{\text{R}})^2} \right) - \frac{\Omega J^e}{R\vartheta} p - \frac{1}{2}\frac{\Omega}{KR\vartheta}(J^e p)^2 \right] c_{\text{test}}\, dv_{\text{R}} = 0.
\end{aligned} \tag{11.1.12}$$

11.2 **Normalization**

Since we plan to solve all the governing equations related to multiple physical fields $\{\mathbf{u}, p, \mu, c_{\text{R}}\}$ simultaneously, it is beneficial to rewrite the four residuals so that they have approximately the same magnitudes. To this end:

1. The species concentration c_{R} mol/m^3 is normalized using the molar volume Ω [m^3/mol] as

$$\hat{c} \stackrel{\text{def}}{=} \Omega c_{\text{R}}. \tag{11.2.1}$$

2. The chemical potential μ [J/mol] is normalized by $R\vartheta$ [J/mol],

$$\hat{\mu} = \frac{\mu}{R\vartheta}. \tag{11.2.2}$$

3. Recall the mass balance equation:

$$\dot{c}_{\mathrm{R}} + \mathrm{Div}\,\mathbf{j}_{\mathrm{R}} = 0.$$

This may be written as,

$$\frac{1}{\Omega}\dot{\hat{c}} + \mathrm{Div}\;\mathbf{j}_{\mathrm{R}} = 0.$$

4. Next, the species flux $\mathbf{j}_{\mathrm{R}}$ is given by the constitutive equation,

$$\mathbf{j}_{\mathrm{R}} = -\mathbf{M}^{\mathrm{mob}}\nabla\mu \qquad \text{with} \qquad \mathbf{M}^{\mathrm{mob}} = \mathbf{C}^{-1}\left(\frac{Dc_{\mathrm{R}}}{R\vartheta}\right). \tag{11.2.3}$$

Using the normalized variables $\hat{\mu}$ and $\hat{c}$ we may rewrite it as follows:

$$\mathbf{j}_{\mathrm{R}} = -\mathbf{M}^{\mathrm{mob}}\nabla\hat{\mu} \times R\vartheta \qquad \text{with} \qquad \mathbf{M}^{\mathrm{mob}} = \mathbf{C}^{-1}\left(\frac{D\hat{c}}{\Omega R\vartheta}\right). \tag{11.2.4}$$

5. Finally we choose to normalize the stress with the shear modulus G_0:

$$\hat{\mathbf{T}}_{\mathrm{R}} \stackrel{\text{def}}{=} \frac{\mathbf{T}_{\mathrm{R}}}{G_0} = \left[\zeta\mathbf{F} - \zeta_0\mathbf{F}^{-\top}\right] - \frac{J\,p}{G_0}\,\mathbf{F}^{-\top}. \tag{11.2.5}$$

6. Summarizing the governing equations may then be expressed as

$$\begin{aligned}
\mathrm{Div}\,\hat{\mathbf{T}}_{\mathrm{R}} &= 0\\
\frac{p}{K} + \frac{\ln J^e}{J^e} &= 0,\\
\dot{\hat{c}} + \mathrm{Div}\,\mathbf{j}_{\mathrm{R}} &= 0,\\
\hat{\mu} - \left(\ln\left(1 - \frac{1}{1+\hat{c}}\right) + \frac{1}{1+\hat{c}} + \frac{\chi}{(1+\hat{c})^2}\right) - \frac{\Omega\,J^e}{R\vartheta}\,p - \frac{1}{2}\frac{\Omega}{K R\vartheta}(J^e p)^2 &= 0.
\end{aligned} \tag{11.2.6}$$

With the normalizations introduced above and the normalized parameters denoted by $\{\hat{\bullet}\}$, the weak forms of the coupled problem may be written as

$$\int_{\mathcal{B}} \hat{\mathbf{T}}_{\mathrm{R}} : \nabla \mathbf{u}_{\text{test}} \, dv_{\mathrm{R}} - \int_{\partial \mathcal{B}_{\mathbf{t}_{\mathrm{R}}}} \bar{\hat{\mathbf{t}}}_{\mathrm{R}} \cdot \mathbf{u}_{\text{test}} \, da_{\mathrm{R}} = 0,$$

$$\int_{\mathcal{B}} \left[\frac{p}{K} + \frac{\ln J^e}{J^e} \right] p_{\text{test}} \, dv_{\mathrm{R}} = 0,$$

$$\int_{\mathcal{B}} \frac{\hat{c} - \hat{c}_{\text{old}}}{\Delta t} \hat{\mu}_{\text{test}} \, dv_{\mathrm{R}} - \int_{\mathcal{B}} \Omega \, \mathbf{j}_{\mathrm{R}} \cdot \nabla \hat{\mu}_{\text{test}} \, dv_{\mathrm{R}} + \int_{\partial \mathcal{B}_{\mathbf{j}_{\mathrm{R}}}} \Omega \, \bar{j}_{\mathrm{R}} \, \hat{\mu}_{\text{test}} \, da_{\mathrm{R}} = 0, \tag{11.2.7}$$

$$\int_{\mathcal{B}} \left[\hat{\mu} - \left(\ln \left(1 - \frac{1}{1+\hat{c}} \right) + \frac{1}{1+\hat{c}} + \frac{\chi}{(1+\hat{c})^2} \right) \right.$$
$$\left. - \frac{\Omega J^e}{R\vartheta} p - \frac{1}{2} \frac{\Omega}{K R \vartheta} (J^e p)^2 \right] \hat{c}_{\text{test}} \, dv_{\mathrm{R}} = 0.$$

11.3 Solution procedure in the FEniCS code

In brief, in FEniCS, we use the following degrees of freedom

$$\{\mathbf{u}, \, p, \, \hat{\mu}, \, \hat{c}\}, \tag{11.3.1}$$

as well as their known values from the previous step

$$\{\mathbf{u}_{\text{old}}, \, p_{\text{old}}, \, \hat{\mu}_{\text{old}}, \, \hat{c}_{\text{old}}\}, \tag{11.3.2}$$

to calculate the *residuals* which correspond to the weak forms of the governing equations, viz.

$$R_{\mathbf{u}} = \int_{\mathcal{B}} \hat{\mathbf{T}}_{\mathrm{R}} : \nabla \mathbf{u}_{\text{test}} \, dv_{\mathrm{R}} - \int_{\partial \mathcal{B}_{\mathbf{t}_{\mathrm{R}}}} \bar{\hat{\mathbf{t}}}_{\mathrm{R}} \cdot \mathbf{u}_{\text{test}} \, da_{\mathrm{R}},$$

$$R_p = \int_{\mathcal{B}} \left[\frac{p}{K} + \frac{\ln J^e}{J^e} \right] p_{\text{test}} \, dv_{\mathrm{R}},$$

$$R_{\hat{\mu}} = \int_{\mathcal{B}} \frac{\hat{c} - \hat{c}_{\text{old}}}{\Delta t} \hat{\mu}_{\text{test}} \, dv_{\mathrm{R}} - \int_{\mathcal{B}} \Omega \, \mathbf{j}_{\mathrm{R}} \cdot \nabla \hat{\mu}_{\text{test}} \, dv_{\mathrm{R}} + \int_{\partial \mathcal{B}_{\mathbf{j}_{\mathrm{R}}}} \Omega \, \bar{j}_{\mathrm{R}} \, \hat{\mu}_{\text{test}} \, da_{\mathrm{R}}, \tag{11.3.3}$$

$$R_{\hat{c}} = \int_{\mathcal{B}} \left[\hat{\mu} - \left(\ln \left(1 - \frac{1}{1+\hat{c}} \right) + \frac{1}{1+\hat{c}} + \frac{\chi}{(1+\hat{c})^2} \right) \right.$$
$$\left. - \frac{\Omega J^e}{R\vartheta} p - \frac{1}{2} \frac{\Omega}{K R \vartheta} (J^e p)^2 \right] \hat{c}_{\text{test}} \, dv_{\mathrm{R}}.$$

FEniCS then differentiates these residuals with respect to the separate degrees of freedom to obtain *tangents*. At each step in time, FEniCS uses the residuals and tangents in a global

Newton–Raphson solver to determine current values of the degrees of freedom $\{\mathbf{u},\ p,\ \hat{\mu},\ \hat{c}\}$ for which the residuals are zero,[1] and which therefore satisfy the governing equations as well as the initial and boundary conditions (in a weak sense).

REMARK

A mixed finite element method is utilized to solve for the displacement $\mathbf{u}$, the pressure p, the chemical potential $\hat{\mu}$, and the concentration field $\hat{c}$, concurrently. To avoid the numerical instabilities we use Taylor–Hood type elements with second-order interpolation for the displacement, while the other variables are interpolated one order lower.

11.4 Algorithm for calculation of the residuals

The algorithm we follow in our code to calculate the residuals in (11.3.3) is summarized below.

1. Begin the step with known values of

$$\left\{\mathbf{u}_{\mathrm{old}},\ p_{\mathrm{old}},\ \hat{\mu}_{\mathrm{old}},\ \hat{c}_{\mathrm{old}}\right\}, \tag{11.4.1}$$

where for the first step these quantities are given by the initial conditions for $\mathbf{u}$, p, $\hat{\mu}$, and $\hat{c}_0$.

2. Calculate necessary kinematical quantities for the current step as,

$$\mathbf{F} = \mathbf{1} + \nabla\,\mathbf{u}, \quad J = \det\mathbf{F}, \quad \mathbf{C} = \mathbf{F}^{\top}\mathbf{F}, \quad \bar{\lambda} = \sqrt{\mathrm{tr}\,\mathbf{C}/3}, \quad \mathbf{C}^{-1} = (\mathbf{F}^{\top}\mathbf{F})^{-1}. \tag{11.4.2}$$

3. Calculate J^e:

$$J^e = \frac{J}{1+\hat{c}}. \tag{11.4.3}$$

4. Calculate the intermediate factors ζ and ζ_0:

$$\begin{aligned}
&\beta \overset{\text{def}}{=} \mathcal{L}^{-1}\left(\frac{\bar{\lambda}}{\lambda_L}\right) \approx z\,\frac{3-z^2}{1-z^2} \qquad \text{where} \qquad z \overset{\text{def}}{=} \frac{\bar{\lambda}}{\lambda_L},\\
&\zeta \overset{\text{def}}{=} \left(\frac{\lambda_L}{3\bar{\lambda}}\right)\beta.
\end{aligned} \tag{11.4.4}$$

$$\begin{aligned}
&\beta_0 \overset{\text{def}}{=} \mathcal{L}^{-1}\left(\frac{1}{\lambda_L}\right) \approx z\,\frac{3-z^2}{1-z^2} \qquad \text{where} \qquad z \overset{\text{def}}{=} \frac{1}{\lambda_L},\\
&\zeta_0 \overset{\text{def}}{=} \left(\frac{\lambda_L}{3}\right)\beta_0.
\end{aligned} \tag{11.4.5}$$

[1] To within some small numerical tolerance.

5. Calculate the Piola stress:

$$\hat{\mathbf{T}}_{\mathrm{R}} = \left[\zeta \mathbf{F} - \zeta_0 \mathbf{F}^{-\top}\right] - \frac{J\,p}{G_0}\,\mathbf{F}^{-\top}, \qquad G_0 = N k_B \vartheta. \tag{11.4.6}$$

6. Calculate the species flux:

$$\mathbf{j}_{\mathrm{R}} = -\mathbf{M}^{\mathrm{mob}} \nabla \hat{\mu} \times R\vartheta \qquad \text{with} \qquad \mathbf{M}^{\mathrm{mob}} = \mathbf{C}^{-1}\left(\frac{D\hat{c}}{\Omega R\vartheta}\right). \tag{11.4.7}$$

7. Use the quantities from the prior steps to compute the residuals:

$$\begin{aligned}
R_{\mathbf{u}} &= \int_{\mathcal{B}} \hat{\mathbf{T}}_{\mathrm{R}} : \nabla \mathbf{u}_{\mathrm{test}}\, dv_{\mathrm{R}} - \int_{\partial \mathcal{B}_{\mathbf{t}_{\mathrm{R}}}} \bar{\hat{\mathbf{t}}}_{\mathrm{R}} \cdot \mathbf{u}_{\mathrm{test}}\, da_{\mathrm{R}}, \\
R_p &= \int_{\mathcal{B}} \left[\frac{p}{K} + \frac{\ln J^e}{J^e}\right] p_{\mathrm{test}}\, dv_{\mathrm{R}}, \\
R_{\hat{\mu}} &= \int_{\mathcal{B}} \frac{\hat{c} - \hat{c}_{\mathrm{old}}}{\Delta t}\, \hat{\mu}_{\mathrm{test}}\, dv_{\mathrm{R}} - \int_{\mathcal{B}} \Omega\, \mathbf{j}_{\mathrm{R}} \cdot \nabla \hat{\mu}_{\mathrm{test}}\, dv_{\mathrm{R}} + \int_{\partial \mathcal{B}_{\mathrm{j}_{\mathrm{R}}}} \Omega\, \bar{\jmath}_{\mathrm{R}}\, \hat{\mu}_{\mathrm{test}}\, da_{\mathrm{R}}, \\
R_{\hat{c}} &= \int_{\mathcal{B}} \left[\hat{\mu} - \left(\ln\left(1 - \frac{1}{1+\hat{c}}\right) + \frac{1}{1+\hat{c}} + \frac{\chi}{(1+\hat{c})^2}\right)\right. \\
&\qquad \left. - \frac{\Omega\, J^e}{R\vartheta}\, p - \frac{1}{2}\frac{\Omega}{K R\vartheta}(J^e p)^2\right] \hat{c}_{\mathrm{test}}\, dv_{\mathrm{R}}.
\end{aligned} \tag{11.4.8}$$

8. FEniCS solves for the current values of $\{\mathbf{u},\, p,\, \hat{\mu},\, \hat{c}\}$.
9. Set these current values as the "old" values for the next step in time,

$$\{\mathbf{u},\, p,\, \hat{\mu},\, \hat{c}\} \implies \{\mathbf{u}_{\mathrm{old}},\, p_{\mathrm{old}},\, \hat{\mu}_{\mathrm{old}},\, \hat{c}_{\mathrm{old}}\}. \tag{11.4.9}$$

10. Repeat for the desired number of steps in time.

12 Representative simulations of poroelasticity of elastomeric gels

In our numerical simulations in FEniCS we will use the basic units listed as SI (mm) in Table 12.1. Representative values of the material parameters for an elastomeric gel are listed in Table 12.2.

Table 12.1 Basic units.

Quantity	SI (m)	SI (mm)
Length	m	mm
Mass	kg	kg
Time	s	s
Mass density	kg/m^3	kg /mm^3
Force	N	mN
Stress	Pa	kPa
Energy	J	μJ
Temperature	K	K
Amount of substance	mol	mol
c_R, species concentration	mol/m^3	mol/mm^3
μ, chemical potential	J/mol	mJ/mol
Ω, molar volume	m^3/mol	mm^3/mol
D, fluid diffusivity	m^2/s	mm^2/s
k_B, Boltzmann's constant	1.38×10^{-23} J/K	1.38×10^{-17} μJ/K
Gas constant R	8.3145 J/(mol K)	8.3145×10^6 μJ/(mol K)

In all the representative simulations shown in this Chapter the underlying dry elastomer is modeled as a Neo-Hookean material, and hence the value of the locking stretch is set to a large number, $\lambda_L = 100$, or greater.[1]

[1] We note that the next part related to thermally responsive gels will include simulations with realistic values of λ_L to exhibit limited chain extensibility.

Introduction to Coupled Theories in Solid Mechanics. Lallit Anand, Eric M. Stewart, Shawn A. Chester, Oxford University Press.
 DOI: 10.1093/oso/9780198986218.003.0013

Table 12.2 Material parameters for a elastomeric gel. Note: to model a Neo-Hookean material set the locking stretch λ_L to a large number, say $\lambda_L = 100$.

Quantity	Value SI (m)	Value SI (mm)
G_0, ground-state shear modulus	1×10^6 Pa	1000 kPa
λ_L, locking stretch	100	100
K, bulk modulus ($1000 \times G_0$)	1000×10^6 Pa	1000×10^3 kPa
Ω, molar volume	1×10^{-4} m^3/mol	1×10^5 mm^3/mol
D, diffusivity	5×10^{-9} m^2/s	5×10^{-3} mm^2/s
χ, Flory–Huggins parameter	0.1	0.1
Gas constant R	8.3145 J/(mol K)	8.3145×10^6 μJ/(mol K)
ϑ_0, reference temperature	298 K	298 K
ϕ_0, initial polymer volume fraction	0.999	0.999

12.1 Free-swelling of an axisymmetric cylinder

As a first example we consider a basic coupled diffusion–deformation problem of free-swelling of an initially dry cylindrical body in which the swelling leads to a large shape change. The cylinder is numerically simulated as an axisymmetric body. Figure 12.1 shows the initial domain of height $H_0 = 5$ mm and radius $R_0 = 2.5$ mm. We make use of appropriate symmetries so that only one half of the axisymmetric slice is modeled, and meshed with 8 triangular elements along each edge.

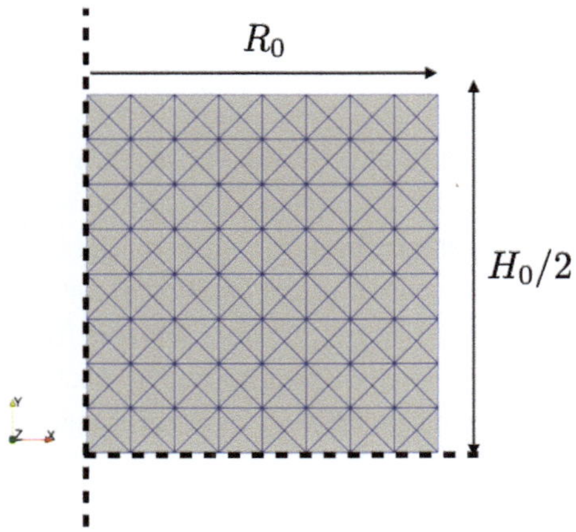

Fig. 12.1 Geometry and mesh used in the free-swelling of an axisymmetric cylinder simulation.

Initial conditions: For an initially dry body the polymer volume fraction is set as $\phi_0 = 0.999$. Equation (10.9.3), when expressed in terms of the normalized concentration $\hat{c} = \Omega c_R$, is

$$\phi = \frac{1}{1+\hat{c}}, \tag{12.1.1}$$

and this gives an initial value of $\hat{c}_0 = 0.001$ for the normalized concentration. Further, equation (10.9.25) for the chemical potential, with $\mu^{\text{ref}} = 0, \chi = 0.1, \phi_0 = 0.999$, and $J^e = 1$, gives the initial value of the normalized chemical potential $\hat{\mu} = \mu/(R\vartheta)$ as

$$\hat{\mu}_0 = \ln(1-\phi_0) + \phi_0 + \chi\phi_0^2 = -5.8. \tag{12.1.2}$$

To these initial conditions we add the initial condition of zero displacement, and take the complete set of initial conditions for the problem at hand as[2]

$$\mathbf{u}_0 = \mathbf{0}, \qquad \hat{\mu}_0 = -5.8, \quad \text{and} \quad \hat{c}_0 = 0.001. \tag{12.1.3}$$

Boundary conditions: The boundary conditions are as follows:

- Mechanical boundary conditions: (i) the left and bottom faces of the domain shown in Fig. 12.1 are prescribed appropriate symmetry conditions, and (ii) the top and right faces of the domain are traction-free.
- Normalized chemical potential boundary conditions: (i) the left and bottom faces of the domain are prescribed zero fluid flux due to symmetry, and (ii) the top and right faces of the domain are prescribed a normalized chemical potential history in the form

$$\bar{\hat{\mu}}(t) = \hat{\mu}_0 \exp(-t/300), \tag{12.1.4}$$

  so that the normalized chemical potential on those faces increases smoothly from its initial value $\hat{\mu}_0 = -5.8$ and approaches $\bar{\hat{\mu}} = 0$. This models the surroundings of the body as a zero-pressure fluid bath.

Figures 12.2(a)–(d) show snapshots of the body at $t = 0$ h, 1.5 h, 3 h, and 6 h, respectively, together with contours of the polymer volume fraction ϕ at these times, and Figs. 12.3(a)–(d) show snapshots of the body at the same times, but with contours of the volume ratio J. These figures clearly show the inhomogeneous transient swelling of the body as it absorbs fluid from the surroundings. The most notable feature that one observes from these results is the large deformations that occur as the gel swells. Specifically, Fig. 12.2(d) and Fig. 12.3(d) show a minimum value of the polymer volume fraction of $\phi \approx 0.18$ and a maximum value of the

[2] In all the subsequent example problems discussed in this section we will use the initial conditions (12.1.3) for an initially dry reference body.

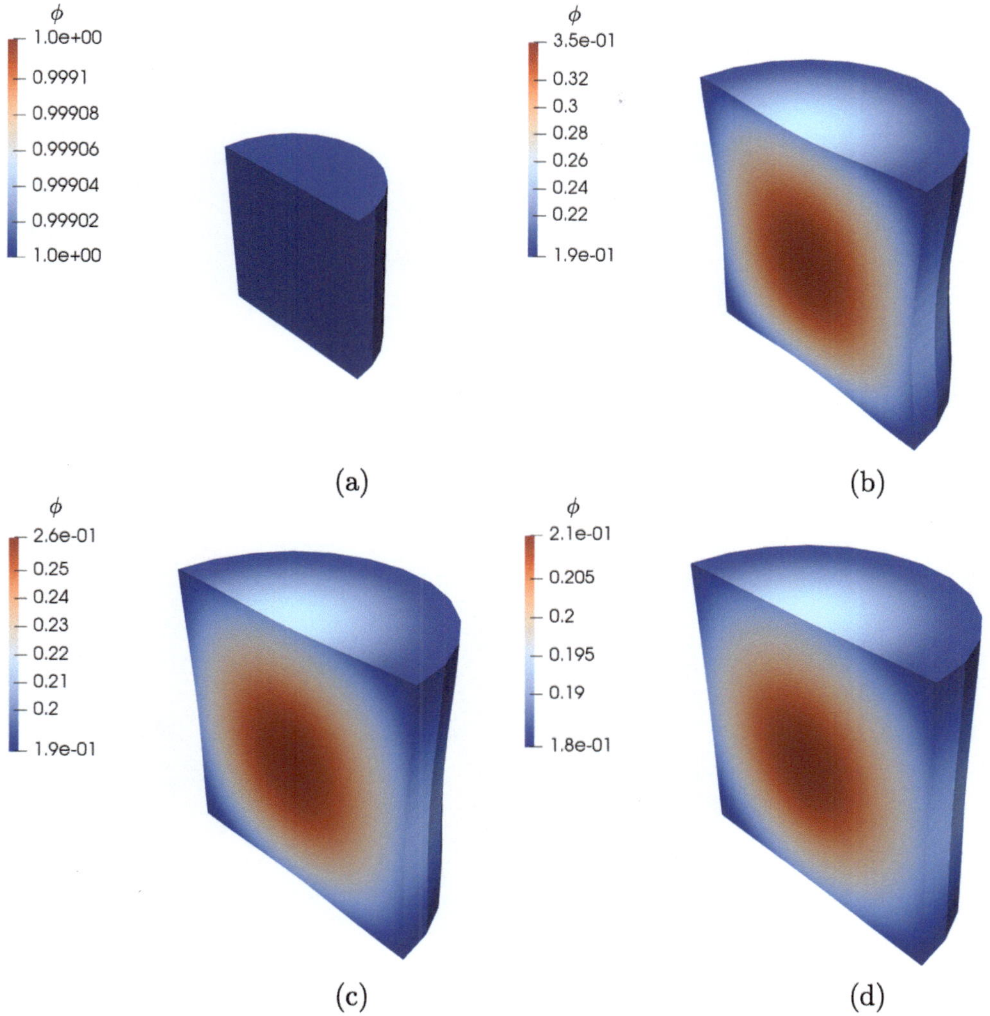

Fig. 12.2 Swollen configurations and contours of the polymer volume fraction, ϕ, at (a) $t = 0$ h, (b) 1.5 h, (c) 3 h, and (d) 6 h. The axisymmetric body has been reflected and rotationally extruded about the axis of symmetry to show one-half of the full geometry. Note the change in scale for the contours of ϕ in each subfigure.

volume ratio of $J \approx 5.5$. This indicates that after 6 hours of free-swelling certain regions of the body are roughly 18% solid and 82% fluid, and in the more fluid-filled regions the volume has increased by ≈ 5.5 times its initial value.

Also note that the edges of the cylinder tend to swell the fastest because of the enhanced flux from both the lateral and top surfaces. Further, the rate of swelling slows with time as the restorative forces due to the elastic deformation of the gel begin to more significantly oppose the swelling forces due to the change in the chemical potential. Equilibrium will eventually be achieved when these two forces balance each other.

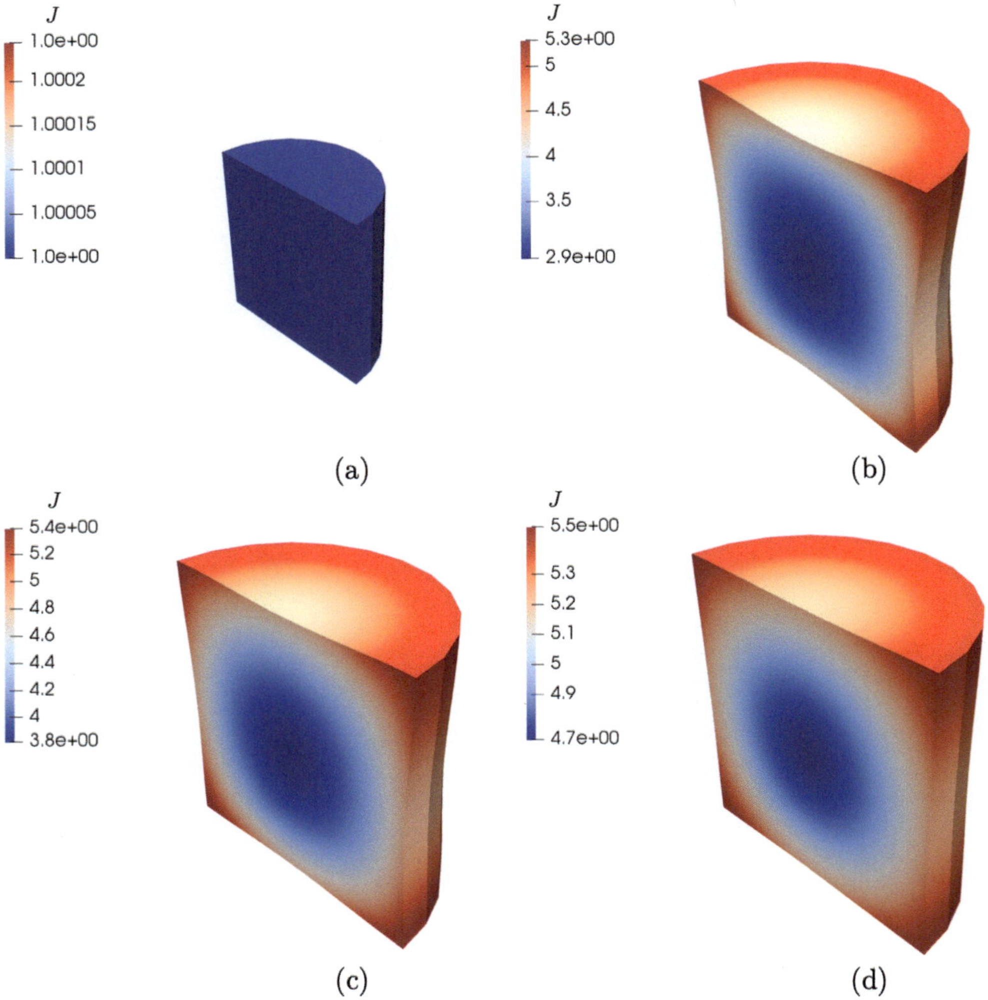

Fig. 12.3 Swollen configurations and contours of the volume ratio, J, at (a) $t = 0$ h, (b) 1.5 h, (c) 3 h, and (d) 6 h. The axisymmetric body has been reflected and rotationally extruded about the axis of symmetry to show one-half of the full geometry. Note the change in scale for the contours of J in each subfigure.

12.2 Free-swelling of a cube

This example mimics the prior example of an axisymmetric free-swelling of a cylinder, but here we consider a full three-dimensional simulation of free-swelling of a cube. The cube has an initial edge length of $L_0 = 5$ mm, and using the symmetries of the problem we model only one eighth of the cube and mesh it with 10 tetrahedral elements along each edge, as shown in Fig. 12.4.

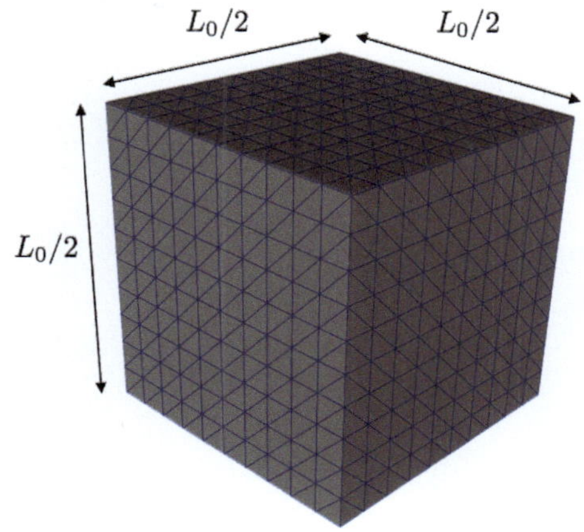

Fig. 12.4 Geometry and mesh used for this 3D free-swelling simulation.

As in the previous example, the initial conditions are

$$\mathbf{u}_0 = \mathbf{0}, \qquad \hat{\mu}_0 = -5.8, \quad \text{and} \quad \hat{c}_0 = 0.001, \tag{12.2.1}$$

and the boundary conditions are as follows:

- Mechanical boundary conditions: (i) the left, back, and bottom faces of the domain are prescribed appropriate symmetry conditions, and (ii) the top, front, and right faces of the domain are traction-free.
- Normalized chemical potential boundary conditions: (i) the left, back, and bottom faces of the domain are prescribed zero fluid due to symmetry, and (ii) the top, front, and right faces of the domain are prescribed a normalized chemical potential history in the form

$$\bar{\hat{\mu}}(t) = \hat{\mu}_0 \exp(-t/400), \tag{12.2.2}$$

so that the normalized chemical potential on those faces increases smoothly and approaches $\bar{\hat{\mu}} = 0$. This models the surroundings as a zero-pressure fluid bath.

Figures 12.5(a)–(d) show snapshots of the polymer volume fraction ϕ at $t = 0\,\text{h}, 0.5\,\text{h}, 1.5\,\text{h}$, and 3 h, respectively, and Figs. 12.6(a)–(d) show snapshots of the body at the same times, but with contours of the volume ratio J. As in the previous example, these figures show the inhomogeneous transient swelling of the body as it absorbs fluid from the surroundings, with similar conclusions drawn.

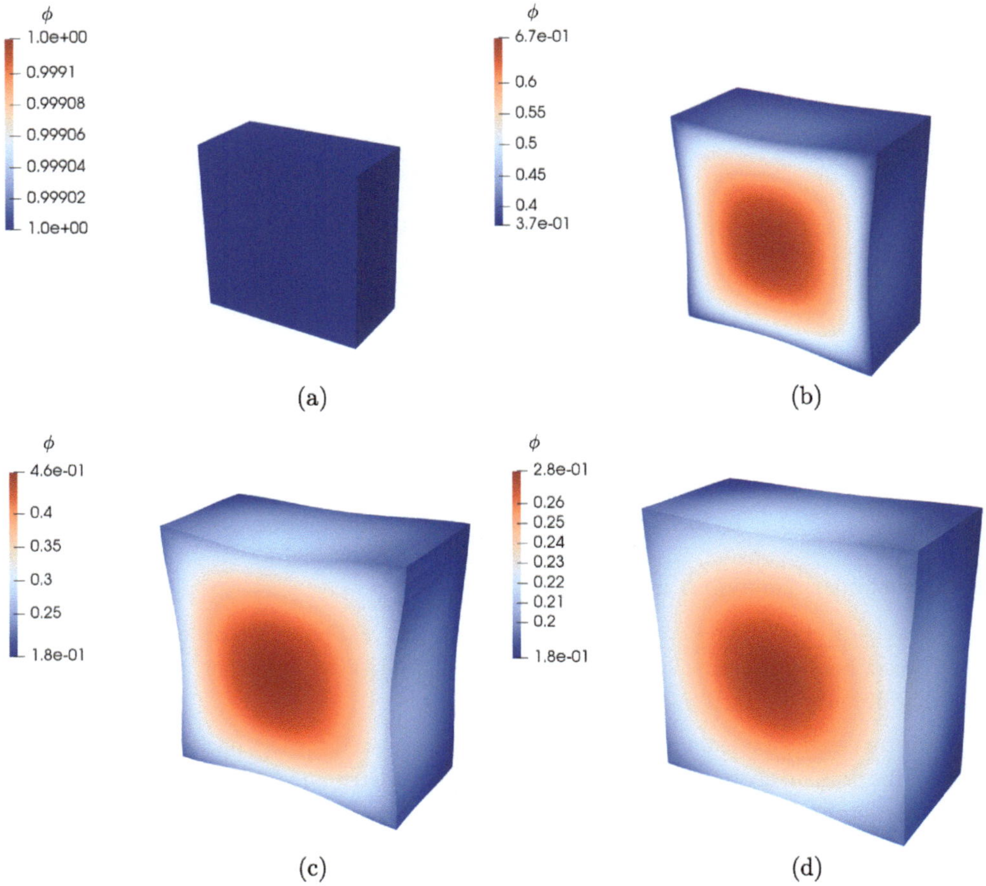

Fig. 12.5 Swollen configurations and contours of the polymer volume fraction, ϕ, at (a) 0 h, (b) 0.5 h, (c) 1 h, and (d) 3 h. The body has been reflected about two planes of symmetry to show one-half of the full geometry. Note the change in scale for the contours of ϕ in each subfigure.

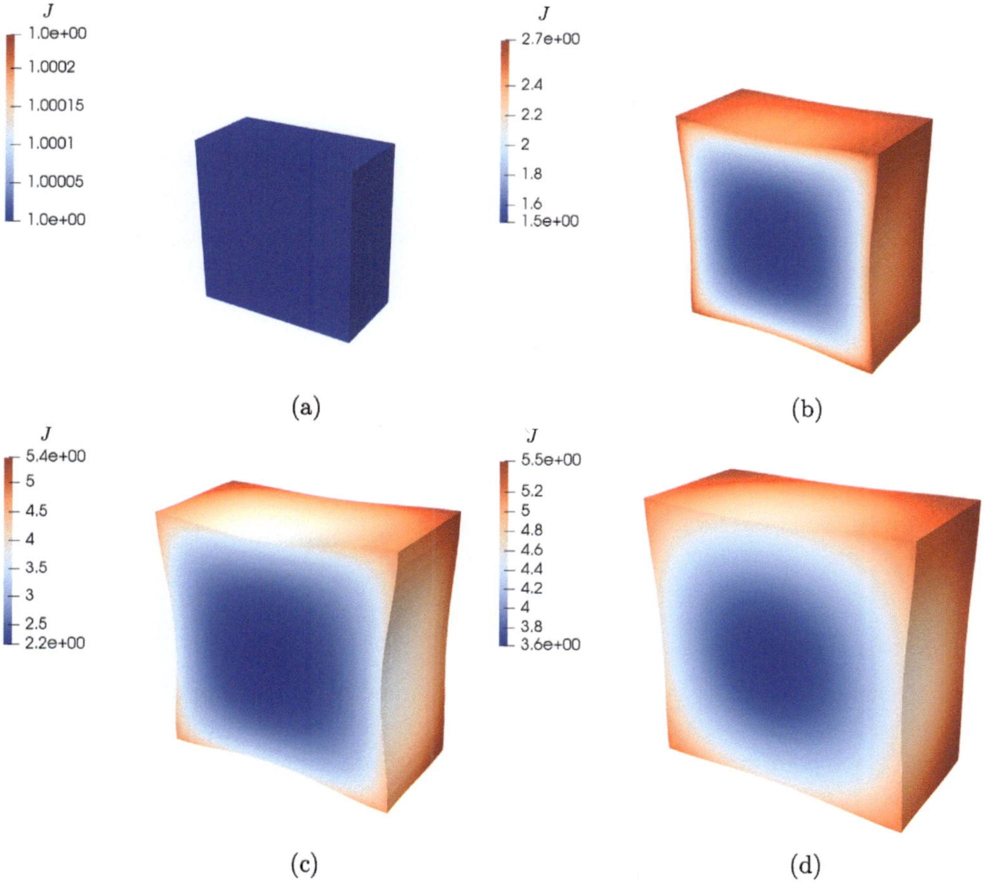

Fig. 12.6 Swollen configurations and contours of the volume ratio, J, at (a) 0 h, (b) 0.5 h, (c) 1 h, and (d) 3 h. The body has been reflected about two planes of symmetry to show one-half of the full geometry. Note the change in scale for the contours of J in each subfigure.

12.3 Plane-strain swelling of a bilayer beam

In this example we simulate the non-uniform swelling of a bilayer beam with different material properties in each layer. The non-uniform swelling gives rise to a pronounced bending of the beam. Such a bilayer beam may be used as an element in the design of soft actuators.

Figure 12.7(a) shows a schematic of the initial configuration of an bilayer beam in which each layer has the same thickness. Here we take the beam to have a length $L_0 = 100$ mm and height $H_0 = 5$ mm. The top and bottom halves of the beam have different material properties:

- The top half is modeled as a *swellable gel* with material parameters listed in Table 12.2.
- The bottom half is modeled as a *non-swellable* elastomer. This region is prescribed a value of $G_0 = 50 \times 10^3$ kPa and a high value of $\chi = 0.9$ which makes it essentially non-swellable; all other material parameters have the same values as those for the top half.

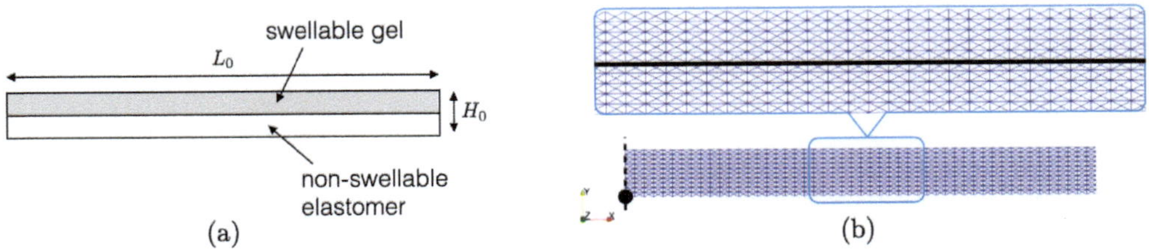

Fig. 12.7 (a) Geometry and (b) finite element mesh used for the plane-strain swelling of a bilayer beam simulation. The dashed line indicates the plane of symmetry.

The beam will be pinned at its midspan. The simulation is carried out under plane-strain conditions, and due to symmetry, only one half of the domain is modeled. A finite element mesh which has 50 triangular elements along the length of the beam, and 10 elements along its height is used. Figure 12.7(b) shows the mesh, and a zoom-in is provided for a clearer picture of the mesh.

The initial conditions for an dry reference body are

$$\mathbf{u}_0 = \mathbf{0}, \qquad \hat{\mu}_0 = -5.8, \quad \text{and} \quad \hat{c}_0 = 0.001, \tag{12.3.1}$$

and the boundary conditions are as follows:

- Mechanical boundary conditions: (i) due to symmetry the left face of the domain is prescribed appropriate symmetry conditions; (ii) the bottom left corner, indicated by the black point in Fig. 12.7(b), is pinned; and (iii) all other faces of the domain are traction-free.
- Normalized chemical potential boundary conditions: (i) the left, right, and bottom faces of the domain are prescribed zero fluid flux, and (ii) the top face of the domain is prescribed a normalized chemical potential history in the form

$$\bar{\hat{\mu}}(t) = \hat{\mu}_0 \exp(-t/300), \tag{12.3.2}$$

so that the normalized chemical potential on that face increases smoothly and approaches $\bar{\hat{\mu}} = 0$. This models the top face of the beam exposed to a fluid bath.

Figures 12.8(a)–(c) show snapshots of the body at $t = 0$ s, 660 s, and 1300 s, respectively, together with contours of the polymer volume fraction ϕ. This figure clearly shows the large bending deformation due to the mismatch in volume expansion between the materials comprising the top and bottom halves of the beam.

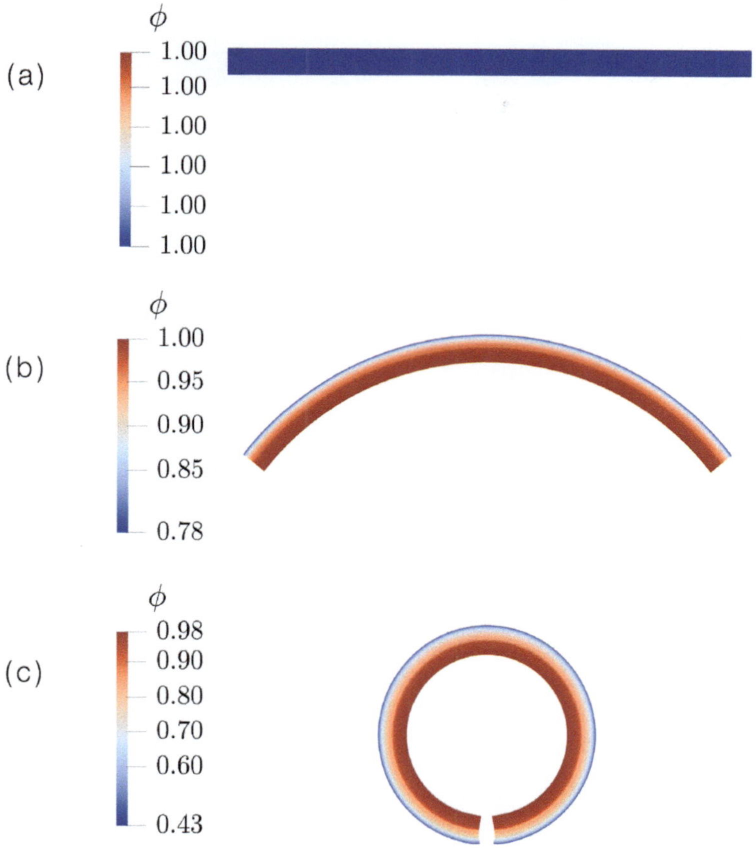

Fig. 12.8 Swollen configurations and contours of the polymer volume fraction ϕ at (a) $t = 0$ s, (b) 660 s, and (c) 1300 s. The modeled domain has been reflected about the plane of symmetry to show the full geometry. Note the change in scale for the contours of ϕ in each subfigure.

12.4 Buckling of constrained cylindrical tubes

In this example we simulate the buckling of cylindrical tubes when one end-face of the tube is geometrically constrained, while the other end-face is subjected to chemical potential boundary conditions which cause the tube to swell. This example is motivated by the experiments of Lee et al. (2012), and highlights how inhomogeneous swelling may cause buckling-type instabilities.

Guided by the experiments of Lee et al. (2012), with reference to Fig. 12.9 we consider two cylindrical tubes:

(a) A "thin" tube with dimensions $H_0 = 0.6$ mm, $D_0 = 4.636$ mm, and $T_0 = 0.206$ mm.

(b) A "thick" tube with dimensions $H_0 = 1.2$ mm, $D_0 = 4.636$ mm, and $T_0 = 0.309$ mm.

As we will see, the thin tube will buckle as it swells, while the thick tube will swell in a more stable manner. Using the symmetry of the problem, only one quarter of the geometry of each tube is meshed.

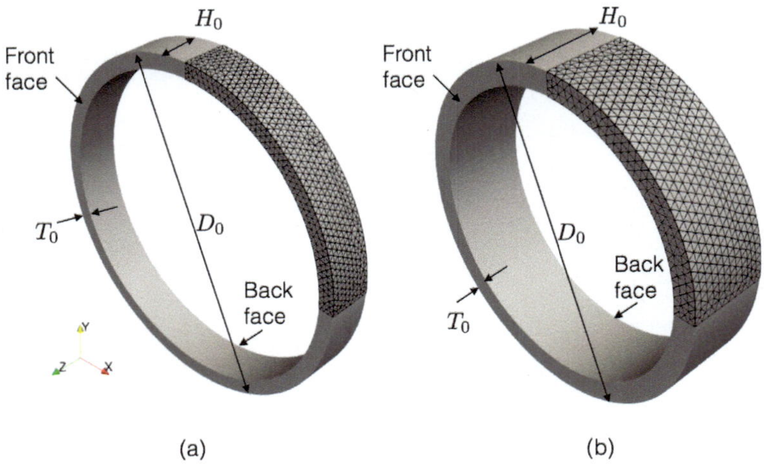

Fig. 12.9 The initial geometry and mesh for: (a) a thin tube with $H_0 = 0.6$ mm, $D_0 = 4.636$ mm, and $T_0 = 0.206$ mm, and (b) a thick tube with $H_0 = 1.2$ mm, $D_0 = 4.636$ mm, and $T_0 = 0.309$ mm. The modeled domain has been reflected about both symmetry planes to show the full geometry.

The initial conditions for these dry reference bodies are

$$\mathbf{u}_0 = \mathbf{0}, \qquad \hat{\mu}_0 = -5.8, \quad \text{and} \quad \hat{c}_0 = 0.001, \tag{12.4.1}$$

and the boundary conditions are as follows:

- Mechanical boundary conditions: (i) the back face of each tube is fully constrained; (ii) the symmetry planes are prescribed appropriate symmetry conditions; and (iii) all other faces of the domain are traction-free.
- Normalized chemical potential boundary conditions: (i) the front face of each tube is prescribed a normalized chemical potential history in the form

$$\bar{\hat{\mu}}(t) = \hat{\mu}_0 \exp(-t/60), \tag{12.4.2}$$

  so that the chemical potential on that face increases smoothly and approaches $\bar{\hat{\mu}} = 0$. (ii) All other faces are prescribed zero fluid flux.

Figures 12.10(a)–(d) for the thin tube show snapshots of the deformed body at $t = 0$ s, 220 s, 250 s, and 900 s, respectively, with contours of the polymer volume fraction ϕ. This figure shows that the constrained thin tube buckles as it swells. Correspondingly, Figs. 12.11(a)–(d) for the thick tube show snapshots of the deformed body at the same times. This figure shows that the constrained thick tube does not buckle (significantly).

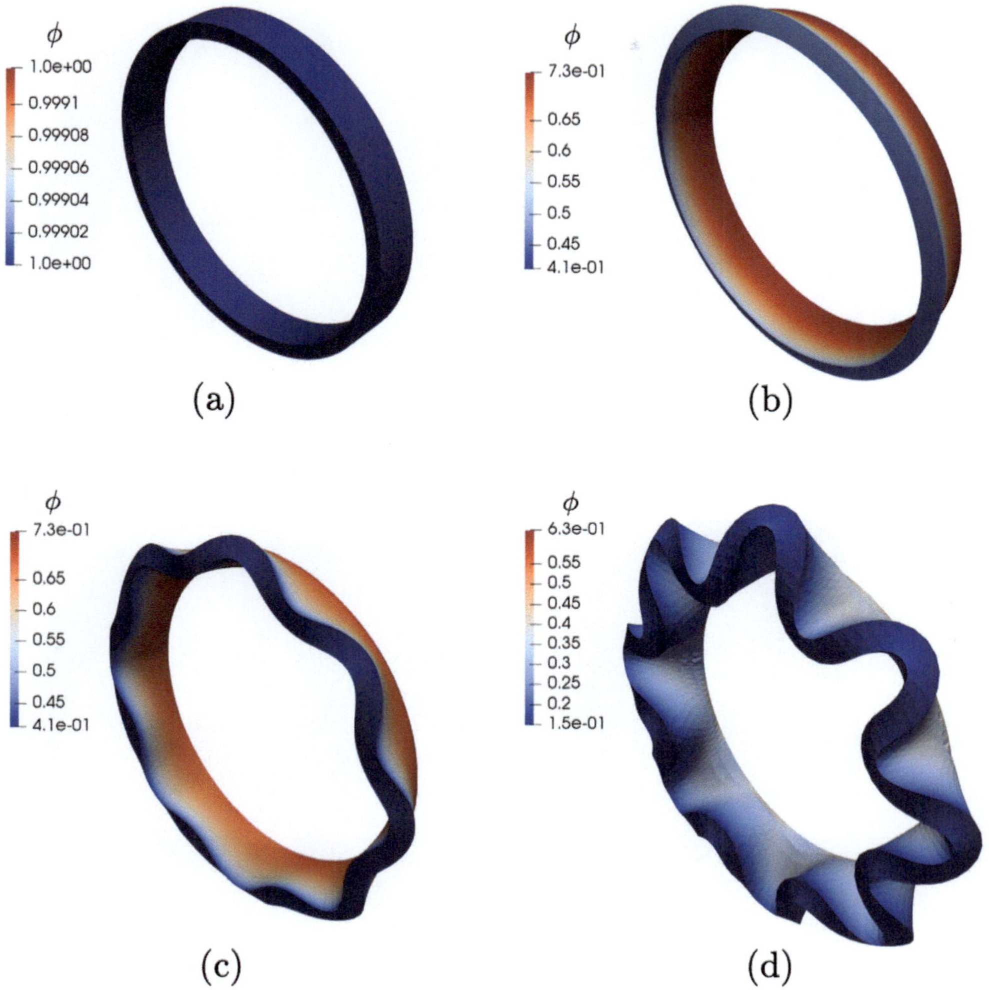

Fig. 12.10 Swollen configurations and contours of the polymer volume fraction, ϕ, at (a) $t = 0\,\mathrm{s}$, (b) 220 s, (c) 250 s, and (c) 900 s for the thin tube. The modeled domain has been reflected about the plane of symmetry to show the full geometry. Note the change in scale for the contours of ϕ in each subfigure.

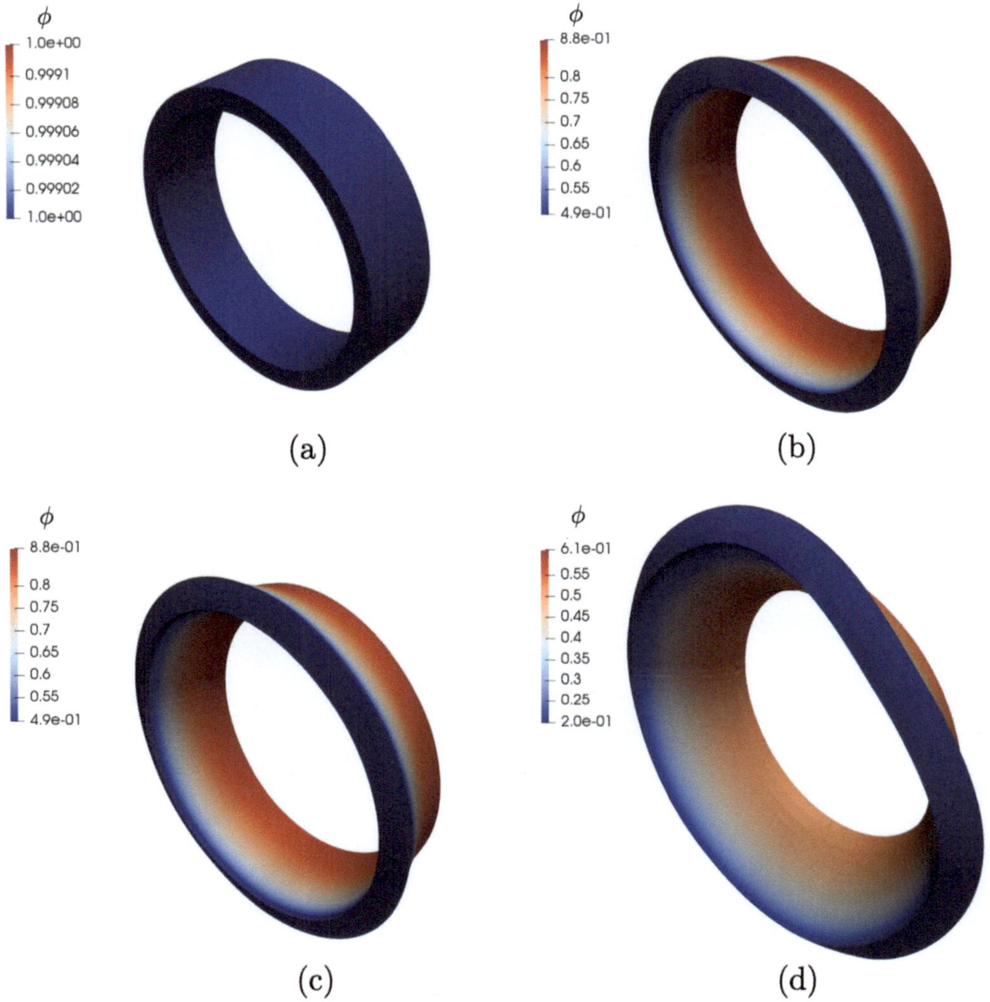

Fig. 12.11 Swollen configurations and contours of the polymer volume fraction, ϕ, at (a) $t = 0\,\text{s}$, (b) 220 s, (c) 250 s, and (c) 900 s for the thick tube. The modeled domain has been reflected about the plane of symmetry to show the full geometry. Note the change in scale for the contours of ϕ in each subfigure.

12.5 **Swell-buckling of a thin disk**

In this example we simulate the out-of-plane buckling of a thin disk when the inner surface of the disk is geometrically constrained, while the top and bottom faces of the disk are subjected to chemical potential boundary conditions which cause the disk to swell. With reference to Fig. 12.12, the disk has dimensions of $D_o = 100\,\text{mm}$, $D_i = 60\,\text{mm}$, and $T_0 = 1\,\text{mm}$. Due to the symmetry of the problem, only one quarter of the geometry is meshed, using three tetrahedral elements through the thickness, and 40 elements across the width.

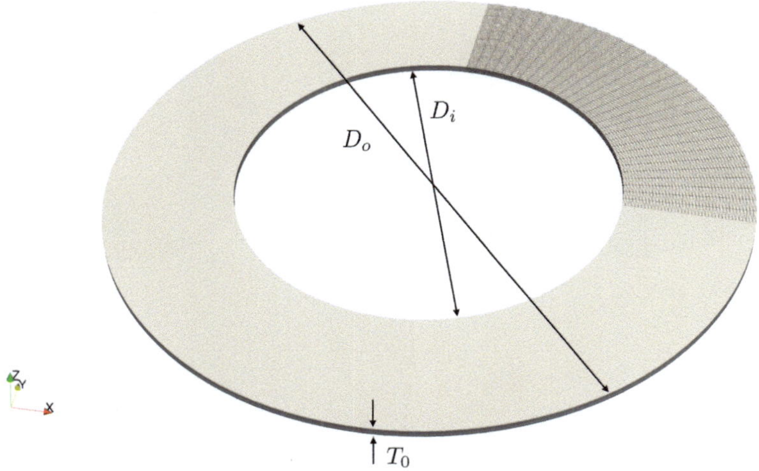

Fig. 12.12 The initial geometry and mesh with dimensions $D_o = 100$ mm, $D_i = 60$ mm, and $T_0 = 1$ mm used for the swell-buckling of a thin disk. Only one-quarter of the disk is modeled. The modeled domain has been reflected about both symmetry planes to show the full geometry.

For a dry reference body the initial conditions are

$$\mathbf{u}_0 = \mathbf{0}, \qquad \hat{\mu}_0 = -5.8, \quad \text{and} \quad \hat{c}_0 = 0.001, \tag{12.5.1}$$

and the boundary conditions in this problem are as follows:

- Mechanical boundary conditions: (i) the face at the inner diameter of the disk is fully constrained; (ii) the symmetry planes are prescribed appropriate symmetry conditions; and (iii) all other faces of the domain are traction-free.
- Normalized chemical potential boundary conditions: (i) the top and bottom faces of the domain are prescribed a normalized chemical potential history in the form

$$\bar{\hat{\mu}}(t) = \hat{\mu}_0 \exp(-t/300), \tag{12.5.2}$$

so that the chemical potential on these faces increases smoothly and approaches $\bar{\hat{\mu}} = 0$, and (ii) all other faces are prescribed zero fluid flux.

Figures 12.13(a)–(d) show snapshots of the body at $t = 0$ s, 575 s, 750 s, and 1000 s, respectively, together with contours of the out-of-plane displacement u_3. It is evident that due to the constraint on the inner surface of the disk, to accommodate the increase in volume due to swelling the disk buckles out of plane, with maximum out-of-plane displacements an order of magnitude larger than the thickness.

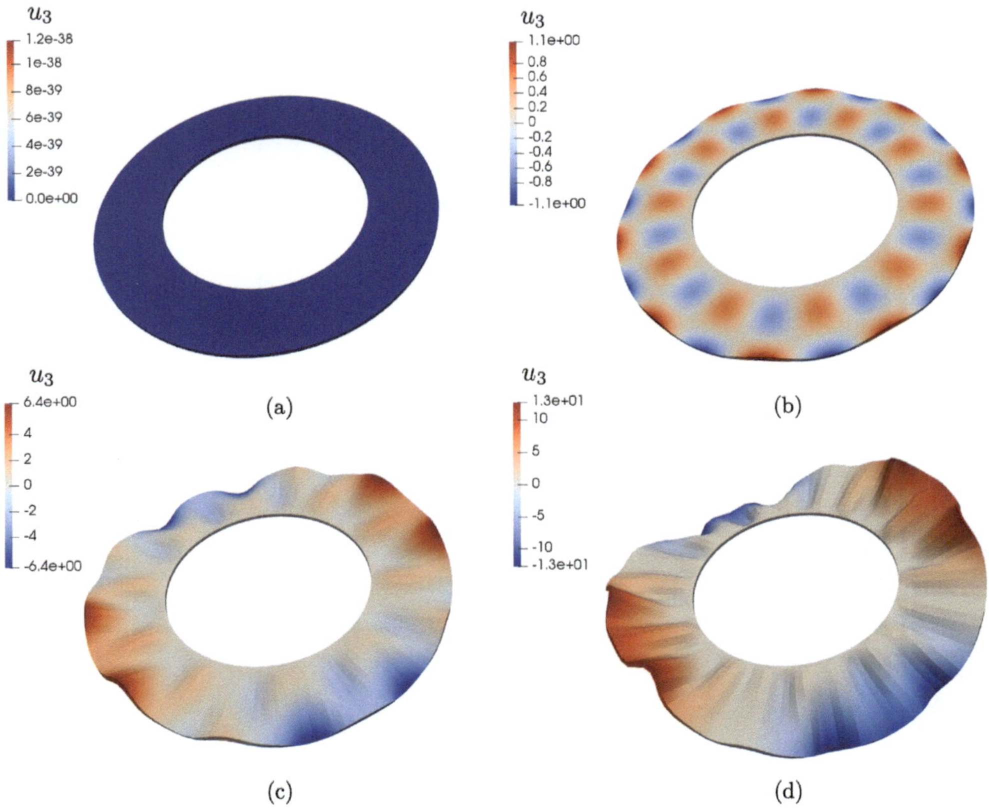

Fig. 12.13 Swollen configurations and contours of the out-of-plane displacement u_3 at (a) $t = 0$ s, (b) 575 s, (c) 750 s, and (d) 1000 s. The modeled domain has been reflected about the planes of symmetry to show the full geometry. Note the change in scale for u_3 in each subfigure.

12.6 **Swelling of an elliptical inclusion in a non-swellable soft matrix**

This example is motivated by the problem of the growth of a tumor embedded in a soft matrix. Specifically, we simulate the swelling of an inclusion embedded in a non-swellable soft matrix, and for simplicity we assume plane-strain conditions. Figure 12.14 shows the geometry: an elliptical inclusion with a major semi-axis $a = 2$ mm and minor semi-axis $b = 1$ mm, which is embedded in a square matrix with edge length $L_0 = 100$ mm.

The material parameters for the matrix and the inclusion used in this simulation are listed in Table 12.3. Also, the Flory–Huggins parameter χ and the molar volume Ω for the matrix are chosen such that the matrix behaves as a non-swellable elastomeric material, while using the same set of constitutive equations as that for the swellable inclusion. For simplicity we do not model diffusion of the fluid in the matrix.

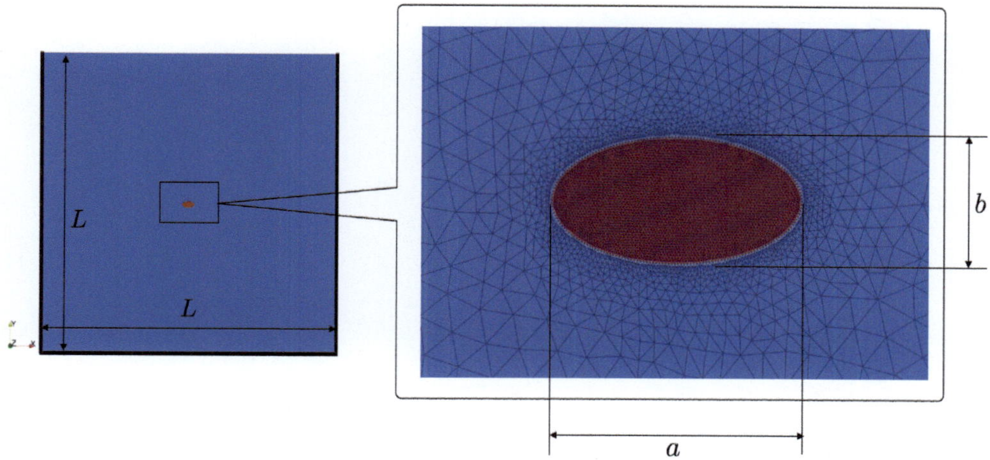

Fig. 12.14 The initial plane-strain geometry and a zoom-in of the mesh around a small elliptical inclusion (red) embedded in a soft matrix (blue). The dimensions are: $L = 100$ mm, $a = 2$ mm, and $b = 1$ mm.

Table 12.3 Material parameters for the matrix and the inclusion. The parameters are chosen such that the matrix approximates a non-swellable elastomeric material, while the inclusion is swellable.

Quantity	Matrix	Inclusion
G_0, ground-state shear modulus	0.3 kPa	0.5 kPa
λ_L, locking stretch	1000	1000
K, bulk modulus $(1000 \times G_0)$	$1000 \times 0.3\,\text{kPa}$	$1000 \times 0.5\,\text{kPa}$
Ω, molar volume	$0.1 \times 10^5\,\text{mm}^3/\text{mol}$	$1 \times 10^5\,\text{mm}^3/\text{mol}$
D, diffusivity	$5 \times 10^{-3}\,\text{mm}^2/\text{s}$	$5 \times 10^{-3}\,\text{mm}^2/\text{s}$
χ, Flory–Huggins parameter	0.9	0.1

As in the previous examples, for a dry reference configuration the initial conditions are

$$\mathbf{u}_0 = \mathbf{0}, \qquad \hat{\mu}_0 = -5.8, \quad \text{and} \quad \hat{c}_0 = 0.001, \tag{12.6.1}$$

and the boundary conditions for this problem are as follows:

- Mechanical boundary conditions: (i) the left, bottom, and right edges of the matrix are fixed; and (ii) the top edge of the matrix is traction-free.
- Normalized chemical potential boundary conditions: (i) the interface between the matrix and inclusion is prescribed a normalized chemical potential history in the form

$$\bar{\hat{\mu}}(t) = \hat{\mu}_0 \exp(-t/300), \tag{12.6.2}$$

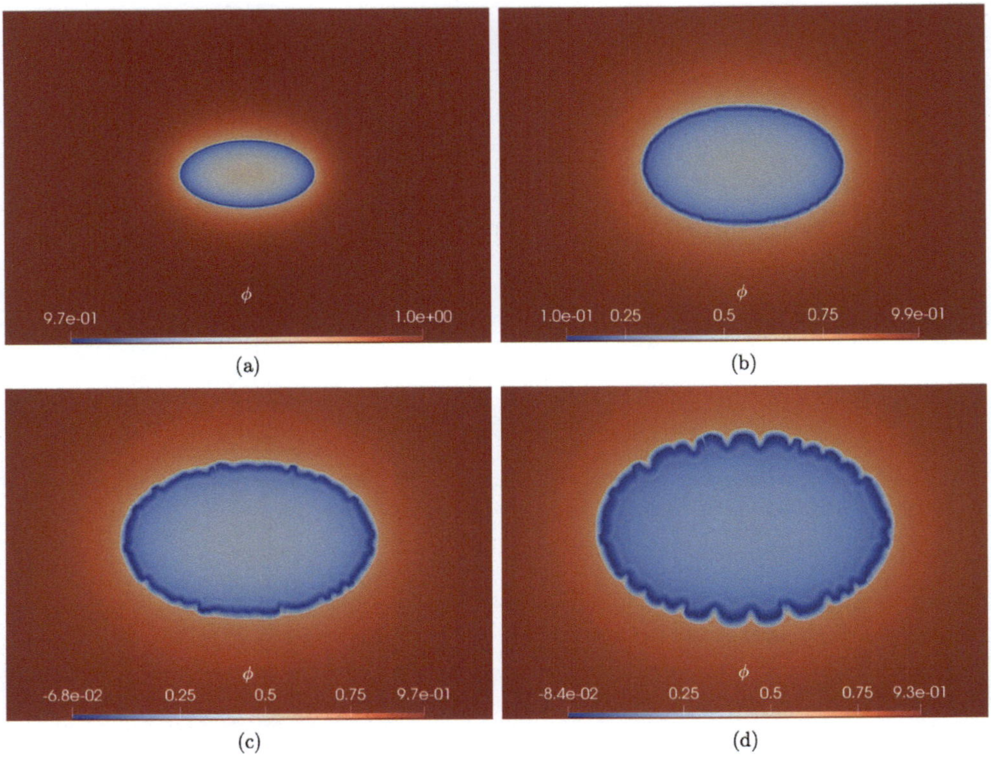

Fig. 12.15 Swollen configurations with contours of the polymer volume fraction ϕ at (a) $t = 250\,\text{s}$, (b) 1800 s, (c) 3600 s, and (d) 7200 s. Note the change in scale for ϕ in each subfigure.

so that the chemical potential on the interface increases smoothly and approaches $\bar{\tilde{\mu}} = 0$, and (ii) all outer edges of the matrix are at zero flux. For simplicity we do not model diffusion of the fluid in the matrix.

Figures 12.15(a)–(d) show snapshots of a region around the inclusion with contours of the polymer volume fraction ϕ at $t = 250$ s, 1800 s, 3600 s, and 7200 s, respectively. As the inclusion grows, the interface region between the inclusion and the matrix starts to roughen and crease. The cause of this creasing is revealed in Figs. 12.16(a)–(d) which show contours of the volume ratio J, the pressure p, and the Piola stress components P_{11} and P_{22}, respectively, at a time of $t = 3600$ s. Note the large compressive values of P_{11} and P_{22} that are generated in the vicinity of the inclusion–matrix interface. As the inclusion swells the matrix retards the volume increase, and this leads to the generation of compressive stresses that cause creasing of the inclusion–matrix interface. Lastly, Fig. 12.17 shows the change in size of the major and minor semi-axis of the elliptical inclusion over time—stretch levels greater than 2 develop in each direction.[3]

[3] Even as the interface between the inclusion and matrix creases.

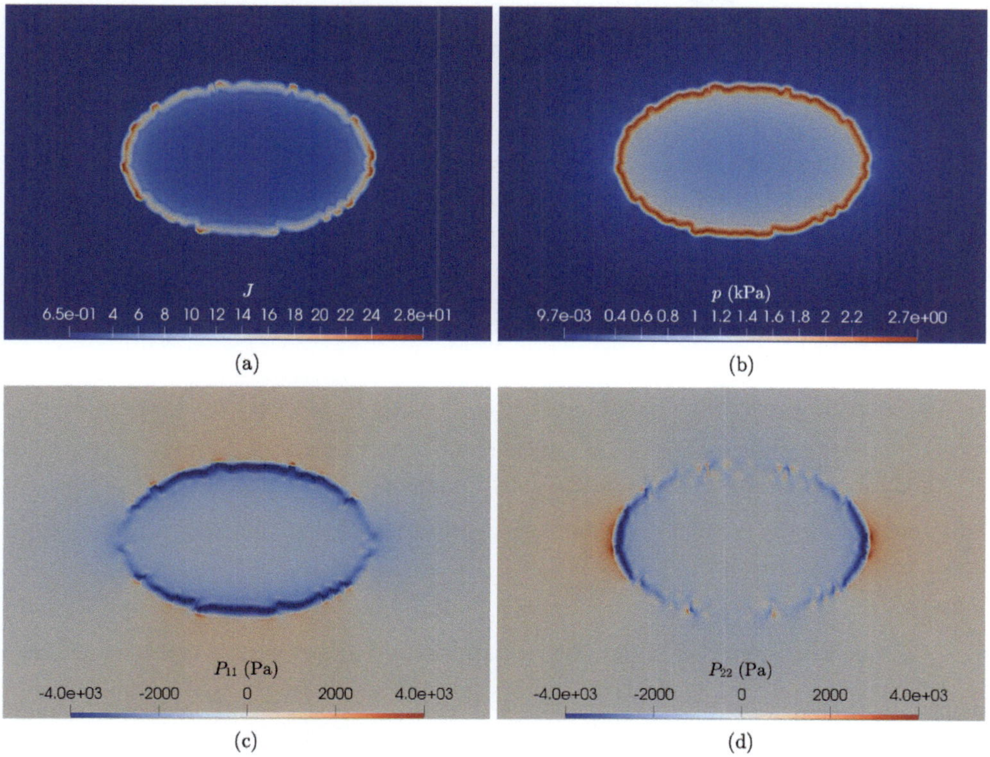

Fig. 12.16 Swollen configuration at $t = 3600$ s with contours of: (a) the volume ratio J; (b) the pressure p; (c) the Piola stress component P_{11}; and (d) the Piola stress component P_{22}. Note the large compressive values of P_{11} and P_{22}, and it is these compressive stresses which lead to the creasing of the inclusion–matrix interface region.

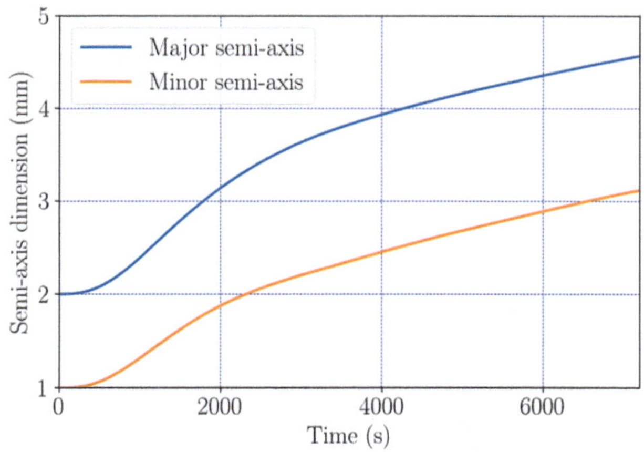

Fig. 12.17 Change in dimensions of the major and minor semi-axes of the elliptical inclusion as a function of time.

12.7 Free-swelling of an axisymmetric cylinder followed by compression

As a final example we consider the problem of free-swelling of an initially dry cylindrical body, after which it is axially compressed so that the fluid exudes from the body over time. We approach the numerical simulation of this problem in two different ways:

1. The first simulation uses a two-step method:
   - Step (i): Start from an initially dry body and allow it to free-swell for 24 hours. Followed by
   - Step (ii): Axial compression.
2. The second simulation uses a one-step method: Start from a homogeneously pre-swollen body at equilibrium and subject it to axial compression.

This example helps to verify the methodology discussed in Section 10.A regarding the modeling of an initially pre-swollen body.

12.7.1 Free-swelling of an initially dry body followed by compression

In this numerical simulation the cylinder is treated as an axisymmetric body of height $H_0 = 2.5\,\text{mm}$ and radius $R_0 = 2.5\,\text{mm}$, cf. Fig. 12.18. The cylinder is compressed between two platens: a fixed impermeable and frictionless bottom platen, and a vertically movable porous and frictionless top platen. We make use of appropriate symmetries so that only an axisymmetric slice is modeled. The material parameters used in the simulation are those shown in Table 12.2, with the exception that $\lambda_L = 10^6$ to model a Neo-Hookean material.

To impose the compression the top platen is initially located at $X_2^{\text{platen}} = 4.5\,\text{mm}$,[4] and it is prescribed a linear displacement ramp of 1.5 mm over 2 hours to compress the body, and is then held fixed for 6 hours to observe relaxation due to fluid leaving the body. The normal contact interaction between the body and the porous platen is modeled using a penalty method similar to what was described in Section 6.2, here using a penalty coefficient of $k_{\text{con}} = 10^6\,\text{kPa/mm}^2$.

For an initially dry reference body the initial conditions are,

$$\mathbf{u}_0 = \mathbf{0}, \qquad p_0 = 0, \qquad \hat{\mu}_0 = -5.8, \quad \text{and} \quad \hat{c}_0 = 0.001, \tag{12.7.1}$$

and the boundary conditions are as follows:

- Mechanical boundary conditions: (i) the left face of the domain shown in Fig. 12.18 is prescribed appropriate symmetry conditions, (ii) the bottom face is prescribed $\bar{u}_2 = 0$, and (iii) the top and right faces of the domain are traction-free, unless in contact with the porous platen.
- Normalized chemical potential boundary conditions: (i) the left and bottom faces of the domain are prescribed zero fluid flux due to symmetry and (ii) the top and right faces of

[4] A location chosen since it is just above the height of the body after the equilibrium free-swelling step.

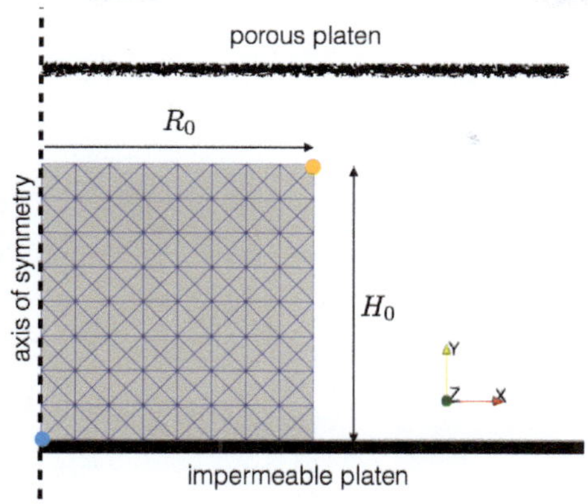

Fig. 12.18 Geometry and mesh used in the free-swelling followed by compression and hold, of an axisymmetric cylinder simulation. The locations marked in blue and orange will be used to extract data for plotting.

the domain are prescribed a normalized chemical potential history in the form

$$\bar{\mu}(t) = \hat{\mu}_0 \exp(-t/300), \tag{12.7.2}$$

so that the normalized chemical potential on those faces increases smoothly from its initial value $\hat{\mu}_0 = -5.8$ and approaches $\bar{\mu} = 0$. This models the surroundings of the body as a zero-pressure fluid bath with a porous platen that allows for unimpeded fluid movement.

12.7.2 Compression of a homogeneously pre-swollen body

For a body which is in a homogeneously pre-swollen equilibrium state, in Section 10.A we discussed how to calculate the initial values of the polymer volume fraction ϕ_0 and the residual pressure p_0 in the body. Thus, solving the implicit equation (10.A.11) we obtain

$$\phi_0 = 0.1817.$$

Then use of (10.9.3) allows for computation of the normalized concentration

$$\hat{c}_0 = \phi_0^{-1} - 1 = 4.5,$$

and the initial swelling stretch

$$\lambda_0 = \phi_0^{-1/3} = 1.765.$$

Next, using (10.A.10) and the value of λ_0 above, the initial pressure is computed as

$$p_0 = 385\,\text{kPa}.$$

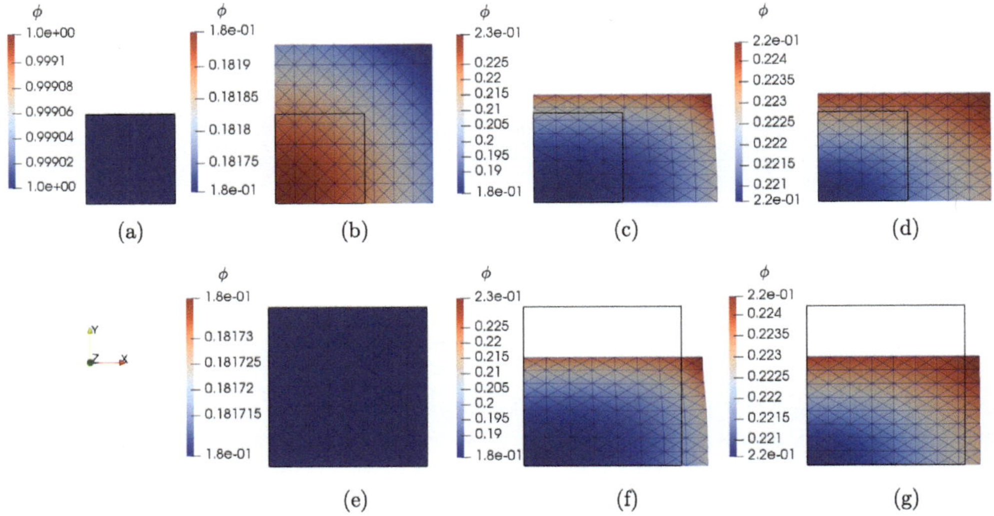

Fig. 12.19 (a)–(d) Deformed configurations at the initial condition $t = 0$ h, 24 h at the end of free swelling, 26 h at the end of the compression, and 32 h at the end of the hold, respectively, with contours of the polymer volume fraction for the initially dry simulation. (e)–(g) Deformed configurations at the initial condition $t = 0$ h which is the freely swollen state, 2 h at the end of the compression, and 8 h at the end of the hold, respectively, with contours of the polymer volume fraction for the initially dry simulation. Note that the thick black line indicates the initial geometry, and the change in scale for the polymer volume fraction in each subfigure.

To these initial conditions we add the initial condition of zero displacement. Thus, the complete set of initial conditions for the problem at hand is,

$$\mathbf{u}_0 = \mathbf{0}, \qquad p_0 = 385\,\text{kPa}, \qquad \hat{\mu}_0 = 0, \quad \text{and} \quad \hat{c}_0 = 4.503. \tag{12.7.3}$$

The boundary conditions are the same as the prior scenario; however, the single difference is here $\bar{\mu} = 0.0$ for the entire simulation.

Using a mesh similar to that shown in Fig. 12.18, the initial homogeneously pre-swollen domain has a height $H_0 = \lambda_0 \times 2.5$ mm and radius $R_0 = \lambda_0 \times 2.5$ mm. As before, we make use of appropriate symmetries so that only one half of the axisymmetric slice is modeled.

12.7.3 **Results**

Figures 12.19(a)–(d) show snapshots of the body that was initially dry, which then swells and is compressed: at times $t = 0$ h, 24 h, 26 h, and 32 h, respectively, together with contours of the polymer volume fraction ϕ at these times.

Figures 12.19(e)–(g) show snapshots of the body that was initially homogeneously pre-swollen then compressed: at times $t = 0$ h, 2 h, and 8 h, respectively, together with contours of the polymer volume fraction ϕ at these times. For ease of comparison these times are aligned from the compression and holding steps of the previous case.

Further,

- Figure 12.20(a) plots the polymer volume fraction at the center and surface—as indicated by the locations marked in blue and orange in Fig. 12.18—as a function of time from the

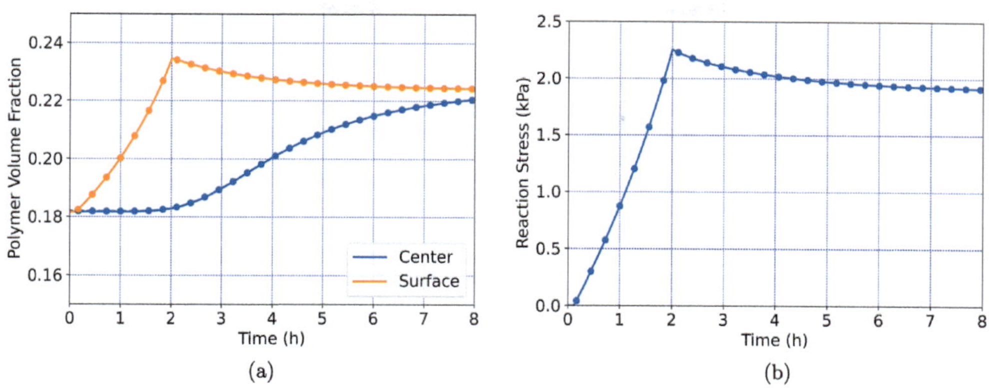

Fig. 12.20 Comparison of (a) the polymer volume fraction at the center and surface locations as indicated in Fig. 12.18, and (b) the reaction stress on the bottom platen, from the start of the compression step through the end of the hold. The markers are results from the initially dry simulation and the solid lines are results from the homogeneously pre-swollen simulation.

compression step onward for both the initially dry and the pre-swollen simulations. The markers are results from the initially dry simulation and the solid lines are results from the homogeneously pre-swollen simulation.

- Figure 12.20(b) plots the reaction stress on the bottom platen as a function of time from the compression step onward for both the initially dry and the pre-swollen simulations.

This example show the correspondence between the results from a simulation which start from an initially dry state, and a simulation that start from an initially pre-swollen state. This is especially useful for situations where the interest lies in simulating deviations from an initially swollen state and not in the swelling process itself.

Part V

Thermally responsive elastomeric gels

13 Thermally responsive elastomeric gels

13.1 Introduction

A subclass of elastomeric gels is known to be *thermally responsive*. A prime example being gels based on poly(N-isopropylacrylamide), often abbreviated as PNIPAM. Hydrogels based on PNIPAM demonstrate distinct thermoresponsive features close to a characteristic temperature known as their *lower critical solution temperature (LCST)* (cf., e.g., Schild, 1992; Wang et al., 2005).[1] When the temperature is below the LCST the hydrogel is hydrophilic and absorbs water; conversely, when the temperature is above the LCST, the hydrogel is hydrophobic and expels water, resulting in volume shrinkage. This volume change is reversible.

PNIPAM is the most studied and widely used thermoresponsive polymer, with applications ranging from tissue engineering to medication delivery and wound dressings. The significant advantage of PNIPAM is that it has an LCST of about $\approx 32^\circ$C and is easily triggered by body temperature of $\approx 37^\circ$C to give a phase transition (Ansari et al., 2022).

In addition to biomedical applications, PNIPAM-based thermally responsive gels have been used for a variety of other applications including valves in microfluidic devices and microscale self-folding "origami" devices, and they show promise for use in a variety of other emerging microscale applications (Harmon et al., 2003; Wang et al., 2005; Yoon et al., 2014; Tang et al., 2021).

In this chapter we extend our theory in Chapter 10 on poroelasticity of gels under isothermal considerations, to account for *non-isothermal conditions*. Many of the considerations remain the same as those discussed in Chapter 10, and we will not repeat them here—we will refer to the pertinent equations in Chapter 10, as necessary.

13.2 Free-energy imbalance under non-isothermal conditions

For the non-isothermal theory under consideration in this chapter, our starting point is the free-energy imbalance (10.6.13) under non-isothermal conditions, viz.

$$\dot{\psi}_{\mathrm{R}} + \eta_{\mathrm{R}}\dot{\vartheta} - \mathbf{T}_{\mathrm{R}} : \dot{\mathbf{F}} - \mu\dot{c}_{\mathrm{R}} + \frac{1}{\vartheta}\,\mathbf{q}_{\mathrm{R}}^{\mathrm{net}} \cdot \nabla\vartheta + \mathbf{j}_{\mathrm{R}} \cdot \nabla\mu \leq 0, \tag{13.2.1}$$

[1] For PNIPAM-based gels the LCST depends on the precise composition of the gel, and may be varied between 30°C and 50°C by suitably varying the composition.

Introduction to Coupled Theories in Solid Mechanics. Lallit Anand, Eric M. Stewart, Shawn A. Chester, Oxford University Press.
 DOI: 10.1093/oso/9780198986218.003.0014

where recall from (10.6.9) that

$$\mu \stackrel{\text{def}}{=} (\mu_\varepsilon - \vartheta\mu_\eta),$$
$$\mathbf{q}_{\mathrm{R}}^{\mathrm{net}} \stackrel{\text{def}}{=} \mathbf{q}_{\mathrm{R}} + \vartheta\mu_\eta \mathbf{j}_{\mathrm{R}}, \tag{13.2.2}$$

represent the *chemical potential* and *net heat flux*, respectively.

Further, as in equation (10.7.10) of Chapter 10, the stress power $\mathbf{T}_{\mathrm{R}} : \dot{\mathbf{F}}$ may be expressed as,

$$\mathbf{T}_{\mathrm{R}} : \dot{\mathbf{F}} = J^s \left(\frac{1}{2}\mathbf{T}^e : \dot{\mathbf{C}}^e\right) + \frac{1}{3}(\operatorname{tr}\mathbf{M}^e)\dot{J}^s, \tag{13.2.3}$$

where

$$\mathbf{T}^e \stackrel{\text{def}}{=} J^e \mathbf{F}^{e-1}\mathbf{T}\mathbf{F}^{e-\top}, \tag{13.2.4}$$

is the elastic second Piola stress, and

$$\mathbf{M}^e \stackrel{\text{def}}{=} J^e \mathbf{F}^{e\top}\mathbf{T}\mathbf{F}^{e-\top} \equiv \mathbf{C}^e\mathbf{T}^e \tag{13.2.5}$$

is the Mandel stress. Hence, using (13.2.3) in (13.2.1), the free-energy imbalance may be written as

$$\dot{\psi}_{\mathrm{R}} + \eta_{\mathrm{R}}\dot{\vartheta} - \frac{1}{2}(J^s\mathbf{T}^e) : \dot{\mathbf{C}}^e - \frac{1}{3}(\operatorname{tr}\mathbf{M}^e)\dot{J}^s - \mu\dot{c}_{\mathrm{R}} + \frac{1}{\vartheta}\mathbf{q}_{\mathrm{R}}^{\mathrm{net}} \cdot \nabla\vartheta + \mathbf{j}_{\mathrm{R}} \cdot \nabla\mu \le 0. \tag{13.2.6}$$

Next, upon enforcing the constitutive assumption (10.5.7), that is $\dot{J}^s = \Omega\dot{c}_{\mathrm{R}}$, the free-energy imbalance (13.2.6) becomes

$$\dot{\psi}_{\mathrm{R}} + \eta_{\mathrm{R}}\dot{\vartheta} - \frac{1}{2}(J^s\mathbf{T}^e) : \dot{\mathbf{C}}^e - \mu^{\mathrm{net}}\,\dot{c}_{\mathrm{R}} + \frac{1}{\vartheta}\mathbf{q}_{\mathrm{R}}^{\mathrm{net}} \cdot \nabla\vartheta + \mathbf{j}_{\mathrm{R}} \cdot \nabla\mu \le 0, \tag{13.2.7}$$

where, as before, we have written

$$\mu^{\mathrm{net}} \stackrel{\text{def}}{=} \mu + \Omega\left(\frac{1}{3}\operatorname{tr}\mathbf{M}^e\right), \tag{13.2.8}$$

for a *net* chemical potential.[2]

We note that under a change in frame

$$\psi_{\mathrm{R}},\ \eta_{\mathrm{R}},\ \vartheta,\ c_{\mathrm{R}},\ \mu,\ \mathbf{q}_{\mathrm{R}}^{\mathrm{net}},\ \nabla\vartheta,\ \mathbf{j}_{\mathrm{R}},\ \text{and } \nabla\mu \text{ are invariant}, \tag{13.2.9}$$

ψ_{R}, η_{R}, ϑ, c_{R}, and μ because they are scalars, and $\mathbf{q}_{\mathrm{R}}^{\mathrm{net}}$, $\nabla\vartheta$, $\mathbf{j}_{\mathrm{R}}$, and $\nabla\mu$ since they are referential vector fields. Using (10.3.6), (10.4.4), the definition $(10.2.8)_3$ of $\mathbf{C}^e$, and the definitions (13.2.4) and (13.2.5) of $\mathbf{T}^e$ and $\mathbf{M}^e$, it is straightforward to show that

[2] Note that $\operatorname{tr}\mathbf{M}^e = J^e \operatorname{tr}\mathbf{T}$.

$$\mathbf{C}^e,\ \mathbf{T}^e,\ \text{and } \mathbf{M}^e \text{ are all invariant under a change in frame.} \tag{13.2.10}$$

Thus, with the invariance properties discussed above, all quantities in the free-energy imbalance (13.2.7) are invariant under a change in frame.

13.3 Constitutive theory

13.3.1 Basic constitutive equations

Guided by the free-energy imbalance (13.2.7), we assume that the free energy ψ_R, the entropy η_R, the stress $\mathbf{T}^e$, the net chemical potential μ^{net}, the fluid flux $\mathbf{j}_R$, and the net heat flux $\mathbf{q}_R^{net}$ are given by the following set of constitutive equations:

$$\begin{aligned}
\psi_R &= \bar{\psi}_R(\mathbf{C}^e, c_R, \vartheta),\\
\eta_R &= \bar{\eta}_R(\mathbf{C}^e, c_R, \vartheta),\\
\mathbf{T}^e &= \bar{\mathbf{T}}^e(\mathbf{C}^e, c_R, \vartheta),\\
\mu^{net} &= \bar{\mu}^{net}(\mathbf{C}^e, c_R, \vartheta),\\
\mathbf{j}_R &= \bar{\mathbf{j}}_R(\mathbf{C}^e, c_R, \vartheta, \nabla\mu),\\
\mathbf{q}_R^{net} &= \bar{\mathbf{q}}_R^{net}(\mathbf{C}^e, c_R, \vartheta, \nabla\vartheta).
\end{aligned} \tag{13.3.1}$$

For simplicity, from the outset we have assumed that the constitutive equations for the free energy, entropy, stress, and net chemical potential are independent of $\nabla\vartheta$ and $\nabla\mu$, and depend only on the mechanical stretch $\mathbf{C}^e$, the fluid content c_R, and the temperature ϑ. We also ignore any cross-coupling Soret-type effects, such as a temperature gradient affecting the fluid flux or the chemical potential gradient affect the heat flux. Using the transformation rules (13.2.9) and (13.2.10) we find that the constitutive equations (13.3.1) are frame-indifferent.

13.3.2 Thermodynamic restrictions

With a view towards determining the restrictions imposed by the local free energy imbalance (13.2.7), note that

$$\dot{\psi}_R = \frac{\partial\bar{\psi}_R}{\partial\mathbf{C}^e} : \dot{\mathbf{C}}^e + \frac{\partial\bar{\psi}_R}{\partial c_R}\,\dot{c}_R + \frac{\partial\bar{\psi}_R}{\partial\vartheta}\,\dot{\vartheta}. \tag{13.3.2}$$

Substituting (13.3.1) and (13.3.2) into (13.2.7) we find that

$$\Big(\frac{\partial\bar{\psi}_R}{\partial\mathbf{C}^e} - \frac{1}{2}J^s\,\bar{\mathbf{T}}^e\Big) : \dot{\mathbf{C}}^e + \Big(\frac{\partial\bar{\psi}_R}{\partial c_R} - \bar{\mu}^{net}\Big)\,\dot{c}_R + \Big(\frac{\partial\bar{\psi}_R}{\partial\vartheta} + \bar{\eta}_R\Big)\,\dot{\vartheta} + \bar{\mathbf{j}}_R\cdot\nabla\mu + \frac{1}{\vartheta}\bar{\mathbf{q}}_R^{net}\cdot\nabla\vartheta \le 0. \tag{13.3.3}$$

This inequality is to hold for all values of $\mathbf{C}^e$, $\dot{\mathbf{C}}^e$, c_{R}, $\dot{c}_{\text{R}}$, ϑ, $\nabla\vartheta$, and $\nabla\mu$. Since $\dot{\mathbf{C}}^e$, $\dot{c}_{\text{R}}$, and $\dot{\vartheta}$ appear linearly, their "coefficients" must vanish, for otherwise $\dot{\mathbf{C}}^e$, $\dot{c}_{\text{R}}$, and $\dot{\vartheta}$ may be chosen to violate (13.3.3). We are therefore led to the thermodynamic restriction that the free energy determines the stress $\mathbf{T}^e$, the chemical potential μ, and the entropy η_{R} through the "state relations"

$$\begin{aligned} \mathbf{T}^e &= 2\,J^{s-1}\,\frac{\partial\bar{\psi}_{\text{R}}(\mathbf{C}^e, c_{\text{R}}, \vartheta)}{\partial\mathbf{C}^e}, \\ \mu &= \frac{\partial\bar{\psi}_{\text{R}}(\mathbf{C}^e, c_{\text{R}}, \vartheta)}{\partial c_{\text{R}}} - \Omega\left(\frac{1}{3}\text{tr}\,\mathbf{M}^e\right), \\ \eta_{\text{R}} &= -\frac{\partial\bar{\psi}_{\text{R}}(\mathbf{C}^e, c_{\text{R}}, \vartheta)}{\partial\vartheta}, \end{aligned} \tag{13.3.4}$$

where in writing $(13.3.4)_2$ we have used the definition (13.2.8) for the net chemical potential. Further, as is standard, we assume that the response function $\bar{\mathbf{j}}_{\text{R}}$ satisfies the *fluid-transport inequality*

$$\bar{\mathbf{j}}_{\text{R}}(\mathbf{C}^e, c_{\text{R}}, \vartheta, \nabla\mu)\cdot\nabla\mu < 0 \quad \text{when} \quad \nabla\mu \neq 0, \tag{13.3.5}$$

while the response function $\bar{\mathbf{q}}_{\text{R}}^{\text{net}}$ satisfies the *heat-conduction inequality*

$$\bar{\mathbf{q}}_{\text{R}}^{\text{net}}(\mathbf{C}^e, c_{\text{R}}, \vartheta, \nabla\vartheta)\cdot\nabla\vartheta < 0 \quad \text{when} \quad \nabla\vartheta \neq 0. \tag{13.3.6}$$

Next, recalling from (13.2.4), (13.2.5), and (10.4.5) that

$$\mathbf{T}^e = J^e\mathbf{F}^{e-1}\mathbf{T}\mathbf{F}^{e-\top}, \quad \mathbf{M}^e = J^e\mathbf{F}^{e\top}\mathbf{T}\mathbf{F}^{e-\top} = \mathbf{C}^e\mathbf{T}^e, \quad \text{and} \quad \mathbf{T}_{\text{R}} = J\,\mathbf{T}\mathbf{F}^{-\top}, \tag{13.3.7}$$

$(13.3.4)_1$ gives the Cauchy stress as

$$\mathbf{T} = J^{-1}\left[2\mathbf{F}^e\frac{\partial\bar{\psi}_{\text{R}}(\mathbf{C}^e, c_{\text{R}})}{\partial\mathbf{C}^e}\mathbf{F}^{e\top}\right], \tag{13.3.8}$$

and the Piola stress as

$$\mathbf{T}_{\text{R}} = 2\mathbf{F}^e\frac{\partial\bar{\psi}_{\text{R}}(\mathbf{C}^e, c_{\text{R}})}{\partial\mathbf{C}^e}\mathbf{F}^{s-\top}. \tag{13.3.9}$$

Also, the mean normal stress $(1/3)\text{tr}\,\mathbf{M}^e$ appearing in $(13.3.4)_2$ is then given by

$$\frac{1}{3}\,\text{tr}\,\mathbf{M}^e = \frac{1}{3}\mathbf{C}^e : \left(2J^{s-1}\,\frac{\partial\bar{\psi}_{\text{R}}(\mathbf{C}^e, c_{\text{R}})}{\partial\mathbf{C}^e}\right). \tag{13.3.10}$$

13.3.3 **Further consequences of thermodynamics**

Gibbs relations

In view of the constitutive equation for the free energy $(13.3.1)_1$ and the state relations (13.3.4) we have the *first Gibbs relation*

$$\dot{\psi}_{\mathrm{R}} = \frac{1}{2}\,(J^s\mathbf{T}^e)\colon\dot{\mathbf{C}}^e - \eta_{\mathrm{R}}\dot{\vartheta} + \left(\mu + \Omega\,\frac{1}{3}\mathrm{tr}\,\mathbf{M}^e\right)\,\dot{c}_{\mathrm{R}}, \tag{13.3.11}$$

which with the definition

$$\psi_{\mathrm{R}} = \varepsilon_{\mathrm{R}} - \vartheta\,\eta_{\mathrm{R}} \tag{13.3.12}$$

for the free energy yields the *second Gibbs relation*

$$\dot{\varepsilon}_{\mathrm{R}} = \vartheta\dot{\eta}_{\mathrm{R}} + \frac{1}{2}\,J^s\,\mathbf{T}^e\colon\dot{\mathbf{C}}^e + \left(\mu + \Omega\,\frac{1}{3}\mathrm{tr}\,\mathbf{M}^e\right)\,\dot{c}_{\mathrm{R}}. \tag{13.3.13}$$

Specific heat

Next, as is standard, the *specific heat* is defined by

$$c_v \stackrel{\text{def}}{=} \frac{\partial\varepsilon_{\mathrm{R}}}{\partial\vartheta}. \tag{13.3.14}$$

Use of (13.3.12) and $(13.3.4)_3$ gives

$$c_v = \vartheta\frac{\partial\eta_{\mathrm{R}}}{\partial\vartheta} = -\vartheta\frac{\partial^2\bar{\psi}_{\mathrm{R}}}{\partial\vartheta^2}. \tag{13.3.15}$$

Evolution equation for temperature

From $(13.3.4)_3$ and (13.3.15)

$$\vartheta\dot{\eta}_{\mathrm{R}} = -\vartheta\frac{\partial^2\bar{\psi}_{\mathrm{R}}}{\partial\vartheta\partial\mathbf{C}^e}\colon\dot{\mathbf{C}}^e - \vartheta\frac{\partial^2\bar{\psi}_{\mathrm{R}}}{\partial\vartheta\partial c_{\mathrm{R}}}\dot{c}_{\mathrm{R}} + c_v\dot{\vartheta}, \tag{13.3.16}$$

which may be rearranged as the following evolution equation for temperature,

$$c_v\dot{\vartheta} = \vartheta\dot{\eta}_{\mathrm{R}} + \vartheta\frac{\partial^2\bar{\psi}_{\mathrm{R}}}{\partial\vartheta\partial\mathbf{C}^e}\colon\dot{\mathbf{C}}^e + \vartheta\frac{\partial^2\bar{\psi}_{\mathrm{R}}}{\partial\vartheta\partial c_{\mathrm{R}}}\dot{c}_{\mathrm{R}}. \tag{13.3.17}$$

Next, we calculate the term $\vartheta\dot{\eta}_{\mathrm{R}}$ in (13.3.17). The second Gibbs relation (13.3.13) gives the entropy balance

$$\vartheta\dot{\eta}_{\mathrm{R}} = \dot{\varepsilon}_{\mathrm{R}} - \frac{1}{2}\,J^s\,\mathbf{T}^e\colon\dot{\mathbf{C}}^e - \left(\mu + \Omega\,\frac{1}{3}\mathrm{tr}\,\mathbf{M}^e\right)\,\dot{c}_{\mathrm{R}}.$$

Recall the balance of energy (10.6.3), viz.,

$$\dot{\varepsilon}_{\mathrm{R}} = \mathbf{T}_{\mathrm{R}} : \dot{\mathbf{F}} + \mu_{\varepsilon}\dot{c}_{\mathrm{R}} - \mathrm{Div}\,\mathbf{q}_{\mathrm{R}} + q_{\mathrm{R}} - \mathbf{j}_{\mathrm{R}} \cdot \nabla\mu_{\varepsilon}. \tag{13.3.18}$$

Using the stress–power relation (13.2.3), viz.

$$\mathbf{T}_{\mathrm{R}} : \dot{\mathbf{F}} = J^s \left(\frac{1}{2}\mathbf{T}^e : \dot{\mathbf{C}}^e\right) + \frac{1}{3}(\mathrm{tr}\,\mathbf{M}^e)\dot{J}^s, \tag{13.3.19}$$

and the constitutive equation $\dot{J}^s = \Omega\dot{c}_{\mathrm{R}}$ this energy balance equation may be written as

$$\dot{\varepsilon}_{\mathrm{R}} = J^s \left(\frac{1}{2}\mathbf{T}^e : \dot{\mathbf{C}}^e\right) + \left(\mu_{\varepsilon} + \Omega\frac{1}{3}(\mathrm{tr}\,\mathbf{M}^e)\right)\dot{c}_{\mathrm{R}} - \mathrm{Div}\,\mathbf{q}_{\mathrm{R}} + q_{\mathrm{R}} - \mathbf{j}_{\mathrm{R}} \cdot \nabla\mu_{\varepsilon}. \tag{13.3.20}$$

Next, recalling the definitions

$$\mu \stackrel{\text{def}}{=} (\mu_{\varepsilon} - \vartheta\mu_{\eta}) \qquad \text{and} \qquad \mathbf{q}_{\mathrm{R}}^{\mathrm{net}} \stackrel{\text{def}}{=} \mathbf{q}_{\mathrm{R}} + \vartheta\mu_{\eta}\mathbf{j}_{\mathrm{R}}, \tag{13.3.21}$$

the energy balance may be written as

$$\begin{aligned}
\dot{\varepsilon}_{\mathrm{R}} &= J^s \left(\frac{1}{2}\mathbf{T}^e : \dot{\mathbf{C}}^e\right) + \left(\mu_{\varepsilon} + \Omega\frac{1}{3}(\mathrm{tr}\,\mathbf{M}^e)\right)\dot{c}_{\mathrm{R}} - \mathrm{Div}\,\mathbf{q}_{\mathrm{R}} + q_{\mathrm{R}} - \mathbf{j}_{\mathrm{R}} \cdot \nabla\mu_{\varepsilon}, \\
&= J^s \left(\frac{1}{2}\mathbf{T}^e : \dot{\mathbf{C}}^e\right) + \left(\mu + \Omega\frac{1}{3}(\mathrm{tr}\,\mathbf{M}^e)\right)\dot{c}_{\mathrm{R}} + \vartheta\mu_{\eta}\dot{c}_{\mathrm{R}} - \mathrm{Div}\,\mathbf{q}_{\mathrm{R}}^{\mathrm{net}} + q_{\mathrm{R}} \\
&\quad + \mathrm{Div}\,(\vartheta\mu_{\eta}\mathbf{j}_{\mathrm{R}}) - \mathbf{j}_{\mathrm{R}} \cdot \nabla\mu - \mathbf{j}_{\mathrm{R}} \cdot \nabla(\vartheta\mu_{\eta}), \\
&= J^s \left(\frac{1}{2}\mathbf{T}^e : \dot{\mathbf{C}}^e\right) + \left(\mu + \Omega\frac{1}{3}(\mathrm{tr}\,\mathbf{M}^e)\right)\dot{c}_{\mathrm{R}} - \mathrm{Div}\,\mathbf{q}_{\mathrm{R}}^{\mathrm{net}} + q_{\mathrm{R}} - \mathbf{j}_{\mathrm{R}} \cdot \nabla\mu \\
&\quad + \left[\vartheta\mu_{\eta}\dot{c}_{\mathrm{R}} + \mathrm{Div}\,(\vartheta\mu_{\eta}\mathbf{j}_{\mathrm{R}}) - \mathbf{j}_{\mathrm{R}} \cdot \nabla(\vartheta\mu_{\eta})\right],
\end{aligned}$$

and since

$$\begin{aligned}
\vartheta\mu_{\eta}\dot{c}_{\mathrm{R}} + \mathrm{Div}\,(\vartheta\mu_{\eta}\mathbf{j}_{\mathrm{R}}) - \mathbf{j}_{\mathrm{R}}\nabla(\vartheta\mu_{\eta}) &= -\vartheta\mu_{\eta}\mathrm{Div}\,\mathbf{j}_{\mathrm{R}} + \vartheta\mu_{\eta}\mathrm{Div}\,\mathbf{j}_{\mathrm{R}} + \mathbf{j}_{\mathrm{R}} \cdot \nabla(\vartheta\mu_{\eta}) \\
&\quad - \mathbf{j}_{\mathrm{R}}\nabla(\vartheta\mu_{\eta}) \equiv 0,
\end{aligned}$$

we obtain the energy balance in the form,

$$\dot{\varepsilon}_{\mathrm{R}} = J^s \left(\frac{1}{2}\mathbf{T}^e : \dot{\mathbf{C}}^e\right) + \left(\mu + \Omega\frac{1}{3}(\mathrm{tr}\,\mathbf{M}^e)\right)\dot{c}_{\mathrm{R}} - \mathrm{Div}\,\mathbf{q}_{\mathrm{R}}^{\mathrm{net}} + q_{\mathrm{R}} - \mathbf{j}_{\mathrm{R}} \cdot \nabla\mu. \tag{13.3.22}$$

Using this energy balance in the second Gibbs relation (13.3.13) gives the entropy balance

$$\vartheta\dot{\eta}_R = -\mathrm{Div}\,\mathbf{q}_R^{\mathrm{net}} + q_R - \mathbf{j}_R \cdot \nabla\mu. \tag{13.3.23}$$

Finally, using this entropy balance in (13.3.17) gives the following partial differential equation for the temperature

$$c_v\dot{\vartheta} = -\mathrm{Div}\,\mathbf{q}_R^{\mathrm{net}} + q_R + \mathcal{H}, \tag{13.3.24}$$

where the term

$$\mathcal{H} \stackrel{\mathrm{def}}{=} \vartheta\frac{\partial^2\bar{\psi}_R}{\partial\vartheta\partial\mathbf{C}^e} : \dot{\mathbf{C}}^e + \vartheta\frac{\partial^2\bar{\psi}_R}{\partial\vartheta\partial c_R}\dot{c}_R - \mathbf{j}_R \cdot \nabla\mu, \tag{13.3.25}$$

characterizes the *thermo-chemo-elastic coupling effect*. We call $\mathcal{H}$ the *thermo-chemo-elastic heating*, in which

$$\vartheta\frac{\partial^2\bar{\psi}_R}{\partial\vartheta\partial\mathbf{C}^e} \tag{13.3.26}$$

represents a "stress–temperature modulus," while the term

$$\vartheta\frac{\partial^2\bar{\psi}_R}{\partial\vartheta\partial c_R} \tag{13.3.27}$$

represents a "chemical-potential-temperature modulus." The partial differential equation (13.3.24) is the central result of this section.

13.4 Specialization of the free-energy function

With a view towards applications of the theory to thermally responsive elastomeric gels we now specialize the theory by imposing additional constitutive assumptions. We confine our attention to materials for which $\bar{\psi}(\mathbf{C}^e, c_R, \vartheta)$ is an isotropic function of $\mathbf{C}^e$, and has the representation

$$\bar{\psi}(\mathbf{C}^e, c_R, \vartheta) = \tilde{\psi}(\mathcal{I}_{\mathbf{C}^e}, c_R, \vartheta), \tag{13.4.1}$$

where

$$\mathcal{I}_{\mathbf{C}^e} = \big(I_1(\mathbf{C}^e), I_2(\mathbf{C}^e), I_3(\mathbf{C}^e)\big)$$

is the list of principal invariants of $\mathbf{C}^e$. Further, we assume that the free energy ψ_R may be written in a separable form as,

$$\tilde{\psi}_R(\mathcal{I}_{\mathbf{C}^e}, c_R, \vartheta) = \mu^{\mathrm{ref}}c_R + \psi^{\mathrm{mix}}(c_R, \vartheta) + \psi^{\mathrm{mech}}(\mathcal{I}_{\mathbf{C}^e}, c_R, \vartheta) + f(\vartheta), \tag{13.4.2}$$

where:

- μ^{ref} is the chemical potential of the unmixed pure solvent;
- $\psi^{\text{mix}}(c_{\text{R}}, \vartheta)$ is the change in free energy due to mixing of the solvent with the polymer network;
- $\psi^{\text{mech}}(\mathcal{I}_{\mathbf{C}^e}, c_{\text{R}}, \vartheta)$ is the contribution to the change in the free energy due to the deformation of the polymer network; and
- $f(\vartheta)$ is a temperature-dependent entropic contribution to the free energy.

Estimate for ψ^{mix}

As in Chapter 10, we introduce the *polymer volume fraction*

$$\phi \stackrel{\text{def}}{=} \frac{1}{1+\Omega c_{\text{R}}} = J^{s-1} = (\lambda^s)^{-3}, \qquad 0 < \phi \le 1, \tag{13.4.3}$$

where *the dry state corresponds to* $\phi = 1$, *and* $\phi < 1$ *represents a swollen state.*

Next, following Doi (1996, 2009), we adopt the following form of the Flory (1942)–Huggins (1942) theory for the contribution to the free energy due to mixing

$$\psi^{\text{mix}} = \frac{R\vartheta}{\Omega}\frac{1}{\phi}\Big((1-\phi)\ln(1-\phi) + \chi(\vartheta)\phi(1-\phi)\Big), \tag{13.4.4}$$

where R is the universal gas constant, and where now the interaction parameter $\chi(\vartheta)$ is a dimensionless temperature-dependent parameter which represents the disaffinity between the polymer and the fluid:

- if $\chi(\vartheta)$ is increased, the fluid molecules are expelled from the gel and the gel shrinks, while if $\chi(\vartheta)$ is decreased, the gel swells.[3]

The expression (13.4.4) for the mixing energy may be written in terms of c_{R} as

$$\psi^{\text{mix}}(c_{\text{R}}, \vartheta) = \frac{R\vartheta}{\Omega}\left(\Omega c_{\text{R}} \ln\left(\frac{\Omega c_{\text{R}}}{1+\Omega c_{\text{R}}}\right) + \chi(\vartheta)\Big(\frac{\Omega c_{\text{R}}}{1+\Omega c_{\text{R}}}\Big)\right). \tag{13.4.5}$$

As reviewed earlier, thermally responsive gels possess a "lower critical solution temperature (LCST)," which we here denote by ϑ_T, across which they exhibit a larger degree of swelling at temperatures below ϑ_T than at temperatures above ϑ_T. To model such experimentally observed phenomena we assume that (Chester and Anand, 2011)

[3] For reference, a typical value of χ which promotes pronounced swelling is $\chi \approx 0.1$, while $\chi \approx 0.5$ is a borderline value for the swelling of a dry polymer. Also, experience shows that the interaction parameter χ may also be a function of the pressure under conditions of highly confined swelling (Stewart et al., 2022).

$$\chi(\vartheta) = \frac{1}{2}(\chi_L + \chi_H) - \frac{1}{2}(\chi_L - \chi_H)\tanh\left(\frac{\vartheta - \vartheta_T}{\Delta}\right), \tag{13.4.6}$$

where ϑ_T is the transition temperature of the gel, χ_L is the value of χ below the transition temperature, $\chi_H > \chi_L$ is the value of χ above the transition temperature, with Δ the width of the transition in temperatures between χ_L and χ_H.

Estimate for ψ^{mech}

In elastomeric materials the major part of ψ^{mech} arises from an entropic contribution. Let

$$\bar{\lambda} \stackrel{\text{def}}{=} \frac{1}{\sqrt{3}}\sqrt{\operatorname{tr}\mathbf{C}} = \frac{1}{\sqrt{3}}\sqrt{(\operatorname{tr}\mathbf{C}^e)\,(1+\Omega c_{\mathrm{R}})^{2/3}} \tag{13.4.7}$$

define an *effective stretch*. Then, classical statistical mechanics models of rubber elasticity (cf., Treloar (1975); Arruda and Boyce (1993)) provide the following estimates for the entropy change due to mechanical stretching:

1. For small to moderate values of $\bar{\lambda}$, based on Gaussian statistics,

$$\eta_{\mathrm{R}}^{\text{mech}} = -\frac{3}{2}Nk_B(\bar{\lambda}^2 - 1) + N\,k_B \ln J, \tag{13.4.8}$$

   where k_B is Boltzmann's constant and N represents the number of polymer chains per unit reference volume.

2. For larger values of $\bar{\lambda}$ it is necessary to use non-Gaussian statistics to account for the limited extensibility of the polymer chains, and for these circumstances the entropy change is given by

$$\begin{aligned}\eta_{\mathrm{R}}^{\text{mech}} = &-Nk_B\lambda_L^2\left[\left(\frac{\bar{\lambda}}{\lambda_L}\right)\beta + \ln\left(\frac{\beta}{\sinh\beta}\right) - \left(\frac{1}{\lambda_L}\right)\beta_0 - \ln\left(\frac{\beta_0}{\sinh\beta_0}\right)\right]\\ &+ Nk_B\left(\frac{\lambda_L}{3}\beta_0\right)\ln J,\end{aligned} \tag{13.4.9}$$

   with

$$\beta \stackrel{\text{def}}{=} \mathcal{L}^{-1}\left(\frac{\bar{\lambda}}{\lambda_L}\right), \qquad \text{and} \qquad \beta_0 \stackrel{\text{def}}{=} \mathcal{L}^{-1}\left(\frac{1}{\lambda_L}\right), \tag{13.4.10}$$

   where $\mathcal{L}^{-1}$ is the inverse of the Langevin function

$$\mathcal{L}(\bullet) = \coth(\bullet) - (\bullet)^{-1}.$$

This functional form for the change in entropy involves two material parameters: N, the number of polymer chains per unit reference volume, and λ_L, the *network locking stretch.*[4]

Next, we take ψ^{mech} to also have a contribution from elastic compressibility and thermal expansion of the material,

$$\varepsilon_{\text{R}}^{\text{mech}} = J^s \left[\tfrac{1}{2}K(\ln J^e)^2 - 3K\alpha(\vartheta - \vartheta_0)(\ln J^e)\right], \tag{13.4.11}$$

where $K > 0$ is the bulk modulus of the gel, and $\alpha > 0$ is a coefficient of thermal expansion.[5] The term within square brackets in (13.4.11) represents a contribution to the free energy measured per unit volume of the intermediate space, and multiplication by J^s converts this to a contribution per unit volume of the reference space.

Then, using (13.4.9) through (13.4.11), for non-Gaussian statistics, and writing

$$G_0 \stackrel{\text{def}}{=} N\, k_B\, \vartheta \tag{13.4.12}$$

for a temperature-dependent *ground-state shear modulus*, we obtain the estimate

$$\begin{aligned}\psi^{\text{mech}}(\mathcal{I}_{\mathbf{C}^e}, c_{\text{R}}, \vartheta) = G_0\lambda_L^2 &\left[\left(\frac{\bar{\lambda}}{\lambda_L}\right)\beta + \ln\left(\frac{\beta}{\sinh\beta}\right) - \left(\frac{1}{\lambda_L}\right)\beta_0 - \ln\left(\frac{\beta_0}{\sinh\beta_0}\right)\right] \\ &- G_0\left(\frac{\lambda_L}{3}\beta_0\right)\ln J + J^s\left[\tfrac{1}{2}K(\ln J^e)^2 - 3K\alpha(\vartheta-\vartheta_0)(\ln J^e)\right].\end{aligned} \tag{13.4.13}$$

Estimate for $f(\vartheta)$

For the entropic contribution $f(\vartheta)$ to the free energy, we assume that (Anand, 1996, eq. (74))

$$f(\vartheta) = c_v(\vartheta - \vartheta_0) - \vartheta c_v(\ln\vartheta - \ln\vartheta_0), \tag{13.4.14}$$

with c_v a constant specific heat.

[4] In the non-Gaussian model, the network locking stretch λ_L is related to the number of links n in a freely jointed chain by $\lambda_L = \sqrt{n}$.

[5] For most elastomeric gels the thermal expansion coefficient α appearing in our theory has a value of approximately 10^{-4} to 10^{-5} /K, and this causes negligibly small dimensional changes for temperature changes of approximately 50 K of interest in modeling thermally responsive gels. It is the variation of the $\chi(\vartheta)$ parameter with temperature, as modeled in equation (13.4.6), which is the cause for the large dimensional changes with temperature variations.

Total free energy

Thus, using (13.4.5), (13.4.13), and (13.4.14) in (13.4.2), a form of the free-energy function which accounts for the combined effects of mixing, swelling, elastic stretching, and thermal effects is

$$
\begin{aligned}
\psi_{\mathrm{R}} = \mu^{\mathrm{ref}} c_{\mathrm{R}} &+ \frac{R\vartheta}{\Omega}\left(\Omega c_{\mathrm{R}} \ln\left(\frac{\Omega c_{\mathrm{R}}}{1+\Omega c_{\mathrm{R}}}\right) + \chi(\vartheta)\left(\frac{\Omega c_{\mathrm{R}}}{1+\Omega c_{\mathrm{R}}}\right)\right) \\
&+ G_0 \lambda_L^2 \left[\left(\frac{\bar{\lambda}}{\lambda_L}\right)\beta + \ln\left(\frac{\beta}{\sinh\beta}\right) - \left(\frac{1}{\lambda_L}\right)\beta_0 - \ln\left(\frac{\beta_0}{\sinh\beta_0}\right)\right] \\
&- G_0\left(\frac{\lambda_L}{3}\beta_0\right)\ln J + J^s\left[\tfrac{1}{2}K(\ln J^e)^2 - 3K\alpha(\vartheta-\vartheta_0)(\ln J^e)\right] \\
&+ c_v(\vartheta-\vartheta_0) - \vartheta c_v(\ln\vartheta - \ln\vartheta_0).
\end{aligned}
\tag{13.4.15}
$$

Stress

Using (13.4.15) and (13.3.8) we find that the Cauchy stress tensor is given by

$$
\mathbf{T} = J^{-1}\left[G\,\phi^{-2/3}\mathbf{B}^e - G_0\zeta_0\mathbf{1}\right] + K\left(\frac{\ln J^e - 3\alpha(\vartheta-\vartheta_0)}{J^e}\right)\mathbf{1}, \tag{13.4.16}
$$

where

$$
G \stackrel{\text{def}}{=} G_0\,\zeta, \quad \text{with} \quad \zeta \stackrel{\text{def}}{=} \left(\frac{\lambda_L}{3\bar{\lambda}}\right)\mathcal{L}^{-1}\left(\frac{\bar{\lambda}}{\lambda_L}\right), \quad \text{and} \quad \zeta_0 \stackrel{\text{def}}{=} \left(\frac{\lambda_L}{3}\right)\mathcal{L}^{-1}\left(\frac{1}{\lambda_L}\right), \tag{13.4.17}
$$

is a *generalized shear modulus*, and $\mathbf{B}^e$ is the elastic left Cauchy–Green tensor.

Next, since

$$
\mathbf{B} = (\lambda^s)^2\mathbf{B}^e = \phi^{-2/3}\mathbf{B}^e, \tag{13.4.18}
$$

(13.4.16) reduces to

$$
\mathbf{T} = J^{-1}\,G_0\,(\zeta\mathbf{B} - \zeta_0\mathbf{1}) - p\,\mathbf{1}. \tag{13.4.19}
$$

where we have introduced the notation

$$
p \stackrel{\text{def}}{=} -K\left(\frac{\ln J^e - 3\alpha(\vartheta-\vartheta_0)}{J^e}\right), \tag{13.4.20}
$$

as a thermoelastic "pressure" associated with the thermo-mechanical deformation of the swollen gel. The Piola stress, $\mathbf{T}_{\mathrm{R}} = J\mathbf{T}\mathbf{F}^{-\top}$ is then given by

$$
\mathbf{T}_{\mathrm{R}} = G_0\left[\zeta\mathbf{F} - \zeta_0\mathbf{F}^{-\top}\right] - J\,p\,\mathbf{F}^{-\top}. \tag{13.4.21}
$$

Chemical potential

Using (13.4.15) and $(13.3.4)_2$ the chemical potential μ is given by

$$\begin{aligned}\mu = \mu^{\text{ref}} + R\vartheta\Bigg(\ln\left(\frac{\Omega c_{\text{R}}}{1+\Omega c_{\text{R}}}\right) + \frac{1}{(1+\Omega c_{\text{R}})} + \chi(\vartheta)\frac{1}{(1+\Omega c_{\text{R}})^2}\Bigg) \\ + \frac{1}{3}\Omega G\,(1+\Omega c_{\text{R}})^{-1/3}\,\text{tr}\,\mathbf{C}^e - \Omega\, G_0\zeta_0\,(1+\Omega c_{\text{R}})^{-1} \\ + \Omega\left[\tfrac{1}{2}K(\ln J^e)^2 - 3K\alpha(\vartheta-\vartheta_0)(\ln J^e)\right] - \Omega\left(\frac{1}{3}\text{tr}\,\mathbf{M}^e\right).\end{aligned} \tag{13.4.22}$$

A little algebra gives

$$\begin{aligned}\mu = \mu^{\text{ref}} + R\vartheta\Big(\ln(1-\phi) + \phi + \chi(\vartheta)\phi^2\Big) + J^e\,p\,\Omega \\ + \Omega\left[\tfrac{1}{2}K(\ln J^e)^2 - 3K\alpha(\vartheta-\vartheta_0)(\ln J^e)\right].\end{aligned} \tag{13.4.23}$$

This may be written equivalently in terms of c_{R} instead of ϕ as

$$\begin{aligned}\mu = \mu^{\text{ref}} + R\vartheta\left[\ln\left(1 - \frac{1}{1+\Omega c_{\text{R}}}\right) + \frac{1}{1+\Omega c_{\text{R}}} + \chi(\vartheta)\left(\frac{1}{1+\Omega c_{\text{R}}}\right)^2\right] \\ + J^e\,p\,\Omega + \Omega\left[\tfrac{1}{2}K(\ln J^e)^2 - 3K\alpha(\vartheta-\vartheta_0)(\ln J^e)\right].\end{aligned} \tag{13.4.24}$$

REMARK

For a theory based on Gaussian statistics for configurational changes in entropy, cf. eq. (13.4.8), the generalized shear modulus is no longer stretch-dependent, so that $G = G_0$, and the expressions (13.4.19) and (13.4.21) for the Cauchy stress and the Piola stress reduce to

$$\begin{aligned}\mathbf{T} &= J^{-1}\left[G_0\,(\mathbf{B} - \mathbf{1})\right] - p\mathbf{1}, \\ \mathbf{T}_{\text{R}} &= G_0\left[\mathbf{F} - \mathbf{F}^{-\top}\right] - Jp\,\mathbf{F}^{-\top},\end{aligned} \tag{13.4.25}$$

where

$$p = -K\left(\frac{\ln J^e - 3\alpha(\vartheta-\vartheta_0)}{J^e}\right) \tag{13.4.26}$$

represents a pressure due to the mechanical deformation $\mathbf{F}^e$, while the expression (13.4.23) for the chemical potential remains unchanged.

Entropy

Using (13.4.15) and $(13.3.4)_3$ we obtain the following expression for the entropy

$$
\begin{aligned}
\eta_{\mathrm{R}} = & -\frac{R}{\Omega}\left(\Omega c_{\mathrm{R}} \ln\left(\frac{\Omega c_{\mathrm{R}}}{1+\Omega c_{\mathrm{R}}}\right) + \chi(\vartheta)\left(\frac{\Omega c_{\mathrm{R}}}{1+\Omega c_{\mathrm{R}}}\right)\right) - \frac{R\vartheta}{\Omega}\frac{\partial\chi(\vartheta)}{\partial\vartheta}\left(\frac{\Omega c_{\mathrm{R}}}{1+\Omega c_{\mathrm{R}}}\right) \\
& - N\,k_B\,\lambda_L^2\left[\left(\frac{\bar{\lambda}}{\lambda_L}\right)\beta + \ln\left(\frac{\beta}{\sinh\beta}\right) - \left(\frac{1}{\lambda_L}\right)\beta_0 - \ln\left(\frac{\beta_0}{\sinh\beta_0}\right)\right] \\
& + Nk_B\left(\frac{\lambda_L}{3}\beta_0\right) + J^s\,(3K\alpha\ln J^e) + c_v(\ln\vartheta - \ln\vartheta_0).
\end{aligned} \tag{13.4.27}
$$

13.4.1 **Specialization of the equations for the fluid flux and heat flux**

For isotropic materials the fluid flux in the *spatial description* is taken as

$$\mathbf{j} = -m\,\mathrm{grad}\,\mu, \tag{13.4.28}$$

in which we assume that the mobility m at a given temperature ϑ is given by

$$m = \frac{Dc}{R\vartheta}, \tag{13.4.29}$$

where $D > 0$, a constant, represents a diffusion coefficient, and $c = c_{\mathrm{R}}/J$ is the fluid concentration measured in moles of fluid per unit deformed volume.

Then, using the standard mapping results

$$\mathbf{F}^{-\top}\nabla\mu = \mathrm{grad}\,\mu \qquad \text{and} \qquad \mathbf{j}_{\mathrm{R}} = J\mathbf{F}^{-1}\mathbf{j}, \tag{13.4.30}$$

we obtain the referential species flux as

$$\mathbf{j}_{\mathrm{R}} = -\mathbf{M}^{\mathrm{mob}}\nabla\mu \qquad \text{with} \qquad \mathbf{M}^{\mathrm{mob}} = \frac{Dc_{\mathrm{R}}}{R\vartheta}\,\mathbf{C}^{-1}, \tag{13.4.31}$$

where $\mathbf{M}^{\mathrm{mob}}$ is the species mobility tensor.

Similarly, and since we ignore cross-coupling terms, for isotropic materials the net heat flux in the *spatial description* is constitutively taken as Fourier's law

$$\mathbf{q}^{\mathrm{net}} = -k\ \mathrm{grad}\,\vartheta, \tag{13.4.32}$$

where

$$\mathbf{q}^{\mathrm{net}} \stackrel{\mathrm{def}}{=} J^{-1}\mathbf{F}\mathbf{q}_{\mathrm{R}}^{\mathrm{net}} \tag{13.4.33}$$

represents the spatial description of the referential heat flux $\tilde{\mathbf{q}}_{\mathrm{R}}$ defined in $(13.2.2)_2$. Then, using the standard mapping relation

$$\mathbf{F}^{-\top}\nabla\vartheta = \mathrm{grad}\,\vartheta \tag{13.4.34}$$

as well as the mapping relation (13.4.33), we obtain the net referential heat flux $\tilde{\mathbf{q}}_{\mathrm{R}}$ as

$$\mathbf{q}_{\mathrm{R}}^{\mathrm{net}} = -\mathbf{K}\nabla\vartheta \qquad \text{with} \qquad \mathbf{K} = Jk\,\mathbf{C}^{-1}, \tag{13.4.35}$$

where $\mathbf{K}$ is the thermal conductivity tensor.

13.4.2 **The heat equation**

The energy balance equation is (13.3.24), viz.

$$c_v\dot{\vartheta} = -\mathrm{Div}\,\mathbf{q}_{\mathrm{R}}^{\mathrm{net}} + q_{\mathrm{R}} + \mathcal{H}, \tag{13.4.36}$$

with the chemo-thermo-mechanical heating term $\mathcal{H}$ given by

$$\mathcal{H} \overset{\text{def}}{=} \vartheta\frac{\partial^2\bar{\psi}_{\mathrm{R}}}{\partial\vartheta\partial\mathbf{C}^e}:\dot{\mathbf{C}}^e + \vartheta\frac{\partial^2\bar{\psi}_{\mathrm{R}}}{\partial\vartheta\partial c_{\mathrm{R}}}\dot{c}_{\mathrm{R}} - \mathbf{j}_{\mathrm{R}}\cdot\nabla\mu. \tag{13.4.37}$$

A little algebra gives that the thermo-chemo-mechanical heating $\mathcal{H}$ may be written as

$$\mathcal{H} \overset{\text{def}}{=} \vartheta\,\mathbf{M}^{\mathcal{H}}:\dot{\mathbf{C}}^e + \vartheta\,M^{\mathcal{H}}\dot{c}_{\mathrm{R}} - \mathbf{j}_{\mathrm{R}}\cdot\nabla\mu, \tag{13.4.38}$$

with

$$\mathbf{M}^{\mathcal{H}} \overset{\text{def}}{=} \frac{1}{2}\left(Nk_B\left(\zeta J^{s\,2/3}\mathbf{1} - \zeta_0\mathbf{C}^{e-1}\right) + \frac{3K\alpha}{J^e}\,\mathbf{C}^{e-1}\right),$$

$$M^{\mathcal{H}} \overset{\text{def}}{=} R\left(\ln\left(\frac{\Omega c_{\mathrm{R}}}{1+\Omega c_{\mathrm{R}}}\right) + \frac{1}{(1+\Omega c_{\mathrm{R}})} + \chi\frac{1}{(1+\Omega c_{\mathrm{R}})^2}\right) + \frac{\Omega N k_B}{1+\Omega c_{\mathrm{R}}}\left(\zeta\,\frac{\mathrm{tr}\,\mathbf{C}}{3} - \zeta_0\right). \tag{13.4.39}$$

Note that, for simplicity, in calculating the term $M^{\mathcal{H}}$ we have neglected the dependence of the χ parameter on the temperature ϑ.

REMARK

In order to avoid *cumbersome notation*, we henceforth drop the superscript on $\mathbf{q}_{\mathrm{R}}^{\mathrm{net}}$ and write the Fourier relation (13.4.35) and the heat equation (13.4.36) simply as

$$\mathbf{q}_{\mathrm{R}} = -\mathbf{K}\nabla\vartheta, \tag{13.4.40}$$

and

$$c_v\dot{\vartheta} = -\mathrm{Div}\,\mathbf{q}_{\mathrm{R}} + q_{\mathrm{R}} + \mathcal{H}, \tag{13.4.41}$$

respectively. The quantity $\mathbf{q}_{\mathrm{R}}^{\mathrm{net}}$ is fully determined by the constitutive equation (13.4.35), and here we are only relabeling it for ease of notation.

13.5 Governing partial differential equations

1. Neglecting body forces and inertia, the local force balance is represented by the pair of equations

$$\begin{aligned} \operatorname{Div} \mathbf{T}_{\mathrm{R}} &= \mathbf{0}, \\ p + K\left(\frac{\ln J^e - 3\alpha(\vartheta - \vartheta_0)}{J^e}\right) &= 0, \end{aligned} \tag{13.5.1}$$

with $\mathbf{T}_{\mathrm{R}}$ given by

$$\begin{aligned} \mathbf{T}_{\mathrm{R}} &= G_0\left[\zeta \mathbf{F} - \zeta_0 \mathbf{F}^{-\top}\right] - Jp\,\mathbf{F}^{-\top}, \\ G_0 &= N k_B \vartheta, \\ \zeta &= \left(\frac{\lambda_L}{3\bar{\lambda}}\right) \mathcal{L}^{-1}\left(\frac{\bar{\lambda}}{\lambda_L}\right), \\ \zeta_0 &= \left(\frac{\lambda_L}{3}\right) \mathcal{L}^{-1}\left(\frac{1}{\lambda_L}\right). \end{aligned} \tag{13.5.2}$$

In writing (13.5.1) we have split the standard force balance equation into two equations because this helps in dealing with the near "elastic" incompressibility of the material. Accordingly, in our numerical finite element implementation in addition to using the displacement $\mathbf{u}$, we will also use p as a separate degree of freedom.

With $\partial\mathcal{B}_{\mathbf{u}}$ and $\partial\mathcal{B}_{\mathbf{t}_{\mathrm{R}}}$ denoting complementary subsurfaces of the boundary $\partial\mathcal{B}$ of the body $\mathcal{B}$, we consider a pair of possible boundary conditions:

$$\begin{aligned} \mathbf{u} &= \bar{\mathbf{u}} \quad \text{on } \partial\mathcal{B}_{\mathbf{u}}, \\ \mathbf{T}_{\mathrm{R}}\mathbf{n}_{\mathrm{R}} &= \bar{\mathbf{t}}_{\mathrm{R}} \quad \text{on } \partial\mathcal{B}_{\mathbf{t}_{\mathrm{R}}}. \end{aligned} \tag{13.5.3}$$

In the boundary conditions above $\bar{\mathbf{u}}$ and $\bar{\mathbf{t}}_{\mathrm{R}}$ are prescribed displacement and tractions, both functions of $\mathbf{X}$ and t.

2. The balance equation (10.5.4) for the fluid content is

$$\dot{c}_{\mathrm{R}} + \operatorname{Div} \mathbf{j}_{\mathrm{R}} = 0, \tag{13.5.4}$$

with the species flux given by

$$\mathbf{j}_{\mathrm{R}} = -\mathbf{M}^{\mathrm{mob}} \nabla \mu, \tag{13.5.5}$$

where

$$\mathbf{M}^{\mathrm{mob}} = \mathbf{C}^{-1}\left(\frac{D c_{\mathrm{R}}}{R\vartheta}\right) \tag{13.5.6}$$

is the mobility tensor, and the chemical potential μ given by

$$\mu = R\vartheta\left[\ln\left(1-\frac{1}{1+\Omega c_{\mathrm{R}}}\right)+\frac{1}{1+\Omega c_{\mathrm{R}}}+\frac{\chi(\vartheta)}{(1+\Omega c_{\mathrm{R}})^2}\right]$$
$$+J^e p\Omega+\Omega\left[\tfrac{1}{2}K(\ln J^e)^2-3K\alpha(\vartheta-\vartheta_0)(\ln J^e)\right], \qquad (13.5.7)$$

where in writing (13.5.7) we have set the reference chemical potential of the fluid to zero, $\mu^{\mathrm{ref}}=0$.

With $\partial\mathcal{B}_\mu$ and $\partial\mathcal{B}_{\mathbf{j}_{\mathrm{R}}}$ denoting complementary subsurfaces of $\partial\mathcal{B}$, we consider a pair of possible boundary conditions:

$$\begin{aligned}\mu &= \bar{\mu} \quad \text{on } \partial\mathcal{B}_\mu,\\ \mathbf{j}_{\mathrm{R}}\cdot\mathbf{n}_{\mathrm{R}} &= \bar{j}_{\mathrm{R}} \quad \text{on } \partial\mathcal{B}_{\mathbf{j}_{\mathrm{R}}},\end{aligned} \qquad (13.5.8)$$

with $\bar{\mu}$ and $\bar{j}_{\mathrm{R}}$ prescribed functions of $\mathbf{X}$ and t.

3. The energy balance equation is (13.3.24), viz.

$$c_v\dot{\vartheta} = -\mathrm{Div}\,\mathbf{q}_{\mathrm{R}}+q_{\mathrm{R}}+\mathcal{H}, \qquad (13.5.9)$$

with the heat flux given by

$$\mathbf{q}_{\mathrm{R}} = -\mathbf{K}\nabla\vartheta \qquad \text{with} \qquad \mathbf{K}=Jk\mathbf{C}^{-1}, \qquad (13.5.10)$$

and the chemo-thermo-mechanical heating term $\mathcal{H}$ given by

$$\begin{aligned}\mathcal{H} &\stackrel{\text{def}}{=} \vartheta\,\mathbf{M}^{\mathcal{H}}:\dot{\mathbf{C}}^e+\vartheta\,M^{\mathcal{H}}\dot{c}_{\mathrm{R}}-\mathbf{j}_{\mathrm{R}}\cdot\nabla\mu, \qquad \text{with}\\ \mathbf{M}^{\mathcal{H}} &\stackrel{\text{def}}{=} \frac{1}{2}\left(Nk_B\,(\zeta J^{s2/3}\mathbf{1}-\zeta_0\mathbf{C}^{e-1})+\frac{3K\alpha}{J^e}\,\mathbf{C}^{e-1}\right),\\ M^{\mathcal{H}} &\stackrel{\text{def}}{=} R\left(\ln\left(\frac{\Omega c_{\mathrm{R}}}{1+\Omega c_{\mathrm{R}}}\right)+\frac{1}{(1+\Omega c_{\mathrm{R}})}+\chi\frac{1}{(1+\Omega c_{\mathrm{R}})^2}\right)\\ &\quad+\frac{\Omega Nk_B}{1+\Omega c_{\mathrm{R}}}\left(\zeta\,\frac{\mathrm{tr}\,\mathbf{C}}{3}-\zeta_0\right).\end{aligned} \qquad (13.5.11)$$

With $\partial\mathcal{B}_\vartheta$ and $\partial\mathcal{B}_{\mathbf{q}_{\mathrm{R}}}$ denoting complementary subsurfaces of $\partial\mathcal{B}$, we consider a pair of possible boundary conditions:

$$\begin{aligned}\vartheta &= \bar{\vartheta} \quad \text{on } \partial\mathcal{B}_\vartheta,\\ \mathbf{q}_{\mathrm{R}}\cdot\mathbf{n}_{\mathrm{R}} &= \bar{q}_{\mathrm{R}} \quad \text{on } \partial\mathcal{B}_{\mathbf{q}_{\mathrm{R}}},\end{aligned} \qquad (13.5.12)$$

with $\bar{\vartheta}$ and $\bar{q}_{\mathrm{R}}$ prescribed functions of $\mathbf{X}$ and t.

Standard initial conditions involve the specifications

$$\mathbf{u}(\mathbf{X},0) = \mathbf{u}_0(\mathbf{X}), \quad \mu(\mathbf{X},0) = \mu_0(\mathbf{X}), \quad \text{and} \quad \vartheta(\mathbf{X},0) = \vartheta_0(\mathbf{X}) \quad \text{in } \mathcal{B}, \tag{13.5.13}$$

with $\mathbf{u}_0$, μ_0, $c_{\mathrm{R}0}$, and ϑ_0 prescribed functions of $\mathbf{X} \in \mathcal{B}$.

REMARK

As discussed in Chapter 10, we also need knowledge of the species concentration c_{R}, which may be obtained by solving

$$\begin{aligned} &\mu - R\vartheta\left(\ln\left(1 - \frac{1}{1+\Omega c_{\mathrm{R}}}\right) + \frac{1}{1+\Omega c_{\mathrm{R}}} + \frac{\chi(\vartheta)}{(1+\Omega c_{\mathrm{R}})^2}\right) \\ &\quad - \Omega\, J^e\, p - \Omega\left[\tfrac{1}{2}K(\ln J^e)^2 - 3K\alpha(\vartheta - \vartheta_0)(\ln J^e)\right] = 0, \end{aligned} \tag{13.5.14}$$

which serves as an implicit equation for c_{R}.

---

14 Numerical implementation of theory for thermally responsive gels

14.1 Weak forms of the governing partial differential equations

1. **Mechanical problem**:
Consider a test field $\mathbf{u}_{\text{test}}(\mathbf{X})$ in $\mathcal{B}$ which satisfies $\mathbf{u}_{\text{test}}(\mathbf{X}) = \mathbf{0}$ on $\partial\mathcal{B}_{\mathbf{u}}$. If the balance equation $\text{Div}\,\mathbf{T}_{\text{R}} = \mathbf{0}$ in $\mathcal{B}$ and the traction condition $(\bar{\mathbf{t}}_{\text{R}} - \mathbf{T}_{\text{R}}\mathbf{n}_{\text{R}}) = \mathbf{0}$ on $\partial\mathcal{B}_{\mathbf{t}_{\text{R}}}$ are satisfied, then it is straightforward to show that the weak form

$$\int_{\mathcal{B}} \mathbf{T}_{\text{R}} : \nabla \mathbf{u}_{\text{test}}\, dv_{\text{R}} - \int_{\partial\mathcal{B}_{\mathbf{t}_{\text{R}}}} \bar{\mathbf{t}}_{\text{R}} \cdot \mathbf{u}_{\text{test}}\, da_{\text{R}} = 0 \tag{14.1.1}$$

holds for all admissible test fields $\mathbf{u}_{\text{test}}$ which vanish on $\partial\mathcal{B}_{\mathbf{u}}$. In addition, we consider a test field $p_{\text{test}}(\mathbf{X})$ in $\mathcal{B}$. Then, the weak form of the pressure equation

$$p + K\left(\frac{\ln J^e - 3\alpha(\vartheta - \vartheta_0)}{J^e}\right) = 0,$$

is written for *numerical convenience* as

$$\int_{\mathcal{B}} \left[\frac{pJ^e}{K} + (\ln J^e - 3\alpha(\vartheta - \vartheta_0))\right] p_{\text{test}}\, dv_{\text{R}} = 0. \tag{14.1.2}$$

2. **Fluid content problem**:
Next, consider a scalar test field $\mu_{\text{test}}(\mathbf{X})$ in $\mathcal{B}$ which satisfies $\mu_{\text{test}}(\mathbf{X}) = \mathbf{0}$ on $\partial\mathcal{B}_{\mu}$. If the balance equation

$$\dot{c}_{\text{R}} + \text{Div}\,\mathbf{j}_{\text{R}} = 0 \qquad \text{in } \mathcal{B},$$

and the boundary condition $\mathbf{j}_{\text{R}} \cdot \mathbf{n} = \bar{j}$ on $\partial\mathcal{B}_{\mathbf{j}_{\text{R}}}$ are satisfied, then

$$\int_{\mathcal{B}} \dot{c}_{\text{R}}\, \mu_{\text{test}}\, dv_{\text{R}} + \int_{\mathcal{B}} (\text{Div}\,\mathbf{j}_{\text{R}})\, \mu_{\text{test}}\, dv_{\text{R}} + \int_{\partial\mathcal{B}_{\mathbf{j}_{\text{R}}}} (\bar{j}_{\text{R}} - \mathbf{j}_{\text{R}} \cdot \mathbf{n})\, \mu_{\text{test}}\, da_{\text{R}} = 0, \tag{14.1.3}$$

Introduction to Coupled Theories in Solid Mechanics. Lallit Anand, Eric M. Stewart, Shawn A. Chester, Oxford University Press.
 DOI: 10.1093/oso/9780198986218.003.0015

or equivalently

$$\int_{\mathcal{B}} \dot{c}_{\mathrm{R}}\, \mu_{\text{test}}\, dv_{\mathrm{R}} + \int_{\mathcal{B}} (\operatorname{Div} \mathbf{j}_{\mathrm{R}})\, \mu_{\text{test}}\, dv_{\mathrm{R}} - \int_{\partial\mathcal{B}_{\mathrm{j_R}}} \mathbf{j}_{\mathrm{R}} \cdot \mathbf{n}_{\mathrm{R}}\, \mu_{\text{test}}\, da_{\mathrm{R}} + \int_{\partial\mathcal{B}_{\mathrm{j_R}}} \bar{j}_{\mathrm{R}}\, \mu_{\text{test}}\, da_{\mathrm{R}} = 0 \quad (14.1.4)$$

holds for all test fields μ_{test}. Thus, since $\mu_{\text{test}} = 0$ on $\partial\mathcal{B}_{\mu}$, the first boundary integral in expression above may be extended to all of $\partial\mathcal{B}$,

$$\int_{\mathcal{B}} \dot{c}_{\mathrm{R}}\, \mu_{\text{test}}\, dv_{\mathrm{R}} + \int_{\mathcal{B}} (\operatorname{Div} \mathbf{j}_{\mathrm{R}})\, \mu_{\text{test}}\, dv_{\mathrm{R}} - \int_{\partial\mathcal{B}} \mathbf{j}_{\mathrm{R}} \cdot \mathbf{n}_{\mathrm{R}}\, \mu_{\text{test}}\, da_{\mathrm{R}} + \int_{\partial\mathcal{B}_{\mathrm{j_R}}} \bar{j}_{\mathrm{R}}\, \mu_{\text{test}}\, da_{\mathrm{R}} = 0 \quad (14.1.5)$$

Next, by the divergence theorem

$$\int_{\partial\mathcal{B}} \mathbf{j}_{\mathrm{R}} \cdot \mathbf{n}_{\mathrm{R}}\, \mu_{\text{test}}\, da_{\mathrm{R}} = \int_{\partial\mathcal{B}} (\mu_{\text{test}}\, \mathbf{j}_{\mathrm{R}}) \cdot \mathbf{n}\, da_{\mathrm{R}} = \int_{\mathcal{B}} \mu_{\text{test}} \operatorname{Div} \mathbf{j}_{\mathrm{R}}\, dv_{\mathrm{R}} + \int_{\mathcal{B}} \mathbf{j}_{\mathrm{R}} \cdot \nabla \mu_{\text{test}}\, dv_{\mathrm{R}}. \quad (14.1.6)$$

Use of (14.1.6) in (14.1.5) gives the weak form

$$\int_{\mathcal{B}} \dot{c}_{\mathrm{R}}\, \mu_{\text{test}}\, dv_{\mathrm{R}} - \int_{\mathcal{B}} \mathbf{j}_{\mathrm{R}} \cdot \nabla \mu_{\text{test}}\, dv_{\mathrm{R}} + \int_{\partial\mathcal{B}_{\mathrm{j_R}}} \bar{j}_{\mathrm{R}}\, \mu_{\text{test}}\, da_{\mathrm{R}} = 0. \quad (14.1.7)$$

We use a first-order finite difference approximation to write

$$\dot{c}_{\mathrm{R}} = \frac{c_{\mathrm{R}} - c_{\mathrm{R\,old}}}{\Delta t}, \quad (14.1.8)$$

where c_{R} is the value of the fluid content at the end of the step, so that the weak form (14.1.7) may be expressed as

$$\int_{\mathcal{B}} \frac{c_{\mathrm{R}} - c_{\mathrm{R\,old}}}{\Delta t}\, \mu_{\text{test}}\, dv_{\mathrm{R}} - \int_{\mathcal{B}} \mathbf{j}_{\mathrm{R}} \cdot \nabla \mu_{\text{test}}\, dv_{\mathrm{R}} + \int_{\partial\mathcal{B}_{\mathrm{j_R}}} \bar{j}_{\mathrm{R}}\, \mu_{\text{test}}\, da_{\mathrm{R}} = 0. \quad (14.1.9)$$

3. **Energy balance problem**:
Next, consider a scalar test field $\vartheta_{\text{test}}(\mathbf{X})$ in $\mathcal{B}$ which satisfies $\vartheta_{\text{test}}(\mathbf{X}) = \mathbf{0}$ on $\partial\mathcal{B}_{\vartheta}$. If the balance equation

$$c_v\, \dot{\vartheta} + \operatorname{Div} \mathbf{q}_{\mathrm{R}} - \left(q_{\mathrm{R}} + \vartheta\, \mathbf{M}^{\mathcal{H}} : \dot{\mathbf{C}}^e + \vartheta\, M^{\mathcal{H}} \dot{c}_{\mathrm{R}} - \mathbf{j}_{\mathrm{R}} \cdot \nabla \mu \right) = 0 \qquad \text{in } \mathcal{B},$$

and the boundary condition $\mathbf{q}_{\mathrm{R}} \cdot \mathbf{n} = \bar{q}_{\mathrm{R}}$ on $\partial\mathcal{B}_{\mathrm{q_R}}$ are satisfied, then

$$\int_{\mathcal{B}} c_v \dot{\vartheta}\, \vartheta_{\text{test}}\, dv_{\mathrm{R}} + \int_{\mathcal{B}} (\operatorname{Div} \mathbf{q}_{\mathrm{R}})\, \vartheta_{\text{test}}\, dv_{\mathrm{R}} + \int_{\partial\mathcal{B}_{\mathrm{q_R}}} (\bar{q}_{\mathrm{R}} - \mathbf{q}_{\mathrm{R}} \cdot \mathbf{n})\, \vartheta_{\text{test}}\, da_{\mathrm{R}}$$
$$- \int_{\mathcal{B}} \left(q_{\mathrm{R}} + \vartheta\, \mathbf{M}^{\mathcal{H}} : \dot{\mathbf{C}}^e + \vartheta\, M^{\mathcal{H}} \dot{c}_{\mathrm{R}} - \mathbf{j}_{\mathrm{R}} \cdot \nabla \mu \right) \vartheta_{\text{test}}\, dv_{\mathrm{R}} = 0, \quad (14.1.10)$$

or equivalently

$$\int_{\mathcal{B}} c_v\,\dot{\vartheta}\,\vartheta_{\text{test}}\,dv_{\mathrm{R}} + \int_{\mathcal{B}} (\operatorname{Div}\mathbf{q}_{\mathrm{R}})\,\vartheta_{\text{test}}\,dv_{\mathrm{R}} - \int_{\partial\mathcal{B}_{\mathbf{q}_{\mathrm{R}}}} \mathbf{q}_{\mathrm{R}}\cdot\mathbf{n}_{\mathrm{R}}\,\vartheta_{\text{test}}\,da_{\mathrm{R}} + \int_{\partial\mathcal{B}_{\mathbf{q}_{\mathrm{R}}}} \bar{q}_{\mathrm{R}}\,\vartheta_{\text{test}}\,da_{\mathrm{R}}$$
$$-\int_{\mathcal{B}} \Big(q_{\mathrm{R}} + \vartheta\,\mathbf{M}^{\mathcal{H}}:\dot{\mathbf{C}}^e + \vartheta\,M^{\mathcal{H}}\dot{c}_{\mathrm{R}} - \mathbf{j}_{\mathrm{R}}\cdot\nabla\mu\Big)\,\vartheta_{\text{test}}\,dv_{\mathrm{R}} = 0 \tag{14.1.11}$$

holds for all test fields ϑ_{test}. Then, since $\vartheta_{\text{test}} = 0$ on $\partial\mathcal{B}_\vartheta$, the first boundary integral in expression above may be extended to all of $\partial\mathcal{B}$,

$$\int_{\mathcal{B}} c_v\,\dot{\vartheta}\,\vartheta_{\text{test}}\,dv_{\mathrm{R}} + \int_{\mathcal{B}} (\operatorname{Div}\mathbf{q}_{\mathrm{R}})\,\vartheta_{\text{test}}\,dv_{\mathrm{R}} - \int_{\partial\mathcal{B}} \mathbf{q}_{\mathrm{R}}\cdot\mathbf{n}_{\mathrm{R}}\,\vartheta_{\text{test}}\,da_{\mathrm{R}} + \int_{\partial\mathcal{B}_{\mathbf{q}_{\mathrm{R}}}} \bar{q}_{\mathrm{R}}\,\vartheta_{\text{test}}\,da_{\mathrm{R}}$$
$$-\int_{\mathcal{B}} \Big(q_{\mathrm{R}} + \vartheta\,\mathbf{M}^{\mathcal{H}}:\dot{\mathbf{C}}^e + \vartheta\,M^{\mathcal{H}}\dot{c}_{\mathrm{R}} - \mathbf{j}_{\mathrm{R}}\cdot\nabla\mu\Big)\,\vartheta_{\text{test}}\,dv_{\mathrm{R}} = 0. \tag{14.1.12}$$

Next, by the divergence theorem

$$\int_{\partial\mathcal{B}} \mathbf{q}_{\mathrm{R}}\cdot\mathbf{n}_{\mathrm{R}}\,\vartheta_{\text{test}}\,da_{\mathrm{R}} = \int_{\partial\mathcal{B}} (\vartheta_{\text{test}}\,\mathbf{q}_{\mathrm{R}})\cdot\mathbf{n}\,da_{\mathrm{R}} = \int_{\mathcal{B}} \vartheta_{\text{test}}\operatorname{Div}\mathbf{q}_{\mathrm{R}}\,dv_{\mathrm{R}} + \int_{\mathcal{B}} \mathbf{q}_{\mathrm{R}}\cdot\nabla\vartheta_{\text{test}}\,dv_{\mathrm{R}}. \tag{14.1.13}$$

Use of (14.1.13) in (14.1.12) gives the weak form

$$\int_{\mathcal{B}} c_v\,\dot{\vartheta}\,\vartheta_{\text{test}}\,dv_{\mathrm{R}} - \int_{\mathcal{B}} \mathbf{q}_{\mathrm{R}}\cdot\nabla\vartheta_{\text{test}}\,dv_{\mathrm{R}} + \int_{\partial\mathcal{B}_{\mathbf{q}_{\mathrm{R}}}} \bar{q}_{\mathrm{R}}\,\vartheta_{\text{test}}\,da_{\mathrm{R}}$$
$$-\int_{\mathcal{B}} \Big(q_{\mathrm{R}} + \vartheta\,\mathbf{M}^{\mathcal{H}}:\dot{\mathbf{C}}^e + \vartheta\,M^{\mathcal{H}}\dot{c}_{\mathrm{R}} - \mathbf{j}_{\mathrm{R}}\cdot\nabla\mu\Big)\,\vartheta_{\text{test}}\,dv_{\mathrm{R}} = 0. \tag{14.1.14}$$

We use first-order finite difference approximations to write

$$\dot{\vartheta} = \frac{\vartheta - \vartheta_{\text{old}}}{\Delta t}, \qquad \dot{\mathbf{C}}^e = \frac{\mathbf{C}^e - \mathbf{C}^e_{\text{old}}}{\Delta t}, \qquad \text{and} \qquad \dot{c}_{\mathrm{R}} = \frac{c_{\mathrm{R}} - c_{\mathrm{R}\text{old}}}{\Delta t}, \tag{14.1.15}$$

so that the weak form (14.1.14) may be expressed as

$$\int_{\mathcal{B}} c_v\left(\frac{\vartheta - \vartheta_{\text{old}}}{\Delta t}\right)\vartheta_{\text{test}}\,dv_{\mathrm{R}} - \int_{\mathcal{B}} \mathbf{q}_{\mathrm{R}}\cdot\nabla\vartheta_{\text{test}}\,dv_{\mathrm{R}} + \int_{\partial\mathcal{B}_{\mathbf{q}_{\mathrm{R}}}} \bar{q}_{\mathrm{R}}\,\vartheta_{\text{test}}\,da_{\mathrm{R}}$$
$$-\int_{\mathcal{B}} \left(q_{\mathrm{R}} + \vartheta\,\mathbf{M}^{\mathcal{H}}:\left(\frac{\mathbf{C}^e - \mathbf{C}^e_{\text{old}}}{\Delta t}\right) + \vartheta\,M^{\mathcal{H}}\left(\frac{c_{\mathrm{R}} - c_{\mathrm{R}\text{old}}}{\Delta t}\right) - \mathbf{j}_{\mathrm{R}}\cdot\nabla\mu\right)\vartheta_{\text{test}}\,dv_{\mathrm{R}} = 0. \tag{14.1.16}$$

4. **Weak form of the implicit equation for c_R:**
As mentioned in the Remark on page 219, we also need knowledge of the species concentration c_R, which may be obtained by solving the following implicit equation,

$$\frac{\mu}{R\vartheta} - \left(\ln\left(1 - \frac{1}{1+\Omega c_R}\right) + \frac{1}{1+\Omega c_R} + \frac{\chi(\vartheta)}{(1+\Omega c_R)^2}\right) - \frac{\Omega J^e}{R\vartheta} p - \frac{\Omega}{R\vartheta}\left[\tfrac{1}{2}K(\ln J^e)^2 - 3K\alpha(\vartheta - \vartheta_0)(\ln J^e)\right] = 0. \tag{14.1.17}$$

On account of the inability of the FEniCS framework to solve nonlinear equations at the Gauss-point level, we solve equation (14.1.17) in a *weak sense* by introducing a test function c_{test} and accompanying weak form

$$\int_{\mathcal{B}} \left[\frac{\mu}{R\vartheta} - \left(\ln\left(1 - \frac{1}{1+\Omega c_R}\right) + \frac{1}{1+\Omega c_R} + \frac{\chi(\vartheta)}{(1+\Omega c_R)^2}\right) - \frac{\Omega J^e}{R\vartheta} p - \frac{\Omega}{R\vartheta}\left[\tfrac{1}{2}K(\ln J^e)^2 - 3K\alpha(\vartheta - \vartheta_0)(\ln J^e)\right]\right] c_{\text{test}}\, dv_R = 0. \tag{14.1.18}$$

Summarizing, the weak forms of the governing equations for the coupled problem are

$$\begin{aligned}
&\int_{\mathcal{B}} \mathbf{T}_R : \nabla \mathbf{u}_{\text{test}}\, dv_R - \int_{\partial\mathcal{B}_{\mathbf{t}_R}} \bar{\mathbf{t}}_R \cdot \mathbf{u}_{\text{test}}\, da_R = 0,\\
&\int_{\mathcal{B}} \left[\frac{pJ^e}{K} + (\ln J^e - 3\alpha(\vartheta - \vartheta_0))\right] p_{\text{test}}\, dv_R = 0,\\
&\int_{\mathcal{B}} \frac{c_R - c_{R\,\text{old}}}{\Delta t}\, \mu_{\text{test}}\, dv_R - \int_{\mathcal{B}} \mathbf{j}_R \cdot \nabla \mu_{\text{test}}\, dv_R + \int_{\partial\mathcal{B}_{\mathbf{j}_R}} \bar{j}_R\, \mu_{\text{test}}\, da_R = 0,\\
&\int_{\mathcal{B}} \left[\frac{\mu}{R\vartheta} - \left(\ln\left(1 - \frac{1}{1+\Omega c_R}\right) + \frac{1}{1+\Omega c_R} + \frac{\chi(\vartheta)}{(1+\Omega c_R)^2}\right)\right.\\
&\qquad \left. - \frac{\Omega J^e}{R\vartheta} p - \frac{\Omega}{R\vartheta}\left[\tfrac{1}{2}K(\ln J^e)^2 - 3K\alpha(\vartheta - \vartheta_0)(\ln J^e)\right]\right] c_{\text{test}}\, dv_R = 0,\\
&\int_{\mathcal{B}} c_v \left(\frac{\vartheta - \vartheta_{\text{old}}}{\Delta t}\right) \vartheta_{\text{test}}\, dv_R - \int_{\mathcal{B}} \mathbf{q}_R \cdot \nabla \vartheta_{\text{test}}\, dv_R + \int_{\partial\mathcal{B}_{\mathbf{q}_R}} \bar{q}_R\, \vartheta_{\text{test}}\, da_R\\
&\qquad - \int_{\mathcal{B}} \left(q_R + \vartheta\, \mathbf{M}^{\mathcal{H}} : \left(\frac{\mathbf{C}^e - \mathbf{C}^e_{\text{old}}}{\Delta t}\right) + \vartheta\, M^{\mathcal{H}} \left(\frac{c_R - c_{R\text{old}}}{\Delta t}\right) - \mathbf{j}_R \cdot \nabla \mu\right) \vartheta_{\text{test}}\, dv_R = 0.
\end{aligned} \tag{14.1.19}$$

14.2 Normalization

Since we plan to solve all the governing equations related to multiple physical fields $\{\mathbf{u}, p, \mu, c_{\text{R}}, \vartheta\}$ simultaneously, it is beneficial to rewrite the residuals so that they have approximately the same magnitudes. To this end:

1. The species concentration c_{R} mol/m^3 is normalized using the molar volume Ω [m^3/mol] as

$$\hat{c} \stackrel{\text{def}}{=} \Omega c_{\text{R}}. \tag{14.2.1}$$

2. The chemical potential μ [J/mol] is normalized by $R\vartheta$ [J/mol],

$$\hat{\mu} = \frac{\mu}{R\vartheta}. \tag{14.2.2}$$

3. Recall the balance of fluid content equation:

$$\dot{c}_{\text{R}} + \operatorname{Div} \mathbf{j}_{\text{R}} = 0.$$

   This may be written as,

$$\frac{1}{\Omega}\dot{\hat{c}} + \operatorname{Div} \mathbf{j}_{\text{R}} = 0.$$

4. Next, the species flux $\mathbf{j}_{\text{R}}$ is given by the constitutive equation,

$$\mathbf{j}_{\text{R}} = -\mathbf{M}\nabla\mu \qquad \text{with} \qquad \mathbf{M} = \mathbf{C}^{-1}\left(\frac{Dc_{\text{R}}}{R\vartheta}\right). \tag{14.2.3}$$

   Using the normalized variables $\hat{\mu}$ and $\hat{c}$ we may rewrite it as follows:

$$\mathbf{j}_{\text{R}} = -\mathbf{M}\nabla\hat{\mu} \times R\vartheta \qquad \text{with} \qquad \mathbf{M} = \mathbf{C}^{-1}\left(\frac{D\hat{c}}{\Omega R\vartheta}\right). \tag{14.2.4}$$

5. Finally we choose to normalize the stress with the shear modulus $G_0 = Nk_B\vartheta$:

$$\hat{\mathbf{T}}_{\text{R}} \stackrel{\text{def}}{=} \frac{\mathbf{T}_{\text{R}}}{G_0} = \left[\zeta\mathbf{F} - \zeta_0\mathbf{F}^{-\top}\right] - \frac{Jp}{G_0}\mathbf{F}^{-\top}. \tag{14.2.5}$$

6. Summarizing, the governing equations may then be expressed as

$$
\begin{aligned}
\operatorname{Div}\hat{\mathbf{T}}_{\mathrm{R}} &= 0\\
\frac{pJ^e}{K} + (\ln J^e - 3\alpha(\vartheta-\vartheta_0)) &= 0,\\
\frac{1}{\Omega}\dot{\hat{c}} + \operatorname{Div}\mathbf{j}_{\mathrm{R}} &= 0,\\
\hat{\mu} - \left(\ln\left(1-\frac{1}{1+\hat{c}}\right) + \frac{1}{1+\hat{c}} + \frac{\chi}{(1+\hat{c})^2}\right) - \frac{\Omega\,J^e}{R\vartheta}\,p &\\
-\frac{\Omega}{R\vartheta}\left[\tfrac{1}{2}K(\ln J^e)^2 - 3K\alpha(\vartheta-\vartheta_0)(\ln J^e)\right] &= 0,\\
c_v\dot{\vartheta} + \operatorname{Div}\mathbf{q}_{\mathrm{R}} - q_{\mathrm{R}} - \mathcal{H} &= 0.
\end{aligned}
\tag{14.2.6}
$$

With the normalizations introduced above and the normalized parameters denoted by $\{\hat{\bullet}\}$, the weak forms of the coupled problem may be written as

$$
\begin{aligned}
\int_{\mathcal{B}} \hat{\mathbf{T}}_{\mathrm{R}} : \nabla\mathbf{u}_{\text{test}}\,dv_{\mathrm{R}} - \int_{\partial\mathcal{B}_{\mathbf{t}_{\mathrm{R}}}} \bar{\hat{\mathbf{t}}}_{\mathrm{R}}\cdot\mathbf{u}_{\text{test}}\,da_{\mathrm{R}} &= 0,\\
\int_{\mathcal{B}} \left[\frac{pJ^e}{K} + (\ln J^e - 3\alpha(\vartheta-\vartheta_0))\right] p_{\text{test}}\,dv_{\mathrm{R}} &= 0,\\
\int_{\mathcal{B}} \frac{\hat{c}-\hat{c}_{\text{old}}}{\Delta t}\,\hat{\mu}_{\text{test}}\,dv_{\mathrm{R}} - \int_{\mathcal{B}} \Omega\,\mathbf{j}_{\mathrm{R}}\cdot\nabla\hat{\mu}_{\text{test}}\,dv_{\mathrm{R}} + \int_{\partial\mathcal{B}_{\mathbf{j}_{\mathrm{R}}}} \Omega\,\bar{j}_{\mathrm{R}}\,\hat{\mu}_{\text{test}}\,da_{\mathrm{R}} &= 0,\\
\int_{\mathcal{B}} \Bigg[\hat{\mu} - \left(\ln\left(1-\frac{1}{1+\hat{c}}\right) + \frac{1}{1+\hat{c}} + \frac{\chi(\vartheta)}{(1+\hat{c})^2}\right) - \frac{\Omega\,J^e}{R\vartheta}\,p &\\
-\frac{\Omega}{R\vartheta}\left[\tfrac{1}{2}K(\ln J^e)^2 - 3K\alpha(\vartheta-\vartheta_0)(\ln J^e)\right]\Bigg]\,\hat{c}_{\text{test}}\,dv_{\mathrm{R}} &= 0,\\
\int_{\mathcal{B}} c_v\,(\vartheta-\vartheta_{\text{old}})\,\vartheta_{\text{test}}\,dv_{\mathrm{R}} - \int_{\mathcal{B}} \Delta t\,\mathbf{q}_{\mathrm{R}}\cdot\nabla\vartheta_{\text{test}}\,dv_{\mathrm{R}} &\\
+\int_{\partial\mathcal{B}_{\mathbf{q}_{\mathrm{R}}}} \Delta t\,\bar{q}_{\mathrm{R}}\,\vartheta_{\text{test}}\,da_{\mathrm{R}} - \int_{\mathcal{B}} \Bigg(\Delta t\,q_{\mathrm{R}} + \vartheta\,\mathbf{M}^{\mathcal{H}} : (\mathbf{C}^e - \mathbf{C}^e_{\text{old}}) &\\
+\vartheta\,M^{\mathcal{H}}\left(\frac{\hat{c}-\hat{c}_{\text{old}}}{\Omega}\right) - \Delta t\,R\vartheta\,\mathbf{j}_{\mathrm{R}}\cdot\nabla\hat{\mu}\Bigg)\,\vartheta_{\text{test}}\,dv_{\mathrm{R}} &= 0.
\end{aligned}
\tag{14.2.7}
$$

REMARK

Note that in writing the weak form for the heat equation above we have multiplied through by Δt. This helps in the numerical implementation.

14.3 Solution procedure in the FEniCS code

In brief, in FEniCS we use the following degrees of freedom

$$\{\mathbf{u},\, p,\, \hat{\mu},\, \hat{c}, \vartheta\}, \tag{14.3.1}$$

as well as their known values from the previous step

$$\{\mathbf{u}_{\text{old}},\, p_{\text{old}},\, \hat{\mu}_{\text{old}},\, \hat{c}_{\text{old}},\, \vartheta_{\text{old}}\}, \tag{14.3.2}$$

to calculate the *residuals* which correspond to the weak forms of the governing equations, viz.

$$\begin{aligned}
R_{\mathbf{u}} &= \int_{\mathcal{B}} \hat{\mathbf{T}}_{\text{R}} \colon \nabla \mathbf{u}_{\text{test}}\, dv_{\text{R}} - \int_{\partial \mathcal{B}_{\mathbf{t}_{\text{R}}}} \bar{\hat{\mathbf{t}}}_{\text{R}} \cdot \mathbf{u}_{\text{test}}\, da_{\text{R}}, \\
R_p &= \int_{\mathcal{B}} \left[\frac{pJ^e}{K} + (\ln J^e - 3\alpha(\vartheta - \vartheta_0)) \right] p_{\text{test}}\, dv_{\text{R}}, \\
R_{\hat{\mu}} &= \int_{\mathcal{B}} \frac{\hat{c} - \hat{c}_{\text{old}}}{\Delta t}\, \hat{\mu}_{\text{test}}\, dv_{\text{R}} - \int_{\mathcal{B}} \Omega\, \mathbf{j}_{\text{R}} \cdot \nabla \hat{\mu}_{\text{test}}\, dv_{\text{R}} + \int_{\partial \mathcal{B}_{\mathbf{j}_{\text{R}}}} \Omega\, \bar{j}_{\text{R}}\, \hat{\mu}_{\text{test}}\, da_{\text{R}}, \\
R_{\hat{c}} &= \int_{\mathcal{B}} \Bigg[\hat{\mu} - \left(\ln\left(1 - \frac{1}{1+\hat{c}}\right) + \frac{1}{1+\hat{c}} + \frac{\chi}{(1+\hat{c})^2} \right) - \frac{\Omega\, J^e}{R\vartheta}\, p \\
&\qquad - \frac{\Omega}{R\vartheta} \left[\tfrac{1}{2} K (\ln J^e)^2 - 3K\alpha(\vartheta - \vartheta_0)(\ln J^e) \right] \Bigg]\, \hat{c}_{\text{test}}\, dv_{\text{R}}, \\
R_{\vartheta} &= \int_{\mathcal{B}} c_v\, (\vartheta - \vartheta_{\text{old}})\, \vartheta_{\text{test}}\, dv_{\text{R}} - \int_{\mathcal{B}} \Delta t\, \mathbf{q}_{\text{R}} \cdot \nabla \vartheta_{\text{test}}\, dv_{\text{R}} + \int_{\partial \mathcal{B}_{\mathbf{q}_{\text{R}}}} \Delta t\, \bar{q}_{\text{R}}\, \vartheta_{\text{test}}\, da_{\text{R}} \\
&\qquad - \int_{\mathcal{B}} \Bigg(\Delta t\, q_{\text{R}} + \vartheta\, \mathbf{M}^{\mathcal{H}} \colon (\mathbf{C}^e - \mathbf{C}^e_{\text{old}}) + \vartheta\, M^{\mathcal{H}} \left(\frac{\hat{c} - \hat{c}_{\text{old}}}{\Omega} \right) \\
&\qquad\qquad - \Delta t\, R\vartheta\, \mathbf{j}_{\text{R}} \cdot \nabla \hat{\mu} \Bigg)\, \vartheta_{\text{test}}\, dv_{\text{R}}.
\end{aligned} \tag{14.3.3}$$

FEniCS then differentiates these residuals with respect to the separate degrees of freedom to obtain *tangents*. At each step in time, FEniCS uses the residuals and tangents in a global Newton–Raphson solver to determine current values of the degrees of freedom $\{\mathbf{u},\, p,\, \hat{\mu},\, \vartheta,\, \hat{c}\}$

for which the residuals are zero,[1] and which therefore satisfy the governing equations as well as the initial and boundary conditions (in a weak sense).

REMARK

A mixed finite element method is utilized to solve for the displacement $\mathbf{u}$, the pressure p, the chemical potential $\hat{\mu}$, the concentration field $\hat{c}$, and the temperature field ϑ concurrently. To avoid the numerical instabilities the interpolation order for the displacement is taken as Lagrange quadratic, while that of the other variables are interpolated one order lower, Lagrange linear.

14.4 Algorithm for calculation of the residuals

The algorithm we follow in our code to calculate the residuals in (11.3.3) is summarized below.

1. Begin the step with known values of

$$\{\mathbf{u}_{\text{old}},\, p_{\text{old}},\, \hat{\mu}_{\text{old}},\, \hat{c}_{\text{old}},\, \vartheta_{\text{old}}\}, \tag{14.4.1}$$

   where for the first step these quantities are given by the initial conditions for $\mathbf{u}$, p, $\hat{\mu}$, and $\hat{c}$.

2. Calculate necessary kinematical quantities for the current step as,

$$\begin{aligned} &\mathbf{F} = \mathbf{1} + \nabla\mathbf{u}, \quad J = \det\mathbf{F}, \quad J^s = 1+\hat{c}, \quad \mathbf{F}^s = (J^s)^{1/3}\mathbf{1}, \\ &\mathbf{F}^e = \mathbf{F}\mathbf{F}^{s-1}, \quad J^e = \det\mathbf{F}^e, \quad \mathbf{C} = \mathbf{F}^\top\mathbf{F}, \quad \mathbf{C}^e = \mathbf{F}^{e\top}\mathbf{F}^e. \end{aligned} \tag{14.4.2}$$

3. Calculate the intermediate factors ζ and ζ_0:

$$\begin{aligned} &\beta \stackrel{\text{def}}{=} \mathcal{L}^{-1}\left(\frac{\bar{\lambda}}{\lambda_L}\right) \approx z\,\frac{3-z^2}{1-z^2} \qquad \text{where} \qquad z \stackrel{\text{def}}{=} \frac{\bar{\lambda}}{\lambda_L}, \\ &\zeta \stackrel{\text{def}}{=} \left(\frac{\lambda_L}{3\bar{\lambda}}\right)\beta. \end{aligned} \tag{14.4.3}$$

$$\begin{aligned} &\beta_0 \stackrel{\text{def}}{=} \mathcal{L}^{-1}\left(\frac{1}{\lambda_L}\right) \approx z\,\frac{3-z^2}{1-z^2} \qquad \text{where} \qquad z \stackrel{\text{def}}{=} \frac{1}{\lambda_L}, \\ &\zeta_0 \stackrel{\text{def}}{=} \left(\frac{\lambda_L}{3}\right)\beta_0. \end{aligned} \tag{14.4.4}$$

4. Calculate $\mathbf{M}^{\mathcal{H}}$ and $M^{\mathcal{H}}$:

$$\begin{aligned} &\mathbf{M}^{\mathcal{H}} \stackrel{\text{def}}{=} \frac{1}{2}\left(Nk_B\,(\zeta J^{s\,2/3}\mathbf{1} - \zeta_0\mathbf{C}^{e-1}) + \frac{3K\alpha}{J^e}\,\mathbf{C}^{e-1}\right), \\ &M^{\mathcal{H}} \stackrel{\text{def}}{=} R\left(\ln\left(\frac{\hat{c}}{1+\hat{c}}\right) + \frac{1}{(1+\hat{c})} + \chi\frac{1}{(1+\hat{c})^2}\right) + \frac{\Omega\,Nk_B}{1+\hat{c}}\left(\zeta\,\frac{\operatorname{tr}\mathbf{C}}{3} - \zeta_0\right). \end{aligned} \tag{14.4.5}$$

[1] To within some small numerical tolerance.

5. Calculate the normalized Piola stress:

$$\hat{\mathbf{T}}_{\mathrm{R}} = \left[\zeta\mathbf{F} - \zeta_0\mathbf{F}^{-\top}\right] - \frac{J\,p}{G_0}\,\mathbf{F}^{-\top}, \qquad G_0 = Nk_B\vartheta. \tag{14.4.6}$$

6. Calculate the species flux:

$$\mathbf{j}_{\mathrm{R}} = -\mathbf{M}^{\mathrm{mob}}\nabla\hat{\mu} \times R\vartheta \qquad \text{with} \qquad \mathbf{M}^{\mathrm{mob}} = \mathbf{C}^{-1}\left(\frac{D\hat{c}}{\Omega R\vartheta}\right). \tag{14.4.7}$$

7. Calculate the heat flux:

$$\mathbf{q}_{\mathrm{R}} = -\mathbf{K}\nabla\vartheta \qquad \text{with} \qquad \mathbf{K} = J\,k\,\mathbf{C}^{-1}. \tag{14.4.8}$$

8. Calculate the χ parameter

$$\chi = \frac{1}{2}(\chi_L + \chi_H) - \frac{1}{2}(\chi_L - \chi_H)\tanh\left(\frac{\vartheta - \vartheta_T}{\Delta}\right). \tag{14.4.9}$$

9. Compute the residuals:

$$\begin{aligned}
R_{\mathbf{u}} &= \int_{\mathcal{B}} \hat{\mathbf{T}}_{\mathrm{R}} : \nabla\mathbf{u}_{\text{test}}\,dv_{\mathrm{R}} - \int_{\partial\mathcal{B}_{\mathbf{t}_{\mathrm{R}}}} \bar{\hat{\mathbf{t}}}_{\mathrm{R}} \cdot \mathbf{u}_{\text{test}}\,da_{\mathrm{R}},\\
R_p &= \int_{\mathcal{B}} \left[\frac{pJ^e}{K} + (\ln J^e - 3\alpha(\vartheta - \vartheta_0))\right] p_{\text{test}}\,dv_{\mathrm{R}},\\
R_{\hat{\mu}} &= \int_{\mathcal{B}} \frac{\hat{c} - \hat{c}_{\text{old}}}{\Delta t}\,\hat{\mu}_{\text{test}}\,dv_{\mathrm{R}} - \int_{\mathcal{B}} \Omega\,\mathbf{j}_{\mathrm{R}} \cdot \nabla\hat{\mu}_{\text{test}}\,dv_{\mathrm{R}} + \int_{\partial\mathcal{B}_{\mathbf{j}_{\mathrm{R}}}} \Omega\,\bar{j}_{\mathrm{R}}\,\hat{\mu}_{\text{test}}\,da_{\mathrm{R}},\\
R_{\hat{c}} &= \int_{\mathcal{B}} \Bigg[\hat{\mu} - \left(\ln\left(1 - \frac{1}{1+\hat{c}}\right) + \frac{1}{1+\hat{c}} + \frac{\chi}{(1+\hat{c})^2}\right) - \frac{\Omega\,J^e}{R\vartheta}\,p\\
&\qquad - \frac{\Omega}{R\vartheta}\left[\tfrac{1}{2}K(\ln J^e)^2 - 3K\alpha(\vartheta - \vartheta_0)(\ln J^e)\right]\Bigg]\,\hat{c}_{\text{test}}\,dv_{\mathrm{R}},\\
R_{\vartheta} &= \int_{\mathcal{B}} c_v\,(\vartheta - \vartheta_{\text{old}})\,\vartheta_{\text{test}}\,dv_{\mathrm{R}} - \int_{\mathcal{B}} \Delta t\,\mathbf{q}_{\mathrm{R}} \cdot \nabla\vartheta_{\text{test}}\,dv_{\mathrm{R}} + \int_{\partial\mathcal{B}_{\mathbf{q}_{\mathrm{R}}}} \Delta t\,\bar{q}_{\mathrm{R}}\,\vartheta_{\text{test}}\,da_{\mathrm{R}}\\
&\quad - \int_{\mathcal{B}} \Bigg(\Delta t\,q_{\mathrm{R}} + \vartheta\,\mathbf{M}^{\mathcal{H}} : (\mathbf{C}^e - \mathbf{C}^e_{\text{old}}) + \vartheta\,M^{\mathcal{H}}\left(\frac{\hat{c} - \hat{c}_{\text{old}}}{\Omega}\right)\\
&\qquad - \Delta t\,R\vartheta\,\mathbf{j}_{\mathrm{R}} \cdot \nabla\hat{\mu}\Bigg)\,\vartheta_{\text{test}}\,dv_{\mathrm{R}}.
\end{aligned} \tag{14.4.10}$$

10. FEniCS solves for the current values of $\{\mathbf{u},\, p,\, \hat{\mu},\, \hat{c},\, \vartheta\}$.
11. Set these current values as the "old" values for the next step in time,

$$\{\mathbf{u},\, p,\, \hat{\mu},\, \hat{c},\, \vartheta\} \implies \{\mathbf{u}_{\text{old}},\, p_{\text{old}},\, \hat{\mu}_{\text{old}},\, \hat{c}_{\text{old}},\, \vartheta\}. \tag{14.4.11}$$

12. Repeat for the desired number of steps in time.

15 Representative simulations of thermally responsive gels

In our numerical simulations in FEniCS we will use the basic units listed as SI (mm) in Table 15.1. Representative values of the material parameters for an elastomeric gel are listed in Table 15.2.

Table 15.1 Basic units.

Quantity	SI (m)	SI (mm)
Length	m	mm
Mass	kg	tonne = 1000 kg
Time	s	s
Mass density	kg/m^3	tonne/mm^3
Force	N	N
Stress	Pa	MPa
Energy	J	mJ
Temperature	K	K
Amount of substance	mol	mol
c_{R}, species concentration	mol/m^3	mol/mm^3
μ, chemical potential	J/mol	mJ/mol
Ω, molar volume	m^3/mol	mm^3/mol
D, species diffusivity	m^2/s	mm^2/s
α, coefficient of thermal expansion	1/K	1/K
c_v, specific heat	J/(m^3 K)	mJ/(mm^3 K)
k, thermal conductivity	W/(m K)	mW/(mm K)
k_B, Boltzmann's constant	1.38×10^{-23} J/K	1.38×10^{-20} mJ/K
R, gas constant	8.3145 J/(mol K)	8.3145×10^3 mJ/(mol K)

Introduction to Coupled Theories in Solid Mechanics. Lallit Anand, Eric M. Stewart, Shawn A. Chester, Oxford University Press.
 DOI: 10.1093/oso/9780198986218.003.0016

Table 15.2 Material parameters for a thermoresponsive elastomeric gel.

Quantity	Value SI (m)	Value SI (mm)
k_B, Boltzmann's constant	1.38×10^{-23} J/K	1.38×10^{-20} mJ/K
R, gas constant	8.3145 J/(mol K)	8.3145×10^3 mJ/(mol K)
ϑ_0, reference temperature	298 K	298 K
G_0, ground-state shear modulus	0.3×10^6 Pa	0.3 MPa
$N = G_0/(k_B\vartheta_0)$, number of chains per unit ref. vol.	#/m^3	#/mm^3
λ_L, locking stretch	5.2	5.2
K, bulk modulus $(1000 \times G_0)$	0.3×10^9 Pa	0.3×10^3 MPa
Ω, molar volume	1×10^{-4} m^3/mol	1×10^5 mm^3/mol
D, diffusivity	5×10^{-9} m^2/s	5×10^{-3} mm^2/s
χ_L, Flory–Huggins parameter, low value	0.1	0.1
χ_H, Flory–Huggins parameter, high value	0.7	0.7
ϑ_T, transition temperature	307 K	307 K
Δ, transition temperature width	5 K	5K
α, coefficient of thermal expansion	70×10^{-6}/K	70×10^{-6}/K
$c_v = \rho_R c$, specific heat per unit volume	4180×10^3 J/(m^3 K)	4.18 mJ/(mm^3 K)
k, thermal conductivity	0.53 W/(m K)	0.53 mW/(mm K)
ϕ_0, initial polymer volume fraction	0.999	0.999

15.1 Swelling, thermal deswelling, and reswelling of a body in plane strain

As a first example we consider a plane-strain analysis of an initially dry square-shaped body which undergoes traction-free thermally activated swelling and deswelling. We consider an initially square domain of edge length $2L_0 = 5$ mm, cf. Fig. 15.1, and make use of appropriate symmetries so that only one quarter of the domain is modeled and meshed with 15 triangular elements along each edge.

Initial conditions: For an initially dry body the polymer volume fraction is set as $\phi_0 = 0.999$, and we set the initial temperature as $\vartheta_0 = 298$ K. Equation (13.4.3), when expressed in terms of the normalized concentration $\hat{c} = \Omega c_R$, is

$$\phi = \frac{1}{1+\hat{c}}, \tag{15.1.1}$$

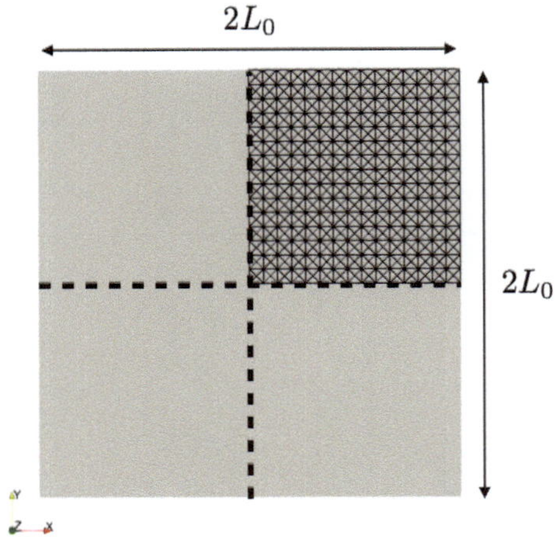

Fig. 15.1 The initial geometry and mesh of a two-dimensional square domain of edge length $2L_0 = 5\,\text{mm}$ used for the plane-strain simulation of thermally activated swelling/deswelling. Only one-quarter of the square domain is modeled. The modeled domain has been reflected about both symmetry planes to show the full geometry.

and this gives an initial value of $\hat{c}_0 = 0.001$ for the normalized concentration. Further, equation (13.4.23) for the chemical potential, with $\mu^{\text{ref}} = 0$, $\chi_L = 0.1$, $\phi_0 = 0.999$, and $J^e = 1$, gives the initial value of the normalized chemical potential $\hat{\mu} = \mu/(R\vartheta)$ as

$$\hat{\mu}_0 = \ln(1 - \phi_0) + \phi_0 + \chi_L \phi_0^2 = -5.8. \tag{15.1.2}$$

To these initial conditions we add the initial condition of zero displacement, and take the complete set of initial conditions for the problem at hand as[1]

$$\mathbf{u}_0 = \mathbf{0}, \qquad \hat{\mu}_0 = -5.8, \qquad \hat{c}_0 = 0.001, \qquad \text{and} \quad \vartheta_0 = 298\,\text{K}. \tag{15.1.3}$$

The simulation is carried out in three steps:

(i) In the first step the body is freely swollen at $\vartheta_0 = 298\,\text{K}$ for 6 h.

(ii) In the second step the surface temperature is increased to $\vartheta = 323\,\text{K}$ leading to deswelling over 8 h.

(iii) In the third step the surface temperature is lowered back to $\vartheta_0 = 298\,\text{K}$ and the body reswells over 6 h.

The prescribed surface temperature history is schematically shown in Fig. 15.2.

Boundary conditions: Referring to Figs. 15.1 and 15.2 the boundary conditions are as follows:

- Step 1 (step time 6 h):
  - Mechanical boundary conditions: (i) the left and bottom faces of the meshed domain shown in Fig. 15.1 are prescribed appropriate symmetry conditions, and (ii) the top and right faces of the domain are traction-free.

[1] In all the subsequent example problems discussed in this section we will use the initial conditions (15.1.3) for an initially-dry reference body.

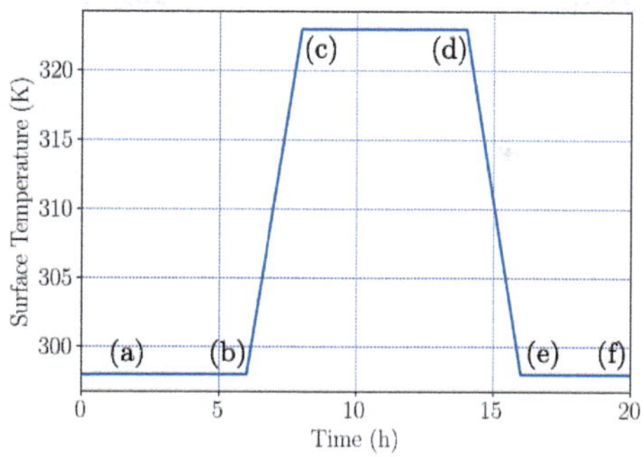

Fig. 15.2 The prescribed surface temperature boundary condition. The points labelled (a) through (f) correspond to the deformed shapes in Fig. 15.3.

- Thermal boundary conditions: (i) the left and bottom faces of the meshed domain are prescribed zero heat flux due to symmetry, and (ii) the top and right faces of the domain are prescribed a constant temperature of $\bar{\vartheta} = 298\,\text{K}$.
- Normalized chemical potential boundary conditions: (i) the left and bottom faces of the domain are prescribed zero fluid flux due to symmetry, and (ii) the top and right faces of the domain are prescribed a normalized chemical potential history in the form

$$\bar{\hat{\mu}}(t) = \hat{\mu}_0 \exp\left(-t/300\right), \tag{15.1.4}$$

so that the normalized chemical potential on those faces increases smoothly from its initial value $\hat{\mu}_0 = -5.8$ and approaches $\bar{\hat{\mu}} = 0$. This models the surroundings of the body as a zero-pressure fluid bath.

- Step 2 (step time 8 h):
  - Mechanical boundary conditions: The mechanical boundary conditions are maintained from the previous step.
  - Thermal boundary conditions: (i) the left and bottom faces of the meshed domain are prescribed zero heat flux due to symmetry, and (ii) the top and right faces of the meshed domain are prescribed a linear temperature increase from $\bar{\vartheta} = 298\,\text{K}$ to $\bar{\vartheta} = 323\,\text{K}$ over a period of 2 h, and then held constant for another 6 h.
  - Normalized chemical potential boundary conditions: The normalized chemical potential boundary conditions are maintained from the previous step.
- Step 3 (step time 6 h):
  - Mechanical boundary conditions: The mechanical boundary conditions are maintained from the previous step.
  - Thermal boundary conditions: (i) the left and bottom faces of the meshed domain are prescribed zero heat flux due to symmetry, and (ii) the top and right faces of the meshed domain are prescribed a linear temperature decrease from $\bar{\vartheta} = 323\,\text{K}$ to $\bar{\vartheta} = 298\,\text{K}$ over a period of 2 h, and then held constant for another 4 h.
  - Normalized chemical potential boundary conditions: The normalized chemical potential boundary conditions are maintained from the previous step.

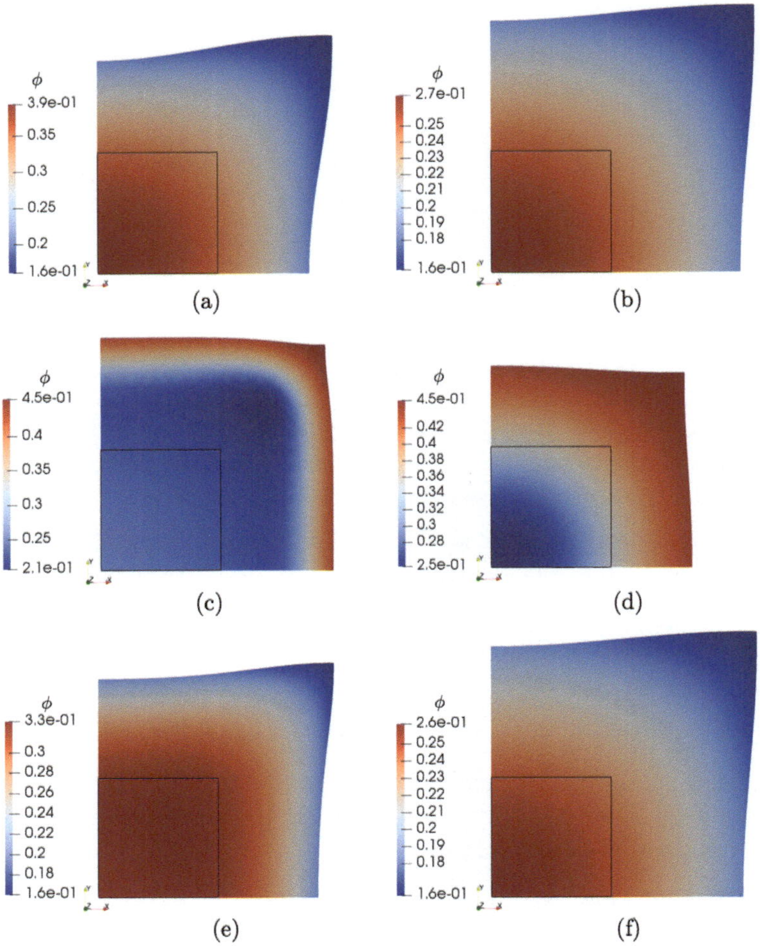

Fig. 15.3 Swollen configurations and contours of the polymer volume fraction ϕ at different times: (a) 2 h, (b) 6 h, (c) 8 h, (d) 14 h, (e) 16 h, and (f) 20 h. The initial dry body is shown by the thick black line. Note the change in scale for the contours of ϕ in each subfigure.

Figures 15.3(a)–(f) show snapshots of the contours of the polymer volume fraction ϕ on the deformed body at times $t = 2\,\text{h}$, $6\,\text{h}$, $8\,\text{h}$, $14\,\text{h}$, $16\,\text{h}$, and $20\,\text{h}$, respectively. Also, Fig. 15.4 shows plots of the polymer volume fraction along the bottom edge of the domain $(x_1 \in [0, L_0], x_2 = 0)$ at these times.

In Step 1, during the initial free swelling shown in Figs. 15.3(a) and (b), the body undergoes an inhomogeneous transient swelling as it absorbs fluid from the surroundings. This is also visible in Fig. 15.4 from the curves for 2 h and 6 h. The polymer volume fraction decreases over time, with the rate of decrease near the surface being higher than that in the interior of the body. Since the polymer volume fraction is not uniform even after 6 h of free swelling, the body is still not in equilibrium.

Next upon heating in Step 2, Figs. 15.3(c) and (d) show that the body reverses course and deswells as the fluid is driven out of the body. The deswelling process is controlled by the temperature-dependence of the χ-parameter, cf. eq. (13.4.6). Once the temperature rises above the "lower critical solution temperature (LCST)" of $\vartheta_T = 307\,\text{K}$, the interaction between the fluid and gel is no longer amenable for swelling and the fluid is driven out of the body. As seen in Fig. 15.4 from the curves for 8 h and 14 h, the polymer volume fraction at the surface increases as fluid leaves the body, with the rate of increase of the polymer volume fraction near the surface being higher than that in the interior of the body. A skin of relatively dry material forms around a wet core which causes the deformed shape in deswelling to be quite different from that during swelling. Also, the fluid diffusion is relatively slower across this drier surface layer, so that even at a time of 14 h there is a relatively large variation in the polymer volume fraction in the body.

Finally upon cooling in Step 3, Figs. 15.3(e) and (f) show that the body once again begins to swell. In this step after a short transient, the behavior begins to resemble that in the initial free-swelling Step 1.

The most notable feature that one observes from these simulations is the large reversible swelling–deswelling that occurs due to the changes in the temperature of the environment of the gel. Also, the deswelling response of the gel is remarkably different from the swelling response because of the formation of the "dry" skin near the surface of the gel during deswelling—this leads to a much reduced rate of volume change during deswelling than during swelling.

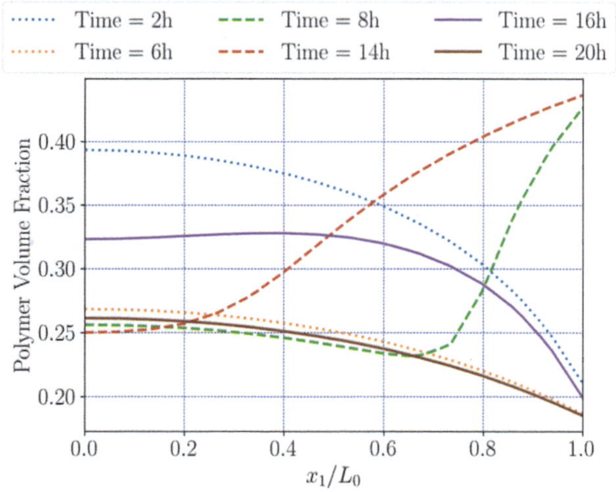

Fig. 15.4 Plots of the polymer volume fraction along the bottom edge of the simulated domain $(x_1 \in [0, L_0], x_2 = 0)$ at various times $t = 2, 6, 8, 14, 16,$ and 20 h corresponding to the deformed configurations in Fig. 15.3.

15.2 Swelling, thermal deswelling, and reswelling of an axisymmetric cylinder

As a second example we consider an axisymmetric simulation of an initially dry cylindrical body which undergoes traction-free thermally activated swelling and deswelling. We consider a cylinder of radius $R_0 = 2.5\,\text{mm}$ and height $H_0 = 5\,\text{mm}$. We make use of appropriate symmetries so that only one half of the axisymmetric slice is modeled and meshed with 15 triangular elements along each edge; cf. Fig. 15.5. The initial conditions are

$$\mathbf{u}_0 = \mathbf{0}, \qquad \hat{\mu}_0 = -5.8, \qquad \hat{c}_0 = 0.001, \qquad \text{and} \quad \vartheta_0 = 298\,\text{K}.$$

Much like the prior example, the simulation is carried out in three steps: (i) in the first step the body is freely swollen at $\vartheta_0 = 298\,\text{K}$ for 2 h; (ii) in the second step the surface temperature is increased to $\vartheta = 323\,\text{K}$ leading to deswelling over 3 h; and (iii) in the third step the surface temperature is lowered back to $\vartheta_0 = 298\,\text{K}$ and the body reswells over 3 h; cf. Fig. 15.6.

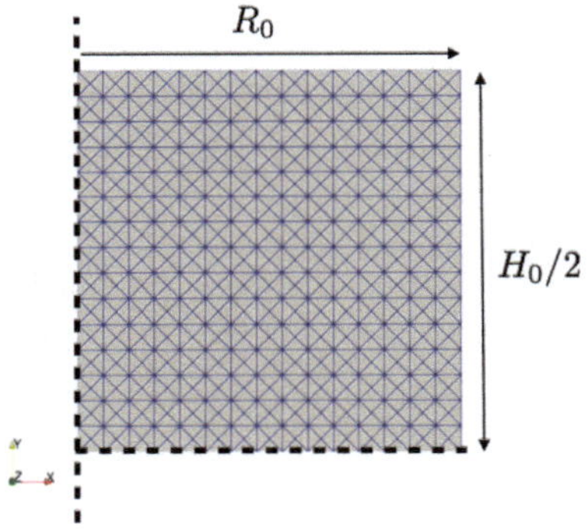

Fig. 15.5 The initial geometry and mesh with dimensions $R_0 = 2.5\,\text{mm}$ and $H_0 = 5\,\text{mm}$ used in the simulation of thermally activated free-swelling of an axisymmetric cylinder.

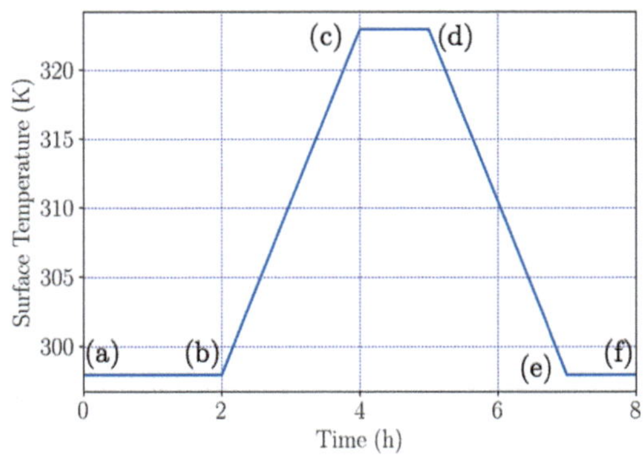

Fig. 15.6 The prescribed surface temperature boundary condition. The points labelled (a) through (f) correspond to the deformed shapes in Fig. 15.7.

Boundary conditions: Referring to Figs. 15.5 and 15.6 the boundary conditions are as follows:

- Step 1 (step time 2 h):
  - Mechanical boundary conditions: (i) the left and bottom faces of the domain shown in Fig. 15.5 are prescribed appropriate symmetry conditions, and (ii) the top and right faces of the domain are traction-free.
  - Thermal boundary conditions: (i) the left and bottom faces of the domain are prescribed zero heat flux due to symmetry, and (ii) the top and right faces of the domain are prescribed a constant temperature of $\bar{\vartheta} = 298\,\mathrm{K}$.
  - Normalized chemical potential boundary conditions: (i) the left and bottom faces of the domain are prescribed zero fluid flux due to symmetry, and (ii) the top and right faces of the domain are prescribed a normalized chemical potential history in the form

$$\bar{\hat{\mu}}(t) = \hat{\mu}_0 \exp\left(-t/300\right),$$

    so that the normalized chemical potential on those faces increases smoothly from its initial value $\hat{\mu}_0 = -5.8$ and approaches $\bar{\hat{\mu}} = 0$. This models the surroundings of the body as a zero-pressure fluid bath.

- Step 2 (step time 3 h):
  - Mechanical boundary conditions: The mechanical boundary conditions are maintained from the previous step.
  - Thermal boundary conditions: (i) the left and bottom faces of the domain are prescribed zero heat flux due to symmetry, and (ii) the top and right faces of the domain are prescribed a linear temperature increase from $\bar{\vartheta} = 298\,\mathrm{K}$ to $\bar{\vartheta} = 323\,\mathrm{K}$ over a period of 2 h, and then held constant for another 1 h.
  - Normalized chemical potential boundary conditions: The normalized chemical potential boundary conditions are maintained from the previous step.

- Step 3 (step time 3 h):
  - Mechanical boundary conditions: The mechanical boundary conditions are maintained from the previous step.
  - Thermal boundary conditions: (i) the left and bottom faces of the domain are prescribed zero heat flux due to symmetry, and (ii) the top and right faces of the domain are prescribed a linear temperature decrease from $\bar{\vartheta} = 323\,\mathrm{K}$ to $\bar{\vartheta} = 298\,\mathrm{K}$ over a period of 2 h, and then held constant for another 1 h.
  - Normalized chemical potential boundary conditions: The normalized chemical potential boundary conditions are maintained from the previous step.

Figures 15.7(a)–(f) show snapshots of the contours of the polymer volume fraction ϕ on the deformed body at times $t = 0\,\mathrm{h}$, 2 h, 4 h, 5 h, 7 h, and 8 h, respectively. Much like the prior example, upon cooling the swollen body begins to deswell as $\chi(\vartheta)$ changes. And upon heating the body reswells and shows a behavior similar to that seen during free-swelling from the dry state.

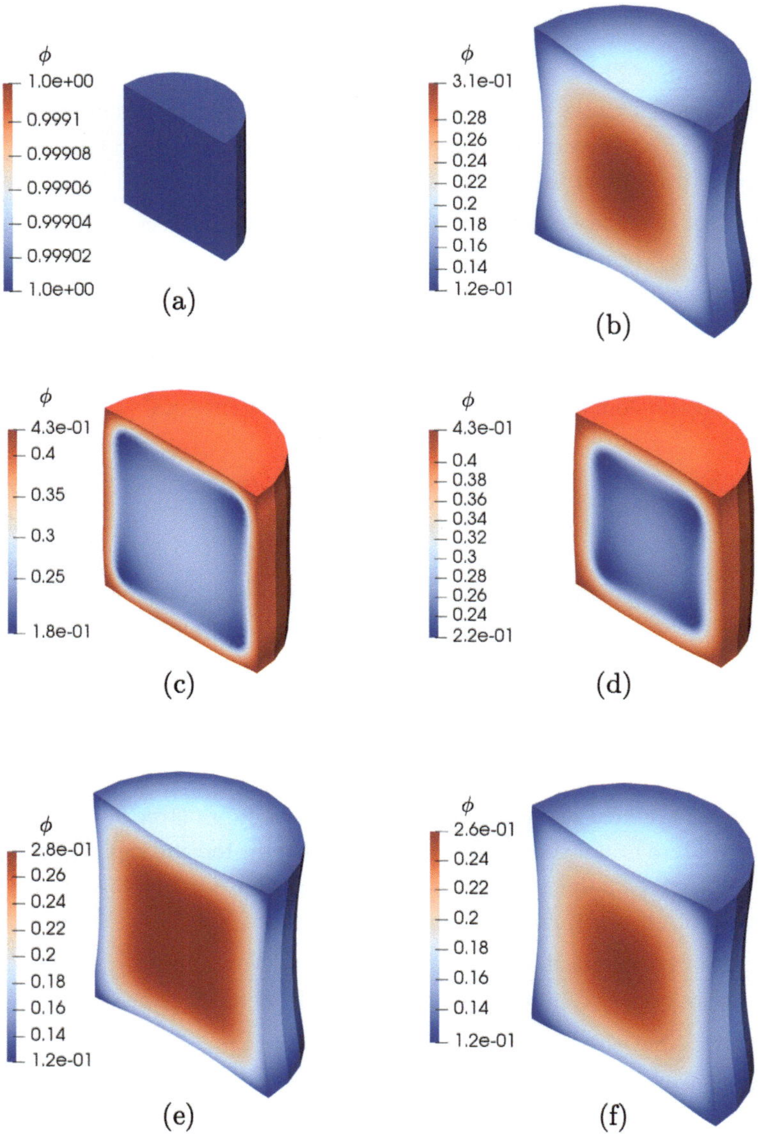

Fig. 15.7 Swollen configurations and contours of the polymer volume fraction ϕ at different times: (a) 0 h, (b) 2 h, (c) 4 h, (d) 5 h, (e) 7 h, and (f) 8 h. Note the change in scale for the contours of ϕ in each subfigure.

15.3 **Swelling, thermal deswelling, and reswelling of a sphere**

As a third example we consider a three-dimensional simulation of an initially dry spherical body which undergoes traction-free thermally activated swelling and deswelling. We consider a sphere of radius $R_0 = 2.5\,\text{mm}$. We make use of appropriate symmetries so that only one eighth of the sphere is modeled and meshed with 10 tetrahedral elements along each edge; cf. Fig. 15.8. The initial conditions are

$$\mathbf{u}_0 = \mathbf{0}, \qquad \hat{\mu}_0 = -5.8, \qquad \hat{c}_0 = 0.001, \quad \text{and} \quad \vartheta_0 = 298\,\text{K}.$$

Much like the prior examples, the simulation is carried out in three steps: (i) in the first step the body is freely swollen at $\vartheta_0 = 298$ K for 2 h; (ii) in the second step the surface temperature is increased to $\vartheta = 323$ K leading to deswelling over 3 h; and (iii) in the third step the surface temperature is lowered back to $\vartheta_0 = 298$ K and the body reswells over 3 h.

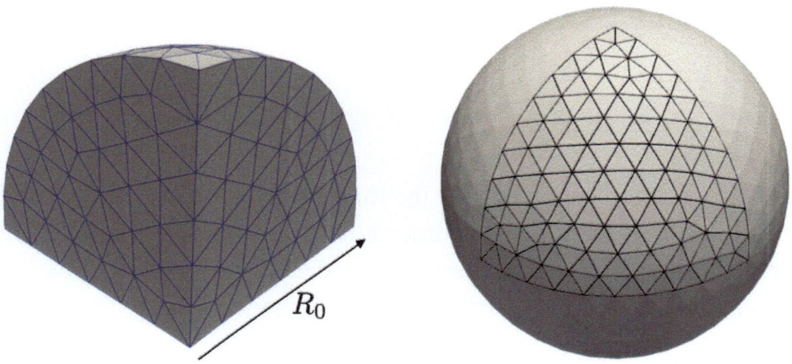

(a) (b)

Fig. 15.8 (a) The initial one eighth geometry and mesh with dimensions $R_0 = 2.5$ mm used in the simulation of thermally activated free-swelling of sphere. (b) The domain reflected about all symmetry planes to show the full geometry, keeping the mesh shown.

Boundary conditions: Referring to Fig. 15.8 the boundary conditions are as follows:

- Step 1 (step time 2 h):
  - Mechanical boundary conditions: (i) the three symmetry faces of the domain shown in Fig. 15.8(a) are prescribed appropriate symmetry conditions, and (ii) the spherical outer surface of the domain is traction-free.
  - Thermal boundary conditions: (i) the three symmetry faces of the domain are prescribed zero heat flux due to symmetry, and (ii) the spherical outer surface of the domain is prescribed a constant temperature of $\bar{\vartheta} = 298$ K.

- Normalized chemical potential boundary conditions: (i) the symmetry faces of the domain are prescribed zero fluid flux due to symmetry, and (ii) the spherical outer surface of the domain is prescribed a normalized chemical potential history in the form

$$\bar{\hat{\mu}}(t) = \hat{\mu}_0 \exp\left(-t/300\right), \tag{15.3.1}$$

so that the normalized chemical potential on that face increases smoothly from its initial value $\hat{\mu}_0 = -5.8$ and approaches $\bar{\hat{\mu}} = 0$. This models the surroundings of the body as a zero-pressure fluid bath.

- Step 2 (step time 3 h):
  - Mechanical boundary conditions: The mechanical boundary conditions are maintained from the previous step.
  - Thermal boundary conditions: (i) the three symmetry faces of the domain are prescribed zero heat flux due to symmetry, and (ii) the spherical outer surface of the domain is prescribed a linear temperature increase from $\bar{\vartheta} = 298\,\text{K}$ to $\bar{\vartheta} = 323\,\text{K}$ over a period of 2 h, and then held constant for another 1 h.
  - Normalized chemical potential boundary conditions: The normalized chemical potential boundary conditions are maintained from the previous step.
- Step 3 (step time 3 h):
  - Mechanical boundary conditions: The mechanical boundary conditions are maintained from the previous step.
  - Thermal boundary conditions: (i) the three symmetry faces of the domain are prescribed zero heat flux due to symmetry, and (ii) the spherical outer surface of the domain is prescribed a linear temperature decrease from $\bar{\vartheta} = 323\,\text{K}$ to $\bar{\vartheta} = 298\,\text{K}$ over a period of 2 h, and then held constant for another 1 h.
  - Normalized chemical potential boundary conditions: The normalized chemical potential boundary conditions are maintained from the previous step.

Figures 15.9(a)–(f) show snapshots of the contours of the polymer volume fraction ϕ on the deformed body at times $t = 0\,\text{h}, 2\,\text{h}, 4\,\text{h}, 5\,\text{h}, 7\,\text{h}$, and $8\,\text{h}$, respectively. Much like the prior example, upon cooling the swollen body begins to deswell as $\chi(\vartheta)$ changes. And upon heating the body reswells and shows a behavior similar to that seen during free-swelling from the dry state. However, distinct from the other examples, this particular geometry does not involve any abrupt surface features, such as a corner; accordingly the swollen configurations are more uniform than the prior examples.

Figure 15.10 shows how the radius of the sphere changes in time as the temperature is varying, and clearly shows the hysteresis between swelling and deswelling.

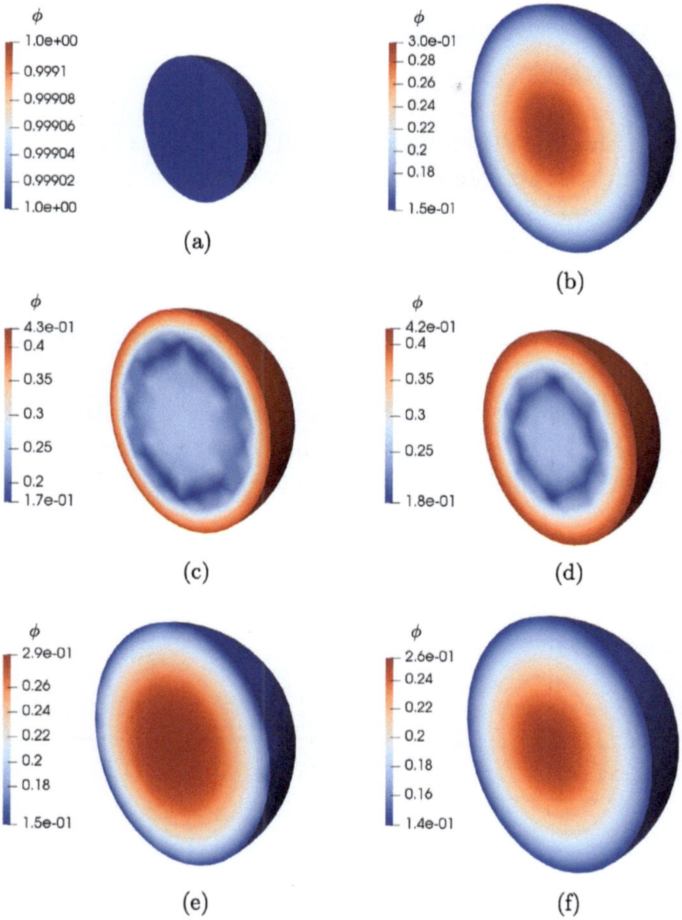

Fig. 15.9 Swollen configurations and contours of the polymer volume fraction ϕ at different times: (a) 0 h, (b) 2 h, (c) 4 h, (d) 5 h, (e)7 h, and (f) 8 h. Note the change in scale for the contours of ϕ in each subfigure.

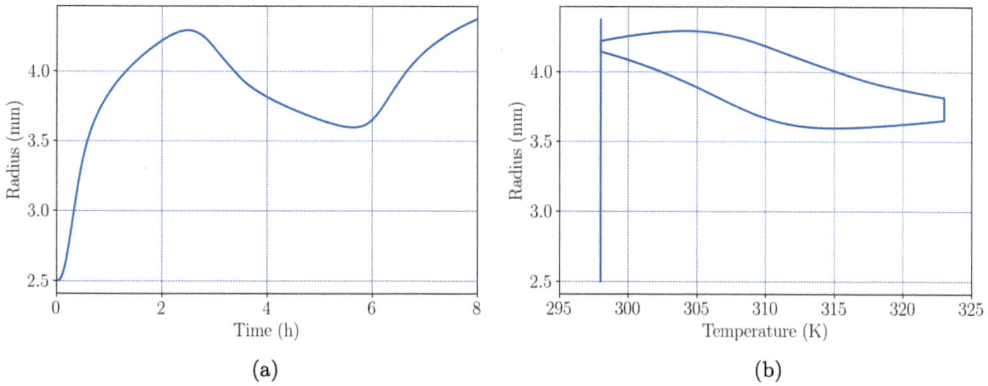

Fig. 15.10 Radius of the sphere as it swells and deswells as a function of (a) time, and (b) prescribed surface temperature.

15.4 **Swelling, thermal deswelling, and reswelling of a bilayer beam**

In this example we simulate the reversible thermal bending of a bilayer beam under plane strain conditions. Such a bilayer beam may be used as a structural element in the design of soft thermal actuators. Figure 15.11(a) shows a schematic of the initial configuration of a bilayer beam of length $L_0 = 100$ mm and height $H_0 = 5$ mm. The top and bottom halves of the beam have the same thickness but different material properties:

- The bottom half is modeled as a *thermally responsive* swellable gel with material parameters listed in Table 15.2.
- The top half is modeled as a *non-thermally responsive* swellable gel. This region is prescribed a constant temperature-independent value of $\chi = 0.1$. All other material parameters have the same values as those for the top half.

The beam is pinned at its midspan, and due to symmetry only one half of the domain is modeled. A finite element mesh which has 50 triangular elements along the length of the beam, and 6 elements along its height is used. Figure 15.11(b) shows the mesh, and a zoom-in is provided for a clearer picture of the mesh.

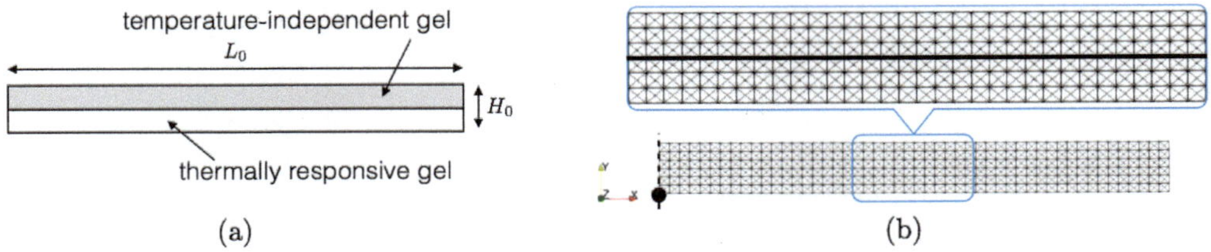

Fig. 15.11 (a) Geometry and (b) finite element mesh used for the plane-strain thermally responsive swelling of a bilayer beam simulation.

The initial conditions for dry reference state for both halves of the beam are

$$\mathbf{u}_0 = \mathbf{0}, \qquad \hat{\mu}_0 = -5.8, \qquad \hat{c}_0 = 0.001, \qquad \text{and} \quad \vartheta_0 = 298\,\text{K}.$$

Much like the prior examples, the simulation is carried out in three steps: (i) in the first step the body is freely swollen at $\vartheta_0 = 298$ K for 3600 s; (ii) in the second step the surface temperature is increased to $\vartheta = 323$ K leading to deswelling over 1500 s; and (iii) in the third step the surface temperature is lowered back to $\vartheta_0 = 298$ K and the body reswells over 7200 s.

Boundary conditions: Referring to Fig. 15.11 the boundary conditions are as follows:

- Step 1 (step time 3600 s):
  - Mechanical boundary conditions: (i) due to symmetry the left face of the domain is prescribed appropriate symmetry conditions; (ii) the bottom left corner, indicated by the black point in Fig. 15.11(b), is pinned; and (iii) all other faces of the domain are traction-free.

- Thermal boundary conditions: (i) due to symmetry the left face is prescribed zero heat flux, and (ii) all other faces are prescribed a constant temperature of $\bar{\vartheta} = 298$ K.
- Normalized chemical potential boundary conditions: (i) due to symmetry the left face is prescribed zero fluid flux, and (ii) all other faces are prescribed a normalized chemical potential history in the form

$$\bar{\hat{\mu}}(t) = \hat{\mu}_0 \exp\left(-t/300\right), \tag{15.4.1}$$

so that the normalized chemical potential on those faces increases smoothly from its initial value $\hat{\mu}_0 = -5.8$ and approaches $\bar{\hat{\mu}} = 0$.

- Step 2 (step time 1500 s):
  - Mechanical boundary conditions: The mechanical boundary conditions are maintained from the previous step.
  - Thermal boundary conditions: (i) due to symmetry the left face is prescribed zero heat flux, and (ii) all other faces are prescribed a linear temperature increase from $\bar{\vartheta} = 298$ K to $\bar{\vartheta} = 323$ K over a period of 1500 s.
  - Normalized chemical potential boundary conditions: The normalized chemical potential boundary conditions are maintained from the previous step.
- Step 3 (step time 7200 s):
  - Mechanical boundary conditions: The mechanical boundary conditions are maintained from the previous step.
  - Thermal boundary conditions: (i) due to symmetry the left face is prescribed zero heat flux, and (ii) all other faces are prescribed a linear temperature decrease from $\bar{\vartheta} = 323$ K to $\bar{\vartheta} = 298$ K over a period of 1500 s, and then held for a total of 2 h.
  - Normalized chemical potential boundary conditions: The normalized chemical potential boundary conditions are maintained from the previous step.

Figures 15.12(a)–(c) show snapshots of the contours of the polymer volume fraction ϕ on the deformed body at times $t = 3600$ s, 6600 s, and 12300 s, respectively.

Here the initial free swelling at a constant temperature of $\vartheta_0 = 298$ K leads to an essentially flat swollen configuration, Fig. 15.12(a). This is because the material properties, in particular the value of χ for both layers, are the same. However, as the temperature is raised above LCST temperature, $\vartheta_T = 307$ K, the difference in values of the χ parameter between the top and bottom layers leads to a deswelling of the bottom half of the beam and consequently the large bending deformation, Fig. 15.12(b). Finally, once the temperature is lowered below ϑ_T the bottom half reswells and the beam reversibly approaches a flat state again, Fig. 15.12(c).

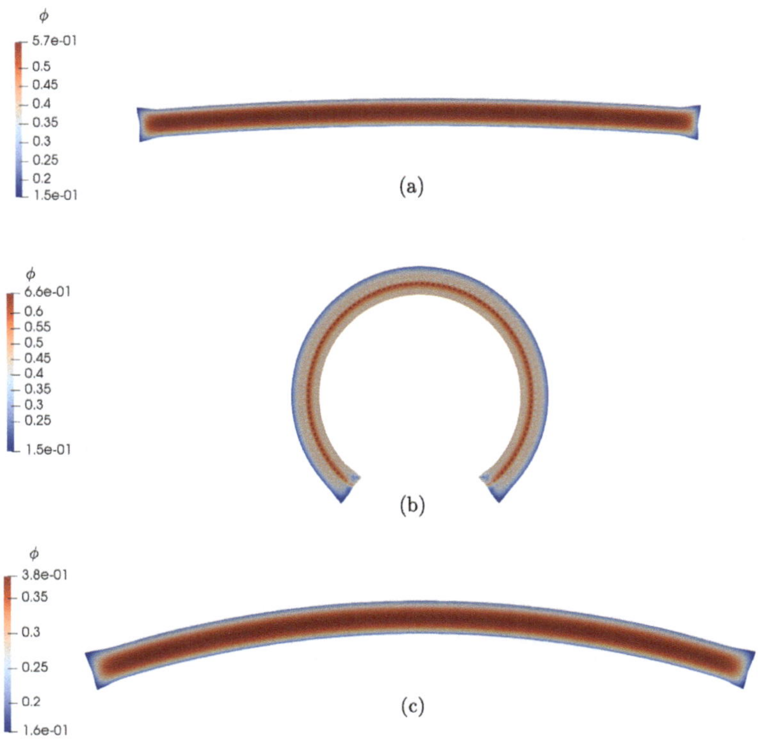

Fig. 15.12 Configurations with contours of the polymer volume fraction ϕ at (a) $t = 3600$ s, (b) 6600 s, and (c) 12300 s. The modeled domain has been reflected about the plane of symmetry to show the full geometry. Note the change in scale for the contours of ϕ in each subfigure.

15.5 Swelling, thermal deswelling, and reswelling of an axisymmetric packer

Swellable elastomeric packers are frequently used for borehole fluid control in oil-fields. Briefly, in such applications an elastomeric packer is mounted on an inner tube and when it is brought into contact with a suitable solvent for a sufficient amount of time, the packer swells and comes into contact with an outer metallic tube and forms a seal (Bhavsar et al., 2008).

As a final example for this chapter we perform an axisymmetric simulation of a packer (on a much smaller scale than packers in oil-fields). Figure 15.13 shows a schematic of the geometry: the rigid inner tube is taken to have a radius of $R_i = 3\,\text{mm}$, the swellable packer is represented as an initially square domain of edge length $L_0 = 6\,\text{mm}$, and the outer rigid tube is taken to have a radius of $R_o = 11\,\text{mm}$, leaving a gap of 2 mm in the dry reference state. In this simulation we used the following values of $\chi_L = 0.7$, $\chi_H = 0.9$, and $D = 5 \times 10^{-2}$ mm^2/s, while the values of all other material parameters are the same as in Table 15.2.

Accounting for these modified material parameters, the initial conditions for the dry swellable elastomeric body are taken as

$$\mathbf{u}_0 = \mathbf{0}, \qquad \hat{\mu}_0 = -5.21, \qquad \hat{c}_0 = 0.001, \qquad \text{and} \quad \vartheta_0 = 298\,\text{K}.$$

As in the prior examples in this chapter, the simulation is carried out in three steps: (i) in the first step the body is freely swollen at $\vartheta_0 = 298\,\text{K}$ for 2 h; (ii) in the second step the surface temperature is increased to $\vartheta = 323\,\text{K}$ leading to deswelling over 3 h; and (iii) in the third

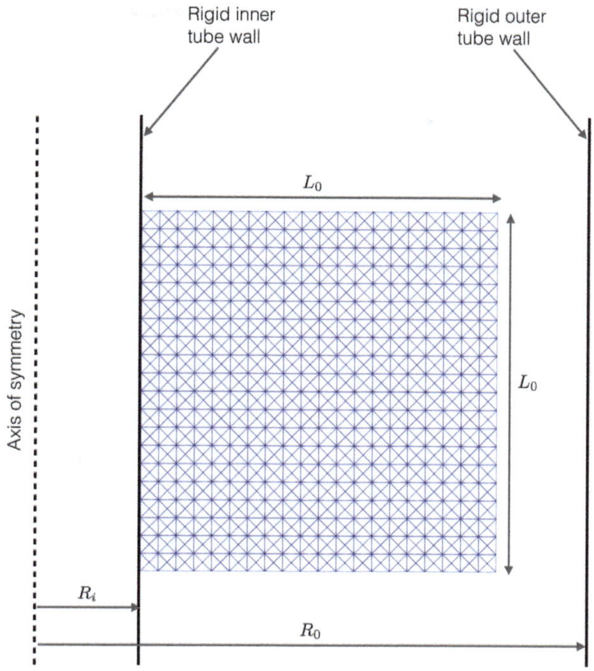

Fig. 15.13 Geometry and finite element mesh used for axisymmetric thermally responsive packer swelling simulation.

step the surface temperature is lowered back to $\vartheta_0 = 298\,\text{K}$ and the body reswells over 3 h. Figure 15.12(a) shows the prescribed surface temperature history.

The inner surface of the packer is modeled as being perfectly adhered to the *inner rigid tube* wall. The contact interaction between the packer and the *outer rigid tube* wall is modeled as follows: tangential contact is modeled as frictionless, and normal contact is modeled as penetrable using a penalty method. Specifically, a linear relationship between penetration $\delta \stackrel{\text{def}}{=} x_1 - R_o$ and contact force f_{con} is used on the domain boundary such that

$$f_{\text{con}} = \begin{cases} -k_{\text{con}}\delta & \text{if} \quad \delta > 0, \\ 0 & \text{if} \quad \delta \leq 0, \end{cases} \tag{15.5.1}$$

in which we set $k_{\text{con}} = 10^3\,\text{kPa/mm}$. In our numerical implementation, with reference to the displacement residual equation $(14.3.3)_1$, the normal contact force is enforced in a weak sense by including it as a traction boundary condition of the form

$$-\int_{\partial\mathcal{B}_{t_{\text{R}}}} (f_{\text{con}}\mathbf{e}_1) \cdot \mathbf{u}_{\text{test}}\, da_{\text{R}}. \tag{15.5.2}$$

Boundary conditions: Referring to Fig. 15.13 the boundary conditions are as follows:

- Step 1 (step time 2 h):
  - Mechanical boundary conditions: (i) the left face is prescribed $u_1 = u_2 = 0$ to model bonded contact with the inner rigid tube, and (ii) all other faces of the domain are traction-free unless in contact with the outer rigid wall.

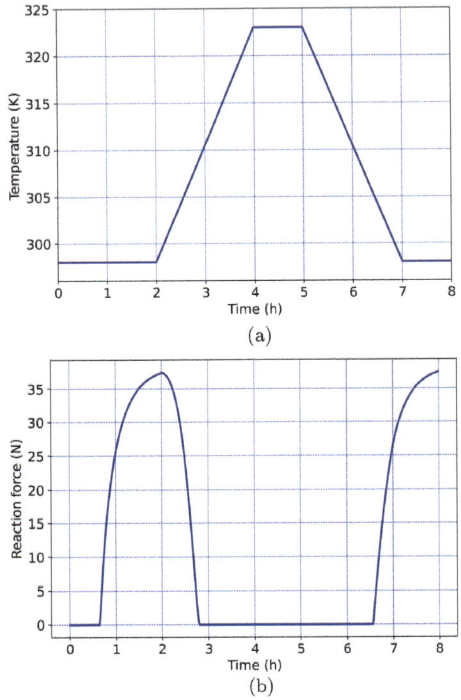

Fig. 15.14 (a) Time history of the prescribed surface temperature boundary condition. (b) The computed time history of the normal force due to the contact interaction of the gel-packer with the wall of the outer tube.

- Thermal boundary conditions: (i) the left face is prescribed zero heat flux, and (ii) all other faces are prescribed a constant temperature of $\bar{\vartheta} = 298\,\text{K}$.
- Normalized chemical potential boundary conditions: (i) the left face is prescribed zero fluid flux, and (ii) all other faces are prescribed a normalized chemical potential history in the form

$$\bar{\hat{\mu}}(t) = \hat{\mu}_0 \exp(-t/300),$$

so that the normalized chemical potential on those faces increases smoothly from its initial value $\hat{\mu}_0 = -5.21$ and approaches $\bar{\hat{\mu}} = 0$. This models the surroundings of the body as a zero-pressure fluid bath.

- Step 2 (step time 3 h):
  - Mechanical boundary conditions: The mechanical boundary conditions are maintained from the previous step.
  - Thermal boundary conditions: (i) the left face is prescribed zero heat flux, and (ii) all other faces are prescribed a linear temperature increase from $\bar{\vartheta} = 298\,\text{K}$ to $\bar{\vartheta} = 323\,\text{K}$ over a period of 2 h, and then held constant for another 1 h.
  - Normalized chemical potential boundary conditions: The normalized chemical potential boundary conditions are maintained from the previous step.

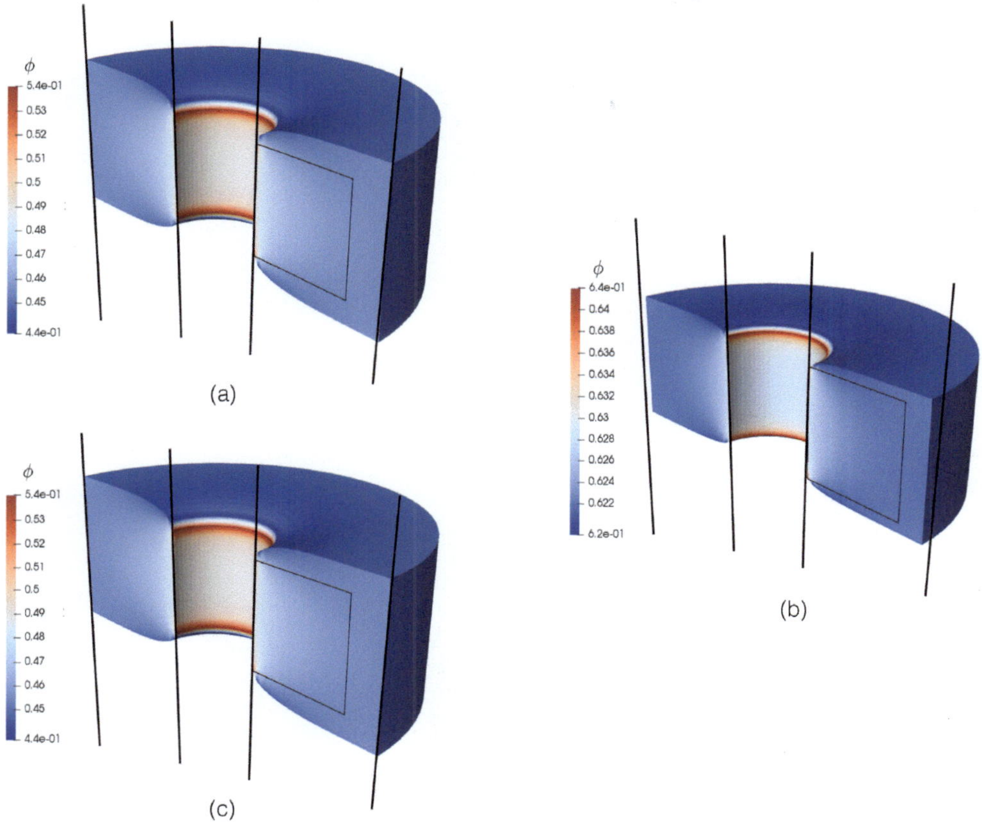

Fig. 15.15 Swollen configurations of the packer with contours of the polymer volume fraction ϕ at times: (a) $t = 2$ h, (b) 5 h, and (c) 8 h. The domain has been revolved about the axis of symmetry to show one-half of the full geometry. The initial dry configuration of the packer is shown in black outline only, and the rigid tube walls are drawn for ease of visualization as lines along the cross-section. Note the change in scale for the contours of ϕ in each subfigure.

- Step 3 (step time 3 h):
  - Mechanical boundary conditions: The mechanical boundary conditions are maintained from the previous step.
  - Thermal boundary conditions: (i) the left face is prescribed zero heat flux, and (ii) all other faces are prescribed a linear temperature decrease from $\bar{\vartheta} = 323\,\text{K}$ to $\bar{\vartheta} = 298\,\text{K}$ over a period of 2 h, and then held for another 1 h.
  - Normalized chemical potential boundary conditions: The normalized chemical potential boundary conditions are maintained from the previous step.

Figures 15.15(a)–(c) show snapshots of contours of the polymer volume fraction ϕ on the swollen packer at times $t = 2$ h, 5 h, and 8 h, respectively, which correspond to the end of each step. Also, Fig. 15.14(b) shows the normal force generated on the wall of the outer tube due to the contact interaction between the gel-packer and the tube. From these simulation results one is able to ascertain the typical time required for closing and opening of the gap between the packer and outer tube wall, and the corresponding contact forces which are generated for sealing purposes.

Part VI

Cahn – Hilliard theory for species diffusion coupled with elastic deformations

16 Cahn–Hilliard theory for species diffusion and phase segregation

16.1 Introduction

We begin in this chapter by reviewing the classical Cahn–Hilliard theory for species diffusion with phase segregation, neglecting any coupling with mechanical deformation (Cahn and Hilliard, 1958; Cahn, 1961). In Chapter 17 we will formulate a *coupled chemo-mechanical theory*.

The **Cahn–Hilliard system of equations** consists of two coupled second-order partial differential equations,

$$\begin{aligned} \dot{\bar{c}}_{\mathrm{R}} - \mathrm{Div}\,(M(\bar{c}_{\mathrm{R}})\nabla\mu) &= 0, \\ \mu - \frac{d\psi^c_{\mathrm{R}}(\bar{c}_{\mathrm{R}})}{d\bar{c}_{\mathrm{R}}} + \lambda\Delta\bar{c}_{\mathrm{R}} &= 0, \end{aligned} \tag{16.1.1}$$

for the fields μ and $\bar{c}_{\mathrm{R}}$ which represent a chemical potential and a normalized concentration, respectively. These two equations can equivalently be combined to give the classical **Cahn–Hilliard equation,**

$$\dot{\bar{c}}_{\mathrm{R}} - \mathrm{Div}\left[M(\bar{c}_{\mathrm{R}})\nabla\left(\frac{d\psi^c_{\mathrm{R}}(\bar{c}_{\mathrm{R}})}{d\bar{c}_{\mathrm{R}}} - \lambda\Delta\bar{c}_{\mathrm{R}}\right)\right] = 0, \tag{16.1.2}$$

which is a fourth-order partial differential equation for the normalized concentration $\bar{c}_{\mathrm{R}}$.

The Cahn–Hilliard equation plays an essential role in materials science since it qualitatively describes important features of two-phase systems related with phase-separation processes. This can be observed, e.g., when a binary alloy is cooled down sufficiently from a high temperature. One then observes a partial nucleation (i.e., the appearance of many very small nuclei in the material) or a total nucleation, known as *spinodal decomposition*: the material quickly becomes inhomogeneous with a very finely dispersed microstructure. In a second stage, which occurs at a slower timescale, the microstructure coarsens. Such phenomena play an essential role in developing strategies for control of microstructures, which eventually govern the mechanical properties of the material, e.g., strength, ductility, and fracture toughness.

Originally derived for phase segregation in a binary alloy system, the Cahn–Hilliard equation has been applied to a wide range of problems, including multiphase fluid flow, image processing, and planet formation. Additionally, viewing a single-species solid as a binary atom–vacancy mixture, the formation of void lattices has been modeled as a process of vacancy segregation by the Cahn–Hilliard treatment coupled with elastic effects. These are just

Introduction to Coupled Theories in Solid Mechanics. Lallit Anand, Eric M. Stewart, Shawn A. Chester, Oxford University Press.
 DOI: 10.1093/oso/9780198986218.003.0017

a few representative examples of a vast body of literature on applications of the Cahn–Hilliard equation (cf., e.g., Miranville, 2017, for a review).

16.2 The Cahn–Hilliard equation

Let

$$\bar{c}_{\mathrm{R}} \quad \text{with} \quad 0 \le \bar{c}_{\mathrm{R}} \le 1,$$

denote a normalized species concentration,

$$\hat{\psi}_{\mathrm{R}}(\bar{c}_{\mathrm{R}}, \nabla\bar{c}_{\mathrm{R}})$$

the free energy per unit volume, and let the functional

$$\Psi(\bar{c}_{\mathrm{R}}) = \int_{\mathcal{B}} \hat{\psi}_{\mathrm{R}}(\bar{c}_{\mathrm{R}}, \nabla\bar{c}_{\mathrm{R}})\, dv_{\mathrm{R}} \tag{16.2.1}$$

denote the total free energy of the region of space occupied by a body $\mathcal{B}$.

Then classically, for theories in which the free energy depends on $\nabla\bar{c}_{\mathrm{R}}$, the chemical potential μ of the chemical species is defined as the *variational derivative*,[1]

$$\mu \stackrel{\text{def}}{=} \frac{\delta\hat{\psi}_{\mathrm{R}}(\bar{c}_{\mathrm{R}}, \nabla\bar{c}_{\mathrm{R}})}{\delta\bar{c}_{\mathrm{R}}} = \frac{\partial\hat{\psi}_{\mathrm{R}}(\bar{c}_{\mathrm{R}}, \nabla\bar{c}_{\mathrm{R}})}{\partial\bar{c}_{\mathrm{R}}} - \mathrm{Div}\left[\frac{\partial\hat{\psi}_{\mathrm{R}}(\bar{c}_{\mathrm{R}}, \nabla\bar{c}_{\mathrm{R}})}{\partial\nabla\bar{c}_{\mathrm{R}}}\right]. \tag{16.2.2}$$

A widely used form for the free energy is the following separable energy

$$\hat{\psi}_{\mathrm{R}}(\bar{c}_{\mathrm{R}}, \nabla\bar{c}_{\mathrm{R}}) = \underbrace{\psi^{c}_{\mathrm{R}}(\bar{c}_{\mathrm{R}})}_{\text{coarse-grain energy}} + \underbrace{\tfrac{1}{2}\lambda|\nabla\bar{c}_{\mathrm{R}}|^2}_{\text{interfacial energy}}, \quad \lambda > 0. \tag{16.2.3}$$

proposed by Cahn and Hilliard. Here,

(i) $\psi^{c}_{\mathrm{R}}(\bar{c}_{\mathrm{R}})$ represents an energy of mixing, a non-convex double-well potential whose wells—the "binodal points"—define the phases, cf. Fig. 16.1.

(ii) The second term $\frac{1}{2}\lambda|\nabla\bar{c}_{\mathrm{R}}|^2$, which depends on the concentration gradient, represents an *interfacial energy*, with λ a positive constant such that $\sqrt{\lambda}$ is proportional to a **length scale** of the phase-segregation problem, related to to the width of the interface between the phases.

Using (16.2.2), the chemical potential corresponding to the free energy (16.2.3) is

$$\mu = \frac{d\psi^{c}_{\mathrm{R}}(\bar{c}_{\mathrm{R}})}{d\bar{c}_{\mathrm{R}}} - \lambda\Delta\bar{c}_{\mathrm{R}}, \tag{16.2.4}$$

[1] Cf. eq. (16.A.15) in the sub-appendix to this chapter for the definition of the variational derivative.

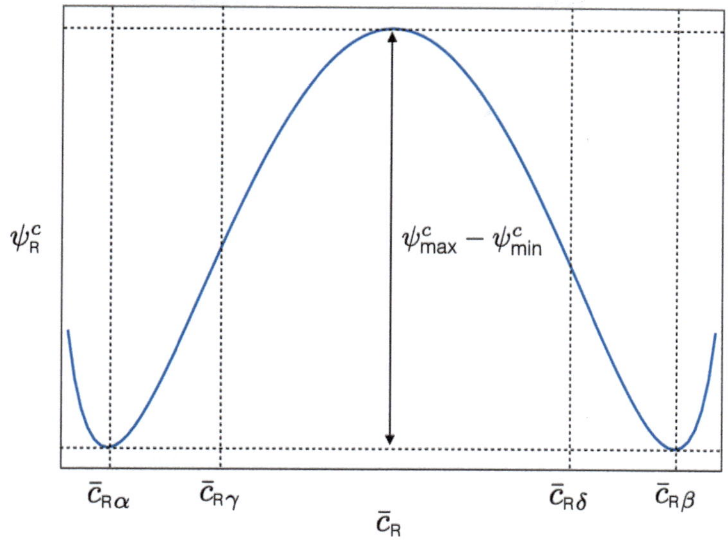

Fig. 16.1 Typical double-well chemical free energy ψ^c. When the initial concentration is in the spinodal region $\bar{c}_R \in (\bar{c}_{R\gamma}, \bar{c}_{R\delta})$, that is regions where $\partial^2\psi^c/\partial\bar{c}_R^2 < 0$, then the system can spinodally decompose into two phases of equilibrium concentration $\bar{c}_{R\alpha}$ and $\bar{c}_{R\beta}$, called the binodal concentrations.

where

$$\Delta(\bullet) = \text{Div}\,(\nabla(\bullet)) \qquad \text{is the Laplace operator.}$$

Next, species balance requires that

$$\dot{\bar{c}}_R = -\text{Div}\,\mathbf{j}_R, \tag{16.2.5}$$

with the species flux $\mathbf{j}_R$ related to the gradient of the chemical potential $\nabla\mu$ through a constitutive equation of the form

$$\mathbf{j}_R = -M(\bar{c}_R)\nabla\mu, \tag{16.2.6}$$

where $M(\bar{c}_R)$ is a nonlinear and positive function, representing the species mobility. Using (16.2.4) and (16.2.6) in (16.2.5) yields the classical **Cahn–Hilliard equation,**

$$\dot{\bar{c}}_R - \text{Div}\left[M(\bar{c}_R)\nabla\left(\frac{d\psi^c_R(\bar{c}_R)}{d\bar{c}_R} - \lambda\Delta\bar{c}_R\right)\right] = 0 \qquad \text{in } \mathcal{B}\times[0,T], \tag{16.2.7}$$

where the unknown field is $\bar{c}_R$. This partial differential equation is typically accompanied by the pair of boundary conditions

$$\left.\begin{aligned} \underbrace{-M(\bar{c}_{\rm R})\nabla\left[\frac{d\psi^c_{\rm R}(\bar{c}_{\rm R})}{d\bar{c}_{\rm R}} - \lambda\Delta\bar{c}_{\rm R}\right]}_{\mathbf{j}_{\rm R}}\cdot\mathbf{n}_{\rm R} &= 0 \quad \text{on } \partial\mathcal{B}\times[0,T], \\ (\lambda\nabla\bar{c}_{\rm R})\cdot\mathbf{n}_{\rm R} &= 0 \quad \text{on } \partial\mathcal{B}\times[0,T], \end{aligned}\right. \tag{16.2.8}$$

where $\mathbf{n}_{\rm R}$ is the outward unit normal to the boundary $\partial\mathcal{B}$ of $\mathcal{B}$. The initial condition for this boundary value problem is

$$\bar{c}_{\rm R}(\mathbf{X},0) = \bar{c}_{\rm R0}(\mathbf{X}) \qquad \text{in } \mathcal{B}. \tag{16.2.9}$$

During the evolution of $\bar{c}_{\rm R}$ a minimization of the chemical energy $\psi^c_{\rm R}(\bar{c}_{\rm R})$ drives the solution to binodal points and separates the phases, and a minimization of the interface energy $(1/2)\lambda|\nabla\bar{c}_{\rm R}|^2$ effectively coarsens the phases.

16.3 The Cahn–Hilliard system of equations

The Cahn–Hilliard equation (16.2.7) is a fourth-order equation, so casting it in a weak form would result in the presence of second-order spatial derivatives, and the problem cannot be solved using a standard linear Lagrange finite element basis. A solution is to rephrase the problem as *two coupled second-order equations*:

$$\left.\begin{aligned} \dot{\bar{c}}_{\rm R} - \text{Div}\,(M(\bar{c}_{\rm R})\nabla\mu) &= 0, \\ \mu - \frac{d\psi^c_{\rm R}(\bar{c}_{\rm R})}{d\bar{c}_{\rm R}} + \lambda\Delta\bar{c}_{\rm R} &= 0, \end{aligned}\right\} \quad \text{in } \mathcal{B}, \tag{16.3.1}$$

where now the unknown fields are μ and $\bar{c}_{\rm R}$.[2] These partial differential equations are accompanied by the pair of boundary conditions

$$\left.\begin{aligned} \underbrace{-M(\bar{c}_{\rm R})\nabla\mu}_{\mathbf{j}_{\rm R}}\cdot\mathbf{n}_{\rm R} &= 0 \quad \text{on } \partial\mathcal{B}\times[0,T], \\ (\lambda\nabla\bar{c}_{\rm R})\cdot\mathbf{n}_{\rm R} &= 0 \quad \text{on } \partial\mathcal{B}\times[0,T], \end{aligned}\right. \tag{16.3.2}$$

[2] Numerical solution techniques for the Cahn–Hilliard equation have spanned finite difference methods, spectral formulations, and finite element methods. Most finite element formulations split the fourth-order Cahn–Hilliard equation into a coupled set of two second-order partial differential equations, which are then solved using mixed finite element methods.

and the initial conditions are

$$\mu(\mathbf{X},0) = \mu_0(\mathbf{X}) \qquad \bar{c}_{\mathrm{R}}(\mathbf{X},0) = \bar{c}_{\mathrm{R}0}(\mathbf{X}) \qquad \text{in } \mathcal{B}. \tag{16.3.3}$$

In Chapter 17 we will formulate a coupled chemo-mechanical theory for species diffusion with phase segregation.

16.A Appendix. First variation of a functional. Variational derivative of a function

- Consider a prototypical problem in which one is asked to find a function $c(\mathbf{x})$ that minimizes the **functional**

$$\Psi\{c(\mathbf{x})\} = \int_\Omega \hat{\psi}(\, c(\mathbf{x}), \nabla c(\mathbf{x}))\, dv \tag{16.A.1}$$

  over a given set of functions, $\mathcal{C}$, for example the set of all continuous functions on the domain Ω.
- In (16.A.1) the integrand $\psi = \hat{\psi}(c(\mathbf{x}), \nabla c(\mathbf{x}))$ is assumed to be a known function of c and ∇c.
- The quantity

$$\delta\Psi\{c; w\} \overset{\text{def}}{=} \frac{d}{d\alpha}\Psi\{c(\mathbf{x}) + \alpha w(\mathbf{x})\}\bigg|_{\alpha=0} \tag{16.A.2}$$

  is known as the **first variation** of Ψ.
- The condition

$$\delta\Psi\{c; w\} = 0 \qquad \text{for all admissible } w \tag{16.A.3}$$

  is a **necessary** condition for $c(\mathbf{x})$ to be a minimizer of the functional Ψ.
- Definition (16.A.2) for the first variation of Ψ closely resembles the conventional directional derivative formula for ordinary functions. In the context of (16.A.3), we then have the interpretation that at a minimizer of Ψ, the value of Ψ does not change for infinitesimally small perturbations in any direction w about c—the functional has "zero slope" at the minimizer.

16.A.1 Application of the necessary condition $\delta\Psi = 0$

- From the functional (16.A.1) we obtain

$$\frac{d}{d\alpha}\Psi\{c+\alpha w\} = \int_\Omega \left[\frac{\partial\hat{\psi}(c+\alpha w, \nabla c+\alpha\nabla w)}{\partial(c+\alpha w)}\, w \right.$$
$$\left. + \frac{\partial\hat{\psi}(c+\alpha w, \nabla c+\alpha\nabla w)}{\partial(\nabla c+\alpha\nabla w)}\cdot\nabla w\right] dv, \tag{16.A.4}$$

and hence the first variation of Ψ is

$$\delta\Psi\{c;w\} = \frac{d}{d\alpha}\Psi\{c+\alpha w\}\bigg|_{\alpha=0} = \int_\Omega \left[\frac{\partial\hat{\psi}(c,\nabla c)}{\partial c}\, w + \frac{\partial\hat{\psi}(c,\nabla c)}{\partial\nabla c}\cdot\nabla w\right] dv. \tag{16.A.5}$$

- Consider the second term in the integrand of (16.A.5), viz.

$$\int_\Omega \frac{\partial\psi}{\partial\nabla c}\cdot\nabla w\, dv. \tag{16.A.6}$$

Using the identity

$$\operatorname{div}(\varphi\mathbf{v}) = \varphi\operatorname{div}\mathbf{v} + \mathbf{v}\cdot\operatorname{grad}\varphi \qquad \Longrightarrow \qquad \mathbf{v}\cdot\operatorname{grad}\varphi = \operatorname{div}(\varphi\mathbf{v}) - \varphi\operatorname{div}\mathbf{v},$$

eq. (16.A.6) may be written as

$$\int_\Omega \frac{\partial\psi}{\partial\nabla c}\cdot\nabla w\, dv = \int_\Omega \operatorname{div}\left(w\left(\frac{\partial\psi}{\partial\nabla c}\right)\right) dv - \int_\Omega w\operatorname{div}\left(\frac{\partial\psi}{\partial\nabla c}\right) dv. \tag{16.A.7}$$

- Next, using the divergence theorem on the first term on the right in (16.A.7) we obtain

$$\int_\Omega \frac{\partial\psi}{\partial\nabla c}\cdot\nabla w\, dv = \int_{\partial\Omega} \left(w\left(\frac{\partial\psi}{\partial\nabla c}\right)\right)\cdot\mathbf{n}_R\, da - \int_\Omega w\operatorname{div}\left(\frac{\partial\psi}{\partial\nabla c}\right) dv. \tag{16.A.8}$$

- Then, using (16.A.8) in (16.A.5) gives

$$\delta\Psi\{c;w\} = \int_\Omega \left[\frac{\partial\psi}{\partial c} - \operatorname{div}\left(\frac{\partial\psi}{\partial\nabla c}\right)\right] w\, dv + \int_{\partial\Omega}\left(\frac{\partial\psi}{\partial\nabla c}\right)\cdot\mathbf{n}_R\, w\, da. \tag{16.A.9}$$

- Hence we find that the necessary condition $\delta\Psi\{c;w\} = 0$, for c to minimize Ψ, requires that

$$\int_\Omega \left[\frac{\partial\psi}{\partial c} - \operatorname{div}\left(\frac{\partial\psi}{\partial\nabla c}\right)\right] w\, dv + \int_{\partial\Omega}\left(\left(\frac{\partial\psi}{\partial\nabla c}\right)\cdot\mathbf{n}_R\right) w\, da = 0 \tag{16.A.10}$$

for all admissible functions w.

- This is true if and only if c satisfies the differential equation

$$\frac{\partial\psi}{\partial c} - \text{div}\left(\frac{\partial\psi}{\partial\nabla c}\right) = 0 \qquad \forall \mathbf{x} \in \Omega, \tag{16.A.11}$$

and the boundary conditions

$$\left(\frac{\partial\psi}{\partial\nabla c}\right) \cdot \mathbf{n}_{\mathrm{R}} = 0 \qquad \text{or} \qquad w = 0 \qquad \forall \mathbf{x} \in \partial\Omega \tag{16.A.12}$$

- **Euler–Lagrange equation**:
  The differential equation (16.A.11) is referred to as the **Euler–Lagrange equation** for the functional Ψ.
- **Essential and natural boundary conditions**: Note that the boundary conditions (16.A.12) may be satisfied in one of two ways:

  (i) Let $\partial\mathcal{B}_1$ and $\partial\mathcal{B}_2$ be complementary subsurfaces of $\partial\Omega$. Suppose that the original minimization problem is performed over the set of functions $\mathcal{C}$ such that they satisfy the boundary condition

$$c(\mathbf{x}) = \hat{c}(\mathbf{x}) \qquad \text{for} \qquad \mathbf{x} \in \partial\mathcal{B}_1. \tag{16.A.13}$$

  where $\hat{c}(\mathbf{x})$ is a prescribed function. Then, we must also require that

$$c(\mathbf{x}) + \alpha w(\mathbf{x}) = \hat{c}(\mathbf{x}) \qquad \text{for} \qquad \mathbf{x} \in \partial\mathcal{B}_1.$$

  This implies that we must only consider those variations for which

$$w(\mathbf{x}) = 0 \qquad \text{for} \qquad \mathbf{x} \in \partial\mathcal{B}_1.$$

  For such variations the boundary condition (16.A.13) is automatically satisfied. A boundary condition of the form (16.A.13) is known as an **essential or Dirichlet boundary condition.**

  (ii) Alternatively, if a boundary condition on $c(\mathbf{x})$ is not imposed for $\mathbf{x} \in \partial\mathcal{B}_2$, then the condition

$$\left(\frac{\partial\psi}{\partial\nabla c}\right) \cdot \mathbf{n}_{\mathrm{R}} = 0 \qquad \text{for} \qquad \mathbf{x} \in \partial\mathcal{B}_2 \tag{16.A.14}$$

  is a necessary condition for c to minimize Ψ. A boundary condition of the form (16.A.14) is known as a **natural or Neumann boundary condition.**

REMARKS

1. The **variational derivative** of the free-energy function $\psi = \hat{\psi}(c, \nabla c)$ is defined as

$$\frac{\delta\hat{\psi}(c,\nabla c)}{\delta c} \overset{\text{def}}{=} \frac{\partial\hat{\psi}(c,\nabla c)}{\partial c} - \operatorname{div}\left[\frac{\partial\hat{\psi}(c,\nabla c)}{\partial\nabla c}\right]. \tag{16.A.15}$$

2. The condition $\delta\Psi = 0$ is only a **necessary** condition for minimizing the functional Ψ, and as such, a function $c(\mathbf{x})$ that satisfies (16.A.11) to (16.A.14) is said to render the functional Ψ **stationary**. In general, a function that satisfies $\delta\Psi = 0$ need not minimize the functional Ψ; it may, in fact, maximize the functional or correspond to a saddle point. Further, a solution $c(\mathbf{x})$ based on this condition may only represent a *local*, rather than a global, minimum.

17 Coupled chemo-mechanical theory for species diffusion and phase segregation

17.1 Introduction

In this chapter we present a theory which couples species diffusion with elastic deformation of a continuum body, accounting for swelling as well as possible phase segregation caused by the diffusing species.[1] A technologically important area of application of the theory is in the chemo-mechanical analysis of the evolution of large stresses which develop because of the volume changes associated with the diffusion of lithium ions in the active electrode particles of lithium-ion batteries during charge–discharge cycles.

As stated above, the objective of this chapter is to formulate a theory which couples species diffusion with elastic deformations of the body—with an accounting for the swelling and phase segregation caused by the diffusing species. In addition to a standard balance for the species concentration, we develop the other balances in our theory by following the *virtual-power* approach of Germain (1973) and Gurtin (2002). This approach leads to "macroforce" and "microforce" balances for the forces associated with the rate-like kinematical descriptors in the theory. The balance law for species concentration and these macro- and microforce balances, when supplemented with a set of thermodynamically consistent constitutive equations provide the governing equations for the theory.

For clarity, in formulating our theory we limit our considerations to *isotropic materials under isothermal conditions*. We neglect any inelastic effects, and since timescales associated with species diffusion are usually considerably longer than those associated with wave propagation, we also neglect inertial effects.

17.2 Basic kinematics

In what follows, the spatially continuous fields that define our continuum theory represent averages meant to apply at length scales which are large compared to the length scales associated with the fine-scale microstructure of the material. We identify such a macroscopically homogeneous body $\mathcal{B}$ with the region of space it occupies in a fixed *reference configuration*, and denote by $\mathbf{X}$ an arbitrary material point of $\mathcal{B}$. A *motion* of $\mathcal{B}$ is then a smooth one-to-one mapping $\mathbf{x} = \boldsymbol{\chi}(\mathbf{X}, t)$ with *deformation gradient*, *velocity*, and *velocity gradient* given by

[1] The theory presented here is essentially that formulated in Anand (2012). The theory of Anand (2012) accounts for not only elastic but also plastic deformation of the host material, but we neglect the effects of plastic deformation here.

Introduction to Coupled Theories in Solid Mechanics. Lallit Anand, Eric M. Stewart, Shawn A. Chester, Oxford University Press.
 DOI: 10.1093/oso/9780198986218.003.0018

$$\mathbf{F} = \nabla\chi, \qquad \mathbf{v} = \dot{\chi}, \qquad \mathbf{L} = \operatorname{grad}\mathbf{v} = \dot{\mathbf{F}}\mathbf{F}^{-1}. \tag{17.2.1}$$

We base the theory on a multiplicative decomposition of the deformation gradient

$$\mathbf{F} = \mathbf{F}^e\mathbf{F}^s, \quad \text{with} \quad \mathbf{F}^s = \lambda^s\mathbf{1}, \quad \lambda^s > 0. \tag{17.2.2}$$

Here, suppressing the argument t:

- $\mathbf{F}^s(\mathbf{X})$ represents the local distortion of the material at $\mathbf{X}$ due to swelling, and λ^s is the *swelling stretch*. This local deformation accounts for the swelling of the material due to absorbed atoms/molecules of the diffusing species which are pinned to the coherent microstructure that resides in the *intermediate* local space at $\mathbf{X}$ (as represented by the range of $\mathbf{F}^s(\mathbf{X})$);
- $\mathbf{F}^e(\mathbf{X})$ represents the subsequent stretching and rotation of this coherent microstructure, and thereby represents a corresponding mechanical elastic distortion.

We refer to $\mathbf{F}^s$ and $\mathbf{F}^e$ as the *swelling and elastic distortions*, respectively.

We write

$$J \stackrel{\text{def}}{=} \det\mathbf{F} > 0, \tag{17.2.3}$$

and hence, using $(17.2.2)_1$,

$$J = J^e J^s, \qquad \text{where} \qquad J^e \stackrel{\text{def}}{=} \det\mathbf{F}^e > 0 \qquad \text{and} \qquad J^s \stackrel{\text{def}}{=} \det\mathbf{F}^s > 0. \tag{17.2.4}$$

Thus, using $(17.2.2)_2$,

$$J^s = (\lambda^s)^3. \tag{17.2.5}$$

As is standard,

$$\mathbf{F} = \mathbf{R}\mathbf{U} \qquad \text{and} \qquad \mathbf{F}^e = \mathbf{R}^e\mathbf{U}^e, \tag{17.2.6}$$

denote the right polar decompositions of $\mathbf{F}$ and $\mathbf{F}^e$, respectively, with $\mathbf{U}$ and $\mathbf{U}^e$ symmetric and positive-definite tensors, and $\mathbf{R}$ and $\mathbf{R}^e$ rotations. Also, the tensors

$$\mathbf{C} = \mathbf{U}^2 = \mathbf{F}^{\top}\mathbf{F}, \quad \text{and} \quad \mathbf{C}^e = \mathbf{U}^{e2} = \mathbf{F}^{e\top}\mathbf{F}^e, \tag{17.2.7}$$

denote the total and elastic right Cauchy–Green tensors.

Next, by $(17.2.1)_3$ and (17.2.2),

$$\mathbf{L} = \mathbf{L}^e + \mathbf{F}^e\mathbf{L}^s\mathbf{F}^{e-1}, \tag{17.2.8}$$

with

$$\mathbf{L}^e = \dot{\mathbf{F}}^e\mathbf{F}^{e-1}, \qquad \mathbf{L}^s = \dot{\mathbf{F}}^s\mathbf{F}^{s-1}. \tag{17.2.9}$$

As is standard, we define the elastic and swelling stretching and spin tensors through

$$\begin{aligned} \mathbf{D}^e &= \operatorname{sym}\mathbf{L}^e, & \mathbf{W}^e &= \operatorname{skw}\mathbf{L}^e, \\ \mathbf{D}^s &= \operatorname{sym}\mathbf{L}^s, & \mathbf{W}^s &= \operatorname{skw}\mathbf{L}^s, \end{aligned} \tag{17.2.10}$$

so that $\mathbf{L}^e = \mathbf{D}^e + \mathbf{W}^e$ and $\mathbf{L}^s = \mathbf{D}^s + \mathbf{W}^s$.

Further from (17.2.2), $(17.2.9)_2$, and (17.2.10)

$$\mathbf{D}^s = (\dot{\lambda}^s\lambda^{s-1})\mathbf{1} \qquad \text{and} \qquad \mathbf{W}^s = \mathbf{0}; \tag{17.2.11}$$

and since

$$\dot{J}^s = J^s \operatorname{tr}\mathbf{D}^s, \tag{17.2.12}$$

we also have

$$\mathbf{D}^s = \frac{1}{3}(\dot{J}^s J^{s-1})\mathbf{1}. \tag{17.2.13}$$

Thus, since $\mathbf{L}^s \equiv \mathbf{D}^s$, we have

$$\dot{\mathbf{F}}^s = \mathbf{D}^s\mathbf{F}^s. \tag{17.2.14}$$

Finally, using (17.2.1), (17.2.9), (17.2.13), and (17.2.14) we may write (17.2.8), for future use, as

$$(\nabla\dot{\boldsymbol{\chi}})\mathbf{F}^{-1} = \dot{\mathbf{F}}^e\mathbf{F}^{e-1} + \frac{1}{3}(\dot{J}^s J^{s-1})\mathbf{1}. \tag{17.2.15}$$

Henceforth, we denote an arbitrary part of $\mathcal{B}$ by $\mathcal{P}$, and by $\mathbf{n}_{\mathrm{R}}$ the outward unit normal on the boundary $\partial\mathcal{P}$ of $\mathcal{P}$. Further, as in previous chapters, we use a subscript "R" to refer to quantities measured in the reference configuration.

17.3 Balance law for the diffusing species

Let $c_{\mathrm{R}}(\mathbf{X}, t)$ denote the *total number of moles of a chemical species per unit reference volume.* Changes in c_{R} in $\mathcal{P}$ are brought about by diffusion of the species across the boundary $\partial\mathcal{P}$, so that

$$-\int_{\partial\mathcal{P}} \mathbf{j}_{\mathrm{R}} \cdot \mathbf{n}_{\mathrm{R}} \, da_{\mathrm{R}}$$

represents the amount of the chemical species entering $\mathcal{P}$ across $\partial\mathcal{P}$ per unit time. The balance law for the chemical species therefore takes the form

$$\overline{\int_{\mathcal{P}} c_{\mathrm{R}} \, dv_{\mathrm{R}}}^{\boldsymbol{\cdot}} = -\int_{\partial\mathcal{P}} \mathbf{j}_{\mathrm{R}} \cdot \mathbf{n}_{\mathrm{R}} \, da_{\mathrm{R}}, \tag{17.3.1}$$

for every part $\mathcal{P}$. Bringing the time derivative in (17.3.1) inside the integral and using the divergence theorem on the integral over $\partial\mathcal{P}$, we find that

$$\int_{\mathcal{P}} (\dot{c}_{\mathrm{R}} + \mathrm{Div}\, \mathbf{j}_{\mathrm{R}}) \, dv_{\mathrm{R}} = 0. \tag{17.3.2}$$

Since $\mathcal{P}$ is arbitrary, this leads to the following local balance,

$$\dot{c}_{\mathrm{R}} = -\mathrm{Div}\, \mathbf{j}_{\mathrm{R}}. \tag{17.3.3}$$

17.4 Constitutive relation for J^s

We assume that changes in J^s arise entirely due to the change in species concentration,

$$J^s(c_{\mathrm{R}}) = 1 + \Omega(c_{\mathrm{R}} - c_{\mathrm{R}0}), \tag{17.4.1}$$

where

$$\Omega > 0 \tag{17.4.2}$$

represents a *molar volume* of the diffusing species. It is convenient to define a related quantity

$$\Theta(c_{\mathrm{R}}) \overset{\text{def}}{=} \frac{\Omega}{J^s(c_{\mathrm{R}})}, \tag{17.4.3}$$

so that

$$\dot{J}^s J^{s-1} = \Theta(c_{\mathrm{R}}) \dot{c}_{\mathrm{R}}. \tag{17.4.4}$$

Using (17.4.4) we can rewrite the swell stretching (17.2.13) as

$$\mathbf{D}^s = \frac{1}{3}\Theta(c_{\mathrm{R}})\dot{c}_{\mathrm{R}}\,\mathbf{1}. \tag{17.4.5}$$

Further, using (17.4.4), the kinematical relation (17.2.15) may be expressed as

$$(\nabla\dot{\chi})\mathbf{F}^{-1} = \dot{\mathbf{F}}^e\mathbf{F}^{e-1} + \frac{1}{3}\Theta(c_{\mathrm{R}})\dot{c}_{\mathrm{R}}\mathbf{1}. \tag{17.4.6}$$

17.5 **Principle of virtual power. Macroscopic and microscopic force balances**

We follow Germain (1973), Gurtin (2002), Gurtin et al. (2010), and Anand (2012) to derive the macroscopic and microscopic force balances in the theory via the method of virtual power.

In developing our theory we take the "rate-like" kinematical descriptors of the theory to be $\dot{\chi}$, $\dot{\mathbf{F}}^e$, and $\dot{c}_{\mathrm{R}}$, and also the gradient $\nabla\dot{c}_{\mathrm{R}}$. In exploiting the method of virtual power we note that the rates $(\dot{\chi}, \dot{\mathbf{F}}^e, \dot{c}_{\mathrm{R}})$ are not independent—they are constrained by (17.4.6), viz.

$$(\nabla\dot{\chi})\mathbf{F}^{-1} = \dot{\mathbf{F}}^e\mathbf{F}^{e-1} + \frac{1}{3}\Theta(c_{\mathrm{R}})\dot{c}_{\mathrm{R}}\mathbf{1}. \tag{17.5.1}$$

With each evolution of the body we associate microscopic and macroscopic force systems. The microscopic system is defined by:

(a) A stress $\mathbf{S}^e$ that expends power over the distortion rate $\dot{\mathbf{F}}^e$.

(b) A scalar microscopic stress ϖ that expends power over the rate $\dot{c}_{\mathrm{R}}$.

(c) A vector microscopic stress $\boldsymbol{\xi}$ that expends power over the gradient $\nabla\dot{c}_{\mathrm{R}}$.

We characterize this microscopic force system through the manner in which these forces expend power. That is, given a part $\mathcal{P}$ we write the internal power in the form

$$\mathcal{W}_{\mathrm{int}}(\mathcal{P}) = \int\limits_{\mathcal{P}} \Big(\mathbf{S}^e : \dot{\mathbf{F}}^e + \varpi\dot{c}_{\mathrm{R}} + \boldsymbol{\xi}\cdot\nabla\dot{c}_{\mathrm{R}}\Big)\, dv_{\mathrm{R}}, \tag{17.5.2}$$

where, $\mathbf{S}^e$, ϖ, and $\boldsymbol{\xi}$, are defined over the body for all time.

This *internal* power must be balanced by power expended *externally* by tractions on $\partial\mathcal{P}$ and body forces acting within $\mathcal{P}$. Accordingly we admit a conventional macroscopic force system defined by:

(i) A traction $\mathbf{t}_{\mathrm{R}}(\mathbf{n}_{\mathrm{R}})$ (for each unit vector $\mathbf{n}_{\mathrm{R}}$) that expends power over the velocity $\dot{\chi}$;

(ii) An external (non-inertial) body force $\mathbf{b}_{0\mathrm{R}}$ that expends power over $\dot{\chi}$.[2]

[2] Regarding the body force, since timescales associated with species diffusion are usually considerably longer than those associated with wave propagation, we *neglect* all inertial effects.

These external forces give rise to the conventional expenditure of external power

$$\int_{\partial\mathcal{P}} \mathbf{t}_{\mathrm{R}}(\mathbf{n}_{\mathrm{R}}) \cdot \dot{\boldsymbol{\chi}}\, da_{\mathrm{R}} + \int_{\mathcal{P}} \mathbf{b}_{0\mathrm{R}} \cdot \dot{\boldsymbol{\chi}}\, dv_{\mathrm{R}}. \tag{17.5.3}$$

However, this power is not sufficiently general to accommodate the internal power (17.5.2); indeed, an additional external power expenditure is needed to balance the internal expenditure $\boldsymbol{\xi} \cdot \nabla \dot{c}_{\mathrm{R}}$. The gradient term $\boldsymbol{\xi} \cdot \nabla \dot{c}_{\mathrm{R}}$ should give rise to a traction term associated with the microscopic stress $\boldsymbol{\xi}$. The following identity, which is based on the divergence theorem, helps us to choose an appropriate form for this traction:

$$\int_{\mathcal{P}} \boldsymbol{\xi} \cdot \nabla \dot{c}_{\mathrm{R}}\, dv_{\mathrm{R}} = -\int_{\mathcal{P}} (\mathrm{Div}\, \boldsymbol{\xi}) \dot{c}_{\mathrm{R}}\, dv_{\mathrm{R}} + \int_{\partial\mathcal{P}} (\boldsymbol{\xi} \cdot \mathbf{n}_{\mathrm{R}}) \dot{c}_{\mathrm{R}}\, da_{\mathrm{R}}. \tag{17.5.4}$$

Guided by the term $(\boldsymbol{\xi} \cdot \mathbf{n}_{\mathrm{R}})\dot{c}_{\mathrm{R}}$, we assume that power is expended externally by a *microscopic traction* $\xi(\mathbf{n})$ conjugate to $\dot{c}_{\mathrm{R}}$. We therefore assume that the *external power* has the form

$$\mathcal{W}_{\mathrm{ext}}(\mathcal{P}) = \int_{\partial\mathcal{P}} \mathbf{t}_{\mathrm{R}}(\mathbf{n}_{\mathrm{R}}) \cdot \dot{\boldsymbol{\chi}}\, da_{\mathrm{R}} + \int_{\mathcal{P}} \mathbf{b}_{0\mathrm{R}} \cdot \dot{\boldsymbol{\chi}}\, dv_{\mathrm{R}} + \int_{\partial\mathcal{P}} \xi(\mathbf{n}_{\mathrm{R}})\, \dot{c}_{\mathrm{R}}\, da_{\mathrm{R}}, \tag{17.5.5}$$

with $\xi(\mathbf{n}_{\mathrm{R}})$ (for each unit vector $\mathbf{n}_{\mathrm{R}}$) defined over the body for all time.

Summarizing: with each evolution of the body we associate macroscopic and microscopic force systems. The macroscopic system is defined by:

(i) A traction $\mathbf{t}_{\mathrm{R}}(\mathbf{n}_{\mathrm{R}})$ (for each unit vector $\mathbf{n}_{\mathrm{R}}$) that expends power over the velocity $\dot{\boldsymbol{\chi}}$;

(ii) An external (non-inertial) body $\mathbf{b}_{0\mathrm{R}}$ force that expends power over $\dot{\boldsymbol{\chi}}$.

The microscopic system, which is non-standard, is defined by:

(a) A stress $\mathbf{S}^e$ that expends power over the distortion rate $\dot{\mathbf{F}}^e$.

(b) A scalar microscopic stress ϖ that expends power over the rate $\dot{c}_{\mathrm{R}}$.

(c) A vector microscopic stress $\boldsymbol{\xi}$ that expends power over the gradient $\nabla \dot{c}_{\mathrm{R}}$.

(d) A scalar microscopic traction $\xi(\mathbf{n}_{\mathrm{R}})$ that expends power over $\dot{c}_{\mathrm{R}}$, acting on the boundary $\partial\mathcal{P}$ of $\mathcal{P}$.

We characterize the force systems through the manner in which these forces expend power. That is, given any part $\mathcal{P}$, $\mathcal{W}_{\mathrm{ext}}(\mathcal{P})$ is the power expended on $\mathcal{P}$ by material external to $\mathcal{P}$, and $\mathcal{W}_{\mathrm{int}}(\mathcal{P})$ is a concomitant expenditure of power within $\mathcal{P}$. Specifically,

$$\left.\begin{aligned} \mathcal{W}_{\mathrm{ext}}(\mathcal{P}) &= \int_{\partial\mathcal{P}} \mathbf{t}_{\mathrm{R}}(\mathbf{n}_{\mathrm{R}}) \cdot \dot{\boldsymbol{\chi}}\, da_{\mathrm{R}} + \int_{\mathcal{P}} \mathbf{b}_{0\mathrm{R}} \cdot \dot{\boldsymbol{\chi}}\, dv_{\mathrm{R}} + \int_{\partial\mathcal{P}} \xi(\mathbf{n}_{\mathrm{R}})\, \dot{c}_{\mathrm{R}}\, da_{\mathrm{R}}, \\ \mathcal{W}_{\mathrm{int}}(\mathcal{P}) &= \int_{\mathcal{P}} \left(\mathbf{S}^e : \dot{\mathbf{F}}^e + \varpi \dot{c}_{\mathrm{R}} + \boldsymbol{\xi} \cdot \nabla \dot{c}_{\mathrm{R}} \right) dv_{\mathrm{R}}, \end{aligned}\right\} \tag{17.5.6}$$

where, $\mathbf{S}^e$, ϖ, $\boldsymbol{\xi}$, and $\xi(\mathbf{n}_{\mathrm{R}})$ are defined over the body for all time.

Assume that, at some arbitrarily chosen but *fixed time*, the fields χ, $\mathbf{F}$, $\mathbf{F}^e$, and c_R are known, and consider the fields $\dot{\chi}$, $\dot{\mathbf{F}}^e$, and $\dot{c}_R$ as virtual velocities to be specified independently in a manner consistent with (17.4.6); that is, denoting the virtual fields by $\tilde{\chi}$, $\tilde{\mathbf{F}}^e$, and $\tilde{c}_R$ to differentiate them from fields associated with the actual evolution of the body, we require that

$$(\nabla\tilde{\chi})\mathbf{F}^{-1} = \tilde{\mathbf{F}}^e\mathbf{F}^{e-1} + \frac{1}{3}\Theta(c_R)\tilde{c}_R\mathbf{1}. \tag{17.5.7}$$

Further, we define a generalized virtual velocity to be a list

$$\mathcal{V} = (\tilde{\chi}, \tilde{\mathbf{F}}^e, \tilde{c}_R),$$

consistent with (17.5.7).

Recall that if the macroscopic motion of a body is rigid, then

$$\dot{\mathbf{F}} = \mathbf{\Omega}\mathbf{F},$$

where at any time t, $\mathbf{\Omega}$ is a spatially constant skew tensor. Accordingly, we refer to a macroscopic virtual field $\mathcal{V}$ as *rigid* if it satisfies

$$(\nabla\tilde{\chi}) = \tilde{\mathbf{F}} = \mathbf{\Omega}\mathbf{F}, \tag{17.5.8}$$

with $\mathbf{\Omega}$ a spatially constant skew tensor, together with

$$\tilde{c}_R = 0 \qquad \text{so that} \qquad \tilde{\mathbf{F}}^e = \mathbf{\Omega}\mathbf{F}^e. \tag{17.5.9}$$

Next, writing

$$\begin{aligned}\mathcal{W}_{\text{ext}}(\mathcal{P},\mathcal{V}) &= \int_{\partial\mathcal{P}} \mathbf{t}_R(\mathbf{n}_R)\cdot\tilde{\chi}\,da_R + \int_{\mathcal{P}} \mathbf{b}_{0R}\cdot\tilde{\chi}\,dv_R + \int_{\partial\mathcal{P}} \xi(\mathbf{n}_R)\,\tilde{c}_R\,da_R,\\ \mathcal{W}_{\text{int}}(\mathcal{P},\mathcal{V}) &= \int_{\mathcal{P}} \Big(\mathbf{S}^e:\tilde{\mathbf{F}}^e + \varpi\tilde{c}_R + \boldsymbol{\xi}\cdot\nabla\tilde{c}_R\Big)\,dv_R,\end{aligned} \tag{17.5.10}$$

respectively, for the external and internal expenditures of virtual power, the *principle of virtual power* consists of two basic requirements:

(V1) Given any part P,

$$\mathcal{W}_{\text{ext}}(\mathcal{P},\mathcal{V}) = \mathcal{W}_{\text{int}}(\mathcal{P},\mathcal{V}) \quad \text{for all generalized virtual velocities } \mathcal{V}. \tag{17.5.11}$$

(V2) Given any part $\mathcal{P}$ and a *rigid* virtual velocity $\mathcal{V}$,

$$\mathcal{W}_{\text{int}}(\mathcal{P},\mathcal{V}) = 0 \quad \text{whenever } \mathcal{V} \text{ is a rigid macroscopic virtual velocity.} \tag{17.5.12}$$

To deduce the consequences of the principle of virtual power, assume that (17.5.11) and (17.5.12) are satisfied. Note that in applying the virtual balance we are at liberty to choose any $\mathcal{V}$ consistent with the constraint (17.5.7).

17.5.1 Macroscopic force and moment balances

Let $\tilde{c}_{\mathrm{R}} = 0$, so that $(\nabla\tilde{\boldsymbol{\chi}})\mathbf{F}^{-1} = \tilde{\mathbf{F}}^e\mathbf{F}^{e-1}$ or equivalently that $(\nabla\tilde{\boldsymbol{\chi}})\mathbf{F}^{s-1} = \tilde{\mathbf{F}}^e$. For this choice of $\mathcal{V}$, (17.5.11) yields

$$\int_{\partial\mathcal{P}} \mathbf{t}_{\mathrm{R}}(\mathbf{n}_{\mathrm{R}})\cdot\tilde{\boldsymbol{\chi}}\,da_{\mathrm{R}} + \int_{\mathcal{P}} \mathbf{b}_{0\mathrm{R}}\cdot\tilde{\boldsymbol{\chi}}\,dv_{\mathrm{R}} = \int_{\mathcal{P}} \mathbf{S}^e:\tilde{\mathbf{F}}^e\,dv_{\mathrm{R}} = \int_{\mathcal{P}} (\mathbf{S}^e\mathbf{F}^{s-\top}):\nabla\tilde{\boldsymbol{\chi}}\,dv_{\mathrm{R}}, \quad (17.5.13)$$

which, by defining

$$\mathbf{T}_{\mathrm{R}} \stackrel{\text{def}}{=} \mathbf{S}^e\mathbf{F}^{s-\top}, \quad (17.5.14)$$

may be rewritten as

$$\int_{\partial\mathcal{P}} \mathbf{t}_{\mathrm{R}}(\mathbf{n}_{\mathrm{R}})\cdot\tilde{\boldsymbol{\chi}}\,da_{\mathrm{R}} = \int_{\mathcal{P}} \Big(\mathbf{T}_{\mathrm{R}}:\nabla\tilde{\boldsymbol{\chi}} - \mathbf{b}_{0\mathrm{R}}\cdot\tilde{\boldsymbol{\chi}}\Big)\,dv_{\mathrm{R}}, \quad (17.5.15)$$

and using the divergence theorem we may conclude that

$$\int_{\partial\mathcal{P}} (\mathbf{t}_{\mathrm{R}}(\mathbf{n}_{\mathrm{R}}) - \mathbf{T}_{\mathrm{R}}\mathbf{n}_{\mathrm{R}})\cdot\tilde{\boldsymbol{\chi}}\,da_{\mathrm{R}} + \int_{\mathcal{P}} (\mathrm{Div}\,\mathbf{T}_{\mathrm{R}} + \mathbf{b}_{0\mathrm{R}})\cdot\tilde{\boldsymbol{\chi}}\,dv_{\mathrm{R}} = 0.$$

Since this relation must hold for all $\mathcal{P}$ and all $\tilde{\boldsymbol{\chi}}$, standard variational arguments yield the traction condition

$$\mathbf{t}_{\mathrm{R}}(\mathbf{n}_{\mathrm{R}}) = \mathbf{T}_{\mathrm{R}}\mathbf{n}_{\mathrm{R}}, \quad (17.5.16)$$

and the local macroscopic force balance

$$\mathrm{Div}\,\mathbf{T}_{\mathrm{R}} + \mathbf{b}_{0\mathrm{R}} = \mathbf{0}, \quad (17.5.17)$$

respectively.

Next, we deduce the consequences of requirement (V2) of the principle of virtual power. Using (17.5.8) and $(17.5.10)_2$, requirement (V2) of the principle of virtual power leads to the requirement that

$$\int_{\mathcal{P}} (\mathbf{T}_{\mathrm{R}}\mathbf{F}^\top):\boldsymbol{\Omega}\,dv_{\mathrm{R}} = 0. \quad (17.5.18)$$

Since $\mathcal{P}$ is arbitrary, we obtain that $(\mathbf{T}_{\mathrm{R}}\mathbf{F}^\top):\boldsymbol{\Omega} = 0$ for all skew tensors $\boldsymbol{\Omega}$, which implies that $\mathbf{T}_{\mathrm{R}}\mathbf{F}^\top$ is *symmetric*:

$$\mathbf{T}_{\mathrm{R}}\mathbf{F}^\top = \mathbf{F}\mathbf{T}_{\mathrm{R}}^\top. \quad (17.5.19)$$

- *In view of* (17.5.17) *and* (17.5.19) *the stress* $\mathbf{T}_{\mathrm{R}}$ *represents the classical Piola stress, with* (17.5.17) *and* (17.5.19) *representing the local macroscopic force and moment balances in the reference body.*

As is standard, the Piola stress $\mathbf{T}_{\mathrm{R}}$ is related to the symmetric Cauchy stress $\mathbf{T}$ in the deformed body by

$$\mathbf{T}_{\mathrm{R}} = J\,\mathbf{T}\,\mathbf{F}^{-\top}, \tag{17.5.20}$$

so that

$$\mathbf{T} = J^{-1}\mathbf{T}_{\mathrm{R}}\mathbf{F}^{\top}. \tag{17.5.21}$$

It is convenient to introduce two new stress measures:

- The elastic second Piola stress,

$$\mathbf{T}^e \overset{\text{def}}{=} J^e\mathbf{F}^{e-1}\mathbf{T}\mathbf{F}^{e-\top}, \tag{17.5.22}$$

  which is *symmetric* on account of the symmetry of the Cauchy stress $\mathbf{T}$.

- The Mandel stress,

$$\mathbf{M}^e \overset{\text{def}}{=} \mathbf{C}^e\mathbf{T}^e = J^e\mathbf{F}^{e\top}\mathbf{T}\mathbf{F}^{e-\top}. \tag{17.5.23}$$

  which in general is *not symmetric.*

Using (17.5.14), (17.5.20), and (17.2.2) we find that

$$\mathbf{S}^e = J\mathbf{T}\mathbf{F}^{e-\top}. \tag{17.5.24}$$

Thus, using the definitions (17.5.22) and (17.5.23) we find that

$$\mathbf{F}^{e-1}\mathbf{S}^e = J^s\mathbf{T}^e, \quad \text{and} \quad \mathbf{F}^{e\top}\mathbf{S}^e = J^s\mathbf{M}^e. \tag{17.5.25}$$

17.5.2 **Microscopic force balance**

Next, consider a generalized virtual velocity with $\tilde{\chi} = \mathbf{0}$, choose the virtual field $\tilde{c}_{\mathrm{R}}$ arbitrarily, and let

$$\tilde{\mathbf{F}}^e = -\frac{1}{3}\Theta(c_{\mathrm{R}})\tilde{c}_{\mathrm{R}}\,\mathbf{F}^e.$$

Then

$$\mathbf{S}^e : \tilde{\mathbf{F}}^e = -\frac{1}{3}\Theta(c_{\mathrm{R}})\tilde{c}_{\mathrm{R}}\,(\mathbf{F}^{e\top}\mathbf{S}^e) : \mathbf{1} = -\,J^s\left(\Theta(c_{\mathrm{R}})\left(\frac{1}{3}\mathrm{tr}\,\mathbf{M}^e\right)\right)\tilde{c}_{\mathrm{R}}, \tag{17.5.26}$$

where we have used $(17.5.25)_2$. Further, recalling the definition (17.4.3) of $\Theta(c_R)$, the expression above may be simplified to read as

$$\mathbf{S}^e : \tilde{\mathbf{F}}^e = -\Omega\left(\frac{1}{3}\mathrm{tr}\,\mathbf{M}^e\right)\tilde{c}_R. \tag{17.5.27}$$

Using (17.5.27) the power balance (17.5.11) then yields the microscopic virtual-power relation

$$\int_{\partial\mathcal{P}} \xi(\mathbf{n}_R)\tilde{c}_R\, da_R = \int_{\mathcal{P}} \left(-\Omega\left(\frac{1}{3}\mathrm{tr}\,\mathbf{M}^e\right)\tilde{c}_R + \varpi\tilde{c}_R + \boldsymbol{\xi}\cdot\nabla\tilde{c}_R\right) dv_R \tag{17.5.28}$$

to be satisfied for all $\tilde{c}_R$ and all $\mathcal{P}$. Equivalently, using the divergence theorem in the form

$$\int_{\mathcal{P}} \boldsymbol{\xi}\cdot\nabla\tilde{c}_R\, dv_R = \int_{\partial\mathcal{P}} \boldsymbol{\xi}\cdot\mathbf{n}_R\,\tilde{c}_R\, da_R - \int_{\mathcal{P}} (\mathrm{Div}\,\boldsymbol{\xi})\tilde{c}_R\, dv_R,$$

we obtain

$$\int_{\partial\mathcal{P}} (\xi(\mathbf{n}_R) - \boldsymbol{\xi}\cdot\mathbf{n}_R)\tilde{c}_R\, da_R + \int_{\mathcal{P}} \left(\mathrm{Div}\,\boldsymbol{\xi} - \varpi + \Omega\left(\frac{1}{3}\mathrm{tr}\,\mathbf{M}^e\right)\right)\tilde{c}_R\, dv_R = 0,$$

and a standard argument yields the microscopic traction condition

$$\xi(\mathbf{n}_R) = \boldsymbol{\xi}\cdot\mathbf{n}_R \tag{17.5.29}$$

and the microscopic force balance

$$\mathrm{Div}\,\boldsymbol{\xi} - \varpi + \Omega\left(\frac{1}{3}\mathrm{tr}\,\mathbf{M}^e\right) = 0. \tag{17.5.30}$$

17.5.3 **Actual external and internal expenditures of power**

Finally, using the traction conditions (17.5.16) and (17.5.29), the actual external expenditure of power is

$$\mathcal{W}_{\mathrm{ext}}(\mathcal{P}) = \int_{\partial\mathcal{P}} (\mathbf{T}_R\mathbf{n}_R)\cdot\dot{\boldsymbol{\chi}}\, da_R + \int_{\mathcal{P}} \mathbf{b}_{0R}\cdot\dot{\boldsymbol{\chi}}\, dv_R + \int_{\partial\mathcal{P}} (\boldsymbol{\xi}\cdot\mathbf{n}_R)\,\dot{c}_R\, da_R. \tag{17.5.31}$$

Also, using $(17.5.25)_1$ and $(17.2.7)_2$ the stress power $\mathbf{S}^e : \dot{\mathbf{F}}^e$ may be alternatively written as

$$\mathbf{S}^e : \dot{\mathbf{F}}^e = (J^s\mathbf{T}^e) : (\mathbf{F}^{e\top}\dot{\mathbf{F}}^e) = \tfrac{1}{2}(J^s\mathbf{T}^e) : \dot{\mathbf{C}}^e. \tag{17.5.32}$$

Thus the corresponding internal expenditure of power may be written as

$$\mathcal{W}_{\text{int}}(\mathcal{P}) = \int\limits_{\mathcal{P}} \Big(\tfrac{1}{2}(J^s\mathbf{T}^e) : \dot{\mathbf{C}}^e + \varpi \dot{c}_{\text{R}} + \boldsymbol{\xi} \cdot \nabla \dot{c}_{\text{R}} \Big)\, dv_{\text{R}}. \tag{17.5.33}$$

And power balance requires that,

$$\mathcal{W}_{\text{ext}}(\mathcal{P}) = \mathcal{W}_{\text{int}}(\mathcal{P}). \tag{17.5.34}$$

17.6 Free-energy imbalance

Under isothermal conditions the two laws of thermodynamics reduce to the statement that the rate of increase in free energy of any part $\mathcal{P}$ is less than or equal to the power expended on $\mathcal{P}$. Precisely, letting ψ_{R} denote the free energy per unit reference volume, this requirement takes the form of a free-energy imbalance

$$\overline{\int_{\mathcal{P}} \psi_{\text{R}}\, dv_{\text{R}}}^{\boldsymbol{\cdot}} \leq \mathcal{W}_{\text{ext}}(\mathcal{P}) + \mathcal{T}(\mathcal{P}), \tag{17.6.1}$$

where:

(i) $\mathcal{W}_{\text{ext}}(\mathcal{P})$ is the external mechanical power given by (17.5.31), viz.

$$\mathcal{W}_{\text{ext}}(\mathcal{P}) = \int\limits_{\partial\mathcal{P}} (\mathbf{T}_{\text{R}}\mathbf{n}_{\text{R}}) \cdot \dot{\boldsymbol{\chi}}\, da_{\text{R}} + \int\limits_{\mathcal{P}} \mathbf{b}_{0\text{R}} \cdot \dot{\boldsymbol{\chi}}\, dv_{\text{R}} + \int\limits_{\partial\mathcal{P}} (\boldsymbol{\xi} \cdot \mathbf{n}_{\text{R}})\, \dot{c}_{\text{R}}\, da_{\text{R}}. \tag{17.6.2}$$

Here, consistent with our omission of inertial effects, we have neglected kinetic energy.

(ii) $\mathcal{T}(\mathcal{P})$ represents the chemical energy carried into part $\mathcal{P}$ by species transport and is given by

$$\mathcal{T}(\mathcal{P}) = -\int_{\partial\mathcal{P}} \mu \mathbf{j}_{\text{R}} \cdot \mathbf{n}_{\text{R}} da_{\text{R}}, \tag{17.6.3}$$

with μ the *chemical potential* of the diffusing species. The chemical potential μ is a primitive quantity that enters the theory through the manner in which it appears in the basic law expressing the free-energy imbalance (Gurtin et al., 2010, p. 65).

Thus, since $\mathcal{W}_{\text{ext}}(\mathcal{P}) = \mathcal{W}_{\text{int}}(\mathcal{P})$, upon recalling (17.5.33) and applying the divergence theorem to the term in (17.6.3) involving the species transport $\mathcal{T}(\mathcal{P})$ with an integral over the boundary $\partial\mathcal{P}$ of $\mathcal{P}$, we obtain

$$\int_{\mathcal{P}} \Big(\dot{\psi}_{\text{R}} - \tfrac{1}{2}(J^s\mathbf{T}^e) : \dot{\mathbf{C}}^e - \varpi \dot{c}_{\text{R}} - \boldsymbol{\xi} \cdot \nabla \dot{c}_{\text{R}} + \mu \operatorname{Div} \mathbf{j}_{\text{R}} + \mathbf{j}_{\text{R}} \cdot \nabla \mu \Big)\, dv_{\text{R}} \leq 0, \tag{17.6.4}$$

which upon use of the balance law (17.3.3), and using the fact that (17.6.4) must hold for all parts $\mathcal{P}$, gives the local form of the free-energy imbalance as

$$\dot{\psi}_{\mathrm{R}} - \tfrac{1}{2}(J^s\mathbf{T}^e):\dot{\mathbf{C}}^e - \mu_{\mathrm{net}}\dot{c}_{\mathrm{R}} - \boldsymbol{\xi}\cdot\nabla\dot{c}_{\mathrm{R}} + \mathbf{j}_{\mathrm{R}}\cdot\nabla\mu \le 0, \tag{17.6.5}$$

where we have written

$$\mu_{\mathrm{net}} \stackrel{\mathrm{def}}{=} \mu + \varpi \tag{17.6.6}$$

for a *net chemical potential.*

For later use we define the dissipation density $\mathcal{D} \geq 0$ per unit volume per unit time by

$$\mathcal{D} = \tfrac{1}{2}(J^s\mathbf{T}^e):\dot{\mathbf{C}}^e + \mu_{\mathrm{net}}\dot{c}_{\mathrm{R}} + \boldsymbol{\xi}\cdot\nabla\dot{c}_{\mathrm{R}} - \mathbf{j}_{\mathrm{R}}\cdot\nabla\mu - \dot{\psi}_{\mathrm{R}} \geq 0. \tag{17.6.7}$$

REMARK

For brevity we have not discussed the transformation properties under a change in frame of the various fields appearing in our theory. Here, we simply note that all quantities in the free-energy imbalance (17.6.5) are invariant under a change in frame (Gurtin et al., 2010).

17.7 Constitutive theory

17.7.1 Energetic constitutive equations

Guided by the free-energy imbalance (17.6.5) we consider constitutive equations for the free energy ψ_{R}, the stress $\mathbf{T}^e$, the net chemical potential μ_{net}, and the microstress $\boldsymbol{\xi}$ of the form

$$\begin{aligned}
\psi_{\mathrm{R}} &= \hat{\psi}_{\mathrm{R}}(\mathbf{C}^e, c_{\mathrm{R}}, \nabla c_{\mathrm{R}}),\\
\mathbf{T}^e &= \hat{\mathbf{T}}^e(\mathbf{C}^e, c_{\mathrm{R}}, \nabla c_{\mathrm{R}}),\\
\mu_{\mathrm{net}} &= \hat{\mu}_{\mathrm{net}}(\mathbf{C}^e, c_{\mathrm{R}}, \nabla c_{\mathrm{R}}),\\
\boldsymbol{\xi} &= \hat{\boldsymbol{\xi}}(\mathbf{C}^e, c_{\mathrm{R}}, \nabla c_{\mathrm{R}}).
\end{aligned} \tag{17.7.1}$$

It is convenient to introduce the notation

$$\boldsymbol{\Lambda} = (\mathbf{C}^e, c_{\mathrm{R}}, \nabla c_{\mathrm{R}}). \tag{17.7.2}$$

Substituting the constitutive equations (17.7.1) into the free-energy imbalance (17.6.5), we find that it may then be written as

$$\begin{aligned}
&\left[\frac{\partial\hat{\psi}_{\mathrm{R}}(\boldsymbol{\Lambda})}{\partial\mathbf{C}^e} - \tfrac{1}{2}(J^s\hat{\mathbf{T}}^e(\boldsymbol{\Lambda}))\right]:\dot{\mathbf{C}}^e + \left[\frac{\partial\hat{\psi}_{\mathrm{R}}(\boldsymbol{\Lambda})}{\partial c_{\mathrm{R}}} - \mu_{\mathrm{net}}\right]\dot{c}_{\mathrm{R}}\\
&\qquad + \left[\frac{\partial\hat{\psi}_{\mathrm{R}}(\boldsymbol{\Lambda})}{\partial\nabla c_{\mathrm{R}}} - \hat{\boldsymbol{\xi}}(\boldsymbol{\Lambda})\right]\cdot\nabla\dot{c}_{\mathrm{R}} + \mathbf{j}_{\mathrm{R}}\cdot\nabla\mu \le 0.
\end{aligned} \tag{17.7.3}$$

Since $\dot{\mathbf{C}}^e$, $\dot{c}_{\mathrm{R}}$, and $\nabla\dot{c}_{\mathrm{R}}$ appear linearly in (17.7.3), their "coefficients" must vanish, for otherwise $\dot{\mathbf{C}}^e$, $\dot{c}_{\mathrm{R}}$, and $\nabla\dot{c}_{\mathrm{R}}$ may be chosen to violate (17.7.3). We are therefore led to the thermodynamic restriction that the free energy determines the stress $\mathbf{T}^e$, the net chemical potential μ_{net}, and the vector microstress $\boldsymbol{\xi}$ through the *state relations*

$$\begin{aligned}\mathbf{T}^e &= J^{s-1}\Big(2\frac{\partial\hat{\psi}_{\mathrm{R}}(\mathbf{C}^e, c_{\mathrm{R}}, \nabla c_{\mathrm{R}})}{\partial\mathbf{C}^e}\Big),\\ \mu_{\mathrm{net}} &= \frac{\partial\hat{\psi}_{\mathrm{R}}(\mathbf{C}^e, c_{\mathrm{R}}, \nabla c_{\mathrm{R}})}{\partial c_{\mathrm{R}}},\\ \boldsymbol{\xi} &= \frac{\partial\hat{\psi}_{\mathrm{R}}(\mathbf{C}^e, c_{\mathrm{R}}, \nabla c_{\mathrm{R}})}{\partial\nabla c_{\mathrm{R}}},\end{aligned} \tag{17.7.4}$$

so that (17.7.3) can be used to write the reduced dissipation inequality,

$$\mathcal{D} = -\mathbf{j}_{\mathrm{R}}\cdot\nabla\mu \geq 0. \tag{17.7.5}$$

17.7.2 Chemical potential

The microforce balance (17.5.30), when combined with the thermodynamically consistent constitutive equations $(17.7.4)_{2,3}$, together with the definition (17.6.6) of μ_{net}, gives the following important expression for the chemical potential in the theory:

$$\mu = \frac{\partial\hat{\psi}_{\mathrm{R}}(\mathbf{C}^e, c_{\mathrm{R}}, \nabla c_{\mathrm{R}})}{\partial c_{\mathrm{R}}} - \mathrm{Div}\left[\frac{\partial\hat{\psi}_{\mathrm{R}}(\mathbf{C}^e, c_{\mathrm{R}}, \nabla c_{\mathrm{R}})}{\partial\nabla c_{\mathrm{R}}}\right] - \Omega\left(\frac{1}{3}\mathrm{tr}\,\mathbf{M}^e\right). \tag{17.7.6}$$

REMARK

In the absence of mechanical deformation, the expression (17.7.6) for the chemical potential,

- *which is deduced from a microforce balance and thermodynamically consistent constitutive equations,*

reduces to an expression for μ which may be derived from an entirely different procedure using a *variational derivative* of an energy functional as in the classical Cahn and Hilliard (1958) theory for species diffusion and phase segregation.

Note that the derivation based on the method of virtual power given here, *accounts for chemo-mechanical coupling in a straightforward manner.*

17.7.3 Dissipative constitutive equation

To the constitutive equations (17.7.1) we append the following relation for the flux of the diffusing species,

$$\mathbf{j}_{\mathrm{R}} = -\mathbf{M}(\tilde{\boldsymbol{\Lambda}})\,\nabla\mu, \tag{17.7.7}$$

where $\mathbf{M}$ is a *mobility tensor*, and where $\tilde{\Lambda}$ denotes the list

$$\tilde{\Lambda} = (\mathbf{C}^e, c_{\mathrm{R}}). \tag{17.7.8}$$

Using (17.7.7) the dissipation inequality (17.7.5) may be written as

$$\mathcal{D} = \nabla\mu \cdot \mathbf{M}(\tilde{\Lambda})\nabla\mu \geq 0. \tag{17.7.9}$$

In order to satisfy (17.7.9), the mobility tensor $\mathbf{M}$ must be positive semi-definite, which we henceforth assume.

17.8 Summary of the constitutive theory

Our chemo-mechanical theory relates the following basic fields:

$\mathbf{x} = \boldsymbol{\chi}(\mathbf{X}, t)$,	motion;
$\mathbf{F} = \nabla\boldsymbol{\chi}, \quad J = \det\mathbf{F} > 0$,	deformation gradient;
$\mathbf{F} = \mathbf{F}^e\mathbf{F}^s$,	multiplicative decomposition of $\mathbf{F}$;
$\mathbf{F}^s = \lambda^s\mathbf{1}, \quad J^s = (\lambda^s)^3 > 0$,	swelling distortion;
$\mathbf{F}^e, \quad J^e = \det\mathbf{F}^e > 0$,	elastic distortion;
$\mathbf{C}^e = \mathbf{F}^{e\top}\mathbf{F}^e$,	elastic right Cauchy–Green tensor;
$\mathbf{T} = \mathbf{T}^\top$,	Cauchy stress;
$\mathbf{T}^e = J^e\mathbf{F}^{e-1}\mathbf{T}\mathbf{F}^{e-\top}$,	Elastic second Piola stress;
$\mathbf{M}^e = \mathbf{C}^e\mathbf{T}^e$,	Mandel stress;
$\mathbf{T}_{\mathrm{R}} = J\mathbf{T}\mathbf{F}^{-\top}$,	Piola stress;
ψ_{R},	free energy density per unit reference volume;
c_{R},	species concentration;
μ,	chemical potential;
$\mathbf{j}_{\mathrm{R}}$,	referential species flux vector;
$\boldsymbol{\xi}$,	vector microstress.

17.8.1 Constitutive equations

1. **Swelling ratio**
   The swelling ratio J^s is related to the species concentration through a constitutive relation

$$J^s = 1 + \Omega(c_{\mathrm{R}} - c_{\mathrm{R}0}), \tag{17.8.1}$$

where

$$\Omega > 0, \tag{17.8.2}$$

represents the *molar volume* of the diffusing species.

2. **Free energy**
The free energy is taken to be a function of the list $(\mathbf{C}^e, c_R, \nabla c_R)$,

$$\psi_R = \hat{\psi}_R(\mathbf{C}^e, c_R, \nabla c_R). \tag{17.8.3}$$

3. **Elastic second Piola stress. Mandel stress. Cauchy stress. Piola stress**
The elastic second Piola stress is given by

$$\mathbf{T}^e = J^{s-1} 2 \frac{\partial \hat{\psi}_R(\mathbf{C}^e, c_R, \nabla c_R)}{\partial \mathbf{C}^e}, \tag{17.8.4}$$

the Mandel stress by

$$\mathbf{M}^e = \mathbf{C}^e \mathbf{T}^e, \tag{17.8.5}$$

the Cauchy stress by

$$\mathbf{T} = J^{e-1} \mathbf{F}^e \mathbf{T}^e \mathbf{F}^{e\top}, \tag{17.8.6}$$

and the Piola stress by

$$\mathbf{T}_R \stackrel{\text{def}}{=} J\mathbf{T}\mathbf{F}^{-\top}. \tag{17.8.7}$$

4. **Chemical potential**
The quantity

$$\mu = \frac{\partial \hat{\psi}_R(\mathbf{C}^e, c_R, \nabla c_R)}{\partial c_R} - \mathrm{Div}\left[\frac{\partial \hat{\psi}_R(\mathbf{C}^e, c_R, \nabla c_R)}{\partial \nabla c_R}\right] - \Omega\left(\frac{1}{3}\mathrm{tr}\,\mathbf{M}^e\right) \tag{17.8.8}$$

represents the chemical potential.

5. **Microstress $\boldsymbol{\xi}$**
The vector microstress $\boldsymbol{\xi}$ is given by

$$\boldsymbol{\xi} = \frac{\partial \hat{\psi}_R(\mathbf{C}^e, c_R, \nabla c_R)}{\partial \nabla c_R}. \tag{17.8.9}$$

6. **Evolution equation for $\mathbf{F}^s$**
The evolution equation for $\mathbf{F}^s$ is

$$\dot{\mathbf{F}}^s = \mathbf{D}^s\,\mathbf{F}^s, \tag{17.8.10}$$

with $\mathbf{D}^s$ given by

$$\mathbf{D}^s = \frac{1}{3}\Theta(c_{\mathrm{R}})\,\dot{c}_{\mathrm{R}}\mathbf{1}, \tag{17.8.11}$$

where

$$\Theta(c_{\mathrm{R}}) \overset{\text{def}}{=} \frac{\Omega}{J^s}. \tag{17.8.12}$$

7. **Species flux**
The species flux $\mathbf{j}_{\mathrm{R}}$ is presumed to obey the constitutive equation

$$\mathbf{j}_{\mathrm{R}} = -\mathbf{M}\,\nabla\mu, \tag{17.8.13}$$

with $\mathbf{M} = \hat{\mathbf{M}}(\mathbf{C}^e, c_{\mathrm{R}})$ a positive semi-definite mobility tensor.

17.9 Governing partial differential equations. Boundary conditions

The governing partial differential equations consist of:

1. The local force balance (17.5.17), viz.

$$\operatorname{Div}\mathbf{T}_{\mathrm{R}} + \mathbf{b}_{0\mathrm{R}} = \mathbf{0}, \tag{17.9.1}$$

where $\mathbf{b}_{0\mathrm{R}}$ is the non-inertial body force, and $\mathbf{T}_{\mathrm{R}}$ given by (17.8.7).

With $\mathbf{u} = \chi - \mathbf{X}$ denoting the displacement, let $\partial\mathcal{B}_{\mathrm{u}}$ and $\partial\mathcal{B}_{\mathrm{t_R}}$ be complementary subsurfaces of the boundary $\partial\mathcal{B}$ of the body $\mathcal{B}$ in the sense $\partial\mathcal{B} = \partial\mathcal{B}_{\mathrm{u}} \cup \partial\mathcal{B}_{\mathrm{t_R}}$ and $\partial\mathcal{B}_{\mathrm{u}} \cap \partial\mathcal{B}_{\mathrm{t_R}} = \emptyset$. Then, as is standard, for a time interval $t \in [0, T]$ we consider a pair of simple boundary conditions in which the displacement is specified on $\partial\mathcal{B}_{\mathrm{u}}$ and the surface traction on $\partial\mathcal{B}_{\mathrm{t_R}}$:

$$\begin{aligned} \mathbf{u} &= \bar{\mathbf{u}} \quad \text{on } \partial\mathcal{B}_{\mathrm{u}} \times [0, T], \\ \mathbf{T}_{\mathrm{R}}\mathbf{n}_{\mathrm{R}} &= \bar{\mathbf{t}}_{\mathrm{R}} \quad \text{on } \partial\mathcal{B}_{\mathrm{t_R}} \times [0, T]. \end{aligned} \tag{17.9.2}$$

2. The local balance equation for the species concentration (17.3.3) together with (17.8.13) gives

$$\dot{c}_{\mathrm{R}} - \mathrm{Div}\,(\mathbf{M}\nabla\mu) = 0, \tag{17.9.3}$$

with the chemical potential given by (17.8.8), viz.

$$\mu = \frac{\partial\hat{\psi}_{\mathrm{R}}(\mathbf{C}^e, c_{\mathrm{R}}, \nabla c_{\mathrm{R}})}{\partial c_{\mathrm{R}}} - \mathrm{Div}\left[\frac{\partial\hat{\psi}_{\mathrm{R}}(\mathbf{C}^e, c_{\mathrm{R}}, \nabla c_{\mathrm{R}})}{\partial\nabla c_{\mathrm{R}}}\right] - \Omega\left(\frac{1}{3}\mathrm{tr}\,\mathbf{M}^e\right). \tag{17.9.4}$$

Next, the flux of the diffusing species from the material in contact with the body changes the energy of the body according to

$$-\int_{\partial\mathcal{B}} \mu\,\mathbf{j}_{\mathrm{R}}\cdot\mathbf{n}_{\mathrm{R}}\,da_{\mathrm{R}}.$$

We use this as a guide for specifying suitable boundary conditions. Let $\partial\mathcal{B}_\mu$ and $\partial\mathcal{B}_{\mathrm{j_R}}$ be complementary subsurfaces of the boundary. We consider a pair of simple boundary conditions in which the chemical potential is specified on $\partial\mathcal{B}_\mu$ and the species flux on $\partial\mathcal{B}_{\mathrm{j_R}}$:

$$\begin{aligned} \mu &= \bar{\mu} \quad \text{on } \partial\mathcal{B}_\mu \times [0,T], \\ \mathbf{j}_{\mathrm{R}}\cdot\mathbf{n}_{\mathrm{R}} &= \bar{j}_{\mathrm{R}} \quad \text{on } \partial\mathcal{B}_{\mathrm{j_R}} \times [0,T], \end{aligned} \tag{17.9.5}$$

with the flux $\mathbf{j}_{\mathrm{R}}$ given by (17.8.13). We associate these boundary conditions with the partial differential equation (17.9.3) for the chemical potential μ.

Finally, the presence of microscopic stresses results in an expenditure of power

$$\int_{\partial\mathcal{B}} (\boldsymbol{\xi}\cdot\mathbf{n}_{\mathrm{R}})\dot{c}_{\mathrm{R}}\,da_{\mathrm{R}}$$

by the material in contact with the body (cf. eq. (17.5.31)), and this necessitates a consideration of boundary conditions on $\partial\mathcal{B}$ involving the microscopic tractions $\boldsymbol{\xi}\cdot\mathbf{n}_{\mathrm{R}}$ and the rate of change of the species concentration $\dot{c}_{\mathrm{R}}$.

- *We restrict attention to boundary conditions that result in a null expenditure of microscopic power in the sense that* $(\boldsymbol{\xi}\cdot\mathbf{n}_{\mathrm{R}})\dot{c}_{\mathrm{R}} = 0$.

A simple boundary condition which satisfies this requirement is[3]

$$\boldsymbol{\xi}\cdot\mathbf{n}_{\mathrm{R}} = 0 \quad \text{on } \partial\mathcal{B} \times [0,T], \tag{17.9.6}$$

with the microforce $\boldsymbol{\xi}$ given by (17.8.9). Thus, (17.9.6) is a boundary condition on the normal derivative of the species concentration. We associate this boundary condition with the

[3] A boundary condition requiring $\dot{c}_{\mathrm{R}} = 0$ on $\partial\mathcal{B}$ for all times would also satisfy the null microscopic power expenditure requirement.

equation (17.9.4) which may be interpreted as a partial differential equation for the species concentration c_R.

The initial data is taken as

$$\mathbf{u}(\mathbf{X},0) = \mathbf{u}_0(\mathbf{X}), \quad c_R(\mathbf{X},0) = c_{R0}(\mathbf{X}) \quad \text{in} \quad \mathcal{B}. \tag{17.9.7}$$

The coupled set of equations (17.9.1) and (17.9.3), together with (17.9.2), (17.9.5), (17.9.6), and (17.9.7) yield an initial/boundary-value problem for the displacement $\mathbf{u}(\mathbf{X},t)$ and the species concentration $c_R(\mathbf{X},t)$.

17.10 Specialization of the constitutive equations

17.10.1 Free energy

We consider a separable free energy of the form

$$\hat{\psi}_R(\mathbf{C}^e, c_R, \nabla c_R) = \psi_R^c(c_R) + \psi_R^s(|\nabla c_R|) + \psi^e(\mathbf{C}^e, c_R), \tag{17.10.1}$$

In the free-energy expression (17.10.1):

(i) ψ_R^c is the change in chemical free energy due to mixing of the species with the host material. As a simple continuum approximation to mixing, we take this to be given by a regular solution model (cf., e.g., DeHoff, 2006)

$$\psi_R^c = \mu^{\text{ref}}\, c_R + R\vartheta\, c_R^{\max}\Big(\bar{c}_R \ln \bar{c}_R + (1-\bar{c}_R)\ln(1-\bar{c}_R)\Big) + R\vartheta\, c_R^{\max}\Big(\chi\, \bar{c}_R(1-\bar{c}_R)\Big), \tag{17.10.2}$$

where

$$\bar{c}_R \overset{\text{def}}{=} \frac{c_R}{c_R^{\max}}, \qquad 0 \le \bar{c}_R \le 1, \tag{17.10.3}$$

is a normalized species concentration, with $c_R^{\max}$ the concentration of the species in moles per unit reference volume when all the accommodating sites in the host material are filled. R is the universal gas constant, and ϑ is the absolute temperature (assumed to be constant). Also,

- μ^{ref} is a reference value of the chemical potential of the diffusing species;
- the second term in (17.10.2) represents the entropy of mixing for an ideal solution; while
- the last term in (17.10.2) expresses a deviation from ideality and represents energetic interactions of mixing where χ is an interaction parameter that controls if there is any phase separation:
  * A value of $\chi = 0$ implies that there is no interaction, and that mixing is ideal and purely entropic in nature.

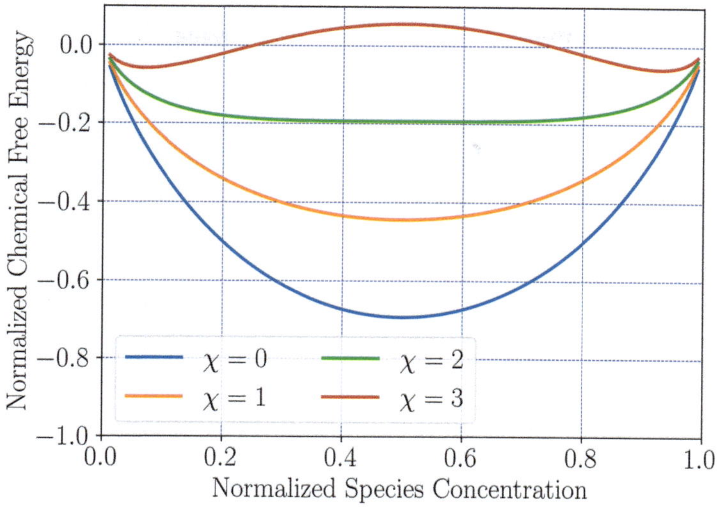

Fig. 17.1 Normalized chemical free energy $(\psi^c_R/c^{max}_R - \mu^{ref}\bar{c}_R)/(R\vartheta)$ versus normalized concentration $\bar{c}_R$ for four different values of the χ-parameter. For $\chi = 3$ the chemical free energy is *non-convex* with two wells.

∗ Positive values of χ correspond to interactions which increase the free energy and oppose mixing.

∗ For $\chi \leq 2$ the chemical free energy has a single well and admits a single phase only.

∗ For $\chi > 2$ the chemical free energy is *non-convex*, with two wells which drives phase segregation into two binodal points—one with a high site fraction of species and another with a lower site fraction; cf. Fig. 17.1.

(ii) ψ^s_R is an interfacial free energy which depends on the concentration gradient. We take this as the following simple quadratic form:

$$\psi^s_R = c^{max}_R \left[\tfrac{1}{2}\lambda\,|\nabla\bar{c}_R|^2\right], \tag{17.10.4}$$

with $\lambda \geq 0$ a gradient energy coefficient that governs the width of interfaces between phases.

REMARK

In order to account for the energy associated with the compositional/structural inhomogeneities occurring at phase-interfaces in the microstructure, following Cahn and Hilliard (1958), we have introduced a free-energy contribution (17.10.4) in which the constant $\lambda \geq 0$ (which has units of energy per mole times length squared) introduces an internal length scale in our continuum-level theory. On physical grounds this length scale should be set by the widths of the interfaces in the underlying microstructure of the material—widths which are probably of the order of a few atomic distances.

An important feature of a gradient theory which accounts for the energy of interfaces using (17.10.4) is that instead of having to deal with *sharp interfaces* across which there is a *jump* in the concentration, a theory using (17.10.4) serves to *regularize* the problem—it allows one to model a phase-interface as a *diffuse interface* in which the concentration field takes distinct values in each of the phases, with a *smooth* change between the values around the interface.

(iii) ψ^e is the contribution to the change in the free energy due to the elastic deformation of the host material, taken to be given by

$$\psi^e = J^s \, \tilde{\psi}^e(\mathbf{C}^e), \tag{17.10.5}$$

in which the term $\tilde{\psi}^e(\mathbf{C}^e)$ represents the free energy per unit volume of the intermediate space defined by the range of $\mathbf{F}^s(\mathbf{X})$, and multiplication of this term by J^s gives us the free energy per unit volume of the reference space.

Next, we introduce the multiplicative decomposition

$$\mathbf{F}^e = \mathbf{F}^{e\,\mathrm{vol}}\bar{\mathbf{F}}^e \tag{17.10.6}$$

of the elastic deformation gradient $\mathbf{F}^e$ into volumetric and isochoric factors $\mathbf{F}^{e\,\mathrm{vol}}$ and $\bar{\mathbf{F}}^e$, where

$$\mathbf{F}^{e\,\mathrm{vol}} = J^{e\,1/3}\mathbf{1} \qquad \text{and} \qquad \bar{\mathbf{F}}^e = J^{e\,-1/3}\mathbf{F}^e. \tag{17.10.7}$$

The decomposition (17.10.6) leads to a multiplicative decomposition of the right elastic Cauchy–Green tensor $\mathbf{C}^e$ in the form

$$\mathbf{C}^e = \mathbf{C}^{e\,\mathrm{vol}}\bar{\mathbf{C}}^e, \qquad \mathbf{C}^{e\,\mathrm{vol}} = J^{e\,2/3}\mathbf{1}, \qquad \bar{\mathbf{C}}^e = J^{e\,-2/3}\mathbf{C}^e. \tag{17.10.8}$$

Similarly, we also have a multiplicative decomposition of the left elastic Cauchy–Green tensor $\mathbf{B}^e$ in the form

$$\mathbf{B}^e = \mathbf{B}^{e\mathrm{vol}}\bar{\mathbf{B}}^e, \qquad \mathbf{B}^{e\mathrm{vol}} = J^{e\,2/3}\mathbf{1}, \qquad \bar{\mathbf{B}}^e = J^{e\,-2/3}\mathbf{B}^e. \tag{17.10.9}$$

Let

$$\bar{I}_1^e = \operatorname{tr}\bar{\mathbf{C}}^e = \operatorname{tr}\bar{\mathbf{B}}^e \tag{17.10.10}$$

denote the first principal invariant of $\bar{\mathbf{C}}^e$, which is equal to the first principal invariant of $\bar{\mathbf{B}}^e$.

Next, we adopt the following compressible Neo-Hookean for the elastic free energy $\tilde{\psi}^e(\mathbf{C}^e)$:

$$\begin{aligned}\tilde{\psi}^e(\mathbf{C}^e) &= \underbrace{\bar{\psi}^e(\bar{I}_1^e)}_{\text{isochoric}} + \underbrace{\psi^{e\text{vol}}(J^e)}_{\text{volumetric}}, \\ &= \tfrac{1}{2}G\left(\bar{I}_1^e - 3\right) + \tfrac{1}{2}K(J^e - 1)^2,\end{aligned} \tag{17.10.11}$$

where the parameters

$$G > 0 \qquad \text{and} \qquad K > 0, \tag{17.10.12}$$

represent the elastic shear and bulk moduli, respectively, of the classical linear elasticity theory.

Thus, using (17.10.2), (17.10.4), and (17.10.11) in (17.10.1), a simple form of the free-energy function which accounts for the combined effects of mixing, swelling, and finite elastic stretching is

$$\begin{aligned}\psi_{\mathrm{R}} = c_{\mathrm{R}}^{\max}\Big[\mu^{\mathrm{ref}}\,\bar{c}_{\mathrm{R}} + R\vartheta\Big(\bar{c}_{\mathrm{R}}\ln\bar{c}_{\mathrm{R}} + (1-\bar{c}_{\mathrm{R}})\ln(1-\bar{c}_{\mathrm{R}}) + \chi\,\bar{c}_{\mathrm{R}}(1-\bar{c}_{\mathrm{R}})\Big)\Big] \\ + c_{\mathrm{R}}^{\max}\left[\tfrac{1}{2}\lambda\,|\nabla\bar{c}_{\mathrm{R}}|^2\right] + J^s\Big[\tfrac{1}{2}G\left(\bar{I}_1^e - 3\right) + \tfrac{1}{2}K(J^e-1)^2\Big].\end{aligned} \tag{17.10.13}$$

17.10.2 **Stress**

Using (17.10.13) and (17.8.4) we find that the elastic second Piola stress is given by

$$\mathbf{T}^e = J^{e-2/3}G\left[\mathbf{1} - \frac{1}{3}\left(\operatorname{tr}\bar{\mathbf{C}}^e\right)\bar{\mathbf{C}}^{e-1}\right] + K\,J^e(J^e-1)\mathbf{C}^{e-1}, \tag{17.10.14}$$

and, since by (17.8.5) $\mathbf{M}^e = \mathbf{C}^e\mathbf{T}^e$, the Mandel stress is given by

$$\mathbf{M}^e = J^{e-2/3}G\left[\mathbf{C}^e - \frac{1}{3}\left(\operatorname{tr}\mathbf{C}^e\right)\mathbf{1}\right] + K\,J^e(J^e-1)\mathbf{1}, \tag{17.10.15}$$

and we find that $\mathbf{M}^e$ is *symmetric*,

$$\mathbf{M}^e = \mathbf{M}^{e\top}. \tag{17.10.16}$$

Further, using the relation

$$\mathbf{T} = J^{e-1}\mathbf{F}^e\mathbf{T}^e\mathbf{F}^{e\top} \quad\Longrightarrow\quad \mathbf{T} = J^{e-1/3}\bar{\mathbf{F}}^e\mathbf{T}^e\bar{\mathbf{F}}^{e\top}, \tag{17.10.17}$$

we obtain that the Cauchy stress is given by

$$\mathbf{T} = J^{e-1}G\left[\bar{\mathbf{B}}^e - \frac{1}{3}\left(\operatorname{tr}\bar{\mathbf{B}}^e\right)\mathbf{1}\right] + K(J^e - 1)\mathbf{1}, \tag{17.10.18}$$

and hence that the Piola stress is given by

$$\mathbf{T}_{\mathrm{R}} = J\mathbf{T}\mathbf{F}^{-\top}. \tag{17.10.19}$$

17.10.3 **Chemical potential**

Also, using (17.8.8) and (17.10.13), the chemical potential μ is given by

$$\begin{aligned}\mu = \mu^{\mathrm{ref}} + R\vartheta\left[\ln\left(\frac{\bar{c}_{\mathrm{R}}}{1-\bar{c}_{\mathrm{R}}}\right) + \chi(1-2\bar{c}_{\mathrm{R}})\right] - \lambda\Delta\bar{c}_{\mathrm{R}} - \Omega\left(\frac{1}{3}\operatorname{tr}\mathbf{M}^e\right)\\ + \Omega\left[\tfrac{1}{2}G\left(\bar{I}_1^e - 3\right) + \tfrac{1}{2}K(J^e-1)^2\right].\end{aligned}$$

Note that the last term in the expression for the chemical potential is quadratic in the elastic strain, and hence small, and negligible relative to the other terms in the expression for the chemical potential. Accordingly we adopt the following slightly simplified expression for the chemical potential,

$$\mu = \mu^{\mathrm{ref}} + R\vartheta\left[\ln\left(\frac{\bar{c}_{\mathrm{R}}}{1-\bar{c}_{\mathrm{R}}}\right) + \chi(1-2\bar{c}_{\mathrm{R}})\right] - \Omega\left(\frac{1}{3}\operatorname{tr}\mathbf{M}^e\right) - \lambda\Delta\bar{c}_{\mathrm{R}}. \tag{17.10.20}$$

17.10.4 **Microstress**

Finally, from (17.10.13) and (17.8.9) the vector microstress $\boldsymbol{\xi}$ is given by

$$\boldsymbol{\xi} = \lambda\nabla\bar{c}_{\mathrm{R}}. \tag{17.10.21}$$

17.10.5 **Species flux**

For isotropic materials the species flux in the *spatial description* is taken as

$$\mathbf{j} = -m\operatorname{grad}\mu, \tag{17.10.22}$$

in which we assume that the mobility $m = \hat{m}(c)$ at a given temperature ϑ is given by

$$m = \frac{Dc}{R\vartheta}\left(1 - \frac{c}{c^{\max}}\right), \tag{17.10.23}$$

where $D > 0$ is a species *diffusivity*, with $c = c_{\mathrm{R}}/J$ and $c^{\max} = c_{\mathrm{R}}^{\max}/J$ the species concentration and its maximum value measured in moles per unit deformed volume. The expression (17.10.23) represents the physical requirement that the pure phases $c = 0$ and $c = 1$ have vanishing mobility.

Then, using the standard mapping results

$$\mathbf{F}^{-\top}\nabla\mu = \operatorname{grad}\mu \qquad \text{and} \qquad \mathbf{j}_{\mathrm{R}} = J\mathbf{F}^{-1}\mathbf{j}, \tag{17.10.24}$$

we obtain the referential species flux as

$$\mathbf{j}_{\mathrm{R}} = -\mathbf{M}\nabla\mu \quad \text{with} \quad \mathbf{M} = \left(\frac{Dc_{\mathrm{R}}}{R\vartheta}\right)(1 - \bar{c}_{\mathrm{R}})\,\mathbf{C}^{-1}, \tag{17.10.25}$$

where $\mathbf{M}$ is the species mobility tensor, and with μ given in (17.10.20).

17.11 Governing equations for the specialized chemo-mechanical theory

(1) Neglecting body forces the macroscopic force balance is,

$$\operatorname{Div}\mathbf{T}_{\mathrm{R}} = \mathbf{0}, \tag{17.11.1}$$

with the Piola stress given by

$$\begin{aligned} \mathbf{T}_{\mathrm{R}} &= J\mathbf{T}\mathbf{F}^{-\top} \qquad \text{where} \\ \mathbf{T} &= J^{e-1}\,\mathbf{F}^{e}\mathbf{T}^{e}\mathbf{F}^{e\top} \qquad \text{and} \\ \mathbf{T}^{e} &= J^{e-2/3}G\left[\mathbf{1} - \frac{1}{3}\left(\operatorname{tr}\bar{\mathbf{C}}^{e}\right)\bar{\mathbf{C}}^{e-1}\right] + K\,J^{e}(J^{e}-1)\mathbf{C}^{e-1}. \end{aligned} \tag{17.11.2}$$

With $\partial\mathcal{B}_{\mathrm{u}}$ and $\partial\mathcal{B}_{\mathrm{t_R}}$ denoting complementary subsurfaces of the boundary $\partial\mathcal{B}$ of the body $\mathcal{B}$, we consider a pair of possible boundary conditions:

$$\begin{aligned} \mathbf{u} &= \bar{\mathbf{u}} \quad \text{on } \partial\mathcal{B}_{\mathrm{u}} \times [0,T], \\ \mathbf{T}_{\mathrm{R}}\mathbf{n}_{\mathrm{R}} &= \bar{\mathbf{t}}_{\mathrm{R}} \quad \text{on } \partial\mathcal{B}_{\mathrm{t_R}} \times [0,T]. \end{aligned} \tag{17.11.3}$$

In the boundary conditions above, $\bar{\mathbf{u}}$ and $\bar{\mathbf{t}}_{\mathrm{R}}$ are prescribed displacement and tractions, both functions of $(\mathbf{X}, t)$.

(2) The species balance equation is,

$$\dot{c}_{\mathrm{R}} + \operatorname{Div}\mathbf{j}_{\mathrm{R}} = 0, \tag{17.11.4}$$

with

$$\mathbf{j}_{\mathrm{R}} = -\mathbf{M}\nabla\mu, \qquad \mathbf{M} = \left(\frac{Dc_{\mathrm{R}}}{R\vartheta}(1-\bar{c}_{\mathrm{R}})\right)\mathbf{C}^{-1}.$$

With $\partial\mathcal{B}_\mu$ and $\partial\mathcal{B}_{\mathrm{j_R}}$ denoting complementary subsurfaces of $\partial\mathcal{B}$, we consider a pair of possible boundary conditions:

$$\begin{aligned} \mu &= \bar{\mu} \quad \text{on } \partial\mathcal{B}_\mu \times [0,T], \\ \mathbf{j}_{\mathrm{R}}\cdot\mathbf{n}_{\mathrm{R}} &= \bar{j}_{\mathrm{R}} \quad \text{on } \partial\mathcal{B}_{\mathrm{j_R}} \times [0,T], \end{aligned} \tag{17.11.5}$$

with $\bar{\mu}$ and $\bar{j}$ prescribed functions of $\mathbf{X}$ and t.

(3) Setting the reference chemical potential for the diffusing species to zero,

$$\mu^{\mathrm{ref}} = 0,$$

the chemical potential from eqn. (17.10.20) is,

$$\mu = R\vartheta\left[\ln\left(\frac{\bar{c}_{\mathrm{R}}}{1-\bar{c}_{\mathrm{R}}}\right) + \chi(1-2\bar{c}_{\mathrm{R}})\right] - \Omega\left(\frac{1}{3}\operatorname{tr}\mathbf{M}^e\right) - \lambda\,\Delta\bar{c}_{\mathrm{R}}. \tag{17.11.6}$$

where

$$\mathbf{M}^e = \mathbf{C}^e\mathbf{T}^e \tag{17.11.7}$$

is the Mandel stress. Let

$$f^{\,c} \overset{\text{def}}{=} R\vartheta\left[\ln\left(\frac{\bar{c}_{\mathrm{R}}}{1-\bar{c}_{\mathrm{R}}}\right) + \chi(1-2\bar{c}_{\mathrm{R}})\right] - \Omega\left(\frac{1}{3}\operatorname{tr}\mathbf{M}^e\right), \tag{17.11.8}$$

then (17.11.6) may be written as

$$\mu - f^{\,c} + \lambda\,\Delta\bar{c}_{\mathrm{R}} = 0. \tag{17.11.9}$$

Equation (17.11.9) represents a partial differential equation for the species concentration $\bar{c}_{\mathrm{R}}$, with boundary condition

$$(\lambda\nabla\bar{c}_{\mathrm{R}})\cdot\mathbf{n}_{\mathrm{R}} = 0 \quad \text{on } \partial\mathcal{B}\times[0,T], \tag{17.11.10}$$

which is a boundary condition on the normal derivative of the species concentration. The physical interpretation of this boundary condition is that phase interface is perpendicular to the outer boundary $\partial\mathcal{B}$ of the body. It is clear that this boundary condition is of no relevance for a theory in which the gradient energy coefficient λ vanishes.

The initial conditions are taken as

$$\mathbf{u}(\mathbf{X},0) = \mathbf{u}_0(\mathbf{X}), \quad \mu(\mathbf{X},0) = \mu_0(\mathbf{X}), \quad \text{and} \quad c_{\mathrm{R}}(\mathbf{X},0) = c_{\mathrm{R}0}(\mathbf{X}) \quad \text{in} \quad \mathcal{B}. \tag{17.11.11}$$

The coupled set of equations (17.11.1), (17.11.4), and (17.11.9), together with (17.11.3), (17.11.5), (17.11.10), and (17.11.11) yield an initial/boundary-value problem for the displacement $\mathbf{u}(\mathbf{X},t)$, the chemical potential $\mu(\mathbf{X},t)$, and the species concentration $c_{\mathrm{R}}(\mathbf{X},t)$.

18 Numerical implementation of the coupled theory chemo-mechanical theory

18.1 Weak forms of the governing partial differential equations

1. **Mechanical problem:**
   Consider a test field $\mathbf{u}_{\text{test}}(\mathbf{X})$ in $\mathcal{B}$ which satisfies $\mathbf{u}_{\text{test}}(\mathbf{X}) = \mathbf{0}$ on $\partial\mathcal{B}_{\mathbf{u}}$. If the balance equation $\operatorname{Div}\mathbf{T}_{\mathrm{R}} = \mathbf{0}$ in $\mathcal{B}$ and the traction condition $(\bar{\mathbf{t}}_{\mathrm{R}} - \mathbf{T}_{\mathrm{R}}\mathbf{n}_{\mathrm{R}}) = \mathbf{0}$ on $\partial\mathcal{B}_{\mathbf{t}_{\mathrm{R}}}$ are satisfied, then it is straightforward to show that the weak form

$$\int_{\mathcal{B}} \mathbf{T}_{\mathrm{R}} : \nabla \mathbf{u}_{\text{test}}\, dv_{\mathrm{R}} - \int_{\partial\mathcal{B}_{\mathbf{t}_{\mathrm{R}}}} \bar{\mathbf{t}}_{\mathrm{R}} \cdot \mathbf{u}_{\text{test}}\, da_{\mathrm{R}} = 0 \tag{18.1.1}$$

   holds for all admissible test fields $\mathbf{u}_{\text{test}}$ which vanish on $\partial\mathcal{B}_{\mathbf{u}}$.

2. **PDE for the chemical potential μ:**
   Next, consider a scalar test field $\mu_{\text{test}}(\mathbf{X})$ in $\mathcal{B}$ which satisfies $\mu_{\text{test}}(\mathbf{X}) = \mathbf{0}$ on $\partial\mathcal{B}_{\mu}$. If the balance equation

$$\dot{c}_{\mathrm{R}} + \operatorname{Div} \mathbf{j}_{\mathrm{R}} = 0 \qquad \text{in } \mathcal{B},$$

   and the boundary condition $\mathbf{j}_{\mathrm{R}} \cdot \mathbf{n}_{\mathrm{R}} = \bar{j}$ on $\partial\mathcal{B}_{\mathbf{j}_{\mathrm{R}}}$ are satisfied, then

$$\int_{\mathcal{B}} \dot{c}_{\mathrm{R}}\, \mu_{\text{test}}\, dv_{\mathrm{R}} + \int_{\mathcal{B}} (\operatorname{Div}\mathbf{j}_{\mathrm{R}})\, \mu_{\text{test}}\, dv_{\mathrm{R}} + \int_{\partial\mathcal{B}_{\mathbf{j}_{\mathrm{R}}}} (\bar{j}_{\mathrm{R}} - \mathbf{j}_{\mathrm{R}} \cdot \mathbf{n}_{\mathrm{R}})\, \mu_{\text{test}}\, da_{\mathrm{R}} = 0, \tag{18.1.2}$$

   or equivalently

$$\int_{\mathcal{B}} \dot{c}_{\mathrm{R}}\, \mu_{\text{test}}\, dv_{\mathrm{R}} + \int_{\mathcal{B}} (\operatorname{Div}\mathbf{j}_{\mathrm{R}})\, \mu_{\text{test}}\, dv_{\mathrm{R}} - \int_{\partial\mathcal{B}_{\mathbf{j}_{\mathrm{R}}}} \mathbf{j}_{\mathrm{R}} \cdot \mathbf{n}_{\mathrm{R}}\, \mu_{\text{test}}\, da_{\mathrm{R}} + \int_{\partial\mathcal{B}_{\mathbf{j}_{\mathrm{R}}}} \bar{j}_{\mathrm{R}}\, \mu_{\text{test}}\, da_{\mathrm{R}} = 0 \tag{18.1.3}$$

   holds for all test fields μ_{test}. Thus, since $\mu_{\text{test}} = 0$ on $\partial\mathcal{B}_{\mu}$, the first boundary integral in the expression above may be extended to all of $\partial\mathcal{B}$,

$$\int_{\mathcal{B}} \dot{c}_{\mathrm{R}}\, \mu_{\text{test}}\, dv_{\mathrm{R}} + \int_{\mathcal{B}} (\operatorname{Div}\mathbf{j}_{\mathrm{R}})\, \mu_{\text{test}}\, dv_{\mathrm{R}} - \int_{\partial\mathcal{B}} \mathbf{j}_{\mathrm{R}} \cdot \mathbf{n}_{\mathrm{R}}\, \mu_{\text{test}}\, da_{\mathrm{R}} + \int_{\partial\mathcal{B}_{\mathbf{j}_{\mathrm{R}}}} \bar{j}_{\mathrm{R}}\, \mu_{\text{test}}\, da_{\mathrm{R}} = 0 \tag{18.1.4}$$

Introduction to Coupled Theories in Solid Mechanics. Lallit Anand, Eric M. Stewart, Shawn A. Chester, Oxford University Press.
 DOI: 10.1093/oso/9780198986218.003.0019

Next, by the divergence theorem

$$\int_{\partial\mathcal{B}} \mathbf{j}_{\mathrm{R}} \cdot \mathbf{n}_{\mathrm{R}}\, \mu_{\text{test}}\, da_{\mathrm{R}} = \int_{\partial\mathcal{B}} (\mu_{\text{test}}\, \mathbf{j}_{\mathrm{R}}) \cdot \mathbf{n}\, da_{\mathrm{R}} = \int_{\mathcal{B}} \mu_{\text{test}}\, \mathrm{Div}\,\mathbf{j}_{\mathrm{R}}\, dv_{\mathrm{R}} + \int_{\mathcal{B}} \mathbf{j}_{\mathrm{R}} \cdot \nabla\mu_{\text{test}}\, dv_{\mathrm{R}}. \tag{18.1.5}$$

Use of (18.1.5) in (18.1.4) gives the weak form

$$\int_{\mathcal{B}} \dot{c}_{\mathrm{R}}\, \mu_{\text{test}}\, dv_{\mathrm{R}} - \int_{\mathcal{B}} \mathbf{j}_{\mathrm{R}} \cdot \nabla\mu_{\text{test}}\, dv_{\mathrm{R}} + \int_{\partial\mathcal{B}_{\mathbf{j}_{\mathrm{R}}}} \bar{j}_{\mathrm{R}}\, \mu_{\text{test}}\, da_{\mathrm{R}} = 0. \tag{18.1.6}$$

We use a first-order finite difference approximation to write

$$\dot{c}_{\mathrm{R}} = \frac{c_{\mathrm{R}} - c_{\mathrm{R}\,\text{old}}}{\Delta t}, \tag{18.1.7}$$

so that the weak form (18.1.6) may be expressed as

$$\int_{\mathcal{B}} \frac{c_{\mathrm{R}} - c_{\mathrm{R}\,\text{old}}}{\Delta t}\, \mu_{\text{test}}\, dv_{\mathrm{R}} - \int_{\mathcal{B}} \mathbf{j}_{\mathrm{R}} \cdot \nabla\mu_{\text{test}}\, dv_{\mathrm{R}} + \int_{\partial\mathcal{B}_{\mathbf{j}_{\mathrm{R}}}} \bar{j}_{\mathrm{R}}\, \mu_{\text{test}}\, da_{\mathrm{R}} = 0. \tag{18.1.8}$$

3. **PDE for $\bar{c}_{\mathrm{R}}$:**
Recall the PDE for $\bar{c}_{\mathrm{R}}$,

$$\mu - f^c + \lambda\, \Delta\bar{c}_{\mathrm{R}} = 0, \tag{18.1.9}$$

with

$$f^c \stackrel{\text{def}}{=} R\vartheta \left[\ln\left(\frac{\bar{c}_{\mathrm{R}}}{1-\bar{c}_{\mathrm{R}}}\right) + \chi(1-2\bar{c}_{\mathrm{R}})\right] - \Omega\left(\frac{1}{3}\mathrm{tr}\,\mathbf{M}^e\right), \tag{18.1.10}$$

and boundary condition

$$(\lambda\nabla\bar{c}_{\mathrm{R}}) \cdot \mathbf{n}_{\mathrm{R}} = 0 \quad \text{on } \partial\mathcal{B} \times [0, T], \tag{18.1.11}$$

which is a boundary condition on the normal derivative of the species concentration.

Let c_{test} be a test function which vanishes on $\partial\mathcal{B}$. Then

$$\int_{\mathcal{B}} \mu\, c_{\text{test}}\, dv_{\mathrm{R}} - \int_{\mathcal{B}} f^{\,c}\, c_{\text{test}}\, dv_{\mathrm{R}} + \int_{\mathcal{B}} (\lambda\Delta\bar{c}_{\mathrm{R}})\; c_{\text{test}}\, dv_{\mathrm{R}} + \int_{\partial\mathcal{B}} (0 - (\lambda\nabla\bar{c}_{\mathrm{R}}) \cdot \mathbf{n}_{\mathrm{R}})\; c_{\text{test}}\, da_{\mathrm{R}} = 0. \tag{18.1.12}$$

Using the divergence theorem we find that

$$\begin{aligned}\int_{\partial\mathcal{B}} c_{\text{test}}\; (\lambda\nabla\bar{c}_{\mathrm{R}}) \cdot \mathbf{n}_{\mathrm{R}}\, da_{\mathrm{R}} &= \int_{\mathcal{B}} \mathrm{Div}\,(c_{\text{test}}\; (\lambda\nabla\bar{c}_{\mathrm{R}}))\; dv_{\mathrm{R}} \\ &= \int_{\mathcal{B}} c_{\text{test}}\; (\lambda\Delta\bar{c}_{\mathrm{R}})\; dv_{\mathrm{R}} + \int_{\mathcal{B}} \lambda\nabla\bar{c}_{\mathrm{R}} \cdot \nabla c_{\text{test}}\, dv_{\mathrm{R}},\end{aligned} \tag{18.1.13}$$

use of which in (18.1.12) gives the weak form

$$\int_{\mathcal{B}} \mu\, c_{\text{test}}\, dv_{\text{R}} - \int_{\mathcal{B}} f^c\, c_{\text{test}}\, dv_{\text{R}} - \int_{\mathcal{B}} \lambda \nabla \bar{c}_{\text{R}} \cdot \nabla c_{\text{test}}\, dv_{\text{R}} = 0, \tag{18.1.14}$$

for all test functions c_{test}.

Summarizing, the weak forms of the governing equations for the coupled problem are

$$\int_{\mathcal{B}} \mathbf{T}_{\text{R}} : \nabla \mathbf{u}_{\text{test}}\, dv_{\text{R}} - \int_{\partial\mathcal{B}_{\mathbf{t}_{\text{R}}}} \bar{\mathbf{t}}_{\text{R}} \cdot \mathbf{u}_{\text{test}}\, da_{\text{R}} = 0,$$

$$\int_{\mathcal{B}} \frac{c_{\text{R}} - c_{\text{R old}}}{\Delta t}\, \mu_{\text{test}}\, dv_{\text{R}} - \int_{\mathcal{B}} \mathbf{j}_{\text{R}} \cdot \nabla \mu_{\text{test}}\, dv_{\text{R}} + \int_{\partial\mathcal{B}_{j_{\text{R}}}} \bar{j}_{\text{R}}\, \mu_{\text{test}}\, da_{\text{R}} = 0, \tag{18.1.15}$$

$$\int_{\mathcal{B}} \mu\, c_{\text{test}}\, dv_{\text{R}} - \int_{\mathcal{B}} f^c\, c_{\text{test}}\, dv_{\text{R}} - \int_{\mathcal{B}} \lambda \nabla \bar{c}_{\text{R}} \cdot \nabla c_{\text{test}}\, dv_{\text{R}} = 0.$$

18.2 Normalization

Since we plan to solve all the governing equations related to multiple physical fields $\{\mathbf{u}, \mu, c_{\text{R}}\}$ simultaneously, it is beneficial to rewrite the residuals so that they have approximately the same magnitudes. To this end:

1. The species concentration c_{R} mol/m^3 is normalized using the molar volume Ω [m^3/mol] as

$$\hat{c} \stackrel{\text{def}}{=} \Omega c_{\text{R}}. \tag{18.2.1}$$

2. The chemical potential μ [J/mol] is normalized by $R\vartheta$ [J/mol],

$$\hat{\mu} = \frac{\mu}{R\vartheta}. \tag{18.2.2}$$

3. Recall the mass balance equation:

$$\dot{c}_{\text{R}} + \text{Div}\, \mathbf{j}_{\text{R}} = 0.$$

This may be written as,

$$\frac{1}{\Omega}\dot{\hat{c}} + \text{Div}\, \mathbf{j}_{\text{R}} = 0.$$

4. Next, the species flux $\mathbf{j}_R$ is given by the constitutive equation,

$$\mathbf{j}_R = -\mathbf{M}\nabla\mu \qquad \text{with} \qquad \mathbf{M} = \left(\frac{Dc_R}{R\vartheta}\right)(1-\bar{c}_R)\,\mathbf{C}^{-1}. \tag{18.2.3}$$

Using the normalized variables $\hat{\mu}$ and $\hat{c}$ we may rewrite it as follows:

$$\mathbf{j}_R = -\mathbf{M}\nabla\hat{\mu} \times R\vartheta \qquad \text{with} \qquad \mathbf{M} = \left(\frac{D\hat{c}}{\Omega R\vartheta}\right)\left(1-\frac{\hat{c}}{\Omega c_R^{\max}}\right)\mathbf{C}^{-1}. \tag{18.2.4}$$

5. Finally we choose to normalize the Piola stress with the shear modulus G:

$$\hat{\mathbf{T}}_R \stackrel{\text{def}}{=} \frac{\mathbf{T}_R}{G}. \tag{18.2.5}$$

6. Summarizing, the governing equations may then be expressed as

$$\begin{aligned} \operatorname{Div}\hat{\mathbf{T}}_R &= 0 \\ \frac{1}{\Omega}\dot{\hat{c}} + \operatorname{Div}\mathbf{j}_R &= 0, \\ \hat{\mu} - \frac{f^c}{R\vartheta} + \frac{\lambda}{R\vartheta}\Delta\bar{c}_R &= 0. \end{aligned} \tag{18.2.6}$$

With the normalizations introduced above and the normalized parameters denoted by $\{\hat{\bullet}\}$, the weak forms of the coupled problem may be written as

$$\begin{aligned} &\int_{\mathcal{B}} \hat{\mathbf{T}}_R : \nabla\mathbf{u}_{\text{test}}\, dv_R - \int_{\partial\mathcal{B}_{\mathbf{t}_R}} \hat{\bar{\mathbf{t}}}_R \cdot \mathbf{u}_{\text{test}}\, da_R = 0, \\ &\int_{\mathcal{B}} \frac{\hat{c}-\hat{c}_{\text{old}}}{\Delta t}\hat{\mu}_{\text{test}}\, dv_R - \int_{\mathcal{B}} \Omega\,\mathbf{j}_R \cdot \nabla\hat{\mu}_{\text{test}}\, dv_R + \int_{\partial\mathcal{B}_{\mathbf{j}_R}} \Omega\,\bar{j}_R\,\hat{\mu}_{\text{test}}\, da_R = 0, \\ &\int_{\mathcal{B}} \hat{\mu}\, c_{\text{test}}\, dv_R - \int_{\mathcal{B}} \frac{f^c}{R\vartheta} c_{\text{test}}\, dv_R - \int_{\mathcal{B}} \frac{\lambda}{R\vartheta}\nabla\bar{c}_R \cdot \nabla c_{\text{test}}\, dv_R = 0. \end{aligned} \tag{18.2.7}$$

18.3 Solution procedure in the FEniCS code

In brief, in FEniCS, we use the following degrees of freedom

$$\{\mathbf{u},\, \hat{\mu},\, \hat{c}\}, \tag{18.3.1}$$

as well as their known values from the previous step

$$\{\mathbf{u}_{\text{old}},\, \hat{\mu}_{\text{old}},\, \hat{c}_{\text{old}}\}, \tag{18.3.2}$$

to calculate the *residuals* which correspond to the weak forms of the governing equations, viz.

$$
\begin{aligned}
R_{\mathbf{u}} &= \int_{\mathcal{B}} \hat{\mathbf{T}}_{\mathrm{R}} : \nabla \mathbf{u}_{\text{test}} \, dv_{\mathrm{R}} - \int_{\partial \mathcal{B}_{\mathbf{t}_{\mathrm{R}}}} \bar{\mathbf{t}}_{\mathrm{R}} \cdot \mathbf{u}_{\text{test}} \, da_{\mathrm{R}}, \\
R_{\hat{\mu}} &= \int_{\mathcal{B}} \frac{\hat{c} - \hat{c}_{\text{old}}}{\Delta t} \hat{\mu}_{\text{test}} \, dv_{\mathrm{R}} - \int_{\mathcal{B}} \Omega \, \mathbf{j}_{\mathrm{R}} \cdot \nabla \hat{\mu}_{\text{test}} \, dv_{\mathrm{R}} + \int_{\partial \mathcal{B}_{\mathbf{j}_{\mathrm{R}}}} \Omega \, \bar{j}_{\mathrm{R}} \, \hat{\mu}_{\text{test}} \, da_{\mathrm{R}}, \qquad (18.3.3) \\
R_{\hat{c}} &= \int_{\mathcal{B}} \hat{\mu} \, c_{\text{test}} \, dv_{\mathrm{R}} - \int_{\mathcal{B}} \frac{f^c}{R\vartheta} c_{\text{test}} \, dv_{\mathrm{R}} - \int_{\mathcal{B}} \frac{\lambda}{R\vartheta} \nabla \bar{c}_{\mathrm{R}} \cdot \nabla c_{\text{test}} \, dv_{\mathrm{R}}.
\end{aligned}
$$

FEniCS then differentiates these residuals with respect to the separate degrees of freedom to obtain *tangents*. At each step in time, FEniCS uses the residuals and tangents in a global Newton–Raphson solver to determine current values of the degrees of freedom $\{\mathbf{u},\ \hat{\mu},\ \hat{c}\}$ for which the residuals are zero,[1] and which therefore satisfy the governing equations as well as the initial and boundary conditions (in a weak sense).

REMARK

A mixed finite element method is utilized to solve for the displacement $\mathbf{u}$, the chemical potential $\hat{\mu}$, and the concentration field $\hat{c}$ concurrently. To avoid the numerical instabilities the interpolation order for the displacement is taken as Lagrange quadratic, while that of the other variables are interpolated one order lower, Lagrange linear.

18.4 **Algorithm for calculation of the residuals**

The algorithm we follow in our code to calculate the residuals in (18.3.3) is summarized below.

1. Begin the step with known values of

$$\{\mathbf{u}_{\text{old}},\ \hat{\mu}_{\text{old}},\ \hat{c}_{\text{old}}\}, \qquad (18.4.1)$$

   where for the first step these quantities are given by the initial conditions for $\mathbf{u}$, $\hat{\mu}$, and $\hat{c}$.

2. Calculate necessary kinematical quantities for the current step as

$$
\begin{aligned}
&\mathbf{F} = \mathbf{1} + \nabla \mathbf{u}, \quad J = \det \mathbf{F}, \quad J^s = 1 + (\hat{c} - \hat{c}_0), \quad \mathbf{F}^s = (J^s)^{1/3}\mathbf{1}, \qquad (18.4.2) \\
&\mathbf{F}^e = \mathbf{F}\mathbf{F}^{s-1}, \quad J^e = \det \mathbf{F}^e, \quad \mathbf{C}^e = \mathbf{F}^{e\top}\mathbf{F}^e, \quad \bar{\mathbf{C}}^e = J^{e-2/3}\mathbf{C}^e.
\end{aligned}
$$

[1] To within some small numerical tolerance.

3. Calculate the elastic second Piola stress:

$$\mathbf{T}^e = J^{e\,-2/3} G \left[\mathbf{1} - \frac{1}{3}\left(\operatorname{tr}\bar{\mathbf{C}}^e\right)\bar{\mathbf{C}}^{e-1} \right] + K\,J^e(J^e - 1)\mathbf{C}^{e-1},$$

with (G, K) material properties—the elastic moduli of the material.

4. Calculate the Mandel stress:

$$\mathbf{M}^e = \mathbf{C}^e\mathbf{T}^e.$$

5. Calculate the Cauchy stress:

$$\mathbf{T} = J^{e-1}\,\mathbf{F}^e\mathbf{T}^e\mathbf{F}^{e-1}.$$

6. Calculate the normalized Piola stress:

$$\hat{\mathbf{T}}_{\mathrm{R}} = \frac{1}{G}\,J\mathbf{T}\mathbf{F}^{-\top}.$$

7. Calculate the normalized concentration and the species flux:
With

$$\bar{c} \stackrel{\text{def}}{=} \frac{\Omega c_{\mathrm{R}}}{\Omega c_{\mathrm{R}}^{\max}} = \frac{\hat{c}}{\Omega c_{\mathrm{R}}^{\max}}, \tag{18.4.3}$$

the species flux is

$$\mathbf{j}_{\mathrm{R}} = -\mathbf{M}\,\nabla\hat{\mu} \times R\vartheta \qquad \text{with} \qquad \mathbf{M} = \left(\frac{D\hat{c}}{\Omega R\vartheta}\right)(1-\bar{c})\,\mathbf{C}^{-1}, \tag{18.4.4}$$

with $(c_{\mathrm{R}}^{\max}, \Omega, D)$ material properties, R the gas constant, and ϑ the absolute temperature.

8. The factor f^c is given by

$$f^c \stackrel{\text{def}}{=} R\vartheta\left[\ln\left(\frac{\bar{c}}{1-\bar{c}}\right) + \chi(1-2\bar{c})\right] - \Omega\left(\frac{1}{3}\operatorname{tr}\mathbf{M}^e\right),$$

where χ is an interaction parameter that controls if there is any phase separation.

9. Compute the residuals:

$$R_{\mathbf{u}} = \int_{\mathcal{B}} \hat{\mathbf{T}}_{\mathrm{R}} : \nabla \mathbf{u}_{\mathrm{test}} \, dv_{\mathrm{R}} - \int_{\partial \mathcal{B}_{\mathbf{t}_{\mathrm{R}}}} \bar{\hat{\mathbf{t}}}_{\mathrm{R}} \cdot \mathbf{u}_{\mathrm{test}} \, da_{\mathrm{R}},$$

$$R_{\hat{\mu}} = \int_{\mathcal{B}} \frac{\hat{c} - \hat{c}_{\mathrm{old}}}{\Delta t} \hat{\mu}_{\mathrm{test}} \, dv_{\mathrm{R}} - \int_{\mathcal{B}} \Omega \, \mathbf{j}_{\mathrm{R}} \cdot \nabla \hat{\mu}_{\mathrm{test}} \, dv_{\mathrm{R}} + \int_{\partial \mathcal{B}_{\mathbf{j}_{\mathrm{R}}}} \Omega \, \bar{j}_{\mathrm{R}} \, \hat{\mu}_{\mathrm{test}} \, da_{\mathrm{R}}, \tag{18.4.5}$$

$$R_{\hat{c}} = \int_{\mathcal{B}} \hat{\mu} \, c_{\mathrm{test}} \, dv_{\mathrm{R}} - \int_{\mathcal{B}} \frac{f^c}{R\vartheta} c_{\mathrm{test}} \, dv_{\mathrm{R}} - \int_{\mathcal{B}} \frac{\lambda}{R\vartheta} \nabla \bar{c}_{\mathrm{R}} \cdot \nabla c_{\mathrm{test}} \, dv_{\mathrm{R}},$$

where λ is an interface parameter which governs the width of the phase boundaries.

10. FEniCS solves for the current values of $\{\mathbf{u},\, \hat{\mu},\, \hat{c}\}$.

11. Set these current values as the "old" values for the next step in time,

$$\{\mathbf{u},\, \hat{\mu},\, \hat{c}\} \implies \{\mathbf{u}_{\mathrm{old}},\, \hat{\mu}_{\mathrm{old}},\, \hat{c}_{\mathrm{old}}\}. \tag{18.4.6}$$

12. Repeat for the desired number of steps in time.

19 Representative simulations of the coupled chemo-mechanical theory

In our numerical simulations in FEniCS we will use the basic units listed as SI (μm) in Table 19.1. Representative values of the material parameters for a cathode particle are listed in Table 19.2.

Table 19.1 Basic units. For example a density of $7800\,\mathrm{kg/m^3}$ would be specified as $7800 \times 10^{-18}\,\mathrm{kg}/\mu\mathrm{m}^3$.

Quantity	SI (m)	SI (μm)	Conversion
Length	m	μm	$10^6\,\mu\mathrm{m} = 1\,\mathrm{m}$
Mass	kg	kg	1
Time	s	s	1
Amount of substance	mol	pmol	$10^{12}\,\mathrm{pmol} = 1\,\mathrm{mol}$
Temperature	K	K	1
Mass density	kg/m^3	kg /μm^3	$10^{-18}\,\mathrm{kg}/\mu\mathrm{m}^3 = 1\ \mathrm{kg/m^3}$
Force	N	μN	$10^6\,\mu\mathrm{N} = 1\mathrm{N}$
Stress	Pa	MPa	$10^{-6}\mathrm{MPa} = 1\mathrm{Pa}$
Energy	J	pJ	$10^{12}\,\mathrm{pJ} = 1\mathrm{J}$
c_R, species concentration	mol/m^3	pmol/μm^3	$10^{-6}\,\mathrm{pmol}/\mu\mathrm{m}^3 = 1\,\mathrm{mol/m^3}$
μ, chemical potential	J/mol	pJ/pmol	1
Ω, molar volume	m^3/mol	μm^3/pmol	$10^6\ \mu\mathrm{m}^3/\mathrm{pmol} = 1\,\mathrm{m^3/mol}$
D, species diffusivity	m^2/s	μm^2/s	$10^{12}\,\mu\mathrm{m}^2/\mathrm{s} = 1\,\mathrm{m^2/s}$
Specific heat	J/(kg K)	pJ/(kg K)	$10^{12}\,\mathrm{pJ/(kg\,K)} = 1\ \mathrm{J/(kg\,K)}$
Thermal conductivity	W/(m K)	pW/(μm K)	$10^6\,\mathrm{pW}/(\mu\mathrm{m\,K}) = 1\ \mathrm{W/(m\,K)}$
k_B, Boltzmann's constant	1.38×10^{-23} J/K	1.38×10^{-11} pJ/K	$10^{12}\,\mathrm{pJ/K} = 1\,\mathrm{J/K}$
R, gas constant	8.3145 J/(mol K)	8.3145 pJ/(pmol K)	1

Introduction to Coupled Theories in Solid Mechanics. Lallit Anand, Eric M. Stewart, Shawn A. Chester, Oxford University Press.
 DOI: 10.1093/oso/9780198986218.003.0020

Table 19.2 Representative material parameters for a cathode particle from Di Leo et al. (2014).

Quantity	Value SI (m)	Value SI (μm)
R, gas constant	8.3145 J/(mol K)	8.3145 pJ/(pmol K)
ϑ, reference temperature	298 K	298 K
E, Young's modulus	124.5×10^9 Pa	124.5×10^3 MPa
ν, Poisson's ratio	0.25	0.25
$G = E/(2(1+\nu))$, shear modulus	49.8×10^9 Pa	49.8×10^3 MPa
$K = E/(3(1-2\nu))$, bulk modulus	83×10^9 Pa	83×10^3 MPa
$c_R^{\max}$	2.29×10^4 mol/m^3	2.29×10^{-2} pmol/μm^3
Ω, molar volume	4.05×10^{-6} m^3/mol	4.05 μm^3/pmol
D, diffusivity	1×10^{-14} m^2/s	1×10^{-2} μm^2/s
χ, phase parameter	3	3
λ, interphase parameter	5.5749×10^{-13} Jm^2/mol	5.5749×10^{-1} (pJ/pmol) μm^2

REMARK

Recall from Section 1.C that the linearized "small-deformation" form of the constitutive equation for the Piola stress for a *compressible Neo-Hookean material, reduces to the classical constitutive equation for the stress in linear isotropic elasticity*. That is, operationally a compressible Neo-Hookean model may be considered a generalization of the classic model of linear isotropic elasticity to moderately large deformations using the ground-state values of the shear and bulk moduli G and K. We have used this constitutive analogy in adopting the particular values of the elastic moduli of a cathode particle listed in Table 19.2.

19.1 Two-dimensional simulation of species diffusion and phase segregation neglecting mechanical effects

In the absence of mechanical coupling, the chemo-mechanically coupled theory in Chapter 17 reduces to the classical Cahn and Hilliard (1958) theory for species diffusion with phase segregation, which may be written as the following two coupled second-order partial differential equations:

(1) The species balance equation (PDE for μ):

$$\dot{c}_R + \mathrm{Div}\,\mathbf{j}_R = 0, \qquad (19.1.1)$$

with

$$\mathbf{j}_{\mathrm{R}} = -M\,\nabla\mu, \qquad M = \left(\frac{Dc_{\mathrm{R}}}{R\vartheta}(1-\bar{c}_{\mathrm{R}})\right).$$

In the two-dimensional phase-separation example problem considered below we will consider a simple no-flux boundary condition,

$$\mathbf{j}_{\mathrm{R}} \cdot \mathbf{n}_{\mathrm{R}} = 0 \quad \text{on } \partial\mathcal{B} \times [0, T]. \tag{19.1.2}$$

(2) The chemical potential equation (PDE for $\bar{c}_{\mathrm{R}}$):

$$\mu - f^{\,c} + \lambda\,\Delta\bar{c}_{\mathrm{R}} = 0, \tag{19.1.3}$$

where

$$f^{\,c} \stackrel{\text{def}}{=} R\vartheta\left[\ln\left(\frac{\bar{c}_{\mathrm{R}}}{1-\bar{c}_{\mathrm{R}}}\right) + \chi(1-2\bar{c}_{\mathrm{R}})\right], \tag{19.1.4}$$

with boundary condition

$$(\lambda\nabla\bar{c}_{\mathrm{R}}) \cdot \mathbf{n}_{\mathrm{R}} = 0 \quad \text{on } \partial\mathcal{B} \times [0, T]. \tag{19.1.5}$$

which is a boundary condition on the normal derivative of the species concentration.

The weak form of these PDEs in terms of the normalized quantities (discussed in Section 18.4) may be read off from equation (18.4.5) as

$$\begin{aligned} R_{\hat{\mu}} &= \int_{\mathcal{B}} \frac{\hat{c} - \hat{c}_{\text{old}}}{\Delta t}\,\hat{\mu}_{\text{test}}\,dv_{\mathrm{R}} - \int_{\mathcal{B}} \Omega\,\mathbf{j}_{\mathrm{R}} \cdot \nabla\hat{\mu}_{\text{test}}\,dv_{\mathrm{R}}. \\ R_{\hat{c}} &= \int_{\mathcal{B}} \hat{\mu}\,c_{\text{test}}\,dv_{\mathrm{R}} - \int_{\mathcal{B}} \frac{f^c}{R\vartheta}\,c_{\text{test}}\,dv_{\mathrm{R}} - \int_{\mathcal{B}} \frac{\lambda}{R\vartheta}\nabla\bar{c}_{\mathrm{R}} \cdot \nabla c_{\text{test}}\,dv_{\mathrm{R}}. \end{aligned} \tag{19.1.6}$$

In this first simulation example we consider:

- Phase separation in a two-dimensional square domain with an edge length $L = 800\,\text{nm}$, which is meshed with 100×100 elements.
- The initial concentration in the domain was given as a random uniform distribution with a mean of $\bar{c}_{\mathrm{R}0} = 0.63$ and a maximum fluctuation of 0.05; cf. Fig. 19.1.

The material properties used in our phase-separation simulation are those shown in Table 19.2, without the elastic moduli.

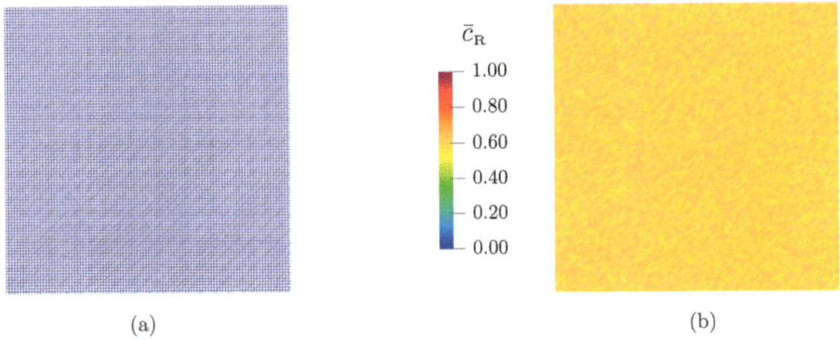

Fig. 19.1 Two-dimensional phase-separation simulation: (a) initial mesh on a square domain with an edge length $L = 800\,\text{nm}$; (b) initial random distribution of normalized concentration $\bar{c}_R$ with a mean of $\bar{c}_{R0} = 0.63$ and a maximum fluctuation of 0.05.

Figure 19.2 illustrates the evolution of phases during spinodal decomposition—from the early stages of the spinodal decomposition, when the components are well-mixed and the concentration is nearly homogenous, until a (near) steady-state is reached, at which time two distinct phases are observed. In this particular simulation, the steady-state morphology consists of a circular concentration-poor region surrounded by a concentration-rich domain.

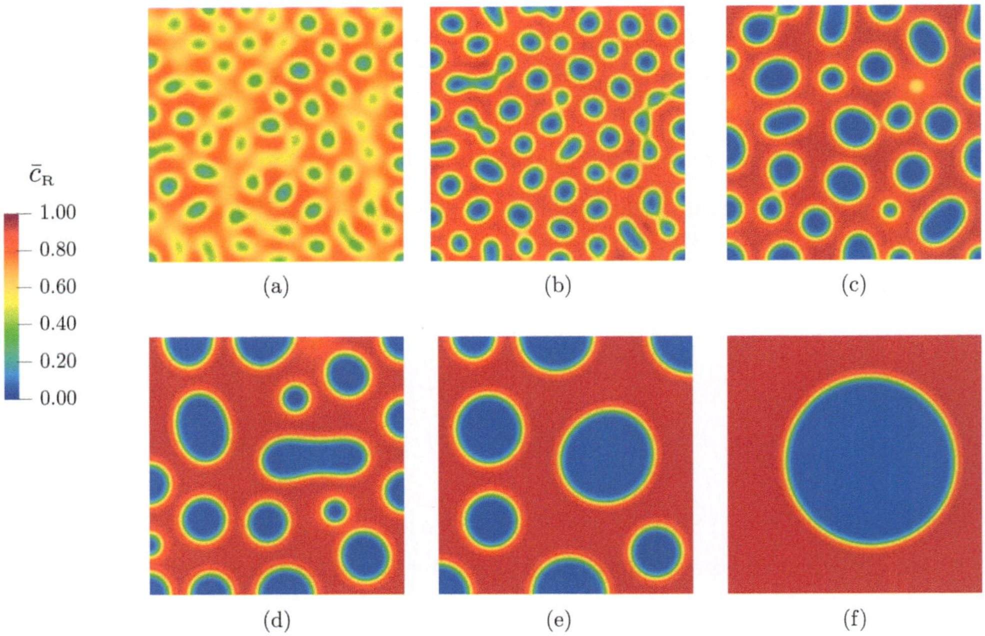

Fig. 19.2 Contours of normalized concentration $\bar{c}_R$ showing the evolution of phase-separation morphology at different times: (a) $t \approx 0.4$s; (b) $t \approx 0.8$s; (c) $t \approx 6.5$s; (d) $t \approx 20$s; (e) $t \approx 100$s; (f) $t \approx 2000$s.

19.2 Coupled chemo-mechanical simulations of lithiation of a spheroidal electrode particle

The two numerical examples shown in this section are those studied and discussed by Di Leo et al. (2014). The discussion below is essentially that given by these authors.

We consider lithiation of a spheroidal particle whose semi-major axis (axis of radial symmetry) is $0.5\,\mu\mathrm{m}$, and semi-minor axis is $0.3\,\mu\mathrm{m}$, see Fig. 19.3. Due to the symmetry of the geometry, we mesh only one quarter of the cross-section of the spheroidal particle. The illustrative material properties used in our simulations are shown in Table 19.2. They represent values for a particle made from an "isotropic" $LiFePO_4$.[1]

As in Di Leo et al. (2014), our interest in this section is to study the effect of stress through the term $(-\Omega(1/3)\mathrm{tr}\,\mathbf{M}^e)$ in the chemical potential (17.10.20) on the lithiation response of the particle. To vary the magnitude of this term we shall—keeping the Poisson's ratio ν fixed—vary the value of the Young's modulus of the particle, which will in turn change the magnitude of the Mandel stress $\mathbf{M}^e$; cf. equation (17.10.15). Thus, introducing a normalized elastic modulus by

$$\bar{E} \stackrel{\text{def}}{=} \frac{E}{E_0}, \tag{19.2.1}$$

where E_0 is the value of the Young's modulus given in Table 19.2, we shall vary the value of $\bar{E}$ in the range

$$\bar{E} \in [0.1, 0.2, 1],$$

and study the effect of stress in the particle on lithiation.

The boundary conditions used in our simulations are as follows:

- **Mechanical boundary conditions**: With reference to Fig. 19.3, the nodes along edge AB (radial symmetry) are constrained to have zero radial displacement, while the nodes along edge BC (mirror symmetry) are constrained to have zero vertical displacement. Further, the exterior boundary AC is taken to be traction-free, $\mathbf{T}_{\mathrm{R}}\mathbf{n}_{\mathrm{R}} = 0$. Also, body forces are neglected.
- **Chemical boundary conditions**: The edges AB and BC are prescribed zero flux conditions, while on the exterior boundary AC the species flux is controlled by the boundary condition

$$\mathbf{j}_{\mathrm{R}} \cdot \mathbf{n}_{\mathrm{R}} = \underbrace{\jmath\,\bar{\jmath}}_{\bar{\jmath}_{\mathrm{R}}}, \tag{19.2.2}$$

where

$$\bar{\jmath} = k(\mu_{\mathrm{Li}}^{\mathrm{electrode}} - \bar{\mu}_{\mathrm{ext}}). \tag{19.2.3}$$

[1] $LiFePO_4$ is an important olivine-type material which is widely used for the positive electrode in modern Li-ion batteries. Actual $LiFePO_4$ particles currently used in practice are highly anisotropic nanoparticles, and not isotropic and micron-dimensioned. So the simulations shown here are of a highly idealized nature.

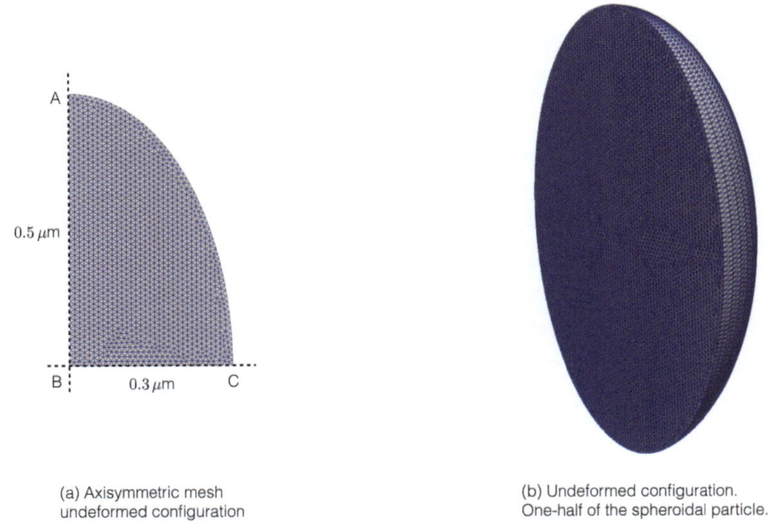

Fig. 19.3 (a) Axisymmetric mesh for simulation of lithiation of a spheroidal particle. (b) One-half of the spheroidal particle after reflection and rotational extrusion.

is the prescribed spatial out-flux, and $\jmath \overset{\text{def}}{=} |\mathbf{F}^{\text{C}}\,\mathbf{n}_{\text{R}}|$ is the areal Jacobian. Here:

- $\mu_{\text{Li}}^{\text{electrode}}$ is the chemical potential at the boundary of the electrode particle—which depends on both the concentration and the stress at the boundary;
- $\bar{\mu}_{\text{ext}}$ is the externally "applied" electrochemical potential; and
- $k > 0$ (units $\text{mol}^2/(\text{m}^2\text{J s})$) is a parameter which characterizes the conductance of the electrode/electrolyte interface.

Thus,

- the condition $\mu_{\text{Li}}^{\text{electrode}} < \bar{\mu}_{\text{ext}}$ corresponds to *insertion* of Li-ions into the electrode, and
- the condition $\mu_{\text{Li}}^{\text{electrode}} > \bar{\mu}_{\text{ext}}$ corresponds to *extraction* of Li-ions from the electrode.

The simple flux boundary condition (19.2.3) is discussed in detail in an Appendix in Section 19.A. In our simulations we use a value of k in equation (19.2.3) of

$$k = 1\,\text{mol}^2/(\text{m}^2\text{J s}) = 1\,\text{pmol}^2/(\mu\text{m}^2\ \text{pJ s}).$$

This value was chosen so that the lithiation of the particle is limited by the diffusion of Li in the bulk of the electrode particle, and not by the conductance of Li across the electrode/electrolyte interface. See the Remark at the end of Appendix 19.A for more details.

Consider the following chemical-loading condition in which $\bar{\mu}_{\text{ext}}$ is ramped linearly from an initial value $\bar{\mu}_i$ to a final value $\bar{\mu}_f$,

$$\bar{\mu}_{\text{ext}} = \bar{\mu}_i + (\bar{\mu}_f - \bar{\mu}_i)(t/T), \tag{19.2.4}$$

over a time $t \in [0, T]$. In (19.2.4) the initial and final values $\bar{\mu}_i$ and $\bar{\mu}_f$ are chosen based on the expression

$$\mu = R\vartheta \left[\ln\left(\frac{\bar{c}_{\text{R}}}{1 - \bar{c}_{\text{R}}} \right) + \chi(1 - 2\bar{c}_{\text{R}}) \right], \tag{19.2.5}$$

for the chemical potential of Li in the stress-free electrode, using initial and final normalized concentrations of $\bar{c}_{\text{R}} = 0.005$ and $\bar{c}_{\text{R}} = 0.995$, respectively.

In the following subsections we show results from two different chemical-loading conditions:

(i) **Slow ramp of** $\bar{\mu}_{\text{ext}}$: In this case $\bar{\mu}_{\text{ext}}$ is ramped from $\bar{\mu}_i$ to $\bar{\mu}_f$ over a time $T = 1800\,\text{s}$. We consider this ramp to be *slow*, since the ramp-time of $T = 1800\,\text{s}$ is much larger than the characteristic timescale for diffusion $\tau = (0.5\,\mu\text{m})^2/D_0 = 25\,\text{s}$ of Li in the particle.

(ii) **Step change in** $\bar{\mu}_{\text{ext}}$: In this case $\bar{\mu}_{\text{ext}}$ is ramped from $\bar{\mu}_i$ to $\bar{\mu}_f$ nearly instantaneously, and then held constant. Numerically, this was realized by ramping $\bar{\mu}_{\text{ext}}$ to its final value in 1 s, and then maintaining it constant for another 29 seconds for a total time of $T = 30\,\text{s}$, until (near) full lithiation was achieved.

Lithiation under a slow ramp of $\bar{\mu}_{\text{ext}}$

When the chemical potential $\bar{\mu}_{\text{ext}}$ is ramped slowly from $\bar{\mu}_i$ to $\bar{\mu}_f$, bulk diffusion of Li in the particle is much faster than the rate of increase of the concentration of Li on the boundary. This leads to a concentration profile inside the particle which is essentially homogeneous up until the point at which phase separation initiates at the boundary of the particle.

A commonly used quantity in the Li-ion battery literature is the *state of charge* (SOC),

$$\text{SOC} \stackrel{\text{def}}{=} \frac{\left(\int_{\mathcal{B}} \bar{c}_{\text{R}}\, dv_{\text{R}}\right)}{\left(\int_{\mathcal{B}} dv_{\text{R}}\right)}, \tag{19.2.6}$$

which represents the volume average of the normalized species concentration in the particle. We use this average quantity as a measure of the extent of lithiation of the particle.

Figure 19.4(a) (left) shows the state-of-charge of the particle as a function of time for the entire simulation span from 0 to 1800 s, while Fig. 19.4(a) (right) focuses in on the time period $1060 < t < 1120\,\text{s}$ in which a Li-rich and a Li-poor phase co-exist within the particle. The effect of varying $\bar{E} \in [0.1, 0.2, 1]$ on the state of charge versus time curves is not very distinguishable in Fig. 19.4(a) (left) but is clearly visible in the time window shown in Fig. 19.4(a) (right) when the particle exhibits phase separation. Fig. 19.4(a) (right) shows that as $\bar{E}$ increases, the time taken to lithiate the particle decreases. That is, the increased level of stress enables a faster lithiation of the particle under the particular chemical-boundary conditions considered here.

The different values of $\bar{E}$ also result in not only different distributions of the normalized Li concentration $\bar{c}_{\text{R}}$, but also of the maximum principal Cauchy stress σ_1 in the particle. The contour plots of these quantities at $t = 1080\text{s}$, indicated by a dashed line in Fig. 19.4(a) (right), are shown in Fig. 19.4(b).

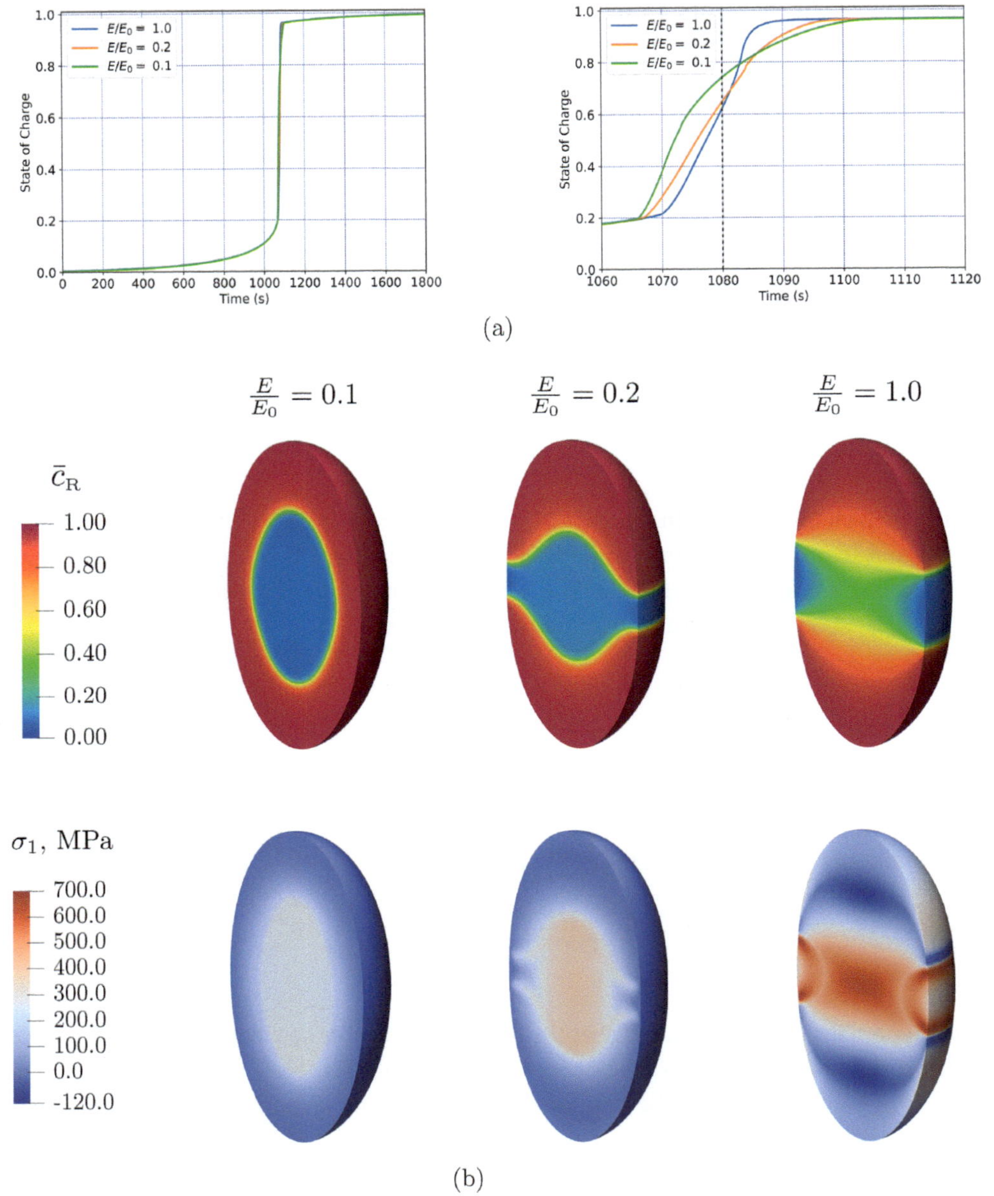

Fig. 19.4 Simulations with varying elastic moduli under an applied *slow ramp* of the external chemical potential. (a) State of charge versus time for the entire lithiation simulation (left), and state of charge versus time focused around the time at which phase separation occurs (right). (b) Contours at time $t = 1080$ s of the normalized concentration $\bar{c}_{\text{R}}$ (top), and of maximum principal stress σ_1 (bottom), for three different values of E/E_0 = 0.1, 0.2, and 1.0.

Figure 19.4(b) (top) shows that at the smallest value of $\bar{E} = 0.1$, one obtains a "core-shell" type of Li distribution in the particle. However, as the value of $\bar{E}$ increases from 0.1 to 1.0, one no longer obtains a core-shell type of lithium distribution—instead, because of the spheroidal geometry, a mildly curved "planar front" separates a high-Li concentration region from the low-Li concentration region. Our numerical simulations show that the high-Li concentration phase nucleates at the poles of the particle and propagates towards the equatorial region of the particle. This is an important result because it indicates that the widely presumed "core-shell" type of lithiation is not the most energetically favorable lithium concentration distribution under the conditions of the problem studied here. The contours in Fig. 19.4(b) (top) also show a widening of the interface separating the two phases as the value of $\bar{E}$ increases—this is another manifestation of the effect of stress in coupled diffusion–deformation problems.

The change in the distribution of the lithium concentration with increasing values of $\bar{E}$ also has an important effect on the resulting stress distribution in the particle. Figure 19.4(b) (bottom) shows contours of the maximum principal stress σ_1. Note that unlike the situation for $\bar{E} = 0.1$, where the maximum principal stress on the periphery of the particle is compressive, $\sigma_1 \lessapprox -100$ MPa, which is what one expects from a core-shell-like distribution of the Li-concentration, for the case $\bar{E} = 1.0$ there are regions on the periphery of the particle with high tensile values of $\sigma_1 \gtrapprox 700$ MPa. Since fracture of brittle electrode particles is usually dictated by large positive values of the maximum principal stress, this result predicts substantially different potential sites for fracture than are predicted by a theory in which the coupling of the diffusion and phase separation with the attendant mechanical swelling is not properly accounted for.

In Fig. 19.5 we show a plot of the normalized externally applied electrochemical potential $\hat{\bar{\mu}}_{\text{ext}} = \bar{\mu}_{\text{ext}}/(R\vartheta)$ as a function of SOC for both Li-insertion and Li-extraction, where the

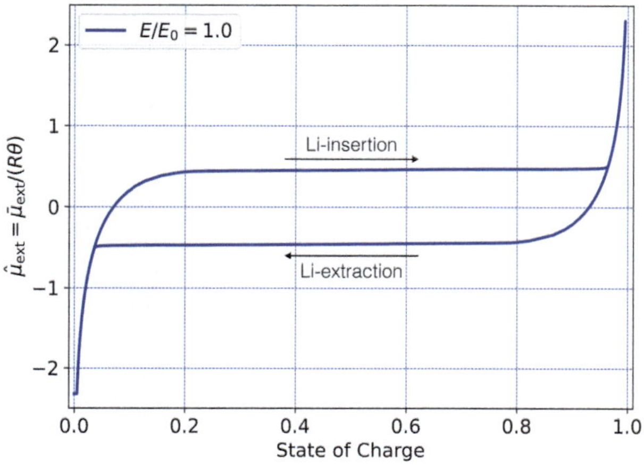

Fig. 19.5 Applied electrochemical potential versus state of charge for both Li-insertion and Li-extraction.

Li-extraction step is simulated by simply reversing the $\bar{\mu}_{\text{ext}}$ ramp to vary from $\bar{\mu}_f$ to $\bar{\mu}_i$. This figure clearly shows the *signature plateau in the applied electrochemical potential versus SOC curves for Li-ion batteries with phase-separating* $LiFePO_4$ *cathodes.*

Lithiation under a step change in $\bar{\mu}_{\text{ext}}$

Next we consider the case in which $\bar{\mu}_{\text{ext}}$ is quickly ramped from $\bar{\mu}_i$ to $\bar{\mu}_f$ in 1 second, and then held constant for another 29 seconds, for a total time of $T = 30\,\text{s}$ until (near) full lithiation was achieved.

Figure 19.6(a) shows the state of charge (SOC) of the particle as a function of time for different values of $\bar{E} = E/E_0$ of 0.1, 0.2, and 1.0. Note first that in contrast to the results in the previous subsection for the slow ramp, for the step-change boundary condition, phase separations occur nearly instantaneously for all values of $\bar{E}$. The SOC versus time curves show different rise-times for different values of $\bar{E}$. Consistent with the results of the simulations in the previous subsection, we observe that an increase of $\bar{E}$ results in *faster lithiation of the particle.*

Figure 19.6(b) shows contours of the normalized concentration $\bar{c}_{\text{R}}$ (top), and of the maximum principal stress σ_1 (bottom). These contours correspond to a time of $t = 3\,\text{s}$. In contrast to the previous simulations, the contours of $\bar{c}_{\text{R}}$ show that in all of the simulations, regardless of the magnitude of $\bar{E}$, the particles lithiate with a "core-shell" type of phase distribution. Such a phase distribution is a direct result of the fast "step"-change of the electrochemical potential on the boundary. Since μ_{ext} is ramped to μ_f in 1s, this amount of time is much shorter than the time for diffusion of $\tau = 25\,\text{s}$ of Li in the particle; there is not enough time for two phases to form and stably co-exist on the surface of the particle, and the result is a core-shell type of phase distribution in the particle. Consistent with the simulations shown previously in Fig. 19.4, an increase in $\bar{E}$ for the present boundary condition also results in a widening of the interface separating the two phases.

In this case the maximum principal Cauchy stress value occurs in the *interior* of the particle during lithiation—it reaches a very high value of $\sigma_1 = 1.2\,\text{GPa}$ for $\bar{E} = 1$, which has obvious implications for particle fracture.

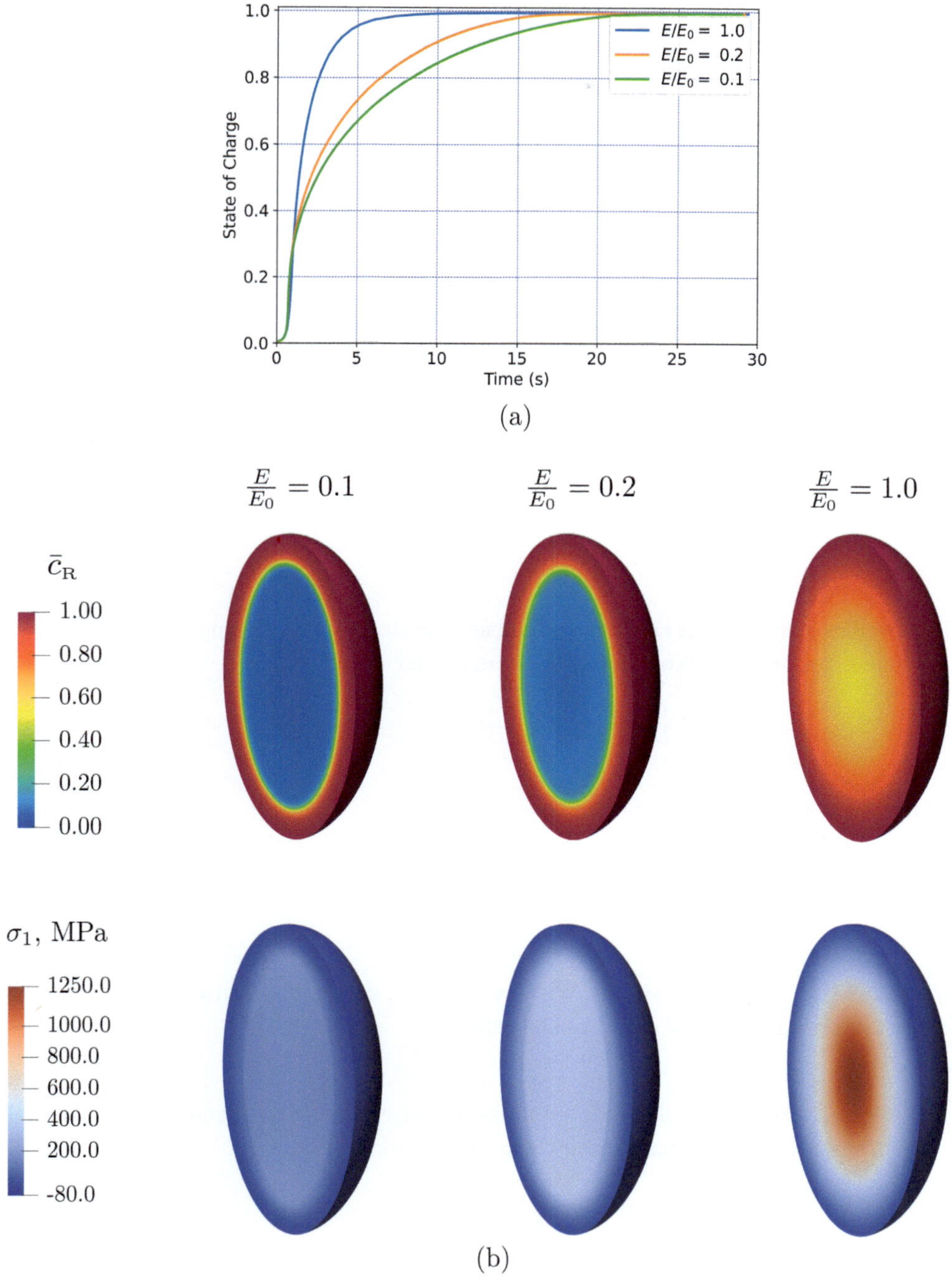

Fig. 19.6 Simulations with varying elastic moduli under an applied *step change* of the external chemical potential. (a) State of charge versus time. (b) Contours at time $t = 3$ s of normalized concentration $\bar{c}_R$ (top), and of maximum principal stress σ_1 (bottom), for three different values of $E/E_0 = 0.1, 0.2$, and 1.0

19.A Appendix. A simple species flux boundary condition

In this section we consider a simple model for specifying $\bar{\jmath}_{\text{R}}(\mathbf{X}, t)$ in the flux boundary condition $(17.11.5)_2$, viz.

$$\mathbf{j}_{\text{R}} \cdot \mathbf{n}_{\text{R}} = \underbrace{\jmath\bar{\jmath}}_{\bar{\jmath}_{\text{R}}}, \tag{19.A.1}$$

where $\bar{\jmath}$ is the prescribed spatial out-flux, and $\jmath \stackrel{\text{def}}{=} |\mathbf{F}^{\text{C}}\mathbf{n}_{\text{R}}|$ is the areal Jacobian.

Consider intercalation of a *monovalent* guest species A into the lattice of the host electrode material $\mathcal{B}$. This electrochemical reaction at the electrode/electrolyte interface may be represented as,

$$A^{\text{electrode}} \rightleftarrows A^{+\text{electrolyte}} + \text{e}^{-\text{electrode}}. \tag{19.A.2}$$

For specificity let A denote Lithium, and consider an electrode containing some Li submerged in a bath of an electrolyte containing Li^+ ions, and let

$$\text{Li}^{\text{electrode}} \rightleftarrows \text{Li}^{+\text{electrolyte}} + \text{e}^{-\text{electrode}} \tag{19.A.3}$$

denote the reaction on the surface of the electrode.

At **equilibrium**, the *electrochemical potentials* of the reacting species obey the balance,

$$\bar{\mu}_{\text{Li}}^{\text{electrode}} = \bar{\mu}_{\text{Li}^+}^{\text{electrolyte}} + \bar{\mu}_{\text{e}^-}^{\text{electrode}}, \tag{19.A.4}$$

where $\bar{\mu}_{\text{Li}}^{\text{electrode}}$, $\bar{\mu}_{\text{Li}^+}^{\text{electrolyte}}$, and $\bar{\mu}_{\text{e}^-}^{\text{electrode}}$, respectively, denote the electrochemical potentials of the Li atoms, Li ions, and electrons respectively.

Next:

- Since Li atoms in the electrode are a neutral species, the electrochemical potential for the Li atoms is simply equal to their chemical potential

$$\bar{\mu}_{\text{Li}}^{\text{electrode}} = \mu_{\text{Li}}^{\text{electrode}}. \tag{19.A.5}$$

- The electrochemical potential for the Li ions in the electrolyte is

$$\bar{\mu}_{\text{Li}^+}^{\text{electrolyte}} = \mu_{\text{Li}^+}^{\text{electrolyte}} + F\phi^{\text{electrolyte}} \tag{19.A.6}$$

  where $\mu_{\text{Li}^+}^{\text{electrolyte}}$ is the chemical potential of the Li ions, F is Faraday's constant, and $\phi^{\text{electrolyte}}$ is the electrostatic potential of the electrolyte.

- The electrochemical potential for the electrons in the electrode is

$$\bar{\mu}_{\text{e}-}^{\text{electrode}} = \mu_{\text{e}-}^{\text{electrode}} - F\phi^{\text{electrode}} \tag{19.A.7}$$

where $\mu_{\text{e}-}^{\text{electrode}}$ is the chemical potential of the electrons, and $\phi^{\text{electrode}}$ is the electrostatic potential of the electrode.

- Substituting (19.A.5), (19.A.6), and (19.A.7) in (19.A.4) gives

$$\mu_{\text{Li}}^{\text{electrode}} = \left(\mu_{\text{Li}^+}^{\text{electrolyte}} + \mu_{\text{e}-}^{\text{electrode}}\right) - F(\phi^{\text{electrode}} - \phi^{\text{electrolyte}}), \tag{19.A.8}$$

where $(\phi^{\text{electrode}} - \phi^{\text{electrolyte}})$ represents the voltage difference between the electrode and the electrolyte.

- Assuming that the activities of the Li^+ ions in the electrolyte and the electrons e^- in the electrode have a value of unity,[2] that is that they may be treated as reservoirs of supply to the chemical reaction, equation (19.A.8) may be simplified as

$$\mu_{\text{Li}}^{\text{electrode}} = \left(\mu_{\text{Li}^+}^0 + \mu_{\text{e}-}^0\right) - F(\phi^{\text{electrode}} - \phi^{\text{electrolyte}}), \tag{19.A.9}$$

where the reference potentials $\mu_{\text{Li}^+}^0$ and $\mu_{\text{e}-}^0$ are constants. The right-hand side of (19.A.9) is thus only a function of the interfacial voltage difference $(\phi^{\text{electrode}} - \phi^{\text{electrolyte}})$.

- For convenience we define

$$\bar{\mu}_{\text{ext}} \stackrel{\text{def}}{=} (\mu_{\text{Li}^+}^0 + \mu_{\text{e}-}^0) - F(\phi^{\text{electrode}} - \phi^{\text{electrolyte}}) \tag{19.A.10}$$

as the *external electrochemical potential.*

Thus, using (19.A.10) the **equilibrium condition** (19.A.4) may be written as

$$\mu_{\text{Li}}^{\text{electrode}} = \bar{\mu}_{\text{ext}}, \tag{19.A.11}$$

where $\mu_{\text{Li}}^{\text{electrode}}$ is the *chemical potential of the neutral lithium atoms in the electrode.*

A **departure from equilibrium** is then due to an imbalance of the electrochemical potentials (19.A.11) and may be quantified by a factor

$$\gamma \stackrel{\text{def}}{=} \mu_{\text{Li}}^{\text{electrode}} - \bar{\mu}_{\text{ext}}, \tag{19.A.12}$$

which measures deviations from equilibrium, and is thus also a driving force for the electrochemical reaction at the surface of the particle.

[2] The chemical potential of a species is usually written as

$$\mu_i = \mu_i^0 + R\vartheta \ln(a_i),$$

where a_i represents the *activity* of the species, and μ_i^0 represents the reference value when $a_i = 1$.

Finally, as a simple flux boundary condition we assume that the spatial surface out-flux $\bar{\jmath}$ in (19.A.1) of Li atoms is given by[3]

$$\bar{\jmath} = k(\mu_{\text{Li}}^{\text{electrode}} - \bar{\mu}_{\text{ext}}), \tag{19.A.13}$$

where

- $k > 0$ (units $\text{mol}^2/(\text{m}^2\text{J s})$) is a parameter that characterizes the conductance of the electrode/electrolyte interface.

The factor k may in general depend on the concentrations of Li and Li^+ at the electrode/electrolyte interface; however, here for simplicity we take it to be a constant. Thus,

- the condition $\mu_{\text{Li}}^{\text{electrode}} < \bar{\mu}_{\text{ext}}$ corresponds to *insertion* of Li-ions into the electrode, and
- the condition $\mu_{\text{Li}}^{\text{electrode}} > \bar{\mu}_{\text{ext}}$ corresponds to *extraction* of Li-ions from the electrode.

REMARK

The relation (19.A.13) for the flux of Li across the surface of the body is similar to *Newton's law of cooling* in heat transfer. The flux of Li through the surface is controlled by the *film coefficient*, k, while the flux of the lithium within the bulk of the electrode particle is controlled by the mobility parameter m (cf. eq. (17.10.23)). The ratio k/m has units of inverse length. Thus, with L denoting a characteristic length dimension of the body and using an analogy with standard ideas in heat transfer, we define a dimensionless "Biot"-number

$$\text{Bi} \stackrel{\text{def}}{=} \frac{kL}{m}. \tag{19.A.14}$$

As in heat transfer,

- if $\text{Bi} \gg 1$ then the surface reaction processes offers little resistance to species transfer, while
- if $\text{Bi} \ll 1$ then the surface reaction processes offer substantial resistance to species transfer.

In our numerical simulations, shown in the next chapter, we shall only consider circumstances in which $\text{Bi} \gg 1$, so that that the surface reaction processes offer little resistance to species transfer.

---

[3] The result (19.A.13) may also be derived as a linearization of existing nonlinear models for electrochemical surface kinetics such as the Butler–Volmer model (cf., e.g. Doyle et al., 1993; Newman and Thomas-Alyea, 2012; Bazant, 2013). A brief summary of the Butler–Volmer model is given in Appendix 19.B.

19.B Appendix. Butler–Volmer equation for species flux boundary condition

In this section we consider a more general model than that considered in Appendix 19.A for specifying $\bar{\jmath}_{\mathrm{R}}(\mathbf{X},t)$ for the flux boundary condition $(17.11.5)_2$, viz.

$$\mathbf{j}_{\mathrm{R}} \cdot \mathbf{n}_{\mathrm{R}} = \bar{\jmath}_{\mathrm{R}} \quad \text{on } \partial\mathcal{B}_{\mathrm{j_R}} \times [0,T], \tag{19.B.1}$$

where $\mathbf{n}_{\mathrm{R}}$ is the outward unit normal to the electrode surface. In the battery literature the normal out-flux, $\bar{\jmath}_{\mathrm{R}}$ [(mol/m^2)/s], is given in terms of

- a local current density, I_{BV} [A/m^2], of positive Li-ions *moving from the electrode to the electrolyte* at a point on the surface of the electrode,

as

$$\bar{\jmath}_{\mathrm{R}} = \jmath \frac{I_{\mathrm{BV}}}{F}, \tag{19.B.2}$$

where $F = 9.64853399(24) \times 10^4\,\mathrm{C\,mol^{-1}}$ is Faraday's constant which is the amount of electric charge per mole of electrons, and $\jmath \overset{\text{def}}{=} |\mathbf{F}^{\mathrm{C}}\mathbf{n}_{\mathrm{R}}|$ is the areal Jacobian which is introduced to account for surface area changes in large deformations.

A model for I_{BV} is given by the *Butler–Volmer equation* which characterizes the kinetics of *reduction–oxidation* reactions on the surface of electrode particles (cf., e.g., Doyle et al., 1993; Newman and Thomas-Alyea, 2012; Bazant, 2013). We elaborate on the Butler–Volmer equation below.

We begin by recalling that the **electrochemical potential** of the ith charged species in a chemical reaction is defined as

$$\bar{\mu}_i \overset{\text{def}}{=} \mu_i + z_i F \phi_\iota, \tag{19.B.3}$$

where μ_i is its **chemical potential**, z_i is its valence, F is Faraday's constant, and ϕ_ι is the electric potential in phase ι in which the charged species is found. Henceforth we distinguish the electrochemical potential with an overbar $\bar{\mu}$, while the chemical potential will not have an overbar, μ.

Next, consider an (insoluble) electrode containing some Li submerged in a bath of a liquid electrolyte containing Li^+ ions. We make the following assumptions regarding this system:

- We assume that within the electrode there are more than enough electrons to combine with the Li ions, so that lithium in the electrode exists as a *neutral species*.
- We denote the electric potential in the electrode by $\phi^{\text{electrode}}$, and the electric potential in the electrolyte by $\phi^{\text{electrolyte}}$.
- We consider the electrical conductivity of the electrode to be large, and the diffusion of Li^+ through the electrolyte to be fast.

Consider now the following *anodic and cathodic reactions* at the interface between an electrode and an electrolyte:

- An *anodic reaction* involves oxidation in which the substance loses electrons, e.g.

$$\underbrace{\mathrm{Li}^{\text{electrode}}}_{\text{reduced state}} \rightarrow \underbrace{\mathrm{Li}^{+\text{electrolyte}}}_{\text{oxidized state}} + \mathrm{e}^{-\text{electrode}}. \tag{19.B.4}$$

- A *cathodic reaction*, on the other hand, is a reduction process in which a substance gains electrons, e.g.

$$\underbrace{\mathrm{Li}^{+\text{electrolyte}}}_{\text{oxidized state}} + \mathrm{e}^{-\text{electrode}} \rightarrow \underbrace{\mathrm{Li}^{\text{electrode}}}_{\text{reduced state}}. \tag{19.B.5}$$

These reactions are often written as

$$\mathrm{Li}^{\text{electrode}} \underset{\text{cathodic}}{\overset{\text{anodic}}{\rightleftarrows}} \mathrm{Li}^{+\text{electrolyte}} + \mathrm{e}^{-\text{electrode}}. \tag{19.B.6}$$

Let,

$$\bar{\mu}_{\mathrm{Li}}^{\text{electrode}}, \quad \bar{\mu}_{\mathrm{Li}^+}^{\text{electrolyte}}, \quad \text{and} \quad \bar{\mu}_{\mathrm{e}^-}^{\text{electrode}}, \tag{19.B.7}$$

respectively, denote the the electrochemical potentials of the Li atoms on the surface of the electrode, Li ions in the electrolyte, and the electrons on the surface of the electrode. Henceforth, whenever convenient, we will omit the superscripts used in (19.B.6) and (19.B.7) to denote the location of each species. Thus, in the ensuing discussion

- the neutral Li as well as the electrons should always be considered as being on the the surface of the electrode,
- while the Li^+ ions should always be considered as being in the electrolyte.

Next, let the difference between the electrochemical potentials of the left and and right sides of (19.B.6) divided by Faraday's constant be denoted by

$$\eta \stackrel{\text{def}}{=} \frac{1}{F}\Big(\bar{\mu}_{\text{Li}} - (\bar{\mu}_{\text{Li}^+} + \bar{\mu}_{\text{e}^-})\Big). \tag{19.B.8}$$

The quantity η is commonly called the interface **overpotential**.[4] It represents a "driving force" for the chemical reaction (19.B.6), in the sense that:

- If $\eta > 0$, then lithium is favored to be expelled form the electrode into the electrolyte, and Li is oxidized to form Li^+ ions,

$$\text{Li}^{\text{electrode}} \rightarrow \text{Li}^{+\text{electrolyte}} + \text{e}^{-\text{electrode}} \qquad \text{(anodic reaction)}. \tag{19.B.9}$$

- If $\eta < 0$, then lithium is favored to be intercalated form the electrolyte into the electrode, and Li^+ ions are reduced to form Li,

$$\text{Li}^{+\text{electrolyte}} + \text{e}^{-\text{electrode}} \rightarrow \text{Li}^{\text{electrode}} \qquad \text{(cathodic reaction)}. \tag{19.B.10}$$

- The condition $\eta = 0$ represents the condition for **equilibrium** of the redox reaction in equation (19.B.6).

The electrochemical potentials in the definition (19.B.8) may be expressed in terms of their chemical potentials and electric potentials as

$$\bar{\mu}_{\text{Li}} = \mu_{\text{Li}}, \qquad \bar{\mu}_{\text{Li}^+} = \mu_{\text{Li}^+} + F\phi^{\text{electrolyte}}, \qquad \bar{\mu}_{\text{e}^-} = \mu_{\text{e}^-} - F\phi^{\text{electrode}}, \tag{19.B.11}$$

where in writing (19.B.11) we have made use of the fact that Li atoms in the electrode are a *neutral species*, that the Li ions are positively charged (valence of $+1$), and that the electrons are negatively charged (valence of -1). Further note that the electric potential acting on the Li^+ ions is that of the electrolyte, $\phi^{\text{electrolyte}}$, and the electric potential acting on the electrons is that of the electrode, $\phi^{\text{electrode}}$.

Next, consistent with our earlier assumptions, we assume further that the activities of the Li^+ ions in the electrolyte and the electrons e^- in the electrode have a value of unity—that is the electrolyte and the electrode may be treated as infinite reservoirs which supply Li^+ ions and electrons e^- to the chemical reaction at the surface of the electrode particle. Then, the electrochemical potentials (19.B.11) may be written as

$$\bar{\mu}_{\text{Li}} = \mu^0_{\text{Li}} + \hat{\mu}_{\text{Li}}(\bar{c}, \mathbf{M}^e, \vartheta), \qquad \bar{\mu}_{\text{Li}^+} = \mu^0_{\text{Li}^+} + F\phi^{\text{electrolyte}}, \qquad \bar{\mu}_{\text{e}^-} = \mu^0_{\text{e}^-} - F\phi^{\text{electrode}}, \tag{19.B.12}$$

where the reference potentials μ^0_{Li}, $\mu^0_{\text{Li}^+}$, and $\mu^0_{\text{e}^-}$ are constants, and where we have recalled that the chemical potential of the Li at the surface of the electrode is a function of the

[4] While in the rest of the book we have used the notation η to denote the entropy density, here, to maintain alignment with the nomenclature in the battery literature, we use η to denote the *overpotential*.

concentration, the Mandel stress, and the temperature (cf. eq. (17.10.20)). Using (19.B.12) we may write the overpotential η defined (19.B.8) as

$$\eta = \Delta\phi - \left(\frac{(\mu^0_{\mathrm{Li}^+} + \mu^0_{\mathrm{e}^-} - \mu^0_{\mathrm{Li}})}{F} - \frac{\hat{\mu}_{\mathrm{Li}}(\bar{c}, \mathbf{M}^e, \vartheta)}{F}\right), \tag{19.B.13}$$

where

$$\Delta\phi \stackrel{\mathrm{def}}{=} \phi^{\mathrm{electrode}} - \phi^{\mathrm{electrolyte}}, \tag{19.B.14}$$

represents the *voltage drop* across the electrode/electroyte interface.

Equilibrium: At equilibrium $\eta = 0$. Thus, using (19.B.13), we may define an *equilibrium potential* as

$$\phi_{\mathrm{eq}} \stackrel{\mathrm{def}}{=} \underbrace{\frac{(\mu^0_{\mathrm{Li}^+} + \mu^0_{\mathrm{e}^-} - \mu^0_{\mathrm{Li}})}{F}}_{\equiv \mathrm{constant}} - \frac{\hat{\mu}_{\mathrm{Li}}(\bar{c}, \mathbf{M}^e, \vartheta)}{F} \equiv \phi_{\mathrm{eq}}(\bar{c}, \mathbf{M}^e, \vartheta). \tag{19.B.15}$$

This is the potential at which no current flows. It is also commonly known as the open-circuit potential (OCP)—it is an experimentally measured quantity.

Deviation from equilibrium: Using (19.B.15), we may rewrite the overpotential η defined in (19.B.13) as

$$\eta = \Delta\phi - \phi_{\mathrm{eq}}. \tag{19.B.16}$$

As is clear from (19.B.16), the overpotential η represents the electric potential drop above (or below) an equilibrium value required to drive the chemical reaction. When the overpotential is non-zero, the chemical reaction (19.B.6) takes place and there is a resulting flow of current I_{BV} [A/m^2], which is a function of the overpotential η [V]. This current is widely taken to be given by the phenomenological Butler–Volmer equation (cf., e.g., Newman and Thomas-Alyea, 2012; Bazant, 2013),[5]

$$I_{\mathrm{BV}} = I_0 \left(\exp\left(\frac{\alpha_a F\eta}{R\vartheta}\right) - \exp\left(-\frac{\alpha_c F\eta}{R\vartheta}\right)\right), \tag{19.B.17}$$

where

- I_0 [A/m^2] is the *exchange current density*,
- α_a is an *anodic transfer coefficient* which represents the fraction of the overpotential that affects the anodic reaction rate, and
- α_c is a *cathodic transfer coefficient* which represents the fraction of the overpotential that affects the cathodic reaction.

[5] The most common derivation of the Butler–Volmer equation is based on Eyring's transition state theory.

Note that the current $I_{\rm BV}$ is *positive* for lithium *de-intercalation*, that is lithium ions leaving the electrode,

$$\mathrm{Li}^{\text{electrode}} \to \mathrm{Li}^{+\text{electrolyte}} + \mathrm{e}^{-\text{electrode}} \qquad \text{(anodic reaction)}. \tag{19.B.18}$$

In a discharging process, as neutral lithium de-intercalates from the electrode to the electrolyte in the form of Li^+ ions, the released electrons flow through the external circuit. For this reason $I_{\rm BV}$ is is called an "anodic current"—it is associated with the oxidation process.

For a reaction involving one electron transfer, the coefficients (α_a, α_c) typically satisfy

$$\alpha_a = \alpha, \qquad \alpha_c = 1 - \alpha, \tag{19.B.19}$$

with α a symmetry factor in the range $0 < \alpha < 1$. In this case the Butler–Volmer equation (19.B.17) becomes,

$$I_{\rm BV} = I_0 \left(\exp\left(\frac{\alpha\, F\eta}{R\vartheta} \right) - \exp\left(-\frac{(1-\alpha)F\eta}{R\vartheta} \right) \right). \tag{19.B.20}$$

Under the assumption that that electrolyte and the electrode are being treated as infinite reservoirs which supply Li^+ ions and electrons e^- to the chemical reaction, the exchange current density I_0 for Li-ion batteries is often modeled as

$$I_0(\bar{c}) = k_0 F\,(1-\bar{c})^{\alpha} \bar{c}^{(1-\alpha)}, \tag{19.B.21}$$

where k_0 a reaction rate constant (units of mol/m^2/s), and $\bar{c}$ is the normalized concentration of lithium in the electrode. Further, the symmetry factor α is often assumed to take a value $\alpha = 1/2$. In this case the Butler–Volmer equation reduces to

$$I_{\rm BV} = 2\, I_0(\bar{c}) \sinh\left(\frac{F\eta}{2R\vartheta} \right) \qquad \text{with} \qquad I_0 = k_0 F\,(1-\bar{c})^{1/2} \bar{c}^{1/2}. \tag{19.B.22}$$

The Butler–Volmer equation (19.B.22) gives the local current density $I_{\rm BV}$ of positive Li-ions *moving from the electrode to the electrolyte* at a point on the surface of the electrode. Thus, as discussed in eqs. (19.B.1) and (19.B.2)$_2$, the species out-flux at a point on the surface of the electrode is specified by specifying normal flux $\bar{\jmath}_{\rm R}$, which is related to the current $I_{\rm BV}$ by

$$\bar{\jmath}_{\rm R} = \jmath\, \frac{I_{\rm BV}}{F}, \tag{19.B.23}$$

where $\jmath \overset{\text{def}}{=} |\mathbf{F}^{\rm c}\, \mathbf{n}_{\rm R}|$ is the areal Jacobian which is introduced to account for surface area changes in large deformations.

REMARKS

1. For the case in which the concentration of Li^+ ions in the electrolyte is variable, a dependence on the normalized concentration of Li^+ ions in the electrolyte in the definition of the exchange current I_0 needs to be included. Thus, with c_R^+ denoting the concentration of Li^+ ions in the electrolyte, $c_{R,\max}^+$ denoting its maximum value, and $\bar{c}_R^+ = c_R^+/c_{R,\max}^+$ denoting the normalized value of the Li^+ concentration in the electrolyte, the expression $(19.B.22)_2$ for the exchange current density is modified to read as

$$I_0(\bar{c}, \bar{c}_R^+) = k_0 F\,(1-\bar{c})^{1/2}\bar{c}^{1/2}(\bar{c}_R^+)^{1/2}. \qquad (19.B.24)$$

2. As reviewed by Deshpande and McMeeking (2023), there are many other empirical modifications to the expression for the exchange current I_0 that have been proposed in the literature. Some of these modifications also account for the mechanical stress difference across the electrode/electrolyte interface.

   For other reviews of the Butler–Volmer model and its many applications in electrochemistry see Dickinson and Wain (2020) and Bard et al. (2022).
3. Since $\sinh(x) \approx x$ for small values of x, the linearized version of the Butler–Volmer equation (19.B.22), recalling (19.B.8), is

$$I_{\mathrm{BV}} \propto \underbrace{\frac{1}{F}\Big(\bar{\mu}_{\mathrm{Li}} - (\bar{\mu}_{\mathrm{Li}^+} + \bar{\mu}_{\mathrm{e}^-})\Big)}_{\eta},$$

   an expression which may be re-arranged and used to arrive at eq. (19.A.13) for $\bar{\jmath}$ discussed in Appendix 19.A.
4. Our considerations in this chapter have focused on modeling the coupled deformation–diffusion response during lithiation of a *single particle* of an electrode in a Li-ion battery. Although such considerations are of fundamental importance, there are numerous other multidimensional and multiphysical issues which arise in when one attempts to model the coupled electrical, chemical, mechanical, and thermal phenomena which occur simultaneously in the discharge or charge of lithium-ion batteries. Such issues need be addressed not only for the ubiquitous lithium-ion batteries, but also next-generation battery technologies such as sodium-ion batteries, and all-solid-state batteries.

Part VII

Electro-elasticity of dielectric elastomers

20 Electroelasticity of dielectric elastomers

20.1 Introduction

A dielectric material is an electrical insulator that can be polarized by an applied electric field. When a dielectric is placed in an electric field, electric charges do not flow through the material as they do in an electrical conductor but only slightly shift from their average equilibrium positions causing dielectric polarization.

When subjected to an electric field, all dielectrics deform. However, the attainable deformation induced by voltage varies markedly from one type of dielectric to another. Elastomeric materials with a ground-state shear modulus in the kPa range can be readily deformed over 100% by a suitably large applied voltage. Such elastomers are called compliant or *soft electroactive dielectrics*.

In the past twenty years or so it has been recognized that the large deformation capacity of these materials may be harnessed for use in actuators and sensors; cf. the classical paper of Pelrine et al. (2000). For recent advancements in new dielectric elastomeric materials and their application in artificial-muscle actuators, see Qiu et al. (2019), Duduta et al. (2019), Hajiesmaili and Clarke (2021) and Shi et al. (2022).

The modeling of the coupled electromechanical response of these materials is based on the theory of *nonlinear electroelasticity*, whose roots go back to a classical paper by Toupin (1956). Since that time there has been substantial work on the formulation of the theory by Eringen (1963), Tiersten (1971), Pao (1978), Hutter and van de Ven (1978), Maugin (1988), Eringen and Maugin (1990), Tiersten (1990), Kovetz (1990, 2000), and Dorfmann and Ogden (cf., e.g., Dorfmann and Ogden, 2005, 2014, 2017), McMeeking and Landis (McMeeking and Landis, 2005; McMeeking et al., 2007), Suo and co-workers (cf., e.g., Suo et al., 2008; Suo, 2010) and Zhao and co-workers (Zhao and Wang, 2014), amongst others. References to the substantial literature on this topic may be found in a review paper by Dorfmann and Ogden (2017).

Papers regarding the numerical implementation of this theory to predict the three-dimensional, large-deformation, coupled electro-mechanical response of dielectric elastomers have also been published in the recent literature. For example, Henann et al. (2013) report on their implementation of the theory in the commercial finite element code Abaqus, and show applications of their numerical implementation in the design of some dielectric-elastomer-based actuators and energy-harvesting devices. References to the other recent publications on the numerical implementation of nonlinear elasticity theory may be found in Henann et al. (2013), and also Vogel et al. (2014) and Wang et al. (2016) who also consider viscoelastic effects. For recent experimental and computational studies on electroactive polymers see Mehnert et al. (2018, 2019) and Kadapa and Hossain (2020).

Introduction to Coupled Theories in Solid Mechanics. Lallit Anand, Eric M. Stewart, Shawn A. Chester, Oxford University Press.
 DOI: 10.1093/oso/9780198986218.003.0021

In this chapter we formulate a thermodynamically consistent theory of nonlinear electroelasticity for dielectric elastomers. We will consider the numerical implementation of the theory in Chapter 21.

20.2 Basic kinematics

We consider a macroscopically homogeneous body $\mathcal{B}$ with the region of space it occupies in a fixed reference configuration, and denote by $\mathbf{X}$ an arbitrary material point of $\mathcal{B}$.[1] A motion of $\mathcal{B}$ is a smooth one-to-one mapping

$$\mathbf{x} = \boldsymbol{\chi}(\mathbf{X}, t), \tag{20.2.1}$$

with displacement

$$\mathbf{u}(\mathbf{X}, t) = \boldsymbol{\chi}(\mathbf{X}, t) - \mathbf{X}, \tag{20.2.2}$$

deformation gradient, velocity, and acceleration given by,

$$\mathbf{F} = \nabla\boldsymbol{\chi} = \mathbf{1} + \nabla\mathbf{u}, \qquad \dot{\mathbf{u}} = \frac{\partial\boldsymbol{\chi}}{\partial t}, \qquad \ddot{\mathbf{u}} = \frac{\partial^2\boldsymbol{\chi}}{\partial t^2}. \tag{20.2.3}$$

We assume that

$$J = \det\mathbf{F} > 0, \tag{20.2.4}$$

so that $\mathbf{F}$ is invertible,

Throughout we denote by $\mathcal{P}$ an arbitrary part of the reference body $\mathcal{B}$ with $\mathbf{n}_{\mathrm{R}}$ the outward unit normal on the boundary $\partial\mathcal{P}$ of $\mathcal{P}$. Also, we denote by $\mathcal{P}_t = \boldsymbol{\chi}(\mathcal{P}, t)$ the image of $\mathcal{P}$ in the deformed body $\mathcal{B}_t = \boldsymbol{\chi}(\mathcal{B}, t)$, with $\mathbf{n}$ the outward unit normal on the boundary $\partial\mathcal{P}_t$ of $\mathcal{P}_t$.

20.3 Balance of forces and moments

Standard considerations of balance of forces and moments, when expressed referentially, give the following results:

(a) There exists a stress tensor $\mathbf{T}_{\mathrm{R}}$, called the Piola stress, such that the surface traction on an oriented element of the surface $\partial\mathcal{P}$ of $\mathcal{P}$, is given by the traction relation

$$\mathbf{t}_{\mathrm{R}}(\mathbf{n}_{\mathrm{R}}) = \mathbf{T}_{\mathrm{R}}\mathbf{n}_{\mathrm{R}}. \tag{20.3.1}$$

[1] Notation: We use standard notation of modern continuum mechanics (Gurtin et al., 2010), cf. Section 1.1. However, as a departure from the usual convention in modern continuum mechanics in which vectors are denoted by lower case bold face letters **a**, **b**, etc., we denote the two important electrical vector fields using bold face sans-serif fonts as follows: we write $\mathsf{\mathbf{e}}$ for the electric field, and $\mathsf{\mathbf{d}}$ for the electric displacement.

(b) $\mathbf{T}_R$ satisfies the macroscopic force balance

$$\operatorname{Div} \mathbf{T}_R + \mathbf{b}_R = \mathbf{0}, \tag{20.3.2}$$

with

$$\mathbf{b}_R = \mathbf{b}_{0R} - \rho_R \ddot{\mathbf{u}}, \tag{20.3.3}$$

where $\mathbf{b}_{0R}$ represents the conventional body force per unit volume of the reference body, and $(-\rho_R \ddot{\mathbf{u}})$ represents the inertial body force, with ρ_R being the mass density of the referential body, and $\ddot{\chi}$ the acceleration.

(c) $\mathbf{T}_R$ obeys the symmetry condition,

$$\mathbf{T}_R \mathbf{F}^{\top} = \mathbf{F} \mathbf{T}_R^{\top}, \tag{20.3.4}$$

which represents a balance of moments.

As is standard, the stress measure

$$\mathbf{T} = J^{-1} \mathbf{T}_R \mathbf{F}^{\top}, \tag{20.3.5}$$

represents the symmetric Cauchy stress, in terms of which:

(a) The surface traction on an oriented element of the surface $\partial \mathcal{P}_t$ of $\mathcal{P}_t$, is given by

$$\mathbf{t}(\mathbf{n}) = \mathbf{T}\mathbf{n}. \tag{20.3.6}$$

(b) And $\mathbf{T}$ satisfies the macroscopic force balance

$$\operatorname{div} \mathbf{T} + \mathbf{b} = \mathbf{0}, \tag{20.3.7}$$

where $\mathbf{b} = \mathbf{b}_0 - \rho \dot{\mathbf{v}}$ represents the body force per unit volume of the deformed body.

REMARK

The Cauchy stress $\mathbf{T}$ is often referred to as the *total* stress in the electroelasticity literature. However, we see no need to distinguish between a purely mechanical stress and an electrostatic, or Maxwell, stress. Such a decomposition appears naturally when we consider specific thermodynamically consistent constitutive equations; cf. Section 20.7.

20.4 Basic equations of electrostatics

The electric field in $\mathcal{B}$ is denoted by $\mathbf{e}_R(\mathbf{X},t)$ (units of V/m). In a polarizable material there is a vector field in $\mathcal{B}$ denoted by $\mathbf{d}_R(\mathbf{X},t)$ called the electric displacement (units of C/m^2). In the theory formulated in this chapter we only consider the electrostatic limit as far as the electrical variables are concerned, so that time t only plays the role of a history parameter.

Under electrostatic conditions there are two governing Maxwell equations that must be satisfied. The first of these is *Faraday's law*, which in the electrostatic limit reads,

$$\operatorname{Curl}\mathbf{e}_R = \mathbf{0}. \tag{20.4.1}$$

Satisfaction of (20.4.1) occurs automatically if we allow the electric field $\mathbf{e}_R(\mathbf{X},t)$ to be represented as the gradient of an electrostatic potential $\phi(\mathbf{X},t)$,[2]

$$\mathbf{e}_R(\mathbf{X},t) \stackrel{\text{def}}{=} -\nabla\phi(\mathbf{X},t). \tag{20.4.2}$$

The negative sign in (20.4.2) follows the convention that the electric field points from a region of high potential to low potential, whereas the vector $\nabla\phi$ by definition points in the direction of increasing ϕ.

The second equation of electrostatics is given by *Gauss's law*. Let the net free charge per unit volume of the reference body be denoted by $q_R(\mathbf{X},t)$. Gauss's law states that the divergence of the electric displacement $\mathbf{d}_R$ must equal q_R,

$$\operatorname{Div}\mathbf{d}_R = q_R. \tag{20.4.3}$$

Equations (20.4.1) and (20.4.3) constitute the governing equations for electrostatics in $\mathcal{B}$.

Further, using the conversion relations

$$\mathbf{e} = \mathbf{F}^{-\top}\mathbf{e}_R \quad \text{and} \quad \mathbf{d} = J^{-1}\mathbf{F}\mathbf{d}_R, \tag{20.4.4}$$

where $\mathbf{e}$ and $\mathbf{d}$ are the spatial counterparts of $\mathbf{e}_R$ and $\mathbf{d}_R$, the spatial counterparts of (20.4.1) and (20.4.3) are

$$\operatorname{curl}\mathbf{e} = \mathbf{0}, \tag{20.4.5}$$

and

$$\operatorname{div}\mathbf{d} = q, \tag{20.4.6}$$

with q the free charge per unit volume of the deformed body. Equations (20.4.5) and (20.4.6) constitute the governing equations for electrostatics in the deformed body $\mathcal{B}_t$.

[2] Recall that for any scalar field ϕ, $\operatorname{Curl}(\nabla\phi) = \mathbf{0}$.

Dielectrics

Dielectric materials are non-conducting and do not possess free charges. Thus for a dielectric

$$q = 0 \qquad \text{and hence} \qquad q_{\mathrm{R}} = 0. \tag{20.4.7}$$

20.5 **Free-energy imbalance**

Under isothermal conditions the two laws of thermodynamics reduce to the statement that *the rate of increase in free energy of any part $\mathcal{P}$ is less than or equal to the power expended on $\mathcal{P}$.* Precisely, letting ψ_{R} denote the *free energy* per unit reference volume, this requirement takes the form of a free-energy imbalance

$$\overline{\int_{\mathcal{P}} \psi_{\mathrm{R}}\, dv_{\mathrm{R}}}^{\boldsymbol{\cdot}} \leq \mathcal{W}(\mathcal{P}) + \mathcal{E}(\mathcal{P}), \tag{20.5.1}$$

where:

(i) The mechanical power $\mathcal{W}(\mathcal{P})$ exerted on $\mathcal{P}$ by the boundary tractions $\mathbf{T}_{\mathrm{R}}\mathbf{n}_{\mathrm{R}}$ and body force $\mathbf{b}_{\mathrm{R}}$ over the velocity $\dot{\mathbf{u}}$ is given by,

$$\mathcal{W}(\mathcal{P}) = \int_{\partial\mathcal{P}} (\mathbf{T}_{\mathrm{R}}\mathbf{n}_{\mathrm{R}}) \cdot \dot{\mathbf{u}}\, da_{\mathrm{R}} + \int_{\mathcal{P}} \mathbf{b}_{\mathrm{R}} \cdot \dot{\mathbf{u}}\, dv_{\mathrm{R}}. \tag{20.5.2}$$

(ii) In electromagnetic field theory the power that is expended on $\mathcal{P}$ is characterized by an electromagnetic energy flux $\mathcal{E}(\mathcal{P})$ which in the **electrostatic limit** for a **dielectric** material is given by

$$\mathcal{E}(\mathcal{P}) = -\int_{\partial\mathcal{P}} \phi\, \dot{\mathbf{d}}_{\mathrm{R}} \cdot \mathbf{n}_{\mathrm{R}}\, da_{\mathrm{R}}. \tag{20.5.3}$$

The genesis of this crucial term comes from the electrostatic limit of the expression for the *flux of electromagnetic energy*, as described by the *Poynting's vector* (see, e.g., Jackson (1999, Sec. 6.7) or Kovetz (2000, Sec. 45 and Sec. 54)). For a brief derivation of equation (20.5.3) see Appendix A.

Thus, using (20.5.2) and (20.5.3), the free-energy imbalance (20.5.1) becomes

$$\overline{\int_{\mathcal{P}} \psi_{\mathrm{R}}\, dv_{\mathrm{R}}}^{\boldsymbol{\cdot}} \leq \int_{\partial\mathcal{P}} (\mathbf{T}_{\mathrm{R}}\mathbf{n}_{\mathrm{R}}) \cdot \dot{\mathbf{u}}\, da_{\mathrm{R}} + \int_{\mathcal{P}} \mathbf{b}_{\mathrm{R}} \cdot \dot{\mathbf{u}}\, dv_{\mathrm{R}} - \int_{\partial\mathcal{P}} \phi\, \dot{\mathbf{d}}_{\mathrm{R}} \cdot \mathbf{n}_{\mathrm{R}}\, da_{\mathrm{R}}. \tag{20.5.4}$$

Using the divergence theorem on the terms involving the integral over $\partial\mathcal{P}$ in (20.5.4) we obtain,

$$\begin{aligned}
\int\limits_{\partial\mathcal{P}} (\mathbf{T}_{\mathrm{R}}\mathbf{n}_{\mathrm{R}}) \cdot \dot{\mathbf{u}}\, da_{\mathrm{R}} &= \int\limits_{\mathcal{P}} (\operatorname{Div}\mathbf{T}_{\mathrm{R}}) \cdot \dot{\mathbf{u}}\, dv + \int\limits_{\mathcal{P}} \mathbf{T}_{\mathrm{R}} : \dot{\mathbf{F}} dv_{\mathrm{R}}, \\
-\int\limits_{\partial\mathcal{P}} (\phi\, \dot{\mathbf{d}}_{\mathrm{R}}) \cdot \mathbf{n}_{\mathrm{R}}\, da_{\mathrm{R}} &= -\int\limits_{\mathcal{P}} \phi\, (\operatorname{Div}\dot{\mathbf{d}}_{\mathrm{R}}) dv_{\mathrm{R}} - \int\limits_{\mathcal{P}} (\nabla\phi) \cdot \dot{\mathbf{d}}_{\mathrm{R}}\, dv_{\mathrm{R}} \\
&= -\int\limits_{\mathcal{P}} \phi\, (\operatorname{Div}\dot{\mathbf{d}}_{\mathrm{R}}) dv_{\mathrm{R}} + \int\limits_{\mathcal{P}} \mathbf{e}_{\mathrm{R}} \cdot \dot{\mathbf{d}}_{\mathrm{R}}\, dv_{\mathrm{R}}.
\end{aligned} \tag{20.5.5}$$

Substituting (20.5.5) in (20.5.4), and using the linear momentum balance (20.3.2), the rate form of Gauss's law (20.4.3), that is Div $\dot{\mathbf{d}}_{\mathrm{R}} = 0$, we obtain,

$$\int_{\mathcal{P}} \left(\dot{\psi}_{\mathrm{R}} - \mathbf{T}_{\mathrm{R}} : \dot{\mathbf{F}} - \mathbf{e}_{\mathrm{R}} \cdot \dot{\mathbf{d}}_{\mathrm{R}} \right) dv_{\mathrm{R}} \leq 0. \tag{20.5.6}$$

Since (20.5.6) must hold for all parts $\mathcal{P}$, we obtain a local free-energy imbalance in the form,

$$\dot{\psi}_{\mathrm{R}} - \mathbf{T}_{\mathrm{R}} : \dot{\mathbf{F}} - \mathbf{e}_{\mathrm{R}} \cdot \dot{\mathbf{d}}_{\mathrm{R}} \leq 0. \tag{20.5.7}$$

Next, the rate of change of $\mathbf{C} = \mathbf{F}^{\top}\mathbf{F}$ is,

$$\dot{\mathbf{C}} = \left(\dot{\mathbf{F}}^{\top}\mathbf{F} + \mathbf{F}^{\top}\dot{\mathbf{F}} \right) = 2\,\mathrm{sym}\,(\mathbf{F}^{\top}\dot{\mathbf{F}}); \tag{20.5.8}$$

hence

$$\mathrm{sym}\,(\mathbf{F}^{\top}\dot{\mathbf{F}}) = \tfrac{1}{2}\dot{\mathbf{C}}. \tag{20.5.9}$$

Using (20.5.9) the "stress power" term $\mathbf{T}_{\mathrm{R}} : \dot{\mathbf{F}}$ in (20.5.12) may be written as

$$\mathbf{T}_{\mathrm{R}} : \dot{\mathbf{F}} = \tfrac{1}{2}\mathbf{T}_{\mathrm{RR}} : \dot{\mathbf{C}}, \tag{20.5.10}$$

where

$$\mathbf{T}_{\mathrm{RR}} \stackrel{\text{def}}{=} J\mathbf{F}^{-1}\mathbf{T}\mathbf{F}^{-\top}, \tag{20.5.11}$$

is the second Piola stress. Thus, we may rewrite the free-energy imbalance (20.5.7) as

$$\dot{\psi}_{\mathrm{R}} - \tfrac{1}{2}\mathbf{T}_{\mathrm{RR}} : \dot{\mathbf{C}} - \mathbf{e}_{\mathrm{R}} \cdot \dot{\mathbf{d}}_{\mathrm{R}} \leq 0. \tag{20.5.12}$$

REMARK
For brevity we have properties under a change in frame of the various fields appearing in our theory. Here, we simply note that all quantities in the free-energy imbalance (20.5.12) are invariant under a change in frame (Gurtin et al., 2010).

20.6 Constitutive theory

Let $\boldsymbol{\Lambda}$ represent the pair

$$\boldsymbol{\Lambda} = (\mathbf{C}, \mathbf{d}_{\mathrm{R}}). \tag{20.6.1}$$

Guided by the free-energy imbalance (20.5.12), we begin by assuming constitutive equations for the free energy ψ_{R}, the stress $\mathbf{T}_{\mathrm{RR}}$, and the electric field $\mathbf{e}_{\mathrm{R}}$ of the form

$$\psi_{\mathrm{R}} = \hat{\psi}_{\mathrm{R}}(\boldsymbol{\Lambda}), \qquad \mathbf{T}_{\mathrm{RR}} = \hat{\mathbf{T}}_{\mathrm{RR}}(\boldsymbol{\Lambda}), \qquad \mathbf{e}_{\mathrm{R}} = \hat{\mathbf{e}}_{\mathrm{R}}(\boldsymbol{\Lambda}), \tag{20.6.2}$$

Then,

$$\dot{\psi}_{\mathrm{R}} = \frac{\partial \hat{\psi}_{\mathrm{R}}(\boldsymbol{\Lambda})}{\partial \mathbf{C}} : \dot{\mathbf{C}} + \frac{\partial \hat{\psi}_{\mathrm{R}}(\boldsymbol{\Lambda})}{\partial \mathbf{d}_{\mathrm{R}}} \cdot \dot{\mathbf{d}}_{\mathrm{R}}. \tag{20.6.3}$$

Using (20.6.3) and substituting the constitutive equations (20.6.2) into the free-energy imbalance (20.5.12), we find that it may be written as

$$\left[\frac{\partial \hat{\psi}_{\mathrm{R}}(\boldsymbol{\Lambda})}{\partial \mathbf{C}} - \tfrac{1}{2}\hat{\mathbf{T}}_{\mathrm{RR}}(\boldsymbol{\Lambda})\right] : \dot{\mathbf{C}} + \left[\frac{\partial \hat{\psi}_{\mathrm{R}}(\boldsymbol{\Lambda})}{\partial \mathbf{d}_{\mathrm{R}}} - \hat{\mathbf{e}}_{\mathrm{R}}(\boldsymbol{\Lambda})\right] \cdot \dot{\mathbf{d}}_{\mathrm{R}} \le 0. \tag{20.6.4}$$

Given a point of the body and any time t, it is possible to find a motion and an electrical displacement field such that $\mathbf{C}$, $\mathbf{d}_{\mathrm{R}}$ and their time derivatives $\dot{\mathbf{C}}$ and $\dot{\mathbf{d}}_{\mathrm{R}}$ have arbitrarily prescribed values at that point and time. Granted this assertion, the coefficients of $\dot{\mathbf{C}}$ and $\dot{\mathbf{d}}_{\mathrm{R}}$ must vanish, for otherwise the rates may be chosen to violate the inequality (20.6.4). We therefore have the *thermodynamic restrictions* that the free energy delivers the stress $\mathbf{T}_{\mathrm{RR}}$ and the electric field $\mathbf{e}_{\mathrm{R}}$ through the state relations:

$$\begin{aligned} \mathbf{T}_{\mathrm{RR}} &= 2\frac{\partial \hat{\psi}_{\mathrm{R}}(\boldsymbol{\Lambda})}{\partial \mathbf{C}}, \\ \mathbf{e}_{\mathrm{R}} &= \frac{\partial \hat{\psi}_{\mathrm{R}}(\boldsymbol{\Lambda})}{\partial \mathbf{d}_{\mathrm{R}}}, \end{aligned} \tag{20.6.5}$$

and that the free-energy imbalance is satisfied with the material exhibiting *no dissipation*.

Using (20.5.11) and $(20.6.5)_1$, the Cauchy stress is given by a constitutive equation of the form

$$\mathbf{T} = 2J^{-1}\mathbf{F}\frac{\partial\hat{\psi}_{\mathrm{R}}(\boldsymbol{\Lambda})}{\partial\mathbf{C}}\mathbf{F}^{\top}. \tag{20.6.6}$$

20.7 Specialization of the constitutive equations

The theory presented thus far is still quite general. We now introduce special constitutive equations which should be useful in applications. We restrict our attention to isotropic materials and consider that the free energy may be expressed in a separable form

$$\psi_{\mathrm{R}} = \psi_{\mathrm{R}}^{\mathrm{mech}}(\mathbf{C}) + \psi_{\mathrm{R}}^{\mathrm{es}}(\mathbf{C}, \mathbf{d}_{\mathrm{R}}). \tag{20.7.1}$$

Where:

(i) The contribution $\psi_{\mathrm{R}}^{\mathrm{mech}}$ is the change in the free energy due to mechanical deformation, which for isotropic materials is expressible as

$$\psi_{\mathrm{R}}^{\mathrm{mech}} = \hat{\psi}_{\mathrm{R}}^{\mathrm{mech}}(I_1, I_2, I_3), \tag{20.7.2}$$

where

$$I_1 = \mathrm{tr}\,\mathbf{C}, \qquad I_2 = \tfrac{1}{2}\left[(\mathrm{tr}\mathbf{C})^2 - \mathrm{tr}\,(\mathbf{C}^2)\right], \qquad I_3 = \det\mathbf{C}, \tag{20.7.3}$$

are the principal invariants of $\mathbf{C}$.

(ii) The contribution $\psi_{\mathrm{R}}^{\mathrm{es}}(\mathbf{C}, \mathbf{d}_{\mathrm{R}})$ is an electrostatic energy. We consider a special form for this energy (Zhao et al., 2007; Suo, 2010),[3]

$$\psi_{\mathrm{R}}^{\mathrm{es}} = J\left(\frac{1}{2\varepsilon}\mathbf{d}\cdot\mathbf{d}\right), \tag{20.7.4}$$

which is *isotropic in the deformed body*, with $\varepsilon = \varepsilon_r\,\varepsilon_0$ representing the *electrical permittivity* of the material, with $\epsilon_0 = 8.854187817 \times 10^{-12}$ F/m the permittivity of free space, and ε_r the relative permittivity of the material.

Since $\mathbf{d} = J^{-1}\mathbf{F}\mathbf{d}_{\mathrm{R}}$, we may express this free energy as

$$\psi_{\mathrm{R}}^{\mathrm{es}} = J^{-1}\left(\frac{1}{2\varepsilon}\mathbf{d}_{\mathrm{R}}\cdot\mathbf{C}\mathbf{d}_{\mathrm{R}}\right). \tag{20.7.5}$$

Note that the free energy (20.7.5) is an isotropic function of $\mathbf{C}$ and $\mathbf{d}_{\mathrm{R}} \otimes \mathbf{d}_{\mathrm{R}}$.

[3] Zhao et al. (2007) motivate this liquid-like choice of electrostatic energy by arguing that when molecular groups in an elastomer can polarize nearly as freely as in liquids, e.g., when the degree of cross-link is low and the deformation is well below the fully extended limit, then the dielectric behavior of the elastomer is expected to be liquid-like.

Thus, the free-energy function for an ideal isotropic dielectric material has the form,

$$\psi_{\mathrm{R}} = \psi_{\mathrm{R}}^{\mathrm{mech}}(I_1, I_2, I_3) + \underbrace{J^{-1}\left(\frac{1}{2\varepsilon}\mathbf{d}_{\mathrm{R}} \cdot \mathbf{C}\mathbf{d}_{\mathrm{R}}\right)}_{\psi_{\mathrm{R}}^{\mathrm{es}}}. \tag{20.7.6}$$

Electric field. Electric displacement

From (20.7.6) and $(20.6.5)_2$ the referential electric field is given by

$$\mathbf{e}_{\mathrm{R}} = \frac{1}{\varepsilon} J^{-1}\mathbf{C}\mathbf{d}_{\mathrm{R}}, \tag{20.7.7}$$

and hence the referential electric displacement is given by

$$\mathbf{d}_{\mathrm{R}} = \varepsilon\, J\mathbf{C}^{-1}\mathbf{e}_{\mathrm{R}}. \tag{20.7.8}$$

Further, using the conversion relations $\mathbf{e} = \mathbf{F}^{-\top}\mathbf{e}_{\mathrm{R}}$ and $\mathbf{d} = J^{-1}\mathbf{F}\mathbf{d}_{\mathrm{R}}$, we obtain that the spatial electric displacement is related to the spatial electric field by

$$\mathbf{d} = \varepsilon\mathbf{e}. \tag{20.7.9}$$

20.7.1 **Stress**

From (20.7.6) and $(20.6.5)_1$ it follows that the second Piola stress is given by,

$$\begin{aligned} \mathbf{T}_{\mathrm{RR}} &= \mathbf{T}_{\mathrm{RR}}^{\mathrm{mech}} + \mathbf{T}_{\mathrm{RR}}^{\mathrm{es}}, \quad \text{where} \\ \mathbf{T}_{\mathrm{RR}}^{\mathrm{mech}} &\stackrel{\text{def}}{=} 2\,\frac{\partial \psi_{\mathrm{R}}^{\mathrm{mech}}(I_1, I_2, I_3)}{\partial \mathbf{C}}, \quad \text{and} \\ \mathbf{T}_{\mathrm{RR}}^{\mathrm{es}} &\stackrel{\text{def}}{=} 2\,\frac{\partial \psi_{\mathrm{R}}^{\mathrm{es}}(\mathbf{C}, \mathbf{d}_{\mathrm{R}})}{\partial \mathbf{C}}. \end{aligned} \tag{20.7.10}$$

Here, $\mathbf{T}_{\mathrm{RR}}^{\mathrm{mech}}$ denotes the stress due to mechanical deformation, and $\mathbf{T}_{\mathrm{RR}}^{\mathrm{es}}$ denotes the stress due to electrostatic effects.

From (20.7.6) and $(20.7.10)_3$

$$\mathbf{T}_{\mathrm{RR}}^{\mathrm{es}} = \frac{1}{\varepsilon}\frac{\partial}{\partial \mathbf{C}}\Big(J^{-1}(\mathbf{d}_{\mathrm{R}} \cdot \mathbf{C}\mathbf{d}_{\mathrm{R}})\Big). \tag{20.7.11}$$

Consider the derivative

$$\frac{\partial}{\partial \mathbf{C}}\Big(J^{-1}(\mathbf{d}_{\mathrm{R}} \cdot \mathbf{C}\mathbf{d}_{\mathrm{R}})\Big) = \frac{\partial J^{-1}}{\partial \mathbf{C}}(\mathbf{d}_{\mathrm{R}} \cdot \mathbf{C}\mathbf{d}_{\mathrm{R}}) + J^{-1}\frac{\partial(\mathbf{d}_{\mathrm{R}} \cdot \mathbf{C}\mathbf{d}_{\mathrm{R}})}{\partial \mathbf{C}}.$$

Using

$$\frac{\partial J^{-1}}{\partial \mathbf{C}} = -\tfrac{1}{2} J^{-1}\mathbf{C}^{-1},$$

and

$$\frac{\partial\, (\mathbf{d}_{\mathrm{R}} \cdot \mathbf{C}\mathbf{d}_{\mathrm{R}})}{\partial \mathbf{C}} = \mathbf{d}_{\mathrm{R}} \otimes \mathbf{d}_{\mathrm{R}},$$

we obtain

$$\begin{aligned}\frac{\partial}{\partial \mathbf{C}}\Big(J^{-1}\,(\mathbf{d}_{\mathrm{R}} \cdot \mathbf{C}\mathbf{d}_{\mathrm{R}})\Big) &= -\tfrac{1}{2} J^{-1}\mathbf{C}^{-1}\,(\mathbf{d}_{\mathrm{R}} \cdot \mathbf{C}\mathbf{d}_{\mathrm{R}}) + J^{-1}\mathbf{d}_{\mathrm{R}} \otimes \mathbf{d}_{\mathrm{R}},\\ &= J^{-1}\Big(\mathbf{d}_{\mathrm{R}} \otimes \mathbf{d}_{\mathrm{R}} - \tfrac{1}{2}\,(\mathbf{d}_{\mathrm{R}} \cdot \mathbf{C}\mathbf{d}_{\mathrm{R}})\,\mathbf{C}^{-1}\Big),\end{aligned}$$

which upon use in (20.7.11) gives

$$\mathbf{T}^{\mathrm{es}}_{\mathrm{RR}} = \frac{1}{\varepsilon}\, J^{-1}\Big(\mathbf{d}_{\mathrm{R}} \otimes \mathbf{d}_{\mathrm{R}} - \tfrac{1}{2}\,(\mathbf{d}_{\mathrm{R}} \cdot \mathbf{C}\mathbf{d}_{\mathrm{R}})\,\mathbf{C}^{-1}\Big). \tag{20.7.12}$$

Thus the total second Piola stress is given by

$$\mathbf{T}_{\mathrm{RR}} = 2\,\frac{\partial \psi_{\mathrm{R}}^{\mathrm{mech}}(I_1, I_2, I_3)}{\partial \mathbf{C}} + \frac{1}{\varepsilon}\, J^{-1}\Big(\mathbf{d}_{\mathrm{R}} \otimes \mathbf{d}_{\mathrm{R}} - \tfrac{1}{2}\,(\mathbf{d}_{\mathrm{R}} \cdot \mathbf{C}\mathbf{d}_{\mathrm{R}})\,\mathbf{C}^{-1}\Big). \tag{20.7.13}$$

Further, using $\mathbf{T} = J^{-1}\mathbf{F}\mathbf{T}_{\mathrm{RR}}\mathbf{F}^{\top}$, $\mathbf{d}_{\mathrm{R}} = \varepsilon\, J\mathbf{C}^{-1}\mathbf{e}_{\mathrm{R}}$, and $\mathbf{e}_{\mathrm{R}} = \mathbf{F}^{\top}\mathbf{e}$ the total Cauchy stress is given by

$$\mathbf{T} = \underbrace{2\,J^{-1}\mathbf{F}\frac{\partial \psi_{\mathrm{R}}^{\mathrm{mech}}(I_1, I_2, I_3)}{\partial \mathbf{C}}\mathbf{F}^{\top}}_{\mathbf{T}^{\mathrm{mech}}} + \underbrace{\varepsilon\,\Big[\mathbf{e} \otimes \mathbf{e} - \tfrac{1}{2}(\mathbf{e} \cdot \mathbf{e})\mathbf{1}\Big]}_{\mathbf{T}^{\mathrm{maxw}}}. \tag{20.7.14}$$

The second term in (20.7.14) is the usual form of the **Maxwell stress**.

As is standard in the theory of large-deformation isotropic elasticity, the mechanical contribution to the Cauchy stress may alternatively be written as (cf., Gurtin et al., 2010, eq. (50.29))

$$\mathbf{T} = \frac{2}{\sqrt{I_3}}\left(\frac{\partial \psi_{\mathrm{R}}^{\mathrm{mech}}(\mathcal{I}_{\mathbf{B}})}{\partial \mathbf{B}}\right)\mathbf{B} + \varepsilon\,\Big[\mathbf{e} \otimes \mathbf{e} - \tfrac{1}{2}(\mathbf{e} \cdot \mathbf{e})\mathbf{1}\Big], \tag{20.7.15}$$

with $\mathcal{I}_{\mathbf{B}}$ the list of principal invariants of $\mathbf{B}$,

$$\mathcal{I}_{\mathbf{B}} \stackrel{\mathrm{def}}{=} (I_1(\mathbf{B}), I_2(\mathbf{B}), I_3(\mathbf{B})). \tag{20.7.16}$$

20.7.2 **Incompressible materials**

Since many elastomeric materials are essentially incompressible, we next consider an incompressible version of the theory considered so far.

An *incompressible body* is one for which only isochoric motions are possible, a requirement manifested in the *constraint*

$$J = \det \mathbf{F} \equiv 1, \tag{20.7.17}$$

in all motions of the body, so that in any such motion, each part $\mathcal{P}$ deforms to a spatial region $\mathcal{P}_t$ with the same volume. An immediate consequence is then that

$$\det \mathbf{C} \equiv 1. \tag{20.7.18}$$

Thus, since

$$\overline{\det \mathbf{C}}^{\,\cdot} = (\det \mathbf{C})\,\mathrm{tr}\,(\dot{\mathbf{C}}\mathbf{C}^{-1}) = (\det \mathbf{C})\mathbf{C}^{-1}\!:\!\dot{\mathbf{C}},$$

for an incompressible body the constraint $\det \mathbf{C} = 1$ is equivalent to

$$\mathbf{C}^{-1}\!:\!\dot{\mathbf{C}} = 0. \tag{20.7.19}$$

Recall that the stress power per unit reference volume for a compressible body is $(1/2)\mathbf{T}_{\text{RR}}:\dot{\mathbf{C}}$, and an incompressible body must satisfy the constraint (20.7.19). Thus, the stress power per unit reference volume for an incompressible body may be written as

$$\text{stress power} = \frac{1}{2}\mathbf{T}_{\text{RR}}\!:\!\dot{\mathbf{C}} + \underbrace{\frac{1}{2}p\,\mathbf{C}^{-1}\!:\!\dot{\mathbf{C}}}_{=0} = \frac{1}{2}\Big(\mathbf{T}_{\text{RR}} + p\mathbf{C}^{-1}\Big):\dot{\mathbf{C}}. \tag{20.7.20}$$

where p is an "arbitrary scalar field" introduced as a Lagrange multiplier to enforce the incompressibility constraint.[4] We may use (20.7.20) to rewrite the referential free-energy imbalance (20.5.12) in the following form for incompressible materials:

$$\dot{\psi}_{\text{R}} - \tfrac{1}{2}\Big(\mathbf{T}_{\text{RR}} + p\mathbf{C}^{-1}\Big):\dot{\mathbf{C}} - \mathbf{e}_{\text{R}}\cdot\dot{\mathbf{d}}_{\text{R}} \le 0. \tag{20.7.21}$$

For incompressible materials, using the energy imbalance (20.7.21), and mimicking the derivations in Section 20.6 we obtain that the second Piola stress and the electric field are given by the state relations:

$$\left.\begin{aligned} \mathbf{T}_{\text{RR}} &= 2\frac{\partial\psi_{\text{R}}(\boldsymbol{\Lambda})}{\partial\mathbf{C}} - p\mathbf{C}^{-1},\\ \mathbf{e}_{\text{R}} &= \frac{\partial\psi_{\text{R}}(\boldsymbol{\Lambda})}{\partial\mathbf{d}_{\text{R}}}. \end{aligned}\right. \tag{20.7.22}$$

[4] The factor $(1/2)$ preceding the term involving p is inconsequential, and is introduced for future convenience.

For incompressible isotropic ideal dielectric materials, $I_3 = 1$, and the free energy (20.7.6) reduces to a separable free energy of the form

$$\psi_{\rm R} = \psi_{\rm R}^{\rm mech}(\tilde{\mathcal{I}}_{\mathbf{B}}) + \underbrace{J^{-1}\left(\frac{1}{2\varepsilon}\mathbf{d}_{\rm R}\cdot\mathbf{C}\mathbf{d}_{\rm R}\right)}_{\psi_{\rm R}^{\rm es}} \tag{20.7.23}$$

with $\tilde{\mathcal{I}}_{\mathbf{B}}$ the pair of principal invariants

$$\tilde{\mathcal{I}}_{\mathbf{B}} \stackrel{\rm def}{=} (I_1(\mathbf{B}), I_2(\mathbf{B})). \tag{20.7.24}$$

Then as is standard in the theory of large-deformation isotropic elasticity for incompressible materials, the mechanical contribution to the Cauchy stress may alternatively be written as (cf., Gurtin et al., 2010, eq. (54.20))

$$\mathbf{T} = -p\mathbf{1} + \beta_1(\tilde{\mathcal{I}}_{\mathbf{B}})\mathbf{B} + \beta_2(\tilde{\mathcal{I}}_{\mathbf{B}})\mathbf{B}^{-1} + \varepsilon\,\mathbf{e}\otimes\mathbf{e}, \tag{20.7.25}$$

with

$$\beta_1(\tilde{\mathcal{I}}_{\mathbf{B}}) = 2\frac{\partial\psi_{\rm R}^{\rm mech}(\tilde{\mathcal{I}}_{\mathbf{B}})}{\partial I_1}, \qquad \beta_2(\tilde{\mathcal{I}}_{\mathbf{B}}) = -2\frac{\partial\psi_{\rm R}^{\rm mech}(\tilde{\mathcal{I}}_{\mathbf{B}})}{\partial I_2}. \tag{20.7.26}$$

Note that the factor

$$\varepsilon\left(-\tfrac{1}{2}(\mathbf{e}\cdot\mathbf{e})\mathbf{1}\right)$$

which appears in the Maxwell stress term in (20.7.14) has been absorbed in the term $-p\mathbf{1}$ in equation (20.7.25).

20.7.3 **Gent elasticity**

Next, as a special case for the mechanical free energy we consider a free energy of the form (Gent, 1996)

$$\psi_{\rm R} = -\tfrac{1}{2}\,G_0\,I_m \ln\left(1 - \frac{I_1 - 3}{I_m}\right), \tag{20.7.27}$$

with G_0 a ground-state shear modulus, and I_m a material parameter which represents the upper limit of $(I_1 - 3)$ associated with the limited extensibility of the polymeric chains. In this case (20.7.25) gives

$$\mathbf{T} = -p\mathbf{1} + G\mathbf{B} + \varepsilon\,\mathbf{e}\otimes\mathbf{e} \qquad \text{with} \qquad G = G_0\left(\frac{I_m}{I_m - (I_1 - 3)}\right). \tag{20.7.28}$$

The spatial electric displacement is still given by (20.7.9), viz.

$$\mathbf{d} = \varepsilon \mathbf{e}. \tag{20.7.29}$$

20.8 **One-dimensional actuation**

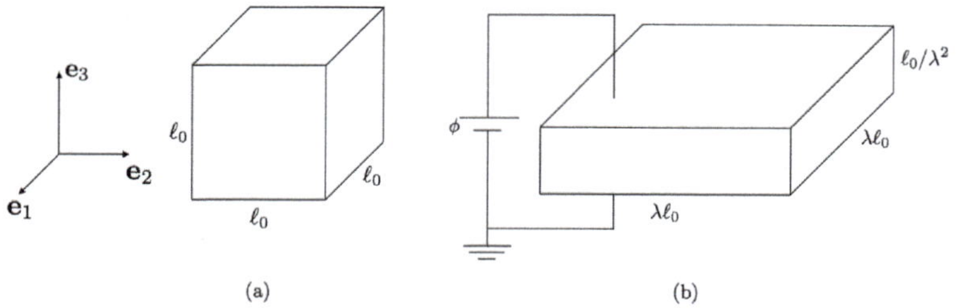

Fig. 20.1 (a) Undeformed reference configuration, a cube of edge length ℓ_0. (b) The cube is subjected to a potential difference ϕ across its top and bottom faces so that it deforms into a rectangular parallelepiped with equi-biaxial stretches λ in the $(\mathrm{e}_1, \mathrm{e}_2)$-plane and a deformed length of ℓ_0/λ^2 in the e_3-direction. The magnitude of the electric field in the e_3-direction in the deformed configuration is $\phi/(\ell_0/\lambda^2)$.

To fix ideas, we consider a dielectric elastomer in the form of a unit cube of initial edge length ℓ_0; cf. Fig. 20.1. The top and bottom faces of the cube are coated with compliant electrodes (not shown) across which an electrical voltage ϕ is applied. Assume that the cube deforms homogeneously into edge lengths $(\lambda\ell_0, \lambda\ell_0, \ell_0/\lambda^2)$ so that

$$[\mathbf{F}] = \begin{bmatrix} \lambda & 0 & 0 \\ 0 & \lambda & 0 \\ 0 & 0 & \lambda^{-2} \end{bmatrix}, \qquad [\mathbf{B}] = \begin{bmatrix} \lambda^2 & 0 & 0 \\ 0 & \lambda^2 & 0 \\ 0 & 0 & \lambda^{-4} \end{bmatrix}, \qquad \operatorname{tr}\mathbf{B} = 2\lambda^2 + \lambda^{-4}, \tag{20.8.1}$$

and that the electric field in the deformed body is also uniform and given by

$$[\mathbf{e}] = \begin{bmatrix} 0 \\ 0 \\ \phi/(\ell_0/\lambda^2) \end{bmatrix}. \tag{20.8.2}$$

Thus, using (20.7.28), the relevant components of the Cauchy stress are,

$$\begin{aligned} T_{11} = T_{22} &= -p + G\lambda^2, \\ T_{33} &= -p + G\lambda^{-4} + \varepsilon\left(\frac{\phi}{\ell_0}\right)^2 \lambda^4, \end{aligned} \tag{20.8.3}$$

with

$$G = G_0\left(\frac{I_m}{I_m - (2\lambda^2 + \lambda^{-4} - 3)}\right). \tag{20.8.4}$$

Note that in this example $\mathbf{T} = \mathbf{0}$, so (20.8.3) gives the pair of equations

$$\begin{aligned} 0 &= -p + G\,\lambda^2, \\ 0 &= -p + G\,\lambda^{-4} + \varepsilon \left(\frac{\phi}{\ell_0}\right)^2 \lambda^4. \end{aligned} \tag{20.8.5}$$

From the first of (20.8.5) we obtain

$$p = G\,\lambda^2,$$

substitution of which in $(20.8.5)_2$ and using (20.8.4) gives

$$0 = \underbrace{-G_0 \left(\frac{I_m}{I_m - (2\lambda^2 + \lambda^{-4} - 3)}\right)(\lambda^{-2} - \lambda^{-8})}_{\text{mechanical stress}} + \underbrace{\varepsilon \left(\frac{\phi}{\ell_0}\right)^2}_{\text{Maxwell stress}}, \tag{20.8.6}$$

or

$$\frac{\phi/\ell_0}{\sqrt{G_0/\varepsilon}} = \sqrt{\left(\frac{I_m}{I_m - (2\lambda^2 + \lambda^{-4} - 3)}\right)(\lambda^{-2} - \lambda^{-8})}. \tag{20.8.7}$$

Figure 20.2(a) shows a plot of the normalized potential $(\phi/\ell_0)/\sqrt{G_0/\varepsilon}$ versus the stretch λ for four different values of $I_m = 7$, 30, 60, and 10^6, with $I_m = 10^6$ approximating a Neo-Hookean material. Note that for $I_m \gtrsim 7$, such a curve exhibits a peak as λ is increased. This peak will result in an *electromechanical instability* under voltage control.

For example, Figure 20.2(b) shows plot normalized potential $(\phi/\ell_0)/\sqrt{G_0/\varepsilon}$ versus λ for $I_m = 60$. This material will exhibit a **snap-through instability** under voltage control. While no snap-through instability will occur for materials with $I_m \lesssim 7$.

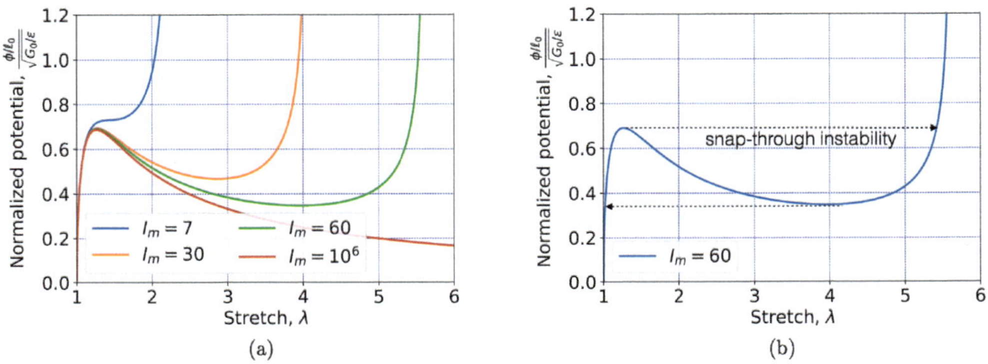

Fig. 20.2 (a) Normalized potential versus stretch for a Gent ideal dielectric material with different values of I_m. (b) Snap-through instability for materials with $I_m \gtrsim 7$.

20.9 **One-dimensional actuation in plane strain**

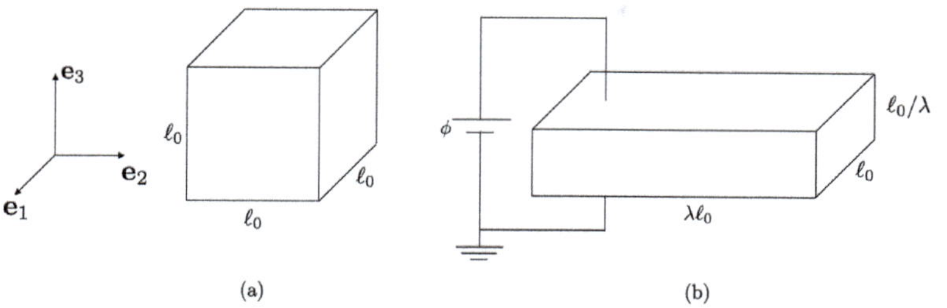

Fig. 20.3 (a) Undeformed reference configuration, a cube of edge length ℓ_0. (b) Under plane-strain conditions in the $\mathbf{e}_1$-direction, the bottom and top faces are subject to a potential difference ϕ so that the deformed length in the $\mathbf{e}_3$-direction is given by ℓ_0/λ and the electric field in the $\mathbf{e}_3$-direction is $\phi/(\ell_0/\lambda)$.

Now consider a dielectric elastomer in the form of a unit cube of initial edge length ℓ_0 as shown in Fig. 20.3. An electrical voltage ϕ is applied across the top and bottom faces of the cube under plane-strain conditions in the $\mathbf{e}_1$-direction. Assume that the cube deforms homogeneously into edge lengths $(\ell_0, \lambda\ell_0, \ell_0/\lambda)$ so that

$$[\mathbf{F}] = \begin{bmatrix} 1 & 0 & 0 \\ 0 & \lambda & 0 \\ 0 & 0 & \lambda^{-1} \end{bmatrix}, \qquad [\mathbf{B}] = \begin{bmatrix} 1 & 0 & 0 \\ 0 & \lambda^2 & 0 \\ 0 & 0 & \lambda^{-2} \end{bmatrix}, \qquad \operatorname{tr}\mathbf{B} = 1 + \lambda^2 + \lambda^{-2}, \tag{20.9.1}$$

and that the electric field in the deformed body is also uniform and given by

$$[\mathbf{e}] = \begin{bmatrix} 0 \\ 0 \\ \phi/(\ell_0/\lambda) \end{bmatrix}. \tag{20.9.2}$$

Thus, using (20.7.28), the relevant components of the Cauchy stress are

$$\begin{aligned} T_{11} &= -p + G, \\ T_{22} &= -p + G\,\lambda^2, \\ T_{33} &= -p + G\,\lambda^{-2} + \varepsilon \left(\frac{\phi}{\ell_0}\right)^2 \lambda^2. \end{aligned} \tag{20.9.3}$$

with

$$G = G_0 \left(\frac{I_m}{I_m - (\lambda^2 + \lambda^{-2} - 2)} \right). \tag{20.9.4}$$

Note that in this example $T_{22} = T_{33} = 0$, so (20.9.3) gives

$$\begin{aligned} 0 &= -p + G\,\lambda^2, \\ 0 &= -p + G\,\lambda^{-2} + \varepsilon \left(\frac{\phi}{\ell_0}\right)^2 \lambda^2. \end{aligned} \tag{20.9.5}$$

From the first of (20.9.5) we obtain

$$p = G\lambda^2,$$

substitution of which in $(20.9.5)_2$ and using (20.9.4) gives

$$0 = \underbrace{-G_0\left(\frac{I_m}{I_m - (\lambda^2 + \lambda^{-2} - 2)}\right)(1 - \lambda^{-4})}_{\text{mechanical stress}} + \underbrace{\varepsilon\left(\frac{\phi}{\ell_0}\right)^2}_{\text{Maxwell stress}}, \tag{20.9.6}$$

or

$$\frac{\phi/\ell_0}{\sqrt{G_0/\varepsilon}} = \sqrt{\left(\frac{I_m}{I_m - (\lambda^2 + \lambda^{-2} - 2)}\right)(1 - \lambda^{-4})}. \tag{20.9.7}$$

Figure 20.4 shows a plot of the normalized potential $(\phi/\ell_0)/\sqrt{G_0/\varepsilon}$ versus the stretch λ for four different values of $I_m = 30$, 60, 90, and 10^6, with $I_m = 10^6$ approximating a Neo-Hookean material. In plane strain, materials with $I_m \lessapprox 90$ will **not** exhibit a snap-through instability. This is *very different* from the 3D case studied in Section 20.8.

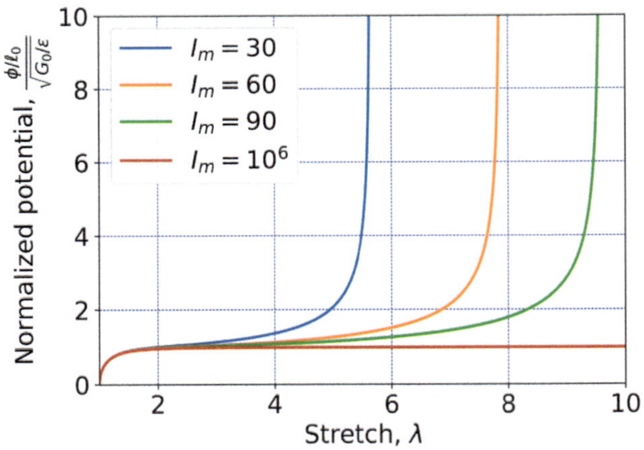

Fig. 20.4 Normalized potential versus stretch for a Gent ideal dielectric material with different values of I_m.

20.10 A sufficiently large pre-stretch can suppress a snap-through instability

In this example we revisit the example of one-dimensional actuation in Section 20.8, but consider the case in which the cube has been biaxially pre-stretched by the application of tractions on its lateral faces before the application of the electric potential across its top and bottom faces. We shall show that a sufficiently large pre-stretch can suppress a snap-through type instability discussed in Section 20.8.

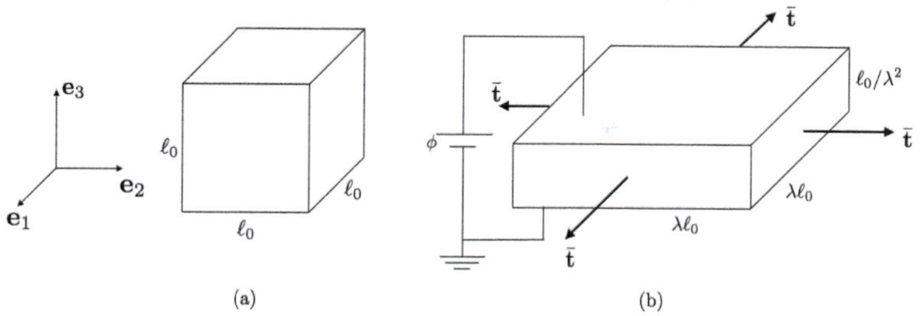

Fig. 20.5 (a) Undeformed reference configuration, a cube of edge length ℓ_0. (b) The cube is subjected to an in-plane biaxial traction $\bar{\mathrm{t}}$ and potential difference ϕ across its top and bottom faces so that it deforms into a rectangular parallelepiped with equi-biaxial stretches λ in the $(\mathrm{e}_1, \mathrm{e}_2)$-plane and a deformed length of ℓ_0/λ^2 in the e_3-direction. The traction acts to impart a pre-stretch while the magnitude of the electric field in the e_3-direction in the deformed configuration is $\phi/(\ell_0/\lambda^2)$ serves to actuate.

Thus, consider a dielectric elastomer in the form of a unit cube of initial edge length ℓ_0 whose top and bottom faces are coated with compliant electrodes (not shown). The lateral faces of the cube have equal tractions $\bar{\mathrm{t}}$ applied to create a pre-stretch, and subsequently an electrical voltage ϕ is applied across the top and bottom faces of the cube. Assume that in its final deformed configuration, Fig. 20.5, the cube has been homogeneously deformed with edge lengths $(\lambda\ell_0, \lambda\ell_0, \ell_0/\lambda^2)$, so that

$$[\mathbf{F}] = \begin{bmatrix} \lambda & 0 & 0 \\ 0 & \lambda & 0 \\ 0 & 0 & \lambda^{-2} \end{bmatrix}, \qquad [\mathbf{B}] = \begin{bmatrix} \lambda^2 & 0 & 0 \\ 0 & \lambda^2 & 0 \\ 0 & 0 & \lambda^{-4} \end{bmatrix}, \qquad \operatorname{tr}\mathbf{B} = 2\lambda^2 + \lambda^{-4}, \tag{20.10.1}$$

and that the electric field in the deformed body is also uniform and given by

$$[\mathbf{e}] = \begin{bmatrix} 0 \\ 0 \\ \phi/(\ell_0/\lambda^2) \end{bmatrix}. \tag{20.10.2}$$

Thus, using (20.7.28), the relevant components of the Cauchy stress are,

$$\begin{aligned} T_{11} = T_{22} &= -p + G\lambda^2, \\ T_{33} &= -p + G\lambda^{-4} + \varepsilon\left(\frac{\phi}{\ell_0}\right)^2 \lambda^4, \end{aligned} \tag{20.10.3}$$

with

$$G = G_0\left(\frac{I_m}{I_m - (2\lambda^2 + \lambda^{-4} - 3)}\right). \tag{20.10.4}$$

Note that in this example $T_{11} = T_{22} = \bar{t}$ and $T_{33} = 0$, so (20.10.3) gives the pair of equations

$$\begin{aligned} \bar{t} &= -p + G\,\lambda^2, \\ 0 &= -p + G\,\lambda^{-4} + \varepsilon \left(\frac{\phi}{\ell_0}\right)^2 \lambda^4. \end{aligned} \tag{20.10.5}$$

From the first of (20.10.5) we obtain

$$p = G\,\lambda^2 - \bar{t},$$

substitution of which in $(20.10.5)_2$ and using (20.10.4) gives

$$0 = \underbrace{-G_0 \left(\frac{I_m}{I_m - (2\lambda^2 + \lambda^{-4} - 3)}\right)(\lambda^{-2} - \lambda^{-8}) + \bar{t}\lambda^{-4}}_{\text{mechanical stress}} + \underbrace{\varepsilon \left(\frac{\phi}{\ell_0}\right)^2}_{\text{Maxwell stress}}. \tag{20.10.6}$$

If one considers for the moment the case of no applied potential, then the magnitude of the traction $\bar{t}$ in order to obtain a desired value of the biaxial pre-stretch λ_{pre} may be obtained by solving,

$$\frac{\bar{t}}{G_0} = \frac{I_m}{I_m - (2\lambda_{\text{pre}}^2 + \lambda_{\text{pre}}^{-4} - 3)} \left(\lambda_{\text{pre}}^2 - \lambda_{\text{pre}}^{-4}\right). \tag{20.10.7}$$

Therefore, under the combined action of an applied biaxial traction followed by the application of a potential difference, one obtains a relation between the deformation and potential in the form

$$\frac{\phi/\ell_0}{\sqrt{G_0/\varepsilon}} = \sqrt{\left(\frac{I_m}{I_m - (2\lambda^2 + \lambda^{-4} - 3)}\right)(\lambda^{-2} - \lambda^{-8}) - \left(\frac{\bar{t}}{G_0}\right)\lambda^{-4}}. \tag{20.10.8}$$

with $\bar{t}$ given by (20.10.7).

Figure 20.6 shows a plot of the normalized potential $(\phi/\ell_0)/\sqrt{G_0/\varepsilon}$ versus the stretch λ for the case of a pre-stretch of $\lambda_{\text{pre}} = 4$ (solid lines), as well as no pre-stretch (dashed lines), for three different values of $I_m = 60$, 90, and 10^6, with $I_m = 10^6$ approximating a Neo-Hookean material. These results indicate that

- by applying a suitably large biaxial pre-stretch and depending on the value of I_m, a snap-through type instability may be suppressed.

This phenomenon is exploited in the design of actuators made from dielectric materials (Pelrine et al., 2000).

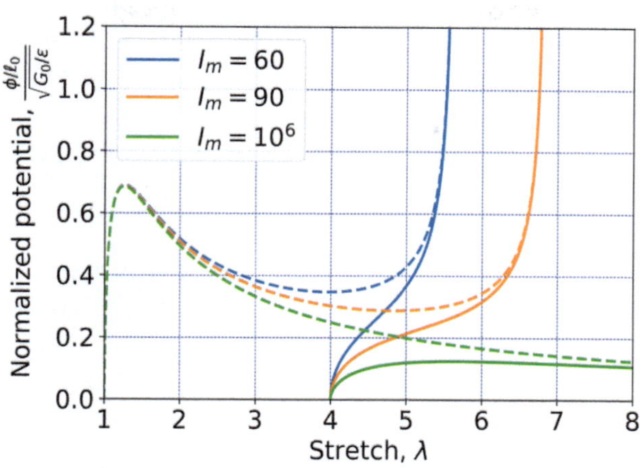

Fig. 20.6 Normalized potential versus stretch for a Gent ideal dielectric material under a biaxial pre-stretch $\lambda_{\text{pre}} = 4$ for different values of I_m. For comparison, the dashed lines are the results from Fig. 20.2 where no pre-stretch is applied.

20.11 **Nearly incompressible materials**

In this section we revert to the more general compressible case, but consider the case when *the compressibility is small.* We begin by decomposing the deformation gradient $\mathbf{F}$ into volumetric and isochoric factors $\mathbf{F}^{\text{vol}}$ and $\bar{\mathbf{F}}$:

$$\mathbf{F} = \mathbf{F}^{\text{vol}}\bar{\mathbf{F}}, \tag{20.11.1}$$

where

$$\mathbf{F}^{\text{vol}} = J^{1/3}\mathbf{1} \qquad \text{and} \qquad \bar{\mathbf{F}} = J^{-1/3}\mathbf{F}. \tag{20.11.2}$$

This leads to a multiplicative decomposition of the left Cauchy–Green tensor in the form

$$\mathbf{C} = \mathbf{C}^{\text{vol}}\bar{\mathbf{C}}, \qquad \mathbf{C}^{\text{vol}} = J^{2/3}\mathbf{1}, \qquad \bar{\mathbf{C}} = \bar{\mathbf{F}}^{\top}\bar{\mathbf{F}}. \tag{20.11.3}$$

We also introduce the invariants,

$$\bar{I}_1 \stackrel{\text{def}}{=} \operatorname{tr}\bar{\mathbf{C}} \qquad \text{and} \qquad \bar{I}_2 \stackrel{\text{def}}{=} \tfrac{1}{2}\left[(\operatorname{tr}\bar{\mathbf{C}})^2 - \operatorname{tr}(\bar{\mathbf{C}}^2)\right]. \tag{20.11.4}$$

For nearly incompressible isotropic ideal dielectric materials, separable free energy (20.7.6) has the form

$$\psi_{\text{R}} = \psi_{\text{R}}^{\text{mech}}(\bar{I}_1, \bar{I}_2, J) + J^{-1}\left(\frac{1}{2\varepsilon}\mathbf{d}_{\text{R}} \cdot \mathbf{C}\mathbf{d}_{\text{R}}\right). \tag{20.11.5}$$

As a special case for the mechanical energy we consider a free energy of the form (Gent, 1996)

$$\psi_{\mathrm{R}}^{\mathrm{mech}} = -\tfrac{1}{2}\, G_0\, I_m \ln\left(1 - \frac{\bar{I}_1 - 3}{I_m}\right) + \tfrac{1}{2} K (\ln J)^2, \tag{20.11.6}$$

with G_0 and K ground-state shear and bulk moduli, respectively, and I_m a material parameter which represents the upper limit of $(\bar{I}_1 - 3)$ associated with the limited extensibility of the polymeric chains.

Thus, the total free energy is taken as

$$\psi_{\mathrm{R}} = -\tfrac{1}{2}\, G_0\, I_m \ln\left(1 - \frac{\bar{I}_1 - 3}{I_m}\right) + \tfrac{1}{2} K (\ln J)^2 + J^{-1}\left(\frac{1}{2\varepsilon}\mathbf{d}_{\mathrm{R}} \cdot \mathbf{C}\mathbf{d}_{\mathrm{R}}\right). \tag{20.11.7}$$

Then, from $(20.6.5)_1$ and (20.11.7) it follows that the second Piola stress is given by,

$$\begin{aligned} \mathbf{T}_{\mathrm{RR}} &= \mathbf{T}_{\mathrm{RR}}^{\mathrm{mech}} + \mathbf{T}_{\mathrm{RR}}^{\mathrm{es}}, \quad \text{where} \\ \mathbf{T}_{\mathrm{RR}}^{\mathrm{mech}} &\stackrel{\text{def}}{=} 2\,\frac{\partial \psi_{\mathrm{R}}^{\mathrm{mech}}(\bar{I}_1, J)}{\partial \mathbf{C}}, \quad \text{and} \\ \mathbf{T}_{\mathrm{RR}}^{\mathrm{es}} &\stackrel{\text{def}}{=} 2\,\frac{\partial \psi_{\mathrm{R}}^{\mathrm{es}}(\mathbf{C}, \mathbf{e}_{\mathrm{R}})}{\partial \mathbf{C}}. \end{aligned} \tag{20.11.8}$$

Then, straightforward calculations mimicking those in Section 1.9 give,

$$\mathbf{T}_{\mathrm{RR}}^{\mathrm{mech}} = \frac{2}{J^{2/3}}\left[\frac{\partial \psi_{\mathrm{R}}^{\mathrm{mech}}(\bar{I}_1, J)}{\partial \bar{\mathbf{C}}} - \frac{1}{3}\left(\bar{\mathbf{C}} : \frac{\partial \psi_{\mathrm{R}}^{\mathrm{mech}}(\bar{I}_1, J)}{\partial \bar{\mathbf{C}}}\right)\bar{\mathbf{C}}^{-1}\right] + J\frac{\partial \psi_{\mathrm{R}}^{\mathrm{mech}}(\bar{I}_1, J)}{\partial J}\mathbf{C}^{-1}, \tag{20.11.9}$$

$$= \frac{2}{J^{2/3}}\frac{\partial \psi_{\mathrm{R}}^{\mathrm{mech}}(\bar{I}_1, J)}{\partial \bar{I}_1}\left[\mathbf{1} - \frac{1}{3}(\mathrm{tr}\,\bar{\mathbf{C}})\bar{\mathbf{C}}^{-1}\right] + K(\ln J)\mathbf{C}^{-1}, \tag{20.11.10}$$

$$= J^{-2/3} G_0 \left(1 - \frac{\bar{I}_1 - 3}{I_m}\right)^{-1}\left[\mathbf{1} - \frac{1}{3}(\mathrm{tr}\,\bar{\mathbf{C}})\bar{\mathbf{C}}^{-1}\right] + K(\ln J)\mathbf{C}^{-1}. \tag{20.11.11}$$

Further, using $\mathbf{T} = J^{-1}\mathbf{F}\mathbf{T}_{\mathrm{RR}}\mathbf{F}^{\top}$ we find that the mechanical part of the Cauchy stress tensor is given by

$$\mathbf{T}^{\mathrm{mech}} = J^{-1}\left[G_0\left(1-\frac{\bar{I}_1-3}{I_m}\right)^{-1}\bar{\mathbf{B}}_0 + K(\ln J)\mathbf{1}\right]. \tag{20.11.12}$$

Hence, from (20.11.12) and (20.7.14) the total Cauchy stress is

$$\begin{aligned}
\mathbf{T} &= \mathbf{T}^{\mathrm{mech}} + \mathbf{T}^{\mathrm{maxw}},\\
\mathbf{T}^{\mathrm{mech}} &= J^{-1}G\bar{\mathbf{B}}_0 - p\mathbf{1} \quad \text{with} \quad G \stackrel{\text{def}}{=} G_0\left(1-\frac{\bar{I}_1-3}{I_m}\right)^{-1},\\
p &\stackrel{\text{def}}{=} -K\frac{\ln J}{J},\\
\mathbf{T}^{\mathrm{maxw}} &= \varepsilon\left[\mathbf{e}\otimes\mathbf{e} - \tfrac{1}{2}(\mathbf{e}\cdot\mathbf{e})\mathbf{1}\right].
\end{aligned} \tag{20.11.13}$$

The spatial electric displacement is still given by (20.7.9), viz.

$$\mathbf{d} = \varepsilon\mathbf{e}. \tag{20.11.14}$$

Finally, since the Piola stress is given by $\mathbf{T}_{\mathrm{R}} = J\mathbf{T}\mathbf{F}^{-\top}$, the referential electric field by $\mathbf{e}_{\mathrm{R}} = \mathbf{F}^{\top}\mathbf{e}$, and referential electric displacement by $\mathbf{d}_{\mathrm{R}} = J\mathbf{F}^{-1}\mathbf{d}$, using (20.11.13) and (20.11.14) we obtain

$$\begin{aligned}
\mathbf{T}_{\mathrm{R}} &= J^{-2/3}G\left(\mathbf{F} - \frac{1}{3}(\operatorname{tr}\mathbf{C})\mathbf{F}^{-\top}\right) - Jp\mathbf{F}^{-\top}\\
&\quad + \varepsilon J\left[(\mathbf{F}^{-\top}\mathbf{e}_{\mathrm{R}})\otimes(\mathbf{C}^{-1}\mathbf{e}_{\mathrm{R}}) - \tfrac{1}{2}(\mathbf{e}_{\mathrm{R}}\cdot\mathbf{C}^{-1}\mathbf{e}_{\mathrm{R}})\mathbf{F}^{-\top}\right],\\
G &\stackrel{\text{def}}{=} G_0\left(1-\frac{\bar{I}_1-3}{I_m}\right)^{-1}, \qquad p \stackrel{\text{def}}{=} -K\frac{\ln J}{J},
\end{aligned} \tag{20.11.15}$$

and

$$\mathbf{d}_{\mathrm{R}} = \varepsilon J\mathbf{C}^{-1}\mathbf{e}_{\mathrm{R}}. \tag{20.11.16}$$

20.12 Governing partial differential equations

1. The governing partial differential equation for the mechanical problem in the absence of body forces and neglect of inertia is the force balance

$$\mathrm{Div}\,\mathbf{T}_{\mathrm{R}} = \mathbf{0}, \tag{20.12.1}$$

where the Piola stress $\mathbf{T}_{\mathrm{R}}$ is given by

$$\begin{aligned}
\mathbf{T}_{\mathrm{R}} &= J\,\mathbf{T}\mathbf{F}^{-\top}\\
\mathbf{T} &= \mathbf{T}^{\mathrm{mech}} + \mathbf{T}^{\mathrm{maxw}},\\
\mathbf{T}^{\mathrm{mech}} &= J^{-1}G\bar{\mathbf{B}}_0 - p\mathbf{1}, \qquad G \overset{\mathrm{def}}{=} G_0\left(1 - \frac{\bar{I}_1 - 3}{I_m}\right)^{-1}, \qquad p \overset{\mathrm{def}}{=} -K\frac{\ln J}{J},\\
\mathbf{T}^{\mathrm{maxw}} &= \varepsilon\left[\mathbf{e}\otimes\mathbf{e} - \tfrac{1}{2}(\mathbf{e}\cdot\mathbf{e})\mathbf{1}\right].
\end{aligned} \tag{20.12.2}$$

With $\partial\mathcal{B}_{\mathbf{u}}$ and $\partial\mathcal{B}_{\mathbf{t}_{\mathrm{R}}}$ denoting complementary subsurfaces of the boundary $\partial\mathcal{B}$ of the body $\mathcal{B}$, we consider a pair of possible boundary conditions:

$$\begin{aligned}
\mathbf{u} &= \bar{\mathbf{u}} && \text{on } \partial\mathcal{B}_{\mathbf{u}},\\
\mathbf{T}_{\mathrm{R}}\mathbf{n}_{\mathrm{R}} &= \bar{\mathbf{t}}_{\mathrm{R}} && \text{on } \partial\mathcal{B}_{\mathbf{t}_{\mathrm{R}}}.
\end{aligned} \tag{20.12.3}$$

In the boundary conditions above $\bar{\mathbf{u}}$ and $\bar{\mathbf{t}}_{\mathrm{R}}$ are prescribed displacement and tractions, both functions of $(\mathbf{X}, t)$.

2. The governing differential equation for electrostatics is Gauss's law,

$$\mathrm{Div}\,\mathbf{d}_{\mathrm{R}} = 0, \tag{20.12.4}$$

where the referential electric displacement $\mathbf{d}_{\mathrm{R}}$ is given by

$$\mathbf{d}_{\mathrm{R}} = \varepsilon J\mathbf{C}^{-1}\mathbf{e}_{\mathrm{R}}. \tag{20.12.5}$$

With $\partial\mathcal{B}_{\phi}$ and $\partial\mathcal{B}_{\mathbf{d}_{\mathrm{R}}}$ denoting complementary subsurfaces of $\partial\mathcal{B}$, we consider a pair of possible boundary conditions:

$$\begin{aligned}
\phi &= \bar{\phi} && \text{on } \partial\mathcal{B}_{\phi},\\
\mathbf{d}_{\mathrm{R}}\cdot\mathbf{n}_{\mathrm{R}} &= \bar{\varpi}_{\mathrm{R}} && \text{on } \partial\mathcal{B}_{\mathbf{d}_{\mathrm{R}}}.
\end{aligned} \tag{20.12.6}$$

In the boundary conditions above $\bar{\phi}$ and $\bar{\varpi}$ are the *prescribed* electrostatic potential and referential surface charge density, respectively, which are both functions of $(\mathbf{X}, t)$.

Standard initial conditions involve the specifications

$$\mathbf{u}(\mathbf{X},0) = \mathbf{u}_0(\mathbf{X}) \qquad \text{and} \qquad \phi(\mathbf{X},0) = \phi_0(\mathbf{X}) \qquad \text{in } \mathcal{B}, \tag{20.12.7}$$

with $\mathbf{u}_0$ and ϕ_0 prescribed functions of $\mathbf{X} \in \mathcal{B}$.

REMARK

In considering the boundary conditions we have neglected the Maxwell stress and the electric displacement in free space.

21 Numerical implementation of the theory for dielectric elastomers

21.1 Weak forms of the governing partial differential equations

1. **Mechanical problem:**
   By following a procedure identical to that laid out in the prior chapters for the mechanical problem, consider a test field $\mathbf{u}_{\text{test}}(\mathbf{X})$ in $\mathcal{B}$ which satisfies $\mathbf{u}_{\text{test}}(\mathbf{X}) = \mathbf{0}$ on $\partial\mathcal{B}_{\mathbf{u}}$. If the balance equation $\text{Div}\,\mathbf{T}_{\text{R}} = \mathbf{0}$ in $\mathcal{B}$ and the traction condition $(\bar{\mathbf{t}}_{\text{R}} - \mathbf{T}_{\text{R}}\mathbf{n}_{\text{R}}) = \mathbf{0}$ on $\partial\mathcal{B}_{\mathbf{t}_{\text{R}}}$ are satisfied, then as shown in the previous chapters, the weak form

$$\int_{\mathcal{B}} \mathbf{T}_{\text{R}} \colon \nabla \mathbf{u}_{\text{test}}\, dv_{\text{R}} - \int_{\partial\mathcal{B}_{\mathbf{t}_{\text{R}}}} \bar{\mathbf{t}}_{\text{R}} \cdot \mathbf{u}_{\text{test}}\, da_{\text{R}} = 0 \tag{21.1.1}$$

   holds for all admissible test fields $\mathbf{u}_{\text{test}}$ which vanish on $\partial\mathcal{B}_{\mathbf{u}}$. Also, the weak form of the pressure equation $(20.12.2)_3$ is

$$\int_{\mathcal{B}} \left[\frac{p}{K} + \frac{\ln J}{J}\right] p_{\text{test}}\, dv_{\text{R}} = 0, \tag{21.1.2}$$

   where $p_{\text{test}}(\mathbf{X})$ is a test field in $\mathcal{B}$.

2. **Gauss's Law:**
   Next, consider a scalar test field $\phi_{\text{test}}(\mathbf{X})$ on $\mathcal{B}$ which satisfies $\phi_{\text{test}}(\mathbf{X}) = \mathbf{0}$ on $\partial\mathcal{B}_{\phi}$. If Gauss's Law

$$\text{Div}\,\mathbf{d}_{\text{R}} = 0 \qquad \text{in } \mathcal{B},$$

   and the boundary condition $\mathbf{d}_{\text{R}} \cdot \mathbf{n}_{\text{R}} = \bar{\varpi}_{\text{R}}$ on $\partial\mathcal{B}_{\mathbf{d}_{\text{R}}}$ are satisfied, then

$$\int_{\mathcal{B}} (\text{Div}\,\mathbf{d}_{\text{R}})\, \phi_{\text{test}}\, dv_{\text{R}} + \int_{\partial\mathcal{B}_{\mathbf{d}_{\text{R}}}} (\bar{\varpi}_{\text{R}} - \mathbf{d}_{\text{R}} \cdot \mathbf{n}_{\text{R}})\, \phi_{\text{test}}\, da_{\text{R}} = 0, \tag{21.1.3}$$

   or equivalently

$$\int_{\mathcal{B}} (\text{Div}\,\mathbf{d}_{\text{R}})\, \phi_{\text{test}}\, dv_{\text{R}} - \int_{\partial\mathcal{B}_{\mathbf{d}_{\text{R}}}} \mathbf{d}_{\text{R}} \cdot \mathbf{n}_{\text{R}}\, \phi_{\text{test}}\, da_{\text{R}} + \int_{\partial\mathcal{B}_{\mathbf{d}_{\text{R}}}} \bar{\varpi}_{\text{R}}\, \phi_{\text{test}}\, da_{\text{R}} = 0, \tag{21.1.4}$$

Introduction to Coupled Theories in Solid Mechanics. Lallit Anand, Eric M. Stewart, Shawn A. Chester, Oxford University Press.
 DOI: 10.1093/oso/9780198986218.003.0022

holds for all test fields ϕ_{test}. Thus, since $\phi_{\text{test}} = 0$ on $\partial\mathcal{B}_\phi$ the first boundary integral in the expression above may be extended to all of $\partial\mathcal{B}$,

$$\int_{\mathcal{B}} (\operatorname{Div}\mathbf{d}_{\text{R}})\,\phi_{\text{test}}\,dv_{\text{R}} - \int_{\partial\mathcal{B}} \mathbf{d}_{\text{R}}\cdot\mathbf{n}_{\text{R}}\,\phi_{\text{test}}\,da_{\text{R}} + \int_{\partial\mathcal{B}_{\mathbf{d}_{\text{R}}}} \bar{\varpi}_{\text{R}}\,\phi_{\text{test}}\,da_{\text{R}} = 0. \tag{21.1.5}$$

Next, by the divergence theorem

$$\begin{aligned}\int_{\partial\mathcal{B}} \mathbf{d}_{\text{R}}\cdot\mathbf{n}_{\text{R}}\,\phi_{\text{test}}\,da_{\text{R}} &= \int_{\partial\mathcal{B}} (\phi_{\text{test}}\,\mathbf{d}_{\text{R}})\cdot\mathbf{n}_{\text{R}}\,da_{\text{R}}\\ &= \int_{\mathcal{B}} \phi_{\text{test}}\operatorname{Div}\mathbf{d}_{\text{R}}\,dv_{\text{R}} + \int_{\mathcal{B}} \mathbf{d}_{\text{R}}\cdot\nabla\phi_{\text{test}}\,dv_{\text{R}}.\end{aligned} \tag{21.1.6}$$

Use of (21.1.6) in (21.1.5) gives the weak form

$$\int_{\mathcal{B}} \mathbf{d}_{\text{R}}\cdot\nabla\phi_{\text{test}}\,dv_{\text{R}} - \int_{\partial\mathcal{B}_{\mathbf{d}_{\text{R}}}} \bar{\varpi}_{\text{R}}\,\phi_{\text{test}}\,da_{\text{R}} = 0. \tag{21.1.7}$$

Summarizing, the weak forms of the governing equations for the coupled problem are

$$\begin{aligned}\int_{\mathcal{B}} \mathbf{T}_{\text{R}}:\nabla\mathbf{u}_{\text{test}}\,dv_{\text{R}} - \int_{\partial\mathcal{B}_{\text{t}}} \bar{\mathbf{t}}_{\text{R}}\cdot\mathbf{u}_{\text{test}}\,da_{\text{R}} &= 0,\\ \int_{\mathcal{B}} \left[\frac{p}{K} + \frac{\ln J}{J}\right] p_{\text{test}}\,dv_{\text{R}} &= 0,\\ \int_{\mathcal{B}} \mathbf{d}_{\text{R}}\cdot\nabla\phi_{\text{test}}\,dv_{\text{R}} - \int_{\partial\mathcal{B}_{\mathbf{d}_{\text{R}}}} \bar{\varpi}_{\text{R}}\,\phi_{\text{test}}\,da_{\text{R}} &= 0,\end{aligned} \tag{21.1.8}$$

for all vector-valued test functions $\mathbf{u}_{\text{test}}(\mathbf{X})$ which vanish on $\partial\mathcal{B}_{\mathbf{u}}$ and all scalar test functions $\phi_{\text{test}}(\mathbf{X})$ which vanish on $\partial\mathcal{B}_\phi$, respectively.

21.2 Solution procedure in the FEniCS code

We use the following degrees of freedom

$$\{\mathbf{u},\, p,\, \phi\}, \tag{21.2.1}$$

as well as their known values from the previous step

$$\{\mathbf{u}_{\text{old}},\, p_{\text{old}},\, \phi_{\text{old}}\}, \tag{21.2.2}$$

to calculate the *residuals* which correspond to the weak forms of the governing equations, viz.

$$
\begin{aligned}
R_{\mathbf{u}} &= \int_{\mathcal{B}} \mathbf{T}_{\mathrm{R}} : \nabla \mathbf{u}_{\text{test}} \, dv_{\mathrm{R}} - \int_{\partial \mathcal{B}_{\mathbf{t}_{\mathrm{R}}}} \bar{\mathbf{t}}_{\mathrm{R}} \cdot \mathbf{u}_{\text{test}} \, da_{\mathrm{R}}, \\
R_p &= \int_{\mathcal{B}} \left[\frac{p}{K} + \frac{\ln J}{J} \right] p_{\text{test}} \, dv_{\mathrm{R}}, \\
R_\phi &= \int_{\mathcal{B}} \mathbf{d}_{\mathrm{R}} \cdot \nabla \phi_{\text{test}} \, dv_{\mathrm{R}} - \int_{\partial \mathcal{B}_{\mathbf{d}_{\mathrm{R}}}} \bar{\varpi}_{\mathrm{R}} \, \phi_{\text{test}} \, da_{\mathrm{R}}.
\end{aligned}
\tag{21.2.3}
$$

FEniCS then automatically differentiates these residuals with respect to the separate degrees of freedom to obtain *tangents*. At each step in time, FEniCS uses the residuals and tangents in a global Newton–Raphson solver to determine current values of the degrees of freedom $\{\mathbf{u},\, p,\, \phi\}$ for which the residuals are zero, to within some small numerical tolerance, and which therefore satisfy the governing equations as well as the initial and boundary conditions (in a weak sense).

REMARK

A mixed finite element method is utilized to solve for the displacement $\mathbf{u}$, pressure p and the electric potential ϕ, concurrently. To avoid the numerical instabilities we use a class of elements known as "Taylor–Hood" elements which combine second-order quadratic interpolation for the displacement, and first-order linear interpolation for the pressure and potential.

21.2.1 **Algorithm for calculation of the residuals**

The algorithm we follow in our code to calculate the residuals in (21.2.3) is summarized below.

1. Begin the step with known values of

$$
\{\mathbf{u}_{\text{old}},\, p_{\text{old}},\, \phi_{\text{old}}\},
\tag{21.2.4}
$$

   where for the first step these quantities are given by the initial conditions for $\mathbf{u}$, p, and ϕ.

2. Calculate necessary kinematical quantities for the current step,

$$
\begin{gathered}
\mathbf{F} = \mathbf{1} + \nabla \mathbf{u}, \quad J = \det \mathbf{F}, \quad \mathbf{C} = \mathbf{F}^{\top}\mathbf{F}, \quad \mathbf{B} = \mathbf{F}\mathbf{F}^{\top}, \\
\bar{\mathbf{C}} = J^{-2/3}\mathbf{C}, \quad \bar{\mathbf{B}} = J^{-2/3}\mathbf{B}, \quad \bar{I}_1 = \operatorname{tr} \bar{\mathbf{C}}.
\end{gathered}
\tag{21.2.5}
$$

3. Calculate electric field for the current step,

$$
\mathbf{e}_{\mathrm{R}} = -\nabla \phi, \quad \mathbf{e} = \mathbf{F}^{-\top}\mathbf{e}_{\mathrm{R}}.
\tag{21.2.6}
$$

4. Calculate the Piola stress,

$$\begin{aligned}
\mathbf{T}_{\mathrm{R}} &= J\,\mathbf{T}\mathbf{F}^{-\top}\\
\mathbf{T} &= \mathbf{T}^{\mathrm{mech}} + \mathbf{T}^{\mathrm{maxw}},\\
\mathbf{T}^{\mathrm{mech}} &= J^{-1}G\bar{\mathbf{B}}_0 - p\mathbf{1}, \qquad G \overset{\mathrm{def}}{=} G_0\left(1 - \frac{\bar{I}_1 - 3}{I_m}\right)^{-1},\\
\mathbf{T}^{\mathrm{maxw}} &= \varepsilon\left[\mathbf{e}\otimes\mathbf{e} - \tfrac{1}{2}(\mathbf{e}\cdot\mathbf{e})\mathbf{1}\right].
\end{aligned} \tag{21.2.7}$$

5. Calculate the referential electric displacement,

$$\mathbf{d}_{\mathrm{R}} = -\varepsilon_r\varepsilon_0\,J\mathbf{C}^{-1}\nabla\phi. \tag{21.2.8}$$

6. Use the quantities from the steps above to compute the residuals,

$$\begin{aligned}
R_{\mathbf{u}} &= \int_{\mathcal{B}} \mathbf{T}_{\mathrm{R}} : \nabla\mathbf{u}_{\mathrm{test}}\,dv_{\mathrm{R}} - \int_{\partial\mathcal{B}_{\mathbf{t}_{\mathrm{R}}}} \bar{\mathbf{t}}_{\mathrm{R}}\cdot\mathbf{u}_{\mathrm{test}}\,da_{\mathrm{R}},\\
R_p &= \int_{\mathcal{B}} \left[\frac{p}{K} + \frac{\ln J}{J}\right] p_{\mathrm{test}}\,dv_{\mathrm{R}},\\
R_\phi &= \int_{\mathcal{B}} \mathbf{d}_{\mathrm{R}}\cdot\nabla\phi_{\mathrm{test}}\,dv_{\mathrm{R}} - \int_{\partial\mathcal{B}_{\mathbf{d}_{\mathrm{R}}}} \bar{\varpi}_{\mathrm{R}}\,\phi_{\mathrm{test}}\,da_{\mathrm{R}}.
\end{aligned} \tag{21.2.9}$$

7. FEniCS solves for the current values of $\{\mathbf{u},\, p,\, \phi\}$.
8. Set these current values as the "old" values for the next step in time,

$$\{\mathbf{u},\, p,\, \phi\} \implies \{\mathbf{u}_{\mathrm{old}},\, p_{\mathrm{old}},\, \phi_{\mathrm{old}}\}. \tag{21.2.10}$$

9. Repeat for the desired number of steps in time.

21.3 **Normalization**

Since we plan to solve all the governing equations related to multiple physical fields simultaneously, it is beneficial to rewrite the three residuals so that their values have approximately the same order of magnitudes. For this particular theory, the normalization of the electrostatic problem is dealt with purely through a prudent selection of the unit system.

Using the units in Table 21.1, the permittivity of the free space becomes $\varepsilon_0 = 8.85\times10^{-3}\,\mathrm{nF/mm}$ rather than $\varepsilon_0 = 8.85\times10^{-12}\,\mathrm{F/m}$, while the numerical value of the mechanical shear modulus changes from Pa to kPa. Thus, the numerical value of the shear modulus increases by three orders of magnitudes and the value of the permittivity of the free space is increased by 10^9. This results in a system of equations which are easier to solve numerically, when compared with those resulting from the use of base SI units.

Unit system in SI (mm) units

Table 21.1 Basic units.

Quantity	SI (m)	SI (mm)
Length	m	mm
Mass	kg	kg
Time	s	s
Electric charge	C	nC
Mass density	kg/m^3	kg /mm^3
Force	N	mN
Energy	J	μJ
Stress	Pa	kPa
ϕ, Electric potential	V (J/C)	kV (μJ/nC)
Capacitance	F (C/V)	pF (nC/kV $= 10^{-12}$ F)
Electric permittivity	F/m	pF/mm (10^{-9} F/m)

22 Representative simulations of dielectric elastomers

In this chapter we present several representative finite element simulations which demonstrate key features of the electroelastic response of dielectric elastomers. The material parameters that we have used in our simulations are listed in Table 22.1.

Table 22.1 Approximate values of material parameters for a dielectric elastomer.

Quantity	Value SI (m)	Value SI (mm)
G, shear modulus	15×10^3 Pa	15 kPa
K, bulk modulus	15×10^6 Pa	15×10^3 kPa
I_m, Gent locking parameter	175	175
ε_0, permittivity of free space	8.85×10^{-12} F/m	8.85×10^{-3} pF/mm
ε_r, relative permittivity for a dielectric elastomer	5	5

22.1 Electrostatic actuation of a three-dimensional cube

As a first example we consider homogeneous electrostatic actuation of a three-dimensional cube in which a pull-in snap-through instability occurs; cf. Section 20.8. We consider a cube of edge length $\ell_0 = 1$ mm, cf. Fig. 22.1. The initial conditions are $\mathbf{u}_0 = \mathbf{0}$, and $\phi_0 = 0$ V, and the boundary conditions are as follows:

- Mechanical boundary conditions: (i) faces ADHE, ABFE, and ADCB are constrained from moving in the directions normal to the coordinate axes; and (ii) all other faces are traction-free.
- Electrostatic boundary conditions: (i) the bottom face ABCD is grounded, $\bar{\phi} = 0$ kV; (ii) the top face EFGH is prescribed a positive normalized electrostatic potential that linearly ramps from 0 to 1 over 10 s; and (iii) all other faces are prescribed zero surface charge, $\bar{\varpi}_{\rm R} = 0$.

The material parameters used in the simulation are those listed in Table 22.1. As discussed in Section 20.8, under such a loading condition a snap-through type instability occurs along the direction of the applied electric field for all values of the Gent locking parameter

Introduction to Coupled Theories in Solid Mechanics. Lallit Anand, Eric M. Stewart, Shawn A. Chester, Oxford University Press.
 DOI: 10.1093/oso/9780198986218.003.0023

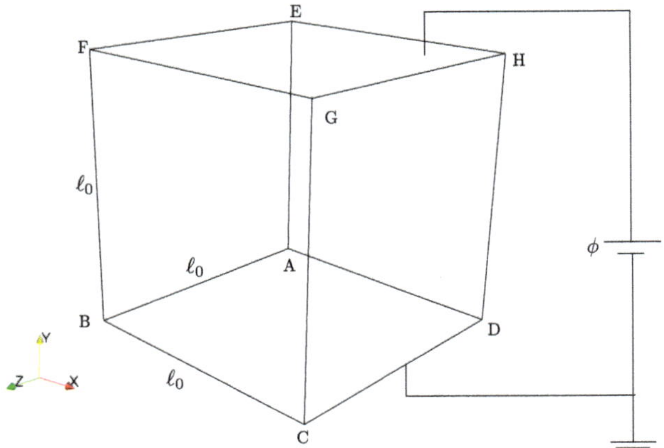

Fig. 22.1 Undeformed reference configuration, a cube of edge length $l_0 = 1\,\text{mm}$, with electrostatic boundary conditions shown.

$I_m \gtrsim 7.3$, while the instability is suppressed for values of $I_m \lesssim 7.3$. Accordingly, we conducted simulations for two different values of the Gent locking parameter, $I_m = 175$ and $I_m = 6$.

Figure 22.2 shows the simulation result for a material with $I_m = 175$. As the electric potential is increased, this material exhibits a "snap-through" thinning instability at a value of $\phi = 12.5\,\text{kV}$.[1] Figure 22.2(a) shows the deformed configuration at the onset of the instability—our numerical simulation fails to converge beyond this point in the simulation. Figure 22.2(b) shows the corresponding curve of the axial stretch versus the normalized potential.

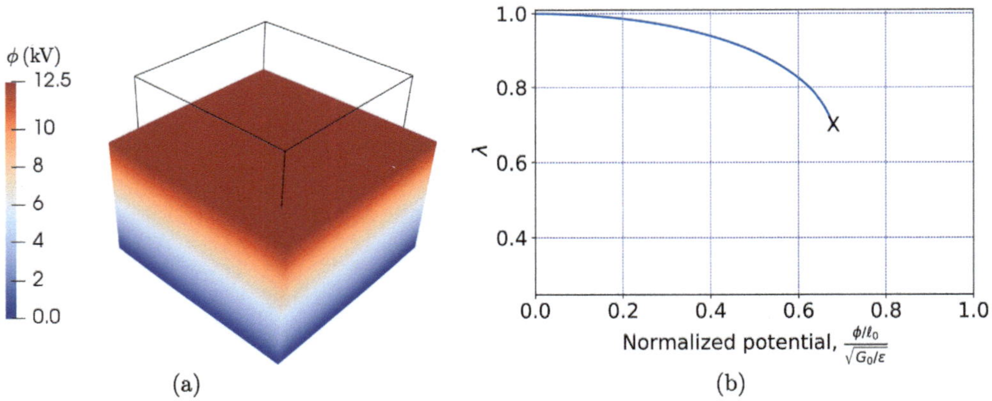

Fig. 22.2 Figure showing snap-through instability for a Gent material with $I_m = 175$ at $\phi = 12.5\,\text{kV}$: (a) deformed configuration at onset of instability; (b) axial stretch versus normalized electrostatic potential curve.

Figure 22.3 shows the simulation result for a Gent material with $I_m = 6$. A material with this value of I_m does not exhibit a snap-through instability. The simulation was carried through to an applied voltage of $\phi = 18.4\,\text{kV}$. Figure 22.3(a) shows the final deformed configuration, and Fig. 22.3(b) shows the corresponding axial stretch versus normalized potential curve.

[1] High voltage can be of concern; however, in actual devices the thickness of the dielectric elastomers are quite small, typically $\approx 100\,\mu\text{m}$, and the voltages required to operate typical dielectric elastomeric actuators are in the one to five to ten kilovolts range.

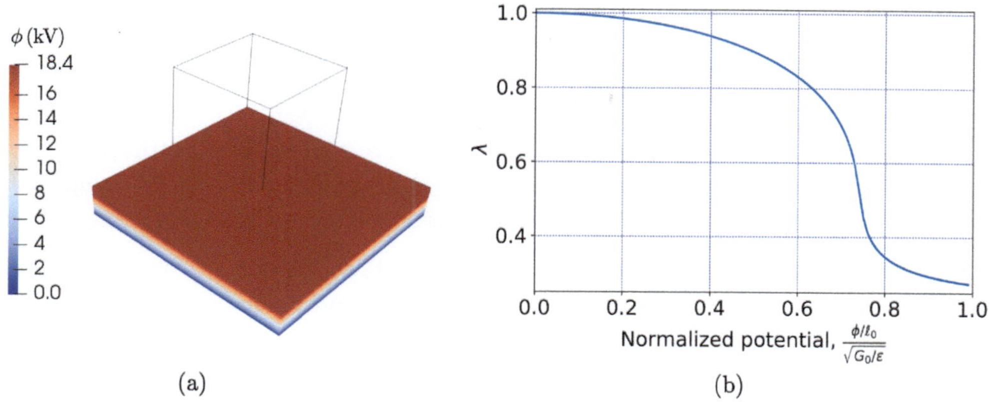

Fig. 22.3 Figure showing actuation of a Gent material with $I_m = 6$: (a) Deformed configuration at 18.4 kV. (b) Axial stretch versus normalized electrostatic potential. A material with $I_m = 6$ does not exhibit a snap-through thinning instability; it gradually thins down as the electrostatic potential is increased.

22.2 Electrostatic actuation of a square domain in plane strain

As discussed in Section 20.9, under plane-strain conditions there is no snap-through instability for any value of the Gent locking parameter I_m. In this example, we consider a square of edge length $l_0 = 1$ mm under plane-strain conditions, and use a finite element mesh shown in Fig. 22.4. The mesh has 10 triangular elements along each edge. The material parameters used in the simulation are those listed in Table 22.1.

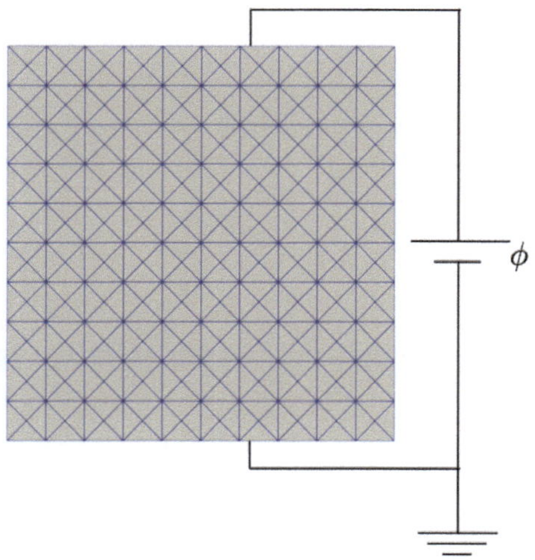

Fig. 22.4 Undeformed reference configuration of a plane-strain square of edge length $\ell_0 = 1$ mm with the corresponding finite element mesh, and a schematic of the applied potential boundary condition.

The initial conditions are $\mathbf{u}_0 = \mathbf{0}$, and $\phi_0 = 0$ V, and the boundary conditions are as follows:

- Mechanical boundary conditions: (i) the left face is prescribed $\bar{u}_1 = 0$; (ii) the bottom face is prescribed $\bar{u}_2 = 0$; and (iii) all other faces of the domain are traction-free.
- Electrostatic boundary conditions: (i) the bottom face is grounded, $\bar{\phi} = 0$ kV; (ii) the top face is prescribed a positive normalized electrostatic potential that linearly ramps from 0 to 1.25 over 1 s; and (iii) all other faces are prescribed zero surface charge, $\bar{\varpi}_R = 0$.

Figure 22.5(a) shows the final deformed configuration, and Fig. 22.5(b) shows the corresponding axial stretch versus normalized potential curve. Under plane-strain conditions the material does not exhibit a snap-through thinning instability; it gradually thins down as the electric potential is increased.

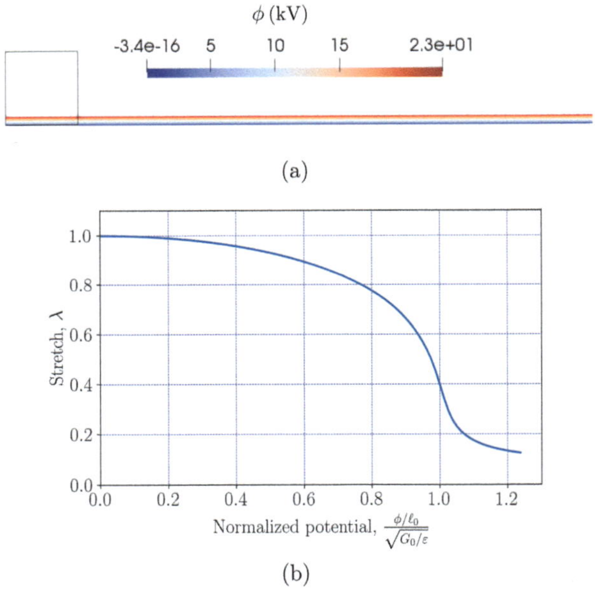

Fig. 22.5 (a) Plane-strain actuation showing the initial body and deformed body. (b) Axial stretch versus normalized electric potential curve. Under plane-strain conditions there is no snap-through thinning instability; the material gradually thins down as the electric potential is increased.

22.3 Suppression of pull-in instabilities in pre-stretched membranes

Pelrine et al. (2000), in their classic paper, showed that pull-in-type instabilities in a membrane of a dielectric elastomer may be suppressed if the membrane is subjected to a large biaxial pre-stretch before electrostatic actuation. We simulate such a response of a dielectric elastomeric membrane in this example.[2]

[2] Cf. Section 20.10 for the homogeneous deformation case.

Figure 22.6(a) shows the initial geometry, which consists of a circular membrane with an initial thickness $T_0 = 0.2\,\text{mm}$ and initial radius $R_0 = 2.5\,\text{mm}$. Central regions on both top and bottom faces of radius $R_0^e = 0.825\,\text{mm}$ (indicated by the region enclosed by the blue line) serve as domains across which the electric field is applied. Due to the symmetry of the problem we only model an axisymmetric slice, using the mesh shown in Fig. 22.6(b). The material parameters used in the simulation for the dielectric elastomer are those listed in Table 22.1 with the exception that $I_m = 90$ following the discussion in Section 20.10. The initial conditions are $\mathbf{u}_0 = \mathbf{0}$, and $\phi_0 = 0\,\text{V}$, and with reference to Fig. 22.6(b), the boundary conditions in this problem are as follows:

- Step 1: Mechanical pre-stretch
  - Mechanical boundary conditions: (i) the left face is prescribed $\bar{u}_1 = 0$ along the axis of symmetry; (ii) the outer right face is prescribed a linear displacement ramp to $\bar{u}_1 = R_0(\lambda_{\text{pre}} - 1)$ at a rate of $10^{-3}\,\text{mm/s}$ such that the applied pre-stretch is $\lambda_{\text{pre}} = 4$; (iii) the left bottom corner is pinned $\bar{\mathbf{u}} = \mathbf{0}$, as indicated by the blue point; and (iv) all other faces are traction-free.
  - Electrostatic boundary conditions: (i) the bottom electrode face is grounded, $\bar{\phi} = 0\,\text{V}$, the top electrode face is also prescribed $\bar{\phi} = 0\,\text{V}$; and (iii) all other faces are prescribed zero surface charge $\bar{\varpi}_{\text{R}} = 0$.
- Step 2: Electrostatic actuation
  - Mechanical boundary conditions: all mechanical boundary conditions from the prior step are maintained; the pre-stretch is held constant.
  - Electrostatic boundary conditions: (i) the bottom electrode face is grounded, $\bar{\phi} = 0\,\text{V}$, the top electrode face is prescribed a normalized electrostatic potential $\frac{\phi/T_0}{\sqrt{G_0/\varepsilon}}$ that linearly ramps from 0 to 0.6 over 1 s; and (iii) all other faces are prescribed zero surface charge $\bar{\varpi}_{\text{R}} = 0$.

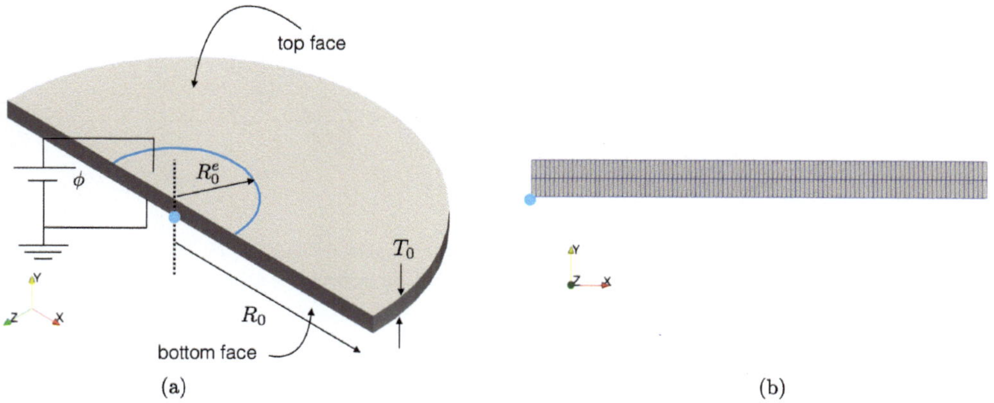

Fig. 22.6 (a) The initial geometry, and (b) finite element mesh used for the pre-stretched membrane simulation. Note that the body is axisymmetric, with only half of the geometry is shown in the schematic, and the mesh is planar.

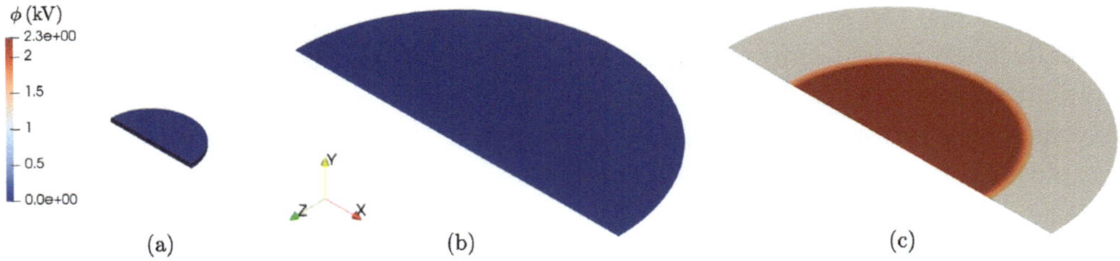

Fig. 22.7 Contours of the electrostatic potential ϕ on (a) the initial body, (b) at the start of the electrostatic actuation, and (c) at the end of the electrostatic actuation. Note that the axisymmetric body has been revolved to show half of the geometry for clarity.

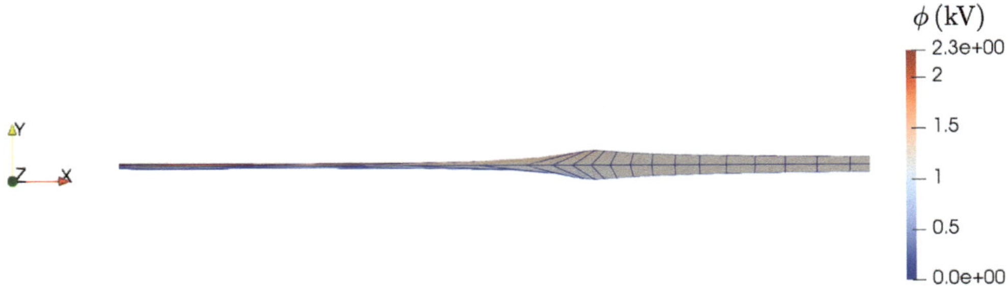

Fig. 22.8 Contours of the electrostatic potential ϕ on the deformed body at the end of electrostatic actuation. Note the dramatic thickness decrease in the electrode region relative to the already thinned pre-stretched membrane around it.

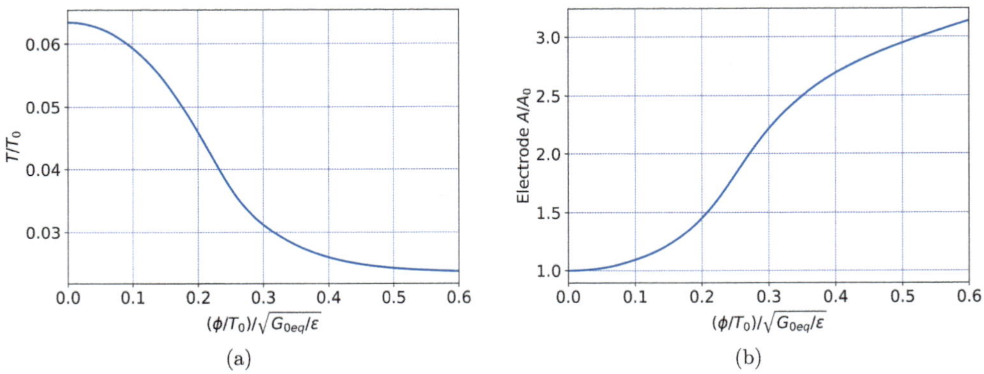

Fig. 22.9 Results showing (a) the ratio of deformed thickness to initial thickness T/T_0, and (b) the change in electrode area A/A_0, as a function of the applied normalized electrostatic potential for a biaxial pre-stretch of $\lambda_{\text{pre}} = 4$.

Figure 22.7 shows the electrostatic potential at the end of pre-stretch, and the deformed configuration at the end of electrostatic actuation. Upon electrostatic actuation the electrode region thins dramatically while the electrode area increases; see Figs. 22.7(b)–(c) and Fig. 22.8. The shape change as a function of applied voltage $\frac{\phi/T_0}{\sqrt{G_0/\varepsilon}}$ is quantified in Fig. 22.9(a) which plots the ratio of deformed thickness to initial thickness T/T_0, and in Fig. 22.9(b) which plots the change in electrode area A/A_0.

Before concluding this example, in Fig. 22.10 we show the difference in response between a membrane which is not pre-stretched, $\lambda_{\text{pre}} = 1$ (orange curve), and a membrane which has been pre-stretched, $\lambda_{\text{pre}} = 4$ (blue curve). The membrane which has been pre-stretched to $\lambda_{\text{pre}} = 4$ shows a substantially larger increase in the ratio A/A_0 upon actuation than the membrane which is not pre-stretched, $\lambda_{\text{pre}} = 1$—as demonstrated experimentally by Pelrine et al. (2000).

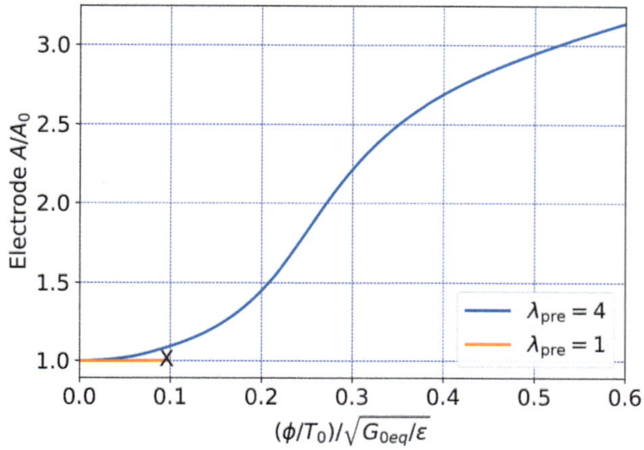

Fig. 22.10 A comparison of the area change after electrostatic actuation between a membrane which has been biaxially pre-stretched to $\lambda_{\text{pre}} = 4$ and a membrane which is not pre-stretched, $\lambda_{\text{pre}} = 1$. These results show the pre-stretched membrane shows dramatically more actuation than the membrane that was not pre-stretched.

22.4 A bilayer beam actuated by a sinusoidal voltage

In this example we simulate the actuation of a bilayer cantilever beam, in which the top half of the dielectric elastomer beam has an electric field applied to it but the bottom half does not. As shown in Fig. 22.11, the beam has a length of $L = 50\,\text{mm}$, a depth $B = 2\,\text{mm}$, and a height $H = 2\,\text{mm}$, and is meshed with 6 tetrahedral elements through the full thickness, 3 for each layer. The material parameters used for the dielectric elastomer are those listed in Table 22.1. The initial conditions are $\mathbf{u}_0 = \mathbf{0}$, and $\phi_0 = 0\,\text{V}$, and with reference to Fig. 22.11, the boundary conditions in this problem are as follows:

- Mechanical boundary conditions: (i) the left face of the beam is fixed, $\bar{\mathbf{u}} = \mathbf{0}$; and (ii) all other faces are traction-free.
- Electrostatic boundary conditions: (i) the mid-plane of the beam which separates the two layers is grounded, $\bar{\phi} = 0\,\text{kV}$; and the top face is prescribed a sinusoidal electrostatic potential in the form $\bar{\phi}(t) = A \sin(2\pi f t)$, where $A = 5\,\text{kV}$ and $f = 1\,\text{Hz}$; and (ii) all other faces are prescribed zero surface charge, $\bar{\varpi}_{\text{R}} = 0$.

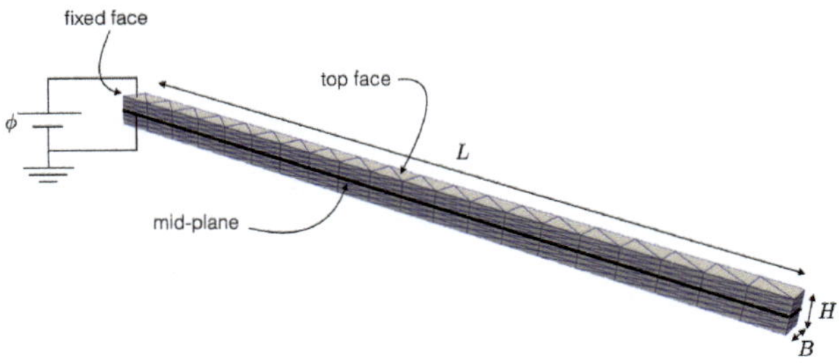

Fig. 22.11 The initial geometry, and finite element mesh used for the bilayer cantilever bending simulation.

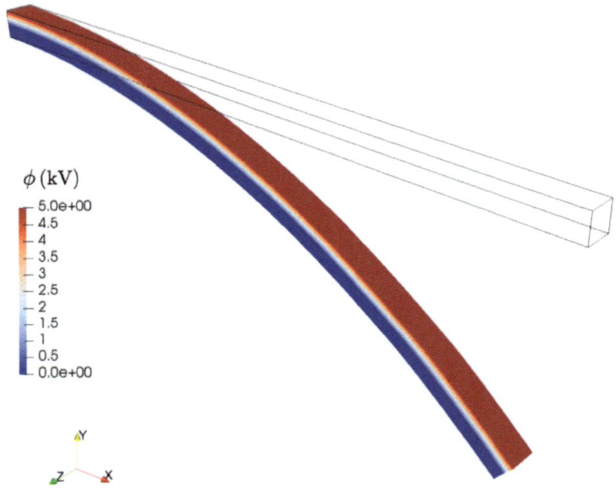

Fig. 22.12 Reference configuration and deformed configuration at maximum deflection, showing contours of the electrostatic potential.

The simulation is performed for 2 cycles of the applied sinusoidal voltage $\bar{\phi}(t)$.

As expected, an application of the potential $\bar{\phi}(t)$ causes the beam to bend because the top layer of the beam expands while the bottom layer prevents overall extension of the beam, leading to an overall bending of the beam. Figure 22.12 shows the deformed configuration of the beam at maximum deflection.

Figure 22.13 shows the time history of the prescribed electrostatic potential and the corresponding displacement of the tip of the beam. *Note that although the input sinusoidal voltage varies between positive and negative values, the tip-displacement is always negative-valued because of the quadratic dependence of the Maxwell stress on the electric field.*

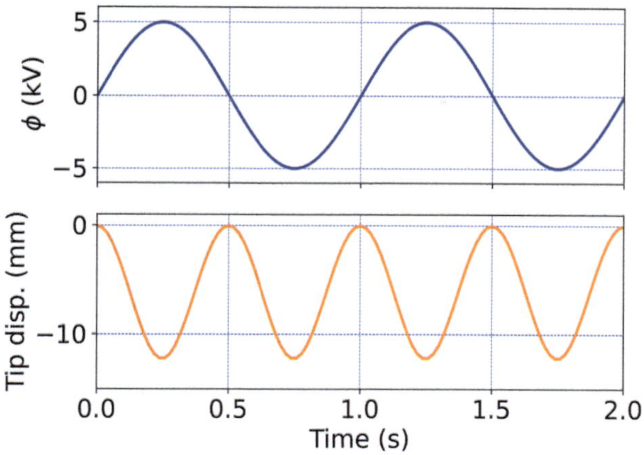

Fig. 22.13 Time history of the prescribed electrostatic potential and corresponding displacement of the tip of the beam.

22.5 A gripper actuated by a sinusoidal voltage

Next we simulate the actuation of a soft "gripper." As shown in Fig. 22.14, the gripper consists of four arms connected at a central square region (marked in blue). The central square region has an edge length of $H = 10\,\text{mm}$, each arm of the gripper has a length of $L = 40\,\text{mm}$, and the total thickness of the gripper is $T = 1\,\text{mm}$. Much like the previous example, we consider a bilayer geometry in which the top half of the gripper across its mid-plane is a dielectric elastomer and the bottom half is a simple non-electroactive elastomer. Due to the symmetry of the geometry, only one quarter of the body is meshed and there are 4 tetrahedral elements through the full thickness, 2 for each layer. The material parameters used for the dielectric elastomer in the top half of the gripper are those listed in Table 22.1, and the non-electroactive elastomer in the bottom half of the gripper is taken to have the same mechanical properties as the top half.

The initial conditions are $\mathbf{u}_0 = \mathbf{0}$ and $\phi_0 = 0\,\text{V}$, and with reference to Fig. 22.14, the boundary conditions in this problem are as follows:

- Mechanical boundary conditions: (i) central square region is fixed by prescribing $\bar{\mathbf{u}} = \mathbf{0}$ on its bottom face; (ii) symmetry planes are prescribed appropriate symmetry conditions; and (iii) all other faces are traction-free.
- Electrostatic boundary conditions: (i) the mid-plane is grounded, $\bar{\phi} = 0$; and the top face is prescribed a sinusoidal electrostatic potential in the form $\bar{\phi}(t) = A\sin(2\pi f t)$, where $A = 4.4\,\text{kV}$ and $f = 1\,\text{Hz}$; and (iii) all other faces are prescribed zero surface charge, $\bar{\varpi}_{\text{R}} = 0$.

The simulation is performed for 1 cycle of the sinusoidal potential.

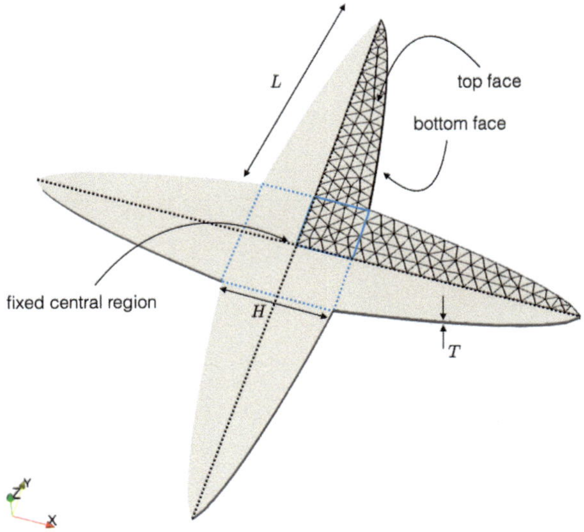

Fig. 22.14 The initial geometry, symmetry planes, and finite element mesh used for the gripper simulation. The modeled domain has been reflected about both symmetry planes to show the full geometry.

Figure 22.15 shows the geometry of the gripper in its initial configuration,[3] and the deformed configurations at two points in time. As in the cantilever beam example, the bending of each arm of the gripper cooperatively creates a gripper-type actuation.

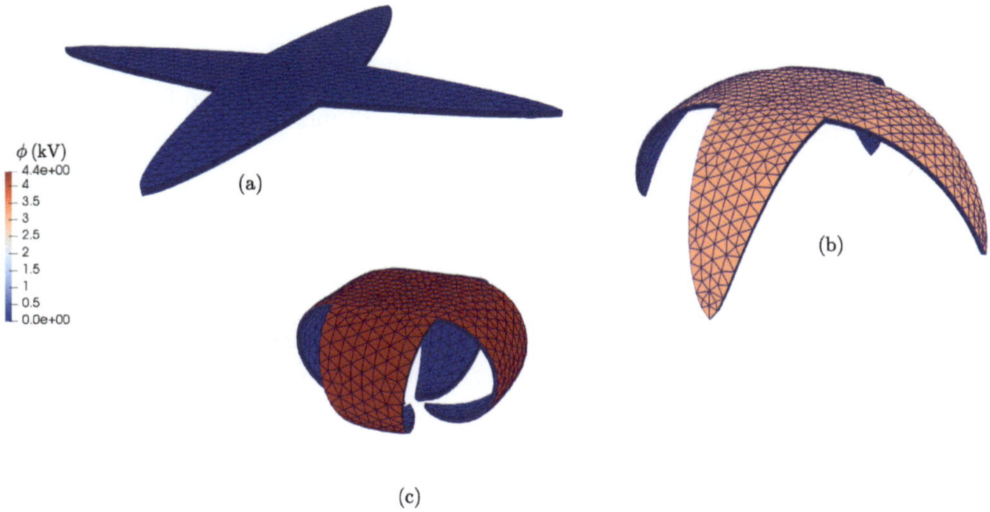

Fig. 22.15 Actuation of a gripper: (a) the reference configuration, and the deformed configurations at (b) $t = 0.13$ s, and (c) $t = 0.26$ s showing the actuation. The modeled domain has been reflected about both symmetry planes to show the full geometry. The contour plots are of the electrostatic potential ϕ.

[3] The modeled domain has been reflected about both symmetry planes to show the full geometry.

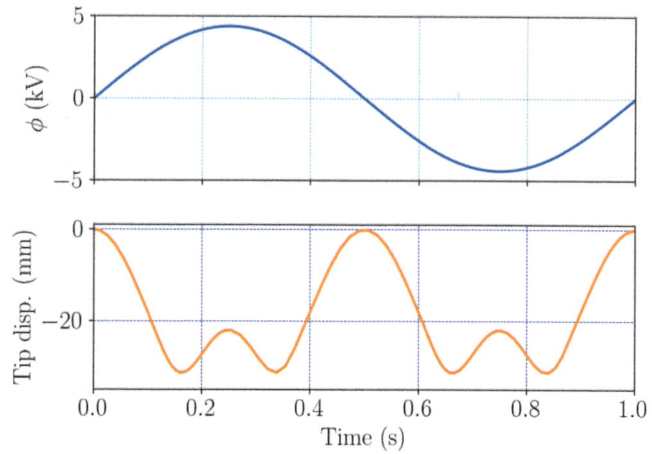

Fig. 22.16 Time history of the prescribed electrostatic potential and corresponding displacement of the tip of one arm.

Lastly, Fig. 22.16 shows the time history of the prescribed electrostatic potential and corresponding displacement of the tip of one of the arms. The non-monotonic tip-displacement is due to the curling of the arm.

22.6 **A helical actuator**

In this last example we simulate the response of a soft helical actuator made from a dielectric elastomer. As shown in Fig. 22.17, the initial geometry consists of a thin helically shaped strip of width $H = 5\,\text{mm}$, and thickness $T = 0.2\,\text{mm}$. The helix has a length of $L = 35\,\text{mm}$, an outer diameter of 5.1 mm, an inner diameter of 4.9 mm, and a pitch of 3 turns per 35 mm. The mesh contains 2 tetrahedral elements through the thickness. The material parameters used in the simulation for the dielectric elastomer are those listed in Table 22.1.

The initial conditions are $\mathbf{u}_0 = \mathbf{0}$, and $\phi_0 = 0\,\text{V}$, and with reference to Fig. 22.17, the boundary conditions in this problem are as follows:

- Mechanical boundary conditions: (i) the lower bottom face of the actuator is fixed; and (ii) all other faces are traction-free.
- Electrostatic boundary conditions: (i) the inner face of the helical actuator is grounded, $\bar{\phi} = 0$, and the outer face is prescribed an electrostatic potential that linearly ramps from $\bar{\phi} = 0\,\text{kV}$ to $\bar{\phi} = 2\,\text{kV}$ over 1 s; and (iii) all other faces are prescribed zero surface charge, $\bar{\varpi}_{\text{R}} = 0$.

Figure 22.15 shows the reference configuration and the deformed configurations at times $t = 0.33,\ 0.66,\ 1\,\text{s}$. An increase of the electrostatic potential across the thickness of the helix causes the thickness to reduce, and the macroscopic shape of the helix to tighten and twist.

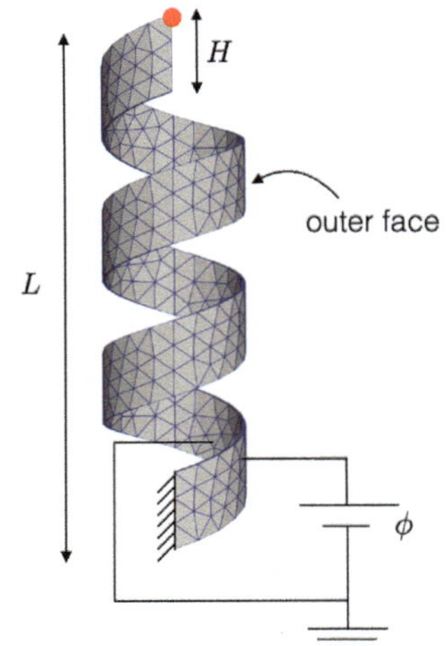

Fig. 22.17 Geometry, mesh, and the boundary conditions used for the helical actuator simulation.

Lastly, Fig. 22.18 shows the time history of the prescribed electrostatic potential and corresponding displacement of the tip of the helix indicated by the red point in Fig. 22.17. The non-monotonic tip-displacement is due to the interplay between the tightening and twisting of the helix during the actuation.

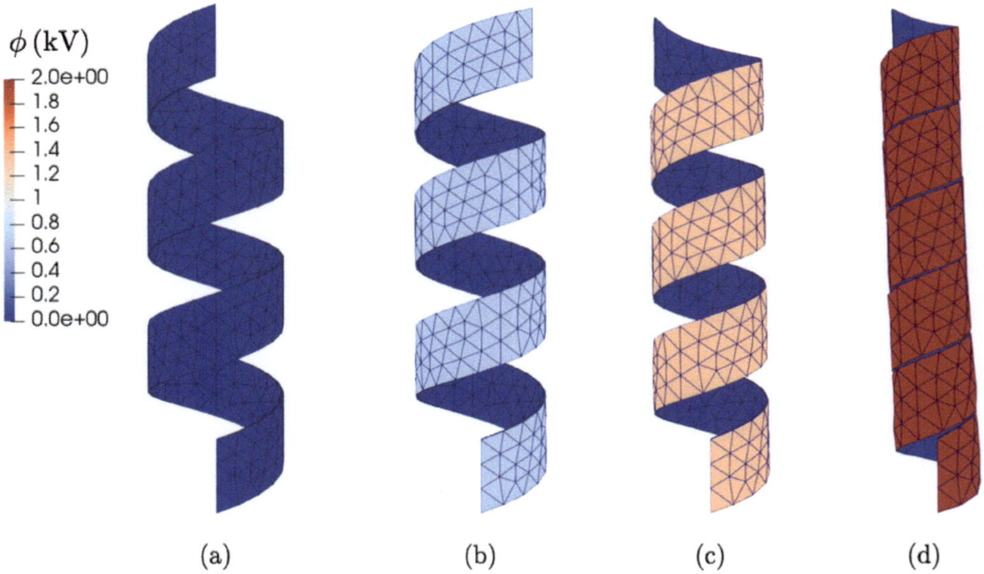

Fig. 22.18 Helical actuator: (a) The reference configuration, and the deformed configurations at times (b) $t = 0.33$ s, (c) $t = 0.66$ s, and (d) $t = 1$ s, showing the actuation response. The contour plots are of the electrostatic potential ϕ.

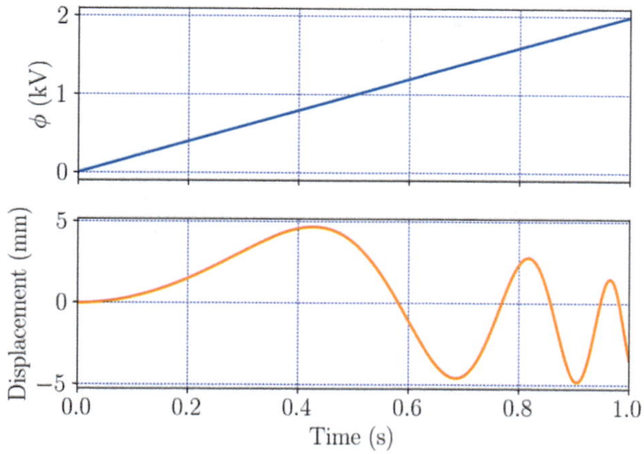

Fig. 22.19 Time history of the prescribed electrostatic potential and corresponding displacement of the tip of the helix indicated by the red point in Fig. 22.17.

Part VIII

ELECTRO-VISCOELASTICITY OF DIELECTRIC ELASTOMERS

23 Electro-viscoelasticity of dielectric elastomers

23.1 Introduction

Soft dielectrics are electrically insulating elastomeric materials, which are capable of large deformation and electrical polarization, and are used as smart transducers for converting between mechanical and electrical energy. In Chapter 20 on "Electro-elasticity of dielectric elastomers" we focused our attention on electro-mechanical coupling, but we *neglected* any viscoelastic characteristics of the elastomer. In this chapter we formulate a constitutive theory for **dielectric viscoelastomers**. The theory is able to account for:

- electro-mechanical coupling,
- large deformations,
- large-stretch chain-locking,
- viscoelasticity, and
- inertial effects.

Most of the theory remains the same as in Chapter 20 on electroelasticity of dielectric elastomers; only the kinematics and the constitutive equations change to account for viscoelasticity effects. These changes will be introduced following our discussion of viscoelasticity and dynamical effects in Chapter 4. Our starting point in this chapter is the free-energy imbalance (20.5.12) of Chapter 20, viz.

$$\dot{\psi}_{\mathrm{R}} - \frac{1}{2}\mathbf{T}_{\mathrm{RR}} : \dot{\mathbf{C}} - \mathbf{e}_{\mathrm{R}} \cdot \dot{\mathbf{d}}_{\mathrm{R}} \le 0, \tag{23.1.1}$$

and the free-energy imbalance (4.3.4) of Chapter 4, viz.

$$\dot{\psi}_{\mathrm{R}} - \sum_{\alpha} \frac{1}{2}\mathbf{T}^{e(\alpha)} : \dot{\mathbf{C}}^{e(\alpha)} - \textstyle\sum_{\alpha} \mathbf{M}_0^{e(\alpha)} : \mathbf{D}^{v(\alpha)} \le 0. \tag{23.1.2}$$

In the case of a dielectric viscoelastomer under isothermal conditions, straightforward calculations will lead to a free-energy imbalance in the form

$$\dot{\psi}_{\mathrm{R}} - \sum_{\alpha} \frac{1}{2}\mathbf{T}^{e(\alpha)} : \dot{\mathbf{C}}^{e(\alpha)} - \textstyle\sum_{\alpha} \mathbf{M}_0^{e(\alpha)} : \mathbf{D}^{v(\alpha)} - \mathbf{e}_{\mathrm{R}} \cdot \dot{\mathbf{d}}_{\mathrm{R}} \le 0. \tag{23.1.3}$$

Introduction to Coupled Theories in Solid Mechanics. Lallit Anand, Eric M. Stewart, Shawn A. Chester, Oxford University Press.
 DOI: 10.1093/oso/9780198986218.003.0024

23.2 Constitutive theory

Let $\boldsymbol{\Lambda}$ represent the list

$$\boldsymbol{\Lambda} = (\mathbf{C}^{e(\alpha)}, \mathbf{d}_{\mathrm{R}}). \tag{23.2.1}$$

Guided by the free-energy imbalance (23.1.1), we begin by assuming constitutive equations for the free energy ψ_{R}, the stresses $\mathbf{T}^{e(\alpha)}$, and the electric field $\mathbf{e}_{\mathrm{R}}$ of the form

$$\psi_{\mathrm{R}} = \hat{\psi}_{\mathrm{R}}(\boldsymbol{\Lambda}), \qquad \mathbf{T}^{e(\alpha)} = \hat{\mathbf{T}}^{e(\alpha)}(\boldsymbol{\Lambda}), \qquad \mathbf{e}_{\mathrm{R}} = \hat{\mathbf{e}}_{\mathrm{R}}(\boldsymbol{\Lambda}), \tag{23.2.2}$$

to which we append viscous flow rules for each $\mathbf{F}^{v(\alpha)}$,

$$\dot{\mathbf{F}}^{v\,(\alpha)} = \mathbf{D}^{v(\alpha)}\,\mathbf{F}^{v(\alpha)}, \tag{23.2.3}$$

with the constraint that viscous flow is isochoric, $\det \mathbf{F}^{v(\alpha)} = 1$. Then,

$$\dot{\psi}_{\mathrm{R}} = \sum_{\alpha} \frac{\partial \hat{\psi}_{\mathrm{R}}(\boldsymbol{\Lambda})}{\partial \mathbf{C}^{e(\alpha)}} : \dot{\mathbf{C}}^{e(\alpha)} + \frac{\partial \hat{\psi}_{\mathrm{R}}(\boldsymbol{\Lambda})}{\partial \mathbf{d}_{\mathrm{R}}} \cdot \dot{\mathbf{d}}_{\mathrm{R}}. \tag{23.2.4}$$

Using (23.2.4) and substituting the constitutive equations (23.2.2) into the free-energy imbalance (23.1.1) we find that it may be written as

$$\sum_{\alpha} \left[\frac{\partial \hat{\psi}_{\mathrm{R}}(\boldsymbol{\Lambda})}{\partial \mathbf{C}^{e(\alpha)}} - \frac{1}{2}\hat{\mathbf{T}}^{e(\alpha)}(\boldsymbol{\Lambda}) \right] : \dot{\mathbf{C}}^{e(\alpha)} + \left[\frac{\partial \hat{\psi}_{\mathrm{R}}(\boldsymbol{\Lambda})}{\partial \mathbf{d}_{\mathrm{R}}} - \hat{\mathbf{e}}_{\mathrm{R}}(\boldsymbol{\Lambda}) \right] \cdot \dot{\mathbf{d}}_{\mathrm{R}} - \sum_{\alpha} \mathbf{M}_0^{e(\alpha)} : \mathbf{D}^{v(\alpha)} \le 0. \tag{23.2.5}$$

Given a point of the body and any time t, it is possible to find a motion and an electrical displacement field such that $\mathbf{C}^{e(\alpha)}$, $\mathbf{d}_{\mathrm{R}}$ and their time derivatives $\dot{\mathbf{C}}^{e(\alpha)}$ and $\dot{\mathbf{d}}_{\mathrm{R}}$ have arbitrarily prescribed values at that point and time. Granted this assertion, the coefficients of $\dot{\mathbf{C}}^{e(\alpha)}$ and $\dot{\mathbf{d}}_{\mathrm{R}}$ must vanish, for otherwise the rates may be chosen to violate the inequality (23.2.5). We therefore have the *thermodynamic restrictions* that the free energy delivers the stresses $\mathbf{T}^{e(\alpha)}$ for each branch and the electric field $\mathbf{e}_{\mathrm{R}}$ through the state relations:

$$\begin{aligned} \mathbf{T}^{e(\alpha)} &= 2\frac{\partial \hat{\psi}_{\mathrm{R}}(\boldsymbol{\Lambda})}{\partial \mathbf{C}^{e(\alpha)}}, \\ \mathbf{e}_{\mathrm{R}} &= \frac{\partial \hat{\psi}_{\mathrm{R}}(\boldsymbol{\Lambda})}{\partial \mathbf{d}_{\mathrm{R}}}. \end{aligned} \tag{23.2.6}$$

We are then left with the reduced dissipation inequality

$$\mathcal{D} = \sum_{\alpha} \mathbf{M}_0^{e(\alpha)} : \mathbf{D}^{v(\alpha)} \ge 0, \tag{23.2.7}$$

which the constitutive equations and the viscous flow rules (23.2.3) for $\mathbf{F}^{v(\alpha)}$ must satisfy.

Using (23.2.6) the Cauchy stress is given by

$$\mathbf{T} = \sum_{\alpha} \mathbf{T}^{(\alpha)} = \sum_{\alpha} \left(J^{e(\alpha)}\right)^{-1} \mathbf{F}^{e(\alpha)} \left(2\frac{\partial \hat{\psi}_{\mathrm{R}}(\boldsymbol{\Lambda})}{\partial \mathbf{C}^{e(\alpha)}}\right) \mathbf{F}^{e(\alpha)\top}, \tag{23.2.8}$$

and as usual, the total Piola stress is given by

$$\mathbf{T}_{\mathrm{R}} = J\mathbf{T}\mathbf{F}^{-\top}. \tag{23.2.9}$$

23.3 Specialization of the constitutive equations

As in Chapter 4, in order to model the response of *solid* viscoelastic elastomers we introduce the following specializations:

(i) We take the first branch, $\alpha = 0$, to represent the *equilibrium elastic response* of the material. We do not associate any viscous relaxation with this network. Accordingly we set $\mathbf{F}^{v(0)} = \mathbf{1}$, and hence

$$\mathbf{F}^{e(0)} = \mathbf{F}. \tag{23.3.1}$$

(ii) The remaining branches, $\alpha = 1, \dots, M$, represent the *non-equilibrium viscoelastic response* of the material.

As before, when it is important to distinguish between the zeroth branch and the other branches of the rheological model, we will use the descriptive terms "equilibrium" for branch $\alpha = 0$ and "non-equilibrium" for branches $\alpha = 1, \dots, M$.

Note that an important feature of the multi-branch decomposition $\mathbf{F} = \mathbf{F}^{e(\alpha)}\mathbf{F}^{v(\alpha)}$, with $\det \mathbf{F}^{v(\alpha)} = 1$, is that

$$J^{e(\alpha)} = J \quad \text{for all} \quad \alpha = 1, \dots, M. \tag{23.3.2}$$

Next, for slightly compressible materials, with $J^{e(\alpha)} = \sqrt{\det \mathbf{C}^{e(\alpha)}}$ denoting the elastic volumetric Jacobian and $\bar{\mathbf{C}}^{e(\alpha)} \stackrel{\text{def}}{=} (J^e)^{-2/3}\mathbf{C}^{e(\alpha)}$ denoting the distortional elastic right Cauchy–Green tensor for each branch, we denote the list of the principal invariants of $\bar{\mathbf{C}}^{e(\alpha)}$ by $\mathcal{I}_{\bar{\mathbf{C}}^{e(\alpha)}}$. And for the special case $\alpha = 0$ where $\mathbf{F}^{v(0)} = \mathbf{1}$, $\mathcal{I}_{\bar{\mathbf{C}}}$ denotes the list of principal invariants of $\bar{\mathbf{C}}$.

We restrict our attention to materials for which the overall free energy of the material may be adequately expressed in a separable form as

$$\psi_{\mathrm{R}} = \psi_{\mathrm{R}}^{\mathrm{eq}}(\mathcal{I}_{\bar{\mathbf{C}}}, J) + \sum_{\alpha=1}^{M} \psi_{\mathrm{R}}^{\mathrm{neq}(\alpha)}(\mathcal{I}_{\bar{\mathbf{C}}^{e(\alpha)}}) + \psi_{\mathrm{R}}^{\mathrm{es}}(\mathbf{C}, \mathbf{d}_{\mathrm{R}}). \tag{23.3.3}$$

Here:

(i) The contribution $\psi_{\rm R}^{\rm eq}$ is the change in free energy due to *equilibrium* elastic response of the elastomer. We assume that this response is compressible, but consider the case when the compressibility is small. As a special case for this equilibrium mechanical energy we consider the Arruda–Boyce form (Arruda and Boyce, 1993; Anand, 1996) in which, with

$$\bar{\lambda} \stackrel{\rm def}{=} \frac{1}{\sqrt{3}}\sqrt{\operatorname{tr}\bar{\mathbf{C}}} \tag{23.3.4}$$

defining an *effective stretch*,

$$\begin{aligned}\psi_{\rm R}^{\rm eq} = G_{\rm eq0}\,\lambda_L^2 \left[\left(\frac{\bar{\lambda}}{\lambda_L}\right)\beta + \ln\left(\frac{\beta}{\sinh\beta}\right) - \left(\frac{1}{\lambda_L}\right)\beta_0 - \ln\left(\frac{\beta_0}{\sinh\beta_0}\right)\right] \\ + \frac{1}{2}K(\ln J)^2,\end{aligned} \tag{23.3.5}$$

with

$$\beta \stackrel{\rm def}{=} \mathcal{L}^{-1}\left(\frac{\bar{\lambda}}{\lambda_L}\right), \qquad \text{and} \qquad \beta_0 \stackrel{\rm def}{=} \mathcal{L}^{-1}\left(\frac{1}{\lambda_L}\right), \tag{23.3.6}$$

where $\mathcal{L}^{-1}$ is the inverse of the Langevin function

$$\mathcal{L}(z) = \coth(z) - (z)^{-1}.$$

This functional form for the free energy involves three material parameters:

- $G_{\rm eq0}$, a ground-state equilibrium shear modulus;
- λ_L, the network locking stretch;[1] and
- K, a bulk modulus.

(ii) To model the non-equilibrium contributions to the free energy we adopt a simple neo-Hookean form

$$\psi_{\rm R}^{{\rm neq}(\alpha)}(\mathcal{I}_{\bar{\mathbf{C}}^{e(\alpha)}}) = \frac{1}{2}\,G_{\rm neq}^{(\alpha)}\left(\bar{I}_1^{e(\alpha)} - 3\right), \tag{23.3.7}$$

where $G_{\rm neq}^{(\alpha)}$ represent shear moduli.

[1] In the non-Gaussian model, the network locking stretch λ_L is related to the number of links n in a freely jointed chain by $\lambda_L = \sqrt{n}$.

(iii) The contribution ψ_R^{es} is an electrostatic energy. We consider a special form for this energy (Zhao et al., 2007; Suo, 2010),

$$\psi_R^{es} = J\left(\frac{1}{2\varepsilon}\mathbf{d}\cdot\mathbf{d}\right), \tag{23.3.8}$$

which is *isotropic in the deformed body*. The scalar parameter $\varepsilon = \varepsilon_r\,\varepsilon_0$ represents the *permittivity* of the material, with ε_0 the *permittivity of free space* and ε_r the *relative permittivity* of the material.

Since $\mathbf{d} = J^{-1}\mathbf{F}\mathbf{d}_R$, we may express this free energy as

$$\psi_R^{es} = J^{-1}\left(\frac{1}{2\varepsilon}\mathbf{d}_R\cdot\mathbf{C}\mathbf{d}_R\right). \tag{23.3.9}$$

Finally, using (23.3.5), (23.3.7), and (23.3.9) the overall free energy is taken as

$$\begin{aligned}\psi_R = {}& \underbrace{G_{eq0}\,\lambda_L^2\left[\left(\frac{\bar\lambda}{\lambda_L}\right)\beta + \ln\left(\frac{\beta}{\sinh\beta}\right) - \left(\frac{1}{\lambda_L}\right)\beta_0 - \ln\left(\frac{\beta_0}{\sinh\beta_0}\right)\right] + \frac{1}{2}K(\ln J)^2}_{\psi_R^{eq}} \\ & + \underbrace{\sum_{\alpha=1}^{M}\frac{1}{2}G_{neq}^{(\alpha)}\left(\bar I_1^{e(\alpha)} - 3\right)}_{\psi_R^{neq}} \\ & + \underbrace{J^{-1}\left(\frac{1}{2\varepsilon}\mathbf{d}_R\cdot\mathbf{C}\mathbf{d}_R\right)}_{\psi_R^{es}}.\end{aligned} \tag{23.3.10}$$

Electric field. Electric displacement

From (23.3.10) and $(23.2.6)_2$ the referential electric field is given by

$$\mathbf{e}_R = \frac{1}{\varepsilon}J^{-1}\mathbf{C}\mathbf{d}_R, \tag{23.3.11}$$

and hence the referential electric displacement is given by

$$\mathbf{d}_R = \varepsilon\,J\mathbf{C}^{-1}\mathbf{e}_R. \tag{23.3.12}$$

Further, using the conversion relations $\mathbf{e} = \mathbf{F}^{-\top}\mathbf{e}_R$ and $\mathbf{d} = J^{-1}\mathbf{F}\mathbf{d}_R$, we obtain that the spatial electric displacement is related to the spatial electric field by

$$\mathbf{d} = \varepsilon\mathbf{e}. \tag{23.3.13}$$

Stress

From (23.3.10) and (23.2.8) it follows that the Cauchy stress is given by,

$$\mathbf{T} = \mathbf{T}^{\mathrm{eq}} + \sum_{\alpha=1}^{M} \mathbf{T}^{\mathrm{neq}(\alpha)} + \mathbf{T}^{\mathrm{es}}, \quad \text{where}$$

$$\mathbf{T}^{\mathrm{eq}} \stackrel{\text{def}}{=} J^{-1}\mathbf{F}\left(2\,\frac{\partial \psi_{\mathrm{R}}^{\mathrm{eq}}}{\partial \mathbf{C}}\right)\mathbf{F}^{\top}, \qquad \mathbf{T}^{\mathrm{neq}(\alpha)} \stackrel{\text{def}}{=} J^{-1}\mathbf{F}^{e(\alpha)}\left(2\,\frac{\partial \psi_{\mathrm{R}}^{\mathrm{neq}}}{\partial \mathbf{C}^{e(\alpha)}}\right)\mathbf{F}^{e(\alpha)\top}, \quad \text{and}$$

$$\mathbf{T}^{\mathrm{es}} \stackrel{\text{def}}{=} J^{-1}\mathbf{F}\left(2\,\frac{\partial \psi_{\mathrm{R}}^{\mathrm{es}}}{\partial \mathbf{C}}\right)\mathbf{F}^{\top}. \tag{23.3.14}$$

- Equilibrium stress: Straightforward calculations, mimicking those in Section 1.10 leading to equation (1.10.11), give

$$\mathbf{T}^{\mathrm{eq}} = J^{-1}\,G_{\mathrm{eq}}\,\bar{\mathbf{B}}_0 - p\,\mathbf{1}, \tag{23.3.15}$$

where

$$p \stackrel{\text{def}}{=} -K\left(\frac{\ln J}{J}\right) \tag{23.3.16}$$

is a pressure, and

$$G_{\mathrm{eq}} \stackrel{\text{def}}{=} G_{\mathrm{eq}0}\left(\frac{\lambda_L}{3\bar{\lambda}}\right)\mathcal{L}^{-1}\left(\frac{\bar{\lambda}}{\lambda_L}\right), \tag{23.3.17}$$

is a *generalized shear modulus*. Note that since $\mathcal{L}^{-1}(\bar{\lambda}/\lambda_L) \to \infty$ as $\bar{\lambda} \to \lambda_L$, the generalized modulus $G_{\mathrm{eq}} \to \infty$ as $\bar{\lambda} \to \lambda_L$, *a situation which reflects the limited chain-extensibility of elastomeric materials*. Note also that as $\lambda_L \to \infty$, it can be shown that $G_{\mathrm{eq}} \to G_{\mathrm{eq}0}$, and we recover the Neo-Hookean result.

- Non-equilibrium stress: Mimicking the calculations in Chapter 4 leading to equation (4.3.31) gives the non-equilibrium Cauchy stress contribution from the αth viscous branch as

$$\mathbf{T}^{\mathrm{neq}(\alpha)} = J^{-1}\,G_{\mathrm{neq}}^{(\alpha)}\,\bar{\mathbf{B}}_0^{e(\alpha)}, \tag{23.3.18}$$

where $\bar{\mathbf{B}}^{e(\alpha)} = J^{e(\alpha)-2/3}\mathbf{B}^{e(\alpha)}$ is the distortional elastic left Cauchy–Green tensor for each branch, and the Mandel stresses are given by

$$\mathbf{M}^{e(\alpha)} = G_{\mathrm{neq}}^{(\alpha)}\,\bar{\mathbf{C}}_0^{e(\alpha)}. \tag{23.3.19}$$

- Electrostatic contribution to Cauchy stress: Mimicking the calculations in Chapter 20 leading to equation (20.7.14) gives the electrostatic contribution to the Cauchy stress as

$$\mathbf{T}^{\text{es}} = \varepsilon\Big(\mathbf{e}\otimes\mathbf{e} - \frac{1}{2}(\mathbf{e}\cdot\mathbf{e})\mathbf{1}\Big), \tag{23.3.20}$$

which is the usual form of the *Maxwell stress.*

Hence the total Cauchy stress in eq. (23.3.14) is given by

$$\mathbf{T} = \mathbf{T}^{\text{eq}} + \mathbf{T}^{\text{neq}} + \mathbf{T}^{\text{es}} \qquad \text{where,}$$

$$\mathbf{T}^{\text{eq}} = J^{-1}G_{\text{eq}}\bar{\mathbf{B}}_0 - p\mathbf{1} \quad \text{with} \quad G_{\text{eq}} = G_{\text{eq}0}\left(\frac{\lambda_L}{3\bar{\lambda}}\right)\mathcal{L}^{-1}\left(\frac{\bar{\lambda}}{\lambda_L}\right) \text{ and } p = -K\left(\frac{\ln J}{J}\right),$$

$$\mathbf{T}^{\text{neq}} = \sum_{\alpha=1}^{M}\mathbf{T}^{\text{neq}(\alpha)} = \sum_{\alpha=1}^{M} J^{-1}G^{(\alpha)}_{\text{neq}}\bar{\mathbf{B}}_0^{e(\alpha)},$$

$$\mathbf{T}^{\text{es}} = \varepsilon\Big(\mathbf{e}\otimes\mathbf{e} - \frac{1}{2}(\mathbf{e}\cdot\mathbf{e})\mathbf{1}\Big). \tag{23.3.21}$$

The total Piola stress is then given by the standard relation

$$\mathbf{T}_{\text{R}} = J\,\mathbf{T}\mathbf{F}^{-\top} \tag{23.3.22}$$

with the Cauchy stress given by (23.3.21).

Viscous flow rules. Evolution equations for the internal variables $\dot{\mathbf{C}}^{v(\alpha)}$

Mimicking the discussion in Chapter 4, Section 4.3.3 and Section 4.3.5, the evolution equation for $\mathbf{F}^{v(\alpha)}$ in the non-equilibrium branches is taken as

$$\dot{\mathbf{F}}^{v(\alpha)} = \mathbf{D}^{v(\alpha)}\,\mathbf{F}^{v(\alpha)}, \tag{23.3.23}$$

with the viscous stretching given by

$$\mathbf{D}^{v(\alpha)} = \dot{\bar{\gamma}}^{v(\alpha)}\left(\frac{\mathbf{M}_0^{e(\alpha)}}{2\,\bar{\tau}^{(\alpha)}}\right), \tag{23.3.24}$$

where

$$\dot{\bar{\gamma}}^{v(\alpha)} \stackrel{\text{def}}{=} \sqrt{2}\,|\mathbf{D}^{v(\alpha)}| \tag{23.3.25}$$

is an *equivalent shear–strain rate* for each branch and

$$\bar{\tau}^{(\alpha)} \stackrel{\text{def}}{=} \frac{1}{\sqrt{2}}\,|\mathbf{M}_0^{e(\alpha)}| \tag{23.3.26}$$

defines an *equivalent shear stress* for each branch. We assume that each $\bar{\tau}^{(\alpha)}$ satisfies a strength relation of the form

$$\bar{\tau}^{(\alpha)} = \mathcal{S}^{(\alpha)}(\dot{\bar{\gamma}}^{v(\alpha)}), \tag{23.3.27}$$

and as a special case we assume that

$$\mathcal{S}^{(\alpha)}(\dot{\bar{\gamma}}^{v(\alpha)}) = \eta^{(\alpha)}\dot{\bar{\gamma}}^{v(\alpha)}, \tag{23.3.28}$$

with $\eta^{(\alpha)} > 0$ a constant *viscosity* parameter for each α, so that

$$\bar{\tau}^{(\alpha)} = \eta^{(\alpha)}\dot{\bar{\gamma}}^{v(\alpha)} \qquad \Longrightarrow \qquad \dot{\bar{\gamma}}^{v(\alpha)} = \frac{\bar{\tau}^{(\alpha)}}{\eta^{(\alpha)}}. \tag{23.3.29}$$

With this choice of the strength relation the dissipation inequality (23.2.7) may be written as

$$\mathcal{D} = \sum_{\alpha} \bar{\tau}^{(\alpha)}\dot{\bar{\gamma}}^{v(\alpha)} = \sum_{\alpha} \eta^{(\alpha)}\,(\dot{\bar{\gamma}}^{v(\alpha)})^2 \geq 0. \tag{23.3.30}$$

which is always satisfied since we have chosen positive viscosities, $\eta^{(\alpha)} > 0$.

Note that using (23.3.19) and (23.3.29) we may rewrite the viscous stretching relation (23.3.24) alternatively as

$$\mathbf{D}^{v(\alpha)} = \frac{\bar{\mathbf{C}}_0^{e(\alpha)}}{2\,\tau_r^{(\alpha)}} \qquad \text{with} \qquad \tau_r^{(\alpha)} \stackrel{\text{def}}{=} \frac{\eta^{(\alpha)}}{G_{\text{neq}}^{(\alpha)}}, \tag{23.3.31}$$

where $\tau_r^{(\alpha)}$ represents a *relaxation time* for the αth non-equilibrium branch.

Next, from $\mathbf{C}^{v(\alpha)} = \mathbf{F}^{v(\alpha)\top}\mathbf{F}^{v(\alpha)}$ we obtain that $\dot{\mathbf{C}}^{v(\alpha)} = \mathbf{F}^{v(\alpha)\top}\left(2\mathbf{D}^{v(\alpha)}\right)\mathbf{F}^{v(\alpha)}$. Hence, using (23.3.31),

$$\dot{\mathbf{C}}^{v(\alpha)} = \mathbf{F}^{v(\alpha)\top}\left(\frac{\bar{\mathbf{C}}_0^{e(\alpha)}}{\tau_r^{(\alpha)}}\right)\mathbf{F}^{v(\alpha)}. \tag{23.3.32}$$

Further, since $\bar{\mathbf{C}}^{e(\alpha)} = J^{-2/3}\left(\mathbf{F}^{v(\alpha)-\top}\mathbf{C}\mathbf{F}^{v(\alpha)-1}\right)$ and $\operatorname{tr}\bar{\mathbf{C}}^{e(\alpha)} = J^{-2/3}\,(\mathbf{C}\!:\!\mathbf{C}^{v(\alpha)-1})$, we can write

$$\bar{\mathbf{C}}_0^{e(\alpha)} = J^{-2/3}\left[\left(\mathbf{F}^{v(\alpha)-\top}\mathbf{C}\mathbf{F}^{v(\alpha)-1}\right) - \frac{1}{3}\left(\mathbf{C}\!:\!\mathbf{C}^{v(\alpha)-1}\right)\mathbf{1}\right], \tag{23.3.33}$$

and use of this in (23.3.32) gives the following evolution equation

$$\dot{\mathbf{C}}^{v(\alpha)} = \frac{1}{\tau_r^{(\alpha)}}\left(\bar{\mathbf{C}} - \frac{1}{3}\left(\bar{\mathbf{C}}\!:\!(\mathbf{C}^{v(\alpha)})^{-1}\right)\mathbf{C}^{v(\alpha)}\right). \tag{23.3.34}$$

Lastly, revisiting (23.3.18), the non-equilibrium stresses in each branch may be written in the form

$$\mathbf{T}^{\text{neq}(\alpha)} = J^{-1}\, G_{\text{neq}}^{(\alpha)} \left(\bar{\mathbf{F}}(\mathbf{C}^{v(\alpha)})^{-1}\bar{\mathbf{F}}^{\top} - \frac{1}{3}(\bar{\mathbf{C}}:(\mathbf{C}^{v(\alpha)})^{-1})\mathbf{1} \right). \tag{23.3.35}$$

Note that in (23.3.35) and (23.3.34) the dependence on the internal variables $\mathbf{F}^{v(\alpha)}$ enters only through the *symmetric tensors* $\mathbf{C}^{v(\alpha)} = (\mathbf{F}^{v(\alpha)})^{\top}\mathbf{F}^{v(\alpha)}$ (cf., e.g., Wijaya et al., 2023; Lefévre et al., 2024).

23.4 Governing partial differential equations

1. In this chapter we account for inertial effects. Accordingly, the governing partial differential equation in the absence of body forces is the momentum balance

$$\left.\begin{aligned} \operatorname{Div}\mathbf{T}_{\mathrm{R}} &= \rho_{\mathrm{R}}\ddot{\mathbf{u}}, \\ \frac{p}{K} + \frac{\ln J}{J} &= 0, \end{aligned}\right\} \quad \text{in } \mathcal{B}, \tag{23.4.1}$$

with the Piola stress $\mathbf{T}_{\mathrm{R}}$ given by (23.3.22). In writing (23.4.1) we have *split* the standard equation of motion into two equations because this helps in numerically dealing with the near-incompressibility of the material. With $\partial\mathcal{B}_{\mathrm{u}}$ and $\partial\mathcal{B}_{\mathrm{t_R}}$ denoting complementary subsurfaces of the boundary $\partial\mathcal{B}$ of the body $\mathcal{B}$, we consider a pair of possible boundary conditions:

$$\begin{aligned} \mathbf{u} &= \bar{\mathbf{u}} \quad \text{on } \partial\mathcal{B}_{\mathrm{u}}, \\ \mathbf{T}_{\mathrm{R}}\mathbf{n}_{\mathrm{R}} &= \bar{\mathbf{t}}_{\mathrm{R}} \quad \text{on } \partial\mathcal{B}_{\mathrm{t_R}}. \end{aligned} \tag{23.4.2}$$

In the boundary conditions above $\bar{\mathbf{u}}$ and $\bar{\mathbf{t}}_{\mathrm{R}}$ are prescribed displacement and tractions, both functions of $(\mathbf{X}, t)$.

2. The governing differential equation for electrostatics is Gauss's Law,

$$\operatorname{Div}\mathbf{d}_{\mathrm{R}} = 0 \quad \text{in } \mathcal{B}, \tag{23.4.3}$$

with the referential electric displacement $\mathbf{d}_{\mathrm{R}}$ given by (23.3.12). With $\partial\mathcal{B}_{\phi}$ and $\partial\mathcal{B}_{\mathbf{d}_{\mathrm{R}}}$ denoting complementary subsurfaces of $\partial\mathcal{B}$, we consider a pair of possible boundary conditions:

$$\begin{aligned} \phi &= \bar{\phi} \quad \text{on } \partial\mathcal{B}_{\phi}, \\ \mathbf{d}_{\mathrm{R}}\cdot\mathbf{n}_{\mathrm{R}} &= \bar{\varpi}_{\mathrm{R}} \quad \text{on } \partial\mathcal{B}_{\mathbf{d}_{\mathrm{R}}}. \end{aligned} \tag{23.4.4}$$

In the boundary conditions above $\bar{\phi}$ and $\bar{\varpi}_{\mathrm{R}}$ are *prescribed* electrostatic potential and referential surface charge density, respectively, which are both functions of $(\mathbf{X}, t)$.

REMARK

In considering the boundary conditions (23.4.4) for the electrostatic problem, we have neglected the Maxwell stress and the electric displacement in free space.

Standard initial conditions involve the specifications

$$\mathbf{u}(\mathbf{X},0) = \mathbf{u}_0(\mathbf{X}), \qquad \dot{\mathbf{u}}(\mathbf{X},0) = \dot{\mathbf{u}}_0(\mathbf{X}), \qquad \text{and} \qquad \phi(\mathbf{X},0) = \phi_0(\mathbf{X}) \qquad \text{in } \mathcal{B}, \tag{23.4.5}$$

with $\mathbf{u}_0$, $\dot{\mathbf{u}}_0$ and ϕ_0 prescribed functions of $\mathbf{X} \in \mathcal{B}$.

24 Numerical implementation of the electro-viscoelasticity theory

24.1 Weak forms of the governing partial differential equations

1. **Balance of linear momentum:**
   Consider a test field $\mathbf{u}_{\text{test}}(\mathbf{X})$ in $\mathcal{B}$ which satisfies $\mathbf{u}_{\text{test}}(\mathbf{X}) = \mathbf{0}$ on $\partial\mathcal{B}_{\mathbf{u}}$. If the balance equation $\text{Div}\,\mathbf{T}_{\text{R}} - \rho_{\text{R}}\,\ddot{\mathbf{u}} = \mathbf{0}$ in $\mathcal{B}$ and the traction condition $(\bar{\mathbf{t}}_{\text{R}} - \mathbf{T}_{\text{R}}\mathbf{n}_{\text{R}}) = \mathbf{0}$ on $\partial\mathcal{B}_{\mathbf{t}_{\text{R}}}$ are satisfied, then it is straightforward to show that the weak form

$$\int_{\mathcal{B}} \mathbf{T}_{\text{R}} : \nabla\mathbf{u}_{\text{test}}\,dv_{\text{R}} + \int_{\mathcal{B}} \rho_{\text{R}}\ddot{\mathbf{u}} \cdot \mathbf{u}_{\text{test}}\,dv_{\text{R}} - \int_{\partial\mathcal{B}_{\mathbf{t}_{\text{R}}}} \bar{\mathbf{t}}_{\text{R}} \cdot \mathbf{u}_{\text{test}}\,da_{\text{R}} = 0 \tag{24.1.1}$$

   holds for all admissible test fields $\mathbf{u}_{\text{test}}$ which vanish on $\partial\mathcal{B}_{\mathbf{u}}$. Also, the weak form of the pressure equation $(23.3.1)_2$ is

$$\int_{\mathcal{B}} \left[\frac{p}{K} + \frac{\ln J}{J}\right] p_{\text{test}}\,dv_{\text{R}} = 0, \tag{24.1.2}$$

   where $p_{\text{test}}(\mathbf{X})$ is a test field in $\mathcal{B}$.

2. **Gauss's Law:**
   Mimicking Chapter 21 Section 21.1, consider a scalar test field $\phi_{\text{test}}(\mathbf{X})$ on $\mathcal{B}$ which satisfies $\phi_{\text{test}}(\mathbf{X}) = \mathbf{0}$ in $\partial\mathcal{B}_{\phi}$. If Gauss's Law

$$\text{Div}\,\mathbf{d}_{\text{R}} = 0 \qquad \text{on } \mathcal{B},$$

   and the boundary condition $\mathbf{d}_{\text{R}} \cdot \mathbf{n}_{\text{R}} = \bar{\varpi}_{\text{R}}$ on $\partial\mathcal{B}_{\mathbf{d}_{\text{R}}}$ are satisfied, then it is straightforward to show that the weak form

$$\int_{\mathcal{B}} \mathbf{d}_{\text{R}} \cdot \nabla\phi_{\text{test}}\,dv_{\text{R}} - \int_{\partial\mathcal{B}_{\mathbf{d}_{\text{R}}}} \bar{\varpi}_{\text{R}}\,\phi_{\text{test}}\,da_{\text{R}} = 0. \tag{24.1.3}$$

   holds for all test fields ϕ_{test}.

Introduction to Coupled Theories in Solid Mechanics. Lallit Anand, Eric M. Stewart, Shawn A. Chester, Oxford University Press.
 DOI: 10.1093/oso/9780198986218.003.0025

Summarizing, the weak forms of the governing equations for the coupled problem are

$$\begin{aligned}\int_{\mathcal{B}} \mathbf{T}_{\mathrm{R}} : \nabla \mathbf{u}_{\text{test}}\, dv_{\mathrm{R}} + \int_{\mathcal{B}} \rho_{\mathrm{R}} \ddot{\mathbf{u}} \cdot \mathbf{u}_{\text{test}}\, dv_{\mathrm{R}} - \int_{\partial \mathcal{B}_{\mathbf{t}_{\mathrm{R}}}} \bar{\mathbf{t}}_{\mathrm{R}} \cdot \mathbf{u}_{\text{test}}\, da_{\mathrm{R}} &= 0,\\ \int_{\mathcal{B}} \left[\frac{p}{K} + \frac{\ln J}{J}\right] p_{\text{test}}\, dv_{\mathrm{R}} &= 0,\\ \int_{\mathcal{B}} \mathbf{d}_{\mathrm{R}} \cdot \nabla \phi_{\text{test}}\, dv_{\mathrm{R}} - \int_{\partial \mathcal{B}_{\mathbf{d}_{\mathrm{R}}}} \bar{\varpi}_{\mathrm{R}}\, \phi_{\text{test}}\, da_{\mathrm{R}} &= 0.\end{aligned} \tag{24.1.4}$$

for all vector-valued test functions $\mathbf{u}_{\text{test}}(\mathbf{X})$ which vanish on $\partial \mathcal{B}_{\mathrm{u}}$, all scalar test functions $p_{\text{test}}(\mathbf{X})$, and scalar test functions $\phi_{\text{test}}(\mathbf{X})$ which vanish on $\partial \mathcal{B}_{\phi}$, respectively.

24.2 Time-integration procedure

With a numerical time-integration strategy in mind, the time domain is discretized in a series of N_t time steps,

$$t_{\mathrm{n}+1} = t_{\mathrm{n}} + \Delta t, \qquad \mathrm{n} = 0, 1, \ldots, N_t, \tag{24.2.1}$$

such that $t_{N_t} = T$. Then, in order to solve a residual based on the weak forms (24.1.4) in a $(\mathbf{u}, p, \phi)$-based finite element solution procedure, given

$$(\mathbf{u}_{\mathrm{n}}, \dot{\mathbf{u}}_{\mathrm{n}}, \ddot{\mathbf{u}}_{\mathrm{n}}, \mathbf{C}^{v(\alpha)}_{\mathrm{n}}, \mathbf{T}_{\mathrm{R}\,\mathrm{n}}, \phi_{\mathrm{n}})$$

at time t_{n}, we need to calculate

$$(\dot{\mathbf{u}}_{\mathrm{n}+1}, \ddot{\mathbf{u}}_{\mathrm{n}+1}, \mathbf{C}^{v(\alpha)}_{\mathrm{n}+1}, \mathbf{T}_{\mathrm{R}\,\mathrm{n}+1})$$

in terms of $(\mathbf{u}_{\mathrm{n}+1}, p_{\mathrm{n}+1}, \phi_{\mathrm{n}+1})$ at time $t_{\mathrm{n}+1} = t_{\mathrm{n}} + \Delta t$.

24.2.1 Calculation of $\mathbf{C}^{v(\alpha)}_{\mathrm{n}+1}$ and $\mathbf{T}_{\mathrm{R}\,\mathrm{n}+1}$

Recall the evolution equation for the internal variable $\mathbf{C}^{v(\alpha)}$ (23.2.35), viz.

$$\dot{\mathbf{C}}^{v(\alpha)} = \frac{1}{\tau_r^{(\alpha)}} \left(\bar{\mathbf{C}} - \frac{1}{3}\left(\bar{\mathbf{C}} : (\mathbf{C}^{v(\alpha)})^{-1}\right) \mathbf{C}^{v(\alpha)}\right). \tag{24.2.2}$$

Following the steps in Chapter 5, Section 5.3, the update formula becomes

$$\mathbf{C}^{v(\alpha)}_{\mathrm{n}+1} = \left(\det\left[\mathbf{C}^{v(\alpha)}_{\mathrm{n}} + \frac{\Delta t}{\tau_r^{(\alpha)}} \bar{\mathbf{C}}_{\mathrm{n}+1}\right]\right)^{-1/3} \left(\mathbf{C}^{v(\alpha)}_{\mathrm{n}} + \frac{\Delta t}{\tau_r^{(\alpha)}} \bar{\mathbf{C}}_{\mathrm{n}+1}\right). \tag{24.2.3}$$

Finally, recalling (23.2.22) and (23.2.21), the total Piola stress at the end of the step is given by

$$
\begin{aligned}
&\mathbf{T}_{\mathrm{R}\,\mathrm{n}+1} = J_{\mathrm{n}+1}\,\mathbf{T}_{\mathrm{n}+1}\mathbf{F}_{\mathrm{n}+1}^{-\top}, \qquad \text{with}\\
&\mathbf{T}_{\mathrm{n}+1} = \mathbf{T}^{\mathrm{eq}}_{\mathrm{n}+1} + \sum_{\alpha}\mathbf{T}^{\mathrm{neq}(\alpha)}_{\mathrm{n}+1} + \mathbf{T}^{\mathrm{es}}_{\mathrm{n}+1}, \qquad \text{where}\\
&\mathbf{T}^{\mathrm{eq}}_{\mathrm{n}+1} = J_{\mathrm{n}+1}^{-1}\,(G_{\mathrm{eq}})_{\mathrm{n}+1}\,(\bar{\mathbf{B}}_0)_{\mathrm{n}+1} - p_{\mathrm{n}+1}\,\mathbf{1} \qquad \text{with}\\
&\qquad (G_{\mathrm{eq}})_{\mathrm{n}+1} = G_{0\mathrm{eq}}\left(\frac{\lambda_L}{3\bar{\lambda}_{\mathrm{n}+1}}\right)\mathcal{L}^{-1}\left(\frac{\bar{\lambda}_{\mathrm{n}+1}}{\lambda_L}\right),\\
&\mathbf{T}^{\mathrm{neq}(\alpha)}_{\mathrm{n}+1} = J_{\mathrm{n}+1}^{-1}\,G^{(\alpha)}_{\mathrm{neq}}\left(\bar{\mathbf{F}}_{\mathrm{n}+1}(\mathbf{C}^{v(\alpha)}_{\mathrm{n}+1})^{-1}\bar{\mathbf{F}}^{\top}_{\mathrm{n}+1} - \frac{1}{3}(\bar{\mathbf{C}}_{\mathrm{n}+1}:(\mathbf{C}^{v(\alpha)}_{\mathrm{n}+1})^{-1})\mathbf{1}\right), \qquad \text{and}\\
&\mathbf{T}^{\mathrm{es}}_{\mathrm{n}+1} = \varepsilon\left(\mathbf{e}_{\mathrm{n}+1}\otimes\mathbf{e}_{\mathrm{n}+1} - \tfrac{1}{2}(\mathbf{e}_{\mathrm{n}+1}\cdot\mathbf{e}_{\mathrm{n}+1})\mathbf{1}\right),
\end{aligned}
\tag{24.2.4}
$$

where $\qquad \mathbf{e}_{\mathrm{n}+1} = \mathbf{F}_{\mathrm{n}+1}^{-\top}\mathbf{e}_{\mathrm{R}\,\mathrm{n}+1} \quad$ with $\quad \mathbf{e}_{\mathrm{R}\,\mathrm{n}+1} = -\nabla\phi_{\mathrm{n}+1}.$

24.2.2 Calculation of $\ddot{\mathbf{u}}_{\mathrm{n}+1}$ and $\dot{\mathbf{u}}_{\mathrm{n}+1}$

Next, to solve $(24.1.4)_1$ a kinematic relationship is required between $(\ddot{\mathbf{u}}_{\mathrm{n}+1}, \dot{\mathbf{u}}_{\mathrm{n}+1})$ and $\mathbf{u}_{\mathrm{n}+1}$ so that the residuals (24.1.4) become a function solely of $(\mathbf{u}_{\mathrm{n}+1}, p_{\mathrm{n}+1}, \phi_{\mathrm{n}+1})$. Using the notation

$$\mathbf{a} \equiv \ddot{\mathbf{u}} \qquad \text{and} \qquad \mathbf{v} \equiv \dot{\mathbf{u}}$$

for the acceleration and velocity, respectively, we approximate these terms using the Newmark time-integration scheme (Newmark, 1959):

$$
\begin{aligned}
\mathbf{a}_{\mathrm{n}+1} &= \frac{\mathbf{u}_{\mathrm{n}+1} - \mathbf{u}_{\mathrm{n}} - \mathbf{v}_{\mathrm{n}}\,\Delta t}{\beta\Delta t^2} - \frac{(1-2\beta)}{2\beta}\,\mathbf{a}_{\mathrm{n}},\\
\mathbf{v}_{\mathrm{n}+1} &= \mathbf{v}_{\mathrm{n}} + \Delta t\,[(1-\gamma)\mathbf{a}_{\mathrm{n}} + \gamma\,\mathbf{a}_{\mathrm{n}+1}],
\end{aligned}
\tag{24.2.5}
$$

where γ and β are parameters which can be chosen in the ranges $\gamma \in (0,1]$ and $\beta \in (0, 0.5]$ to yield different time-integration schemes.

In fact, since time-discrete solutions often contain small amplitude high frequency oscillations, one can use an enhanced version of the Newmark method called the α-method (Hilber et al., 1977) to dampen such high-frequency oscillations. In this method the Newmark kinematic approximations (24.2.5) are retained for evaluating $\mathbf{a}_{\mathrm{n}+1}$ and $\mathbf{v}_{\mathrm{n}+1}$. However, one introduces a positive-valued parameter $\alpha \in [0, 1)$, and replaces $\mathbf{u}_{\mathrm{n}+1}$ and $\mathbf{v}_{\mathrm{n}+1}$ with $\mathbf{u}$ and $\mathbf{v}$ evaluated at an intermediate time

$$t_\alpha = (1-\alpha)t_{\mathrm{n}+1} + \alpha t_{\mathrm{n}} = t_{\mathrm{n}+1} - \alpha\Delta t$$

between time steps $\mathrm{n}+1$ and n, that is

$$\begin{aligned} \mathbf{u}_\alpha &= (1-\alpha)\mathbf{u}_{\mathrm{n}+1} + \alpha\mathbf{u}_{\mathrm{n}}, \\ \mathbf{v}_\alpha &= (1-\alpha)\mathbf{v}_{\mathrm{n}+1} + \alpha\mathbf{v}_{\mathrm{n}}, \end{aligned} \tag{24.2.6}$$

so that displacement and velocity are linearly interpolated to an *intermediate* time t_α.

In principle, the algorithmic parameters α, β, and γ of the α-method may be independent. However, care must be taken to achieve consistency, stability, and favorable dissipation properties of the time-integration procedure; cf. Erlicher et al. (2002). Note that:

1. If $\alpha = 0$, the method reverts back to the Newmark method.
2. If

   $$\alpha = 0, \qquad \gamma = 1/2, \qquad \beta = 1/4, \tag{24.2.7}$$

   then one obtains the trapezoidal time-integration scheme which is known to be unconditionally stable, and hence is preferred in many numerical simulations.
3. If the parameters α, γ, and β are selected such that,

   $$\alpha \le \tfrac{1}{2}, \qquad \gamma = \tfrac{1}{2} + \alpha, \qquad \beta = \frac{1}{4}(\gamma + \tfrac{1}{2})^2, \tag{24.2.8}$$

   then this results in an unconditionally stable and second-order accurate scheme, provided the time step remains constant.

For further detailed discussions regarding the choices for (α, γ, β)—and also other time-integration schemes—see Erlicher et al. (2002) and Bonet et al. (2021).

REMARK

In the representative simulations for dielectric visco-elastomers shown in Chapter 25, the presence of viscoelasticity provides sufficient damping for numerical stability and we therefore use $\alpha = 0$. However, for completeness the numerical implementation in this chapter and the accompanying FEniCS codes still include the α-method relations.

24.3 Solution procedure in the FEniCS code

For ease of notation we will write

$$(\bullet) \equiv (\bullet)_{\mathrm{n}+1} \qquad \text{and} \qquad (\bullet)_{\mathrm{old}} \equiv (\bullet)_{\mathrm{n}},$$

for all time-varying quantities $(\bullet)$ evaluated at times $\mathrm{n}+1$ and n, respectively.

In FEniCS we use the following degrees of freedom

$$\{\mathbf{u},\, p, \phi\}, \tag{24.3.1}$$

as well as their known values from the previous step

$$\{\mathbf{u}_{\mathrm{old}},\, p_{\mathrm{old}},\, \phi_{\mathrm{old}}\}, \tag{24.3.2}$$

to calculate the residuals resulting from the weak forms of the governing equations, viz.

$$\begin{aligned}
R_{\mathbf{u}} &= \int_{\mathcal{B}} \mathbf{T}_{\mathrm{R}} : \nabla \mathbf{u}_{\mathrm{test}}\, dv_{\mathrm{R}} + \int_{\mathcal{B}} \rho_{\mathrm{R}} \ddot{\mathbf{u}} \cdot \mathbf{u}_{\mathrm{test}}\, dv_{\mathrm{R}} - \int_{\partial\mathcal{B}_{\mathbf{t}_{\mathrm{R}}}} \bar{\mathbf{t}}_{\mathrm{R}} \cdot \mathbf{u}_{\mathrm{test}}\, da_{\mathrm{R}}, \\
R_p &= \int_{\mathcal{B}} \left[\frac{p}{K} + \frac{\ln J}{J} \right] p_{\mathrm{test}}\, dv_{\mathrm{R}}, \\
R_\phi &= \int_{\mathcal{B}} \mathbf{d}_{\mathrm{R}} \cdot \nabla \phi_{\mathrm{test}}\, dv_{\mathrm{R}} - \int_{\partial\mathcal{B}_{\mathbf{d}}} \bar{\varpi}_{\mathrm{R}}\, \phi_{\mathrm{test}}\, da_{\mathrm{R}}.
\end{aligned} \tag{24.3.3}$$

FEniCS then automatically differentiates these residuals with respect to the separate degrees of freedom to obtain *tangents*. At each step in time, FEniCS uses the residuals and tangents in a global Newton–Raphson solver to determine current values of the degrees of freedom $\{\mathbf{u},\, p, \phi\}$ for which the residuals are zero (to within some small numerical tolerance) and which therefore satisfy the governing equations as well as the initial and boundary conditions (in a weak sense).

REMARK

A mixed finite element method is utilized to solve for the displacement $\mathbf{u}$, the pressure p, and the electric potential ϕ, concurrently. To avoid the numerical instabilities we use a class of elements known as "Taylor–Hood" elements which combine second-order quadratic interpolation for the displacement, and first-order linear interpolation for the scalar pressure and the potential.

24.3.1 **Algorithm for calculation of the residuals**

The algorithm we follow in our code to calculate the residuals in (24.3.3) is summarized below.

1. Begin the step with known values of

$$\{\mathbf{u}_{\mathrm{old}},\, \dot{\mathbf{u}}_{\mathrm{old}},\, \ddot{\mathbf{u}}_{\mathrm{old}},\, p_{\mathrm{old}},\, \mathbf{C}^{v(\alpha)}_{\mathrm{old}}, \phi_{\mathrm{old}}\}. \tag{24.3.4}$$

   where for the first step these quantities are given by the initial conditions for $\mathbf{u}$, $\dot{\mathbf{u}}$, $\ddot{\mathbf{u}}$, p, $\mathbf{C}^{v(\alpha)}$, and ϕ.

2. Calculate the acceleration and velocity using the Newmark scheme:

$$\begin{aligned}\ddot{\mathbf{u}} &= \frac{\mathbf{u} - \mathbf{u}_{\text{old}} - \dot{\mathbf{u}}_{\text{old}}\,\Delta t}{\beta \Delta t^2} - \frac{(1-2\beta)}{2\beta}\,\ddot{\mathbf{u}}_{\text{old}},\\ \dot{\mathbf{u}} &= \dot{\mathbf{u}}_{\text{old}} + \Delta t\,[(1-\gamma)\ddot{\mathbf{u}}_{\text{old}} + \gamma\ddot{\mathbf{u}}].\end{aligned} \tag{24.3.5}$$

3. Calculate the intermediate step values of $\mathbf{u}$ and p, respectively denoted by $\mathbf{u}_{\text{avg}}$ and p_{avg}, for use in the α-method of time integration

$$\begin{aligned}\mathbf{u}_{\text{avg}} &= \alpha\mathbf{u}_{\text{old}} + (1-\alpha)\mathbf{u},\\ p_{\text{avg}} &= \alpha p_{\text{old}} + (1-\alpha)p.\end{aligned} \tag{24.3.6}$$

4. Calculate necessary kinematical quantities for the current step:

$$\begin{aligned}&\mathbf{F} = \mathbf{1} + \nabla\,\mathbf{u}_{\text{avg}}, \quad J = \det\mathbf{F}, \quad \mathbf{C} = \mathbf{F}^{\top}\mathbf{F},\\ &\bar{\mathbf{F}} = J^{-1/3}\mathbf{F}, \quad \bar{\mathbf{C}} = \bar{\mathbf{F}}^{\top}\bar{\mathbf{F}}, \quad \bar{\mathbf{B}} = \bar{\mathbf{F}}\bar{\mathbf{F}}^{\top}, \quad \bar{\lambda} = \sqrt{(\operatorname{tr}\bar{\mathbf{C}})/3}.\end{aligned} \tag{24.3.7}$$

5. Calculate the electric field for the current step:

$$\mathbf{e}_{\text{R}} = -\nabla\phi, \quad \mathbf{e} = \mathbf{F}^{-\top}\mathbf{e}_{\text{R}}.$$

6. Update the internal variables $\mathbf{C}^{v(\alpha)}$:

$$\mathbf{C}^{v(\alpha)} = \left(\det\left[\mathbf{C}^{v(\alpha)}_{\text{old}} + \frac{\Delta t}{\tau_r^{(\alpha)}}\bar{\mathbf{C}}\right]\right)^{-1/3}\left(\mathbf{C}^{v(\alpha)}_{\text{old}} + \frac{\Delta t}{\tau_r^{(\alpha)}}\bar{\mathbf{C}}\right). \tag{24.3.8}$$

7. Calculate the referential electric displacement,

$$\mathbf{d}_{\text{R}} = \varepsilon_r\varepsilon_0\,J\,\mathbf{C}^{-1}\mathbf{e}_{\text{R}}. \tag{24.3.9}$$

8. Calculate the Piola stress,

$$
\begin{aligned}
\mathbf{T}_{\mathrm{R}} &= J\,\mathbf{T}\mathbf{F}^{-\top}, \qquad \text{with} \\
\mathbf{T} &= \mathbf{T}^{\mathrm{eq}} + \sum_{\alpha} \mathbf{T}^{\mathrm{neq}(\alpha)} + \mathbf{T}^{\mathrm{es}}, \qquad \text{where} \\
\mathbf{T}^{\mathrm{eq}} &= J^{-1} G_{\mathrm{eq}}\, \bar{\mathbf{B}}_0 - p\,\mathbf{1}, \\
& \quad G_{\mathrm{eq}} = G_{0\,\mathrm{eq}} \left(\frac{\lambda_L}{3\bar{\lambda}}\right) \mathcal{L}^{-1}\left(\frac{\bar{\lambda}}{\lambda_L}\right), \\
& \quad p = -K\left(\frac{\ln J}{J}\right), \\
\mathbf{T}^{\mathrm{neq}(\alpha)} &= J^{-1} G^{(\alpha)}_{\mathrm{neq}} \left(\bar{\mathbf{F}}(\mathbf{C}^{v(\alpha)})^{-1}\bar{\mathbf{F}}^{\top} - \frac{1}{3}(\bar{\mathbf{C}} : (\mathbf{C}^{v(\alpha)})^{-1})\mathbf{1}\right), \quad \text{and} \\
\mathbf{T}^{\mathrm{es}} &= \varepsilon\left[\mathbf{e}\otimes\mathbf{e} - \tfrac{1}{2}(\mathbf{e}\cdot\mathbf{e})\mathbf{1}\right].
\end{aligned}
\tag{24.3.10}
$$

In our numerical implementation we compute the inverse of the Langevin function $\mathcal{L}^{-1}(x)$ in equation (24.3.10) using a Padé approximation (Cohen, 1991)

$$
\mathcal{L}^{-1}(x) \approx x\,\frac{3 - x^2}{1 - x^2}. \tag{24.3.11}
$$

9. Compute the residuals:

$$
\begin{aligned}
R_{\mathrm{u}} &= \int_{\mathcal{B}} \mathbf{T}_{\mathrm{R}} : \nabla \mathbf{u}_{\text{test}}\, dv_{\mathrm{R}} + \int_{\mathcal{B}} \rho_{\mathrm{R}} \ddot{\mathbf{u}} \cdot \mathbf{u}_{\text{test}}\, dv_{\mathrm{R}} - \int_{\partial\mathcal{B}_{\mathbf{t}_{\mathrm{R}}}} \bar{\mathbf{t}}_{\mathrm{R}} \cdot \mathbf{u}_{\text{test}}\, da_{\mathrm{R}}, \\
R_p &= \int_{\mathcal{B}} \left[\frac{p}{K} + \frac{\ln J}{J}\right] p_{\text{test}}\, dv_{\mathrm{R}}, \\
R_{\phi} &= \int_{\mathcal{B}} \mathbf{d}_{\mathrm{R}} \cdot \nabla \phi_{\text{test}}\, dv_{\mathrm{R}} - \int_{\partial\mathcal{B}_{\mathbf{d}}} \bar{\varpi}_{\mathrm{R}}\, \phi_{\text{test}}\, da_{\mathrm{R}}.
\end{aligned}
\tag{24.3.12}
$$

10. FEniCS solves for the current values of $\{\mathbf{u},\, p,\, \phi\}$.
11. Set the current values as the "old" values for the next step in time,

$$
\{\mathbf{u},\, \dot{\mathbf{u}},\, \ddot{\mathbf{u}},\, p,\, \mathbf{C}^{v(\alpha)}, \phi\} \implies \{\mathbf{u}_{\text{old}},\, \dot{\mathbf{u}}_{\text{old}},\, \ddot{\mathbf{u}}_{\text{old}},\, p_{\text{old}},\, \mathbf{C}^{v(\alpha)}_{\text{old}}\, \phi_{\text{old}}\}. \tag{24.3.13}
$$

12. Repeat the steps above for the desired number of steps in time.

25 Representative simulations of dielectric viscoelastomers

In this chapter we present several representative finite element simulations which demonstrate key features of the response of dielectric viscoelastomers as well as their interaction with inertial effects. The material parameters that we have used for our simulations are listed in Table 25.1.[1]

Table 25.1 Material parameters for VHB-4910 from Wang et al. (2016). Note that for the non-equilibrium branches the relation times $\tau_r^{(\alpha)}$ may be converted to viscosities $\eta^{(\alpha)}$ using $\eta^{(\alpha)} = \tau_r^{(\alpha)} \times G_{\rm neq}^{(\alpha)}$.

Quantity	Value SI (m)	Value SI (mm)
$\rho_{\rm R}$, mass density	$1000\ \text{kg/m}^3$	$1 \times 10^{-6}\ \text{kg/mm}^3$
$G_0^{\rm eq}$, ground-state equilibrium shear modulus	15.36×10^3 Pa	15.36 kPa
K, bulk modulus	15.36×10^6 Pa	15.36×10^3 kPa
λ_l, locking stretch	5.99	5.99
$G_{\rm neq}^{(1)}$, non-equilibrium shear modulus	26.06×10^3 Pa	26.06 kPa
$\tau^{(1)}$, relaxation time	0.6074 s	0.6074 s
$G_{\rm neq}^{(2)}$, non-equilibrium shear modulus	26.53×10^3 Pa	26.53 kPa
$\tau^{(2)}$, relaxation time	6.56 s	6.56 s
$G_{\rm neq}^{(3)}$, non-equilibrium shear modulus	10.83×10^3 Pa	10.83 kPa
$\tau^{(3)}$, relaxation time	61.25 s	61.25 s
ε_0, permittivity of free space	8.85×10^{-12} F/m	8.85×10^{-3} pF/mm
ε_r, relative permittivity	4.8	4.8

25.1 Electrostatic actuation of a three-dimensional cube

We first examine the same basic homogeneous one-dimensional actuation problem discussed in Section 22.1, where a thinning "pull-in" instability is known to occur. As in Fig. 22.1, we consider a cube of edge length $\ell_0 = 1$ mm. However, in contrast to our considerations in

[1] We use the unit system listed in Table 21.1 of Section 21.3.

Introduction to Coupled Theories in Solid Mechanics. Lallit Anand, Eric M. Stewart, Shawn A. Chester, Oxford University Press.
 DOI: 10.1093/oso/9780198986218.003.0026

Section 22.1, here multiple rates of electrostatic loading are used in order to delineate the contribution due to viscoelastic effects.

The material parameters used in the simulation are those listed in Table 25.1. The initial conditions are $\mathbf{u}_0 = \mathbf{0}$, and $\phi_0 = 0$ V, and with reference to Fig. 22.1 the boundary conditions are as follows:

- Mechanical boundary conditions: (i) faces ADHE, ABFE, and ADCB are constrained from moving in the directions normal to the coordinate axes; and (ii) all other faces are traction-free.
- Electrostatic boundary conditions: (i) the bottom face is grounded, i.e., $\bar{\phi} = 0$ kV; (ii) the top face is prescribed a positive normalized electrostatic potential that linearly ramps at the following rates,

$$\frac{\dot{\phi}/\ell_0}{\sqrt{G_0/\varepsilon}} = \{10^{-3},\ 10^{-2},\ 10^{-1},\ 10^{0},\ 10^{1}\}/\text{s};$$

  and (iii) all other faces are prescribed zero surface charge, i.e., $\bar{\varpi}_{\text{R}} = 0$.

The simulations are run until convergence is not achieved at the onset of the instability.

Figure 25.1 shows the simulation results in the form of axial stretch versus normalized potential. As expected, this viscoelastic dielectric material also exhibits a "pull-in" thinning instability; however, the onset is now clearly rate-dependent. An increased loading rate $\dot{\phi}$ delays the onset of instability, while a decreased loading rate tends toward the equilibrium response studied in in Section 22.1.

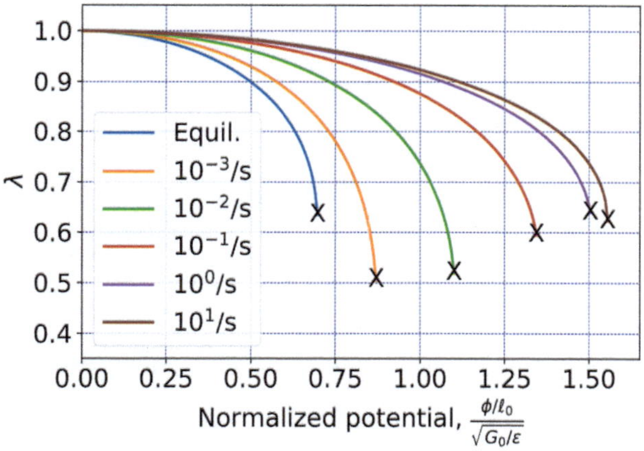

Fig. 25.1 Axial stretch λ (in the $\mathbf{e}_2$-direction) as a function of the normalized electric potential, $\frac{\phi/\ell_0}{\sqrt{G_0/\varepsilon}}$, for electric potential ramp rates of $\frac{\dot{\phi}/\ell_0}{\sqrt{G_0/\varepsilon}} = 10^{-3}$/s, 10^{-2}/s, 10^{-1}/s, 10^{0}/s, and 10^{1}/s; the equilibrium response is also shown. The onset of the instability is indicated by an "X", and it is delayed as the loading rate $\dot{\phi}$ increases.

25.2 Electrostatic actuation of a cantilever bilayer beam by a step electrostatic potential

In this example we simulate the actuation of a bilayer cantilever beam, in which the top half of the dielectric viscoelastomer beam has an electric field applied to it but the bottom half does not. As shown in Fig. 25.2, the beam has a length of $L = 50\,\text{mm}$, a depth $B = 2\,\text{mm}$, and a height $H = 2\,\text{mm}$ and is meshed with 6 tetrahedral elements through the full thickness, 3 for each layer. The material parameters used for the dielectric elastomer beam are those listed in Table 25.1.

The initial conditions are $\mathbf{u}_0 = \mathbf{0}$, and $\phi_0 = 0\,\text{V}$. With reference to Fig. 25.2, the boundary conditions in this problem are as follows:

- Mechanical boundary conditions: (i) the left face of the beam is fixed, i.e., $\bar{\mathbf{u}} = \mathbf{0}$; and (ii) all other faces are traction-free.
- Electrostatic boundary conditions: (i) the mid-plane of the beam separating the two layers is grounded, i.e., $\bar{\phi} = 0\,\text{kV}$; the top face is prescribed a step input in electrostatic potential with a finite ramp-up time, $\bar{\phi}(t) = A(1 - \exp(-t/\tau))$, with amplitude $A = 5\,\text{kV}$ and ramp-up time constant $\tau = 0.25\,\text{s}$; and (ii) all other faces are prescribed zero surface charge $\bar{\varpi}_{\text{R}} = 0$.

The simulation is performed for a total of 5 seconds.

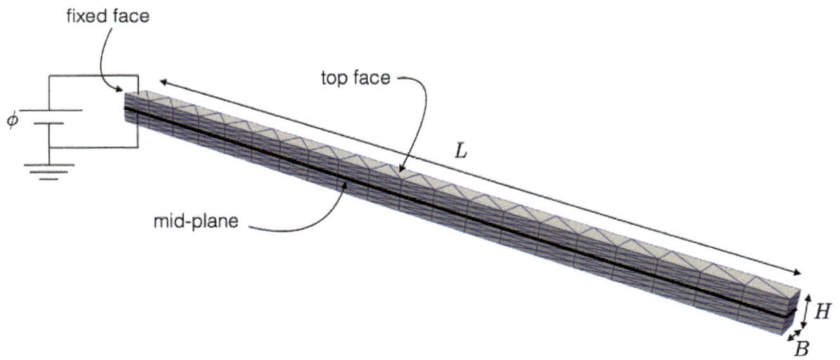

Fig. 25.2 The initial geometry, and finite element mesh used for the bilayer cantilever bending simulation.

As expected, an application of the potential $\bar{\phi}(t)$ causes the beam to bend because the top layer of the beam expands while the bottom layer prevents overall extension of the beam, leading to an overall bending of the beam. Figure 25.3 shows the deformed configuration of the beam at maximum deflection.

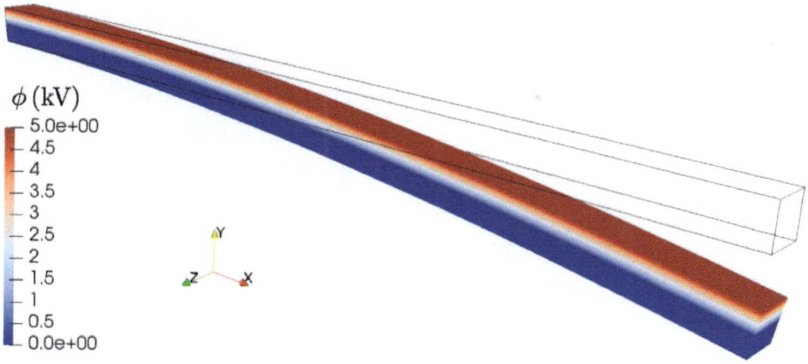

Fig. 25.3 Reference configuration and deformed configuration at maximum deflection, showing contours of the electrostatic potential.

Figure 25.4 shows the time history of the prescribed electrostatic potential and the corresponding tip-displacement of the beam. Note that for an input electrostatic potential which is essentially a step loading, the tip-displacement oscillates due to the interplay between electroelastic, inertial, and viscoelastic effects.

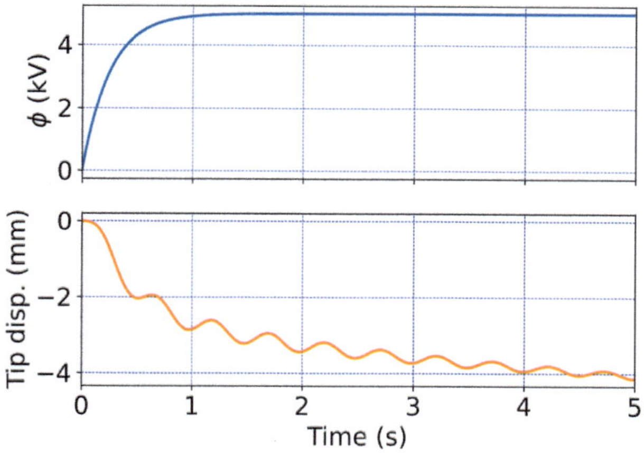

Fig. 25.4 Time history of the prescribed electrostatic potential and corresponding displacement of the tip of the beam.

25.3 Blistering of an electrostatically actuated membrane

As discussed previously in Section 22.1 and Section 25.1, electrostatic actuation can lead to a "pull-in" instability, and as described in Section 22.3, that instability may be suppressed with an adequate pre-stretch. However, simulations in the literature have shown that when the potential is applied to only a portion of a mechanically constrained membrane, first an out-of-plane "blistering" deformation occurs, which eventually culminates in a "pull-in" type instability (cf., e.g., Wang et al., 2016).

In this example, we simulate the response of a square dielectric viscoelastomer membrane that is mechanically fixed along its outer edges, and subjected to electrostatic actuation only in a central circular region, as shown in Fig. 25.5. The membrane has an initial thickness

$T_0 = 0.612\,\text{mm}$ and initial edge length $L_0 = 150\,\text{mm}$. The central region which serves as the electrode has radius $R_0^e = 90\,\text{mm}$, and is indicated by the region enclosed by the blue circle, on both top and bottom faces. Due to symmetry of the problem we only model a quarter of the geometry, with a mesh shown in Fig. 25.5 which contains 2 tetrahedral elements through the thickness. The material parameters used in the simulation for the dielectric elastomer are those listed in Table 25.1.

The initial conditions are $\mathbf{u}_0 = \mathbf{0}$, and $\phi_0 = 0\,\text{V}$. With reference to Fig. 25.5, the boundary conditions in this problem are as follows:

- Mechanical boundary conditions: (i) the external faces are prescribed $\bar{\mathbf{u}} = \mathbf{0}$; (ii) appropriate symmetry conditions are prescribed along the planes of symmetry; and (iii) the top and bottom faces are traction-free.
- Electrostatic boundary conditions: (i) the top electrode face is prescribed a positive normalized electrostatic potential that linearly ramps up at rates of

$$\frac{\dot{\phi}/T_0}{\sqrt{G_0/\varepsilon}} = \{10^{-3},\, 10^{-2},\, 10^{-1}\}/\text{s};$$

(ii) the bottom face is grounded, $\bar{\phi} = 0\,\text{V}$, and (iii) all other faces are prescribed zero surface charge $\bar{\varpi}_{\text{R}} = 0$.

The simulations shown are run until non-convergence occurs, which indicates the occurrence of a "pull-in" instability.

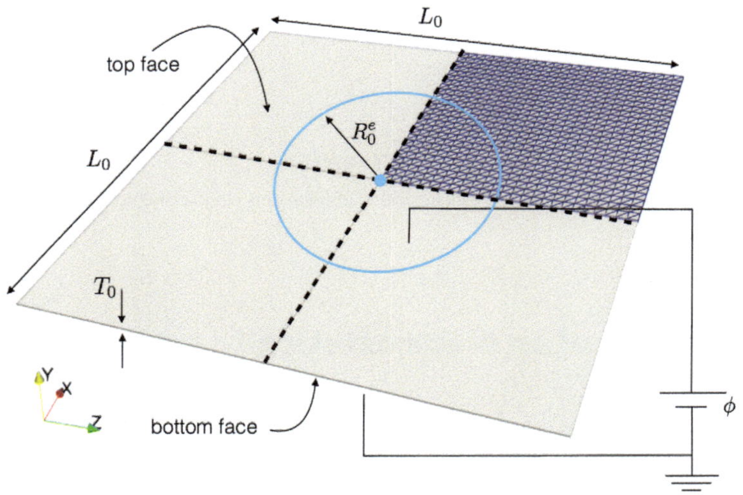

Fig. 25.5 The initial geometry, finite element mesh, and schematic of boundary conditions used for the membrane blistering simulations. The area enclosed by the blue circle is the electrode area. Because of symmetry only one quarter of the geometry is meshed; the planes of symmetry are denoted by the dashed lines.

Figure 25.6 shows the displacement component u_2 at three different points in time for the case $\frac{\dot{\phi}/T_0}{\sqrt{G_0/\varepsilon}} = 10^{-1}$/s. Upon electrostatic actuation the electrode region dramatically thins, while the electrode area increases. Due to the mechanical constraint around the outer faces of the membrane, an out-of-plane "blistering" deformation ensues, which culminates in a "pull-in" instability.

The out-of-plane blistering deformation is easily explained: since the outer edges of the membrane are fixed, an out-of-plane deformation is necessary to accommodate the thinning and areal expansion of the electrode area. Additionally, after the formation of the initial smooth blister seen in Fig. 25.6(b), the transition region between the two nominally flat regions of the membrane (analogous to a cliff face) undergoes a wrinkling instability, as seen in Fig. 25.6(c). Since the transition region is within the electrode area it is continually thinning, and this puts it in circumferential compression which leads to the wrinkling.

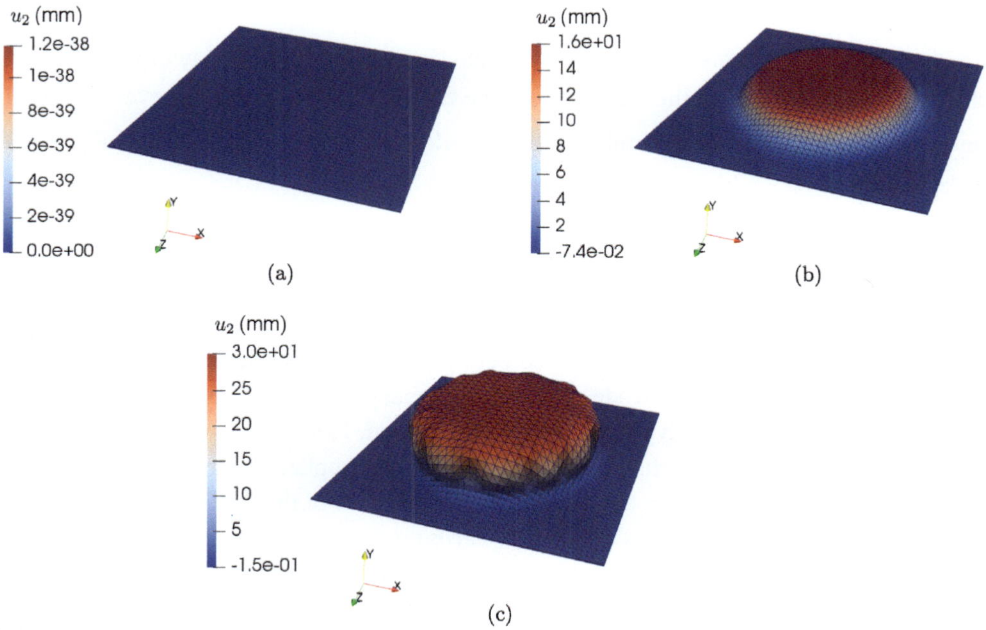

Fig. 25.6 Contours of the displacement component u_2 on (a) the initial body, (b) at 100 s, and (c) at the onset of instability at 121 s for case of $\frac{\dot{\phi}/\ell_0}{\sqrt{G_0/\varepsilon}} = 10^{-1}$. Note that the body has been mirrored about both symmetry planes to show the full geometry.

Figure 25.7 shows plots of the normalized displacement component (u_2/T_0) at the top center of the blister, as a function of the applied normalized electrostatic potential $\frac{\phi/T_0}{\sqrt{G_0/\varepsilon}}$ at the three different $\frac{\dot{\phi}/T_0}{\sqrt{G_0/\varepsilon}}$ rates. The points "X" in these plots indicate non-convergence of

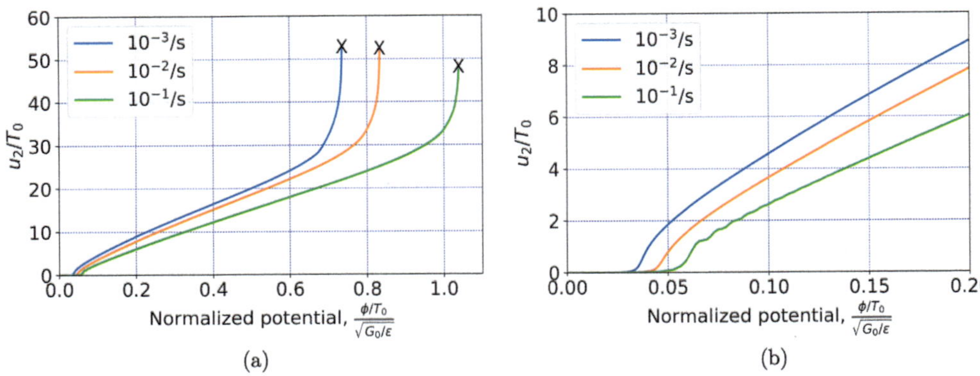

Fig. 25.7 (a) The normalized displacement component u_2/T_0 as a function of the applied normalized electrostatic potential. (b) Detailed view of the same data near zero applied potential, showing the rate-dependence of the onset of blistering.

the numerical simulations. From the plots we can clearly observe that higher rates of electrostatic loading $\frac{\dot{\phi}/T_0}{\sqrt{G_0/\varepsilon}}$ delay the onset of both the blistering behavior and the eventual pull-in instability of the membrane, similar to the rate-dependent delay of the "pull-in" behavior discussed in Section 25.1.

25.4 Electrocreasing

When a soft dielectric layer is sandwiched between a rigid electrode on the bottom side of the layer and a compliant electrode on the top side, and lateral expansion of the dielectric is prevented, then a pull-in type instability is suppressed. However, experiments conducted by Wang et al. (2011) found that at a critical value of the applied electric field, creases suddenly form at imperfections on the compliant electrode side of the dielectric layer. This type of electromechanical instability is called *electrocreasing*.

In this example we simulate the phenomenon of electrocreasing instability based on (Wang et al., 2016). For simplicity, we adopt a two-dimensional plane-strain idealization, which assumes that the creases will be aligned and regularly spaced along the $\mathbf{e}_1$-direction, and long in the $\mathbf{e}_3$-direction relative to the inter-crease spacing in the $\mathbf{e}_1$-direction, as shown schematically in Fig. 25.8(a).

The periodic nature of the assumed creasing pattern enables modeling only a single cell, indicated by the dashed box in Fig. 25.8(a). Then, due to the symmetry of the cell geometry, we can simulate one half of the cell and approximate the effect of the other half by symmetry boundary conditions, as shown in Fig. 25.8(b). The half cell geometry has initial height l_0 and width $l_0/6$, where in this case $l_0 = 120\,\text{mm}$. The material parameters used for the dielectric viscoelastomer are those listed in Table 25.1.

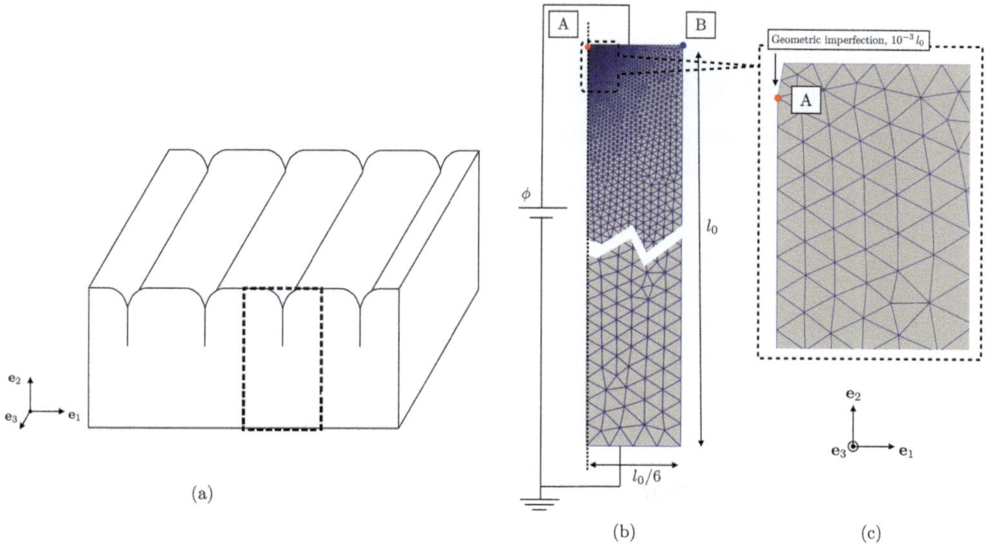

Fig. 25.8 (a) Schematic of a soft dielectric with periodic creases. In our simulations we model a single plane-strain cell as indicated by the dashed box. (b) Finite element mesh and electromechanical boundary conditions used in the two-dimensional, plane-strain electrocreasing simulations. The dashed line along the symmetry plane represents a fixed, frictionless, rigid surface that serves to model self-contact once the crease is formed. (c) Detailed view of the upper-left corner of the mesh, showing the geometric imperfection at point A.

The mesh, shown in in Fig. 25.8(b), consists of 3195 triangular elements with increased refinement toward the region where the crease will form.[2] A small imperfection of $10^{-3}\, l_0$ is introduced to the top face at point A to serve as a geometric inhomogeneity which triggers the electrocreasing instability, Fig. 25.8(c). This geometric imperfection reduces the distance between point A and the lower electrode, resulting in higher electric fields at point A relative to the rest of edge AB for a given potential difference across the electrodes. The locally higher electric fields at point A, in turn, lead to locally higher Maxwell stresses, which drive crease formation.

The initial conditions are $\mathbf{u}_0 = \mathbf{0}$, and $\phi_0 = 0\,\mathrm{V}$, and with reference to Fig. 25.8, the boundary conditions in this problem are as follows:

- Mechanical boundary conditions: (i) The bottom face is fixed by prescribing $\bar{\mathbf{u}} = \mathbf{0}$; (ii) the symmetry planes along the left and right faces are prescribed appropriate symmetry conditions; (iii) frictionless self-contact of the top face AB with the symmetry plane is modeled using the penalty method described in Section 15.5 using a penalty stiffness of 10^6 kPa/mm; and (iv) all other faces are traction-free.

[2] Due to the extreme deformations, the elements in this simulation use a linear interpolation for all degrees of freedom.

- Electrostatic boundary conditions: (i) the bottom face is grounded, i.e., $\bar{\phi} = 0$; and the top face AB is prescribed a positive normalized electrostatic potential that linearly ramps up at a rate of $\frac{\dot{\phi}/l_0}{\sqrt{G_0/\varepsilon}} = 10^{-3}$/s to a value of 1.25, and in one case then reverses at the same rate, or in another case is held fixed; and (ii) all other faces are prescribed zero surface charge $\bar{\varpi}_{\mathrm{R}} = 0$.

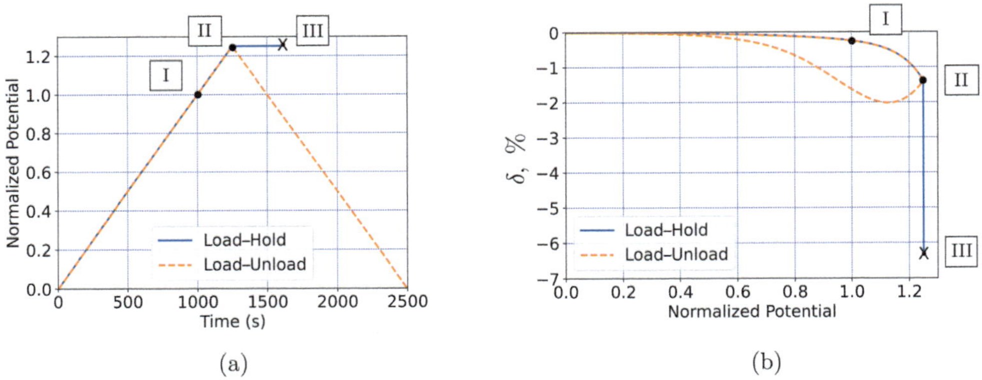

Fig. 25.9 (a) Normalized electrostatic potential for the two loading cases of load–unload (dashed red line) and load–hold (blue line). The deformed configurations corresponding to the points marked I, II, and III are shown in Fig. 25.10. (b) Normalized crease depth δ for the two loading cases. The "X" indicates the onset of numerical non-convergence for the load–hold simulation.

Fig. 25.9(a) shows the time history of the prescribed electrostatic potential for the two loading cases, of load-unload (dashed red line) and load-hold (blue line). Then, we define

$$\delta \stackrel{\text{def}}{=} \left(u_2^{\mathrm{A}} - u_2^{\mathrm{B}}\right) / \, l_0, \tag{25.4.1}$$

where u_2^{A} and u_2^{B} are the vertical displacements of points A and B in Fig. 25.8. Thus δ measures the depth of the crease at point A relative to the overall displacement of the top surface, normalized by the initial height l_0. Figure 25.9(b) plots δ as a function of normalized electrostatic potential for the two loading cases. During loading the rate-dependent stiffening of the soft dielectric delays the onset of the electrocreasing instability.

- In the load–unload case, gradual reduction of the electrostatic potential back to zero leads to a corresponding reduction in crease depth, with some hysteresis.
- However, in the load–hold case viscoelastic relaxation during the holding phase of the simulation leads to runaway deepening of the crease. As the crease deepens the local deformation in the crease becomes severe, which eventually leads to numerical non-convergence indicated by an "X" in Figs. 25.9(a)–(b).

Figures 25.10(a), (b), and (c) show contours of the displacement component u_2 plotted on the deformed configuration during electrocreasing at times 1000 s, 1250 s, and 1612 s, respectively, for the "load–hold" case. The deformation associated with the crease formation is quite severe, as shown in Fig. 25.10(c).

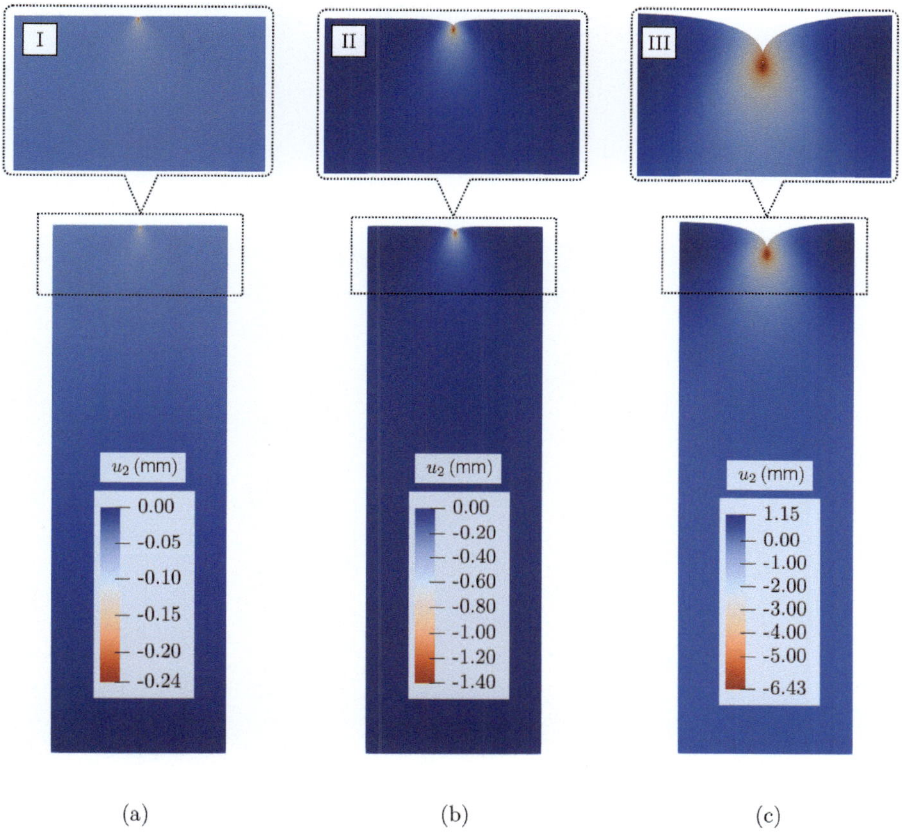

Fig. 25.10 Contour plots of the displacement component u_2 shown on the deformed configuration and a zoomed-in region near the crease, at (a) $t = 1000$ s, (b) $t = 1250$ s, and (c) $t = 1612$ s. These times correspond to the points marked I, II, and III, respectively in Fig. 25.9.

Part IX

Electro-chemo-elasticity of ionic polymers

26 Electro-chemo-elasticity of ionic polymers

26.1 Introduction

An ionic polymer consists of a polymer network with charged functional side-groups, and a solvent with counter-ions filling the interstitial spaces in the network. When a membrane of such a material is stimulated with an electrical potential across its thickness, the mobile ions migrate and redistribute in the material. The redistribution of the mobile ions causes a differential swelling of the membrane, which in turn causes the membrane to deform. Such ionically conductive polymers represent an important class of soft active materials for use in a variety of soft-robotics and biomimetic applications.

In this chapter we formulate a theory which couples

- electrostatics,
- transport of an ionic species, and
- large elastic deformations,

for modeling the response of ionic polymers. For specificity, we will formulate and apply the theory for modeling the response of a class of devices based on ionic polymers known as *ionic polymer–metal composites* (IPMCs), which we describe in the subsection below.

However, we note that the theory presented in this chapter is more general and it should also be useful for other applications involving coupled electro-chemo-mechanical phenomena. For example, recall that *electrolytes* in energy conversion and storage devices are materials that are ion-conductive but electron-insulating—the basic function of an electrolyte is to mediate ion transport between the electrodes in such a device. The theory should also be useful for modeling the response of soft-solid-electrolytes for flexible lithium-ion batteries and fuel-cells.

26.1.1 Ionic polymer–metal composites

Ionic polymer–metal composites (IPMCs) consist of membranes of ionic polymers ($\approx 200\,\mu$m thick), with noble metal electrodes (≈ 5 to $10\,\mu$m thick) plated on both faces of the membrane. The matrix of an IPMC is solvated with a suitable solvent, and neutralized with an amount of cations (e.g., Li^+, Na^+, K^+, or H^+) which balance the charge of the anions that are covalently fixed to the polymer chains.

A prominent example of an ionic polymer used is Nafion.[1] It is made up of cross-linked polymer chains which have a backbone of a perfluorinated copolymer of

[1] Nafion is a brand name of the Chemours company.

Introduction to Coupled Theories in Solid Mechanics. Lallit Anand, Eric M. Stewart, Shawn A. Chester, Oxford University Press.
 DOI: 10.1093/oso/9780198986218.003.0027

polytetrafluoroethylene, with regularly spaced long perfluorovinyl ether pendant side chains that terminate in negatively charged ionic sulfonate groups (anions). Nafion membranes are permeable to water or other polar solvents and contain mobile positively charged ions (cations) that move within the perfluorinated backbone.

When a solvated IPMC membrane is stimulated with a suddenly applied small step potential (≈ 1 to 3 V), both the anions and cations are subjected to an electric field. While the anions remain fixed to the polymer chains, the mobile cations migrate toward the cathode, transporting with them some solvent molecules due to hydration effects. The redistribution of hydrated cations causes a differential swelling of the membrane, and the resulting gradient in the volumetric deformation across the thickness of the membrane causes the IPMC to bend. A schematic of this actuation mechanism is shown in Fig. 26.1. Under an alternating potential, a cantilevered strip of an IPMC performs bending oscillations at the frequency of the applied voltage, usually no more than a few to a few tens of hertz. When an IPMC membrane is suddenly bent, a small voltage of the order of millivolts is produced across its faces. Hence, IPMCs can serve as soft actuators and sensors (cf., e.g., Shahinpoor and Kim, 2001; Nemat-Nasser, 2008; Jo et al., 2013).

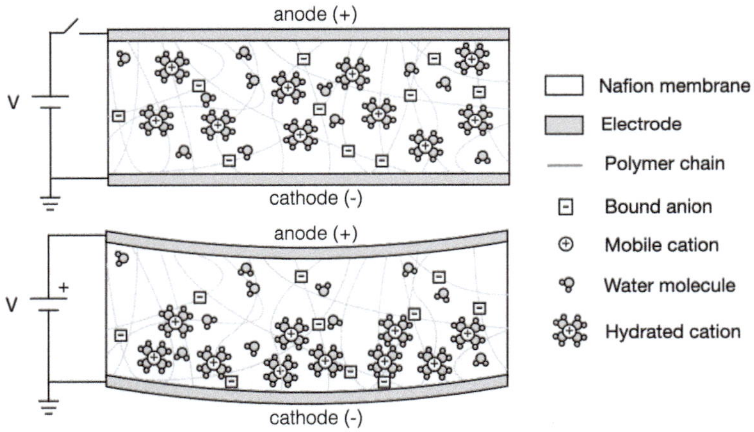

Fig. 26.1 Schematic of the actuation mechanism of an IPMC membrane due to the migration of hydrated cations under an applied voltage.

The low actuation voltage and high compliance of IPMCs make them excellent candidates for use as actuators in both wet and dry environments. They can serve as "artificial muscles" for use in a wide range of soft-robotics and biomimetic applications (Minaian et al., 2021). The use of ionic polymers as actuators was first proposed by Oguro et al. (1992). Subsequent efforts towards understanding the underlying physical mechanisms of IPMCs were led by the experiments performed and models proposed by Shahinpoor and Kim (Shahinpoor and Kim, 2001; Kim and Shahinpoor, 2003; Shahinpoor and Kim, 2004), and Nemat-Nasser and co-workers (Nemat-Nasser and Li, 2000; Nemat-Nasser, 2002). For a more comprehensive discussion of the early experimental and modeling literature on IPMCs, see the book chapters by Nemat-Nasser and Thomas (2004) and Nemat-Nasser (2008), and the references cited therein.

While the major underlying physical mechanism behind the actuation and sensing response of IPMCs is reasonably well understood and has been agreed upon for some time, viz.

migration of mobile hydrated cations under the gradient of an electric potential, the continuum-mechanical modeling of the coupled chemo-electro-mechanical response of IPMCs in a thermodynamically consistent fashion has been a more recent effort in the literature (cf., e.g., Nardinocchi et al., 2011; Galante et al., 2013; Nardinocchi and Pezzulla, 2013; Nardinocchi et al., 2014; Rossi and Wallmersperger, 2018; Rossi et al., 2018; Cha and Porfiri, 2014; Porfiri et al., 2017; Leronni and Bardella, 2021; Narayan et al., 2021).

In this chapter we follow Narayan et al. (2021) and formulate a continuum-mechanical theory for coupled electro-chemo-mechanics. In formulating the theory we make the following major modeling assumptions:

- The polymer–solvent–ions mixture is treated as a single homogenized continuum body allowing for a mass flux to be carried by the mobile hydrated cations.
- The body is under isothermal conditions at a fixed absolute temperature ϑ.
- Since the neutral solvent (water) molecules diffuse slowly, we neglect the effects of solvent diffusion in the membrane.[2]
- Further, since the motion of the cations occurs at timescales considerably longer than those associated with wave propagation in solids, all inertial effects are neglected.

26.2 Theory

26.2.1 Basic kinematics

Consider a macroscopically homogeneous body $\mathcal{B}$ with the region of space it occupies in a fixed *reference configuration*, and denote by $\mathbf{X}$ an arbitrary material point of $\mathcal{B}$.[3] A motion of $\mathcal{B}$ is a smooth one-to-one mapping $\mathbf{x} = \boldsymbol{\chi}(\mathbf{X}, t)$, with $\mathbf{u} = \mathbf{x} - \mathbf{X}$ the displacement, and deformation gradient and velocity given by

$$\mathbf{F} = \nabla\boldsymbol{\chi} = \mathbf{1} + \nabla\mathbf{u}, \qquad \mathbf{v} = \dot{\mathbf{u}}, \qquad \mathbf{L} = \operatorname{grad}\mathbf{v} = \dot{\mathbf{F}}\mathbf{F}^{-1}. \tag{26.2.1}$$

We write

$$J = \det\mathbf{F} > 0, \tag{26.2.2}$$

so that $\mathbf{F}$ is invertible.

Throughout we denote by $\mathcal{P}$ an arbitrary part of the reference body $\mathcal{B}$ with $\mathbf{n}_R$ the outward unit normal on the boundary $\partial\mathcal{P}$ of $\mathcal{P}$. Also, we denote by $\mathcal{P}_t = \boldsymbol{\chi}(\mathcal{P}, t)$ the image of $\mathcal{P}$ in the deformed body $\mathcal{B}_t = \boldsymbol{\chi}(\mathcal{B}, t)$, with $\mathbf{n}$ the outward unit normal on the boundary $\partial\mathcal{P}_t$ of $\mathcal{P}_t$.

[2] Nafion-IPMCs invariably first bend towards the anode under a step voltage, and then relax back towards the cathode while the applied voltage is being maintained. A detailed physical understanding and modeling of this complex phenomena of "back-relaxation"—which occurs at long timescales—is at present incomplete, and the literature on this aspect of the response of IPMCs is mixed (cf., e.g., Arnold et al., 2022).

[3] Notation: We use standard notation of modern continuum mechanics (Gurtin et al., 2010), cf. Section 1.1. However, as a departure from the usual convention in modern continuum mechanics in which vectors are denoted by lower case bold face letters **a**, **b**, etc., we denote the two important electrical vector fields using bold face sans serif fonts as follows: we write $\boldsymbol{\mathsf{e}}$ for the electric field, and $\boldsymbol{\mathsf{d}}$ for the electric displacement.

26.2.2 **Balance of forces and moments**

Standard considerations of balance of forces and moments, when expressed referentially, give the following results:

(a) There exists a stress tensor $\mathbf{T}_{\mathrm{R}}$, called the Piola stress, such that the surface traction on an oriented element of the surface $\partial\mathcal{P}$ of $\mathcal{P}$, is given by the traction relation

$$\mathbf{t}_{\mathrm{R}}(\mathbf{n}_{\mathrm{R}}) = \mathbf{T}_{\mathrm{R}}\mathbf{n}_{\mathrm{R}}. \tag{26.2.3}$$

(b) $\mathbf{T}_{\mathrm{R}}$ satisfies the macroscopic force balance

$$\mathrm{Div}\,\mathbf{T}_{\mathrm{R}} + \mathbf{b}_{\mathrm{R}} = \mathbf{0}, \tag{26.2.4}$$

where $\mathbf{b}_{\mathrm{R}}$ represents the conventional body force per unit volume of the reference body.

(c) $\mathbf{T}_{\mathrm{R}}$ obeys the symmetry condition,

$$\mathbf{T}_{\mathrm{R}}\mathbf{F}^{\top} = \mathbf{F}\mathbf{T}_{\mathrm{R}}^{\top}, \tag{26.2.5}$$

which represents a balance of moments.

As is standard, the stress measure

$$\mathbf{T} = J^{-1}\mathbf{T}_{\mathrm{R}}\mathbf{F}^{\top}, \tag{26.2.6}$$

represents the symmetric Cauchy stress, in terms of which:

(a) The surface traction on an oriented element of the surface $\partial\mathcal{P}_t$ of $\mathcal{P}_t$, is given by

$$\mathbf{t}(\mathbf{n}) = \mathbf{T}\mathbf{n}. \tag{26.2.7}$$

(b) And $\mathbf{T}$ satisfies the macroscopic force balance

$$\mathrm{div}\,\mathbf{T} + \mathbf{b} = \mathbf{0}, \tag{26.2.8}$$

where $\mathbf{b}$ represents the conventional body force per unit volume of the deformed body.

REMARK

The Cauchy stress $\mathbf{T}$ is often referred to as the *total* stress in the electroelasticity literature. However, we see no need to distinguish between a purely mechanical stress and an electrostatic, or Maxwell, stress. Such a decomposition appears naturally when we consider specific thermodynamically consistent constitutive equations; cf. Section 26.2.12.

26.2.3 **Balance of concentration and charge of the migrating cations**

Balance of concentration

Let $c_{\mathrm{R}}(\mathbf{X},t)$ [mol/m^3] denote the total number of moles of mobile cations per unit volume of the reference body. Changes in ion concentration c_{R} in a part $\mathcal{P}$ of the reference body $\mathcal{B}$ are brought about by the migration of the species across its boundary $\partial\mathcal{P}$. The migration is characterized by a referential species flux $\mathbf{j}_{\mathrm{R}}(\mathbf{X},t)$, measured as the number of moles of the species per unit area per unit time, so that

$$-\int_{\partial\mathcal{P}} \mathbf{j}_{\mathrm{R}}\cdot\mathbf{n}_{\mathrm{R}}\,da_{\mathrm{R}}$$

represents the number of moles of cations entering $\mathcal{P}$ across $\partial\mathcal{P}$ per unit time.

Thus the rate of change of the number of cations in $\mathcal{P}$ (assuming no internal sources) is given by

$$\overline{\int_{\mathcal{P}} c_{\mathrm{R}}\,dv_{\mathrm{R}}}^{\,\boldsymbol{\cdot}} = -\int_{\partial\mathcal{P}} \mathbf{j}_{\mathrm{R}}\cdot\mathbf{n}_{\mathrm{R}}\,da_{\mathrm{R}} \tag{26.2.9}$$

for every part $\mathcal{P}$. Bringing the time derivative inside the integral and using the divergence theorem in (26.2.9) on the integral over $\partial\mathcal{P}$, we find that

$$\int_{\mathcal{P}} (\dot{c}_{\mathrm{R}} + \mathrm{Div}\,\mathbf{j}_{\mathrm{R}})\,dv_{\mathrm{R}} = 0, \tag{26.2.10}$$

and since $\mathcal{P}$ is arbitrary, this leads to the following local balance law for referential cation concentration c_{R},

$$\dot{c}_{\mathrm{R}} = -\mathrm{Div}\,\mathbf{j}_{\mathrm{R}}. \tag{26.2.11}$$

This balance may also be expressed in the deformed body as,

$$\dot{c}_{\mathrm{R}} = -J\,\mathrm{div}\mathbf{j} \quad\Longrightarrow\quad \overline{(J\,c)}^{\,\boldsymbol{\cdot}} = -J\,\mathrm{div}\mathbf{j} \quad\Longrightarrow\quad \dot{c} + c\,\mathrm{div}\mathbf{v} = -\mathrm{div}\mathbf{j}, \tag{26.2.12}$$

where $c = J^{-1}c_{\mathrm{R}}$ is the spatial concentration and $\mathbf{j} = J^{-1}\mathbf{F}\mathbf{j}_{\mathrm{R}}$ is the spatial flux of the cations.

Balance of charge

Initially, that is at time $t = 0$, the charge of the mobile univalent cations is

$$F\,\underbrace{c_{\mathrm{R}}^{+}(\mathbf{X},0)}_{\equiv c_{\mathrm{R}}(\mathbf{X},0)}, \tag{26.2.13}$$

and that of the fixed univalent anions is

$$-\,F\,c_{\mathrm{R}}^{-}(\mathbf{X},0), \tag{26.2.14}$$

measured in Coulombs per unit reference volume [C/m^3], where $F = 96,485$ C/mol is the Faraday constant. *Electroneutrality* at time $t = 0$ requires that,

$$F\, c_{\mathrm{R}}^{+}(\mathbf{X},0) - F\, c_{\mathrm{R}}^{-}(\mathbf{X},0) = 0, \tag{26.2.15}$$

or

$$c_{\mathrm{R}}^{+}(\mathbf{X},0) = c_{\mathrm{R}}^{-}(\mathbf{X},0) \equiv \underbrace{c_{\mathrm{R}0}}_{\text{fixed}}. \tag{26.2.16}$$

Then at time $t > 0$ the charge of the mobile cations is

$$F\, c_{\mathrm{R}}(\mathbf{X},t), \tag{26.2.17}$$

and that of the fixed anions is

$$-\, F\, c_{\mathrm{R}0}. \tag{26.2.18}$$

Thus the *net charge* per unit reference volume of the body at times $t > 0$ is

$$q_{\mathrm{R}} = F\,(c_{\mathrm{R}} - c_{\mathrm{R}0}). \tag{26.2.19}$$

Let,

$$\mathbf{i}_{\mathrm{R}} \stackrel{\text{def}}{=} F\mathbf{j}_{\mathrm{R}} \tag{26.2.20}$$

denote the referential *current density* or flux of positive charge due to the migration of the cations. Then the balance law (26.2.11) gives the following balance law for q_{R},

$$\dot{q}_{\mathrm{R}} = -\mathrm{Div}\,\mathbf{i}_{\mathrm{R}}. \tag{26.2.21}$$

26.2.4 **Basic equations of electrostatics**

The referential *electric field* in $\mathcal{B}$ is denoted by $\mathbf{e}_{\mathrm{R}}(\mathbf{X},t)$. In a polarizable material there is a vector field in $\mathcal{B}$ denoted by $\mathbf{d}_{\mathrm{R}}(\mathbf{X},t)$ called the *electric displacement*. In the theory formulated in this chapter we only consider the *electrostatic* (quasi-static) limit as far as the electrical variables are concerned, so that time t only plays the role of a history parameter.

Under electrostatic conditions there are two governing Maxwell equations that must be satisfied. The first of these is *Faraday's law*, which in the electrostatic limit reads,

$$\mathrm{Curl}\,\mathbf{e}_{\mathrm{R}} = \mathbf{0}. \tag{26.2.22}$$

Satisfaction of (26.2.22) occurs automatically if we consider the electric field $\mathbf{e}_R(\mathbf{X}, t)$ to be represented as the gradient of an *electrostatic potential* $\phi(\mathbf{X}, t)$,[4]

$$\mathbf{e}_R(\mathbf{X}, t) \stackrel{\text{def}}{=} -\nabla\phi(\mathbf{X}, t). \tag{26.2.23}$$

The negative sign in (26.2.23) follows the convention that the electric field points from a region of high potential to low potential, whereas the vector $\nabla\phi$ by definition points in the direction of increasing ϕ.

The second equation of electrostatics is given by *Gauss's law*. Gauss's law states that the divergence of the electric displacement $\mathbf{d}_R$ must be equal to the density of the net charge per unit volume q_R,

$$\text{Div}\,\mathbf{d}_R = q_R. \tag{26.2.24}$$

Equations (26.2.22) and (26.2.24) constitute the governing equations for electrostatics in $\mathcal{B}$. Using the conversion relations

$$\mathbf{e} = \mathbf{F}^{-\top}\mathbf{e}_R \qquad \text{and} \qquad \mathbf{d} = J^{-1}\mathbf{F}\mathbf{d}_R, \tag{26.2.25}$$

the spatial counterparts of (26.2.22) and (26.2.24) are

$$\text{curl}\,\mathbf{e} = \mathbf{0}, \tag{26.2.26}$$

and

$$\text{div}\,\mathbf{d} = q \qquad \text{with} \qquad q = F\,(c - c_0), \tag{26.2.27}$$

where

$$c = J^{-1}c_R, \qquad c_0 = J^{-1}c_{R0} \tag{26.2.28}$$

denote the concentration of the cations and fixed anions per unit volume of the deformed body. Equations (26.2.26) and (26.2.27) constitute the governing equations for electrostatics in the deformed body $\mathcal{B}_t$.

26.2.5 **Free-energy imbalance**

Under isothermal conditions the two laws of thermodynamics reduce to the statement that *the rate of increase in free energy of any part $\mathcal{P}$ is less than or equal to the power expended on $\mathcal{P}$.*

[4] Recall that for any scalar field ϕ, $\text{Curl}\,(\nabla\phi) = \mathbf{0}$.

Precisely, letting ψ_R denote the *free energy* per unit reference volume, this requirement takes the form of a free-energy imbalance

$$\overline{\int_{\mathcal{P}} \psi_R \, dv_R}^{\,\cdot} \leq \mathcal{W}(\mathcal{P}) + \mathcal{E}(\mathcal{P}) + \mathcal{T}(\mathcal{P}), \tag{26.2.29}$$

where:

(i) The mechanical power $\mathcal{W}(\mathcal{P})$ exerted on $\mathcal{P}$ by the boundary tractions $\mathbf{T}_R\mathbf{n}_R$ and body force $\mathbf{b}_R$ over the velocity $\dot{\mathbf{u}}$ is given by,

$$\mathcal{W}(\mathcal{P}) = \int_{\partial\mathcal{P}} (\mathbf{T}_R\mathbf{n}_R) \cdot \dot{\mathbf{u}} \, da_R + \int_{\mathcal{P}} \mathbf{b}_R \cdot \dot{\mathbf{u}} \, dv_R. \tag{26.2.30}$$

(ii) In electromagnetic field theory the electric power that is expended on $\mathcal{P}$ by charges external to $\mathcal{P}$ is characterized by an electromagnetic energy flux $\mathcal{E}(\mathcal{P})$ which in the electrostatic limit is given by (cf., eq. (A.3.14)),

$$\mathcal{E}(\mathcal{P}) = -\int_{\partial\mathcal{P}} \phi \, \dot{\mathbf{d}}_R \cdot \mathbf{n}_R \, da_R - \int_{\partial\mathcal{P}} F\phi \, \mathbf{j}_R \cdot \mathbf{n}_R \, da_R. \tag{26.2.31}$$

(iii) Finally to characterize the chemical energy carried into part $\mathcal{P}$ by cation transport we introduce

$$\mathcal{T}(\mathcal{P}) = -\int_{\partial\mathcal{P}} \mu \, \mathbf{j}_R \cdot \mathbf{n}_R \, da_R, \tag{26.2.32}$$

with μ the *chemical potential* of the diffusing species (cf., e.g., Gurtin et al., 2010, p. 65).

Thus, using (26.2.30), (26.2.31), and (26.2.32), the free-energy imbalance (26.2.29) becomes

$$\overline{\int_{\mathcal{P}} \psi \, dv_R}^{\,\cdot} \leq \int_{\partial\mathcal{P}} (\mathbf{T}_R\mathbf{n}_R) \cdot \dot{\mathbf{u}} \, da_R + \int_{\mathcal{P}} \mathbf{b}_R \cdot \dot{\mathbf{u}} \, dv_R - \int_{\partial\mathcal{P}} \phi \, \dot{\mathbf{d}}_R \cdot \mathbf{n}_R \, da_R - \int_{\partial\mathcal{P}} \omega \mathbf{j}_R \cdot \mathbf{n}_R \, da_R, \tag{26.2.33}$$

where we have introduced the notation

$$\omega \stackrel{\text{def}}{=} \mu + F\phi, \tag{26.2.34}$$

for an *electrochemical potential* of the cations.[5]

[5] The electrochemical potential is also often denoted by $\bar{\mu}$ instead of ω in the electrochemistry literature (e.g., equation (19.B.3) in Section 19.B).

Using the divergence theorem on the terms involving the integral over $\partial\mathcal{P}$, (26.2.23), and the species balance (26.2.11) we obtain,

$$
\left.
\begin{aligned}
\int_{\partial\mathcal{P}} (\mathbf{T}_{\mathrm{R}}\mathbf{n}_{\mathrm{R}}) \cdot \dot{\boldsymbol{\chi}}\, da_{\mathrm{R}} &= \int_{\mathcal{P}} (\mathrm{Div}\, \mathbf{T}_{\mathrm{R}}) \cdot \dot{\boldsymbol{\chi}}\, dv + \int_{\mathcal{P}} \mathbf{T}_{\mathrm{R}} : \dot{\mathbf{F}} dv_{\mathrm{R}}, \\
-\int_{\partial\mathcal{P}} (\phi\, \dot{\mathbf{d}}_{\mathrm{R}}) \cdot \mathbf{n}_{\mathrm{R}}\, da_{\mathrm{R}} &= -\int_{\mathcal{P}} \phi\, (\mathrm{Div}\, \dot{\mathbf{d}}_{\mathrm{R}}) dv_{\mathrm{R}} - \int_{\mathcal{P}} (\nabla\phi) \cdot \dot{\mathbf{d}}_{\mathrm{R}}\, dv_{\mathrm{R}} \\
&= -\int_{\mathcal{P}} \phi\, (\mathrm{Div}\, \dot{\mathbf{d}}_{\mathrm{R}}) dv_{\mathrm{R}} + \int_{\mathcal{P}} \mathbf{e}_{\mathrm{R}} \cdot \dot{\mathbf{d}}_{\mathrm{R}}\, dv_{\mathrm{R}}, \\
-\int_{\partial\mathcal{P}} \omega\, \mathbf{j}_{\mathrm{R}} \cdot \mathbf{n}_{\mathrm{R}}\, da_{\mathrm{R}} &= -\left[\int_{\mathcal{P}} \omega\, \mathrm{Div}\, \mathbf{j}_{\mathrm{R}}\, dv_{\mathrm{R}} + \int_{\mathcal{P}} \mathbf{j}_{\mathrm{R}} \cdot \nabla\omega\, dv_{\mathrm{R}} \right] \\
&= \int_{\mathcal{P}} \Big(\omega \dot{c}_{\mathrm{R}} - \mathbf{j}_{\mathrm{R}} \cdot \nabla\omega \Big) dv_{\mathrm{R}}.
\end{aligned}
\right\} \qquad (26.2.35)
$$

Substituting (26.2.35) in (26.2.33), and using (26.2.4), (26.2.34), and the rate form of (26.2.24) we obtain,

$$
\int_{\mathcal{P}} \Big(\dot{\psi}_{\mathrm{R}} - \mathbf{T}_{\mathrm{R}} : \dot{\mathbf{F}} - \mu \dot{c}_{\mathrm{R}} - \mathbf{e}_{\mathrm{R}} \cdot \dot{\mathbf{d}}_{\mathrm{R}} + \mathbf{j}_{\mathrm{R}} \cdot \nabla\omega \Big)\, dv_{\mathrm{R}} \le 0. \qquad (26.2.36)
$$

Since (26.2.36) has to hold for all parts $\mathcal{P}$, we obtain a local free-energy imbalance in the form,

$$
\dot{\psi}_{\mathrm{R}} - \mathbf{T}_{\mathrm{R}} : \dot{\mathbf{F}} - \mu \dot{c}_{\mathrm{R}} - \mathbf{e}_{\mathrm{R}} \cdot \dot{\mathbf{d}}_{\mathrm{R}} + \mathbf{j}_{\mathrm{R}} \cdot \nabla\omega \le 0. \qquad (26.2.37)
$$

For later use we define the dissipation density $\mathcal{D} \ge 0$ per unit volume per unit time by

$$
\mathcal{D} = \mathbf{T}_{\mathrm{R}} : \dot{\mathbf{F}} + \mu \dot{c}_{\mathrm{R}} + \mathbf{e}_{\mathrm{R}} \cdot \dot{\mathbf{d}}_{\mathrm{R}} - \mathbf{j}_{\mathrm{R}} \cdot \nabla\omega - \dot{\psi}_{\mathrm{R}} \ge 0. \qquad (26.2.38)
$$

26.2.6 The decomposition $\mathbf{F} = \mathbf{F}^e\mathbf{F}^s$

Next, to account for coupling between diffusion of the cations and mechanical swelling, we base our theory on a multiplicative decomposition of the deformation gradient (Chester and Anand, 2010),[6]

$$
\mathbf{F} = \mathbf{F}^e\mathbf{F}^s, \qquad (26.2.39)
$$

where,

- $\mathbf{F}^s$ represents the local distortion of the material neighborhood of $\mathbf{X}$ due to the insertion/extraction of migrating cations and the associated volume changes, and

[6] Cf. Chapter 10.

- $\mathbf{F}^e$ represents the subsequent stretching and rotation of this coherent species–distorted material neighborhood, and thereby represents a corresponding elastic mechanical distortion.

We refer to $\mathbf{F}^s$ and $\mathbf{F}^e$ as the swelling and mechanical distortions, respectively. Hence we have

$$J = J^e J^s, \quad \text{and we assume that} \quad J^e = \det \mathbf{F}^e > 0, \quad \text{and} \quad J^s = \det \mathbf{F}^s > 0, \tag{26.2.40}$$

so that $\mathbf{F}^e$ and $\mathbf{F}^s$ are invertible.

The total and elastic right Cauchy–Green deformation tensors are given by

$$\mathbf{C} = \mathbf{F}^\top \mathbf{F} \qquad \text{and} \qquad \mathbf{C}^e = \mathbf{F}^{e\top} \mathbf{F}^e, \tag{26.2.41}$$

respectively. Further, by the definition $(26.2.1)_3$, we have

$$\mathbf{L} = \mathbf{L}^e + \mathbf{F}^e \mathbf{L}^s \mathbf{F}^{e-1}, \quad \text{with} \quad \mathbf{L}^e = \dot{\mathbf{F}}^e \mathbf{F}^{e-1}, \quad \text{and} \quad \mathbf{L}^s = \dot{\mathbf{F}}^s \mathbf{F}^{s-1}, \tag{26.2.42}$$

where $\mathbf{L}^s$ represents a distortion rate due to insertion/extraction of the migrating cations.

Finally, we assume that the swelling distortion $\mathbf{F}^s$ is isotropic and given by,

$$\mathbf{F}^s = \lambda^s \mathbf{1}, \quad \lambda^s = (J^s)^{1/3} > 0; \tag{26.2.43}$$

we call λ^s the swell-stretch. Further, from the definition of $\mathbf{L}^s$ in (26.2.42), and the specific expression for $\mathbf{F}^s$ in (26.2.43), we have,

$$\mathbf{L}^s = (\dot{\lambda}^s \lambda^{s-1}) \mathbf{1}; \tag{26.2.44}$$

and since $\dot{J}^s = J^s \operatorname{tr} \mathbf{L}^s$, we have

$$\mathbf{L}^s = \frac{1}{3} (\dot{J}^s J^{s-1}) \mathbf{1}. \tag{26.2.45}$$

26.2.7 Constitutive equation for J^s

Now, $(J^s - 1)$ represents the change in volume per unit reference volume due to swelling caused by the migrating cations. We assume that this change arises entirely due to the change in the concentration of the hydrated cations, so that we have the important swelling constraint,

$$J^s = 1 + \Omega (c_R - c_{R0}), \tag{26.2.46}$$

where

- Ω is the molar volume of the mobile hydrated cations, and c_{R0} is the initial concentration of the cations which is equal to that of the fixed anions. Thus note that,

$$\dot{J}^s = \Omega \dot{c}_R. \tag{26.2.47}$$

26.2.8 Free-energy imbalance revisited

Next, using (26.2.39), (26.2.42)$_2$, and (26.2.6), we may write the stress power $\mathbf{T}_R : \dot{\mathbf{F}}$ appearing in the free-energy imbalance (26.2.37) as

$$\mathbf{T}_R : \dot{\mathbf{F}} = J^s \tfrac{1}{2} (J^e \mathbf{F}^{e\,-1} \mathbf{T} \mathbf{F}^{e-\top}) : \dot{\mathbf{C}}^e + J^s (J^e \mathbf{F}^{e\top} \mathbf{T} \mathbf{F}^{e-\top}) : \mathbf{L}^s.$$

For convenience, we define two new stress measures

$$\mathbf{T}^e \stackrel{\text{def}}{=} J^e \mathbf{F}^{e\,-1} \mathbf{T} \mathbf{F}^{e\,-\top} \qquad \text{and} \qquad \mathbf{M}^e \stackrel{\text{def}}{=} \mathbf{C}^e \mathbf{T}^e = J^e \mathbf{F}^{e\top} \mathbf{T} \mathbf{F}^{e-\top}, \tag{26.2.48}$$

which we call the *elastic second Piola stress* and the *Mandel stress*, respectively. Then

$$\mathbf{T}_R : \dot{\mathbf{F}} = J^s \tfrac{1}{2} \mathbf{T}^e : \dot{\mathbf{C}}^e + J^s \mathbf{M}^e : \mathbf{L}^s. \tag{26.2.49}$$

Further, recalling (26.2.45) we may write (26.2.49) as

$$\mathbf{T}_R : \dot{\mathbf{F}} = J^s \tfrac{1}{2} \mathbf{T}^e : \dot{\mathbf{C}}^e + \frac{1}{3} (\operatorname{tr} \mathbf{M}^e) \dot{J}^s. \tag{26.2.50}$$

Let

$$p \stackrel{\text{def}}{=} -\frac{1}{3} (\operatorname{tr} \mathbf{M}^e) = J^e \left(-\frac{1}{3} \operatorname{tr} \mathbf{T} \right), \tag{26.2.51}$$

define a *mean normal pressure*. Then, using (26.2.51) and (26.2.47), the stress power (26.2.50) may be written as

$$\mathbf{T}_R : \dot{\mathbf{F}} = \tfrac{1}{2} J^s \, \mathbf{T}^e : \dot{\mathbf{C}}^e - p \, \Omega \, \dot{c}_R. \tag{26.2.52}$$

Introducing a net chemical potential, defined by

$$\mu_{\text{net}} \stackrel{\text{def}}{=} \mu - p \Omega, \tag{26.2.53}$$

we may rewrite the free-energy imbalance (26.2.37) under isothermal conditions as

$$\dot{\psi}_R - \tfrac{1}{2} J^s \, \mathbf{T}^e : \dot{\mathbf{C}}^e - \mu_{\text{net}} \dot{c}_R - \mathbf{e}_R \cdot \dot{\mathbf{d}}_R + \mathbf{j}_R \cdot \nabla \omega \leq 0. \tag{26.2.54}$$

Also the dissipation density (26.2.38) becomes

$$\mathcal{D} = \tfrac{1}{2} J^s \, \mathbf{T}^e : \dot{\mathbf{C}}^e + \mu_{\text{net}} \dot{c}_R + \mathbf{e}_R \cdot \dot{\mathbf{d}}_R - \mathbf{j}_R \cdot \nabla \omega - \dot{\psi}_R \geq 0. \tag{26.2.55}$$

REMARK

For brevity we have not discussed the transformation properties under a change in frame of the various fields appearing in our theory. Here, we simply note that all quantities in the free-energy imbalance (26.2.54) are invariant under a change in frame (Gurtin et al., 2010).

26.2.9 Constitutive theory

Let $\boldsymbol{\Lambda}$ represent the list

$$\boldsymbol{\Lambda} = \{\mathbf{C}^e, c_{\mathrm{R}}, \mathbf{d}_{\mathrm{R}}\}. \tag{26.2.56}$$

Guided by the free-energy imbalance (26.2.54), we begin by assuming constitutive equations for the free energy ψ_{R}, the stress $\mathbf{T}^e$, the net chemical potential μ_{net}, and the electric field $\mathbf{e}_{\mathrm{R}}$ of the form

$$\psi_{\mathrm{R}} = \hat{\psi}_{\mathrm{R}}(\boldsymbol{\Lambda}), \qquad \mathbf{T}^e = \hat{\mathbf{T}}^e(\boldsymbol{\Lambda}), \qquad \mu_{\mathrm{net}} = \hat{\mu}_{\mathrm{net}}(\boldsymbol{\Lambda}), \quad \mathbf{e}_{\mathrm{R}} = \hat{\mathbf{e}}_{\mathrm{R}}(\boldsymbol{\Lambda}). \tag{26.2.57}$$

Then,

$$\dot{\psi}_{\mathrm{R}} = \frac{\partial \hat{\psi}_{\mathrm{R}}(\boldsymbol{\Lambda})}{\partial \mathbf{C}^e} : \dot{\mathbf{C}}^e + \frac{\partial \hat{\psi}_{\mathrm{R}}(\boldsymbol{\Lambda})}{\partial c_{\mathrm{R}}} \dot{c}_{\mathrm{R}} + \frac{\partial \hat{\psi}_{\mathrm{R}}(\boldsymbol{\Lambda})}{\partial \mathbf{d}_{\mathrm{R}}} \cdot \dot{\mathbf{d}}_{\mathrm{R}}. \tag{26.2.58}$$

Using (26.2.58) and substituting the constitutive equations (26.2.57) into the free-energy imbalance (26.2.54), we find that it may be written as

$$\left[\frac{\partial \hat{\psi}_{\mathrm{R}}(\boldsymbol{\Lambda})}{\partial \mathbf{C}^e} - \tfrac{1}{2} J^s\, \hat{\mathbf{T}}^e(\boldsymbol{\Lambda})\right] : \dot{\mathbf{C}}^e + \left[\frac{\partial \hat{\psi}_{\mathrm{R}}(\boldsymbol{\Lambda})}{\partial c_{\mathrm{R}}} - \hat{\mu}_{\mathrm{net}}(\boldsymbol{\Lambda})\right] \dot{c}_{\mathrm{R}} + \left[\frac{\partial \hat{\psi}_{\mathrm{R}}(\boldsymbol{\Lambda})}{\partial \mathbf{d}_{\mathrm{R}}} - \hat{\mathbf{e}}_{\mathrm{R}}(\boldsymbol{\Lambda})\right] \cdot \dot{\mathbf{d}}_{\mathrm{R}} + \mathbf{j}_{\mathrm{R}} \cdot \nabla\omega \le 0. \tag{26.2.59}$$

We assume constitutively that the free energy delivers the stress $\mathbf{T}^e$, the net chemical potential μ_{net}, and the electric field $\mathbf{e}_{\mathrm{R}}$ through the state relations

$$\mathbf{T}^e = J^{s-1}\left(2\frac{\partial \hat{\psi}_{\mathrm{R}}(\boldsymbol{\Lambda})}{\partial \mathbf{C}^e}\right), \qquad \mu_{\mathrm{net}} = \frac{\partial \hat{\psi}_{\mathrm{R}}(\boldsymbol{\Lambda})}{\partial c_{\mathrm{R}}}, \qquad \mathbf{e}_{\mathrm{R}} = \frac{\partial \hat{\psi}_{\mathrm{R}}(\boldsymbol{\Lambda})}{\partial \mathbf{d}_{\mathrm{R}}}. \tag{26.2.60}$$

Using (26.2.60) in (26.2.59) leads to the following *dissipation inequality*

$$\mathcal{D} = -\mathbf{j}_{\mathrm{R}} \cdot \nabla\omega \ge 0. \tag{26.2.61}$$

26.2.10 **Chemical potential and electrochemical potential**

The thermodynamically consistent constitutive equation $(26.2.60)_2$, together with the definition (26.2.53) of μ_{net}, gives the *chemical potential* for the cations,

$$\mu = \frac{\partial \hat{\psi}_{\text{R}}(\boldsymbol{\Lambda})}{\partial c_{\text{R}}} + p\,\Omega, \tag{26.2.62}$$

and recalling (26.2.34) we have that the *electrochemical potential* ω is given by

$$\omega = \frac{\partial \hat{\psi}(\boldsymbol{\Lambda})}{\partial c_{\text{R}}} + p\,\Omega + F\,\phi. \tag{26.2.63}$$

26.2.11 **Dissipative constitutive equations**

We assume that the flux $\mathbf{j}_{\text{R}}$ depends *linearly* on the gradient $\nabla\omega$ of electrochemical potential gradient, by a *Fick*-type relation,

$$\mathbf{j}_{\text{R}} = -\mathbf{M}(\boldsymbol{\Lambda})\nabla\omega, \tag{26.2.64}$$

with $\mathbf{M}$ the **mobility tensor**. A consequence of the dissipation inequality (26.2.61) is then that the mobility tensor $\mathbf{M}$ is positive-semidefinite,

$$\mathbf{a}\cdot\mathbf{M}(\boldsymbol{\Lambda})\mathbf{a} \geq 0 \qquad \text{for all vectors } \mathbf{a} \neq \mathbf{0}. \tag{26.2.65}$$

26.2.12 **Specialization of the constitutive equations**

The theory presented thus far is quite general. We now introduce special constitutive equations which should be useful in applications. We limit our attention to materials which may be idealized to be isotropic.

Free energy

Henceforth we restrict our attention to materials for which the response function $\hat{\psi}(\boldsymbol{\Lambda})$ may be taken in the separable form

$$\hat{\psi}_{\text{R}}(\mathbf{C}^e, c_{\text{R}}, \mathbf{d}_{\text{R}}) = \psi_{\text{R}}^{\text{mech}}(\mathbf{C}^e, c_{\text{R}}) + \psi_{\text{R}}^{\text{es}}(\mathbf{C}^e, c_{\text{R}}, \mathbf{d}_{\text{R}}) + \psi_{\text{R}}^{\text{mix}}(c_{\text{R}}). \tag{26.2.66}$$

Here:

(i) $\psi_{\text{R}}^{\text{mech}}$ is the contribution to the change in the free energy due to the elastic deformation of the host material, taken to be given by

$$\psi_{\text{R}}^{\text{mech}} = J^s\,\tilde{\psi}^{\text{mech}}(\mathbf{C}^e), \tag{26.2.67}$$

in which the term $\tilde{\psi}^{\text{mech}}(\mathbf{C}^e)$ represents the free energy per unit volume of the intermediate space defined by the range of $\mathbf{F}^s(\mathbf{X})$, and multiplication of this term by J^s gives us the free energy per unit volume of the reference space.

Next, we introduce the multiplicative decomposition

$$\mathbf{F}^e = \mathbf{F}^{e\,\mathrm{vol}}\bar{\mathbf{F}}^e \tag{26.2.68}$$

of the elastic deformation gradient $\mathbf{F}^e$ into volumetric and isochoric factors $\mathbf{F}^{e\,\mathrm{vol}}$ and $\bar{\mathbf{F}}^e$, where

$$\mathbf{F}^{e\,\mathrm{vol}} = J^{e\,1/3}\mathbf{1} \qquad \text{and} \qquad \bar{\mathbf{F}}^e = J^{e\,-1/3}\mathbf{F}^e. \tag{26.2.69}$$

The decomposition (26.2.68) leads to a multiplicative decomposition of the right elastic Cauchy–Green tensor $\mathbf{C}^e$ in the form

$$\mathbf{C}^e = \mathbf{C}^{e\,\mathrm{vol}}\bar{\mathbf{C}}^e, \qquad \mathbf{C}^{e\,\mathrm{vol}} = J^{e\,2/3}\mathbf{1}, \qquad \bar{\mathbf{C}}^e = J^{e\,-2/3}\mathbf{C}^e. \tag{26.2.70}$$

Similarly, we also have a multiplicative decomposition of the left elastic Cauchy–Green tensor $\mathbf{B}^e$ in the form

$$\mathbf{B}^e = \mathbf{B}^{e\mathrm{vol}}\bar{\mathbf{B}}^e, \qquad \mathbf{B}^{e\mathrm{vol}} = J^{e\,2/3}\mathbf{1}, \qquad \bar{\mathbf{B}}^e = J^{e\,-2/3}\mathbf{B}^e. \tag{26.2.71}$$

Let

$$\bar{I}_1^e = \mathrm{tr}\,\bar{\mathbf{C}}^e = \mathrm{tr}\,\bar{\mathbf{B}}^e \tag{26.2.72}$$

denote the first principal invariant of $\bar{\mathbf{C}}^e$, which is equal to the first principal invariant of $\bar{\mathbf{B}}^e$.

Next, we adopt the following compressible Neo-Hookean for the elastic free energy $\tilde{\psi}^{\mathrm{mech}}(\mathbf{C}^e)$:

$$\begin{aligned}\tilde{\psi}^{\mathrm{mech}}(\mathbf{C}^e) &= \underbrace{\bar{\psi}^{\mathrm{mech}}(\bar{I}_1^e)}_{\text{isochoric}} + \underbrace{\psi^{\mathrm{mechvol}}(J^e)}_{\text{volumetric}},\\ &= \tfrac{1}{2}G\left(\bar{I}_1^e - 3\right) + \tfrac{1}{2}K(J^e - 1)^2,\end{aligned} \tag{26.2.73}$$

where the parameters

$$G > 0 \qquad \text{and} \qquad K > 0, \tag{26.2.74}$$

represent the elastic shear and bulk moduli, respectively, of the classical linear elasticity theory. Thus,

$$\psi_{\mathrm{R}}^{\mathrm{mech}} = J^s\left[\tfrac{1}{2}G\left(\bar{I}_1^e - 3\right) + \tfrac{1}{2}K(J^e - 1)^2\right]. \tag{26.2.75}$$

(ii) $\psi_{\mathrm{R}}^{\mathrm{es}}(\mathbf{C}^e, c_{\mathrm{R}}, \mathbf{d}_{\mathrm{R}})$ is an electrostatic or polarization energy. We consider a special form for this energy (Zhao et al., 2007; Suo, 2010),

$$\psi_{\mathrm{R}}^{\mathrm{es}} = J\left(\frac{1}{2\varepsilon}\mathbf{d}\cdot\mathbf{d}\right), \tag{26.2.76}$$

which is *isotropic in the deformed body*, with $\varepsilon > 0$ a scalar constant representing the *electrical permittivity* of the material, which in general could depend on the cation concentration, but we take it to be a constant here. Since $\mathbf{d} = J^{-1}\mathbf{F}\mathbf{d}_R$, we may express this free energy as

$$\psi_R^{es} = J^{-1}\left(\frac{1}{2\varepsilon}\mathbf{d}_R \cdot \mathbf{C}\mathbf{d}_R\right). \tag{26.2.77}$$

The electrical permittivity ε is often expressed as

$$\varepsilon = \varepsilon_r\,\varepsilon_0, \tag{26.2.78}$$

where $\varepsilon_0 = 8.85 \times 10^{-12}$ F/m is a scalar representing the electrical permittivity of vacuum, and ε_r is a *relative electrical permittivity* of the material.

(iii) $\hat{\psi}_R^{mix}(\mathbf{C}^e, c_R)$ is the change in free energy due to mixing of the migrating cations in the host material. As a simple continuum approximation of this mixing energy we take it to be given by a concentrated solution model of the form

$$\psi_R^{mix}(c_R) = \mu^{ref} c_R + R\vartheta\, c_R^{max}\left[\bar{c}_R \ln \bar{c}_R + (1 - \bar{c}_R)\ln(1 - \bar{c}_R)\right], \tag{26.2.79}$$

where

$$\bar{c}_R \overset{\text{def}}{=} \frac{c_R}{c_R^{max}}, \qquad 0 \le \bar{c}_R \le 1,$$

is a normalized species concentration, with c_R^{max} the concentration of the species in moles per unit reference volume when all the accommodating sites in the host material are filled. Also R is the universal gas constant, and ϑ is the absolute temperature, assumed to be constant in our isothermal theory.

Thus, using (26.2.75), (26.2.77), and (26.2.79) in (26.2.66), a simple form of the free-energy function which accounts for the combined effects of finite elastic stretching, electrostatic interactions, mixing, and swelling, is

$$\begin{aligned}\psi_R = J^s\left[\tfrac{1}{2}G\,(\bar{I}_1^e - 3) + \tfrac{1}{2}K(J^e - 1)^2\right] + J^{-1}\left(\frac{1}{2\varepsilon}\mathbf{d}_R \cdot \mathbf{C}\mathbf{d}_R\right)\\ + \mu^{ref} c_R + R\vartheta\, c_R^{max}\left[\bar{c}_R \ln \bar{c}_R + (1 - \bar{c}_R)\ln(1 - \bar{c}_R)\right].\end{aligned} \tag{26.2.80}$$

Electric field

From $(26.2.60)_3$ and (26.2.80) the referential electric field is given by

$$\mathbf{e}_R = \frac{1}{\varepsilon}J^{-1}\mathbf{C}\mathbf{d}_R. \tag{26.2.81}$$

Thus,

$$\mathbf{d}_R = \varepsilon\, J\mathbf{C}^{-1}\mathbf{e}_R. \tag{26.2.82}$$

Further, using the conversion relations $\mathbf{e} = \mathbf{F}^{-\top}\mathbf{e}_{\mathrm{R}}$ and $\mathbf{d} = J^{-1}\mathbf{F}\mathbf{d}_{\mathrm{R}}$, we obtain that the spatial electric displacement is given by

$$\mathbf{d} = \varepsilon\mathbf{e}. \tag{26.2.83}$$

Stress

Mechanical contribution to the stress

Using (26.2.80) and $(26.2.60)_1$ we find that the mechanical contribution to elastic second Piola stress is given by

$$\mathbf{T}^{e\,\mathrm{mech}} = J^{e-2/3}G\left[\mathbf{1} - \frac{1}{3}\left(\operatorname{tr}\bar{\mathbf{C}}^{e}\right)\bar{\mathbf{C}}^{e-1}\right] + K\,J^{e}(J^{e}-1)\mathbf{C}^{e-1}. \tag{26.2.84}$$

Further, using the relation

$$\mathbf{T} = J^{e-1}\mathbf{F}^{e}\mathbf{T}^{e}\mathbf{F}^{e\top} \quad\Longrightarrow\quad \mathbf{T} = J^{e-1/3}\bar{\mathbf{F}}^{e}\mathbf{T}^{e}\bar{\mathbf{F}}^{e\top}, \tag{26.2.85}$$

we obtain that the mechanical contribution to the Cauchy stress is given by

$$\mathbf{T}^{\mathrm{mech}} = J^{e-1}G\left[\bar{\mathbf{B}}^{e} - \frac{1}{3}\left(\operatorname{tr}\bar{\mathbf{B}}^{e}\right)\mathbf{1}\right] + K(J^{e}-1)\mathbf{1}. \tag{26.2.86}$$

Electrostatic contribution to the stress

Since

$$J^{-1} = J^{e-1}J^{s-1} = J^{e-1}(\lambda^{s})^{-3} \qquad \text{and} \qquad \mathbf{C} = (\lambda^{s})^{2}\mathbf{C}^{e},$$

we may rewrite (26.2.77) as

$$\psi_{\mathrm{R}}^{\mathrm{es}} = \frac{(\lambda^{s})^{-1}}{2\varepsilon}J^{e-1}\left(\mathbf{d}_{\mathrm{R}}\cdot\mathbf{C}^{e}\mathbf{d}_{\mathrm{R}}\right), \tag{26.2.87}$$

From $(26.2.60)_1$ and (26.2.87) the electrostatic contribution to the elastic second Piola stress is

$$\mathbf{T}^{e\,\mathrm{es}} = J^{s-1}\left(2\frac{(\lambda^{s})^{-1}}{2\varepsilon}\left[\frac{\partial J^{e-1}}{\partial\mathbf{C}^{e}}\left(\mathbf{d}_{\mathrm{R}}\cdot\mathbf{C}^{e}\mathbf{d}_{\mathrm{R}}\right) + J^{e-1}\frac{\partial}{\partial\mathbf{C}^{e}}\left(\mathbf{d}_{\mathrm{R}}\cdot\mathbf{C}^{e}\mathbf{d}_{\mathrm{R}}\right)\right]\right).$$

Using,

$$\frac{\partial J^{e-1}}{\partial\mathbf{C}^{e}} = -\tfrac{1}{2}J^{e-1}\mathbf{C}^{e-1}, \tag{26.2.88}$$

and

$$\frac{\partial}{\partial\mathbf{C}^{e}}\left(\mathbf{d}_{\mathrm{R}}\cdot\mathbf{C}^{e}\mathbf{d}_{\mathrm{R}}\right) = \mathbf{d}_{\mathrm{R}}\otimes\mathbf{d}_{\mathrm{R}}, \tag{26.2.89}$$

we obtain

$$\mathbf{T}^{e\text{es}} = J^{-1}(\lambda^s)^{-1}\left(\frac{1}{\varepsilon}\left[\mathbf{d}_{\text{R}} \otimes \mathbf{d}_{\text{R}} - \tfrac{1}{2}\mathbf{C}^{e-1}\left(\mathbf{d}_{\text{R}} \cdot \mathbf{C}^e \mathbf{d}_{\text{R}}\right)\right]\right). \tag{26.2.90}$$

Hence the electrostatic contribution to the Cauchy stress, $\mathbf{T}^{\text{es}} = J^{e-1}\mathbf{F}^e\mathbf{T}^{e\text{es}}\mathbf{F}^{e\top}$, after a bit of algebra is,

$$\mathbf{T}^{\text{es}} = \varepsilon\left[\mathbf{e} \otimes \mathbf{e} - \tfrac{1}{2}(\mathbf{e} \cdot \mathbf{e})\mathbf{1}\right], \tag{26.2.91}$$

which is the usual form of the Maxwell stress.

Total stress

Using (26.2.86) and (26.2.91), the total Cauchy stress is then given by

$$\mathbf{T} = \underbrace{J^{e-1}G\left(\bar{\mathbf{B}}^e - \frac{1}{3}(\operatorname{tr}\bar{\mathbf{B}}^e)\mathbf{1}\right) + K(J^e - 1)\mathbf{1}}_{\text{mechanical stress}} + \underbrace{\varepsilon\left[\mathbf{e} \otimes \mathbf{e} - \tfrac{1}{2}(\mathbf{e} \cdot \mathbf{e})\,\mathbf{1}\right]}_{\text{Maxwell stress}}, \tag{26.2.92}$$

and since $\mathbf{T}_{\text{R}} = J\mathbf{T}\mathbf{F}^{-\top}$, the corresponding total Piola stress is

$$\mathbf{T}_{\text{R}} = \underbrace{J^s\left[G\left(\bar{\mathbf{B}}^e - \frac{1}{3}(\operatorname{tr}\bar{\mathbf{B}}^e)\mathbf{1}\right) + K(J^e - 1)\mathbf{1}\right]\mathbf{F}^{-\top}}_{\text{mechanical stress}} + \underbrace{\varepsilon J\left[\mathbf{e} \otimes \mathbf{e} - \tfrac{1}{2}(\mathbf{e} \cdot \mathbf{e})\,\mathbf{1}\right]\mathbf{F}^{-\top}}_{\text{Maxwell stress}}. \tag{26.2.93}$$

Electrochemical potential

Further, using $(26.2.60)_2$, (26.2.63), and (26.2.80), the electrochemical potential ω is given by

$$\begin{aligned}\omega = {} & \mu^{\text{ref}} + R\vartheta \ln\left(\frac{\bar{c}_{\text{R}}}{1 - \bar{c}_{\text{R}}}\right) \\ & + \Omega \underbrace{\tilde{p}}_{\text{mechanical pressure}} + F\phi + \Omega \underbrace{\left[\tfrac{1}{2}G\left(\bar{I}_1^e - 3\right) + \tfrac{1}{2}K(J^e - 1)^2\right]}_{\text{energy quadratic in elastic deformation measures}},\end{aligned} \tag{26.2.94}$$

where we have introduced a mechanical pressure $\tilde{p}$ defined by

$$\tilde{p} \stackrel{\text{def}}{=} -K(J^e - 1). \tag{26.2.95}$$

Henceforth, we set the reference chemical potential μ^{ref} to zero, and we neglect the effects of the last term in (26.2.94), which is anticipated to be much smaller in magnitude than the other terms. Under these approximations (26.2.94) reduces to

$$\omega = R\vartheta \ln\left(\frac{\bar{c}_{\text{R}}}{1 - \bar{c}_{\text{R}}}\right) + \Omega\tilde{p} + F\phi. \tag{26.2.96}$$

Species flux

We assume a spherical *spatial* mobility tensor

$$m(c)\mathbf{1}, \tag{26.2.97}$$

in which

$$m(c) = \frac{Dc(1-\bar{c})}{R\vartheta} \geq 0 \tag{26.2.98}$$

is a positive-valued scalar mobility at a given temperature ϑ, with $D > 0$ representing a cation diffusion coefficient with units of [m^2/s], and

$$c = J^{-1}c_{\mathrm{R}}, \qquad c^{\max} = J^{-1}c_{\mathrm{R}}^{\max}, \qquad \bar{c} = \frac{c}{c^{\max}} \tag{26.2.99}$$

is the species concentration and its normalized version, per unit volume of the deformed body. We also take the *spatial flux* $\mathbf{j}$ of the hydrated cations to depend only on the *spatial gradient* of its electrochemical potential, such that it is given by

$$\mathbf{j} = -m(c)\,\mathrm{grad}\,\omega. \tag{26.2.100}$$

Next, from (26.2.96),

$$\mathrm{grad}\,\omega = \frac{R\vartheta}{c(1-\bar{c})}\,\mathrm{grad}\,c + \Omega\,\mathrm{grad}\,\tilde{p} + F\mathrm{grad}\,\phi, \tag{26.2.101}$$

and hence using (26.2.100), (26.2.98), and (26.2.101) the flux of the mobile species in the deformed body is given by

$$\mathbf{j} = -D\left[\mathrm{grad}\,c + \left(\frac{Fc(1-\bar{c})}{R\vartheta}\right)\mathrm{grad}\,\phi + \left(\frac{\Omega c(1-\bar{c})}{R\vartheta}\right)\mathrm{grad}\,\tilde{p}\right]. \tag{26.2.102}$$

REMARK

Equation (26.2.102) for the flux of the mobile cations has a form similar to the **Nernst–Planck equation** commonly used to describe the migration of a chemical species in a fluid under an electric field, but here it also includes the effects of mechanical pressure. It describes the flux of ions as the sum of the diffusive component which is related to the gradient of the concentration, $\mathrm{grad}\,c$, and a migrative component associated with the gradient of the electric potential, $\mathrm{grad}\,\phi$, and another migrative component associated with the gradient of the mechanical pressure, $\mathrm{grad}\,\tilde{p}$.

Finally, using the standard mapping results

$$\mathbf{F}^{-\top}\nabla\omega = \operatorname{grad}\omega \qquad \text{and} \qquad \mathbf{j}_{\mathrm{R}} = J\mathbf{F}^{-1}\mathbf{j}, \tag{26.2.103}$$

we obtain the referential species flux as

$$\mathbf{j}_{\mathrm{R}} = -\mathbf{M}\nabla\omega \qquad \text{with} \qquad \mathbf{M} = \mathbf{C}^{-1}\left(\frac{Dc_{\mathrm{R}}}{R\vartheta}\right)(1-\bar{c}_{\mathrm{R}}), \tag{26.2.104}$$

where $\mathbf{M}$ is the species mobility tensor, and with the electrochemical potential ω given in (26.2.96), viz.

$$\omega = R\vartheta\ln\left(\frac{\bar{c}_{\mathrm{R}}}{1-\bar{c}_{\mathrm{R}}}\right) + \Omega\,\tilde{p} + F\,\phi. \tag{26.2.105}$$

We can evaluate $\nabla\omega$ as

$$\nabla\omega = \frac{R\vartheta}{\bar{c}_{\mathrm{R}}(1-\bar{c}_{\mathrm{R}})}\nabla\bar{c}_{\mathrm{R}} + \Omega\,\nabla\tilde{p} + F\,\nabla\phi, \tag{26.2.106}$$

so that the referential species flux is

$$\begin{aligned}
\mathbf{j}_{\mathrm{R}} &= -\mathbf{M}\,\nabla\omega, \\
&= -\mathbf{C}^{-1}\left(\frac{Dc_{\mathrm{R}}(1-\bar{c}_{\mathrm{R}})}{R\vartheta}\right)\left(\frac{R\vartheta}{\bar{c}_{\mathrm{R}}(1-\bar{c}_{\mathrm{R}})}\nabla\bar{c}_{\mathrm{R}} + \Omega\,\nabla\,\tilde{p} + F\,\nabla\phi\right) \\
&= -D\mathbf{C}^{-1}\left(\frac{c_{\mathrm{R}}}{\bar{c}_{\mathrm{R}}}\nabla\bar{c}_{\mathrm{R}} + \frac{\Omega c_{\mathrm{R}}(1-\bar{c}_{\mathrm{R}})}{R\vartheta}\nabla\,\tilde{p} + \frac{Fc_{\mathrm{R}}(1-\bar{c}_{\mathrm{R}})}{R\vartheta}\nabla\phi\right) \\
&= -D\mathbf{C}^{-1}\left(\nabla c_{\mathrm{R}} + \frac{\Omega c_{\mathrm{R}}(1-\bar{c}_{\mathrm{R}})}{R\vartheta}\nabla\,\tilde{p} + \frac{Fc_{\mathrm{R}}(1-\bar{c}_{\mathrm{R}})}{R\vartheta}\nabla\phi\right),
\end{aligned} \tag{26.2.107}$$

or

$$\mathbf{j}_{\mathrm{R}} = -D\mathbf{C}^{-1}\left(\nabla c_{\mathrm{R}} + \frac{\Omega c_{\mathrm{R}}(1-\bar{c}_{\mathrm{R}})}{R\vartheta}\nabla\,\tilde{p} + \frac{Fc_{\mathrm{R}}(1-\bar{c}_{\mathrm{R}})}{R\vartheta}\nabla\phi\right) \tag{26.2.108}$$

which is a Nernst–Planck-type relation for the referential species flux induced by gradients in the species concentration c_{R}, mechanical pressure $\tilde{p}$, and electrostatic potential ϕ.

26.3 Governing partial differential equations and boundary conditions

1. The governing partial differential equation for the mechanical problem in the absence of body forces and neglect of inertia is the force balance

$$\operatorname{Div}\mathbf{T}_{\mathrm{R}} = \mathbf{0}, \tag{26.3.1}$$

where the Piola stress $\mathbf{T}_{\mathrm{R}}$ is given by

$$\begin{aligned}\mathbf{T}_{\mathrm{R}} &= \mathbf{T}_{\mathrm{R}}^{\mathrm{mech}} + \mathbf{T}_{\mathrm{R}}^{\mathrm{es}}, \quad \text{with}\\ \mathbf{T}_{\mathrm{R}}^{\mathrm{mech}} &= J^s\left[G\left(\bar{\mathbf{B}}^e - \frac{1}{3}(\operatorname{tr}\bar{\mathbf{B}}^e)\mathbf{1}\right) + K(J^e-1)\mathbf{1}\right]\mathbf{F}^{-\top}\\ \mathbf{T}_{\mathrm{R}}^{\mathrm{es}} &= J\,\varepsilon\left[\mathbf{e}\otimes\mathbf{e} - \frac{1}{2}(\mathbf{e}\cdot\mathbf{e})\,\mathbf{1}\right]\mathbf{F}^{-\top}.\end{aligned} \tag{26.3.2}$$

Next, with $\partial\mathcal{B}_{\mathrm{u}}$ and $\partial\mathcal{B}_{\mathrm{t_R}}$ denoting complementary subsurfaces of the boundary $\partial\mathcal{B}$ of the body $\mathcal{B}$, we consider a pair of possible boundary conditions:

$$\begin{aligned}\mathbf{u} &= \bar{\mathbf{u}} \quad \text{on } \partial\mathcal{B}_{\mathrm{u}},\\ \mathbf{T}_{\mathrm{R}}\mathbf{n}_{\mathrm{R}} &= \bar{\mathbf{t}}_{\mathrm{R}} \quad \text{on } \partial\mathcal{B}_{\mathrm{t_R}}.\end{aligned} \tag{26.3.3}$$

In the boundary conditions above $\bar{\mathbf{u}}$ and $\bar{\mathbf{t}}_{\mathrm{R}}$ are prescribed displacement and tractions, both functions of $(\mathbf{X}, t)$.

2. The governing partial differential equation for the cation concentration is

$$\dot{c}_{\mathrm{R}} + \operatorname{Div}\mathbf{j}_{\mathrm{R}} = 0, \tag{26.3.4}$$

where the referential species flux is given by

$$\mathbf{j}_{\mathrm{R}} = -D\mathbf{C}^{-1}\left(\nabla c_{\mathrm{R}} + \frac{\Omega c_{\mathrm{R}}(1-\bar{c}_{\mathrm{R}})}{R\vartheta}\nabla\tilde{p} + \frac{Fc_{\mathrm{R}}(1-\bar{c}_{\mathrm{R}})}{R\vartheta}\nabla\phi\right). \tag{26.3.5}$$

Regarding boundary conditions for the PDE (26.3.5) we will only consider flux boundary conditions,

$$\mathbf{j}_{\mathrm{R}}\cdot\mathbf{n}_{\mathrm{R}} = \bar{j}_{\mathrm{R}} \quad \text{on } \partial\mathcal{B}, \tag{26.3.6}$$

in the application of the theory to modeling IPMCs in the next chapter.

3. The governing differential equation for electrostatics is Gauss's law,

$$\text{Div}\, \mathbf{d}_{\text{R}} = q_{\text{R}}, \tag{26.3.7}$$

where the referential electric displacement $\mathbf{d}_{\text{R}}$ is given by

$$\mathbf{d}_{\text{R}} = \varepsilon\, J\mathbf{C}^{-1}\mathbf{e}_{\text{R}}, \tag{26.3.8}$$

and the referential net free charge q_{R} is given by

$$q_{\text{R}} = F\, (c_{\text{R}} - c_{\text{R}0}). \tag{26.3.9}$$

With $\partial\mathcal{B}_{\phi}$ and $\partial\mathcal{B}_{\mathbf{d}_{\text{R}}}$ denoting complementary subsurfaces of $\partial\mathcal{B}$, we consider a pair of possible boundary conditions:

$$\left.\begin{aligned} \phi &= \bar{\phi} \quad \text{on } \partial\mathcal{B}_{\phi}, \\ \mathbf{d}_{\text{R}}\cdot\mathbf{n}_{\text{R}} &= \bar{\varpi}_{\text{R}} \quad \text{on } \partial\mathcal{B}_{\mathbf{d}_{\text{R}}}. \end{aligned}\right. \tag{26.3.10}$$

In the boundary conditions above $\bar{\phi}$ and $\bar{\varpi}_{\text{R}}$ are prescribed electrostatic potential and referential surface charge density, respectively, which are both functions of $(\mathbf{X}, t)$.

Standard initial conditions involve the specifications

$$\mathbf{u}(\mathbf{X},0) = \mathbf{u}_0(\mathbf{X}), \quad c_{\text{R}}(\mathbf{X},0) = c_{\text{R}0}(\mathbf{X}), \quad \text{and} \quad \phi(\mathbf{X},0) = \phi_0(\mathbf{X}) \quad \text{in } \mathcal{B}, \tag{26.3.11}$$

with $\mathbf{u}_0$, $c_{\text{R}0}$, and ϕ_0 prescribed functions of $\mathbf{X} \in \mathcal{B}$.

27 Numerical implementation of theory for ionic polymers

27.1 Weak forms of the governing partial differential equations

1. **Mechanical problem**:
   Consider a test field $\mathbf{u}_{\text{test}}(\mathbf{X})$ in $\mathcal{B}$ which satisfies $\mathbf{u}_{\text{test}}(\mathbf{X}) = \mathbf{0}$ on $\partial\mathcal{B}_{\mathbf{u}}$. If the balance equation $\text{Div}\,\mathbf{T}_{\text{R}} = \mathbf{0}$ on $\mathcal{B}$ and the traction condition $(\bar{\mathbf{t}}_{\text{R}} - \mathbf{T}_{\text{R}}\mathbf{n}_{\text{R}}) = \mathbf{0}$ on $\partial\mathcal{B}_{\mathbf{t}_{\text{R}}}$ are satisfied, then it is straightforward to show that the weak form

$$\int_{\mathcal{B}} \mathbf{T}_{\text{R}} : \nabla\mathbf{u}_{\text{test}}\, dv_{\text{R}} - \int_{\partial\mathcal{B}_{\mathbf{t}_{\text{R}}}} \bar{\mathbf{t}}_{\text{R}} \cdot \mathbf{u}_{\text{test}}\, da_{\text{R}} = 0 \tag{27.1.1}$$

   holds for all admissible test fields $\mathbf{u}_{\text{test}}$ which vanish on $\partial\mathcal{B}_{\mathbf{u}}$.

2. **Species concentration balance problem**:
   Next, consider a scalar test field $c_{\text{test}}(\mathbf{X})$ in $\mathcal{B}$ which satisfies $c_{\text{test}}(\mathbf{X}) = \mathbf{0}$ on $\partial\mathcal{B}_{c_{\text{R}}}$. If the balance equation

$$\dot{c}_{\text{R}} + \text{Div}\,\mathbf{j}_{\text{R}} = 0 \qquad \text{in } \mathcal{B},$$

   and the boundary condition $\mathbf{j}_{\text{R}} \cdot \mathbf{n} = \bar{j}_{\text{R}}$ in $\partial\mathcal{B}_{\mathbf{j}_{\text{R}}}$ are satisfied, then

$$\int_{\mathcal{B}} \dot{c}_{\text{R}}\, c_{\text{test}}\, dv_{\text{R}} + \int_{\mathcal{B}} (\text{Div}\,\mathbf{j}_{\text{R}})\, c_{\text{test}}\, dv_{\text{R}} + \int_{\partial\mathcal{B}_{\mathbf{j}_{\text{R}}}} (\bar{j}_{\text{R}} - \mathbf{j}_{\text{R}} \cdot \mathbf{n}_{\text{R}})\, c_{\text{test}}\, da_{\text{R}} = 0, \tag{27.1.2}$$

   or equivalently

$$\int_{\mathcal{B}} \dot{c}_{\text{R}}\, c_{\text{test}}\, dv_{\text{R}} + \int_{\mathcal{B}} (\text{Div}\,\mathbf{j}_{\text{R}})\, c_{\text{test}}\, dv_{\text{R}} - \int_{\partial\mathcal{B}_{\mathbf{j}_{\text{R}}}} \mathbf{j}_{\text{R}} \cdot \mathbf{n}_{\text{R}}\, c_{\text{test}}\, da_{\text{R}} + \int_{\partial\mathcal{B}_{\mathbf{j}_{\text{R}}}} \bar{j}_{\text{R}}\, c_{\text{test}}\, da_{\text{R}} = 0 \tag{27.1.3}$$

   holds for all test fields c_{test}. Thus, since $c_{\text{test}} = 0$ on $\partial\mathcal{B}_{c_{\text{R}}}$, the first boundary integral in expression above may be extended to all of $\partial\mathcal{B}$,

$$\int_{\mathcal{B}} \dot{c}_{\text{R}}\, c_{\text{test}}\, dv_{\text{R}} + \int_{\mathcal{B}} (\text{Div}\,\mathbf{j}_{\text{R}})\, c_{\text{test}}\, dv_{\text{R}} - \int_{\partial\mathcal{B}} \mathbf{j}_{\text{R}} \cdot \mathbf{n}_{\text{R}}\, c_{\text{test}}\, da_{\text{R}} + \int_{\partial\mathcal{B}_{\mathbf{j}_{\text{R}}}} \bar{j}_{\text{R}}\, c_{\text{test}}\, da_{\text{R}} = 0. \tag{27.1.4}$$

Introduction to Coupled Theories in Solid Mechanics. Lallit Anand, Eric M. Stewart, Shawn A. Chester, Oxford University Press.
 DOI: 10.1093/oso/9780198986218.003.0028

Next, by the divergence theorem

$$\int_{\partial\mathcal{B}} \mathbf{j}_{\mathrm{R}} \cdot \mathbf{n}_{\mathrm{R}}\, c_{\mathrm{test}}\, da_{\mathrm{R}} = \int_{\partial\mathcal{B}} (c_{\mathrm{test}}\, \mathbf{j}_{\mathrm{R}}) \cdot \mathbf{n}\, da_{\mathrm{R}} = \int_{\mathcal{B}} c_{\mathrm{test}}\, \mathrm{Div}\, \mathbf{j}_{\mathrm{R}}\, dv_{\mathrm{R}} + \int_{\mathcal{B}} \mathbf{j}_{\mathrm{R}} \cdot \nabla c_{\mathrm{test}}\, dv_{\mathrm{R}}. \tag{27.1.5}$$

Use of (27.1.5) in (27.1.4) gives the weak form

$$\int_{\mathcal{B}} \dot{c}_{\mathrm{R}}\, c_{\mathrm{test}}\, dv_{\mathrm{R}} - \int_{\mathcal{B}} \mathbf{j}_{\mathrm{R}} \cdot \nabla c_{\mathrm{test}}\, dv_{\mathrm{R}} + \int_{\partial\mathcal{B}_{\mathbf{j}_{\mathrm{R}}}} \bar{j}_{\mathrm{R}}\, c_{\mathrm{test}}\, da_{\mathrm{R}} = 0. \tag{27.1.6}$$

We use a first-order finite difference approximation to write

$$\dot{c}_{\mathrm{R}} = \frac{c_{\mathrm{R}} - c_{\mathrm{R\,old}}}{\Delta t}, \tag{27.1.7}$$

so that the weak form (27.1.6) may be expressed as

$$\int_{\mathcal{B}} \frac{c_{\mathrm{R}} - c_{\mathrm{R\,old}}}{\Delta t}\, c_{\mathrm{test}}\, dv_{\mathrm{R}} - \int_{\mathcal{B}} \mathbf{j}_{\mathrm{R}} \cdot \nabla c_{\mathrm{test}}\, dv_{\mathrm{R}} + \int_{\partial\mathcal{B}_{\mathbf{j}_{\mathrm{R}}}} \bar{j}_{\mathrm{R}}\, c_{\mathrm{test}}\, da_{\mathrm{R}} = 0. \tag{27.1.8}$$

3. **Gauss's law**:
Finally, consider a scalar test field $\phi_{\mathrm{test}}(\mathbf{X})$ in $\mathcal{B}$ which satisfies $\phi_{\mathrm{test}}(\mathbf{X}) = \mathbf{0}$ in $\partial\mathcal{B}_{\phi}$. If Gauss's law

$$\mathrm{Div}\, \mathbf{d}_{\mathrm{R}} = q_{\mathrm{R}} \qquad \text{in } \mathcal{B},$$

and the boundary condition $\mathbf{d}_{\mathrm{R}} \cdot \mathbf{n}_{\mathrm{R}} = \bar{\varpi}_{\mathrm{R}}$ on $\partial\mathcal{B}_{\mathbf{d}_{\mathrm{R}}}$ are satisfied, then

$$\int_{\mathcal{B}} (\mathrm{Div}\, \mathbf{d}_{\mathrm{R}})\, \phi_{\mathrm{test}}\, dv_{\mathrm{R}} - \int_{\mathcal{B}} q_{\mathrm{R}}\, \phi_{\mathrm{test}}\, dv_{\mathrm{R}} + \int_{\partial\mathcal{B}_{\mathbf{d}_{\mathrm{R}}}} (\bar{\varpi}_{\mathrm{R}} - \mathbf{d}_{\mathrm{R}} \cdot \mathbf{n}_{\mathrm{R}})\, \phi_{\mathrm{test}}\, da_{\mathrm{R}} = 0, \tag{27.1.9}$$

or equivalently

$$\int_{\mathcal{B}} (\mathrm{Div}\, \mathbf{d}_{\mathrm{R}})\, \phi_{\mathrm{test}}\, dv_{\mathrm{R}} - \int_{\mathcal{B}} q_{\mathrm{R}}\, \phi_{\mathrm{test}}\, dv_{\mathrm{R}} - \int_{\partial\mathcal{B}_{\mathbf{d}_{\mathrm{R}}}} \mathbf{d}_{\mathrm{R}} \cdot \mathbf{n}_{\mathrm{R}}\, \phi_{\mathrm{test}}\, da_{\mathrm{R}} + \int_{\partial\mathcal{B}_{\mathbf{d}_{\mathrm{R}}}} \bar{\varpi}_{\mathrm{R}}\, \phi_{\mathrm{test}}\, da_{\mathrm{R}} = 0, \tag{27.1.10}$$

holds for all test fields ϕ_{test}. Thus, since $\phi_{\mathrm{test}} = 0$ on $\partial\mathcal{B}_{\phi}$ the first boundary integral in expression above may be extended to all of $\partial\mathcal{B}$,

$$\int_{\mathcal{B}} (\mathrm{Div}\, \mathbf{d}_{\mathrm{R}})\, \phi_{\mathrm{test}}\, dv_{\mathrm{R}} - \int_{\mathcal{B}} q_{\mathrm{R}}\, \phi_{\mathrm{test}}\, dv_{\mathrm{R}} - \int_{\partial\mathcal{B}} \mathbf{d}_{\mathrm{R}} \cdot \mathbf{n}_{\mathrm{R}}\, \phi_{\mathrm{test}}\, da_{\mathrm{R}} + \int_{\partial\mathcal{B}_{\mathbf{d}_{\mathrm{R}}}} \bar{\varpi}_{\mathrm{R}}\, \phi_{\mathrm{test}}\, da_{\mathrm{R}} = 0, \tag{27.1.11}$$

Next, by the divergence theorem

$$\int_{\partial\mathcal{B}} \mathbf{d}_R \cdot \mathbf{n}_R\, \phi_{\text{test}}\, da_R = \int_{\partial\mathcal{B}} (\phi_{\text{test}}\, \mathbf{d}_R) \cdot \mathbf{n}_R\, da_R = \int_{\mathcal{B}} \phi_{\text{test}}\, \text{Div}\, \mathbf{d}_R\, dv_R + \int_{\mathcal{B}} \mathbf{d}_R \cdot \nabla\phi_{\text{test}}\, dv_R. \tag{27.1.12}$$

Use of (27.1.12) in (27.1.11) and recalling $q_R = F\,(c_R - c_{R0})$ we have the weak form

$$\int_{\mathcal{B}} \mathbf{d}_R \cdot \nabla\phi_{\text{test}}\, dv_R + \int_{\mathcal{B}} F\,(c_R - c_{R0})\, \phi_{\text{test}}\, dv_R - \int_{\partial\mathcal{B}_{\mathbf{d}_R}} \bar{\varpi}_R\, \phi_{\text{test}}\, da_R = 0. \tag{27.1.13}$$

Summarizing, the weak forms of the governing equations for the coupled problem are

$$\begin{gathered}
\int_{\mathcal{B}} \mathbf{T}_R : \nabla\mathbf{u}_{\text{test}}\, dv_R - \int_{\partial\mathcal{B}_{\mathbf{t}_R}} \bar{\mathbf{t}}_R \cdot \mathbf{u}_{\text{test}}\, da_R = 0, \\
\int_{\mathcal{B}} \frac{c_R - c_{R\,\text{old}}}{\Delta t}\, c_{\text{test}}\, dv_R - \int_{\mathcal{B}} \mathbf{j}_R \cdot \nabla c_{\text{test}}\, dv_R + \int_{\partial\mathcal{B}_{\mathbf{j}_R}} \bar{j}_R\, c_{\text{test}}\, da_R = 0, \\
\int_{\mathcal{B}} \mathbf{d}_R \cdot \nabla\phi_{\text{test}}\, dv_R + \int_{\mathcal{B}} F\,(c_R - c_{R0})\, \phi_{\text{test}}\, dv_R - \int_{\partial\mathcal{B}_{\mathbf{d}_R}} \bar{\varpi}_R\, \phi_{\text{test}}\, da_R = 0.
\end{gathered} \tag{27.1.14}$$

for all vector-valued test functions $\mathbf{u}_{\text{test}}(\mathbf{X})$ which vanish on $\partial\mathcal{B}_{\mathbf{u}}$ and all scalar test functions $c_{\text{test}}(\mathbf{X})$ and $\phi_{\text{test}}(\mathbf{X})$ which vanish on $\partial\mathcal{B}_{c_R}$ and $\partial\mathcal{B}_{\phi}$, respectively.

27.2 Normalization

Since we plan to solve all the governing equations related to multiple physical fields simultaneously, it is beneficial to rewrite the three residuals so that they have approximately the same magnitudes, and to use degrees of freedom which take on values on the order of unity. For this particular theory, the normalization of the electrostatic problem is dealt with through a combination of a prudent selection of the unit system (see Table 28.1), and:

1. The species concentration c_R [mol/m^3] is normalized by the saturation concentration $c_R^{\max}$ [mol/m^3] as

$$\bar{c}_R \stackrel{\text{def}}{=} \frac{c_R}{c_R^{\max}}, \qquad 0 \le \bar{c}_R \le 1, \tag{27.2.1}$$

so that $\bar{c}_R \in [0, 1]$. Recall the species concentration balance equation,

$$\dot{c}_R + \text{Div}\,\mathbf{j}_R = 0.$$

This may be written as,

$$c_{\mathrm{R}}^{\max}\,\dot{\bar{c}}_{\mathrm{R}} + \operatorname{Div}\mathbf{j}_{\mathrm{R}} = 0.$$

2. We choose to normalize the Piola stress with the shear modulus G,

$$\hat{\mathbf{T}}_{\mathrm{R}} \overset{\text{def}}{=} \frac{\mathbf{T}_{\mathrm{R}}}{G}. \tag{27.2.2}$$

3. Summarizing, the normalized governing equations may then be expressed as

$$\begin{aligned}\operatorname{Div}\hat{\mathbf{T}}_{\mathrm{R}} &= 0,\\ c_{\mathrm{R}}^{\max}\,\dot{\bar{c}}_{\mathrm{R}} + \operatorname{Div}\mathbf{j}_{\mathrm{R}} &= 0,\\ \operatorname{Div}\mathbf{d}_{\mathrm{R}} &= q_{\mathrm{R}}.\end{aligned} \tag{27.2.3}$$

The weak forms of the coupled problem may be written as:

$$\begin{aligned}&\int_{\mathcal{B}} \hat{\mathbf{T}}_{\mathrm{R}} : \nabla\mathbf{u}_{\text{test}}\,dv_{\mathrm{R}} - \int_{\partial\mathcal{B}_{\mathbf{t}}} \hat{\bar{\mathbf{t}}}_{\mathrm{R}}\cdot\mathbf{u}_{\text{test}}\,da_{\mathrm{R}} = 0,\\ &\int_{\mathcal{B}} \frac{\bar{c}_{\mathrm{R}} - \bar{c}_{\mathrm{R}\text{old}}}{\Delta t}\,c_{\text{test}}\,dv_{\mathrm{R}} - \int_{\mathcal{B}} \frac{1}{c_{\mathrm{R}}^{\max}}\,\mathbf{j}_{\mathrm{R}}\cdot\nabla c_{\text{test}}\,dv_{\mathrm{R}} + \int_{\partial\mathcal{B}_{\mathbf{j}}} \frac{1}{c_{\mathrm{R}}^{\max}}\,\bar{j}_{\mathrm{R}}\,c_{\text{test}}\,da_{\mathrm{R}} = 0,\\ &\int_{\mathcal{B}} \mathbf{d}_{\mathrm{R}}\cdot\nabla\phi_{\text{test}}\,dv_{\mathrm{R}} + \int_{\mathcal{B}} F\,c_{\mathrm{R}}^{\max}\left(\bar{c}_{\mathrm{R}} - \bar{c}_{\mathrm{R}0}\right)\phi_{\text{test}}\,dv_{\mathrm{R}} - \int_{\partial\mathcal{B}_{\mathbf{d}}} \bar{\varpi}_{\mathrm{R}}\,\phi_{\text{test}}\,da_{\mathrm{R}} = 0.\end{aligned} \tag{27.2.4}$$

REMARK

Note that in writing the second of the weak forms above we have also divided through by $c_{\mathrm{R}}^{\max}$. This helps in the numerical implementation.

27.3 **Solution procedure in the FEniCS code**

In FEniCS we use the following degrees of freedom

$$\{\mathbf{u},\,\bar{c}_{\mathrm{R}},\,\phi\}, \tag{27.3.1}$$

as well as their known values from the previous step

$$\{\mathbf{u}_{\text{old}},\,\bar{c}_{\mathrm{R}\text{old}},\,\phi_{\text{old}}\}, \tag{27.3.2}$$

to calculate the *residuals* which correspond to the weak forms of the governing equations, viz.

$$
\begin{aligned}
R_{\mathbf{u}} &= \int_{\mathcal{B}} \hat{\mathbf{T}}_{\mathrm{R}} : \nabla \mathbf{u}_{\text{test}} \, dv_{\mathrm{R}} - \int_{\partial\mathcal{B}_{\mathbf{t}_{\mathrm{R}}}} \bar{\mathbf{t}}_{\mathrm{R}} \cdot \mathbf{u}_{\text{test}} \, da_{\mathrm{R}}, \\
R_{\bar{c}_{\mathrm{R}}} &= \int_{\mathcal{B}} \frac{\bar{c}_{\mathrm{R}} - \bar{c}_{\mathrm{R}\text{old}}}{\Delta t} c_{\text{test}} \, dv_{\mathrm{R}} - \int_{\mathcal{B}} \frac{1}{c_{\mathrm{R}}^{\max}} \mathbf{j}_{\mathrm{R}} \cdot \nabla c_{\text{test}} \, dv_{\mathrm{R}} + \int_{\partial\mathcal{B}_{\mathbf{j}_{\mathrm{R}}}} \frac{1}{c_{\mathrm{R}}^{\max}} \bar{j}_{\mathrm{R}} \, c_{\text{test}} \, da_{\mathrm{R}}, \\
R_{\phi} &= \int_{\mathcal{B}} \mathbf{d}_{\mathrm{R}} \cdot \nabla \phi_{\text{test}} \, dv_{\mathrm{R}} + \int_{\mathcal{B}} F \, c_{\mathrm{R}}^{\max} \left(\bar{c}_{\mathrm{R}} - \bar{c}_{\mathrm{R}0}\right) \phi_{\text{test}} \, dv_{\mathrm{R}} - \int_{\partial\mathcal{B}_{\mathbf{d}_{\mathrm{R}}}} \bar{\varpi}_{\mathrm{R}} \, \phi_{\text{test}} \, da_{\mathrm{R}}
\end{aligned}
\tag{27.3.3}
$$

FEniCS then automatically differentiates these residuals with respect to the separate degrees of freedom to obtain *tangents*. At each step in time, FEniCS uses the residuals and tangents in a global Newton–Raphson solver to determine current values of the degrees of freedom $\{\mathbf{u}, \bar{c}_{\mathrm{R}}, \phi\}$ for which the residuals are zero, and which therefore satisfy the governing equations as well as the initial and boundary conditions (in a weak sense).

REMARK

A mixed finite element method is utilized to solve for the displacement $\mathbf{u}$, normalized concentration $\bar{c}_{\mathrm{R}}$, and the electrostatic potential ϕ fields, concurrently. To avoid numerical instabilities the basis functions for the displacement are taken as Lagrange quadratic, while those of the concentration and electrostatic potential are one order lower, Lagrange linear.

27.3.1 Algorithm for calculation of the residuals

1. Begin the step with known values of

$$
\{\mathbf{u}_{\text{old}}, \bar{c}_{\mathrm{R}\text{old}}, \phi_{\text{old}}\}, \tag{27.3.4}
$$

where for the first step these quantities are given by the initial conditions for $\mathbf{u}$, $\bar{c}_{\mathrm{R}0}$, and ϕ.

2. Calculate necessary kinematical and electric field quantities for the current step as,

$$
\begin{aligned}
&\mathbf{F} = \mathbf{1} + \nabla \mathbf{u}, \quad J = \det \mathbf{F}, \quad \mathbf{C}^{-1} = (\mathbf{F}^{\top}\mathbf{F})^{-1}, \\
&\mathbf{e}_{\mathrm{R}} = -\nabla \phi, \qquad \mathbf{e} = \mathbf{F}^{-\top}\mathbf{e}_{\mathrm{R}}.
\end{aligned}
\tag{27.3.5}
$$

3. Calculate the volumetric swelling,

$$
J^{s} = 1 + \Omega \, c_{\mathrm{R}}^{\max} \left(\bar{c}_{\mathrm{R}} - \bar{c}_{\mathrm{R}0}\right), \tag{27.3.6}
$$

and use J^{s} to calculate the elastic deformation measures

$$
\mathbf{F}^{e} = \mathbf{F}(J^{s})^{-1/3}, \quad J^{e} = \det \mathbf{F}^{e}, \quad \mathbf{B}^{e} = \mathbf{F}^{e}\mathbf{F}^{e\top}, \quad \bar{\mathbf{B}}^{e} = (J^{e})^{-2/3}\mathbf{B}^{e}. \tag{27.3.7}
$$

4. Calculate the Piola stress,

$$\begin{aligned} \mathbf{T}_{\mathrm{R}} &= \mathbf{T}_{\mathrm{R}}^{\mathrm{mech}} + \mathbf{T}_{\mathrm{R}}^{\mathrm{es}}, \qquad \text{where} \\ \mathbf{T}_{\mathrm{R}}^{\mathrm{mech}} &= J^s \left[G\left(\bar{\mathbf{B}}^e - \frac{1}{3}(\operatorname{tr}\bar{\mathbf{B}}^e)\mathbf{1} \right) + K(J^e - 1)\mathbf{1} \right] \mathbf{F}^{-\top}, \\ \mathbf{T}_{\mathrm{R}}^{\mathrm{es}} &= J\,\varepsilon_r \varepsilon_0 \left[\mathbf{e} \otimes \mathbf{e} - \frac{1}{2}(\mathbf{e}\cdot\mathbf{e})\,\mathbf{1} \right] \mathbf{F}^{-\top}, \end{aligned} \tag{27.3.8}$$

and associated mechanical pressure,

$$\tilde{p} = -K(J^e - 1). \tag{27.3.9}$$

5. Calculate the referential species flux,

$$\mathbf{j}_{\mathrm{R}} = -D\mathbf{C}^{-1} \left(\nabla\,(c_{\mathrm{R}}^{\max}\,\bar{c}_{\mathrm{R}}) + \frac{\Omega c_{\mathrm{R}}^{\max}\,\bar{c}_{\mathrm{R}}(1-\bar{c}_{\mathrm{R}})}{R\vartheta}\,\nabla\,\tilde{p} + \frac{F c_{\mathrm{R}}^{\max}\,\bar{c}_{\mathrm{R}}(1-\bar{c}_{\mathrm{R}})}{R\vartheta}\,\nabla\phi \right). \tag{27.3.10}$$

6. Calculate the referential electric displacement,

$$\mathbf{d}_{\mathrm{R}} = -\varepsilon_0\,\varepsilon_r\,J\mathbf{C}^{-1}\nabla\phi. \tag{27.3.11}$$

7. Use the quantities from steps 1–6 to compute the residuals as

$$\begin{aligned} R_{\mathbf{u}} &= \int_{\mathcal{B}} \hat{\mathbf{T}}_{\mathrm{R}} : \nabla \mathbf{u}_{\mathrm{test}}\, dv_{\mathrm{R}} - \int_{\partial\mathcal{B}_{\mathrm{t}}} \bar{\hat{\mathbf{t}}}_{\mathrm{R}} \cdot \mathbf{u}_{\mathrm{test}}\, da_{\mathrm{R}}, \\ R_{\bar{c}_{\mathrm{R}}} &= \int_{\mathcal{B}} \frac{\bar{c}_{\mathrm{R}} - \bar{c}_{\mathrm{R}\mathrm{old}}}{\Delta t}\, c_{\mathrm{test}}\, dv_{\mathrm{R}} - \int_{\mathcal{B}} \frac{1}{c_{\mathrm{R}}^{\max}}\, \mathbf{j}_{\mathrm{R}} \cdot \nabla c_{\mathrm{test}}\, dv_{\mathrm{R}} + \int_{\partial\mathcal{B}_{\mathbf{j}}} \frac{1}{c_{\mathrm{R}}^{\max}}\, \bar{j}_{\mathrm{R}}\, c_{\mathrm{test}}\, da_{\mathrm{R}}, \\ R_{\phi} &= \int_{\mathcal{B}} \mathbf{d}_{\mathrm{R}} \cdot \nabla\phi_{\mathrm{test}}\, dv_{\mathrm{R}} + \int_{\mathcal{B}} F\, c_{\mathrm{R}}^{\max}\,(\bar{c}_{\mathrm{R}} - \bar{c}_0)\,\phi_{\mathrm{test}}\, dv_{\mathrm{R}} - \int_{\partial\mathcal{B}_{\mathbf{d}}} \bar{\varpi}_{\mathrm{R}}\,\phi_{\mathrm{test}}\, da_{\mathrm{R}} \end{aligned} \tag{27.3.12}$$

8. FEniCS solves for the current values of $\{\mathbf{u},\, \bar{c}_{\mathrm{R}},\, \phi\}$.
9. Set these current values as the "old" values for the next step in time,

$$\{\mathbf{u},\, \bar{c}_{\mathrm{R}},\, \phi\} \implies \{\mathbf{u}_{\mathrm{old}},\, \bar{c}_{\mathrm{R}\mathrm{old}},\, \phi_{\mathrm{old}}\}. \tag{27.3.13}$$

10. Repeat the steps above for the desired number of steps in time.

27.4 A numerical issue regarding charge separation in boundary layers

Recall the spatial version of Gauss's law (26.2.26), viz,

$$\operatorname{div}\mathbf{d} = q \qquad \text{with} \qquad q = F(c - c_0). \tag{27.4.1}$$

With spatial electric displacement given by $\mathbf{d} = \varepsilon\mathbf{e}$, equation (27.4.1) for a homogeneous material may be written as

$$\Delta\phi = -\frac{F}{\varepsilon}\,(c - c_0) \qquad \text{with} \qquad \varepsilon = \varepsilon_r\,\varepsilon_0. \tag{27.4.2}$$

Dimensional analysis of the Poisson equation (27.4.2) leads us to certain conclusions. Normalizing the potential ϕ by the "thermal volt" $R\vartheta/F$ and the spatial concentration of cations c by the initial concentration c_0,

$$\phi^* \stackrel{\text{def}}{=} \frac{\phi}{(R\vartheta/F)}, \qquad c^* \stackrel{\text{def}}{=} \frac{c}{c_0}, \tag{27.4.3}$$

we may rewrite (27.4.2) as

$$2\,\lambda_{\text{Debye}}^2 \Delta\phi^* + (c^* - 1) = 0, \tag{27.4.4}$$

where the length scale

$$\lambda_{\text{Debye}} \stackrel{\text{def}}{=} \sqrt{\frac{\varepsilon_r\varepsilon_0 R\vartheta}{2F^2 c_0}} \tag{27.4.5}$$

is commonly called the *Debye length*.

Since the electrodes in IPMCs are impermeable to the mobile cations,

- *electric double layers* of charge form in the beam at the interfaces with the electrodes, where charge neutrality is *not* satisfied.

The thickness of the electric double layers for a typical IPMC is on the order of a few Debye lengths,

$$\lambda_{\text{debye}} \stackrel{\text{def}}{=} \sqrt{\frac{\varepsilon_r\varepsilon_0 R\vartheta}{2F^2 c_0}} \approx 1.5\,\text{nm}, \tag{27.4.6}$$

while the thickness of a typical IPMC is about $\sim 300\,\mu$m.

- *It is extremely difficult to numerically resolve such thin boundary layers in beams which are several hundred microns thick using standard finite element procedures.*

In order to deal with this numerical issue, in our simulations—as an approximation—we will introduce a positive-valued *numerical permittivity factor* $\varepsilon_{\rm num}$ such that the electrical permittivity is given by (Narayan et al., 2021)

$$\varepsilon = \varepsilon_{\rm num}\varepsilon_r\varepsilon_0. \tag{27.4.7}$$

With this modification of the permittivity,

$$\lambda_{\rm debye,num} \stackrel{\rm def}{=} \sqrt{\frac{\varepsilon_{\rm num}\varepsilon_r\varepsilon_0 R\vartheta}{2F^2 c_0}} \tag{27.4.8}$$

represents a *numerically enhanced* Debye length. We will use a rather large value of

$$\varepsilon_{\rm num} \approx 2.5 \times 10^7$$

in our simulations, and in this case the thickness of boundary layers is increased to $\sim 3\,\mu$m, which is about 1% of the thickness of $\sim$ 300 μm of an IPMC film. We emphasize that:

- We will use the $\varepsilon_{\rm num}$ factor only in the expression for the electric displacement.
- We *will not* use this factor in the expression for the Cauchy stress, which would artificially enhance the effects of the Maxwell stress.

REMARK

Before closing this section we note that Narayan et al. (2021) have shown that if one adopts this assumption, then the effect of the Maxwell stress on the actuation response of IPMCs is found to be *negligible* in comparison to the effect of the chemo-mechanical stress; see the Remark on p. 17 and Figure 8 in the paper by Narayan et al. (2021). They justified the use of this assumption by showing that the results of their numerical calculations based on this assumption compare well against results from physical experiments available in the literature.

---

28 Representative simulations of ionic polymers

In this chapter we present some representative finite element simulations which demonstrate key features of the response of Nafion-IPMCs. The unit system and material parameters that we have used for our simulations are listed in Table 28.1 and Table 28.2.

Table 28.1 Basic units.

Quantity	SI (m)	SI (mm)
Length	m	mm
Mass	kg	tonne (10^3 kg)
Time	s	s
Amount of substance	mol	mol
Electric charge	C	mC
Temperature	K	K
Force	N	N
Energy	J	mJ (10^{-3} J)
c_R, species concentration	mol/m^3	mol/mm^3
Stress	Pa	MPa (N/mm^2)
Ω, molar volume	m^3/mol	mm^3/mol
D, species diffusivity	m^2/s	mm^2/s
ϕ, electric potential	V (J/C)	V (mJ/mC)
Electric permittivity	F/m	mF/mm
ε_0, permittivity of free space	8.85×10^{-12} F/m	8.85×10^{-12} mF/mm
R, gas constant	8.3145 J/(mol K)	8.3145×10^3 mJ/(mol K)
F, Faraday constant	96.485×10^3 C/mol	96.485×10^6 mC/mol
Capacitance	F (C/V)	mF (mC/V)

Introduction to Coupled Theories in Solid Mechanics. Lallit Anand, Eric M. Stewart, Shawn A. Chester, Oxford University Press.
 DOI: 10.1093/oso/9780198986218.003.0029

Table 28.2 Approximate values of material parameters for a Nafion-based IPMC.

Quantity	Value SI (m)	Value SI (mm)
E, Young's modulus	100×10^6 Pa	100 MPa
ν, Poisson's ratio	0.4	0.4
$G = E/(2(1+\nu)$, shear modulus	35.71×10^6 Pa	35.71 MPa
$K = E/(3(1-2\nu))$, bulk modulus	166.67×10^6 Pa	166.67 MPa
Ω, molar volume of hydrated lithium cations	1.04×10^{-4} m^3/mol	1.04×10^5 mm^3/mol
D, diffusivity of solvated lithium cation	1.3×10^{-11} m^2/s	1.3×10^{-5} mm^2/s
c_{R0}, initial concentration of solvated lithium cations	1500 mol/m^3	1500×10^{-9} mol/mm^3
c_R^{max}, saturation concentration of solvated lithium cations	4750 mol/m^3	4750×10^{-9} mol/mm^3
ϑ, reference temperature	298 K	298 K
ε_0, permittivity of free space	8.85×10^{-12} F/m	8.85×10^{-12} mF/mm
ε_r, relative permittivity of ionic polymer	50	50

28.1 Curling response of a cantilevered Nafion-IPMC beam subjected to a step voltage

Nguyen et al. (2007) have reported on several experiments on Nafion-IPMCs with mobile Li^+ cations, and platinum-infused electrodes. Figure 28.1 shows a result from one of their experiments in which a cantilevered beam was subjected to a step voltage of 3 V. Figure 28.1(a) shows the initial configuration at time $t = 0$, and Fig. 28.1(b) shows the deformed configuration at $t = 20$ s. Clearly the migration of the cations under the action of the applied potential leads to a large bending deformation of the IPMC beam.

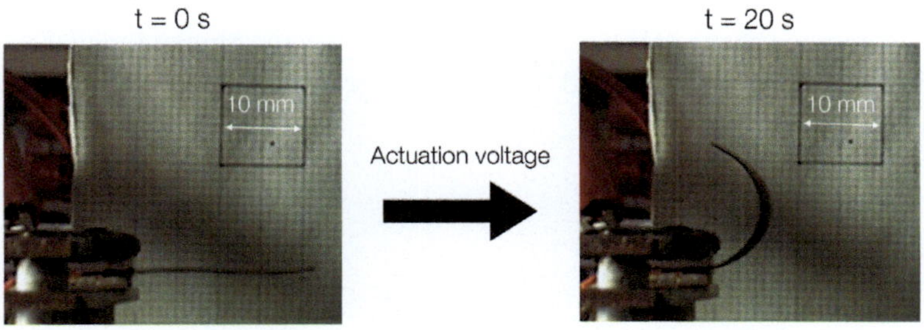

Fig. 28.1 (a) Initial cantilever beam geometry at $t = 0$ s, and (b) deformed configuration at $t = 20$ s.

Figure 28.2 shows a schematic of their experiment. The beam has a length of 20 mm, a height of 300 μm, and a depth of 5 mm (into the plane of the paper). In our simulations we model the beam under plane-strain conditions. The finite element mesh (not shown) consists of 100 uniformly spaced elements across its length, and 120 elements across its thickness with a biased spacing. The mesh-density is progressively higher towards the top and bottom surfaces of the beam, with successive element decreasing in size by a ratio 1.09. The higher mesh-density towards the top and bottom surfaces of the beam is needed to resolve the large gradients in electrostatic potential and cation concentration near the outer surfaces of the beam. The material parameters used for the IPMC cantilever beam are those listed in Table 28.2.

With reference to Fig. 28.2, the boundary conditions in this simulation are as follows:

- Mechanical boundary conditions: (i) the left face is fixed by prescribing $\bar{\mathbf{u}} = \mathbf{0}$; and (ii) all other faces are traction-free.
- Electrostatic boundary conditions: (i) the bottom face of the beam is grounded, $\bar{\phi} = 0$; (ii) the top face is prescribed a "step" electrostatic potential of 3 V using

$$\bar{\phi}(t) = 3(1 - \exp(-t/0.2)),$$

  as shown in Fig. 28.3; and (iii) all other faces are prescribed zero surface charge, $\bar{\varpi}_{\mathrm{R}} = 0$.
- Cation flux boundary conditions: We do not permit any flux of cations at the boundaries of the beam, $\bar{j}_{\mathrm{R}} = 0$.

The simulation is performed for a total time of 20 seconds.

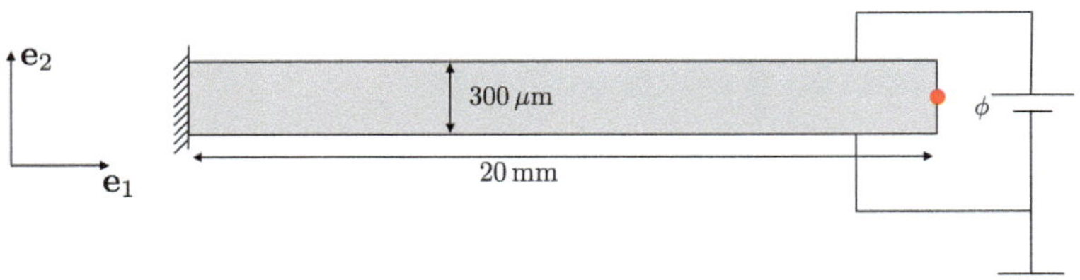

Fig. 28.2 Schematic of the initial IPMC cantilever beam geometry and boundary conditions.

Figure 28.3 shows the applied electric potential on the top face and the resulting displacement of the tip of the beam in the $\mathbf{e}_2$-direction as a function of time. The tip-displacement is extracted from the simulation results using the point indicated by the red dot in Fig. 28.2. The numerically computed shape of the curled cantilevered beam is shown in Fig. 28.4. The black outline shows the initial geometry, and the deformed geometry is shown with contours of the electrostatic potential ϕ at $t = 20$ s.

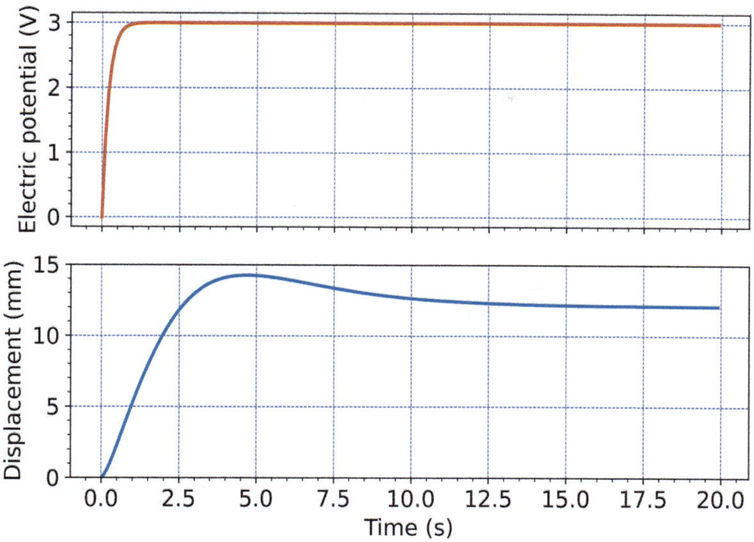

Fig. 28.3 Applied electric potential, and the corresponding tip-displacement of the beam in the e_2-direction as a function of time.

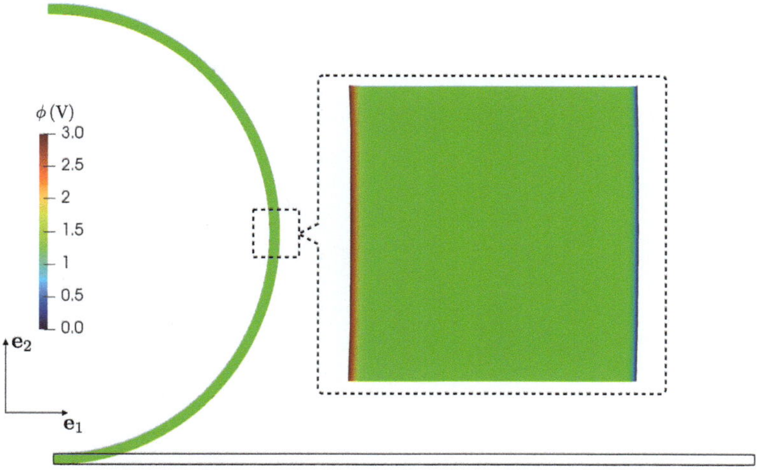

Fig. 28.4 Numerically calculated shape of a curled cantilevered beam. The black outline shows the initial geometry and the deformed geometry at $t = 20$ s is shown in green. The contours of the voltage ϕ on the deformed geometry are nearly uniform, except for sharp gradients of ϕ near the top and bottom surfaces of the beam.

Figure 28.5 shows a comparison of the numerically calculated profile of the beam against the profile measured experimentally by Nguyen et al. (2007). The bright green line in this figure shows the results from the simulation, which are overlayed atop the experimentally observed shape, at $t = 20$ s. The numerical simulation is in good qualitative agreement with the experiment.

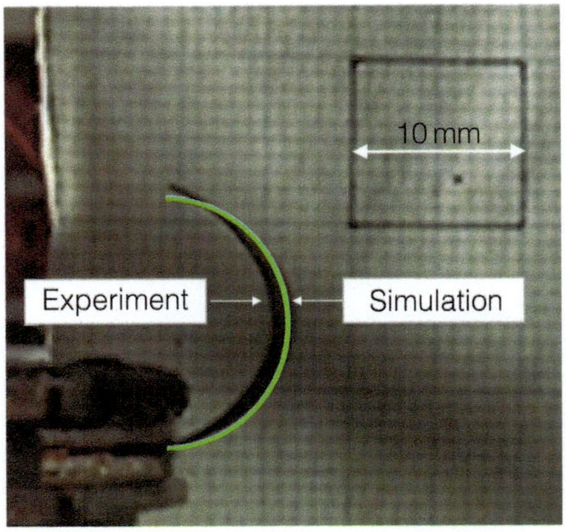

Fig. 28.5 The simulated deformed shape overlayed on the experimentally observed shape of a Nafion-IPMC cantilever beam at $t = 20$ s from Nguyen et al. (2007).

As described earlier, the concentration of the hydrated cations in the beam evolves with time in response to the applied electrostatic potential. As the concentration of the cations increases near the bottom of the beam it is accompanied by chemical volume expansion, and at the same time the upper part of the beam experiences compressive strains since the cation concentration in that region decreases; cf. Fig. 28.6. This differential swelling leads to a marked bending of the beam.

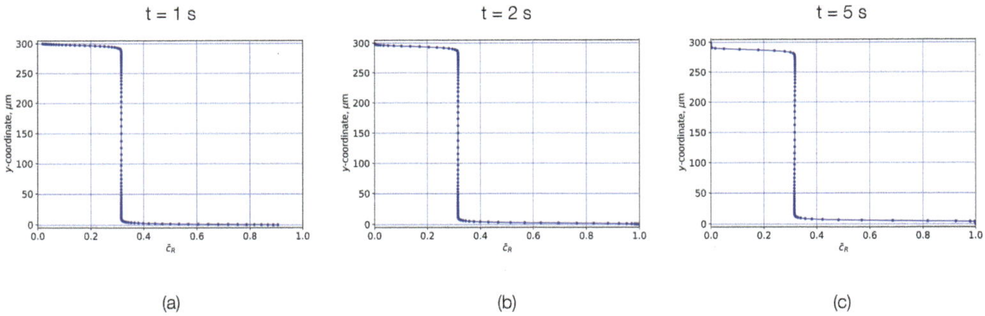

Fig. 28.6 Figure showing the variation of the normalized cation concentration $\bar{c}_R$ through the thickness of the beam at (a) $t = 1$ s, (b) $t = 2$ s, and (c) $t = 5$ s.

28.2 Blocking force response of a constrained Nafion-IPMC beam subjected to a step voltage

Using a Nafion-IPMC specimen similar to that discussed in the previous subsection, Nguyen et al. (2007) also measured the blocking force generated by the beam under the same step potential of 3 V, but using a constrained geometry shown schematically in Fig. 28.7. As before, the material parameters used for this simulation are those listed in Table 28.2. With reference to Fig. 28.7, the boundary conditions in this simulation are as follows:

- Mechanical boundary conditions: (i) the left face is fixed by prescribing $\bar{\mathbf{u}} = \mathbf{0}$; (ii) the midpoint of the right face is prescribed $\bar{u}_2 = 0$, and (iii) all other faces are traction-free.
- Electrostatic boundary conditions: (i) the bottom face of the beam is grounded, $\bar{\phi} = 0$; (ii) the top face is prescribed a "step" electrostatic potential of 3 V using

$$\bar{\phi}(t) = 3(1 - \exp(-t/0.5)),$$

  as shown in Fig. 28.9(a); and (iii) all other faces are prescribed zero surface charge, $\bar{\varpi}_{\mathrm{R}} = 0$.
- Cation flux boundary conditions: We do not permit any flux of cations at the boundaries of the beam, $\bar{j}_{\mathrm{R}} = 0$.

The simulation is performed for a total time of 30 seconds. The deformed geometry of the constrained beam from our simulation at $t = 30$ s is shown in Fig. 28.8.

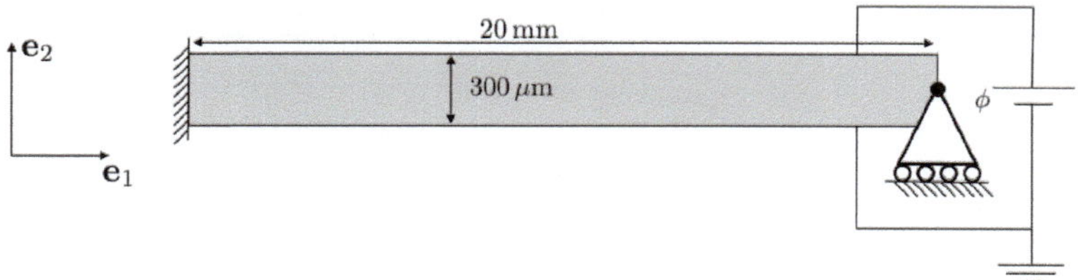

Fig. 28.7 A schematic of the geometry and boundary conditions for the constrained beam for the blocking force simulation.

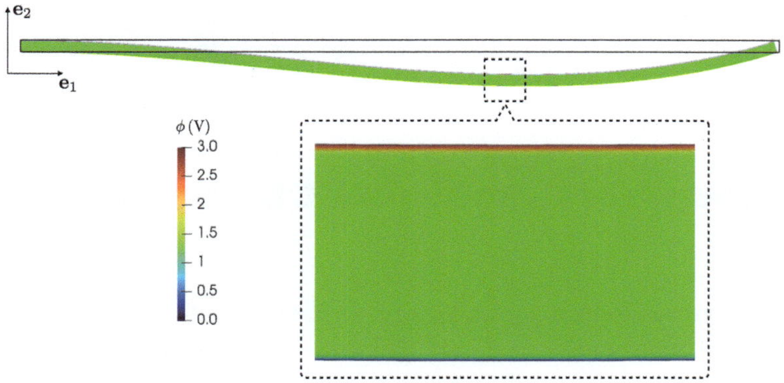

Fig. 28.8 Numerically predicted deformed shape of the IPMC beam at $t = 30$ s. The black outline shows the initial geometry and the deformed geometry is shown with contours of the electrostatic potential ϕ. The contours of the voltage ϕ on the deformed geometry are nearly uniform, except for sharp gradients of ϕ near the top and bottom surfaces of the beam as shown in the inset.

Figure 28.9(a) shows the applied electric potential as function of time, and the solid blue line in Fig. 28.9(b) shows the corresponding numerically calculated reaction force exerted on the support to maintain the constraint in the vertical direction.[1] Figure 28.9(b) also shows the

[1] See Section 28.A for a note on the method that we used to numerically calculate the reaction force in FEniCS.

experimentally measured reaction force by Nguyen et al. (2007) as open black circles. Starting from zero, the blocking force increases during the time over which the cation redistribution occurs. In our simulations the blocking force increases and then saturates at a constant value once steady state is reached. Up to around $t = 15$ s, the numerically calculated reaction force versus time curve is in very good agreement with the experimental measurement. However, in their experiments, Nguyen et al. observed some relaxation at long times, which they attributed to a loss of solvent molecules from the hydrated beam to the surrounding environment. In the theory considered here such a long-time loss of water molecules to the surroundings is not accounted for.

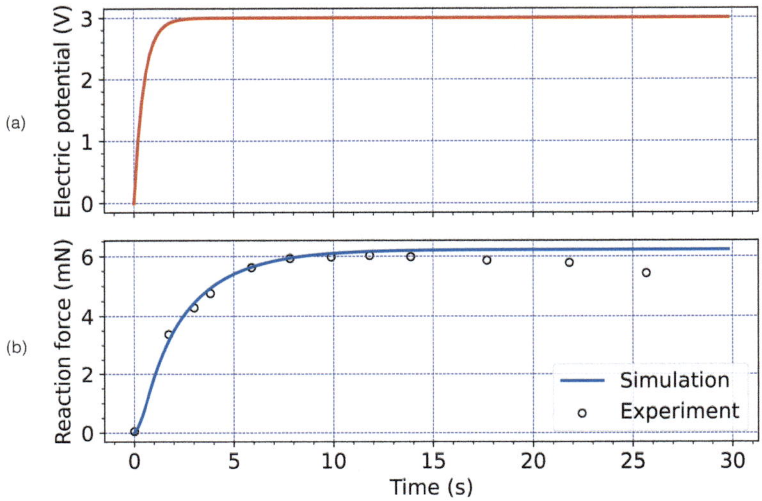

Fig. 28.9 (a) Applied electric potential, and (b) the corresponding reaction force at the tip of the constrained beam in the $\mathbf{e}_2$-direction as a function of time. The result from the simulation is shown as the solid blue line and the experimental result from Nguyen et al. (2007) is shown as open black circles.

28.3 Response of an Nafion-IPMC beam to a sinusoidal voltage

In this simulation we model the response of a cantilever IPMC beam to a sinusoidal electrostatic potential and compare the result against the corresponding experimental response reported in Nguyen et al. (2007). Specifically, we simulate the response of an unconstrained cantilevered beam as in the first example shown in Fig. 28.2(a). As before, the material parameters used for the IPMC are those listed in Table 28.2. With reference to Fig. 28.2(a), the boundary conditions in this problem are as follows:

- Mechanical boundary conditions: (i) the left face is fixed by prescribing $\bar{\mathbf{u}} = \mathbf{0}$; and (ii) all other faces are traction-free.
- Electrostatic boundary conditions: (i) the bottom face of the beam is grounded, $\bar{\phi} = 0$; (ii) the top face is prescribed a sinusoidal electrostatic potential at a frequency of 1 Hz and amplitude of 3 V using

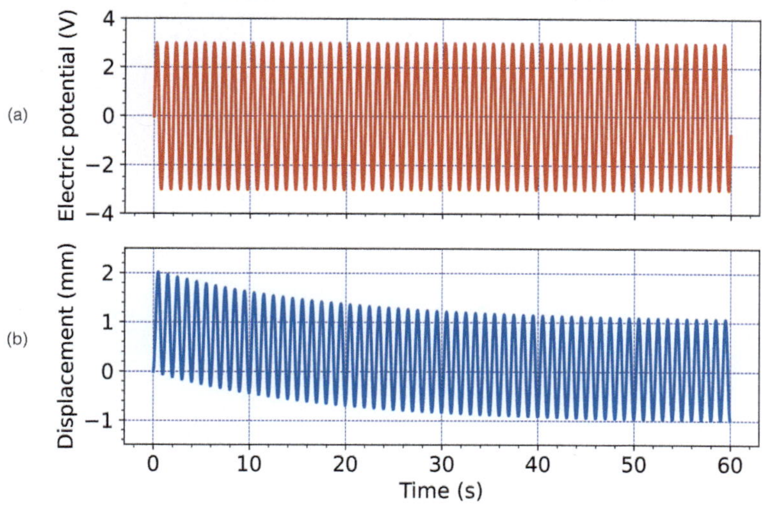

Fig. 28.10 (a) Applied 1 Hz sinusoidal electrostatic potential with an amplitude of 3 V. (b) Tip-displacement of the beam for the first 60 cycles.

$$\bar{\phi}(t) = 3\sin(2\pi t),$$

as shown in Fig. 28.10(a); and (iii) all other faces are prescribed zero surface charge, $\bar{\varpi}_{\mathrm{R}} = 0$.

- Cation flux boundary conditions: We do not permit any flux of cations at the boundaries of the beam, $\bar{j}_{\mathrm{R}} = 0$.

The simulation is performed for a total time of 60 seconds, which corresponds to 60 cycles.

Figure 28.10(a) shows the applied electrostatic potential as a function of time for the first 60 cycles, and Fig. 28.10(b) shows the corresponding tip-displacement of the beam. The displacement in the positive and negative directions over the first few cycles is evidently not symmetric about the zero tip-displacement state. This is because the initial polarity of the applied voltage biases the deformation. With an increasing number of cycles the asymmetry gradually decreases as the influence of the initial state decays out. After about 50 cycles the tip-displacement has reached a steady-state response, which is shown in Fig. 28.11. At steady state the displacements are sinusoidal with the same frequency as the input voltage. However, since there is a time delay associated with the migration of the cations following an applied voltage and the subsequent bending deformation (which is a result of the cation redistribution), Fig. 28.11 shows that there is a phase lag between the output displacements and the input voltage.

In Fig. 28.12 we show a comparison of the numerically predicted tip-displacement response at steady state against the corresponding experimental measurement reported in Nguyen et al. (2007). The agreement is excellent.

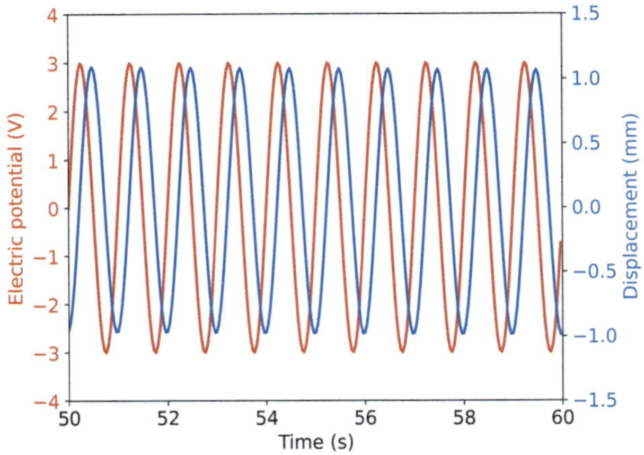

Fig. 28.11 Tip-displacement response from 50 to 60 cycles. The applied electrostatic potential is plotted in red and the resulting tip-displacement is plotted in blue.

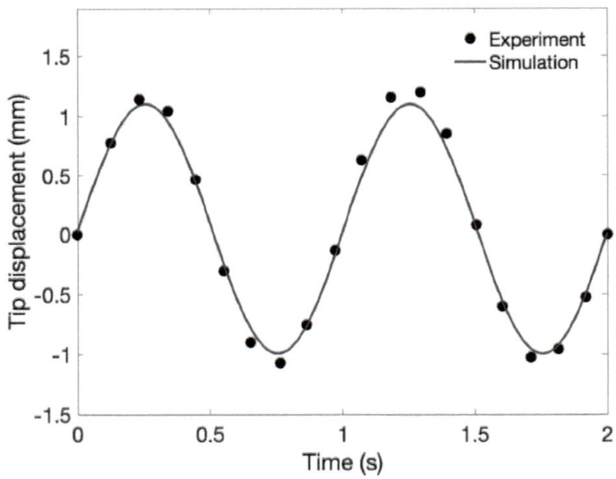

Fig. 28.12 Comparison of predicted tip-displacement at steady state against the experimentally measured response from Nguyen et al. (2007).

REMARK

For geometrically more complex applications of the theory we refer the reader to the paper by Narayan et al. (2021) who formulated a theory for ionic polymers which is essentially equivalent to the one considered here, except they used a logarithmic elastic strain-based description of the elastic response, rather than a Neo-Hookean type elastic response used here. They implemented their theory in the commercial finite element program Abaqus/Standard (Abaqus, 2017) and showed applications of their theory for two soft-robotics applications with some geometric complexity: (i) aquatic fins, and (ii) grippers. Such simulations cannot at present be performed within the FEniCS environment because of lack of a robust capability to account for contact and friction.

28.A **Computing reaction forces at boundaries with applied displacement boundary conditions**

As part of post-processing of a numerical solution one often needs to compute a reaction force at a boundary where displacements are prescribed. Quite often such a reaction force will be computed using the stress field associated with the computed displacement, but for certain meshes—like the specially biased mesh used in this chapter to resolve the boundary layers near the electrodes of the IPMC beam—this can lead to an inaccurate result. Another approach is to calculate reaction forces using the virtual power principle, which gives more accurate results (Bleyer, 2024). We describe this virtual power approach for calculating reaction forces below.

Let $\tilde{\mathbf{u}}$ denote a virtual velocity field and let $\nabla\tilde{\mathbf{u}} = \tilde{\mathbf{H}} \equiv \tilde{\mathbf{F}}$ denote the corresponding virtual variation $\tilde{\mathbf{H}}$ in the displacement gradient $\mathbf{H}$, which is equivalent to a virtual variation $\tilde{\mathbf{F}}$ in the deformation gradient $\mathbf{F}$. Then with $\mathbf{b}_{\mathrm{R}}$ denoting a generalized body force per unit reference volume defined on $\mathcal{B}$, $\mathbf{t}_{\mathrm{R}}(\mathbf{n}_{\mathrm{R}})$ a traction field defined on $\partial\mathcal{B}$, and $\mathbf{T}_{\mathrm{R}}$ a stress field conjugate to $\tilde{\mathbf{F}}$ defined on $\mathcal{B}$, let

$$\begin{aligned} \mathcal{W}_{\text{ext}}(\mathcal{B},\tilde{\mathbf{u}}) &= \int_{\mathcal{B}} \mathbf{b}_{\mathrm{R}}\cdot\tilde{\mathbf{u}}\,dv_{\mathrm{R}} + \int_{\partial\mathrm{B}} \mathbf{t}_{\mathrm{R}}(\mathbf{n}_{\mathrm{R}})\cdot\tilde{\mathbf{u}}\,da_{\mathrm{R}}, \\ \mathcal{W}_{\text{int}}(\mathcal{B},\tilde{\mathbf{u}}) &= \int_{\mathcal{B}} \mathbf{T}_{\mathrm{R}}:\nabla\tilde{\mathbf{u}}, \end{aligned} \tag{28.A.1}$$

denote the external and internal virtual powers expended on $\mathcal{B}$. Then, the principle of virtual power (when applied to the whole body) requires that the balance

$$\mathcal{W}_{\text{ext}}(\mathcal{B},\tilde{\mathbf{u}}) = \mathcal{W}_{\text{int}}(\mathcal{B},\tilde{\mathbf{u}}) \tag{28.A.2}$$

hold for all virtual velocities $\tilde{\mathbf{u}}$.

Now consider a boundary-value problem for a body $\mathcal{B}$ with boundary

$$\partial\mathcal{B} = \partial\mathcal{B}_{\mathrm{u}} \cup \partial\mathcal{B}_{\mathrm{t_R}} \quad \text{with} \quad \partial\mathcal{B}_{\mathrm{u}} \cap \partial\mathcal{B}_{\mathrm{t_R}} = \emptyset,$$

and boundary conditions

(i) $\mathbf{u} = \bar{\mathbf{u}}$ prescribed on $\partial\mathcal{B}_{\mathrm{u}}$; and

(ii) $\mathbf{t}_{\mathrm{R}}(\mathbf{n}_{\mathrm{R}}) = \tilde{\mathbf{t}}_{\mathrm{R}}$ prescribed on $\partial\mathcal{B}_{\mathrm{t_R}}$.

Also, let the Piola stress field in $\mathcal{B}$ be constitutively prescribed in terms of the actual displacement $\mathbf{u}$, that is

$$\mathbf{T}_{\mathrm{R}} = \mathbf{T}_{\mathrm{R}}(\mathbf{u}).$$

Then the virtual power balance (28.A.2) requires that

$$\int_{\mathcal{B}} \mathbf{T}_{\mathrm{R}}(\mathbf{u}):\nabla\tilde{\mathbf{u}}\,dv_{\mathrm{R}} - \int_{\mathcal{B}} \mathbf{b}_{\mathrm{R}}\cdot\tilde{\mathbf{u}}\,dv_{\mathrm{R}} - \int_{\partial\mathrm{B}_{\mathrm{t_R}}} \tilde{\mathbf{t}}_{\mathrm{R}}\cdot\tilde{\mathbf{u}}\,da_{\mathrm{R}} = \int_{\partial\mathrm{B}_{\mathrm{u}}} \mathbf{t}_{\mathrm{R}}(\mathbf{n}_{\mathrm{R}})\cdot\tilde{\mathbf{u}}\,da_{\mathrm{R}} \tag{28.A.3}$$

hold for all virtual velocities $\tilde{\mathbf{u}}$. Thus, defining a **residual** by

$$\underbrace{R(\mathbf{u}, \tilde{\mathbf{u}})}_{\text{residual}} \stackrel{\text{def}}{=} \int_{\mathcal{B}} \mathbf{T}_{\mathrm{R}}(\mathbf{u}) : \nabla \tilde{\mathbf{u}} \, dv_{\mathrm{R}} - \int_{\mathcal{B}} \mathbf{b}_{\mathrm{R}} \cdot \tilde{\mathbf{u}} \, dv_{\mathrm{R}} - \int_{\partial \mathbf{B}_{\mathbf{t}_{\mathrm{R}}}} \tilde{\mathbf{t}}_{\mathrm{R}} \cdot \tilde{\mathbf{u}} \, da_{\mathrm{R}}, \tag{28.A.4}$$

the virtual power balance (28.A.3) may be written as the requirement that

$$R(\mathbf{u}, \tilde{\mathbf{u}}) = \int_{\partial \mathbf{B}_{\mathbf{u}}} \mathbf{t}_{\mathrm{R}}(\mathbf{n}_{\mathrm{R}}) \cdot \tilde{\mathbf{u}} \, da_{\mathrm{R}} \tag{28.A.5}$$

hold for all virtual velocities $\tilde{\mathbf{u}}$. As is standard, a weak finite element solution for the displacement field $\mathbf{u}$ in the boundary-value problem is obtained by solving

$$R(\mathbf{u}, \tilde{\mathbf{u}}) = 0 \quad \text{for } \mathbf{u} \qquad \text{for all virtual velocities which vanish, } \tilde{\mathbf{u}} = \mathbf{0}, \text{ on } \partial \mathbf{B}_{\mathbf{u}}.$$

Now consider that we wish to calculate the reaction force N_{11} in the $\mathbf{e}_1$-direction on a portion $\mathcal{S}_1$ of $\partial \mathbf{B}_{\mathbf{u}}$ which has an outward unit normal $\mathbf{n}_{\mathrm{R}} = \mathbf{e}_1$,

$$N_{11} \stackrel{\text{def}}{=} \int_{\mathcal{S}_1} \mathbf{e}_1 \cdot \mathbf{t}_{\mathrm{R}}(\mathbf{e}_1) \, da_{\mathrm{R}} = \int_{\mathcal{S}_1} \mathbf{e}_1 \cdot \mathbf{T}_{\mathrm{R}} \mathbf{e}_1 \, da_{\mathrm{R}} = \int_{\mathcal{S}_1} (\mathbf{T}_{\mathrm{R}})_{11} \, da_{\mathrm{R}}. \tag{28.A.6}$$

This reaction force may be numerically calculated by *using a virtual velocity field* $\tilde{\mathbf{u}}$ *which does not vanish on* $\partial \mathcal{S}_1$, *but has a unit value* $\tilde{\mathbf{u}} = 1\, \mathbf{e}_1$ *on* $\partial \mathcal{S}_1$ in the virtual power balance (28.A.5), as

$$N_{11} = \int_{\mathcal{S}_1} \mathbf{e}_1 \cdot \mathbf{t}_{\mathrm{R}}(\mathbf{e}_1) \, da_{\mathrm{R}} = R(\mathbf{u}, \tilde{\mathbf{u}}) \qquad \text{with} \quad \tilde{\mathbf{u}} = 1\, \mathbf{e}_1 \text{ on } \mathcal{S}_1. \tag{28.A.7}$$

That is,

$$N_{11} = \int_{\mathcal{B}} \mathbf{T}_{\mathrm{R}}(\mathbf{u}) : \nabla \tilde{\mathbf{u}} \, dv_{\mathrm{R}} - \int_{\mathcal{B}} \mathbf{b}_{\mathrm{R}} \cdot \tilde{\mathbf{u}} \, dv_{\mathrm{R}} - \int_{\partial \mathbf{B}_{\mathbf{t}_{\mathrm{R}}}} \tilde{\mathbf{t}}_{\mathrm{R}} \cdot \tilde{\mathbf{u}} \, da_{\mathrm{R}} \qquad \text{with} \quad \tilde{\mathbf{u}} = 1\, \mathbf{e}_1 \text{ on } \mathcal{S}_1. \tag{28.A.8}$$

We have used this method in FEniCS to calculate the reaction force for the problem considered in Section 28.2.

Part X

Magneto-elasticity of hard-magnetic soft-elastomers

29 Magneto-viscoelasticity of hard-magnetic soft-elastomers

29.1 Introduction

Magnetorheological elastomers (MREs) are particulate composites made of a non-magnetic elastomeric matrix (such as a silicone rubber) dispersed with micron-sized magnetic particles. When the particles are made from a "hard-magnetic" material, such as NdFeB, the composites are called "hard"-MREs or h-MREs, and when the particles are made from a "soft-magnetic" material, such as carbonyl iron powders, the composites are called "soft"-MREs or s-MREs.

There are many aspects of the mechanical response of both h-MREs and s-MREs which are influenced by an externally applied magnetic field. A distinguishing characteristic of MREs is that when they are subjected to an externally applied magnetic field, they change shape. The macroscopic shape change is due to the microscopic rearrangement of the magnetic particles under the action of the applied magnetic field, which in turn deforms the surrounding non-magnetic soft polymer matrix. The shape-changing response of MREs under the action of an applied magnetic field is known as *magnetostriction*, which may be harnessed to develop novel magnetic actuators for engineering applications in several fields, including soft robotics and biomedical engineering.

In this chapter we formulate a theory for modeling the coupled magneto-mechanical response of h-MREs. A corresponding coupled theory for s-MREs will be considered in Chapter 32.

But before embarking on the details of the formulation of a theory for this special class of materials, we begin by recalling a subset of Maxwell's equation for magnetostatics (with no free currents).[1]

- Under magnetostatic conditions there is
  - a vector field denoted by

$$\mathbf{h}(\mathbf{x}, t)$$

    called the **magnetic field** [Amperes/meter, A/m], and
  - another field

$$\mathbf{b}(\mathbf{x}, t)$$

    called the **magnetic flux density** [Tesla, T].

[1] See Appendix A.

Introduction to Coupled Theories in Solid Mechanics. Lallit Anand, Eric M. Stewart, Shawn A. Chester, Oxford University Press.
 DOI: 10.1093/oso/9780198986218.003.0030

- In vacuum, under non-relativistic conditions, an inertial frame exists in which the field **h** is related to the field **b** by

$$\mathbf{h} = \mu_0^{-1}\mathbf{b}, \tag{29.1.1}$$

  where

$$\mu_0 = 4\pi \times 10^{-7}\ \mathrm{T/(A/m)}$$

  is the magnetic permeability of free space.
- In a material, the field **h** is related to the field **b** by

$$\mathbf{h} = \mu_0^{-1}\mathbf{b} - \mathbf{m}, \tag{29.1.2}$$

  where $\mathbf{m}(\mathrm{x}, t)$ is the bound **magnetization** [A/m] induced in the material.
- Under magnetostatic conditions there are two governing Maxwell equations:
  - **Gauss's law for magnetism**:

$$\mathrm{div}\mathbf{b} = 0. \tag{29.1.3}$$

  - **Maxwell–Ampère's law** (with no free current):

$$\mathrm{curl}\mathbf{h} = \mathbf{0}. \tag{29.1.4}$$

As described above, h-MREs are composites made of a "mechanically soft" elastomeric matrix phase with a dispersion of small particles of a "hard-magnetic" material, such as neodymium–iron–boron (NdFeB). Typically the magnetizable particles have a small diameter ($\approx$ 35 μm) and they make up a small volume fraction ($\approx$ 10%) of the overall composite material, so that the composite material is magnetically responsive but it remains mechanically flexible and low-weight. When a strong external magnetic field **h** of 1–5 MA/m (or equivalently, an external magnetic flux density $\mathbf{b} = \mu_0\mathbf{h}$ of 1–6 Teslas) is applied to a hard-magnetic soft-elastomeric composite, it causes a magnetization **m** to develop in the magnetizable particles according to a complicated **h**–**m** hysteresis loop, as shown schematically in Fig. 29.1, while leaving the elastomeric matrix un-magnetized.

- A major characteristic feature of a hysteresis loop for a "hard-magnetic" material is that it exhibits a high coercivity $\mathrm{h_c}$, and therefore it can retain a high remanent magnetization $\mathbf{m}^{\mathrm{rem}}$ when a large magnetic field **h** is applied and then subsequently removed.

Thus, the composite material after being subjected to a strong external magnetic field has a remanent magnetization field—on the order of $|\mathbf{m}^{\mathrm{rem}}| \approx 100\ \mathrm{kA/m}$—which is oriented in the direction of the applied magnetic field.

After this initial magnetization of the composite material, application of an external magnetic field with strength substantially below the coercivity limit of the hard-magnetic particles *essentially* does not change the magnetization of the particles. Instead, any misalignment

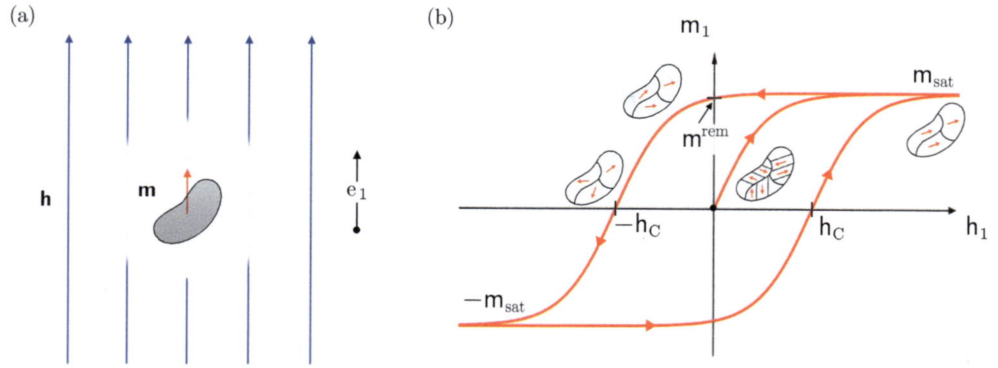

Fig. 29.1 (a) A hard-magnetic particle immersed in a magnetic field. (b) The accompanying magnetic hysteresis loop. Arrows in the inset schematics indicate the microstructure of magnetized ferromagnetic domains, which align under the action of the imposed magnetic field. Figure adapted from Coey (2010).

between the remanent magnetization $\mathbf{m}^{\text{rem}}$ and applied magnetic flux density $\mathbf{b}^{\text{app}} = \mu_0 \mathbf{h}^{\text{app}}$ generates a body torque

$$\boldsymbol{\tau}^{\text{mag}} \stackrel{\text{def}}{=} \mathbf{m}^{\text{rem}} \times \mathbf{b}^{\text{app}}, \tag{29.1.5}$$

which deforms the material so that $\mathbf{m}^{\text{rem}}$ tends to align with $\mathbf{b}^{\text{app}}$. Owing to the flexibility of the elastomeric matrix, this body torque is capable of driving large deformations in structures made from magnetized hard-magnetic soft-materials when they are subjected to magnetic flux densities as low as a few tens of milli-Teslas.

Magneto-elasticity has long been studied in the continuum mechanics literature, and pioneering contributions to the field include the works of Truesdell and Toupin (1960), Brown (1966), Pao (1978), Kovetz (2000), Dorfmann and Ogden (2003; 2004; 2014), amongst others. With a specific focus on hard-magnetic soft-materials, Kim et al. (2018) and Zhao et al. (2019) proposed a continuum theory for modeling the response of such materials, and they showed that the simulation results based on their theory compare well against results from corresponding experiments.

In this chapter we describe a theory for hard-magnetic soft-elastomers (h-MREs) proposed by Stewart and Anand (2023), which extends the theory of Zhao et al. (2019) to account for *viscoelasticity and inertial effects*—phenomena which are of importance in modeling the response of actuators made from such materials.

29.2 Basic kinematics

We consider a macroscopically homogeneous body $\mathcal{B}$ with the region of space it occupies in a fixed reference configuration, and denote by $\mathbf{X}$ an arbitrary material point of $\mathcal{B}$. A motion of $\mathcal{B}$ is a smooth one-to-one mapping

$$\mathbf{x} = \boldsymbol{\chi}(\mathbf{X}, t), \tag{29.2.1}$$

with displacement

$$\mathbf{u}(\mathbf{X},t) = \boldsymbol{\chi}(\mathbf{X},t) - \mathbf{X}, \tag{29.2.2}$$

and the deformation gradient, velocity, acceleration, and velocity gradient are given by,

$$\mathbf{F} = \nabla\boldsymbol{\chi} = \mathbf{1} + \nabla\mathbf{u}, \qquad \dot{\mathbf{u}} = \frac{\partial\boldsymbol{\chi}}{\partial t}, \qquad \ddot{\mathbf{u}} = \frac{\partial^2\boldsymbol{\chi}}{\partial t^2}, \qquad \mathbf{L} = \operatorname{grad}\dot{\mathbf{u}} = \dot{\mathbf{F}}\mathbf{F}^{-1}. \tag{29.2.3}$$

We assume that

$$J = \det\mathbf{F} > 0, \tag{29.2.4}$$

so that $\mathbf{F}$ is invertible.

Basic to our discussion of the constitutive equations for large deformations of hard-magnetic soft-materials is the polar decomposition

$$\mathbf{F} = \mathbf{R}\mathbf{U}, \tag{29.2.5}$$

of the deformation gradient into a rotation tensor $\mathbf{R}$, together with a symmetric and positive-definite right stretch tensor $\mathbf{U}$, satisfying

$$\mathbf{U} = \sqrt{\mathbf{F}^\top\mathbf{F}}. \tag{29.2.6}$$

The symmetric and positive-definite right Cauchy–Green tensor is defined as,

$$\mathbf{C} \stackrel{\text{def}}{=} \mathbf{F}^\top\mathbf{F}. \tag{29.2.7}$$

For later use we note that the deformation gradient $\mathbf{F}$ may be decomposed multiplicatively into volumetric and isochoric factors $\mathbf{F}^{\text{vol}}$ and $\bar{\mathbf{F}}$ as,

$$\mathbf{F} = \mathbf{F}^{\text{vol}}\bar{\mathbf{F}}, \tag{29.2.8}$$

where

$$\mathbf{F}^{\text{vol}} = J^{1/3}\mathbf{1} \qquad \text{and} \qquad \bar{\mathbf{F}} = J^{-1/3}\mathbf{F}. \tag{29.2.9}$$

This leads to a multiplicative decomposition of the Cauchy–Green tensor $\mathbf{C} = \mathbf{F}^\top\mathbf{F}$ of the form,

$$\mathbf{C} = \mathbf{C}^{\text{vol}}\bar{\mathbf{C}}, \qquad \mathbf{C}^{\text{vol}} = J^{2/3}\mathbf{1}, \qquad \bar{\mathbf{C}} = \bar{\mathbf{F}}^\top\bar{\mathbf{F}}. \tag{29.2.10}$$

We also introduce the invariants,

$$\bar{I}_1 \stackrel{\text{def}}{=} \operatorname{tr}\bar{\mathbf{C}}, \qquad \bar{I}_2 \stackrel{\text{def}}{=} \tfrac{1}{2}\left[(\operatorname{tr}\bar{\mathbf{C}})^2 - \operatorname{tr}(\bar{\mathbf{C}}^2)\right], \qquad \text{and} \qquad I_3 = J. \tag{29.2.11}$$

Throughout we shall denote an arbitrary part of the reference body by $\mathcal{P}$, with $\mathbf{n}_{\text{R}}$ the outward unit normal on the boundary $\partial\mathcal{P}$ of $\mathcal{P}$.

Kinematics of viscoelasticity based on the decomposition $\mathbf{F} = \mathbf{F}^{e(\alpha)}\mathbf{F}^{v(\alpha)}$

To account for viscoelasticity of h-MREs we introduce a multi-branch multiplicative decomposition of the deformation gradient of the form (Sidoroff, 1974),[2]

$$\mathbf{F} = \mathbf{F}^{e(\alpha)}\mathbf{F}^{v(\alpha)} \qquad \alpha = 0, 1, \ldots, M, \tag{29.2.12}$$

which allows for $(M+1)$ viscoelastic relaxation mechanisms. We refer to $\mathbf{F}^{v(\alpha)}$ and $\mathbf{F}^{e(\alpha)}$ as *viscous and elastic distortions*, respectively. Using (29.2.4) and (29.2.12) the volumetric Jacobian J can be decomposed as

$$J = J^{e(\alpha)}J^{v(\alpha)}. \tag{29.2.13}$$

We assume that

$$J^{e(\alpha)} \stackrel{\text{def}}{=} \det \mathbf{F}^{e(\alpha)} > 0, \quad \text{and} \quad J^{v(\alpha)} \stackrel{\text{def}}{=} \det \mathbf{F}^{v(\alpha)} > 0, \tag{29.2.14}$$

so that $\mathbf{F}^{e(\alpha)}$ and $\mathbf{F}^{v(\alpha)}$ are invertible.

The elastic distortions $\mathbf{F}^{e(\alpha)}$ may be decomposed into volumetric and isochoric factors $\mathbf{F}^{e(\alpha)_{\text{vol}}}$ and $\bar{\mathbf{F}}^{e(\alpha)}$:

$$\mathbf{F}^{e(\alpha)} = \mathbf{F}^{e(\alpha)_{\text{vol}}}\bar{\mathbf{F}}^{e(\alpha)}, \tag{29.2.15}$$

where

$$\mathbf{F}^{e(\alpha)_{\text{vol}}} = (J^{e(\alpha)})^{1/3}\mathbf{1} \qquad \text{and} \qquad \bar{\mathbf{F}}^{e(\alpha)} = (J^{e(\alpha)})^{-1/3}\mathbf{F}^{e(\alpha)}. \tag{29.2.16}$$

The right and left elastic Cauchy–Green tensors are given by

$$\mathbf{C}^{e(\alpha)} = \mathbf{F}^{e(\alpha)\top}\mathbf{F}^{e(\alpha)}, \qquad \mathbf{B}^{e(\alpha)} = \mathbf{F}^{e(\alpha)}\mathbf{F}^{e(\alpha)\top}, \tag{29.2.17}$$

and these may be decomposed into volumetric and isochoric forms as,

$$\begin{aligned} \mathbf{C}^{e(\alpha)} &= \mathbf{C}^{e(\alpha)_{\text{vol}}}\bar{\mathbf{C}}^{e(\alpha)}, \qquad \mathbf{C}^{e(\alpha)_{\text{vol}}} = (J^{e(\alpha)})^{2/3}\mathbf{1}, \qquad \bar{\mathbf{C}}^{e(\alpha)} = \bar{\mathbf{F}}^{e(\alpha)\top}\bar{\mathbf{F}}^{e(\alpha)} \\ \mathbf{B}^{e(\alpha)} &= \mathbf{B}^{e(\alpha)_{\text{vol}}}\bar{\mathbf{B}}^{e(\alpha)}, \qquad \mathbf{B}^{e(\alpha)_{\text{vol}}} = (J^{e(\alpha)})^{2/3}\mathbf{1}, \qquad \bar{\mathbf{B}}^{e(\alpha)} = \bar{\mathbf{F}}^{e(\alpha)}\bar{\mathbf{F}}^{e(\alpha)\top}. \end{aligned} \tag{29.2.18}$$

We also introduce the invariants,

$$\bar{I}_1^e \stackrel{\text{def}}{=} \operatorname{tr}\bar{\mathbf{C}}^{e(\alpha)}, \qquad \bar{I}_2^e \stackrel{\text{def}}{=} \tfrac{1}{2}\left((\operatorname{tr}\bar{\mathbf{C}}^{e(\alpha)})^2 - \operatorname{tr}\left((\bar{\mathbf{C}}^{e(\alpha)})^2\right)\right), \qquad \text{and} \qquad I_3^e = J^{e(\alpha)}. \tag{29.2.19}$$

Next, by $(29.2.3)_4$ and (29.2.12), the velocity gradient $\mathbf{L} = \dot{\mathbf{F}}\mathbf{F}^{-1}$ may be decomposed as

$$\mathbf{L} = \mathbf{L}^{e(\alpha)} + \mathbf{F}^{e(\alpha)}\mathbf{L}^{v(\alpha)}\mathbf{F}^{e(\alpha)-1}, \tag{29.2.20}$$

[2] We follow our development of the viscoelasticity theory in Chapter 4.

with

$$\mathbf{L}^{e(\alpha)} = \dot{\mathbf{F}}^{e(\alpha)}\mathbf{F}^{e(\alpha)-1}, \qquad \mathbf{L}^{v(\alpha)} = \dot{\mathbf{F}}^{v(\alpha)}\mathbf{F}^{v(\alpha)-1}. \tag{29.2.21}$$

As is standard, we define the total, elastic, and viscous stretching and spin tensors through

$$\begin{aligned} \mathbf{D} &= \operatorname{sym}\mathbf{L}, & \mathbf{W} &= \operatorname{skw}\mathbf{L}, \\ \mathbf{D}^{e(\alpha)} &= \operatorname{sym}\mathbf{L}^{e(\alpha)}, & \mathbf{W}^{e(\alpha)} &= \operatorname{skw}\mathbf{L}^{e(\alpha)}, \\ \mathbf{D}^{v(\alpha)} &= \operatorname{sym}\mathbf{L}^{v(\alpha)}, & \mathbf{W}^{v(\alpha)} &= \operatorname{skw}\mathbf{L}^{v(\alpha)}, \end{aligned} \tag{29.2.22}$$

so that $\mathbf{L} = \mathbf{D} + \mathbf{W}$, $\mathbf{L}^{e(\alpha)} = \mathbf{D}^{e(\alpha)} + \mathbf{W}^{e(\alpha)}$, and $\mathbf{L}^{v(\alpha)} = \mathbf{D}^{v(\alpha)} + \mathbf{W}^{v(\alpha)}$.

As in Chapter 4, we make two basic kinematical assumptions concerning viscous flow:

(i) First, we make the standard assumption that viscous flow is incompressible, so that

$$J^{v(\alpha)} = \det \mathbf{F}^{v(\alpha)} = 1 \qquad \text{and} \qquad \operatorname{tr}\mathbf{L}^{v(\alpha)} = \operatorname{tr}\mathbf{D}^{v(\alpha)} = 0. \tag{29.2.23}$$

Hence, using (29.2.13), we note that

$$J^{e(\alpha)} = J \qquad \text{for all } \alpha. \tag{29.2.24}$$

(ii) Next, from the outset we constrain the theory by limiting our discussion to circumstances under which the material may be idealized as *isotropic*. For isotropic elastic-viscous theories utilizing the multiplicative decomposition of $\mathbf{F}$, it is widely assumed that the viscous flow is *irrotational* (cf., e.g., Gurtin et al., 2010, Section 97), in the sense that

$$\mathbf{W}^{v(\alpha)} = \mathbf{0}. \tag{29.2.25}$$

Then, $\mathbf{L}^{v(\alpha)} \equiv \mathbf{D}^{v(\alpha)}$, using which the evolution equation for $\mathbf{F}^{v(\alpha)}$ is

$$\dot{\mathbf{F}}^{v(\alpha)} = \mathbf{D}^{v(\alpha)}\mathbf{F}^{v(\alpha)}. \tag{29.2.26}$$

29.3 Balance of forces

In the multibranch theory for viscoelasticity based on $\mathbf{F} = \mathbf{F}^{e(\alpha)}\mathbf{F}^{v(\alpha)}$, the Cauchy stress admits the additive decomposition

$$\mathbf{T} = \sum_{\alpha} \mathbf{T}^{(\alpha)}, \tag{29.3.1}$$

where $\mathbf{T}^{(\alpha)}$ is the Cauchy stress in each branch (cf., e.g., Mao et al., 2017). Since $\mathbf{T}$ is symmetric, we assume that each $\mathbf{T}^{(\alpha)}$ is *symmetric*.

Next, as is standard, we introduce the Piola stress $\mathbf{T}_R$ which is related to the Cauchy stress $\mathbf{T}$ by,

$$\mathbf{T}_R = J\,\mathbf{T}\mathbf{F}^{-\top}, \tag{29.3.2}$$

in terms of which:

(a) The surface traction on an oriented element of the boundary $\partial\mathcal{P}$ of $\mathcal{P}$, is given by the traction relation

$$\mathbf{t}_R(\mathbf{n}_R) = \mathbf{T}_R\mathbf{n}_R. \tag{29.3.3}$$

(b) $\mathbf{T}_R$ satisfies the force balance

$$\mathrm{Div}\,\mathbf{T}_R + \mathbf{f}_R = \mathbf{0}, \tag{29.3.4}$$

with

$$\mathbf{f}_R = \mathbf{f}_{0R} - \rho_R\ddot{\mathbf{u}}, \tag{29.3.5}$$

where $\mathbf{f}_{0R}$ represents the conventional body force per unit volume of the reference body, and $(-\rho_R\ddot{\mathbf{u}})$ represents the inertial body force, with ρ_R being the mass density of the referential body and $\ddot{\mathbf{u}}$ the acceleration.[3]

29.4 Free-energy imbalance

Under isothermal conditions the two laws of thermodynamics reduce to the statement that *the rate of increase in free energy of any part $\mathcal{P}$ is less than or equal to the power expended on $\mathcal{P}$.* Precisely, letting ψ_R denote the *free energy* per unit reference volume, this requirement takes the form of a free-energy imbalance

$$\overline{\int_{\mathcal{P}} \psi_R\,dv_R}^{\boldsymbol{\cdot}} \le \mathcal{W}(\mathcal{P}) \tag{29.4.1}$$

where the mechanical power $\mathcal{W}(\mathcal{P})$ exerted on $\mathcal{P}$ by the boundary tractions $\mathbf{T}_R\mathbf{n}_R$ and body force $\mathbf{f}_R$ over the velocity $\dot{\mathbf{u}}$ is given by,

$$\mathcal{W}(\mathcal{P}) = \int_{\partial\mathcal{P}} (\mathbf{T}_R\mathbf{n}_R)\cdot\dot{\mathbf{u}}\,da_R + \int_{\mathcal{P}} \mathbf{f}_R\cdot\dot{\mathbf{u}}\,dv_R. \tag{29.4.2}$$

[3] In the previous chapters we have denoted the referential conventional and generalized body forces by $\mathbf{b}_{0R}$ and $\mathbf{b}_R$, respectively, but here—in order not to cause confusion with notation for the magnetic flux density—we use the notations $\mathbf{f}_{0R}$ and $\mathbf{f}_R$ for the body forces.

Thus, using (29.4.2), the free-energy imbalance (29.4.1) becomes

$$\overline{\int_{\mathcal{P}} \psi_{\mathrm{R}}\, dv_{\mathrm{R}}}^{\,\boldsymbol{\cdot}} \le \int_{\partial\mathcal{P}} (\mathbf{T}_{\mathrm{R}}\mathbf{n}_{\mathrm{R}}) \cdot \dot{\mathbf{u}}\, da_{\mathrm{R}} + \int_{\mathcal{P}} \mathbf{f}_{\mathrm{R}} \cdot \dot{\mathbf{u}}\, dv_{\mathrm{R}}. \tag{29.4.3}$$

Using the divergence theorem on the term involving the integral over $\partial\mathcal{P}$ in (29.4.3) we obtain

$$\int_{\partial\mathcal{P}} (\mathbf{T}_{\mathrm{R}}\mathbf{n}_{\mathrm{R}}) \cdot \dot{\mathbf{u}}\, da_{\mathrm{R}} = \int_{\mathcal{P}} (\mathrm{Div}\,\mathbf{T}_{\mathrm{R}}) \cdot \dot{\mathbf{u}}\, dv_{\mathrm{R}} + \int_{\mathcal{P}} \mathbf{T}_{\mathrm{R}} : \dot{\mathbf{F}} dv_{\mathrm{R}}. \tag{29.4.4}$$

Substituting (29.4.4) in (29.4.3), and using the linear momentum balance (29.3.4) we obtain,

$$\int_{\mathcal{P}} \left(\dot{\psi}_{\mathrm{R}} - \mathbf{T}_{\mathrm{R}} : \dot{\mathbf{F}} \right) dv_{\mathrm{R}} \le 0. \tag{29.4.5}$$

Since (29.4.5) must hold for all parts $\mathcal{P}$, we obtain a local free-energy imbalance in the form,

$$\dot{\psi}_{\mathrm{R}} - \mathbf{T}_{\mathrm{R}} : \dot{\mathbf{F}} \le 0. \tag{29.4.6}$$

Using (29.2.12), $(29.2.21)_2$, (29.2.26), $(29.2.23)_1$, and (29.3.2), the stress power $\mathbf{T}_{\mathrm{R}} : \dot{\mathbf{F}}$ admits the decomposition,

$$\begin{aligned}
\mathbf{T}_{\mathrm{R}} : \dot{\mathbf{F}} &= \sum_{\alpha} \mathbf{T}_{\mathrm{R}} : (\dot{\mathbf{F}}^{e(\alpha)}\mathbf{F}^{v(\alpha)} + \mathbf{F}^{e(\alpha)}\dot{\mathbf{F}}^{v(\alpha)}) \\
&= \sum_{\alpha} (\mathbf{T}_{\mathrm{R}}\mathbf{F}^{v(\alpha)\top}) : \dot{\mathbf{F}}^{e(\alpha)} + \sum_{\alpha} (\mathbf{F}^{e(\alpha)\top}\mathbf{T}_{\mathrm{R}}) : \dot{\mathbf{F}}^{v(\alpha)}, \\
&= \sum_{\alpha} (\mathbf{T}_{\mathrm{R}}\mathbf{F}^{v(\alpha)\top}) : \dot{\mathbf{F}}^{e(\alpha)} + \sum_{\alpha} (\mathbf{F}^{e(\alpha)\top}\mathbf{T}_{\mathrm{R}}\mathbf{F}^{v(\alpha)\top}) : \mathbf{L}^{v(\alpha)}, \\
&= \sum_{\alpha} (J\mathbf{T}^{(\alpha)}\mathbf{F}^{e(\alpha)-\top}) : \dot{\mathbf{F}}^{e(\alpha)} + \sum_{\alpha} (J\mathbf{F}^{e(\alpha)\top}\mathbf{T}^{(\alpha)}\mathbf{F}^{e(\alpha)-\top}) : \mathbf{D}^{v(\alpha)}, \\
&= \sum_{\alpha} (J^{e(\alpha)}\mathbf{T}^{(\alpha)}\mathbf{F}^{e(\alpha)-\top}) : \dot{\mathbf{F}}^{e(\alpha)} + \sum_{\alpha} (J^{e(\alpha)}\mathbf{F}^{e(\alpha)\top}\mathbf{T}^{(\alpha)}\mathbf{F}^{e(\alpha)-\top}) : \mathbf{D}^{v(\alpha)}, \\
&= \sum_{\alpha} (J^{e(\alpha)}\mathbf{F}^{e(\alpha)-1}\mathbf{T}^{(\alpha)}\mathbf{F}^{e(\alpha)-\top}) : \mathbf{F}^{e(\alpha)\top}\dot{\mathbf{F}}^{e(\alpha)} + \sum_{\alpha} (J^{e(\alpha)}\mathbf{F}^{e(\alpha)\top}\mathbf{T}^{(\alpha)}\mathbf{F}^{e(\alpha)-\top}) : \mathbf{D}^{v(\alpha)}.
\end{aligned} \tag{29.4.7}$$

In view of (29.4.7) we introduce two new stress measures:

- The *elastic second Piola stress* for each α,

$$\mathbf{T}^{e(\alpha)} \stackrel{\text{def}}{=} J^{e(\alpha)}\mathbf{F}^{e(\alpha)-1}\mathbf{T}^{(\alpha)}\mathbf{F}^{e(\alpha)-\top}, \tag{29.4.8}$$

which is *symmetric* on account of the symmetry of $\mathbf{T}$.

- The *Mandel stress* for each α,

$$\mathbf{M}^{e(\alpha)} \stackrel{\text{def}}{=} J^{e(\alpha)}\mathbf{F}^{e(\alpha)\top}\mathbf{T}^{(\alpha)}\mathbf{F}^{e(\alpha)-\top} \equiv \mathbf{C}^{e(\alpha)}\mathbf{T}^{e(\alpha)}, \tag{29.4.9}$$

  which in general is not symmetric.

Then, the stress power (29.4.7) may be written as

$$\mathbf{T}_{\text{R}}:\dot{\mathbf{F}} = \sum_{\alpha}\mathbf{T}^{e(\alpha)}:\mathbf{F}^{e(\alpha)\top}\dot{\mathbf{F}}^{e(\alpha)} + \sum_{\alpha}\mathbf{M}_0^{e(\alpha)}:\mathbf{D}^{v(\alpha)}, \tag{29.4.10}$$

where we have used the fact that $\operatorname{tr}\mathbf{D}^{v(\alpha)} = 0$ (cf. eq. (29.2.14)) so that it is the deviator of the Mandel stress that appears in (29.4.10). Noting that the material derivative of the elastic right Cauchy–Green tensor $\mathbf{C}^{e(\alpha)}$ is

$$\dot{\mathbf{C}}^{e(\alpha)} = \mathbf{F}^{e(\alpha)\top}\dot{\mathbf{F}}^{e(\alpha)} + \dot{\mathbf{F}}^{e(\alpha)\top}\mathbf{F}^{e(\alpha)}, \tag{29.4.11}$$

and recalling the symmetry of $\mathbf{T}^{e(\alpha)}$, gives the stress-power decomposition

$$\mathbf{T}_{\text{R}}:\dot{\mathbf{F}} = \sum_{\alpha}\frac{1}{2}\mathbf{T}^{e(\alpha)}:\dot{\mathbf{C}}^{e(\alpha)} + \sum_{\alpha}\mathbf{M}_0^{e(\alpha)}:\mathbf{D}^{v(\alpha)}. \tag{29.4.12}$$

Hence, using (29.4.12) in (29.4.6), the local free-energy imbalance may be written as

$$\dot{\psi}_{\text{R}} - \sum_{\alpha}\frac{1}{2}\mathbf{T}^{e(\alpha)}:\dot{\mathbf{C}}^{e(\alpha)} - \sum_{\alpha}\mathbf{M}_0^{e(\alpha)}:\mathbf{D}^{v(\alpha)} \le 0. \tag{29.4.13}$$

29.5 Constitutive theory

In a magnetizable material there is a vector field in the deformed $\mathcal{B}_t = \chi(\mathcal{B},t)$ denoted by $\mathbf{h}(\mathbf{x},t)$ called the *magnetic field*, and another field $\mathbf{b}(\mathbf{x},t)$ in $\mathcal{B}_t$ called the *magnetic flux density*. The field $\mathbf{h}$ is related to the field $\mathbf{b}$ by

$$\mathbf{h} = \mu_0^{-1}\mathbf{b} - \mathbf{m}, \tag{29.5.1}$$

where $\mathbf{m}(\mathbf{x},t)$ is a magnetization field, and $\mu_0 = 4\pi \times 10^{-7}$ Tesla/(A/m) is the magnetic permeability of free space. The referential counterparts of $\mathbf{h}$ and $\mathbf{b}$ are

$$\mathbf{h}_{\text{R}} = \mathbf{F}^{\top}\mathbf{h} \qquad \text{and} \qquad \mathbf{b}_{\text{R}} = J\mathbf{F}^{-1}\mathbf{b}. \tag{29.5.2}$$

The focus of this chapter is on materials composed of particles of hard-magnetic materials—that is, particles of high-coercivity ferromagnetic materials such as neodymium–iron–boron

(NdFeB)—which are embedded in soft elastomeric matrices. A major characteristic of hard-magnetic materials is that they retain a high residual magnetization $\mathbf{m}_{\mathrm{R}}^{\mathrm{rem}}$ after they have been magnetically saturated. In developing the theory we make the following central assumptions:

(A1) The remanent magnetization $\mathbf{m}_{\mathrm{R}}^{\mathrm{rem}}$ per unit volume in the reference configuration is known a priori. The vector field $\mathbf{m}_{\mathrm{R}}^{\mathrm{rem}}$, being an internal variable defined in the reference body, is invariant under a change in frame.

(A2) Next, consider for the moment that the soft matrix and hard particles are both completely incompressible, and that at a material point in the reference body the macroscopically averaged magnetization vector field per unit volume is $\mathbf{m}_{\mathrm{R}}^{\mathrm{rem}}$. A central question in the formulation of the theory is how $\mathbf{m}_{\mathrm{R}}^{\mathrm{rem}}$ convects when the body deforms. Based on their computational homogenization studies of such composites, Mukherjee et al. (2021) found that the direction of $\mathbf{m}_{\mathrm{R}}^{\mathrm{rem}}$ is essentially *unaltered* when the neighborhood of a material point is simply stretched, that is with $\mathbf{F} = \mathbf{R}\mathbf{U}$ the vector $\mathbf{m}_{\mathrm{R}}^{\mathrm{rem}}$ is *unaltered* when $\mathbf{F} = \mathbf{U}$ and $\mathbf{R} = \mathbf{1}$. However, the direction of $\mathbf{m}_{\mathrm{R}}^{\mathrm{rem}}$ is *significantly altered* when $\mathbf{R} \neq \mathbf{1}$; see Fig. 29.2 for a schematic.

Thus, allowing for a slight compressibility of the composite, and recalling that an elemental volume dv_{R} in the reference body is related to an elemental volume dv in the deformed body by $dv_{\mathrm{R}} = J^{-1}dv$, we assume that the remanent magnetization per unit volume in the deformed body $\mathbf{m}^{\mathrm{rem}}$ is given by

$$\mathbf{m}^{\mathrm{rem}} = J^{-1}\mathbf{R}\mathbf{m}_{\mathrm{R}}^{\mathrm{rem}}. \tag{29.5.3}$$

REMARK

In their paper Zhao et al. (2019) used the mapping relation

$$\mathbf{m}^{\mathrm{rem}} = J^{-1}\mathbf{F}\mathbf{m}_{\mathrm{R}}^{\mathrm{rem}} \tag{29.5.4}$$

for the remanent magnetization. Use of the $\mathbf{R}$-based mapping (29.5.3) for $\mathbf{m}^{\mathrm{rem}}$, rather than $\mathbf{F}$-based mapping (29.5.4), is a major point of departure of the theory considered here relative to the theory in Zhao et al. (2019). Also see Yan et al. (2023), in the context of a plate theory, for a discussion regarding the use of $\mathbf{R}$-based rather than $\mathbf{F}$-based mapping of the residual magnetization.

(A3) The permeability of saturated hard-magnetic particles as well as that of the elastomeric material is close to that of vacuum, μ_0.

(A4) The remanent magnetization $\mathbf{m}_{\mathrm{R}}^{\mathrm{rem}}$ and the externally applied magnetic flux density $\mathbf{b}^{\mathrm{app}}$ are assumed to be constants in so far as the development of the constitutive theory.[4]

[4] The applied magnetic flux density $\mathbf{b}^{\mathrm{app}}$ can of course be varied when considering actual boundary-value problems.

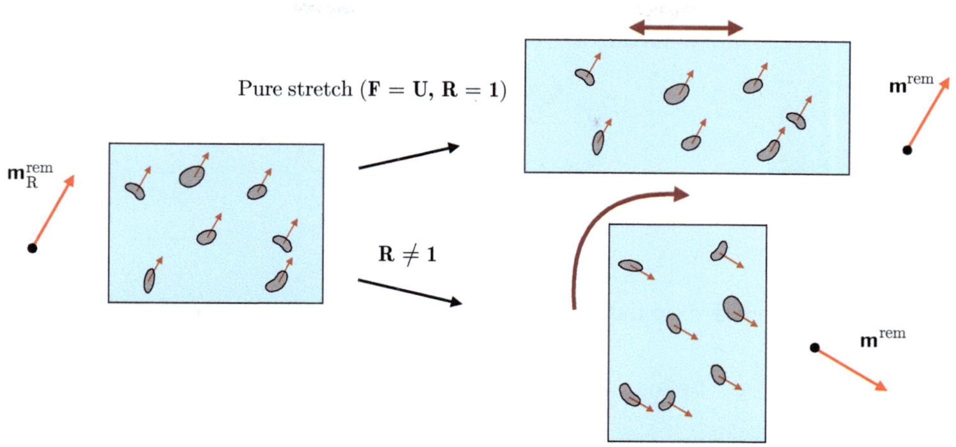

Fig. 29.2 The deformation of a material neighborhood in an h-MRE material under cases of a pure stretch ($\mathbf{F} = \mathbf{U}$, $\mathbf{R} = \mathbf{1}$) and a pure rotation ($\mathbf{F} = \mathbf{R} \neq \mathbf{1}$, $\mathbf{U} = \mathbf{1}$). Annotated arrows indicate the change in direction (or lack thereof) for the spatial remanent magnetization $\mathbf{m}^{\mathrm{rem}}$ relative to the referential remanent magnetization $\mathbf{m}_{\mathrm{R}}^{\mathrm{rem}}$.

(A5) The free energy of the material per unit volume depends on the remanent magnetization $\mathbf{m}^{\mathrm{rem}}$ and its interaction with the externally applied magnetic flux density $\mathbf{b}^{\mathrm{app}}$.

Then, guided by the free-energy imbalance (29.4.13) and with $\mathbf{\Lambda}$ denoting the list

$$\mathbf{\Lambda} \stackrel{\mathrm{def}}{=} \left\{ \mathbf{C}^{e(\alpha)}, \mathbf{m}_{\mathrm{R}}^{\mathrm{rem}}, \mathbf{b}^{\mathrm{app}} \right\}, \tag{29.5.5}$$

we assume constitutive equations for the free-energy density ψ_{R} and the elastic second Piola stress $\mathbf{T}^{e(\alpha)}$ of the form

$$\psi_{\mathrm{R}} = \hat{\psi}_{\mathrm{R}}(\mathbf{\Lambda}), \qquad \mathbf{T}^{e(\alpha)} = \hat{\mathbf{T}}^{e(\alpha)}(\mathbf{\Lambda}). \tag{29.5.6}$$

Then, since the magnetic variables $(\mathbf{m}_{\mathrm{R}}^{\mathrm{rem}}, \mathbf{b}^{\mathrm{app}})$ are *fixed*,

$$\dot{\psi}_{\mathrm{R}} = \sum_{\alpha} \frac{\partial \hat{\psi}_{\mathrm{R}}(\mathbf{\Lambda})}{\partial \mathbf{C}^{e(\alpha)}} : \dot{\mathbf{C}}^{e(\alpha)}, \tag{29.5.7}$$

and the free-energy imbalance (29.4.13) then requires that the constitutive equations satisfy the inequality

$$\sum_{\alpha} \left(\frac{\partial \hat{\psi}_{\mathrm{R}}(\mathbf{\Lambda})}{\partial \mathbf{C}^{e(\alpha)}} - \frac{1}{2} \hat{\mathbf{T}}^{e(\alpha)}(\mathbf{\Lambda}) \right) : \dot{\mathbf{C}}^{e(\alpha)} - \sum_{\alpha} \mathbf{M}^{e(\alpha)} : \mathbf{D}^{v(\alpha)} \leq 0. \tag{29.5.8}$$

Satisfaction of the free-energy imbalance (29.5.8) for all $\dot{\mathbf{C}}^{e(\alpha)}$ requires that the elastic second Piola stress for each α is given by

$$\mathbf{T}^{e(\alpha)} = 2 \frac{\partial \hat{\psi}_{\mathrm{R}}(\mathbf{\Lambda})}{\partial \mathbf{C}^{e(\alpha)}}, \tag{29.5.9}$$

and we are then left with the reduced dissipation inequality

$$\mathcal{D} = \sum_{\alpha} \mathbf{M}_0^{e(\alpha)} : \mathbf{D}^{v(\alpha)} \geq 0. \tag{29.5.10}$$

Also, recalling (29.4.9), the Mandel stresses are given by

$$\mathbf{M}^{e(\alpha)} = \mathbf{C}^{e(\alpha)}\mathbf{T}^{e(\alpha)}, \tag{29.5.11}$$

and the contributions $\mathbf{T}^{(\alpha)}$ to the Cauchy stress $\mathbf{T}$ are given by

$$\mathbf{T}^{(\alpha)} = (J^{e(\alpha)})^{-1}\left(2\mathbf{F}^{e(\alpha)}\frac{\partial\hat{\psi}_{\mathrm{R}}(\mathbf{\Lambda})}{\partial\mathbf{C}^{e(\alpha)}}\mathbf{F}^{e(\alpha)\top}\right). \tag{29.5.12}$$

The total Cauchy stress $\mathbf{T} = \sum_{\alpha} \mathbf{T}^{(\alpha)}$ is then given by

$$\mathbf{T} = \sum_{\alpha} (J^{e(\alpha)})^{-1}\left(2\mathbf{F}^{e(\alpha)}\frac{\partial\hat{\psi}_{\mathrm{R}}(\mathbf{\Lambda})}{\partial\mathbf{C}^{e(\alpha)}}\mathbf{F}^{e(\alpha)\top}\right), \tag{29.5.13}$$

and hence the total Piola stress $\mathbf{T}_{\mathrm{R}}$ is

$$\mathbf{T}_{\mathrm{R}} = J\mathbf{T}\mathbf{F}^{-\top}. \tag{29.5.14}$$

Evolution equations for $\mathbf{F}^{v(\alpha)}$

The evolution equation for $\mathbf{F}^{v(\alpha)}$ in each of the branches is taken as[5]

$$\dot{\mathbf{F}}^{v(\alpha)} = \mathbf{D}^{v(\alpha)}\,\mathbf{F}^{v(\alpha)}, \tag{29.5.15}$$

with the viscous stretching given by

$$\mathbf{D}^{v(\alpha)} = \dot{\bar{\gamma}}^{v(\alpha)}\left(\frac{\mathbf{M}_0^{e(\alpha)}}{2\,\bar{\tau}^{(\alpha)}}\right), \tag{29.5.16}$$

with the constraint that viscous flow is isochoric, $\det \mathbf{F}^{v(\alpha)} = 1$, and where

$$\dot{\bar{\gamma}}^{v(\alpha)} \stackrel{\text{def}}{=} \sqrt{2}\,|\mathbf{D}^{v(\alpha)}| \tag{29.5.17}$$

is an *equivalent shear-strain rate* for each branch and

$$\bar{\tau}^{(\alpha)} \stackrel{\text{def}}{=} \frac{1}{\sqrt{2}}\,|\mathbf{M}_0^{e(\alpha)}| \tag{29.5.18}$$

defines an *equivalent shear stress* for each branch. We assume that each $\bar{\tau}^{(\alpha)}$ satisfies a strength relation of the form

$$\bar{\tau}^{(\alpha)} = \mathcal{S}^{(\alpha)}(\dot{\bar{\gamma}}^{v(\alpha)}). \tag{29.5.19}$$

[5] Mimicking the discussion in Chapter 4, Section 4.3.3 and Section 4.3.5.

29.6 Specialization of the theory to model hard-magnetic soft-elastomeric composites

We now introduce special constitutive equations which should be useful in applications. In order to model the viscoelastic response of h-MREs we introduce the following specializations:

(i) We take the first branch, $\alpha = 0$, to represent the *equilibrium elastic response* of the material. We do not associate any viscous relaxation with this network. Accordingly we set $\mathbf{F}^{v(0)} = \mathbf{1}$, and hence

$$\mathbf{F}^{e(0)} = \mathbf{F}. \tag{29.6.1}$$

(ii) The remaining branches, $\alpha = 1, \ldots, M$, represent the *non-equilibrium viscoelastic response* of the material.

When it is important to distinguish between the zeroth branch and the other branches of the rheological model, we will use the descriptive terms "equilibrium" for branch $\alpha = 0$ and "non-equilibrium" for branches $\alpha = 1, \ldots, M$.

Note that an important feature of the multi-branch decomposition $\mathbf{F} = \mathbf{F}^{e(\alpha)}\mathbf{F}^{v(\alpha)}$, with $\det \mathbf{F}^{v(\alpha)} = 1$, is that

$$J^{e(\alpha)} = J \quad \text{for all} \quad \alpha = 1, \ldots, M. \tag{29.6.2}$$

We restrict our attention to materials for which the overall free energy of the composite may be adequately expressed in a separable form as

$$\psi_{\mathrm{R}} = \psi_{\mathrm{R}}^{\mathrm{eq}}(\bar{I}_1, J) + \sum_{\alpha=1}^{M} \psi_{\mathrm{R}}^{\mathrm{neq}(\alpha)}(\bar{I}_1^e) + \psi_{\mathrm{R}}^{\mathrm{ms}}(\mathbf{R}, \mathbf{m}_{\mathrm{R}}^{\mathrm{rem}}, \mathbf{b}^{\mathrm{app}}). \tag{29.6.3}$$

Here:

(i) The contribution $\psi_{\mathrm{R}}^{\mathrm{eq}}$ is the change in the free energy due to *equilibrium* elastic response of the matrix. We assume that this response is compressible, but consider the case when *the compressibility is small*. As a special case for this equilibrium mechanical energy we consider a simple slightly compressible Neo-Hookean free energy of the form

$$\psi_{\mathrm{R}}^{\mathrm{eq}} = \tfrac{1}{2}\, G(\bar{I}_1 - 3) + \tfrac{1}{2} K (J - 1)^2, \tag{29.6.4}$$

with G and K shear and bulk moduli, respectively.[6]

[6] The shear modulus G here represents the *equilibrium shear modulus*. But for economy of notation, we do not decorate it with a subscript "eq."

(ii) To model the non-equilibrium contributions to the free energy we adopt a simple Neo-Hookean form

$$\psi_{\mathrm{R}}^{\mathrm{neq}(\alpha)}(\bar{I}_1^e) = \tfrac{1}{2}\, G_{\mathrm{neq}}^{(\alpha)} \left(\bar{I}_1^{e(\alpha)} - 3\right), \tag{29.6.5}$$

where $G_{\mathrm{neq}}^{(\alpha)}$ represent non-equilibrium shear moduli.

(iii) We assume that the magnetic part of the Helmholtz free energy *per unit volume of the deformed body*, ψ^{ms}, is given by the Zeeman energy which may be calculated as the work per unit volume required to rotate the remanent magnetization $\mathbf{m}^{\mathrm{rem}}$ relative to a magnetic flux density $\mathbf{b}^{\mathrm{app}}$,

$$\psi^{\mathrm{ms}} = -\mathbf{m}^{\mathrm{rem}} \cdot \mathbf{b}^{\mathrm{app}}. \tag{29.6.6}$$

Next, since the free energy per unit volume of the reference body is

$$\psi_{\mathrm{R}}^{\mathrm{ms}} = J\, \psi^{\mathrm{ms}}, \tag{29.6.7}$$

and using the mapping relation (29.5.3) for $\mathbf{m}^{\mathrm{rem}}$, the magnetostatic free energy per unit reference volume is[7]

$$\psi_{\mathrm{R}}^{\mathrm{ms}} = -\mathbf{R}\mathbf{m}_{\mathrm{R}}^{\mathrm{rem}} \cdot \mathbf{b}^{\mathrm{app}}. \tag{29.6.8}$$

– Note that $\psi_{\mathrm{R}}^{\mathrm{ms}}$ is presumed to depend on $\mathbf{R}$ which transforms under a change in frame as $\mathbf{R}^* = \mathbf{Q}\mathbf{R}$, where $\mathbf{Q}$ denotes the frame rotation. Further because the applied field $\mathbf{b}^{\mathrm{app}}$ also transforms as $\mathbf{b}^{\mathrm{app}*} = \mathbf{Q}\mathbf{b}^{\mathrm{app}}$ under a frame change, the free energy $\psi_{\mathrm{R}}^{\mathrm{ms}}$ is invariant under a change in frame (as is also the free energy ψ^{ms}).

Finally, using (29.6.4), (29.6.5), and (29.6.8) the overall free energy per unit reference volume is taken as

$$\psi_{\mathrm{R}} = \underbrace{\tfrac{1}{2}\, G(\bar{I}_1 - 3) + \tfrac{1}{2}K(J-1)^2}_{\psi_{\mathrm{R}}^{\mathrm{eq}}} + \underbrace{\sum_{\alpha=1}^{m} \tfrac{1}{2}\, G_{\mathrm{neq}}^{(\alpha)} \left(\bar{I}_1^{e(\alpha)} - 3\right)}_{\psi_{\mathrm{R}}^{\mathrm{neq}}} + \underbrace{\left(-\mathbf{R}\mathbf{m}_{\mathrm{R}}^{\mathrm{rem}} \cdot \mathbf{b}^{\mathrm{app}}\right)}_{\psi_{\mathrm{R}}^{\mathrm{ms}}}. \tag{29.6.9}$$

[7] Since $\mathbf{b}^{\mathrm{app}}$ is *fixed* in space and external to the body, it is not mapped to the reference body using the mapping relation $(29.5.2)_2$.

REMARK

For the physical basis behind the assumption (29.6.6), consider a magnet with uniform magnetization $\mathbf{m}^{\text{rem}}$ in a uniform magnetic flux density field $\mathbf{b}^{\text{app}}$. Then, with reference to Fig. 29.3, let $\theta = 0$ be the reference state when $\mathbf{m}^{\text{rem}}$ is aligned with $\mathbf{b}^{\text{app}}$. When $\mathbf{m}^{\text{rem}}$ is not aligned with $\mathbf{b}^{\text{app}}$, a torque

$$\boldsymbol{\tau} = \mathbf{m}^{\text{rem}} \times \mathbf{b}^{\text{app}} = |\mathbf{m}^{\text{rem}}|\,|\mathbf{b}^{\text{app}}| \sin\theta\, \mathbf{e}_3$$

will tend to align the dipoles in the magnet so that they are parallel to the field $\mathbf{b}^{\text{app}}$; this torque acts counter-clockwise about the $\mathbf{e}_3$-axis.

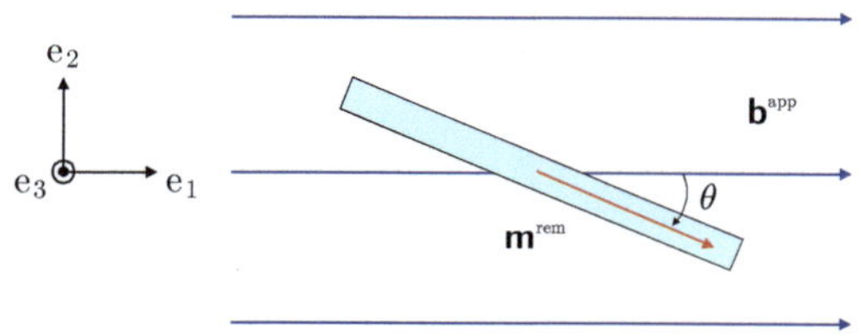

Fig. 29.3 A magnet with magetization $\mathbf{m}^{\text{rem}}$ in a magnetic flux density $\mathbf{b}^{\text{app}}$.

The energy which needs to be expended to rotate the magnet clockwise by an angle θ about the $\mathbf{e}_3$-axis is then given by,

$$\int_0^\theta |\mathbf{m}^{\text{rem}}|\,|\mathbf{b}^{\text{app}}| \sin\bar{\theta}\, d\bar{\theta} = |\mathbf{m}^{\text{rem}}|\,|\mathbf{b}^{\text{app}}| \left[-\cos\bar{\theta}\right]_0^\theta = -|\mathbf{m}^{\text{rem}}|\,|\mathbf{b}^{\text{app}}| \cos\theta + \text{constant}.$$

Thus, to within a constant, the energy of a magnet with magnetization $\mathbf{m}^{\text{rem}}$ relative to an applied magnetic field $\mathbf{b}^{\text{app}}$ is given by

$$\text{magnetostatic energy} = -\mathbf{m}^{\text{rem}} \cdot \mathbf{b}^{\text{app}},$$

and is known as the *Zeeman energy*. This energy is negative because the torque is counter-clockwise while the rotation θ is clockwise about the $\mathbf{e}_3$-axis. This is the physical basis for the expression (29.6.6) for the magnetostatic energy ψ^{ms}.

Finally, for the viscous flow rule we assume a strength relation (29.5.19) in the simple form

$$S^{(\alpha)}(\dot{\bar{\gamma}}^{v(\alpha)}) = \eta^{(\alpha)} \dot{\bar{\gamma}}^{v(\alpha)}, \tag{29.6.10}$$

with $\eta^{(\alpha)} > 0$ a constant *viscosity* parameter for each α, so that

$$\bar{\tau}^{(\alpha)} = \eta^{(\alpha)} \dot{\bar{\gamma}}^{v(\alpha)} \quad \Longrightarrow \quad \dot{\bar{\gamma}}^{v(\alpha)} = \frac{\bar{\tau}^{(\alpha)}}{\eta^{(\alpha)}}. \tag{29.6.11}$$

With this choice of the strength relation the dissipation inequality (29.5.10) may be written as

$$\mathcal{D} = \sum_{\alpha} \bar{\tau}^{(\alpha)} \dot{\bar{\gamma}}^{v(\alpha)} = \sum_{\alpha} \eta^{(\alpha)} (\dot{\bar{\gamma}}^{v(\alpha)})^2 \geq 0. \tag{29.6.12}$$

which is always satisfied since we have chosen positive viscosities, $\eta^{(\alpha)} > 0$.

29.7 Stress

Using (29.5.12), (29.5.14), and (29.5.13) we can write the Piola stress $\mathbf{T}_{\mathrm{R}}$ as

$$\mathbf{T}_{\mathrm{R}} = \mathbf{T}_{\mathrm{R}}^{\mathrm{eq}} + J\left(\sum_{\alpha=1}^{M} \mathbf{T}^{(\alpha)}\right)\mathbf{F}^{-\top} + \mathbf{T}_{\mathrm{R}}^{\mathrm{ms}}, \quad \text{where}$$

$$\mathbf{T}_{\mathrm{R}}^{\mathrm{eq}} \stackrel{\text{def}}{=} \mathbf{F}\left(2\frac{\partial \psi_{\mathrm{R}}^{\mathrm{eq}}}{\partial \mathbf{C}}\right), \quad \mathbf{T}^{(\alpha)} = (J^{e(\alpha)})^{-1}\left(2\mathbf{F}^{e(\alpha)}\frac{\partial \psi_{\mathrm{R}}^{\mathrm{neq}}}{\partial \mathbf{C}^{e(\alpha)}}\mathbf{F}^{e(\alpha)\top}\right), \quad \text{and} \quad \mathbf{T}_{\mathrm{R}}^{\mathrm{ms}} \stackrel{\text{def}}{=} \frac{\partial \psi_{\mathrm{R}}^{\mathrm{ms}}}{\partial \mathbf{F}}. \tag{29.7.1}$$

(i) First, straightforward calculations give (cf., Gurtin et al., 2010, Chapter 55, eq. (55.7)):

$$2\frac{\partial \psi_{\mathrm{R}}^{\mathrm{eq}}}{\partial \mathbf{C}} = \frac{2}{J^{2/3}}\left[\frac{\partial \psi_{\mathrm{R}}^{\mathrm{eq}}(\bar{I}_1, J)}{\partial \bar{\mathbf{C}}} - \frac{1}{3}\left(\bar{\mathbf{C}} : \frac{\partial \psi_{\mathrm{R}}^{\mathrm{eq}}(\bar{I}_1, J)}{\partial \bar{\mathbf{C}}}\right)\bar{\mathbf{C}}^{-1}\right] + J\frac{\partial \psi_{\mathrm{R}}^{\mathrm{eq}}(\bar{I}_1, J)}{\partial J}\mathbf{C}^{-1}, \tag{29.7.2}$$

$$= \frac{2}{J^{2/3}}\frac{\partial \psi_{\mathrm{R}}^{\mathrm{eq}}(\bar{I}_1, J)}{\partial \bar{I}_1}\left[\mathbf{1} - \frac{1}{3}(\operatorname{tr}\bar{\mathbf{C}})\bar{\mathbf{C}}^{-1}\right] + KJ(J-1)\mathbf{C}^{-1}, \tag{29.7.3}$$

$$= J^{-2/3}G\left[\mathbf{1} - \frac{1}{3}(\operatorname{tr}\bar{\mathbf{C}})\bar{\mathbf{C}}^{-1}\right] + KJ(J-1)\mathbf{C}^{-1}. \tag{29.7.4}$$

Hence the mechanical equilibrium part of the Piola stress is

$$\mathbf{T}_{\mathrm{R}}^{\mathrm{eq}} = J^{-2/3}G\left[\mathbf{F} - \frac{1}{3}(\operatorname{tr}\mathbf{C})\mathbf{F}^{-\top}\right] - Jp\mathbf{F}^{-\top}, \tag{29.7.5}$$

where for later convenience we have introduced the notation

$$p \stackrel{\text{def}}{=} -K(J-1) \tag{29.7.6}$$

for a pressure.

(ii) Next we calculate the contributions to the Piola stress from the viscoelastic mechanisms. Mimicking the calculations in Chapter 4 leading to equation (4.3.31) gives the non-equilibrium Cauchy stress contribution from the αth viscous branch as

$$\mathbf{T}^{(\alpha)} = J^{-1} G_{\text{neq}}^{(\alpha)} \bar{\mathbf{B}}_0^{e(\alpha)}, \tag{29.7.7}$$

and the Mandel stress for each branch α is then given by

$$\mathbf{M}^{e(\alpha)} = G_{\text{neq}}^{(\alpha)} \bar{\mathbf{C}}_0^{e(\alpha)}. \tag{29.7.8}$$

(iii) Next, with

$$\psi_{\text{R}}^{\text{ms}} = -\mathbf{R}\mathbf{m}_{\text{R}}^{\text{rem}} \cdot \mathbf{b}^{\text{app}} \qquad \text{with} \quad \mathbf{R} = \mathbf{R}(\mathbf{F}), \tag{29.7.9}$$

the magnetostatic part of the Piola stress is

$$\mathbf{T}_{\text{R}}^{\text{ms}} = -\frac{\partial}{\partial \mathbf{F}}\Big(\mathbf{R}(\mathbf{F})\mathbf{m}_{\text{R}}^{\text{rem}} \cdot \mathbf{b}^{\text{app}}\Big). \tag{29.7.10}$$

The derivative on the right-hand side of (29.7.10) is easiest to calculate using components. Thus, temporarily writing $\mathbf{T} = \mathbf{T}_{\text{R}}^{\text{ms}}$, $\mathbf{m} = \mathbf{m}_{\text{R}}^{\text{rem}}$, and $\mathbf{b} = \mathbf{b}^{\text{app}}$, we obtain

$$\begin{aligned} T_{kl} &= -\frac{\partial}{\partial F_{kl}}\left(R_{ij}(F_{pq})m_j b_i\right) = -\frac{\partial R_{ij}}{\partial F_{pq}}\frac{\partial F_{pq}}{\partial F_{kl}} m_j b_i = -\frac{\partial R_{ij}}{\partial F_{pq}}\delta_{pk}\delta_{ql} m_j b_i \\ &= -\frac{\partial R_{ij}}{\partial F_{kl}} m_j b_i, = -\left(\frac{\partial \mathbf{R}}{\partial \mathbf{F}}\right)^{\top}_{klij} b_i m_j, \end{aligned}$$

and hence reinstating the original notation for $\mathbf{T} = \mathbf{T}_{\text{R}}^{\text{ms}}$, $\mathbf{m} = \mathbf{m}_{\text{R}}^{\text{rem}}$, and $\mathbf{b} = \mathbf{b}^{\text{app}}$, and expressing the result above in direct notation we obtain

$$\mathbf{T}_{\text{R}}^{\text{ms}} = -\left(\frac{\partial \mathbf{R}}{\partial \mathbf{F}}\right)^{\top} [\mathbf{b}^{\text{app}} \otimes \mathbf{m}_{\text{R}}^{\text{rem}}], \tag{29.7.11}$$

where $(\partial \mathbf{R}/\partial \mathbf{F})^{\top}$ is a fourth-order tensor which is the transpose of the derivative of the rotation tensor $\mathbf{R} = \mathbf{R}(\mathbf{F})$ with respect to $\mathbf{F}$.

Thus, the total Piola stress is given by

$$\begin{aligned} \mathbf{T}_{\text{R}} &= J^{-2/3} G\left[\mathbf{F} - \frac{1}{3}(\operatorname{tr}\mathbf{C})\mathbf{F}^{-\top}\right] - J\,p\,\mathbf{F}^{-\top} + \sum_{\alpha=1}^{M} G_{\text{neq}}^{(\alpha)} \bar{\mathbf{B}}_0^{e(\alpha)} \mathbf{F}^{-\top} \\ &\quad - \left(\frac{\partial \mathbf{R}}{\partial \mathbf{F}}\right)^{\top} [\mathbf{b}^{\text{app}} \otimes \mathbf{m}_{\text{R}}^{\text{rem}}], \qquad \text{with} \end{aligned} \tag{29.7.12}$$

$$p = -K(J-1),$$

and the total Cauchy stress is given by

$$\mathbf{T} = J^{-1} G \bar{\mathbf{B}}_0 - p\mathbf{1} + \sum_{\alpha=1}^{M} J^{-1} G_{\text{neq}}^{(\alpha)} \bar{\mathbf{B}}_0^{e(\alpha)} - J^{-1} \left(\frac{\partial \mathbf{R}}{\partial \mathbf{F}}\right)^{\top} [\mathbf{b}^{\text{app}} \otimes \mathbf{m}_{\text{R}}^{\text{rem}}]\mathbf{F}^{\top}, \qquad \text{with}$$

$$p = -K(J-1). \tag{29.7.13}$$

29.8 Viscous flow

Note that using (29.7.8) and (29.6.11) we may rewrite the viscous stretching relation (29.5.16) alternatively as

$$\mathbf{D}^{v(\alpha)} = \frac{\bar{\mathbf{C}}_0^{e(\alpha)}}{2\,\tau_r^{(\alpha)}} \qquad \text{with} \qquad \tau_r^{(\alpha)} \stackrel{\text{def}}{=} \frac{\eta^{(\alpha)}}{G_{\text{neq}}^{(\alpha)}}, \tag{29.8.1}$$

where $\tau_r^{(\alpha)}$ represents a *relaxation time* for the αth non-equilibrium branch.

Next, from $\mathbf{C}^{v(\alpha)} = \mathbf{F}^{v(\alpha)\top}\mathbf{F}^{v(\alpha)}$ we obtain that $\dot{\mathbf{C}}^{v(\alpha)} = \mathbf{F}^{v(\alpha)\top}\left(2\mathbf{D}^{v(\alpha)}\right)\mathbf{F}^{v(\alpha)}$. Hence, using (29.8.1),

$$\dot{\mathbf{C}}^{v(\alpha)} = \mathbf{F}^{v(\alpha)\top}\left(\frac{\bar{\mathbf{C}}_0^{e(\alpha)}}{\tau_r^{(\alpha)}}\right)\mathbf{F}^{v(\alpha)}. \tag{29.8.2}$$

Next, since $\bar{\mathbf{C}}^{e(\alpha)} = J^{-2/3}\left(\mathbf{F}^{v(\alpha)-\top}\mathbf{C}\mathbf{F}^{v(\alpha)-1}\right)$ and $\operatorname{tr}\bar{\mathbf{C}}^{e(\alpha)} = J^{-2/3}\left(\mathbf{C}\!:\!\mathbf{C}^{v(\alpha)-1}\right)$, we can write

$$\bar{\mathbf{C}}_0^{e(\alpha)} = J^{-2/3}\left[\left(\mathbf{F}^{v(\alpha)-\top}\mathbf{C}\mathbf{F}^{v(\alpha)-1}\right) - \frac{1}{3}\left(\mathbf{C}\!:\!\mathbf{C}^{v(\alpha)-1}\right)\mathbf{1}\right], \tag{29.8.3}$$

and use of this in (29.8.2) gives the following evolution equation

$$\dot{\mathbf{C}}^{v(\alpha)} = \frac{1}{\tau_r^{(\alpha)}}\left(\bar{\mathbf{C}} - \frac{1}{3}\left(\bar{\mathbf{C}}\!:\!(\mathbf{C}^{v(\alpha)})^{-1}\right)\mathbf{C}^{v(\alpha)}\right). \tag{29.8.4}$$

Lastly, revisiting (29.7.7), the non-equilibrium Cauchy stresses in each branch may be written in the form

$$\mathbf{T}^{(\alpha)} = J^{-1} G_{\text{neq}}^{(\alpha)}\left(\bar{\mathbf{F}}(\mathbf{C}^{v(\alpha)})^{-1}\bar{\mathbf{F}}^{\top} - \frac{1}{3}(\bar{\mathbf{C}}\!:\!(\mathbf{C}^{v(\alpha)})^{-1})\mathbf{1}\right). \tag{29.8.5}$$

Note that in (29.8.5) and (29.8.4) the dependence on the internal variables $\mathbf{F}^{v(\alpha)}$ enters only through the *symmetric tensors* $\mathbf{C}^{v(\alpha)} = (\mathbf{F}^{v(\alpha)})^{\top}\mathbf{F}^{v(\alpha)}$ (cf., e.g., Wijaya et al., 2023; Lefévre et al., 2024).

29.9 Governing partial differential equation

Neglecting body forces, we express the balance of linear momentum by the pair of equations

$$\begin{aligned} \operatorname{Div}\mathbf{T}_{\mathrm{R}} &= \rho_{\mathrm{R}}\ddot{\mathbf{u}}, \\ \frac{p}{K} + (J-1) &= 0, \end{aligned} \tag{29.9.1}$$

with $\mathbf{T}_{\mathrm{R}}$ given by $\mathbf{T}_{\mathrm{R}} = J\mathbf{T}\mathbf{F}^{-\top}$, where the total Cauchy stress is

$$\begin{aligned} \mathbf{T} = J^{-1}G\bar{\mathbf{B}}_0 - p\mathbf{1} + \sum_{\alpha=1}^{M} J^{-1}G_{\mathrm{neq}}^{(\alpha)}\left(\bar{\mathbf{F}}(\mathbf{C}^{v(\alpha)})^{-1}\bar{\mathbf{F}}^{\top} - \frac{1}{3}(\bar{\mathbf{C}}:(\mathbf{C}^{v(\alpha)})^{-1})\mathbf{1}\right) \\ - J^{-1}\left(\frac{\partial\mathbf{R}}{\partial\mathbf{F}}\right)^{\top}[\mathbf{b}^{\mathrm{app}}\otimes\mathbf{m}_{\mathrm{R}}^{\mathrm{rem}}]\mathbf{F}^{\top}. \end{aligned} \tag{29.9.2}$$

In writing (29.9.1) we have *split* the standard equation of motion into two equations—one for the balance of linear momentum with the Piola stress expressed as a function of $(\mathbf{u}, p)$, and another equation for the pressure p expressed in terms of $\mathbf{u}$; this helps in numerically dealing with the near-incompressibility of the material. In our numerical finite element implementation of the theory rather than just using the displacement $\mathbf{u}$, as in standard displacement-based finite element procedures, we will use $(\mathbf{u}, p)$ as separate degrees of freedom.

The only governing partial differential equation in this particular theory for hard-magnetic soft-elastomers is the (split) equation of motion $(29.9.1)_1$, with the Piola stress $\mathbf{T}_{\mathrm{R}}$ given by $\mathbf{T}_{\mathrm{R}} = J\mathbf{T}\mathbf{F}^{-\top}$ and the Cauchy stress $\mathbf{T}$ by (29.9.2). It is in the Cauchy stress expression (29.9.2) that we explicitly see the coupling contribution from the magnetization with the mechanics. In this *particular* theory for magnetoelasticity there is no separate partial differential equation which needs to be solved for the magnetic field.

Finally, let $\partial\mathcal{B}_{\mathrm{u}}$ and $\partial\mathcal{B}_{\mathrm{t}_{\mathrm{R}}}$ be complementary subsurfaces of the boundary $\partial\mathcal{B}$ of the body $\mathcal{B}$. Then for a time interval $t \in (0, T]$ we consider a pair of boundary conditions:

$$\begin{aligned} \mathbf{u} &= \bar{\mathbf{u}} \quad \text{on } \partial\mathcal{B}_{\mathrm{u}} \times (0, T], \\ \mathbf{T}_{\mathrm{R}}\mathbf{n}_{\mathrm{R}} &= \bar{\mathbf{t}}_{\mathrm{R}} \quad \text{on } \partial\mathcal{B}_{\mathrm{t}_{\mathrm{R}}} \times (0, T]. \end{aligned} \tag{29.9.3}$$

In the boundary conditions above $\bar{\mathbf{u}}$ and $\bar{\mathbf{t}}_{\mathrm{R}}$ are prescribed functions of $\mathbf{X}$ and t. Standard initial conditions involve the specifications:

$$\mathbf{u}(\mathbf{X}, 0) = \mathbf{u}_0(\mathbf{X}), \quad \text{and} \quad \dot{\mathbf{u}}(\mathbf{X}, 0) = \dot{\mathbf{u}}_0(\mathbf{X}) \tag{29.9.4}$$

with $\mathbf{u}_0$ and $\dot{\mathbf{u}}_0$ prescribed functions on $\mathcal{B}$.

30 Numerical implementation of the theory for hard-magnetic soft-elastomers

The numerical implementation of the theory for hard-magnetic soft materials is essentially identical to that detailed for viscoelasticity with inertial effects considered in Chapter 5. Accordingly here we summarize the procedure considered there, but highlight the differences which are peculiar to the theory under consideration.

30.1 Weak form of the governing partial differential equation

Consider a test field $\mathbf{u}_{\text{test}}(\mathbf{X})$ in $\mathcal{B}$ which satisfies $\mathbf{u}_{\text{test}}(\mathbf{X}) = \mathbf{0}$ on $\partial\mathcal{B}_{\mathbf{u}}$. If the balance equation $\operatorname{Div}\mathbf{T}_{\text{R}} - \rho_{\text{R}}\,\ddot{\mathbf{u}} = \mathbf{0}$ on $\mathcal{B}$ and the traction condition $(\bar{\mathbf{t}}_{\text{R}} - \mathbf{T}_{\text{R}}\mathbf{n}_{\text{R}}) = \mathbf{0}$ on $\partial\mathcal{B}_{\mathbf{t}_{\text{R}}}$ are satisfied, then the weak form

$$\int_{\mathcal{B}} \mathbf{T}_{\text{R}} : \nabla\mathbf{u}_{\text{test}}\, dv_{\text{R}} + \int_{\mathcal{B}} \rho_{\text{R}}\ddot{\mathbf{u}}\cdot\mathbf{u}_{\text{test}}\, dv_{\text{R}} - \int_{\partial\mathcal{B}_{\mathbf{t}_{\text{R}}}} \bar{\mathbf{t}}_{\text{R}}\cdot\mathbf{u}_{\text{test}}\, da_{\text{R}} = 0 \tag{30.1.1}$$

holds for all admissible test fields $\mathbf{u}_{\text{test}}$ which vanish on $\partial\mathcal{B}_{\mathbf{u}}$. Also, the weak form of the pressure equation $p/K + (J-1) = 0$ is

$$\int_{\mathcal{B}} \left[\frac{p}{K} + (J-1)\right] p_{\text{test}}\, dv_{\text{R}} = 0, \tag{30.1.2}$$

where $p_{\text{test}}(\mathbf{X})$ is a test field on $\mathcal{B}$.

Recall that the Piola stress appearing in the weak form (30.1.1) in this theory is given by

$$\mathbf{T}_{\text{R}} = J^{-2/3}G\left[\mathbf{F} - \frac{1}{3}(\operatorname{tr}\mathbf{C})\mathbf{F}^{-\top}\right] - J\,p\mathbf{F}^{-\top} + \sum_{\alpha=1}^{M} G_{\text{neq}}^{(\alpha)}\Big(\bar{\mathbf{F}}(\mathbf{C}^{v(\alpha)})^{-1}\bar{\mathbf{F}}^{\top}$$
$$-\frac{1}{3}(\bar{\mathbf{C}} : (\mathbf{C}^{v(\alpha)})^{-1})\mathbf{1}\Big)\mathbf{F}^{-\top} - \left(\frac{\partial\mathbf{R}}{\partial\mathbf{F}}\right)^{\top}[\mathbf{b}^{\text{app}}\otimes\mathbf{m}_{\text{R}}^{\text{rem}}]. \tag{30.1.3}$$

The only real difficulty in the numerical implementation is the calculation of the rotation $\mathbf{R}$ and its derivative $\left(\dfrac{\partial\mathbf{R}}{\partial\mathbf{F}}\right)^{\top}$ appearing in the magnetic contribution to the Piola stress. We discuss the calculation of these terms below.

Introduction to Coupled Theories in Solid Mechanics. Lallit Anand, Eric M. Stewart, Shawn A. Chester, Oxford University Press.
 DOI: 10.1093/oso/9780198986218.003.0031

30.1.1 Calculation of the rotation $\mathrm{R} = \mathrm{R}(\mathrm{F})$ and its derivative $(\partial\mathrm{R}(\mathrm{F})/\partial\mathrm{F})^{\top}$

(i) To calculate the rotation $\mathbf{R} = \mathbf{R}(\mathbf{F})$, we first analytically compute $\mathbf{U}^{-1}$ in terms of the invariants of $\mathbf{C} = \mathbf{F}^{\top}\mathbf{F}$, and then compute the rotation using $\mathbf{R} = \mathbf{F}\mathbf{U}^{-1}$.[1] The details of this method for computing the right polar decomposition $\mathbf{F} = \mathbf{R}\mathbf{U}$ in FEniCS are provided in Appendix 30.A, and here we present a summary of the results.

With the three principal invariants of $\mathbf{C}$ denoted by

$$I_1 = \mathrm{tr}\,\mathbf{C}, \qquad I_2 = \frac{1}{2}\left[(\mathrm{tr}\,\mathbf{C})^2 - \mathrm{tr}\,(\mathbf{C}^2)\right], \qquad I_3 = \det\mathbf{C}, \tag{30.1.4}$$

and the three principal invariants of $\mathbf{U}$ denoted by

$$i_1 = \mathrm{tr}\,\mathbf{U}, \qquad i_2 = \frac{1}{2}\left[(\mathrm{tr}\,\mathbf{U})^2 - \mathrm{tr}\,(\mathbf{U}^2)\right], \qquad i_3 = \det\mathbf{U}, \tag{30.1.5}$$

the tensor $\mathbf{U}^{-1}$ may expressed in terms of the invariants of $\mathbf{C}$ as follows (c.f., e.g., Hoger and Carlson, 1984; Sawyers, 1986; Guan-Suo, 1998; Norris, 2007; Scott, 2020):

$$\mathbf{U}^{-1} = ((i_1 i_2 - i_3)i_3)^{-1}\left[\left(i_1 i_2^2 - i_3 i_1^2 - i_3 i_2\right)\mathbf{1} - \left(i_3 + i_1^3 - 2i_1 i_2\right)\mathbf{C} + i_1\mathbf{C}^2\right], \tag{30.1.6}$$

with

$$\begin{aligned}
i_1 &= \lambda + \sqrt{I_1 - \lambda^2 + 2\sqrt{I_3}/\lambda},\\
i_2 &= \sqrt{I_3}/\lambda + \sqrt{I_1\lambda^2 - \lambda^4 + 2\sqrt{I_3}\,\lambda},\\
i_3 &= \sqrt{I_3},
\end{aligned} \tag{30.1.7}$$

where

$$\lambda = \frac{1}{\sqrt{3}}\sqrt{I_1 + 2\left(I_1^2 - 3I_2\right)^{1/2}\cos\theta} \quad \text{with} \quad \theta = \frac{1}{3}\cos^{-1}\left(\frac{2I_1^3 - 9I_1I_2 + 27I_3}{2\left(I_1^2 - 3I_2\right)^{3/2}}\right), \tag{30.1.8}$$

is *an* eigenvalue of $\mathbf{U}$.

The rotation tensor $\mathbf{R}$ and stretch tensor $\mathbf{U}$ are then calculated using

$$\mathbf{R} = \mathbf{F}\mathbf{U}^{-1} \quad \text{and} \quad \mathbf{U} = \mathbf{R}^{\top}\mathbf{F}. \tag{30.1.9}$$

[1] In the present version of FEniCS it is not possible to calculate $\mathbf{U}^{-1}$ by using standard methods to first calculate $\mathbf{U} = \sqrt{\mathbf{C}}$ using the spectral decomposition of $\mathbf{C}$, and then taking its inverse to calculate $\mathbf{U}^{-1}$ and $\mathbf{R} = \mathbf{F}\mathbf{U}^{-1}$.

(ii) In the expression (29.7.12) for the Piola stress the fourth-order tensor $(\partial \mathbf{R}/\partial \mathbf{F})^{\top}$ acts on the second-order tensor $\mathbf{b}^{\text{app}} \otimes \mathbf{m}_{\text{R}}^{\text{rem}}$. We calculate this term using the formula given by Chen and Wheeler (1993), eq. $(11)_1$,

$$\left(\frac{\partial \mathbf{R}}{\partial \mathbf{F}}\right)^{\top} [\mathbf{b}^{\text{app}} \otimes \mathbf{m}_{\text{R}}^{\text{rem}}] = \det(\mathbf{Y})^{-1}\mathbf{R}\mathbf{Y}\,(\mathbf{R}^{\top}[\mathbf{b}^{\text{app}} \otimes \mathbf{m}_{\text{R}}^{\text{rem}}] - [\mathbf{b}^{\text{app}} \otimes \mathbf{m}_{\text{R}}^{\text{rem}}]^{\top}\mathbf{R})\,\mathbf{Y}, \tag{30.1.10}$$

where

$$\mathbf{Y} \stackrel{\text{def}}{=} (\operatorname{tr}\mathbf{U})\,\mathbf{1} - \mathbf{U}. \tag{30.1.11}$$

REMARK

Note that if instead of the $\mathbf{R}$-based mapping for $\mathbf{m}_{\text{R}}^{\text{rem}}$,

$$\mathbf{m}^{\text{rem}} = J^{-1}\mathbf{R}\mathbf{m}_{\text{R}}^{\text{rem}}, \tag{30.1.12}$$

one uses the $\mathbf{F}$-based mapping,

$$\mathbf{m}^{\text{rem}} = J^{-1}\mathbf{F}\mathbf{m}_{\text{R}}^{\text{rem}}, \tag{30.1.13}$$

then the term $\left(\dfrac{\partial \mathbf{R}}{\partial \mathbf{F}}\right)^{\top}$ in the expression (30.1.3) for the Piola stress reduces to the fourth-order identity tensor $\mathbb{I}$, and the Piola stress expression simplifies to

$$\begin{aligned}\mathbf{T}_{\text{R}} &= J^{-2/3}G\left[\mathbf{F} - \frac{1}{3}(\operatorname{tr}\mathbf{C})\mathbf{F}^{-\top}\right] - J\,p\,\mathbf{F}^{-\top} \\ &\quad + \sum_{\alpha=1}^{M} G_{\text{neq}}^{(\alpha)}\left(\bar{\mathbf{F}}(\mathbf{C}^{v(\alpha)})^{-1}\bar{\mathbf{F}}^{\top} - \frac{1}{3}(\bar{\mathbf{C}}:(\mathbf{C}^{v(\alpha)})^{-1})\mathbf{1}\right)\mathbf{F}^{-\top} - \mathbf{b}^{\text{app}} \otimes \mathbf{m}_{\text{R}}^{\text{rem}}, \\ &\text{with}\quad p = -K(J-1).\end{aligned} \tag{30.1.14}$$

This considerably simplifies numerical calculations, and this is the approach used in Zhao et al. (2019).

- In situations where the stretch is *small*, or equivalently when the Green strain $\mathbf{E}^G = 1/2\,(\mathbf{C} - \mathbf{1})$ is small, $|\mathbf{E}^G| \ll 1$, the results from an $\mathbf{R}$-based theory and those from an $\mathbf{F}$-based theory are almost indistinguishable from each other. Thus in the context of many simple actuators where the strains are small, the error induced by using an $\mathbf{F}$-based theory is small. However, when the components of the Green strain are large ($\gtrapprox 10\%$), the error induced by using the $\mathbf{F}$-based theory can be significant, and such cases the $\mathbf{R}$-based theory should be used.
- We emphasize that in the numerical simulations that we show below, *we have used the* $\mathbf{R}$*-based form of the theory.*

30.2 Time-integration procedure

With a numerical time-integration strategy in mind, the time domain is discretized in a series of N_t time steps,

$$t_{n+1} = t_n + \Delta t, \qquad n = 0, 1, \ldots, N_t, \tag{30.2.1}$$

such that $t_{N_t} = T$. Then, in order to solve a residual based on the weak form (30.1.1) in a $(\mathbf{u}, p)$-based finite element solution procedure, given

$$(\mathbf{u}_n, \dot{\mathbf{u}}_n, \ddot{\mathbf{u}}_n, \mathbf{C}^{v(\alpha)}_n, \mathbf{T}_{Rn})$$

at time t_n, we need to calculate

$$(\dot{\mathbf{u}}_{n+1}, \ddot{\mathbf{u}}_{n+1}, \mathbf{C}^{v(\alpha)}_{n+1}, \mathbf{T}_{Rn+1})$$

in terms of $(\mathbf{u}_{n+1}, p_{n+1})$ at time $t_{n+1} = t_n + \Delta t$.

30.2.1 Calculation of $\mathbf{C}^{v(\alpha)}_{n+1}$ and $\mathbf{T}_{Rn+1}$

As detailed in Section 5.3 the evolution equations for the internal variables $\mathbf{C}^{v(\alpha)}$ (29.8.4) are implicitly integrated to obtain

$$\mathbf{C}^{v(\alpha)}_{n+1} = \left(\det\left[\mathbf{C}^{v(\alpha)}_n + \frac{\Delta t}{\tau_r^{(\alpha)}}\bar{\mathbf{C}}_{n+1}\right]\right)^{-1/3}\left(\mathbf{C}^{v(\alpha)}_n + \frac{\Delta t}{\tau_r^{(\alpha)}}\bar{\mathbf{C}}_{n+1}\right), \tag{30.2.2}$$

and hence the non-equilibrium Cauchy stress at the end of the step is,

$$\mathbf{T}^{(\alpha)}_{n+1} = J^{-1}_{n+1}\, G^{(\alpha)}_{\text{neq}}\left(\bar{\mathbf{F}}_{n+1}(\mathbf{C}^{v(\alpha)}_{n+1})^{-1}\bar{\mathbf{F}}^{\top}_{n+1} - \frac{1}{3}(\bar{\mathbf{C}}_{n+1} : (\mathbf{C}^{v(\alpha)}_{n+1})^{-1})\mathbf{1}\right). \tag{30.2.3}$$

Thus, with

$$\begin{aligned} \mathbf{T}^{\text{eq}}_{n+1} &= J^{-1}_{n+1}\, G(\bar{\mathbf{B}}_0)_{n+1} - p_{n+1}\mathbf{1}, \\ p_{n+1} &= -K\left(J_{n+1} - 1\right), \end{aligned} \tag{30.2.4}$$

and

$$\mathbf{T}^{\text{ms}}_{n+1} = -J^{-1}_{n+1}\left(\frac{\partial \mathbf{R}}{\partial \mathbf{F}}\right)^{\top}_{n+1}\left[\mathbf{b}^{\text{app}}_{n+1} \otimes \mathbf{m}^{\text{rem}}_{R}\right]\mathbf{F}^{\top}_{n+1}, \tag{30.2.5}$$

the total Cauchy stress at the end of the step is,

$$\mathbf{T}_{n+1} = \mathbf{T}^{\text{eq}}_{n+1} + \sum_{\alpha=1}^{M}\mathbf{T}^{(\alpha)}_{n+1} + \mathbf{T}^{\text{ms}}_{n+1}, \tag{30.2.6}$$

and the total Piola stress is

$$\mathbf{T}_{\mathrm{R}\,\mathrm{n}+1} = J_{\mathrm{n}+1}\,\mathbf{T}_{\mathrm{n}+1}\mathbf{F}_{\mathrm{n}+1}^{-\top}. \tag{30.2.7}$$

30.2.2 Calculation of $\ddot{\mathbf{u}}_{\mathrm{n}+1}$ and $\dot{\mathbf{u}}_{\mathrm{n}+1}$

At each time step in the numerical implementation, the velocity and acceleration are computed according to the Newmark kinematic relations, cf. Section 5.4 (cf., e.g., Bonet et al., 2021)

$$\begin{aligned}
\ddot{\mathbf{u}} &= \frac{\mathbf{u} - \mathbf{u}_{\mathrm{old}} - \Delta t\,\dot{\mathbf{u}}_{\mathrm{old}}}{(1/4)\Delta t^2} - \ddot{\mathbf{u}}_{\mathrm{old}},\\
\dot{\mathbf{u}} &= \dot{\mathbf{u}}_{\mathrm{old}} + \frac{1}{2}\Delta t\,(\ddot{\mathbf{u}}_{\mathrm{old}} + \ddot{\mathbf{u}}),
\end{aligned} \tag{30.2.8}$$

where Δt is the size of the increment in time, and the subscript $(\cdot)_{\mathrm{old}}$ denotes a quantity from the previous time step. We enhance the Newmark method by also using the α-method of time integration discussed in Section 5.4.

30.3 Solution procedure in the FEniCS code

For ease of notation we will write

$$(\bullet) \equiv (\bullet)_{\mathrm{n}+1} \qquad \text{and} \qquad (\bullet)_{\mathrm{old}} \equiv (\bullet)_{\mathrm{n}},$$

for all time-varying quantities $(\bullet)$ evaluated at times $\mathrm{n}+1$ and n, respectively. In FEniCS we use the following degrees of freedom

$$\{\mathbf{u},\, p\}, \tag{30.3.1}$$

as well as their known values from the previous step

$$\{\mathbf{u}_{\mathrm{old}},\, p_{\mathrm{old}}\}, \tag{30.3.2}$$

to calculate the *residuals* resulting from the weak forms of the governing equations, viz.

$$\begin{aligned}
R_{\mathbf{u}} &= \int_{\mathcal{B}} \mathbf{T}_{\mathrm{R}} : \nabla\mathbf{u}_{\mathrm{test}}\, dv_{\mathrm{R}} + \int_{\mathcal{B}} \rho_{\mathrm{R}}\ddot{\mathbf{u}}\cdot\mathbf{u}_{\mathrm{test}}\, dv_{\mathrm{R}} - \int_{\partial\mathcal{B}_{\mathbf{t}_{\mathrm{R}}}} \bar{\mathbf{t}}_{\mathrm{R}}\cdot\mathbf{u}_{\mathrm{test}}\, da_{\mathrm{R}},\\
R_p &= \int_{\mathcal{B}} \left[\frac{p}{K} + (J-1)\right] p_{\mathrm{test}}\, dv_{\mathrm{R}}.
\end{aligned} \tag{30.3.3}$$

FEniCS then automatically differentiates these residuals with respect to the separate degrees of freedom to obtain *tangents*. At each step in time, FEniCS uses the residuals and tangents in a global Newton–Raphson solver to determine current values of the degrees of freedom $\{\mathbf{u},\, p\}$ for which the residuals are zero (to within some small numerical tolerance) and which therefore satisfy the governing equations as well as the initial and boundary conditions (in a weak sense).

REMARK

A mixed finite element method is utilized to solve for the displacement $\mathbf{u}$ and the pressure p, concurrently. To avoid numerical instabilities we use a class of elements known as "Taylor–Hood" elements which combine second-order quadratic interpolation for the displacement, and first-order linear interpolation for the scalar pressure. Additionally, the remanent magnetization $\mathbf{m}_{\mathrm{R}}^{\mathrm{rem}}$ uses a discontinuous Galerkin vector function space such that different regions of the body may have different remanent magnetizations, and the applied magnetic field is a spatially uniform vector which may vary in time.

30.3.1 Algorithm for calculation of the residuals

The algorithm we follow in our code to calculate the residuals in (30.3.3) is summarized below.

1. Begin the step with known values of

$$\{\mathbf{u}_{\mathrm{old}},\, \dot{\mathbf{u}}_{\mathrm{old}},\, \ddot{\mathbf{u}}_{\mathrm{old}},\, p_{\mathrm{old}},\, \mathbf{C}^{v(\alpha)}_{\mathrm{old}}\}. \tag{30.3.4}$$

   where for the first step these quantities are given by the initial conditions for $\mathbf{u}$, $\dot{\mathbf{u}}$, $\ddot{\mathbf{u}}$, p, and $\mathbf{C}^{v(\alpha)}$.

2. Calculate the acceleration and velocity using the Newmark scheme:

$$\begin{aligned}\ddot{\mathbf{u}} &= \frac{\mathbf{u}-\mathbf{u}_{\mathrm{old}}-\dot{\mathbf{u}}_{\mathrm{old}}\,\Delta t}{(1/4)\,\Delta t^2} - \ddot{\mathbf{u}}_{\mathrm{old}},\\ \dot{\mathbf{u}} &= \dot{\mathbf{u}}_{\mathrm{old}} + \frac{1}{2}\Delta t\,(\ddot{\mathbf{u}}_{\mathrm{old}}+\ddot{\mathbf{u}})\,.\end{aligned} \tag{30.3.5}$$

3. Calculate the intermediate step values of $\mathbf{u}$ and p, respectively denoted by $\mathbf{u}_{\mathrm{avg}}$ and p_{avg}, for use in the α-method of time integration

$$\begin{aligned}\mathbf{u}_{\mathrm{avg}} &= \alpha\mathbf{u}_{\mathrm{old}} + (1-\alpha)\mathbf{u},\\ p_{\mathrm{avg}} &= \alpha p_{\mathrm{old}} + (1-\alpha)p.\end{aligned} \tag{30.3.6}$$

4. Calculate the necessary kinematical quantities for the current step:

$$\mathbf{F} = \mathbf{1} + \nabla\mathbf{u}_{\mathrm{avg}}, \qquad J = \det\mathbf{F}, \qquad \bar{\mathbf{F}} = J^{-1/3}\mathbf{F}, \qquad \bar{\mathbf{C}} = \bar{\mathbf{F}}^{\top}\bar{\mathbf{F}}. \tag{30.3.7}$$

5. Update the internal variables $\mathbf{C}^{v(\alpha)}$:

$$\mathbf{C}^{v(\alpha)} = \left(\det\left[\mathbf{C}^{v(\alpha)}_{\text{old}} + \frac{\Delta t}{\tau_r^{(\alpha)}}\bar{\mathbf{C}}\right]\right)^{-1/3}\left(\mathbf{C}^{v(\alpha)}_{\text{old}} + \frac{\Delta t}{\tau_r^{(\alpha)}}\bar{\mathbf{C}}\right). \tag{30.3.8}$$

6. Calculate the Piola stress as,

$$\begin{aligned}\mathbf{T}_{\mathrm{R}} &= J^{-2/3}G\left[\mathbf{F} - \frac{1}{3}(\operatorname{tr}\mathbf{C})\mathbf{F}^{-\top}\right] - J\,p\mathbf{F}^{-\top}\\ &\quad + J\left[\sum_{\alpha=1}^{M} J^{-1}G^{(\alpha)}_{\text{neq}}\left(\bar{\mathbf{F}}(\mathbf{C}^{v(\alpha)})^{-1}\bar{\mathbf{F}}^{\top} - \frac{1}{3}(\bar{\mathbf{C}}:(\mathbf{C}^{v(\alpha)})^{-1})\mathbf{1}\right)\right]\mathbf{F}^{-\top}\\ &\quad - \left(\frac{\partial\mathbf{R}}{\partial\mathbf{F}}\right)^{\top}[\mathbf{b}^{\text{app}}\otimes\mathbf{m}^{\text{rem}}_{\mathrm{R}}].\end{aligned} \tag{30.3.9}$$

7. Compute the residuals:

$$\begin{aligned}R_{\mathbf{u}} &= \int_{\mathcal{B}}\mathbf{T}_{\mathrm{R}}:\nabla\mathbf{u}_{\text{test}}\,dv_{\mathrm{R}} + \int_{\mathcal{B}}\rho_{\mathrm{R}}\ddot{\mathbf{u}}\cdot\mathbf{u}_{\text{test}}\,dv_{\mathrm{R}} - \int_{\partial\mathcal{B}_{\mathbf{t}_{\mathrm{R}}}}\bar{\mathbf{t}}_{\mathrm{R}}\cdot\mathbf{u}_{\text{test}}\,da_{\mathrm{R}},\\ R_p &= \int_{\mathcal{B}}\left[\frac{p}{K} + (J-1)\right]p_{\text{test}}\,dv_{\mathrm{R}}.\end{aligned} \tag{30.3.10}$$

8. FEniCS solves for the current values of $\{\mathbf{u},\, p\}$.
9. Set the current values as the "old" values for the next step in time,

$$\{\mathbf{u},\,\dot{\mathbf{u}},\,\ddot{\mathbf{u}},\,p,\,\mathbf{C}^{v(\alpha)}\} \implies \{\mathbf{u}_{\text{old}},\,\dot{\mathbf{u}}_{\text{old}},\,\ddot{\mathbf{u}}_{\text{old}},\,p_{\text{old}},\,\mathbf{C}^{v(\alpha)}_{\text{old}}\}. \tag{30.3.11}$$

10. Repeat for the desired number of steps in time.

30.A Appendix: Computing the right polar decomposition $\mathbf{F} = \mathbf{RU}$ in FEniCS

In this Appendix we present brief derivations of analytical methods for computing the right polar decomposition $\mathbf{F} = \mathbf{RU}$ which we use in FEniCS. The advantage of such methods is that FEniCS can then take automatic derivatives of the analytical expressions for $\mathbf{R}$ and $\mathbf{U}$ with respect to the numerical degrees of freedom to automatically obtain analytical tangent or "Jacobian" expressions.

This Appendix is organized as follows. First, in Section 30.A.1 we derive analytical expressions for the eigenvalues of $\mathbf{C}$. Then, in Section 30.A.2 we derive an analytical expression for $\mathbf{U}^{-1}$ in terms of the invariants of $\mathbf{C}$. The derivation for the calculation of $\mathbf{U}^{-1}$ is based on the papers by Hoger and Carlson (1984), Sawyers (1986), Guan-Suo (1998), Norris (2007), and Scott (2020). With $\mathbf{U}^{-1}$ in hand, we can easily compute

$$\mathbf{R} = \mathbf{F}\mathbf{U}^{-1} \quad \text{and} \quad \mathbf{U} = \mathbf{R}^{\top}\mathbf{F}. \tag{30.A.1}$$

30.A.1 Analytical expressions for the eigenvalues of $\mathbf{C}$

The eigenvalues of $\mathbf{C}$ are the squared stretches $\{\lambda_1^2, \lambda_2^2, \lambda_3^2\}$, and we may assume the ordering

$$\lambda_1 \geq \lambda_2 \geq \lambda_3. \tag{30.A.2}$$

The three principal invariants of $\mathbf{C}$ are

$$I_1 = \operatorname{tr}\mathbf{C}, \qquad I_2 = \frac{1}{2}\left[I_1^2 - \operatorname{tr}(\mathbf{C}^2)\right], \qquad I_3 = \det\mathbf{C}, \tag{30.A.3}$$

which in terms of the squared stretches, become

$$I_1 = \lambda_1^2 + \lambda_2^2 + \lambda_3^2, \qquad I_2 = \lambda_1^2\lambda_2^2 + \lambda_2^2\lambda_3^2 + \lambda_3^2\lambda_1^2, \qquad I_3 = \lambda_1^2\lambda_2^2\lambda_3^2. \tag{30.A.4}$$

The characteristic equation for $\mathbf{C}$ is

$$\det(\mathbf{C} - \lambda^2\mathbf{1}) = 0, \tag{30.A.5}$$

which on expansion, and writing

$$x = \lambda^2,$$

becomes

$$x^3 - I_1 x^2 + I_2 x - I_3 = 0. \tag{30.A.6}$$

By introducing the substitution

$$x = \frac{1}{3}I_1 + y \tag{30.A.7}$$

one can show that the cubic equation (30.A.6) is reduced to

$$y^3 = py + q, \tag{30.A.8}$$

in which

$$\begin{aligned} p &= \frac{1}{3}\left(I_1^2 - 3I_2\right), \\ q &= \frac{1}{27}\left(2I_1^3 - 9I_1I_2 + 27I_3\right). \end{aligned} \tag{30.A.9}$$

From the definitions of the invariants $\{I_1, I_2, I_3\}$, cf. (30.A.4), it can be shown that

$$p = \frac{1}{6}\left[(\lambda_1^2 - \lambda_2^2)^2 + (\lambda_2^2 - \lambda_3^2)^2 + (\lambda_3^2 - \lambda_1^2)^2\right] \geq 0, \tag{30.A.10}$$

which vanishes only when

$$\lambda_1 = \lambda_2 = \lambda_3.$$

Also, q admits the factorization

$$q = \frac{1}{27}\left(2\lambda_1^2 - \lambda_2^2 - \lambda_3^2\right)\left(2\lambda_2^2 - \lambda_3^2 - \lambda_1^2\right)\left(2\lambda_3^2 - \lambda_1^2 - \lambda_2^2\right). \tag{30.A.11}$$

It can be shown that

$$p = \frac{1}{2}\text{tr}\left(\mathbf{C}_0^2\right) \quad \text{and} \quad q = \det \mathbf{C}_0, \tag{30.A.12}$$

where $\mathbf{C}_0$ is the deviatoric part of $\mathbf{C}$.

The roots of (30.A.8) can be obtained by a trigonometric solution to the cubic equation

$$y^3 - py - q = 0, \tag{30.A.13}$$

as follows. Let

$$y = m\cos\theta \qquad \text{and choose} \qquad m = 2\sqrt{\frac{p}{3}}.$$

Substituting into (30.A.13) and dividing by $m^3/4$, we have

$$4\cos^3\theta - 3\cos\theta - \frac{q/2}{(p/3)^{3/2}} = 0. \tag{30.A.14}$$

Considering the trigonometric identity

$$4\cos^3\theta - 3\cos\theta - \cos(3\theta) = 0, \tag{30.A.15}$$

we see that a solution θ_1 exists which satisfies

$$\cos(3\theta_1) = \frac{q/2}{(p/3)^{3/2}} \qquad \text{or} \qquad \theta_1 = \frac{1}{3}\cos^{-1}\left(\frac{q/2}{(p/3)^{3/2}}\right),$$

which also has solutions

$$\theta_2 = \theta_1 - \frac{2\pi}{3} \qquad \text{and} \qquad \theta_3 = \theta_1 + \frac{2\pi}{3}.$$

Choosing the branch of $\cos^{-1}$ with $0 \le \theta_1 \le \pi$ we find that the squared stretches, namely, the three roots (30.A.7) of the characteristic equation (30.A.6) are given by

$$\begin{aligned}
\lambda_i^2 &= \frac{I_1}{3} + 2\left(\frac{p}{3}\right)^{1/2}\cos\theta_i, \qquad i = 1,2,3\\
\theta_1 &= \frac{1}{3}\cos^{-1}\left(\frac{q/2}{(p/3)^{3/2}}\right)\\
\theta_2 &= \theta_1 - \frac{2\pi}{3},\\
\theta_3 &= \theta_1 + \frac{2\pi}{3}.
\end{aligned} \tag{30.A.16}$$

The angles θ_i have been chosen so that roots (30.A.16) satisfy the ordering

$$\lambda_1 \ge \lambda_2 \ge \lambda_3. \tag{30.A.17}$$

Accounting for eigenvalue multiplicity

- **Accounting for eigenvalue multiplicity three:**
  In the case of a spherical tensor $\mathbf{C}$ which has a single eigenvalue with multiplicity three, it is immediately clear that

  $$\mathbf{C}_0 = \mathbf{0},$$

  and therefore

  $$p = \frac{1}{2}\text{tr}\left(\mathbf{C}_0^2\right) = 0.$$

  Allowing $p = 0$, however, would introduce division by zero in $(30.A.16)_2$. To prevent division by zero in our numerical implementation, we therefore introduce the following conditional statement

$$\theta_1 = \begin{cases} \dfrac{1}{3}\cos^{-1}\left(\dfrac{q/2}{(p/3)^{3/2}}\right) & \text{if} \quad p \neq 0 \\ \dfrac{1}{3}\cos^{-1}(0) \quad \text{if} \quad p = 0, \end{cases} \tag{30.A.18}$$

with the latter case resulting in $\lambda_1 = \lambda_2 = \lambda_3 = I_1/3$ as required when $p = 0$.

- **Accounting for eigenvalue multiplicity two**:
  In the case of an eigenvalue of **C** having multiplicity two, for example if $\lambda_1 = \lambda_2$, it follows from (30.A.10) and (30.A.11) that

$$\begin{aligned} p &= -\frac{1}{3}(\lambda_1^2 - \lambda_3^2)^2, \quad \text{and} \\ q &= -\frac{2}{27}(\lambda_1^2 - \lambda_3^2)^3. \end{aligned} \tag{30.A.19}$$

In this case, the argument of $\cos^{-1}$ in $(30.A.16)_2$ becomes

$$\frac{q/2}{(p/3)^{3/2}} = 1. \tag{30.A.20}$$

More generally, if

$$\frac{q/2}{(p/3)^{3/2}} = \pm 1, \tag{30.A.21}$$

then two of the eigenvalues of **C** are equal to each other. However, the condition (30.A.21) for the argument of $\cos^{-1}$ results in a divide-by-zero in the derivative of $\cos^{-1}$ since

$$\frac{\partial}{\partial x}\cos^{-1}(x) = -\frac{1}{\sqrt{1 - x^2}}. \tag{30.A.22}$$

Therefore, in the FEniCS code we restrict the argument of $\cos^{-1}$ to be within $(-1, 1)$.

Summary of algorithm for computing the eigenvalues of C

In our algorithm for computing the eigenvalues of **C**, we use the notation of Ang et al. (2022) to write

$$\begin{aligned} \lambda_i^2 &= d_{par} + 2\,e_{par}\,\cos\theta_i, \qquad i = 1, 2, 3 \\ \theta_1 &= h_{par} + \frac{6\pi}{3} \\ \theta_2 &= h_{par} + \frac{4\pi}{3}, \\ \theta_3 &= h_{par} + \frac{2\pi}{3}, \end{aligned} \tag{30.A.23}$$

where the intermediate quantities are defined as follows:

$$
\begin{aligned}
d_{par} &\overset{\text{def}}{=} \frac{1}{3} I_1 \\
e_{par} &\overset{\text{def}}{=} \sqrt{p/3} = \sqrt{\frac{1}{6}\text{tr}((\mathbf{C} - d_{par}\mathbf{1})^2)} \\
f_{par} &\overset{\text{def}}{=} \begin{cases} \dfrac{\mathbf{C} - d_{par}\mathbf{1}}{e_{par}} & \text{if} \quad e_{par} \neq 0 \\ \mathbf{0} \quad \text{if} \quad e_{par} = 0, \end{cases} \\
g_{par} & \begin{cases} \dfrac{\det(f_{par})}{2} & \text{if} \quad \dfrac{\det(f_{par})}{2} \in (-1 + 3 \times 10^{-16}, 1 - 3 \times 10^{-16}) \\ 1 - 3 \times 10^{-16} & \text{if} \quad \dfrac{\det(f_{par})}{2} \geq 1 - 3 \times 10^{-16}, \\ -1 + 3 \times 10^{-16} & \text{if} \quad \dfrac{\det(f_{par})}{2} \leq -1 + 3 \times 10^{-16}, \end{cases} \\
h_{par} &= \frac{\cos^{-1}(g_{par})}{3}.
\end{aligned}
\tag{30.A.24}
$$

Note that the tensor $\mathbf{C}$ in this algorithm can be replaced with any general three-dimensional symmetric tensor $\mathbf{A}$, and the method will return an ordered list of real eigenvalues of $\mathbf{A}$.

30.A.2 An analytical expression for $\mathbf{U}^{-1}$

We consider relations among some basic tensors of three-dimensional continuum kinematics,

$$\mathbf{F} = \mathbf{RU}, \qquad \mathbf{C} = \mathbf{F}^{\top}\mathbf{F}, \tag{30.A.25}$$

where since $\mathbf{C}$ is symmetric and positive-definite,

$$\mathbf{U} = \sqrt{\mathbf{C}}. \tag{30.A.26}$$

Although the square root of a second-order positive-definite symmetric tensor is unique and unambiguous it is not, however, a simple matter to calculate it algebraically. We first express $\mathbf{U}$ as a polynomial in $\mathbf{C}$ starting from the Cayley–Hamilton equation

$$\mathbf{U}^3 - i_1\mathbf{U}^2 + i_2\mathbf{U} - i_3\mathbf{1} = \mathbf{0}, \tag{30.A.27}$$

where $\{i_1, i_2, i_3\}$ are the invariants of $\mathbf{U}$,

$$i_1 = \text{tr}\,\mathbf{U}, \qquad i_2 = \frac{1}{2}\left[(\text{tr}\,\mathbf{U})^2 - \text{tr}(\mathbf{U}^2)\right], \qquad i_3 = \det\mathbf{U}. \tag{30.A.28}$$

Multiplying (30.A.27) by $(\mathbf{U} + i_1\mathbf{1})$ and replacing $\mathbf{U}^2$ and $\mathbf{U}^4$ by $\mathbf{C}$ and $\mathbf{C}^2$, respectively, gives

$$\mathbf{C}^2 + (i_2 - i_1^2)\mathbf{C} + (i_1 i_2 - i_3)\mathbf{U} - i_1 i_3\mathbf{1} = \mathbf{0},$$

which can be solved for $\mathbf{U}$ to give

$$\mathbf{U} = (i_1 i_2 - i_3)^{-1}\left(i_1 i_3 \mathbf{1} + (i_1^2 - i_2)\mathbf{C} - \mathbf{C}^2\right). \tag{30.A.29}$$

To obtain the inverse of $\mathbf{U}$, multiply (30.A.29) with $\mathbf{C}^{-1}$ to get

$$\mathbf{U}^{-1} = (i_1 i_2 - i_3)^{-1}\left(i_1 i_3 \mathbf{C}^{-1} + (i_1^2 - i_2)\mathbf{1} - \mathbf{C}\right). \tag{30.A.30}$$

Further, the Cayley–Hamilton theorem for $\mathbf{C}$ gives,

$$\mathbf{C}^3 - I_1\mathbf{C}^2 + I_2\mathbf{C} - I_3\mathbf{1} = \mathbf{0}, \tag{30.A.31}$$

with

$$I_I = \operatorname{tr}\mathbf{C}, \qquad I_2 = \frac{1}{2}\left[(\operatorname{tr}\mathbf{C})^2 - \operatorname{tr}(\mathbf{C}^2)\right], \qquad I_3 = \det\mathbf{C}\,; \tag{30.A.32}$$

multiplying (30.A.31) with $\mathbf{C}^{-1}$ yields

$$\mathbf{C}^2 - I_1\mathbf{C} + I_2\mathbf{1} - I_3\mathbf{C}^{-1} = \mathbf{0}, \tag{30.A.33}$$

or

$$\mathbf{C}^{-1} = I_3^{-1}\left(I_2\mathbf{1} - I_1\mathbf{C} + \mathbf{C}^2\right). \tag{30.A.34}$$

Substituting (30.A.34) in (30.A.30) and using $I_3^{-1} = i_3^{-2}$ gives

$$\mathbf{U}^{-1} = ([(i_1 i_2 - i_3) i_3]^{-1}\left[(i_1 I_2 + i_3(i_1^2 - i_2))\,\mathbf{1} - (i_1 I_1 + i_3)\,\mathbf{C} + i_1\mathbf{C}^2\right]. \tag{30.A.35}$$

It can be shown that,

$$\begin{aligned} I_1 &= i_1^2 - 2i_2, \\ I_2 &= i_2^2 - 2i_1 i_3, \\ I_3 &= i_3^2. \end{aligned} \tag{30.A.36}$$

Substitution of I_1 and I_2 from $(30.A.36)_{1,2}$ in (30.A.35) gives

$$\mathbf{U}^{-1} = i_3^{-1}(i_1 i_2 - i_3)^{-1}\left[(i_1 i_2^2 - i_3(i_1^2 + i_2))\,\mathbf{1} - (i_3 + i_1(i_1^2 - 2i_2))\,\mathbf{C} + i_1\mathbf{C}^2\right]. \tag{30.A.37}$$

It still remains to express the invariants (i_1, i_2, i_3) of $\mathbf{U}$ in terms of the known invariants (I_1, I_2, I_3) of $\mathbf{C}$.

Invariants i_i of $\mathbf{U}$ in terms of the invariants I_i of $\mathbf{C}$

A solution for i_1 and i_2 may be obtained by starting with the *ansatz*

$$i_1 = \lambda + \rho, \tag{30.A.38}$$

where λ is *any one* of the triplet of the eigenvalues of $\{\lambda_1, \lambda_2, \lambda_3\}$ of $\mathbf{U}$. For instance if $\lambda = \lambda_1$, then $\rho = \lambda_2 + \lambda_3$, and the invariant

$$i_2 = \lambda_1(\lambda_2 + \lambda_3) + \lambda_2\lambda_3$$

may be written as

$$i_2 = \rho\lambda + \frac{i_3}{\lambda}. \tag{30.A.39}$$

Note that this holds no matter which value $\{\lambda_1, \lambda_2, \lambda_3\}$ that λ takes. Substitution of (30.A.38) and (30.A.39) in $(30.A.36)_1$ implies that

$$I_1 = (\lambda + \rho)^2 - 2\left(\rho\lambda + \frac{i_3}{\lambda}\right),$$

which can be solved for ρ to find

$$\rho = \sqrt{I_1 - \lambda^2 + 2\frac{i_3}{\lambda}}. \tag{30.A.40}$$

Using (30.A.40) in (30.A.38) and (30.A.39) and recalling $i_3 = I_3^{1/2}$ gives

$$\begin{aligned} i_1 &= \lambda + \sqrt{I_1 - \lambda^2 + 2\sqrt{I_3}/\lambda}, \\ i_2 &= \sqrt{I_3}/\lambda + \sqrt{I_1\lambda^2 - \lambda^4 + 2\sqrt{I_3}\,\lambda}, \\ i_3 &= \sqrt{I_3}, \end{aligned} \tag{30.A.41}$$

where λ is *any* positive eigenvalue of $\mathbf{U}$, obtainable using the methods of Section 30.A.1.

Summary of expressing $\mathbf{U}^{-1}$ in terms of the invariants of $\mathbf{C}$

$$\mathbf{U}^{-1} = i_3^{-1}(i_1 i_2 - i_3)^{-1}\left[\left(i_1 i_2^2 - i_3(i_1^2 + i_2)\right)\mathbf{1} - \left(i_3 + i_1(i_1^2 - 2i_2)\right)\mathbf{C} + i_1\mathbf{C}^2\right], \tag{30.A.42}$$

with

$$\begin{aligned} i_1 &= \lambda + \sqrt{I_1 - \lambda^2 + 2\sqrt{I_3}/\lambda} \\ i_2 &= \sqrt{I_3}/\lambda + \sqrt{I_1\lambda^2 - \lambda^4 + 2\sqrt{I_3}\,\lambda} \\ i_3 &= \sqrt{I_3}, \end{aligned} \tag{30.A.43}$$

where, using the methods of Section 30.A.1, the eigenvalue λ of $\mathbf{U}$ is given by

$$\begin{aligned} \lambda &= \sqrt{\frac{I_1}{3} + 2\left(\frac{p}{3}\right)^{1/2}\cos\theta} \qquad \text{with} \qquad \theta = \frac{1}{3}\cos^{-1}\left(\frac{q/2}{(p/3)^{3/2}}\right), \\ p &= \frac{1}{3}\left(I_1^2 - 3I_2\right), \\ q &= \frac{1}{27}\left(2I_1^3 - 9I_1 I_2 + 27 I_3\right). \end{aligned} \tag{30.A.44}$$

31 Representative simulations of hard-magnetic soft-elastomers

In the applications of the theory shown in this chapter we will use the unit system shown in Table 31.1. Representative material parameters are shown in Table 31.2.

Table 31.1 Units.

Quantity	SI (m)	SI (mm)	Conversion factor C_f
Length	m	mm	$1\,\text{m} = 10^3\,\text{mm}$
Mass	kg	kg	1
Time	s	s	1
Electric charge	C	kC	$1\,\text{C} = 10^{-3}\,\text{kC}$
Force	N	mN	$1\,\text{N} = 10^3\,\text{mN}$
Energy	J	mJ	$1\,\text{J} = 10^3\,\text{mJ}$
Mass density	kg/m^3	kg/mm^3	$1\,\text{kg/m}^3 = 10^{-9}\,\text{kg/mm}^3$
Stress	Pa	kPa (mN/mm^2)	$1\,\text{Pa} = 10^{-3}\,\text{kPa}$
Current	A (C/s)	kA (kC/s)	$1\,\text{A} = 10^{-3}\,\text{kA}$
Magnetic field strength	A/m	kA/ mm	$1\,\text{A/m} = 10^{-6}\,\text{kA/mm}$
Magnetic flux density	T (N s / C m)	mT (mN s / kC mm)	$1\,\text{T} = 10^3\,\text{mT}$

Table 31.2 Representative values of material parameters for a hard-magnetic soft-elastomeric material from Stewart and Anand (2023). We only use one Maxwell branch with material parameters $G^{(1)}_{\text{neq}}$ and $\tau^{(1)}_r$ to model the viscoelasticity of the material.

Quantity	Value SI (m)	Value SI (mm)
ρ_{R}, mass density	$2000\ \text{kg/m}^3$	$2 \times 10^{-6}\ \text{kg/mm}^3$
G, equilibrium shear modulus	$1.4 \times 10^6\ \text{Pa}$	$1400\ \text{kPa}$
K, bulk modulus	$1.4 \times 10^9\ \text{Pa}$	$1400 \times 10^3\ \text{kPa}$
$G^{(1)}_{\text{neq}}$, non-equilibrium shear modulus	$0.5 \times 10^6\ \text{Pa}$	$500\ \text{kPa}$
$\tau^{(1)}_r$, relaxation time	$0.01\,\text{s}$	$0.01\,\text{s}$
$\lvert\mathbf{m}^{\text{rem}}\rvert$, remanent magnetization strength	$53.7 \times 10^3\ \text{A/m}$	$53.7 \times 10^{-3}\ \text{kA/mm}$
μ_0, permeability of free space	$4\,\pi \times 10^{-7}\ \text{T / (A/m)}$	$4\,\pi \times 10^2\ \text{mT / (kA/mm)}$

Introduction to Coupled Theories in Solid Mechanics. Lallit Anand, Eric M. Stewart, Shawn A. Chester, Oxford University Press.
 DOI: 10.1093/oso/9780198986218.003.0032

31.1 Bending of a beam with 90°-actuation

Zhao et al. (2019) conducted several beam-bending experiments in which the remanent magnetization $\mathbf{m}_{\mathrm{R}}^{\mathrm{rem}}$ was aligned with the longitudinal $\mathbf{e}_1$-direction of the beam, Fig. 31.1(a). In the experiments the left end of the beam was clamped while the other end was free to move under the action of a uniform magnetic flux $\mathbf{b}^{\mathrm{app}}$ applied in the $\mathbf{e}_2$-direction, which is perpendicular to the length direction of the beam—a geometry which we call 90°-actuation. Here we simulate one of their experiments in which they used a beam of length $L = 17.2\,\mathrm{mm}$, height $H = 0.84\,\mathrm{mm}$, and width $B = 5\,\mathrm{mm}$, which corresponds to a beam with an aspect ratio of $L/H = 20.5$. They report an average value of shear modulus and remanent magnetization for their beams of

$$G = 303\,\mathrm{kPa} \quad \text{and} \quad |\mathbf{m}_{\mathrm{R}}^{\mathrm{rem}}| = 114\,\mathrm{kA}\,/\,\mathrm{m}\,.$$

We adopt these material parameters, and take the bulk modulus K to be 1000 times larger than the shear modulus G, as is typical for nearly incompressible elastomers.

Zhao et al. (2019) reported the results from their experiments for the normalized deflection δ/L of the tip of the beam in the $\mathbf{e}_2$-direction as a function of $\mathbf{b}^{\mathrm{app}}$ whose magnitude was varied from 0 to 50 mT. Since this experiment of Zhao et al. (2019) was quasistatic in nature, inertial as well as viscoelastic effects in our simulation were suppressed. In our simulation, the initial conditions are $\mathbf{u}_0 = \mathbf{0}$ and, referring to Fig. 31.1(a), the boundary and prescribed conditions are as follows:

- Mechanical boundary conditions: (i) the left face of the beam is fully constrained, $\bar{\mathbf{u}} = \mathbf{0}$; and (ii) all other faces are traction-free.
- Magnetic prescribed conditions: a uniform magnetic flux is prescribed $\mathbf{b}^{\mathrm{app}}$ in the $\mathbf{e}_2$-direction with a magnitude that linearly ramps from 0 mT to 50 mT.

Figure 31.1(b) shows the experimental data of Zhao et al. (2019) as black open circles. The δ/L versus $|\mathbf{b}^{\mathrm{app}}|$ curve from our numerical simulation is also shown in this figure—the simulation result compares well with the experimental result from Zhao et al. (2019). Figure 31.1(c) shows the deformed configuration of the beam when $|\mathbf{b}^{\mathrm{app}}| = 50\,\mathrm{mT}$. The red arrows in this figure show a vector plot of the remanent magnetization $\mathbf{m}^{\mathrm{rem}}$ in the body. This figure clearly shows that it is the tendency of the $\mathbf{m}^{\mathrm{rem}}$ field to line up with the applied $\mathbf{b}^{\mathrm{app}}$ field which causes the bending of the beam.

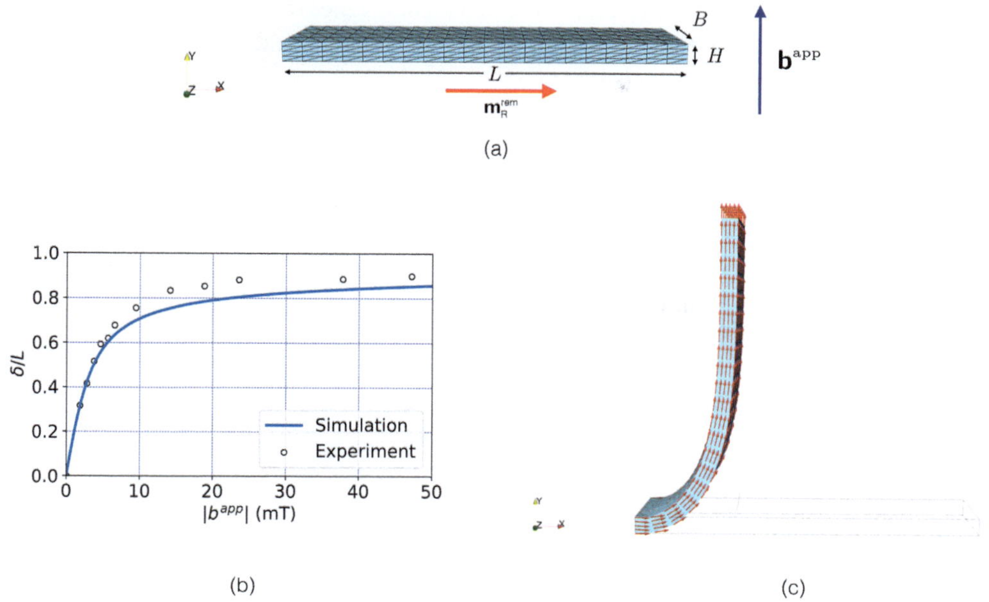

Fig. 31.1 (a) Initial configuration and finite element mesh for 90°-actuation of a beam with an aspect ratio $L/H = 20.5$. (b) Normalized deflection of the tip of the beam, δ/L, as a function of the applied $\mathbf{b}^{\mathrm{app}}$ field whose magnitude varies from 0 to 50 mT. The blue solid line is the simulation result and the open black circles are experimental data from Zhao et al. (2019). (c) Deformed configuration of the beam when $|\mathbf{b}^{\mathrm{app}}| = 50$ mT. The red arrows in this figure represent a vector plot of the remanent magnetization $\mathbf{m}^{\mathrm{rem}}$ in the deformed body.

31.2 Bending of a beam with 180°-actuation

In another experiment, aimed at creating extreme bending, Zhao et al. (2019) using a beam with the same dimensions and material properties and remanent magnetization as in the previous example discussed in Section 31.1, used a $\mathbf{b}^{\mathrm{app}}$ field which was applied in a direction *opposite* to the initial $\mathbf{m}_{\mathrm{R}}^{\mathrm{rem}}$ field, as shown in Fig. 31.2(a), a geometry and loading which we call 180°-actuation.

For this simulation we used the same mesh, but introduced a small misalignment of 5° between the applied $\mathbf{b}^{\mathrm{app}}$ field and the remanent $\mathbf{m}_{\mathrm{R}}^{\mathrm{rem}}$ field—such a misalignment is needed so that the two fields are not initially perfectly anti-parallel to each other, for otherwise there would be no magnetic torque which would cause bending of the beam. And as before, since the experiment was quasi-static in nature, viscoelasticity and internal effects are neglected in this simulation.

Zhao et al. (2019) reported the result from their experiment for the normalized deflection δ/L of the tip of the beam in the $\mathbf{e}_2$-direction as a function of the applied $\mathbf{b}^{\mathrm{app}}$ field whose magnitude was varied from 0 to 40 mT. Figure 31.1(b) shows their experimental results as black open circles. The δ/L versus $|\mathbf{b}^{\mathrm{app}}|$ curve from our numerical simulation is also shown in this figure—the simulation result compares very well with the experimental result. Figure 31.2(c) shows the deformed configuration of the beam when $|\mathbf{b}^{\mathrm{app}}| = 40$ mT. The red arrows in this figure show a vector plot of the remanent magnetization $\mathbf{m}^{\mathrm{rem}}$ in the body. This figure again shows that it is the tendency of the $\mathbf{m}^{\mathrm{rem}}$ field to line up with the applied $\mathbf{b}^{\mathrm{app}}$ field which causes the extreme bending of the beam.

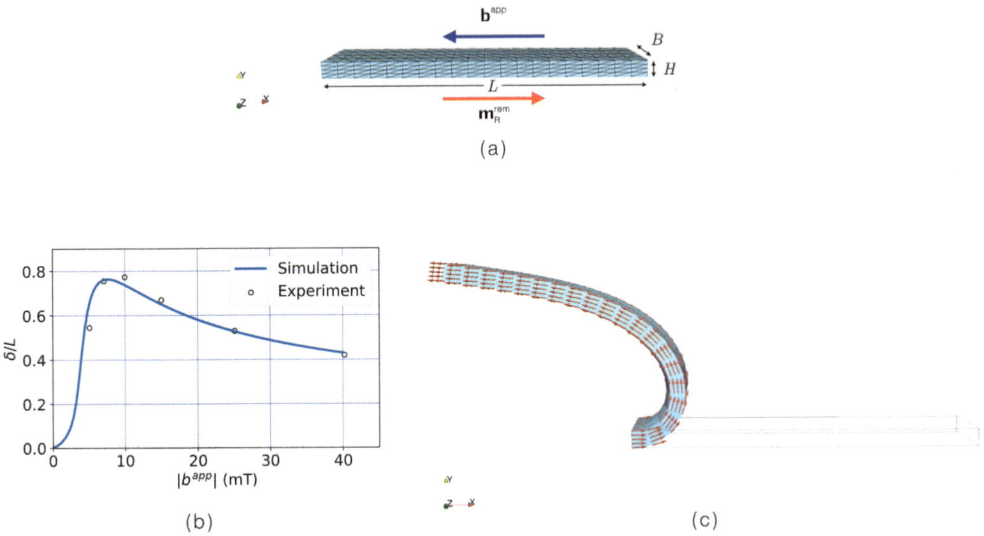

Fig. 31.2 (a) Initial configuration and finite element mesh for $180°$-actuation of a beam with an aspect ratio $L/H = 20.5$. (b) Normalized deflection of the tip of the beam, δ/L, as a function of the applied $\mathbf{b}^{\mathrm{app}}$ field whose magnitude varies from 0 to 40 mT. The blue solid line is the simulation result and the open black circles are experimental data from Zhao et al. (2019). (c) Deformed configuration of the beam when $|\mathbf{b}^{\mathrm{app}}| = 40\,\mathrm{mT}$. The red arrows in this figure represent a vector plot of the remanent magnetization $\mathbf{m}^{\mathrm{rem}}$ in the deformed body.

31.3 Simulation of a patterned planar domain morphing into a complex 3-D shape

Kim et al. (2018) developed a new 3-D-printing technique to create complex shape-morphing structures based on intricately patterned 2-D domains in soft materials with different orientations of the magnetization field $\mathbf{m}_{\mathrm{R}}^{\mathrm{rem}}$ in different sub-domains of the body. One of these patterned shape-morphing structures is shown in Fig. 31.3(a). Kim et al. (2018) reported their experimental observation of the morphed shape for this initial geometry after it had been subjected to an applied $\mathbf{b}^{\mathrm{app}}$ field of magnitude $|\mathbf{b}^{\mathrm{app}}| = 200\,\mathrm{mT}$ which was perpendicular to the initial planar geometry. Figure 31.3(b) shows the experimentally observed morphed geometry. In this section we show the results from our simulation of this shape-morphing experiment of Kim et al. (2018).

The planar dimensions of the geometry are shown in Fig. 31.3(a), and the body has a thickness of 0.84 mm in the out-of-plane $\mathbf{e}_3$-direction. In our numerical simulation we model one-quarter of the full geometry, and approximate the remaining three-quarters of the body by using symmetry boundary conditions. The quarter-geometry mesh with 2 tetrahedral elements through the thickness is shown in Fig. 31.4, with red arrows indicating the orientation of the (uniform strength) remanent magnetization $\mathbf{m}_{\mathrm{R}}^{\mathrm{rem}}$ in each sub-domain. Since the experiment was quasi-static in nature, viscoelasticity and internal effects are neglected

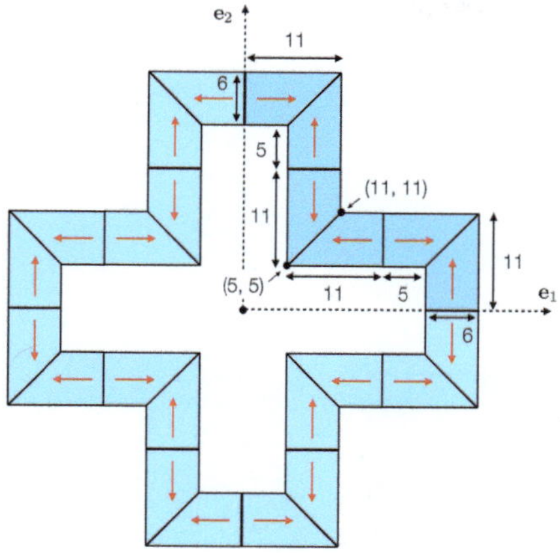

(a)

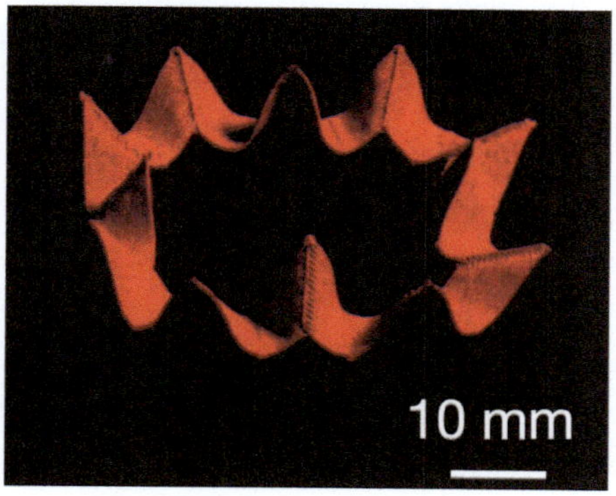

(b)

Fig. 31.3 (a) Initial planar domain pattern with different initial $\mathbf{m}_R^{rem}$ fields indicated by arrows and dimensions given in mm. In the experiment the applied $\mathbf{b}^{app}$ field points out of the plane of the paper. (b) Deformed configuration observed in the experiment when $|\mathbf{b}^{app}| = 200$ mT. From Kim et al. (2018).

in this simulation. We adopt the material properties for this material given by Zhao et al. (2019):

$$G = 330\,\text{kPa} \quad \text{and} \quad |\mathbf{m}_R^{rem}| = 81\,\text{kA}\,/\,\text{m},$$

and take the bulk modulus K to be 1000 times larger than the shear modulus G, as is typical for nearly incompressible elastomers.

The initial conditions are $\mathbf{u}_0 = \mathbf{0}$, and with reference to Fig. 31.4, the boundary and prescribed conditions are:

- Mechanical boundary conditions: (i) faces AB and CD are prescribed appropriate symmetry conditions; (ii) point E is fully constrained $\bar{\mathbf{u}} = \mathbf{0}$, and (iii) all other faces are traction-free.
- Magnetic prescribed conditions: a uniform magnetic flux $\mathbf{b}^{app}$ is prescribed along the $\mathbf{e}_3$-direction with a magnitude that linearly ramps from 0 mT to 200 mT.

Snapshots of the deformed configuration of the cross-shaped structure at various applied magnetic flux densities are shown in Fig. 31.5. When $|\mathbf{b}^{app}| = 20$ mT, Fig. 31.5(a), there is a small amount of bending as the magnetization of the different sub-domains deforms the structure to bring it in greater alignment with $\mathbf{b}^{app}$. As the applied magnetic flux density is ramped up the extent of bending becomes greater and the structure transforms into a contracting ring-shape, as shown in the snapshots for $|\mathbf{b}^{app}| = 50$ mT and $|\mathbf{b}^{app}| = 100$ mT in Figs. 31.5(b)–(c). Finally, when $|\mathbf{b}^{app}| = 200$ mT, Fig. 31.5(d), the initially flat structure has taken on a complicated wavy ring-like conformation which resembles the one observed in the experiments of Kim et al. (2018); cf. Fig. 31.3(b).

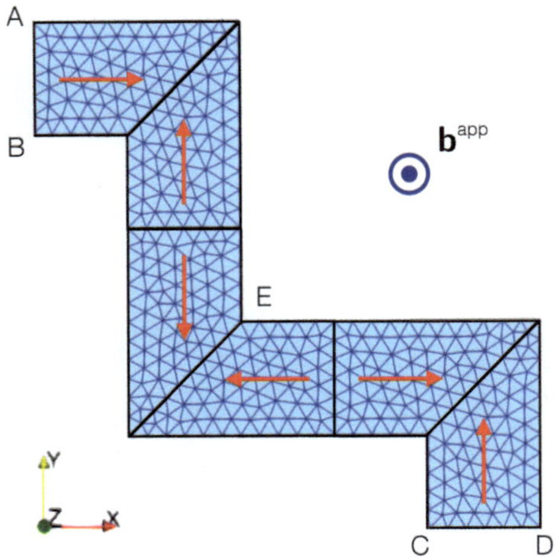

Fig. 31.4 The mesh used to model one quarter of the shape-morphing structure. Red arrows indicate the orientation of $\mathbf{m}_{\mathrm{R}}^{\mathrm{rem}}$ in each sub-domain, and the applied magnetic flux density $\mathbf{b}^{\mathrm{app}}$ is oriented in the out-of-plane e_3-direction as shown.

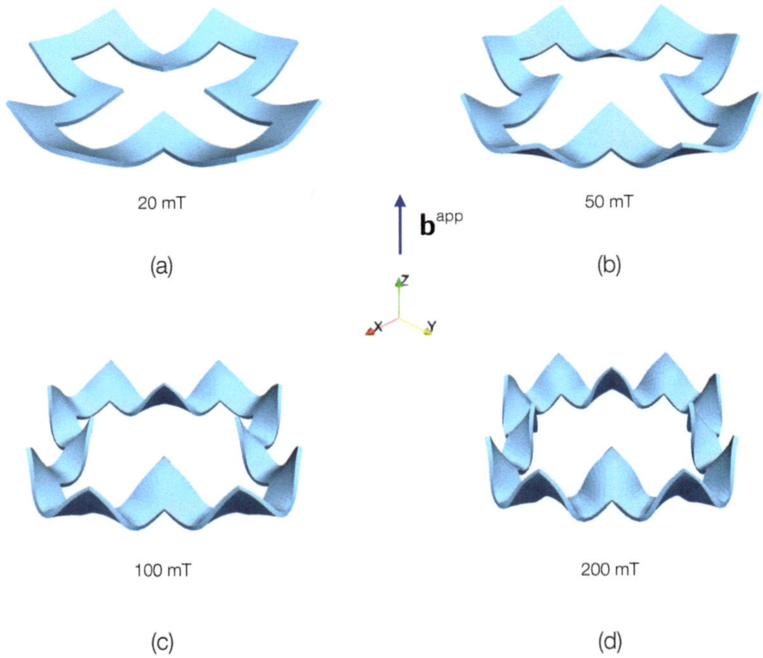

Fig. 31.5 Snapshots of the deformed configuration of the cross-shaped structure for various strengths of the applied magnetic flux density $\mathbf{b}^{\mathrm{app}}$ (a) 20 mT, (b) 50 mT, (c) 100 mT, and (d) 200 mT.

31.4 Snap-through buckling of a bistable beam

A promising application of hard-magnetic soft-materials is as a magnetic actuator which uses a "bistable beam." The beam is made such that it is comprised of two segments with antiparallel magnetization across its mid-length, and it is set into a bistable curved configuration by imposing an end-to-end compression. Reversible dynamic snapping is possible between the two stable states of such a beam (cf., e.g., Tan et al., 2022). Schematic figures showing the magnetization process, magnetized domains, stable configurations, and magnetic actuation of such a bistable magnetic actuator are presented in Fig. 31.6.

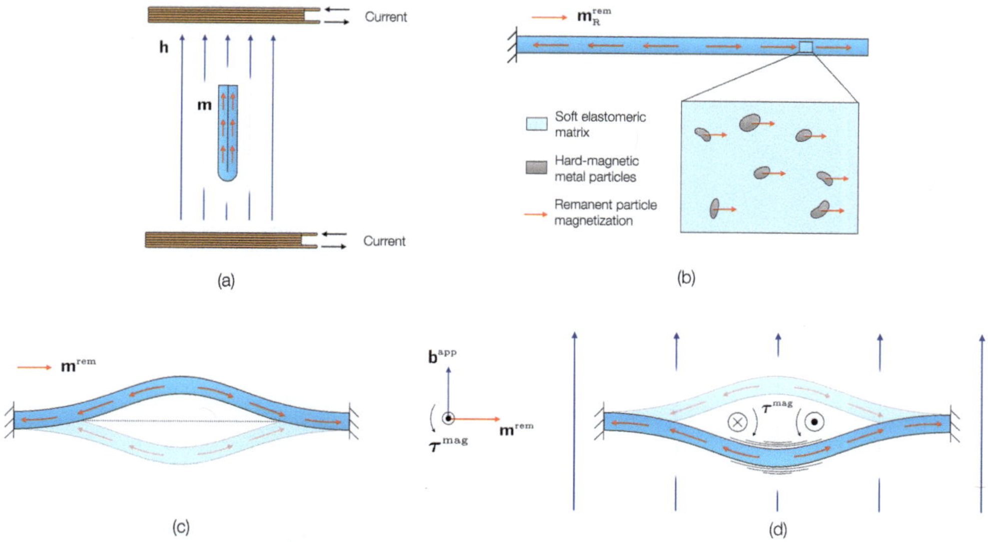

Fig. 31.6 (a) Magnetization of a hard-magnetic soft-elastomeric composite beam which has been bent upon itself in a uniform magnetic field generated by a pair of Helmholtz coils. (b) The two resulting anti-parallel domains of equal-strength remanent magnetization in the beam, with an inset showing a schematic of the microstructure of the material. (c) The two possible buckled configurations of the beam which has been subjected to an end-to-end shortening. (d) When a magnetic flux density $\mathbf{b}^{app}$ in the orientation shown is applied to the buckled beam, a rapid dynamic transition or "snap-through" between the two bistable states is triggered by the magnetically generated body torque.

In their experiments on such a bistable arch Tan et al. (2022) observed that the dynamic snap-through event occurs quite rapidly in $\approx 50\,\text{ms}$ when a magnetic flux density of 7 mT is applied to their device. Experimental observations of the displacement of the mid-span of the arch during such a dynamic snap-through event show clear overshoot, followed by oscillations about the second stable configuration which dissipate out over time. Clearly, viscoelastic and inertial effects both play a significant role in the snap-through process of this magnetically actuated device.

In this section we discuss the simulations carried out by Stewart and Anand (2023) of the bistable arch experiments of Tan et al. (2022).[1] The experiments of Tan et al. (2022) were conducted on a beam which was made from a composite of a polydimethylsiloxane (PDMS) matrix with a dispersion of magnetizable NdFeB particles prepared such that the resulting weight percentage of the metallic particles and elastomeric matrix are equal. They magnetized this composite beam in a uniform 1.5 T magnetic flux density in such a way as to obtain two domains of remanent magnetization which are oriented anti-parallel to one another across its mid-length and in the $\mathbf{e}_1$-direction in the reference configuration, Fig 31.7(a). The magnetized composite beam is 2.5 mm thick, 20 mm wide, and around 120.5 mm long. When the beam is

[1] The rest of this chapter is taken, almost verbatim, from Stewart and Anand (2023).

axially compressed to an end-to-end length of 120 mm the beam buckles and its mid-span undergoes a lateral deflection δ of around 4 mm in the $\mathbf{e}_2$-direction, Fig 31.7(b). A photograph of the bistable arch of Tan et al. (2022) is shown in Fig. 31.7(c).

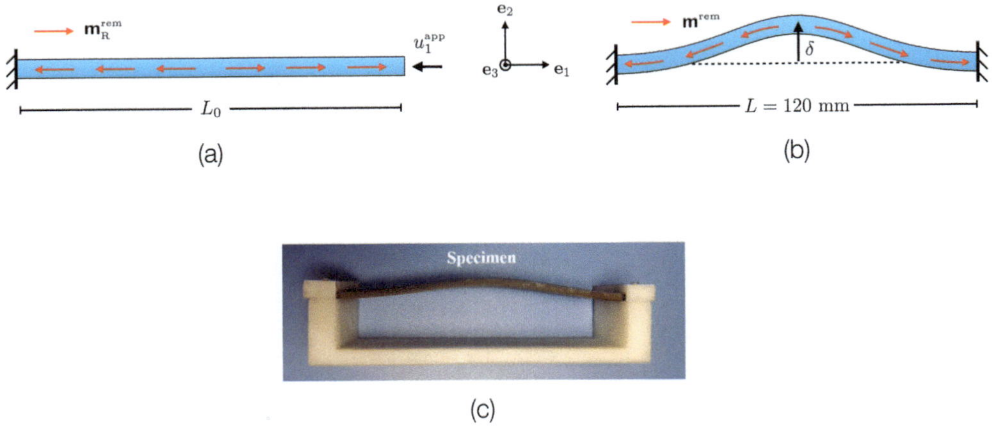

Fig. 31.7 (a) The composite beam of Tan et al. (2022), containing two domains of equal-strength remanent magnetization $\mathbf{m}_R^{rem}$ oriented as shown. The beam has a length of around 120.5 mm, thickness of 2.5 mm, and an out-of-page width of 20 mm. (b) When the beam is compressed to an end-to-end length of 120 mm it buckles and a lateral deflection $\delta \approx 4$ mm develops in the $\mathbf{e}_2$-direction. (c) A photograph of the bistable hard-magnetic soft-elastomeric arch studied experimentally by Tan et al. (2022).

Tan et al. (2022) conducted several experiments on their magnetically responsive bistable arch, reporting corresponding mid-span displacement data for:

(i) damped free oscillations of the arch in response to a small impulsive mechanical stimulus;

(ii) a dynamic snap-through event in response to an applied magnetic field which is linearly ramped up over time and exceeds a critical value at which snap-through occurs;

(iii) an instantly applied magnetic field above the critical value at which snap-through occurs; and

(iv) an applied magnetic field in the form of a triangle-wave which causes snap-through in a "forward" and a "reverse" direction.

The material parameters used to model the experiments of Tan et al. (2022) are listed in Table 31.2. Briefly, the rationale for the choice of the values of material parameters is as follows:

- The value for the shear modulus G of the composite material was estimated by using the Young's modulus of the composite material reported by Tan et al. (4.2 MPa) and assuming incompressibility so that $G = E/3$.
- The bulk modulus K for the material was taken to be much larger than the shear modulus, $K = 1000\,G$, which is representative of nearly incompressible elastomers like the PDMS employed in the experiments.
- The mass density of composite material ρ_R is an estimate which results from using the rule of mixtures to find the mass density of the composite material.
- The viscoelastic parameters $G^{(1)}_{\text{neq}}$ and $\tau^{(1)}$ were calibrated to match the damped oscillations observed in their experiment (i) of Tan et al. (2022).[2]
- The uniform strength of the remanent magnetization $|\mathbf{m}^{\text{rem}}|$ in the composite arch was calibrated to match experimental data from Tan et al. (2022) concerning the critical strength of applied magnetic flux density $|\mathbf{b}^{\text{app}}_{\text{cr}}|$ required to induce a snap-through event in their experiment (ii).

31.4.1 Initial buckling of the beam into an arch configuration

The numerical simulations start with a flat pre-magnetized beam of length $L_0 = 120.47$ mm, as shown in Fig. 31.8(a), and the beam is then compressed axially until the end-to-end length of the beam matches the length $L = 120$ mm reported by Tan et al. (2022). During this axial compression the beam buckles, and a significant lateral deflection develops as shown in Fig 31.8(b).[3] The lateral displacement of the mid-span of the beam in the $\mathbf{e}_2$-direction is plotted as a function of decreasing axial length L in Fig. 31.8(c). Onset of buckling in the beam occurs when $L \approx 120.3$ mm. The beam is compressed over a long time period of 1000 seconds so that inertial and viscoelastic effects are negligible and the observed buckling behavior is essentially quasi-static.

The numerical buckling process was carried out as an "initial" step prior to all of the numerical simulations of bistable arches. When subsequent mechanical or magnetic prescribed conditions are applied to the arch, the end-to-end length of the device is held fixed at 120 mm, and both ends have all mechanical displacement degrees of freedom fixed—that is the ends are "built-in".

[2] Stewart and Anand (2023) found that only one Maxwell branch was sufficient to model the viscoelastic dissipation observed in these experiments.

[3] A small initial imperfection of 0.01 mm in the $\mathbf{e}_2$-direction is introduced at the mid-span of the beam for all the simulations shown. This imperfection is visually imperceptible in Fig. 31.8(a), but was sufficiently large to ensure that initial buckling always occurs in the positive $\mathbf{e}_2$-direction.

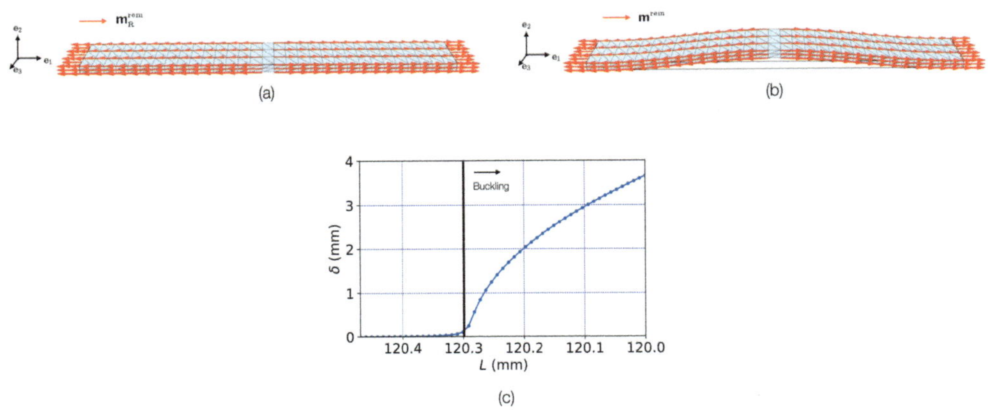

Fig. 31.8 (a) The computational reference configuration and mesh used in the numerical simulations of the bistable arch, with annotated arrows indicating the orientation of uniform-strength remanent magnetization at each node. (b) The deformed configuration after the beam is compressed axially to an end-to-end length of 120 mm, during which the beam buckles and deflects laterally. (c) A plot of the lateral deflection δ of the mid-span of the arch in the $\mathbf{e}_2$-direction versus the total end-to-end length L.

31.4.2 **Calibration of the viscoelasticity parameters and the remanent magnetization**

Calibration of the viscoelastic material parameters

The open black circles in Fig. 31.9(b) show the experimentally measured time signal of mid-span deflection, $u_2^{\text{mid-span}}$, of the arch in the $\mathbf{e}_2$-direction when Tan et al. subjected the beam to a small mechanical impulse (but with no applied magnetic field), and damped free oscillations of the arch were observed. Stewart and Anand (2023) used this data to calibrate the material parameters related to viscoelasticity in the theory which are reproduced here in Table 31.2.

The mechanical "impulse" in this experiment was modeled by linearly ramping up the mid-span displacement of the arch to 0.33 mm in the negative $\mathbf{e}_2$-direction over a short period of 70 ms, and then suddenly removing this displacement boundary condition. Upon removal of the imposed displacement, damped free oscillations of the mid-span of the arch are observed.

The free oscillation characteristics of the arch in terms of amplitude, frequency, and damping over time are driven by a combination of the mechanical material properties, namely the equilibrium shear modulus G of the material, its mass density ρ_{R}, and the non-equilibrium shear moduli $G_{\text{neq}}^{(\alpha)}$ with corresponding relaxation times $\tau^{(\alpha)}$. By fixing the values of G and ρ_{R} at the known values in Table 31.2, and by varying the viscoelastic parameters of the viscoelastic Maxwell elements, Stewart and Anand (2023) found that using a *single* Maxwell branch with

- $G_{\text{neq}}^{(1)} = 500\,\text{kPa}$ and $\tau^{(1)} = 0.01\,\text{s}$

yielded an excellent fit between the numerically predicted response and the experimentally observed response of Tan et al., as shown in Fig. 31.9(b).

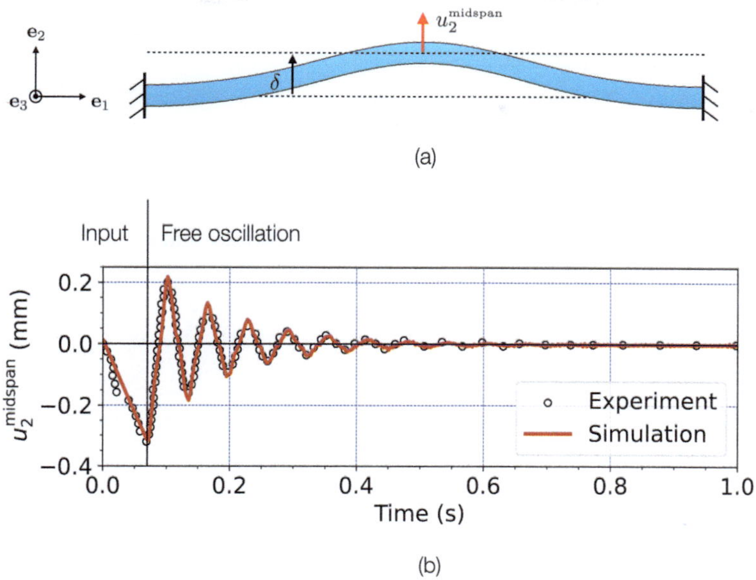

Fig. 31.9 (a) A diagram showing the mid-span displacement signal, $u_2^{\text{mid-span}}$, which is measured relative to the initial lateral deflection δ of the arch. (b) Comparison of the time signals of $u_2^{\text{mid-span}}$ from the numerical simulation and the experiment of Tan et al. (2022), showing excellent agreement in terms of frequency, amplitude, and damping over time.

Calibration of the initial magnetization

In order to simulate the response of the arch to an applied **b**-field one needs to know the magnitude of the remanent magnetization in the arch. Stewart and Anand (2023) determined a value of the remanent magnetization by attempting to numerically match the experimentally observed response of the arch to an applied **b**-field. Tan et al. have reported the results from another experiment in which they linearly ramp up the applied magnetic flux density as shown in Figs. 31.10(a) and (b), until a snap-through event occurs. They report the following "critical" value of the applied **b**-field at which snap-through is first observed in their experiment,

$$|\mathbf{b}_{\text{cr}}^{\text{app}}| = 6.3 \text{ mT}. \tag{31.4.1}$$

Stewart and Anand (2023) adjusted the value of the remanent magnetization $|\mathbf{m}^{\text{rem}}|$ in their simulations to match the experimentally observed value of $|\mathbf{b}_{\text{cr}}^{\text{app}}|$ from Tan et al., and found that

- a remanent magnetization strength of $|\mathbf{m}^{\text{rem}}| = 53.7 \text{ kA/m}$

resulted in an excellent agreement between the value of the critical magnetic flux density $|\mathbf{b}_{\text{cr}}^{\text{app}}| = 6.3\,\text{mT}$ for inducing snap-through in the numerical simulation and that measured by Tan et al., as shown in Fig. 31.10(c). Beyond the good match between the time at which the snap-through initiates, note the ability of the numerical simulation to predict the mid-span displacement values at all major stages of the simulation: (i) prior to the snap-through instability while the field strength is increasing; (ii) during the overshoot and damped oscillations subsequent to the snap-through instability (cf. the inset of Fig. 31.10(c)); and (iii) in the steady-state ramp period after the oscillations from snap-through have died down and the field strength continues to increase.

31.4.3 **Predictive capabilities of the theory**

Having calibrated the material parameters and the remanent magnetization, we now evaluate the predictive capabilities of the theory. To that end, two additional experiments on the same arch reported by Tan et al. (2022) are simulated:

(i) An instantly applied magnetic field above the critical value $|\mathbf{b}_{\mathrm{cr}}^{\mathrm{app}}| = 6.3\,\mathrm{mT}$ at which snap-through occurs.

(ii) An applied magnetic field in the form of a triangle-wave which causes snap-through in a "forward" and a "reverse" direction.

Instantaneously applied magnetic field

As shown in Fig. 31.11(a), a uniform e_2-directed **b**-field with a strength of $|\mathbf{b}^{\mathrm{app}}| = 7\,\mathrm{mT}$, just above $|\mathbf{b}_{\mathrm{cr}}^{\mathrm{app}}| = 6.3\,\mathrm{mT}$, was applied to the arch and the numerically predicted mid-span deflection of the arch during and after the resulting snap-through event is reported in Fig. 31.11(b). The predicted mid-span deflection matches the corresponding experimental data quite well in

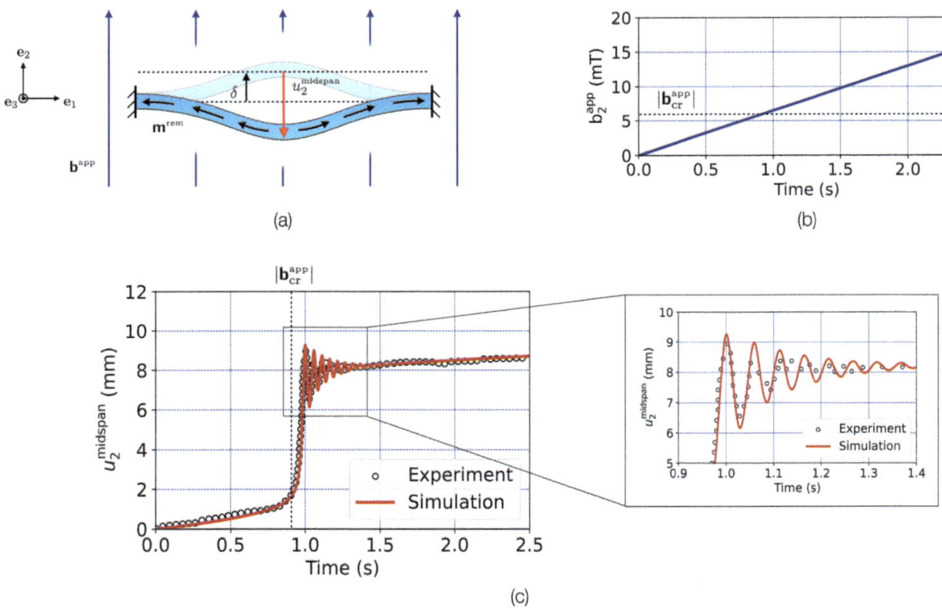

Fig. 31.10 (a) Schematic of the bistable arch studied by Tan et al. (2022) showing an applied uniform **b**-field in the e_2-direction. (b) Prescribed time history of the strength of the magnetic flux density used in the linear ramp experiment, with the annotated dashed line indicating the critical value $|\mathbf{b}_{\mathrm{cr}}^{\mathrm{app}}| = 6.3\,\mathrm{mT}$ for inducing snap-through. (c) Comparison of the numerically simulated mid-span displacement of the arch in response to the applied **b**-field with the corresponding experimental data from Tan et al. The inset plot shows details of the overshoot and subsequent damped oscillations about the second stable configuration.

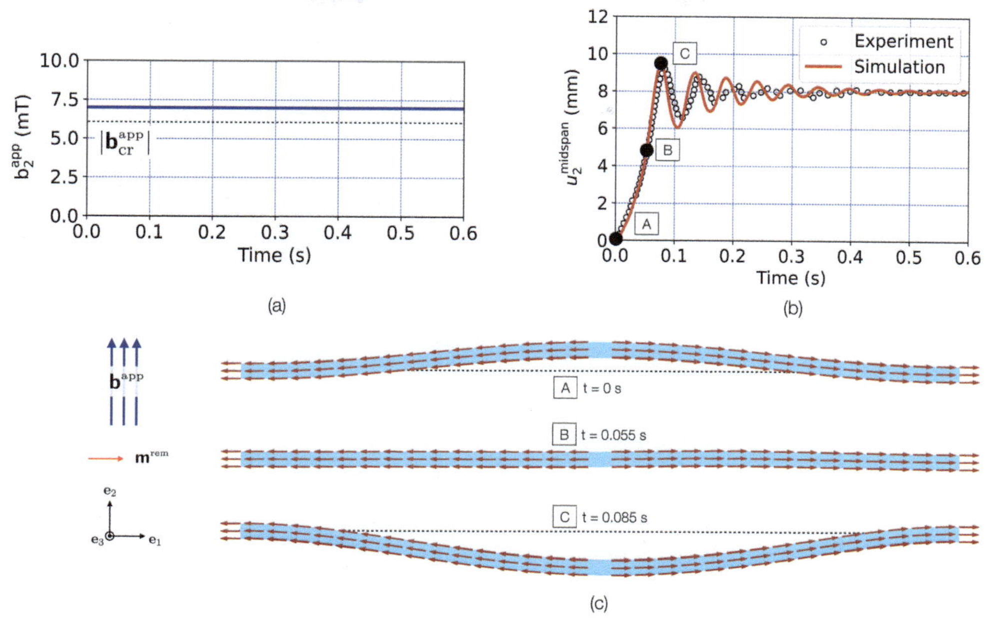

Fig. 31.11 (a) Prescribed time history of the instantaneously applied constant magnetic flux density of 7 mT, slightly above the critical value for snap-through $|\mathbf{b}_{cr}^{app}| = 6.3\,\mathrm{mT}$. (b) Comparison of the numerically predicted mid-span deflection against the experimental measurement from Tan et al. (2022). (c) Snapshots of the simulated deformed configuration at selected times during the snap-through process: A shows the initial arch configuration at time $t = 0$ s; B shows the snap-through event in progress at time $t = 0.055$ s; and C shows the arch has fully reversed due to the applied magnetic field at time $t = 0.085$ s.

terms of response time, magnitude of overshoot, and eventual steady-state deflection. Snapshots of the simulated deformed configuration at selected times during the snap-through process are shown in Fig. 31.11(c).

Triangle-wave magnetic field

Next, we predict the results from an experiment in which two successive snap-through events occur in response to an applied **b**-field whose strength varies according to a triangle-wave signal with an amplitude greater than $|\mathbf{b}_{cr}^{app}| = 6.3\,\mathrm{mT}$, as shown in Fig. 31.12(a).

During the ramp-up phase of the triangle wave the strength of the applied **b**-field exceeds $|\mathbf{b}_{cr}^{app}|$ after about 1 second, and an initial snap-through is observed. Then, at 2.5 seconds the magnetic flux density starts to decrease in magnitude and at 5 seconds changes orientation from the positive $\mathbf{e}_2$-direction to the negative $\mathbf{e}_2$-direction. Eventually, the strength of the **b**-field drops below $-|\mathbf{b}_{cr}^{app}|$ at around 6 seconds, and a second snap-through event occurs which returns the arch to its initial configuration. The numerically predicted mid-span deflection of the arch during the application of the triangle wave **b**-field is shown in Fig. 31.12(b). The prediction matches the corresponding experimental data quite well. Snapshots of the simulated configuration of the arch at selected times, $t = 0$ s, $t = 2.5$ s, and $t = 7.5$ s are shown in Fig. 31.12(c).

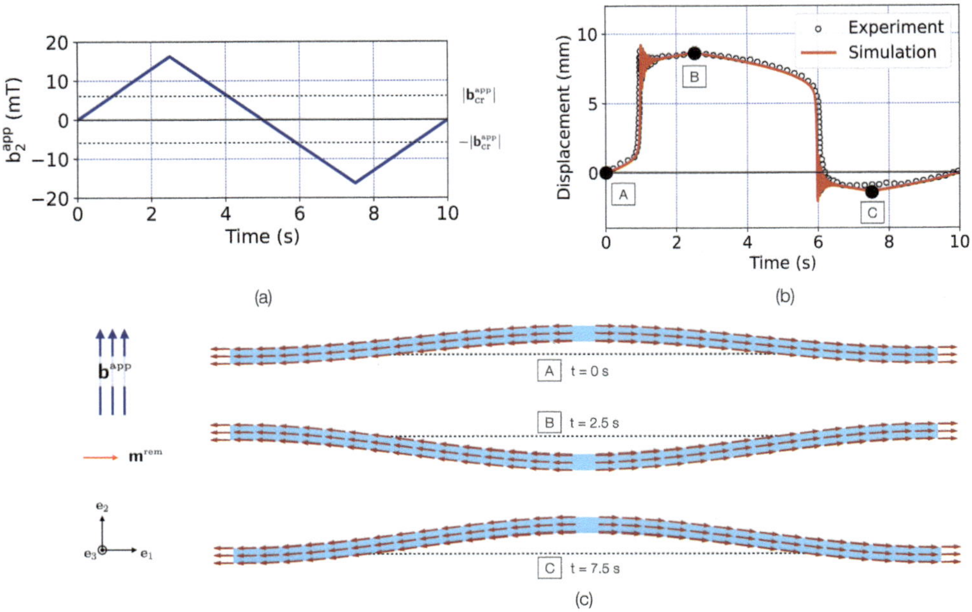

Fig. 31.12 (a) Prescribed time history of the triangle-wave b-field applied to the bistable arch, which exceeds $|\mathbf{b}_{\mathrm{cr}}^{\mathrm{app}}| = 6.3\,\mathrm{mT}$ in the positive direction before reversing direction and passing $-|\mathbf{b}_{\mathrm{cr}}^{\mathrm{app}}| = -6.3\,\mathrm{mT}$ in the negative direction. (b) Comparison of numerically predicted mid-span deflection against that measured in the experiment of Tan et al. (2022). (c) Snapshots of the bistable arch simulation's deformed configuration from selected times during the experiment: A shows the initial arch configuration which is buckled in the positive $\mathbf{e}_2$-direction at time $t = 0\,\mathrm{s}$; B shows the direction of buckling has reversed at time $t = 2.5\,\mathrm{s}$; and in C, reversing the direction of the magnetic field has returned the arch to its initial direction of buckling at time $t = 7.5\,\mathrm{s}$.

31.5 Magnetic eversion of a hemispherical shell

In this section the broader capabilities of the theory and its numerical implementation are demonstrated by simulating the magnetically induced eversion of a hemispherical shell made from a hard-magnetic soft-material. It is also demonstrated that this shape-change is reversible—by reversing the direction of the applied field, the shell reverts to its initial configuration.

The dimensions, finite element mesh, magnetic fields, undeformed and deformed shapes are shown in Fig. 31.13. The hemispherical shell has a uniform-strength remanent magnetization which is oriented in the $\mathbf{e}_2$-direction, while the initial applied **b**-field is oriented nearly opposite to the $\mathbf{e}_2$-direction but with a slight 1 degree misalignment as shown in Fig. 31.13(c). Representative values of the material parameters used by Stewart and Anand (2023) for this simulation are listed in Table 31.3.[4]

In the simulation a 200 mT **b**-field is applied instantly at time $t = 0$ in the direction shown in Fig. 31.14(a) to trigger a snap-through event which everts the hemisphere.[5] The time

[4] This simulation uses a shear modulus value of $G = 150$ kPa, which is significantly lower than that used in modeling the arch of Tan et al. (2022), but still within the range of possible values for PDMS (cf., e.g. Huang and Anand, 2005).

[5] A 200 mT **b**-field is relatively large but still well below the coercivity limit of NdFeB alloys, which is commonly ≈ 1 T (cf., e.g. Fuerst and Brewer, 1993). A similar 200 mT **b**-field was used in the shape-changing hard-magnetic soft-material experiments of Kim et al. (2018); cf. Section 31.3.

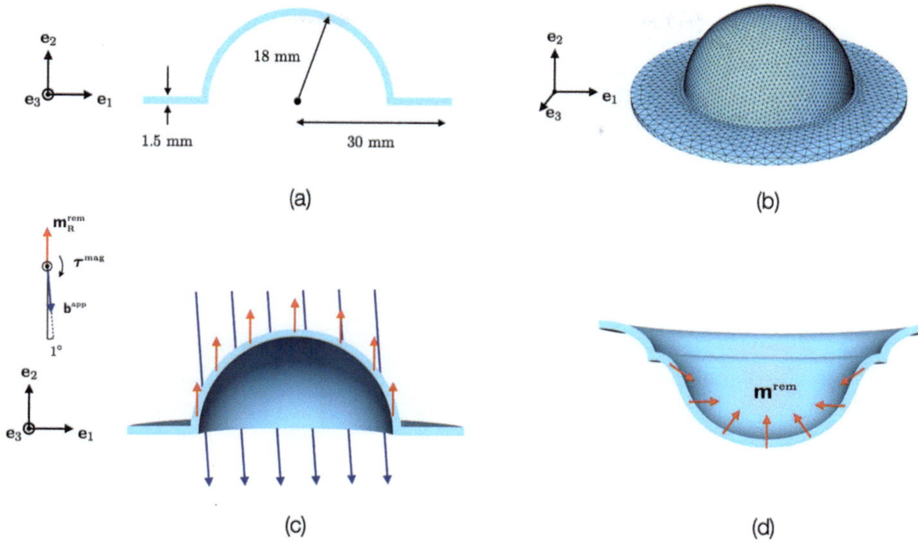

Fig. 31.13 (a) The dimensions of the hemispherical shell. (b) The 3-D mesh of tetrahedral elements used in the simulation. (c) A section view schematic of the hemispherical shell with arrows indicating the remanent magnetization $\mathbf{m}_R^{\text{rem}}$ and the actuation field $\mathbf{b}^{\text{app}}$. (d) The stable everted shape of the hemispherical shell with arrows indicating $\mathbf{m}^{\text{rem}}$ after eversion.

Table 31.3 Material parameters used for modeling the magnetic eversion of a hemispherical shell.

G	K/G	ρ_R	$G_{\text{neq}}^{(1)}$	$\tau^{(1)}$	$\lvert\mathbf{m}^{\text{rem}}\rvert$
kPa	–	kg/m^3	kPa	s	kA/m
150	10^3	2000	100	0.01	80

history of the prescribed strength of the actuation field $\mathbf{b}^{\text{app}}$ and the corresponding vertical displacement of the apex point $\boxed{A}$ of the hemisphere—marked in Fig. 31.14(a)—are shown in Figs. 31.14(b) and (c), respectively. After eversion, at $t = 0.1$ s the field is shut off to demonstrate that the hemisphere is stable in its everted configuration without an applied field. Then in order to revert the hemisphere back to its initial configuration, at $t = 0.2$ s an 80 mT $\mathbf{b}$-field is applied in the opposite direction to the direction shown in Fig. 31.14(a). Finally at $t = 0.3$ s the field is again shut off to demonstrate that the hemisphere eventually stabilizes in its reverted configuration.

We now describe the complex and dynamic snap-through eversion and reversion events in some detail. At time $t = 0$, the magnetic body torque induced in the hemispherical shell by the 200 mT $\mathbf{b}$-field is directed in the negative $\mathbf{e}_3$-direction, so that the shell initially rotates about the negative $\mathbf{e}_3$-direction as shown in the snapshots in Fig. 31.15 for $t = 0$ ms, 30 ms, and 55 ms. Then at ≈ 67.5 ms the far corner of the hemisphere buckles and inverts locally, which kicks off a cascading global snap-through process that takes $\approx$10 ms and is shown in Fig. 31.15 for $t = 67.5$ ms, 72 ms, and 75 ms. After this rapid snap-through process, the hemisphere is in its everted state and exhibits some damped oscillations about the second stable configuration. The oscillations have mostly dissipated by $t = 200$ ms, which is shown in the final snapshot of Fig. 31.15.

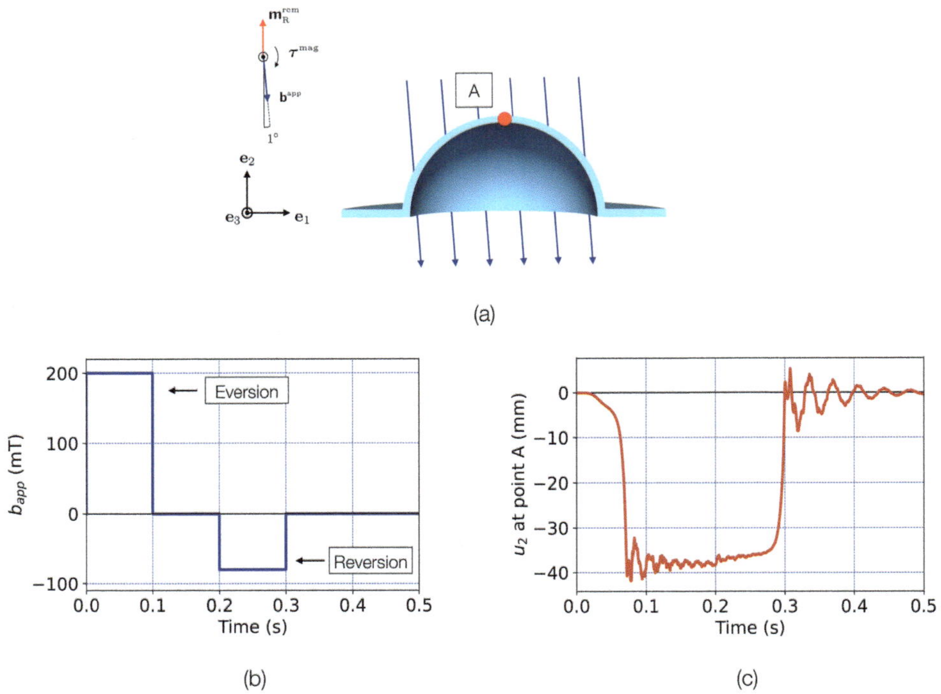

Fig. 31.14 (a) A section view schematic of the hemispherical shell with blue arrows indicating the actuation field $\mathbf{b}^{app}$ and a red dot indicating the material point A whose displacement is tracked in part (c) of this figure. (b) Prescribed time history of the strength of the magnetic flux density in the orientation of $\mathbf{b}^{app}$ shown during the eversion and reversion of the hemispherical shell. (c) Time history of the displacement of point A in the $\mathbf{e}_2$-direction during eversion and reversion.

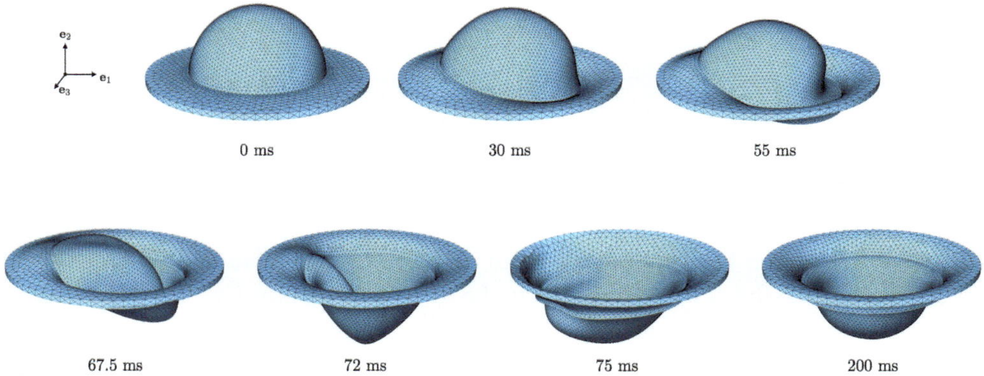

Fig. 31.15 Sequential snapshots of the deformed geometry of the hemispherical shell before, during, and after its magnetically induced eversion. After application of the 200 mT magnetic flux density at time $t = 0$, at times $t = 30$ ms and $t = 55$ ms the hemispherical shell rotates about the negative $\mathbf{e}_3$-direction. Around $t = 67.5$ ms the hemisphere suddenly buckles, starting a dramatic and dynamic snap-through process which is shown at $t = 72$ ms and $t = 75$ ms. After time $t = 200$ ms, inertial oscillations from the snap-through have largely dissipated and the hemisphere remains in its second stable configuration.

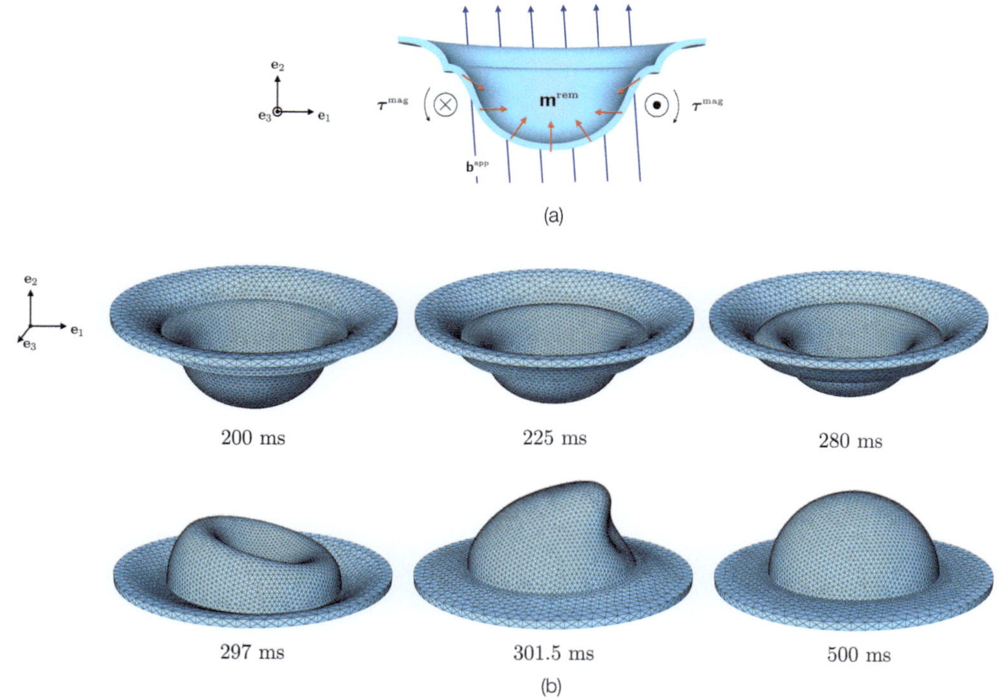

Fig. 31.16 (a) Schematic of the 80 mT magnetic flux density applied at $t = 200\,\text{ms}$ to revert the shell back to its original configuration. (b) Sequential snapshots of the deformed geometry of the shell before, during, and after its magnetically induced reversion. After application of the reversed 80 mT magnetic flux density at time $t = 200\,\text{ms}$, at times $t = 225$ ms and $t = 280$ ms the inner ring of the hemispherical shell bulges upwards in the $\mathbf{e}_2$-direction. Around $t = 297.5$ ms the hemisphere suddenly buckles, starting a dramatic and dynamic snap-through process which is almost complete by $t = 301.5$ ms. By time $t = 500$ ms the inertial oscillations have dissipated and the hemisphere is back to its initial stable configuration.

In order to revert the shell back to its initial configuration, at $t = 200$ ms the direction of the applied **b**-field is reversed, as shown in Fig. 31.16(a). The distribution of remanent magnetization in the everted shape of the shell is already quite misaligned with this reversed **b**-field, so only an 80 mT field is used to trigger the reversion. Sequential snapshots from the simulation of the reversion process are shown in Fig. 31.16(b). After application of the reversed magnetic flux density, at times $t = 225$ ms and $t = 280$ ms the inner ring of the hemispherical shell bulges upwards in the $\mathbf{e}_2$-direction. At $t \approx 297$ ms the hemisphere suddenly buckles, starting a dramatic and dynamic snap-through process which is almost complete by $t \approx 301.5$ ms, and then followed by inertial oscillations of the shell. By time $t = 500$ ms the inertial oscillations have mostly dissipated, and the hemisphere is back in its initial stable configuration.

Part XI

Magneto-elasticity of soft-magnetic soft-elastomers

32 Magneto-viscoelasticity of soft-magnetic soft-elastomers

32.1 Introduction

In Chapter 29 we considered "hard"-magnetorheological elastomers (h-MREs) which are particulate composites made of a non-magnetic elastomeric matrix dispersed with micron-sized magnetic particles of a "hard-magnetic" material, such as NdFeB. In this chapter we turn our attention to modeling the magnetostriction response of "soft"-magnetorheological elastomers (s-MREs) in which the particles are made from a "soft-magnetic" material such as carbonyl iron powders. Both major classes of MREs display actuation behavior under an applied external magnetic field; however, the underlying mechanism of the magnetic-actuation behavior for s-MREs is quite different from the magnetic-actuation mechanism for h-MREs, and will be discussed in detail in Chapter 34.

The continuum mechanical theory for s-MREs that we consider in this chapter is rooted in the foundational works on magnetoelasticity by Brown (1966), Pao (1978), Kovetz (2000), and Dorfmann and Ogden (2003; 2004; 2014), amongst others. In recent years several authors have considered specializations of the general theory to model the magneto-mechanical response of s-MREs (cf., e.g., Kankanala and Triantafyllidis, 2004; Danas et al., 2012; Lefèvre et al., 2017; Garcia-Gonzalez and Hossain, 2021; Rambausek et al., 2022; Danas, 2024). In this chapter, following Stewart and Anand (2025), we develop a coupled magneto-mechanical theory for s-MREs which also accounts for (i) the effects of viscoelasticity; (ii) the effects of the magnetic field in the air-space surrounding s-MRE specimens; and (iii) also inertial effects, because inertia plays a vital role in applications of s-MREs as actuators.

32.2 Theory

32.2.1 Basic kinematics

We consider a macroscopically homogeneous body $\mathcal{B}$ identified with the region of space it occupies in a fixed reference configuration, and denote by $\mathbf{X}$ an arbitrary material point of $\mathcal{B}$. A motion of $\mathcal{B}$ is a smooth one-to-one mapping

$$\mathbf{x} = \boldsymbol{\chi}(\mathbf{X}, t), \tag{32.2.1}$$

with displacement

$$\mathbf{u}(\mathbf{X}, t) = \boldsymbol{\chi}(\mathbf{X}, t) - \mathbf{X}. \tag{32.2.2}$$

Introduction to Coupled Theories in Solid Mechanics. Lallit Anand, Eric M. Stewart, Shawn A. Chester, Oxford University Press.
 DOI: 10.1093/oso/9780198986218.003.0033

Throughout we denote by $\mathcal{P}$ an arbitrary part of the reference body $\mathcal{B}$ with $\mathbf{n}_{\mathrm{R}}$ the outward unit normal on the boundary $\partial\mathcal{P}$ of $\mathcal{P}$. Also, we denote by $\mathcal{P}_t = \chi(\mathcal{P}, t)$ the image of $\mathcal{P}$ in the deformed body $\mathcal{B}_t = \chi(\mathcal{B}, t)$, with $\mathbf{n}$ the outward unit normal on the boundary $\partial\mathcal{P}_t$ of $\mathcal{P}_t$.

From the motion (32.2.1), the deformation gradient, velocity, acceleration, and velocity gradient are given by

$$\mathbf{F} = \nabla\chi = \mathbf{1} + \nabla\mathbf{u}, \quad \dot{\mathbf{u}} = \frac{\partial\chi}{\partial t}, \quad \ddot{\mathbf{u}} = \frac{\partial^2\chi}{\partial t^2}, \quad \mathbf{L} = \mathrm{grad}\,\dot{\mathbf{u}} = \dot{\mathbf{F}}\mathbf{F}^{-1}. \tag{32.2.3}$$

We assume that

$$J = \det\mathbf{F} > 0, \tag{32.2.4}$$

so that $\mathbf{F}$ is invertible. The spatial descriptions of the velocity $\dot{\mathbf{u}}$ and the acceleration $\ddot{\mathbf{u}}$ are denoted by $\mathbf{v}$ and $\dot{\mathbf{v}}$, respectively.

The symmetric and positive definite right and left Cauchy–Green tensors are defined by,

$$\mathbf{C} = \mathbf{F}^{\top}\mathbf{F}, \qquad \mathbf{B} = \mathbf{F}\mathbf{F}^{\top}, \tag{32.2.5}$$

respectively.

For later use we note that the deformation gradient $\mathbf{F}$ may be decomposed into volumetric and isochoric factors $\mathbf{F}^{\mathrm{vol}}$ and $\bar{\mathbf{F}}$:

$$\mathbf{F} = \mathbf{F}^{\mathrm{vol}}\bar{\mathbf{F}}, \tag{32.2.6}$$

where

$$\mathbf{F}^{\mathrm{vol}} = J^{1/3}\mathbf{1} \quad \text{and} \quad \bar{\mathbf{F}} = J^{-1/3}\mathbf{F}. \tag{32.2.7}$$

This leads to decompositions of the Cauchy–Green tensors $\mathbf{C} = \mathbf{F}^{\top}\mathbf{F}$ and $\mathbf{B} = \mathbf{F}\mathbf{F}^{\top}$ of the form,

$$\begin{aligned} \mathbf{C} &= \mathbf{C}^{\mathrm{vol}}\bar{\mathbf{C}}, \qquad \mathbf{C}^{\mathrm{vol}} = J^{2/3}\mathbf{1}, \qquad \bar{\mathbf{C}} = \bar{\mathbf{F}}^{\top}\bar{\mathbf{F}}, \\ \mathbf{B} &= \mathbf{B}^{\mathrm{vol}}\bar{\mathbf{B}}, \qquad \mathbf{B}^{\mathrm{vol}} = J^{2/3}\mathbf{1}, \qquad \bar{\mathbf{B}} = \bar{\mathbf{F}}\bar{\mathbf{F}}^{\top}. \end{aligned} \tag{32.2.8}$$

We also introduce the invariants of $\bar{\mathbf{C}}$,

$$\bar{I}_1 \stackrel{\text{def}}{=} \mathrm{tr}\,\bar{\mathbf{C}}, \qquad \bar{I}_2 \stackrel{\text{def}}{=} \tfrac{1}{2}\left((\mathrm{tr}\,\bar{\mathbf{C}})^2 - \mathrm{tr}(\bar{\mathbf{C}}^2)\right), \qquad \text{and} \qquad I_3 = J. \tag{32.2.9}$$

The invariants of $\mathbf{B}$ are similarly defined, and these are equal in value to the invariants of $\mathbf{C}$.

Kinematics of viscoelasticity based on the decomposition $\mathbf{F} = \mathbf{F}^{e(\alpha)}\mathbf{F}^{v(\alpha)}$

We base our treatment of viscoelasticity on a multi-branch multiplicative decomposition of the deformation gradient of the form (Sidoroff, 1974),

$$\mathbf{F} = \mathbf{F}^{e(\alpha)}\mathbf{F}^{v(\alpha)} \qquad \alpha = 0, 1, \ldots, M, \tag{32.2.10}$$

which allows for $(M+1)$ viscoelastic relaxation mechanisms. We refer to $\mathbf{F}^{v(\alpha)}$ and $\mathbf{F}^{e(\alpha)}$ as *viscous, and elastic distortions*, respectively. Using (32.2.4) and (32.2.10) the volumetric Jacobian J can be decomposed as

$$J = J^{e(\alpha)} J^{v(\alpha)}. \tag{32.2.11}$$

We assume that

$$J^{e(\alpha)} \overset{\text{def}}{=} \det \mathbf{F}^{e(\alpha)} > 0, \quad \text{and} \quad J^{v(\alpha)} \overset{\text{def}}{=} \det \mathbf{F}^{v(\alpha)} > 0, \tag{32.2.12}$$

so that $\mathbf{F}^{e(\alpha)}$ and $\mathbf{F}^{v(\alpha)}$ are invertible.

The elastic distortions $\mathbf{F}^{e(\alpha)}$ may be decomposed into volumetric and isochoric factors $\mathbf{F}^{e(\alpha)\text{vol}}$ and $\bar{\mathbf{F}}^{e(\alpha)}$:

$$\mathbf{F}^{e(\alpha)} = \mathbf{F}^{e(\alpha)\text{vol}}\bar{\mathbf{F}}^{e(\alpha)}, \tag{32.2.13}$$

where

$$\mathbf{F}^{e(\alpha)\text{vol}} = (J^{e(\alpha)})^{1/3}\mathbf{1} \qquad \text{and} \qquad \bar{\mathbf{F}}^{e(\alpha)} = (J^{e(\alpha)})^{-1/3}\mathbf{F}^{e(\alpha)}. \tag{32.2.14}$$

The right and left elastic Cauchy–Green tensors are given by

$$\mathbf{C}^{e(\alpha)} = \mathbf{F}^{e(\alpha)\top}\mathbf{F}^{e(\alpha)}, \qquad \mathbf{B}^{e(\alpha)} = \mathbf{F}^{e(\alpha)}\mathbf{F}^{e(\alpha)\top}, \tag{32.2.15}$$

and these may be decomposed into volumetric and isochoric forms as,

$$\begin{aligned} &\mathbf{C}^{e(\alpha)} = \mathbf{C}^{e(\alpha)\text{vol}}\bar{\mathbf{C}}^{e(\alpha)}, \qquad \mathbf{C}^{e(\alpha)\text{vol}} = (J^{e(\alpha)})^{2/3}\mathbf{1}, \qquad \bar{\mathbf{C}}^{e(\alpha)} = \bar{\mathbf{F}}^{e(\alpha)\top}\bar{\mathbf{F}}^{e(\alpha)} \\ &\mathbf{B}^{e(\alpha)} = \mathbf{B}^{e(\alpha)\text{vol}}\bar{\mathbf{B}}^{e(\alpha)}, \qquad \mathbf{B}^{e(\alpha)\text{vol}} = (J^{e(\alpha)})^{2/3}\mathbf{1}, \qquad \bar{\mathbf{B}}^{e(\alpha)} = \bar{\mathbf{F}}^{e(\alpha)}\bar{\mathbf{F}}^{e(\alpha)\top}. \end{aligned} \tag{32.2.16}$$

We also introduce the invariants,

$$\bar{I}_1^e \overset{\text{def}}{=} \operatorname{tr}\bar{\mathbf{C}}^{e(\alpha)}, \qquad \bar{I}_2^e \overset{\text{def}}{=} \tfrac{1}{2}\left((\operatorname{tr}\bar{\mathbf{C}}^{e(\alpha)})^2 - \operatorname{tr}((\bar{\mathbf{C}}^{e(\alpha)})^2)\right), \qquad \text{and} \qquad I_3^e = J^{e(\alpha)}. \tag{32.2.17}$$

Next, by $(32.2.3)_4$ and (32.2.10), the velocity gradient $\mathbf{L} = \dot{\mathbf{F}}\mathbf{F}^{-1}$ may be decomposed as

$$\mathbf{L} = \mathbf{L}^{e(\alpha)} + \mathbf{F}^{e(\alpha)}\mathbf{L}^{v(\alpha)}\mathbf{F}^{e(\alpha)-1}, \tag{32.2.18}$$

with

$$\mathbf{L}^{e(\alpha)} = \dot{\mathbf{F}}^{e(\alpha)}\mathbf{F}^{e(\alpha)-1}, \qquad \mathbf{L}^{v(\alpha)} = \dot{\mathbf{F}}^{v(\alpha)}\mathbf{F}^{v(\alpha)-1}. \tag{32.2.19}$$

As is standard, we define the total, elastic, and viscous stretching and spin tensors through

$$\begin{aligned} \mathbf{D} &= \operatorname{sym}\mathbf{L}, & \mathbf{W} &= \operatorname{skw}\mathbf{L}, \\ \mathbf{D}^{e(\alpha)} &= \operatorname{sym}\mathbf{L}^{e(\alpha)}, & \mathbf{W}^{e(\alpha)} &= \operatorname{skw}\mathbf{L}^{e(\alpha)}, \\ \mathbf{D}^{v(\alpha)} &= \operatorname{sym}\mathbf{L}^{v(\alpha)}, & \mathbf{W}^{v(\alpha)} &= \operatorname{skw}\mathbf{L}^{v(\alpha)}, \end{aligned} \tag{32.2.20}$$

so that $\mathbf{L} = \mathbf{D} + \mathbf{W}$, $\mathbf{L}^{e(\alpha)} = \mathbf{D}^{e(\alpha)} + \mathbf{W}^{e(\alpha)}$, and $\mathbf{L}^{v(\alpha)} = \mathbf{D}^{v(\alpha)} + \mathbf{W}^{v(\alpha)}$.

We make two basic kinematical assumptions concerning viscous flow:

(i) First, we make the standard assumption that viscous flow is incompressible, so that

$$J^{v(\alpha)} = \det \mathbf{F}^{v(\alpha)} = 1 \qquad \text{and} \qquad \operatorname{tr}\mathbf{L}^{v(\alpha)} = \operatorname{tr}\mathbf{D}^{v(\alpha)} = 0. \tag{32.2.21}$$

Hence, using (32.2.11), we note that

$$J^{e(\alpha)} = J \qquad \text{for all } \alpha. \tag{32.2.22}$$

(ii) Next, from the outset we constrain the theory by limiting our discussion to circumstances under which the material may be idealized as *isotropic*. For isotropic elastic-viscous theories utilizing the multiplicative decomposition of $\mathbf{F}$, it is widely assumed that the viscous flow is *irrotational* (cf., e.g., Gurtin et al., 2010, Section 97), in the sense that

$$\mathbf{W}^{v(\alpha)} = \mathbf{0}. \tag{32.2.23}$$

Then, $\mathbf{L}^{v(\alpha)} \equiv \mathbf{D}^{v(\alpha)}$, using which the evolution equation for $\mathbf{F}^{v(\alpha)}$ is

$$\dot{\mathbf{F}}^{v(\alpha)} = \mathbf{D}^{v(\alpha)}\mathbf{F}^{v(\alpha)}. \tag{32.2.24}$$

32.2.2 Balance of forces and moments

Standard considerations of balance of forces, when expressed spatially, give the following results:

(a) There exists a symmetric stress tensor $\mathbf{T}(\mathbf{x}, t)$, called the Cauchy stress, such that the surface traction on an oriented element of the surface $\partial\mathcal{P}_t$ of $\mathcal{P}_t$, is given by

$$\mathbf{t}(\mathbf{n}) = \mathbf{T}\mathbf{n}. \tag{32.2.25}$$

(b) The stress $\mathbf{T}$ satisfies the force balance

$$\mathrm{div}\mathbf{T} + \mathbf{f} = \mathbf{0}, \tag{32.2.26}$$

with

$$\mathbf{f} = \mathbf{f}_0 - \rho\dot{\mathbf{v}}, \tag{32.2.27}$$

where $\mathbf{f}_0$ represents the conventional body force per unit volume of the deformed body, and $(-\rho\dot{\mathbf{v}})$ represents the inertial body force, with ρ the mass density of the deformed body and $\dot{\mathbf{v}}$ the acceleration.[1]

The stress tensor $\mathbf{T}$ is the *total* Cauchy stress. This stress accounts for the stress caused by pure mechanical deformation (if the magnetic field were to be absent) and also the stress induced by a magnetic field.

In the multibranch theory for viscoelasticity based on $\mathbf{F} = \mathbf{F}^{e(\alpha)}\mathbf{F}^{v(\alpha)}$, the Cauchy stress admits the additive decomposition

$$\mathbf{T} = \sum_{\alpha} \mathbf{T}^{(\alpha)}, \tag{32.2.28}$$

where $\mathbf{T}^{(\alpha)}$ is the Cauchy stress in each branch (cf., e.g., Mao et al., 2017). Since $\mathbf{T}$ is symmetric, we assume that each $\mathbf{T}^{(\alpha)}$ is *symmetric.*

Next, as is standard, the stress measure

$$\mathbf{T}_{\mathrm{R}} = J\,\mathbf{T}\mathbf{F}^{-\top} \tag{32.2.29}$$

represents the Piola stress, in terms of which:

(a) The surface traction on an oriented element of the surface $\partial\mathcal{P}$ of $\mathcal{P}$, is given by the traction relation

$$\mathbf{t}_{\mathrm{R}}(\mathbf{n}_{\mathrm{R}}) = \mathbf{T}_{\mathrm{R}}\mathbf{n}_{\mathrm{R}}. \tag{32.2.30}$$

(b) $\mathbf{T}_{\mathrm{R}}$ satisfies the force balance

$$\mathrm{Div}\,\mathbf{T}_{\mathrm{R}} + \mathbf{f}_{\mathrm{R}} = \mathbf{0}, \tag{32.2.31}$$

with

$$\mathbf{f}_{\mathrm{R}} = \mathbf{f}_{0\mathrm{R}} - \rho_{\mathrm{R}}\ddot{\mathbf{u}}, \tag{32.2.32}$$

where $\mathbf{f}_{0\mathrm{R}}$ represents the conventional body force per unit volume of the reference body, and $(-\rho_{\mathrm{R}}\ddot{\mathbf{u}})$ represents the inertial body force, with ρ_{R} the mass density of the referential body and $\ddot{\mathbf{u}}$ the acceleration.

[1] In previous chapters we have denoted the referential conventional and generalized body forces expressed referentially by $\mathbf{b}_{0\mathrm{R}}$ and $\mathbf{b}_{\mathrm{R}}$, respectively, but here—in order not to cause confusion with notation for the magnetic flux density—we use the notations $\mathbf{f}_{0\mathrm{R}}$ and $\mathbf{f}_{\mathrm{R}}$ for the body forces.

32.2.3 Magnetostatics. Ampere's and Gauss's laws

In a magnetizable material there is a vector field in $\mathcal{B}_t$ denoted by $\mathbf{h}(\mathbf{x}, t)$ called the *magnetic field*, and another field $\mathbf{b}(\mathbf{x}, t)$ in $\mathcal{B}_t$ called the *magnetic flux density*. The field $\mathbf{h}$ is related to the field $\mathbf{b}$ by

$$\mathbf{h} = \mu_0^{-1}\mathbf{b} - \mathbf{m}, \tag{32.2.33}$$

where $\mathbf{m}(\mathbf{x}, t)$ is a magnetization field in the material, and μ_0 is the magnetic permeability of free space.

The governing equations of magnetostatics in $\mathcal{B}_t$, *in the absence of steady currents*, are Ampere's law

$$\operatorname{curl}\mathbf{h} = \mathbf{0} \quad \text{in } \mathcal{B}_t, \tag{32.2.34}$$

and Gauss's law

$$\operatorname{div}\mathbf{b} = 0 \qquad \text{in } \mathcal{B}_t. \tag{32.2.35}$$

The referential counterparts of $\mathbf{h}$ and $\mathbf{b}$ are

$$\mathbf{h}_{\mathrm{R}} = \mathbf{F}^{\top}\mathbf{h} \qquad \text{and} \qquad \mathbf{b}_{\mathrm{R}} = J\mathbf{F}^{-1}\mathbf{b}. \tag{32.2.36}$$

Also, when expressed referentially, Ampere's law and Gauss's law read as

$$\operatorname{Curl}\mathbf{h}_{\mathrm{R}} = \mathbf{0}, \tag{32.2.37}$$

and

$$\operatorname{Div}\mathbf{b}_{\mathrm{R}} = 0. \tag{32.2.38}$$

Satisfaction of Ampere's law (32.2.37) occurs automatically if we allow the field $\mathbf{h}_{\mathrm{R}}(\mathbf{X}, t)$ to be represented as the gradient of a magnetostatic potential $\phi(\mathbf{X}, t)$,[2]

$$\mathbf{h}_{\mathrm{R}}(\mathbf{X}, t) \overset{\text{def}}{=} -\nabla\phi(\mathbf{X}, t). \tag{32.2.39}$$

While the variable $\phi(\mathbf{X}, t)$ does not have physical significance, it is widely used as a primary magnetic variable in finite element formulations of magnetostatics in the absence of free currents.

[2] Recall that for any scalar field ϕ, $\operatorname{Curl}(\nabla\phi) = \mathbf{0}$.

Shortly we will prescribe a constitutive equation for $\mathbf{b}_R$ in terms of $\mathbf{h}_R$ and the elastic Cauchy–Green tensors $\vec{\mathbf{C}}^{e(\alpha)}$,

$$\mathbf{b}_R = \hat{\mathbf{b}}_R(\vec{\mathbf{C}}^{e(\alpha)}, \mathbf{h}_R), \qquad \text{with} \qquad \vec{\mathbf{C}}^{e(\alpha)} = \{\mathbf{C}^{e(\alpha)} | \alpha = 0, \ldots, M\}. \tag{32.2.40}$$

Using (32.2.39) this may be written as

$$\mathbf{b}_R = \hat{\mathbf{b}}_R(\vec{\mathbf{C}}^{e(\alpha)}, \phi). \tag{32.2.41}$$

Then Gauss's law (32.2.38) requires that

$$\operatorname{Div} \hat{\mathbf{b}}_R(\vec{\mathbf{C}}^{e(\alpha)}, \phi) = 0 \tag{32.2.42}$$

be satisfied. This is the partial differential equation that one has to solve to determine ϕ, and through that the field $\mathbf{h}_R$.

32.2.4 **Free-energy imbalance**

Under isothermal conditions the two laws of thermodynamics reduce to the statement that *the rate of increase in free energy of any part $\mathcal{P}$ is less than or equal to the power expended on $\mathcal{P}$.* Precisely, letting ψ_R denote the free energy per unit reference volume, this requirement takes the form of a free-energy imbalance

$$\overline{\int_{\mathcal{P}} \psi_R \, dv_R}^{\cdot} \leq \mathcal{W}(\mathcal{P}) + \mathcal{E}(\mathcal{P}), \tag{32.2.43}$$

where:

(i) The mechanical power $\mathcal{W}(\mathcal{P})$ exerted on $\mathcal{P}$ by the boundary tractions $\mathbf{T}_R\mathbf{n}_R$ and body force $\mathbf{f}_R$ over the velocity $\dot{\mathbf{u}}$ is given by,

$$\mathcal{W}(\mathcal{P}) = \int_{\partial\mathcal{P}} (\mathbf{T}_R\mathbf{n}_R) \cdot \dot{\mathbf{u}} \, da_R + \int_{\mathcal{P}} \mathbf{f}_R \cdot \dot{\mathbf{u}} \, dv_R. \tag{32.2.44}$$

(ii) In electromagnetic field theory the power that is expended on $\mathcal{P}$ is characterized by an electromagnetic energy flux $\mathcal{E}(\mathcal{P})$ which in the *magnetostatic limit* is given by (cf. Appendix A, eq. (A.4.28))

$$\mathcal{E}(\mathcal{P}) = -\int_{\partial\mathcal{P}} \phi \, \dot{\mathbf{b}}_R \cdot \mathbf{n}_R \, da_R. \tag{32.2.45}$$

The genesis of this crucial term comes from the magnetostatic limit of the expression for the *flux of electromagnetic energy*, as described by the *Poynting's vector* (see also, e.g., Jackson (1999, Section 6.7) or Kovetz (2000, Sections 45 and 54)).

Thus, using (32.2.44) and (32.2.45), the free energy imbalance (32.2.43) becomes

$$\overline{\int_{\mathcal{P}} \psi_{\mathrm{R}}\, dv_{\mathrm{R}}}^{\boldsymbol{\cdot}} \leq \int_{\partial\mathcal{P}} (\mathbf{T}_{\mathrm{R}}\mathbf{n}_{\mathrm{R}}) \cdot \dot{\mathbf{u}}\, da_{\mathrm{R}} + \int_{\mathcal{P}} \mathbf{f}_{\mathrm{R}} \cdot \dot{\mathbf{u}}\, dv_{\mathrm{R}} - \int_{\partial\mathcal{P}} \phi\, \dot{\mathbf{b}}_{\mathrm{R}} \cdot \mathbf{n}_{\mathrm{R}}\, da_{\mathrm{R}}. \tag{32.2.46}$$

Using the divergence theorem on the terms involving the integral over $\partial\mathcal{P}$ in (32.2.46) we obtain,

$$\begin{aligned}
\int_{\partial\mathcal{P}} (\mathbf{T}_{\mathrm{R}}\mathbf{n}_{\mathrm{R}}) \cdot \dot{\mathbf{u}}\, da_{\mathrm{R}} &= \int_{\mathcal{P}} (\operatorname{Div} \mathbf{T}_{\mathrm{R}}) \cdot \dot{\mathbf{u}}\, dv + \int_{\mathcal{P}} \mathbf{T}_{\mathrm{R}} \colon \dot{\mathbf{F}}\, dv_{\mathrm{R}}, \\
-\int_{\partial\mathcal{P}} (\phi\, \dot{\mathbf{b}}_{\mathrm{R}}) \cdot \mathbf{n}_{\mathrm{R}}\, da_{\mathrm{R}} &= -\int_{\mathcal{P}} \phi\, (\operatorname{Div} \dot{\mathbf{b}}_{\mathrm{R}}) dv_{\mathrm{R}} - \int_{\mathcal{P}} (\nabla\phi) \cdot \dot{\mathbf{b}}_{\mathrm{R}}\, dv_{\mathrm{R}} \\
&= -\int_{\mathcal{P}} \phi\, (\operatorname{Div} \dot{\mathbf{b}}_{\mathrm{R}}) dv_{\mathrm{R}} + \int_{\mathcal{P}} \mathbf{h}_{\mathrm{R}} \cdot \dot{\mathbf{b}}_{\mathrm{R}}\, dv_{\mathrm{R}}.
\end{aligned} \tag{32.2.47}$$

Substituting (32.2.47) in (32.2.46), and using the linear momentum balance (32.2.26) and the rate form of Gauss's law (32.2.38), that is Div $\dot{\mathbf{b}}_{\mathrm{R}} = 0$, we obtain,

$$\int_{\mathcal{P}} \left(\dot{\psi}_{\mathrm{R}} - \mathbf{T}_{\mathrm{R}} \colon \dot{\mathbf{F}} - \mathbf{h}_{\mathrm{R}} \cdot \dot{\mathbf{b}}_{\mathrm{R}} \right) dv_{\mathrm{R}} \leq 0. \tag{32.2.48}$$

Since (32.2.48) must hold for all parts $\mathcal{P}$, we obtain a local free-energy imbalance in the form

$$\dot{\psi}_{\mathrm{R}} - \mathbf{T}_{\mathrm{R}} \colon \dot{\mathbf{F}} - \mathbf{h}_{\mathrm{R}} \cdot \dot{\mathbf{b}}_{\mathrm{R}} \leq 0. \tag{32.2.49}$$

The term $\mathbf{T}_{\mathrm{R}} \colon \dot{\mathbf{F}}$ in (32.2.49) represents the stress power per unit reference volume. Using (32.2.10), (32.2.15), (32.2.28), (32.2.29), (32.2.22), and (32.2.24) this stress power may be written as the sum of an elastic stress power and a viscous stress power,

$$\begin{aligned}
\mathbf{T}_{\mathrm{R}} \colon \dot{\mathbf{F}} = &\underbrace{\frac{1}{2} \sum_{\alpha} \left(J^{e(\alpha)} \mathbf{F}^{e(\alpha)-1} \mathbf{T}^{(\alpha)} \mathbf{F}^{e(\alpha)-\top} \right) \colon \dot{\mathbf{C}}^{e(\alpha)}}_{\text{elastic stress power}} \\
&+ \underbrace{\sum_{\alpha} \mathbf{C}^{e(\alpha)} \left(J^{e(\alpha)} \mathbf{F}^{e(\alpha)-1} \mathbf{T}^{(\alpha)} \mathbf{F}^{e(\alpha)-\top} \right) \colon \mathbf{D}^{v(\alpha)}}_{\text{viscous stress power}}.
\end{aligned} \tag{32.2.50}$$

In view of (32.2.50) we introduce two sets of new stress measures:

- The elastic second Piola stresses

$$\mathbf{T}^{e(\alpha)} \stackrel{\text{def}}{=} J^{e(\alpha)} \mathbf{F}^{e(\alpha)-1} \mathbf{T}^{(\alpha)} \mathbf{F}^{e(\alpha)-\top}, \tag{32.2.51}$$

  which are *symmetric* on account of the symmetry of the stresses $\mathbf{T}^{(\alpha)}$.

- The Mandel stresses

$$\mathbf{M}^{e(\alpha)} \stackrel{\text{def}}{=} \mathbf{C}^{e(\alpha)}\mathbf{T}^{e(\alpha)} = J^{e(\alpha)}\mathbf{F}^{e(\alpha)\top}\mathbf{T}^{(\alpha)}\mathbf{F}^{e(\alpha)-\top}, \tag{32.2.52}$$

  which in general are *not symmetric.*

Using these new stress measures the stress power (32.2.50) may be written as

$$\mathbf{T}_{\text{R}}:\dot{\mathbf{F}} = \underbrace{\frac{1}{2}\sum_{\alpha}\mathbf{T}^{e(\alpha)}:\dot{\mathbf{C}}^{e(\alpha)}}_{\text{elastic stress power}} + \underbrace{\sum_{\alpha}\mathbf{M}_0^{e(\alpha)}:\mathbf{D}^{v(\alpha)}}_{\text{viscous stress power}}, \tag{32.2.53}$$

and using this the free-energy imbalance (32.2.49) may be expressed as

$$\dot{\psi}_{\text{R}} - \frac{1}{2}\sum_{\alpha}\mathbf{T}^{e(\alpha)}:\dot{\mathbf{C}}^{e(\alpha)} - \mathbf{h}_{\text{R}}\cdot\dot{\mathbf{b}}_{\text{R}} - \sum_{\alpha}\mathbf{M}_0^{e(\alpha)}:\mathbf{D}^{v(\alpha)} \le 0. \tag{32.2.54}$$

REMARK

For brevity we have not discussed the transformation properties under a change in frame of the various fields appearing in our theory. Here, we simply note that all quantities in the free-energy imbalance (32.2.54) are invariant under a change in frame.

32.2.5 Constitutive theory

With $\boldsymbol{\Lambda}$ denoting the list

$$\boldsymbol{\Lambda} \stackrel{\text{def}}{=} \left(\vec{\mathbf{C}}^{e(\alpha)}, \mathbf{b}_{\text{R}}\right) \qquad \text{where} \qquad \vec{\mathbf{C}}^{e(\alpha)} = \{\mathbf{C}^{e(\alpha)} | \alpha = 0, 1, \ldots, M\}, \tag{32.2.55}$$

we assume constitutive equations for ψ_{R}, $\mathbf{T}^{e(\alpha)}$, and $\mathbf{h}_{\text{R}}$ of the form

$$\psi_{\text{R}} = \sum_{\alpha}\psi_{\text{R}}^{(\alpha)}(\boldsymbol{\Lambda}), \qquad \mathbf{T}^{e(\alpha)} = \hat{\mathbf{T}}^{e\,(\alpha)}(\boldsymbol{\Lambda}), \qquad \mathbf{h}_{\text{R}} = \hat{\mathbf{h}}_{\text{R}}(\boldsymbol{\Lambda}). \tag{32.2.56}$$

From $(32.2.56)_1$ we have,

$$\dot{\psi}_{\text{R}} = \sum_{\alpha}\frac{\partial\psi_{\text{R}}^{(\alpha)}(\boldsymbol{\Lambda})}{\partial\mathbf{C}^{e(\alpha)}}:\dot{\mathbf{C}}^{e(\alpha)} + \frac{\partial\psi_{\text{R}}(\boldsymbol{\Lambda})}{\partial\mathbf{b}_{\text{R}}}\cdot\dot{\mathbf{b}}_{\text{R}}. \tag{32.2.57}$$

Using (32.2.57) and substituting the constitutive equations (32.2.56) into the free-energy imbalance (32.2.54), we find that it may be written as

$$\sum_{\alpha}\left(\frac{\partial\psi_{\text{R}}^{(\alpha)}(\boldsymbol{\Lambda})}{\partial\mathbf{C}^{e(\alpha)}} - \frac{1}{2}\hat{\mathbf{T}}^{e\,(\alpha)}(\boldsymbol{\Lambda})\right):\dot{\mathbf{C}}^{e(\alpha)} + \left(\frac{\partial\psi_{\text{R}}(\boldsymbol{\Lambda})}{\partial\mathbf{b}_{\text{R}}} - \hat{\mathbf{h}}_{\text{R}}(\boldsymbol{\Lambda})\right)\cdot\dot{\mathbf{b}}_{\text{R}} - \sum_{\alpha}\mathbf{M}_0^{e(\alpha)}:\mathbf{D}^{v(\alpha)} \le 0. \tag{32.2.58}$$

We assume constitutively that the free energy delivers the elastic second Piola stresses $\mathbf{T}^{e(\alpha)}$ and the magnetic field $\mathbf{h}_{\mathrm{R}}$ through the state relations,

$$\begin{aligned} \mathbf{T}^{e(\alpha)} &= 2\,\frac{\partial \hat{\psi}_{\mathrm{R}}^{(\alpha)}(\boldsymbol{\Lambda})}{\partial \mathbf{C}^{e(\alpha)}}, \\ \mathbf{h}_{\mathrm{R}} &= \frac{\partial \psi_{\mathrm{R}}(\boldsymbol{\Lambda})}{\partial \mathbf{b}_{\mathrm{R}}}, \end{aligned} \tag{32.2.59}$$

and we are then left with the reduced dissipation inequality

$$\mathcal{D} = \sum_{\alpha} \mathbf{M}_0^{e(\alpha)} : \mathbf{D}^{v(\alpha)} \geq 0, \tag{32.2.60}$$

which represents the dissipation due to viscous flow.

Magnetic field $\mathbf{h}_{\mathrm{R}}$ as an independent variable

In applications it is often preferable to replace constitutive dependence upon the magnetic flux density $\mathbf{b}_{\mathrm{R}}$ by constitutive dependence upon the magnetic field $\mathbf{h}_{\mathrm{R}}$. To effect this change we assume that for fixed $\vec{\mathbf{C}}^{e(\alpha)}$ the relation

$$\mathbf{h}_{\mathrm{R}} = \hat{\mathbf{h}}_{\mathrm{R}}(\vec{\mathbf{C}}^{e(\alpha)}, \mathbf{b}_{\mathrm{R}}) \tag{32.2.61}$$

is *smoothly invertible* in $\mathbf{b}_{\mathrm{R}}$, so that

$$\mathbf{b}_{\mathrm{R}} = \breve{\mathbf{b}}_{\mathrm{R}}(\vec{\mathbf{C}}^{e(\alpha)}, \mathbf{h}_{\mathrm{R}}). \tag{32.2.62}$$

Thus, let

$$\Omega_{\mathrm{R}} \stackrel{\text{def}}{=} \psi_{\mathrm{R}} - \mathbf{h}_{\mathrm{R}} \cdot \mathbf{b}_{\mathrm{R}} \tag{32.2.63}$$

define a partially Legendre-transformed energy function.

REMARK

The Legendre-transformed energy function Ω_{R} is sometimes called the *magnetic enthalpy*. However, for brevity, here we will refer to Ω_{R} as the free energy.

Then,

$$\Omega_R = \breve{\Omega}_R(\vec{\mathbf{C}}^{e(\alpha)}, \mathbf{h}_R)$$
$$= \hat{\psi}_R(\vec{\mathbf{C}}^{e(\alpha)}, \breve{\mathbf{b}}_R(\vec{\mathbf{C}}^{e(\alpha)}, \mathbf{h}_R)) - \mathbf{h}_R \cdot \breve{\mathbf{b}}_R(\vec{\mathbf{C}}^{e(\alpha)}, \mathbf{h}_R), \tag{32.2.64}$$

while $(32.2.59)_1$ yields

$$\mathbf{T}^{e(\alpha)} = \hat{\mathbf{T}}^{e(\alpha)}(\vec{\mathbf{C}}^{e(\alpha)}, \mathbf{b}_R) \tag{32.2.65}$$
$$= \breve{\mathbf{T}}^{e(\alpha)}(\vec{\mathbf{C}}^{e(\alpha)}, \breve{\mathbf{b}}_R(\vec{\mathbf{C}}^{e(\alpha)}, \mathbf{h}_R)). \tag{32.2.66}$$

Thus, bearing in mind that a "breve" denotes a function of $(\vec{\mathbf{C}}^{e(\alpha)}, \mathbf{h}_R)$ while a "hat" denotes a function of $(\vec{\mathbf{C}}^{e(\alpha)}, \mathbf{b}_R)$, we find, using (32.2.59) that

$$\frac{\partial \breve{\Omega}_R}{\partial \mathbf{C}^{e(\alpha)}} = \frac{\partial \hat{\psi}_R}{\partial \mathbf{C}^{e(\alpha)}} + \underbrace{\left(\frac{\partial \hat{\psi}_R}{\partial \mathbf{b}_R} - \mathbf{h}_R\right)}_{=0} \frac{\partial \breve{\mathbf{b}}_R}{\partial \mathbf{C}^{e(\alpha)}} = \frac{1}{2}\breve{\mathbf{T}}^{e(\alpha)},$$

and

$$\frac{\partial \breve{\Omega}_R}{\partial \mathbf{h}_R} = \underbrace{\left(\frac{\partial \hat{\psi}_R}{\partial \mathbf{b}_R} - \mathbf{h}_R\right)}_{=0} \frac{\partial \breve{\mathbf{b}}_R}{\partial \mathbf{h}_R} - \breve{\mathbf{b}}_R = -\breve{\mathbf{b}}_R.$$

The elastic second Piola stresses and the magnetic flux density are therefore determined by the free-energy function $\Omega_R = \breve{\Omega}_R(\vec{\mathbf{C}}^{e(\alpha)}, \mathbf{h}_R)$ via the relations

$$\mathbf{T}^{e(\alpha)} = 2\,\frac{\partial \breve{\Omega}_R(\vec{\mathbf{C}}^{e(\alpha)}, \mathbf{h}_R)}{\partial \mathbf{C}^{e(\alpha)}},$$
$$\mathbf{b}_R = -\frac{\partial \breve{\Omega}_R(\vec{\mathbf{C}}^{e(\alpha)}, \mathbf{h}_R)}{\partial \mathbf{h}_R}. \tag{32.2.67}$$

Also, recalling (32.2.52) the Mandel stresses are given by

$$\mathbf{M}^{e(\alpha)} = \mathbf{C}^{e(\alpha)}\mathbf{T}^{e(\alpha)}, \tag{32.2.68}$$

and the contributions $\mathbf{T}^{(\alpha)}$ to the Cauchy stress $\mathbf{T}$ are given by

$$\mathbf{T}^{(\alpha)} = (J^{e(\alpha)})^{-1}\left(2\mathbf{F}^{e(\alpha)}\frac{\partial \breve{\Omega}_R(\vec{\mathbf{C}}^{e(\alpha)}, \mathbf{h}_R)}{\partial \mathbf{C}^{e(\alpha)}}\mathbf{F}^{e(\alpha)\top}\right). \tag{32.2.69}$$

The total Cauchy stress $\mathbf{T} = \sum_\alpha \mathbf{T}^{(\alpha)}$ is then given by

$$\mathbf{T} = \sum_\alpha (J^{e(\alpha)})^{-1}\Big(2\mathbf{F}^{e(\alpha)}\frac{\partial \breve{\Omega}_{\scriptscriptstyle R}(\vec{\mathbf{C}}^{e(\alpha)},\mathbf{h}_{\scriptscriptstyle R})}{\partial \mathbf{C}^{e(\alpha)}}\mathbf{F}^{e(\alpha)\top}\Big), \tag{32.2.70}$$

and hence the total Piola stress $\mathbf{T}_{\scriptscriptstyle R}$ is

$$\mathbf{T}_{\scriptscriptstyle R} = J\mathbf{T}\mathbf{F}^{-\top}. \tag{32.2.71}$$

Evolution equations for $\mathbf{F}^{v(\alpha)}$

The evolution equation for $\mathbf{F}^{v(\alpha)}$ in the viscous branches is taken as[3]

$$\dot{\mathbf{F}}^{v(\alpha)} = \mathbf{D}^{v(\alpha)}\,\mathbf{F}^{v(\alpha)}, \tag{32.2.72}$$

with the viscous stretching given by

$$\mathbf{D}^{v(\alpha)} = \dot{\bar{\gamma}}^{v(\alpha)}\Big(\frac{\mathbf{M}_0^{e(\alpha)}}{2\,\bar{\tau}^{(\alpha)}}\Big), \tag{32.2.73}$$

with the constraint that viscous flow is isochoric, $\det \mathbf{F}^{v(\alpha)} = 1$, and where

$$\dot{\bar{\gamma}}^{v(\alpha)} \stackrel{\text{def}}{=} \sqrt{2}\,|\mathbf{D}^{v(\alpha)}| \tag{32.2.74}$$

is an *equivalent shear strain rate* for each branch and

$$\bar{\tau}^{(\alpha)} \stackrel{\text{def}}{=} \frac{1}{\sqrt{2}}\,|\mathbf{M}_0^{e(\alpha)}| \tag{32.2.75}$$

defines an *equivalent shear stress* for each branch. We assume that each $\bar{\tau}^{(\alpha)}$ satisfies a strength relation of the form

$$\bar{\tau}^{(\alpha)} = \mathcal{S}^{(\alpha)}(\dot{\bar{\gamma}}^{v(\alpha)}). \tag{32.2.76}$$

[3] Mimicking the discussion in Chapter 4 Section 4.3.3 and Section 4.3.5.

32.3 Summary of the constitutive theory, governing partial differential equations, and boundary conditions

The theory relates the following basic fields:

$\mathbf{x} = \boldsymbol{\chi}(\mathbf{X}, t)$,	motion;
$\mathbf{F} = \nabla\boldsymbol{\chi}, \quad J = \det \mathbf{F} > 0$,	deformation gradient;
$\mathbf{h}$,	spatial magnetic field;
$\mathbf{b}$,	spatial magnetic flux density;
$\mathbf{m} = \mu_0^{-1}\mathbf{b} - \mathbf{h}$,	spatial magnetization;
$\mathbf{h}_{\scriptscriptstyle R} = \mathbf{F}^{\top}\mathbf{h}$,	referential magnetic field;
$\mathbf{b}_{\scriptscriptstyle R} = J\mathbf{F}^{-1}\mathbf{b}$,	referential magnetic flux density;
ϕ,	magnetostatic potential;
$\mathbf{C} = \mathbf{F}^{\top}\mathbf{F}$,	right Cauchy–Green tensor;
$\mathbf{B} = \mathbf{F}\mathbf{F}^{\top}$,	left Cauchy–Green tensor;
$\bar{\mathbf{F}} = J^{-1/3}\mathbf{F}, \quad \det \bar{\mathbf{F}} = 1$,	isochoric part of $\mathbf{F}$;
$\bar{\mathbf{C}} = \bar{\mathbf{F}}^{\top}\bar{\mathbf{F}}, \quad \det \bar{\mathbf{C}} = 1$,	isochoric part of $\mathbf{C}$;
$\bar{I}_1 = \operatorname{tr} \bar{\mathbf{C}}$,	first invariant of $\bar{\mathbf{C}}$;
$\bar{\mathbf{B}} = \bar{\mathbf{F}}\bar{\mathbf{F}}^{\top}, \quad \det \bar{\mathbf{B}} = 1$,	isochoric part of $\mathbf{B}$;
$\mathbf{F} = \mathbf{F}^{e(\alpha)}\mathbf{F}^{v(\alpha)} \quad \alpha = 0, 1, \ldots, M$,	multi-branch multiplicative decomposition of $\mathbf{F}$;
$\mathbf{F}^{v(\alpha)}, \ J^{v(\alpha)} = \det \mathbf{F}^{v(\alpha)} = 1$,	isochoric viscous distortion for the αth branch;
$\mathbf{C}^{v(\alpha)} = \mathbf{F}^{v(\alpha)\top}\mathbf{F}^{v(\alpha)}$,	viscous right Cauchy–Green tensors;
$\mathbf{F}^{e(\alpha)}, \ J^{e\,(\alpha)} = \det \mathbf{F}^{e(\alpha)} > 0$,	elastic distortion for the αth branch;
$\mathbf{C}^{e(\alpha)} = \mathbf{F}^{e(\alpha)\top}\mathbf{F}^{e(\alpha)}$,	elastic right Cauchy–Green tensors;
$\mathbf{B}^{e(\alpha)} = \mathbf{F}^{e(\alpha)}\mathbf{F}^{e(\alpha)\top}$,	elastic left Cauchy–Green tensors;
$\bar{\mathbf{F}}^{e(\alpha)} = \bar{\mathbf{F}}(\mathbf{F}^{v(\alpha)})^{-1} \ \det \bar{\mathbf{F}}^{e(\alpha)} = 1$,	isochoric part of $\mathbf{F}^{e(\alpha)}$;
$\bar{\mathbf{C}}^{e(\alpha)} = \bar{\mathbf{F}}^{e(\alpha)\top}\bar{\mathbf{F}}^{e(\alpha)}, \ \det \bar{\mathbf{C}}^{e(\alpha)} = 1$,	isochoric elastic right Cauchy–Green tensors;
$\bar{I}_1^e = \operatorname{tr} \bar{\mathbf{C}}^{e(\alpha)}$,	first invariant of $\bar{\mathbf{C}}^{e(\alpha)}$;
$\bar{\mathbf{B}}^{e(\alpha)} = \bar{\mathbf{F}}^{e(\alpha)}\bar{\mathbf{F}}^{e(\alpha)\top}, \ \det \bar{\mathbf{B}}^{e(\alpha)} = 1$,	isochoric elastic left Cauchy–Green tensors;
$\mathbf{T}^{(\alpha)}$,	Cauchy stress for the αth branch;
$\mathbf{T}^{e(\alpha)} = J^{e(\alpha)}\,\mathbf{F}^{e(\alpha)-1}\mathbf{T}^{(\alpha)}\mathbf{F}^{e(\alpha)-\top}$,	elastic second Piola stress for the αth branch;
$\mathbf{M}^{e(\alpha)} = \mathbf{C}^{e(\alpha)}\mathbf{T}^{e(\alpha)}$,	Mandel stress for the αth branch;
$\mathbf{T} = \sum_{\alpha} \mathbf{T}^{(\alpha)}$,	Cauchy stress;
$\mathbf{T}_{\scriptscriptstyle R} = J\mathbf{T}\mathbf{F}^{-\top}$,	Piola stress;
$\psi_{\scriptscriptstyle R}$,	free energy density per unit reference volume;
$\Omega_{\scriptscriptstyle R} = \psi_{\scriptscriptstyle R} - \mathbf{h}_{\scriptscriptstyle R} \cdot \mathbf{b}_{\scriptscriptstyle R}$	Legendre-transformed energy per unit reference volume.

- For brevity we will refer to both $\psi_{\scriptscriptstyle R}$ and the Legendre-transformed energy $\Omega_{\scriptscriptstyle R}$ as free energies.
- We limit our attention to situations under which the material may be idealized to be *isotropic*. Accordingly, all constitutive functions considered below are presumed to be isotropic in character.

- Also, when convenient, we shall use the shorthand

$$\sum_{\alpha} = \sum_{\alpha=0}^{M}, \tag{32.3.1}$$

32.3.1 **Constitutive relations**

1. **Free energy**:
   The free energy Ω_R is given by

$$\Omega_R = \breve{\Omega}_R(\vec{\mathbf{C}}^{e(\alpha)}, \mathbf{h}_R) \qquad \text{with} \qquad \vec{\mathbf{C}}^{e(\alpha)} = \{\mathbf{C}^{e(\alpha)} | \alpha = 0, 1, \ldots, M\}. \tag{32.3.2}$$

2. **Elastic second Piola stress. Mandel Stress. Piola stress. Cauchy stress**:
   The contribution of each branch α to the elastic second Piola stress is given by

$$\mathbf{T}^{e(\alpha)} = 2\,\frac{\partial \breve{\Omega}_R(\vec{\mathbf{C}}^{e(\alpha)}, \mathbf{h}_R)}{\partial \mathbf{C}^{e(\alpha)}}, \tag{32.3.3}$$

   the symmetric Mandel stress by

$$\mathbf{M}^{e(\alpha)} = \mathbf{C}^{e(\alpha)}\mathbf{T}^{e(\alpha)}, \tag{32.3.4}$$

   and the contribution $\mathbf{T}^{(\alpha)}$ to the Cauchy stress $\mathbf{T}$ by

$$\mathbf{T}^{(\alpha)} = (J^{e(\alpha)})^{-1}\left(2\mathbf{F}^{e(\alpha)}\frac{\partial \breve{\Omega}_R(\vec{\mathbf{C}}^{e(\alpha)}, \mathbf{h}_R)}{\partial \mathbf{C}^{e(\alpha)}}\mathbf{F}^{e(\alpha)\top}\right). \tag{32.3.5}$$

   The total Cauchy stress $\mathbf{T} = \sum_{\alpha} \mathbf{T}^{(\alpha)}$ is then given by

$$\mathbf{T} = \sum_{\alpha} (J^{e(\alpha)})^{-1}\left(2\mathbf{F}^{e(\alpha)}\frac{\partial \breve{\Omega}_R(\vec{\mathbf{C}}^{e(\alpha)}, \mathbf{h}_R)}{\partial \mathbf{C}^{e(\alpha)}}\mathbf{F}^{e(\alpha)\top}\right), \tag{32.3.6}$$

   and the total Piola stress $\mathbf{T}_R$ is

$$\mathbf{T}_R = J\mathbf{T}\mathbf{F}^{-\top}. \tag{32.3.7}$$

3. **Magnetic flux density**:
   The magnetic flux density is given by

$$\mathbf{b}_R = -\frac{\partial \breve{\Omega}_R(\vec{\mathbf{C}}^{e(\alpha)}, \mathbf{h}_R)}{\partial \mathbf{h}_R}. \tag{32.3.8}$$

4. **Viscous flow rules:**
The evolution equation for each $\mathbf{F}^{v(\alpha)}$ is

$$\dot{\mathbf{F}}^{v\,(\alpha)} = \mathbf{D}^{v(\alpha)}\,\mathbf{F}^{v(\alpha)}, \tag{32.3.9}$$

with the viscous stretching given by

$$\mathbf{D}^{v(\alpha)} = \dot{\bar{\gamma}}^{v(\alpha)}\left(\frac{\mathbf{M}_0^{e(\alpha)}}{2\,\bar{\tau}^{(\alpha)}}\right), \tag{32.3.10}$$

where

$$\bar{\tau}^{(\alpha)} \stackrel{\text{def}}{=} \frac{1}{\sqrt{2}}\,|\mathbf{M}_0^{e(\alpha)}| \tag{32.3.11}$$

defines an *equivalent shear stress* for each α, and

$$\dot{\bar{\gamma}}^{v(\alpha)} \stackrel{\text{def}}{=} \sqrt{2}\,|\mathbf{D}^{v(\alpha)}|, \tag{32.3.12}$$

is an *equivalent viscous shear strain rate.*
We assume that $\bar{\tau}^{(\alpha)}$ satisfies the strength relation

$$\bar{\tau}^{(\alpha)} = \mathcal{S}^{(\alpha)}(\dot{\bar{\gamma}}^{v(\alpha)}), \tag{32.3.13}$$

with $\mathcal{S}^{(\alpha)}(\dot{\bar{\gamma}}^{v(\alpha)})$ a positive-valued strictly increasing function of $\dot{\bar{\gamma}}^{v(\alpha)}$, so that the function $\mathcal{S}^{(\alpha)}(\dot{\bar{\gamma}}^{v(\alpha)})$ is *invertible*, with

$$\dot{\bar{\gamma}}^{v(\alpha)} = \mathcal{F}^{(\alpha)}(\bar{\tau}^{(\alpha)}) \geq 0. \tag{32.3.14}$$

5. **Total dissipation rate:**
The total rate of viscous dissipation per unit reference volume is

$$\mathcal{D} = \sum_{\alpha} \mathcal{S}^{(\alpha)}(\dot{\bar{\gamma}}^{v(\alpha)})\,\dot{\bar{\gamma}}^{v(\alpha)} \geq 0. \tag{32.3.15}$$

32.3.2 Governing partial differential equations

The governing partial differential equations (PDEs) consist of:

1. **The equation of motion:**
This is given by

$$\operatorname{Div}\mathbf{T}_{\scriptscriptstyle R} + \mathbf{f}_{\scriptscriptstyle R} = \mathbf{0}, \tag{32.3.16}$$

with

$$\mathbf{f}_{\scriptscriptstyle R} = \mathbf{f}_{0\scriptscriptstyle R} - \rho_{\scriptscriptstyle R}\ddot{\mathbf{u}}, \tag{32.3.17}$$

where $\mathbf{f}_{0R}$ represents the conventional body force per unit volume of the reference body, and $(-\rho_R\ddot{\mathbf{u}})$ represents the inertial body force, with ρ_R the mass density of the referential body and $\ddot{\mathbf{u}}$ the acceleration.

2. **Gauss's law for magnetism**:
   This is given by

$$\mathrm{Div}\,\mathbf{b}_R = 0. \tag{32.3.18}$$

3. **Ampere's law**:
   In the magnetostatic limit and in the absence of steady currents this is given by,

$$\mathrm{Curl}\,\mathbf{h}_R = 0. \tag{32.3.19}$$

   We take

$$\mathbf{h}_R \stackrel{\text{def}}{=} -\nabla\phi, \tag{32.3.20}$$

   with ϕ the magnetostatic potential, so that (32.3.19) is automatically satisfied.

32.3.3 Boundary and initial conditions

We also need boundary and initial conditions for the equation of motion and Gauss's law for magnetism in order to complete our theory. None are necessary for Ampere's law, since it is automatically satisfied.

1. **Boundary conditions for the PDE governing the displacement $\mathbf{u}$**:
   For a time interval $t \in [0, T]$, let $\partial\mathcal{B}_{\mathbf{u}}$ and $\partial\mathcal{B}_{\mathbf{t}_R}$ denote complementary subsurfaces of the boundary $\partial\mathcal{B}$ of the body $\mathcal{B}$. We consider a pair of possible boundary conditions:

$$\begin{aligned} \mathbf{u} &= \bar{\mathbf{u}} \quad \text{on } \partial\mathcal{B}_{\mathbf{u}} \times [0, T], \\ \mathbf{T}_R\mathbf{n}_R &= \bar{\mathbf{t}}_R \quad \text{on } \partial\mathcal{B}_{\mathbf{t}_R} \times [0, T]. \end{aligned} \tag{32.3.21}$$

   In the boundary conditions above $\bar{\mathbf{u}}$ and $\bar{\mathbf{t}}_R$ are prescribed displacement and tractions, both functions of $(\mathbf{X}, t)$.

2. **Boundary conditions for the PDE governing the magnetic potential ϕ**:
   For a time interval $t \in [0, T]$, let $\partial\mathcal{B}_\phi$ and $\partial\mathcal{B}_{\mathbf{b}_R}$ denote complementary subsurfaces of the boundary $\partial\mathcal{B}$ of the body $\mathcal{B}$. We consider a pair of possible boundary conditions:

$$\begin{aligned} \phi &= \bar{\phi} \quad \text{on } \partial\mathcal{B}_\phi \times [0, T], \\ \mathbf{b}_R \cdot \mathbf{n}_R &= \bar{b}_R \quad \text{on } \partial\mathcal{B}_{\mathbf{b}_R} \times [0, T]. \end{aligned} \tag{32.3.22}$$

In the boundary conditions above $\bar{\phi}$ and $\bar{\mathrm{b}}_{\mathrm{R}}$ are prescribed magnetostatic potentials and normal components of the referential magnetic flux density, both functions of $(\mathbf{X}, t)$.

Standard initial conditions involve the specifications

$$\begin{aligned} \mathbf{u}(\mathbf{X},0) = \mathbf{u}_0(\mathbf{X}), \qquad \dot{\mathbf{u}}(\mathbf{X},0) &= \dot{\mathbf{u}}_0(\mathbf{X}), \qquad \phi(\mathbf{X},0) = \phi_0(\mathbf{X}), \\ \mathbf{F}^{v(\alpha)}(\mathbf{X},0) = \mathbf{1} \qquad &\text{in } \mathcal{B}, \end{aligned} \tag{32.3.23}$$

with $\mathbf{u}_0$, $\dot{\mathbf{u}}_0$, and ϕ_0 prescribed functions on $\mathcal{B}$.

32.4 Specialization of the constitutive equations

The theory presented thus far is quite general. We now introduce special constitutive equations which should be useful in applications:

- To fix ideas, in the first subsection below we *neglect viscoelastic effects* and consider a relatively simple set of constitutive equations which we consider as a model for an *ideal* magnetoelastic s-MRE material which exhibits Neo-Hookean elasticity and linear paramagnetism.

  For this special case we only consider one branch $\alpha = 0$ in which

  $$\mathbf{F}^{v(0)} = \mathbf{1}, \tag{32.4.1}$$

  and therefore,

  $$\mathbf{F}^{e(0)} = \mathbf{F}, \qquad \mathbf{C}^{e(0)} = \mathbf{C}. \tag{32.4.2}$$

- In the second subsection we will extend our considerations of the first subsection to also account for:
  - viscoelastic effects, and
  - nonlinear paramagnetism.

Throughout we restrict our attention to isotropic materials under isothermal conditions.

32.4.1 "Ideal" magnetoelastic s-MREs

We begin by assuming that the total energy Ω_{R} is given by the sum of a mechanical energy and a magnetic energy:

$$\Omega_{\mathrm{R}} = \underbrace{\Omega_{\mathrm{R}}^{\mathrm{mech}}(\mathbf{C})}_{\text{mechanical energy}} + \underbrace{\Omega_{\mathrm{R}}^{\mathrm{totmag}}(\mathbf{C}, \mathbf{h}_{\mathrm{R}})}_{\text{total magnetic energy}} . \tag{32.4.3}$$

The contribution $\Omega_{\rm R}^{\rm mech}(\mathbf{C})$ in (32.4.3) is the mechanical energy due to the nonlinear elastic deformation of the material, which for isotropic slightly compressible materials is taken in the simple additive form,

$$\Omega_{\rm R}^{\rm mech}(\mathbf{C}) = \underbrace{\Omega_{\rm R}^{\rm dis}(\bar{I}_1)}_{\text{distortional energy}} + \underbrace{\Omega_{\rm R}^{\rm vol}(J)}_{\text{volumetric energy}}, \tag{32.4.4}$$

with the distortional energy taken as a Neo-Hookean energy,

$$\Omega_{\rm R}^{\rm dis}(\bar{I}_1) = \tfrac{1}{2}G\left(\bar{I}_1 - 3\right), \tag{32.4.5}$$

with $G > 0$ a shear modulus, and the volumetric energy taken as

$$\Omega_{\rm R}^{\rm vol}(J) = \tfrac{1}{2}K(J-1)^2, \tag{32.4.6}$$

with $K > 0$ a bulk modulus of the material.

Next, for a *linear* paramagnetic material the magnetic contribution to the energy per unit volume of the *deformed body* is taken to be given by

$$\begin{aligned}\Omega^{\rm totmag}(\mathbf{h}) &= \Omega^{\rm maxw}(\mathbf{h}) + \Omega^{\rm mag}(\mathbf{h})\\ &= \underbrace{\left(-\mu_0\frac{1}{2}\mathbf{h}\cdot\mathbf{h}\right)}_{\text{field energy}} + \underbrace{\left(-\mu_0\,\chi\,\frac{1}{2}\mathbf{h}\cdot\mathbf{h}\right)}_{\text{magnetization energy}},\end{aligned} \tag{32.4.7}$$

where the first term in (32.4.7) represents the magnetic field energy, and the second term is the energy due to magnetization of the body. Here,

$$\mu_0 = 4\pi \times 10^{-7}\,\mathrm{T/(A/m)}$$

is the *permeability of free space*, and

$$\chi > 0 \qquad \text{(dimensionless)}$$

is the *magnetic susceptibility* of the paramagnetic material. Thus the total magnetic energy may be written as

$$\Omega^{\rm totmag}(\mathbf{h}) = -\mu\,\frac{1}{2}\mathbf{h}\cdot\mathbf{h}, \tag{32.4.8}$$

with

$$\mu \stackrel{\rm def}{=} \mu_0(1+\chi) \tag{32.4.9}$$

the *magnetic permeability* of the material. Next, recall that the spatial magnetic field $\mathbf{h}$ and the referential magnetic field $\mathbf{h}_{\rm R}$ are related by

$$\mathbf{h}_{\rm R} = \mathbf{F}^{\top}\mathbf{h}, \tag{32.4.10}$$

so that

$$\mathbf{h} \cdot \mathbf{h} = \mathbf{h}_{\mathrm{R}} \cdot \mathbf{C}^{-1}\mathbf{h}_{\mathrm{R}}. \tag{32.4.11}$$

Use of (32.4.11) allows us to write the total magnetic energy per unit volume of the reference body for a linear paramagnetic material as

$$\Omega_{\mathrm{R}}^{\mathrm{totmag}}(\mathbf{C}, \mathbf{h}_{\mathrm{R}}) = J\left(-\mu\,\frac{1}{2}\mathbf{h}_{\mathrm{R}} \cdot \mathbf{C}^{-1}\mathbf{h}_{\mathrm{R}}\right). \tag{32.4.12}$$

Summarizing, the total energy for an ideal linear paramagnetic s-MRE material is taken to be given by

$$\Omega_{\mathrm{R}} = \tfrac{1}{2}G\left(\bar{I}_1 - 3\right) + \tfrac{1}{2}K(J-1)^2 + J\left(-\mu\,\frac{1}{2}\mathbf{h}_{\mathrm{R}} \cdot \mathbf{C}^{-1}\mathbf{h}_{\mathrm{R}}\right), \tag{32.4.13}$$

in which there are only three material parameters (G, K, χ) which need to be determined from experiments.

Stress

Using (32.4.13), $(32.2.59)_1$, and (32.2.51), the Cauchy stress is given by[4]

$$\begin{aligned}
\mathbf{T} &= J^{-1}G\bar{\mathbf{B}}_0 - p\mathbf{1} + \underbrace{\mu\left[\mathbf{h}\otimes\mathbf{h} - \frac{1}{2}(\mathbf{h}\cdot\mathbf{h})\mathbf{1}\right]}_{\mathbf{T}^{\mathrm{maxw}} \equiv \text{Maxwell stress}}, \qquad \text{with} \\
p &\stackrel{\text{def}}{=} -K(J-1),
\end{aligned} \tag{32.4.14}$$

where $\bar{\mathbf{B}}_0$ is the deviatoric part of the distortional left Cauchy–Green stretch tensor $\bar{\mathbf{B}} = J^{-2/3}\mathbf{F}\mathbf{F}^{\top}$, and p is a mechanical pressure. The last term in $(32.4.14)_1$ is the usual form of the *Maxwell stress* in the material.

Magnetic flux density

Also, using (32.4.13) and $(32.2.67)_2$ the referential magnetic flux density is given by

$$\mathbf{b}_{\mathrm{R}} = (\mu\,J\,\mathbf{C}^{-1})\,\mathbf{h}_{\mathrm{R}}. \tag{32.4.15}$$

Using $\mathbf{h}_{\mathrm{R}} = \mathbf{F}^{\top}\mathbf{h}$ and $\mathbf{b}_{\mathrm{R}} = J\mathbf{F}^{-1}\mathbf{b}$ the corresponding spatial form of the magnetic flux is the familiar linear relation

$$\mathbf{b} = \mu\mathbf{h}. \tag{32.4.16}$$

[4] Detailed calculations for the expressions for the stress and magnetic flux density in an "ideal" s-MRE are provided in Appendix 32.A.1

Magnetization

Using (32.4.16) and (32.2.37), viz. $\mathbf{b} = \mu_0(\mathbf{h} + \mathbf{m})$, we obtain that for a linear paramagnetic material the magnetization is given by

$$\mathbf{m} = \chi\, \mathbf{h}, \tag{32.4.17}$$

which implies that for isotropic materials the magnetization $\mathbf{m}$ is collinear with the magnetic field, and that

$$|\mathbf{m}| = \chi\, |\mathbf{h}|. \tag{32.4.18}$$

32.4.2 Constitutive equations which account for additional physical effects

In this subsection we extend our considerations of the previous subsection to account for:

- viscoelastic effects, and
- nonlinear paramagnetism.

We begin by making the following specializations for modeling the viscoelastic response of the material. Recalling the multi-branch multiplicative decomposition of the deformation gradient $\mathbf{F} = \mathbf{F}^{e(\alpha)}\mathbf{F}^{v(\alpha)}$ $(\alpha = 0, 1, \ldots, M)$:

(i) We take the first branch, $\alpha = 0$, to represent the *equilibrium elastic response* of the material. We do not associate any viscous relaxation with this network. Accordingly we set $\mathbf{F}^{v(0)} = \mathbf{1}$, and we therefore have

$$\mathbf{F}^{e\,(0)} = \mathbf{F}, \qquad \mathbf{C}^{e\,(0)} = \mathbf{C}. \tag{32.4.19}$$

(ii) The remaining branches, $\alpha = 1, \ldots, M$, represent the *non-equilibrium viscoelastic response* of the material.

(iii) A feature of the multi-branch decomposition $\mathbf{F} = \mathbf{F}^{e(\alpha)}\mathbf{F}^{v(\alpha)}$, with $\det \mathbf{F}^{v(\alpha)} = 1$, is that

$$J^{e(\alpha)} = J \quad \text{for all } \alpha.$$

Henceforth, when it is important to distinguish between the zeroth branch and the other branches of the rheological model, we will use the descriptive terms "equilibrium" for branch $\alpha = 0$ and "non-equilibrium" for branches $\alpha = 1, \ldots, M$.

32.4.3 Free energy

We assume that the total free energy Ω_R is given by a sum of mechanical and magnetic energies:[5]

$$\Omega_R = \breve{\Omega}_R(\vec{\mathbf{C}}^{e(\alpha)}, \mathbf{h}_R) = \underbrace{\Omega_R^{\text{mech}}(\bar{I}_1^{e(\alpha)}, J)}_{\text{mechanical energy}} + \underbrace{\Omega_R^{\text{totmag}}(\mathbf{C}, \mathbf{h}_R)}_{\text{total magnetic energy}} \quad . \tag{32.4.20}$$

1. The mechanical contribution to the free energy is taken as

$$\Omega_R^{\text{mech}}(\bar{I}_1^{e(\alpha)}, J) = \underbrace{\Omega_R^{\text{eq dis}}(\bar{I}_1) + \Omega_R^{\text{eq vol}}(J)}_{\text{equilibrium energy}} + \underbrace{\sum_{\alpha=1}^{M} \Omega_R^{\text{neq}(\alpha)}(\bar{I}_1^{e(\alpha)})}_{\text{non-equilibrium energy}}. \tag{32.4.21}$$

Note that in writing (32.4.21), we have implicitly neglected any time-dependence of the volumetric response of the material, and associated all volume changes with equilibrium response of the material.

Special forms for the two terms in (32.4.21) are given below:

(i) *Equilibrium energy*: For the equilibrium distortional energy we consider a Neo-Hookean form,

$$\Omega_R^{\text{eq dis}}(\bar{I}_1) = \frac{G_{\text{eq}}}{2}(\bar{I}_1 - 3), \tag{32.4.22}$$

and we take equilibrium volumetric energy to be given by

$$\Omega_R^{\text{eq vol}}(J) = \tfrac{1}{2}K(J-1)^2. \tag{32.4.23}$$

Here:
- The parameter $G_{\text{eq}} > 0$ is an equilibrium shear modulus.
- The parameter $K > 0$ is a bulk modulus of the material.

(ii) *Non-equilibrium energy*: To model the non-equilibrium contributions to the free energy we again adopt simple Neo-Hookean forms,

$$\Omega_R^{\text{neq}(\alpha)}(\bar{I}_1^{e(\alpha)}) = \tfrac{1}{2}\, G_{\text{neq}}^{(\alpha)} \left(\bar{I}_1^{e(\alpha)} - 3\right), \tag{32.4.24}$$

where $G_{\text{neq}}^{(\alpha)}$ represent non-equilibrium shear moduli.

[5] Here for simplicity, the notation $\bar{I}_1^{e(\alpha)}$ represents the list of all invariants $\bar{I}_1^{e(\alpha)}$ for $\alpha = 0, \dots, M$.

REMARK

It has been observed in experiments that the magnetic field $\mathbf{h}_R$ has the effect of increasing the equilibrium shear modulus of s-MREs, and that it can also influence the viscoelastic material parameters (cf., e.g., Haldar et al., 2016; Pelteret et al., 2018; Haldar, 2021; Moreno et al., 2021; Wang et al., 2023). However, the precise mathematical forms for expressing such physical dependencies are not yet firmly established in the literature—for simplicity we neglect them here.

2. The total magnetic contribution to the free energy is taken as

$$\Omega_R^{\text{totmag}} = \underbrace{\Omega_R^{\text{maxw}}(\mathbf{C}, \mathbf{h}_R)}_{\text{field energy}} + \underbrace{\Omega_R^{\text{mag}}(\mathbf{C}, \mathbf{h}_R)}_{\text{magnetization energy}} \quad . \tag{32.4.25}$$

Here:

(i) The magnetic field energy per unit volume of the *deformed body* is taken to be given by

$$\Omega^{\text{maxw}}(\mathbf{h}) = -\mu_0 \frac{1}{2}\mathbf{h} \cdot \mathbf{h}, \tag{32.4.26}$$

where,

$$\mu_0 = 4\pi \times 10^{-7}\, \text{T}/(\text{A}/\text{m})$$

is the *permeability of free space*. Use of the mapping $\mathbf{h} = \mathbf{F}^{-\top}\mathbf{h}_R$ allows us to write the magnetic field energy per unit volume of the reference body as

$$\Omega_R^{\text{maxw}}(\mathbf{C}, \mathbf{h}_R) = J\left(-\mu_0 \frac{1}{2}\mathbf{h}_R \cdot \mathbf{C}^{-1}\mathbf{h}_R\right). \tag{32.4.27}$$

(ii) Next, we consider the nonlinear magnetization contribution to the free energy. Many paramagnetic materials exhibit a nonlinear magnetization as $|\mathbf{h}|$ increases, and indeed exhibit a **saturation magnetization** limit m_s. A simple phenomenological model for such a magnetization response is given by (Danas, 2017)

$$\frac{|\mathbf{m}|}{m_s} = \tanh\left(\chi \frac{|\mathbf{h}|}{m_s}\right). \tag{32.4.28}$$

Figure 32.1 shows a plot of a $|\mathbf{m}|/m_s$ versus $|\mathbf{h}|/(m_s)$ for a value of the magnetic susceptibility of $\chi = 2.0$.

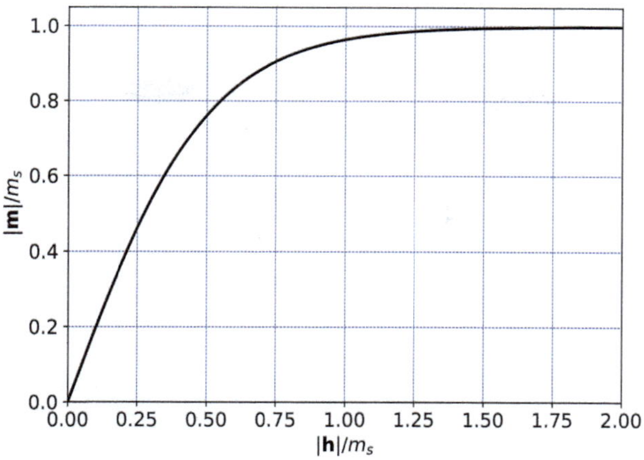

Fig. 32.1 A plot of a $|\mathbf{m}|/m_s$ versus $|\mathbf{h}|/(m_s)$ response from eq. (32.4.28) for $\chi = 2.0$.

This form of saturation magnetization is obtained from a spatial magnetization energy of the form,

$$\Omega^{\text{mag}}(\mathbf{h}) = -\mu_0 \left[\frac{m_s^2}{\chi} \ln\left(\cosh\left(\frac{\chi}{m_s} \sqrt{\mathbf{h}\cdot\mathbf{h}} \right) \right) \right]. \tag{32.4.29}$$

Use of the mapping $\mathbf{h} = \mathbf{F}^{-\top}\mathbf{h}_{\text{R}}$ allows us to write the nonlinear magnetization energy per unit volume of the reference body as

$$\Omega_{\text{R}}^{\text{mag}}(\mathbf{C}, \mathbf{h}_{\text{R}}) = J \left(-\mu_0 \left[\frac{m_s^2}{\chi} \ln\left(\cosh\left(\frac{\chi}{m_s} \sqrt{\mathbf{h}_{\text{R}}\cdot\mathbf{C}^{-1}\mathbf{h}_{\text{R}}} \right) \right) \right] \right). \tag{32.4.30}$$

Summarizing, we assume that the total free energy is given by,

$$\begin{aligned}
\Omega_{\text{R}} = & \underbrace{\frac{G_{\text{eq}}}{2}(\bar{I}_1 - 3) + \tfrac{1}{2}K(J-1)^2}_{\text{equilibrium energy}} + \underbrace{\sum_{\alpha=1}^{M} \tfrac{1}{2} G_{\text{neq}}^{(\alpha)} \left(\bar{I}_1^{e(\alpha)} - 3 \right)}_{\text{non-equilibrium energy}} \\
& + \underbrace{J\left(-\mu_0 \frac{1}{2} \mathbf{h}_{\text{R}}\cdot\mathbf{C}^{-1}\mathbf{h}_{\text{R}} \right)}_{\text{field energy}} \\
& + \underbrace{J\left(-\mu_0 \left[\frac{m_s^2}{\chi} \ln\left(\cosh\left(\frac{\chi}{m_s} \sqrt{\mathbf{h}_{\text{R}}\cdot\mathbf{C}^{-1}\mathbf{h}_{\text{R}}} \right) \right) \right] \right)}_{\text{magnetization energy}}.
\end{aligned} \tag{32.4.31}$$

32.4.4 **Stress**

Using the free energy (32.4.31) and the stress relation (32.3.3), the elastic second Piola stress from each non-equilibrium branch is given by

$$\mathbf{T}^{e(\alpha)} = (J^{e(\alpha)})^{-2/3} G^{(\alpha)}_{\text{neq}} \left[\mathbf{1} - \frac{1}{3} (\operatorname{tr} \mathbf{C}^{e(\alpha)}) \mathbf{C}^{e(\alpha)-1} \right], \tag{32.4.32}$$

so that the Mandel stresses $\mathbf{M}^{e(\alpha)} = \mathbf{C}^{e(\alpha)} \mathbf{T}^{e(\alpha)}$ associated with the non-equilibrium branches are

$$\mathbf{M}^{e(\alpha)} = G^{(\alpha)}_{\text{neq}} \left[\bar{\mathbf{C}}^{e(\alpha)} - \frac{1}{3} (\operatorname{tr} \bar{\mathbf{C}}^{e(\alpha)}) \mathbf{1} \right], \tag{32.4.33}$$

which shows that it is deviatoric, and can therefore be written as

$$\mathbf{M}_0^{e(\alpha)} = G^{(\alpha)}_{\text{neq}} \bar{\mathbf{C}}_0^{e(\alpha)}. \tag{32.4.34}$$

Then, using the free energy (32.4.31) in the stress relation (32.3.6) yields (after a lengthy calculation included in Appendix 32.A) that the total Cauchy stress is given by

$$\mathbf{T} = \mathbf{T}^{\text{eq}} + \sum_{\alpha=1}^{M} \mathbf{T}^{(\alpha)} + \mathbf{T}^{\text{maxw}} + \mathbf{T}^{\text{mag}}, \tag{32.4.35}$$

where

$$\begin{aligned}
\mathbf{T}^{\text{eq}} &= \underbrace{J^{-1} G_{\text{eq}} \bar{\mathbf{B}}_0 + K (J-1) \mathbf{1}}_{\text{equilibrium stress}}, \\
\mathbf{T}^{(\alpha)} &= \underbrace{J^{-1} G_{\text{neq}} \bar{\mathbf{B}}_0^{e(\alpha)}}_{\text{non-equilibrium stress}}, \\
\mathbf{T}^{\text{maxw}} &= \underbrace{\mu_0 \left[\mathbf{h} \otimes \mathbf{h} - \frac{1}{2} (\mathbf{h} \cdot \mathbf{h}) \mathbf{1} \right]}_{\text{Maxwell stress}}, \qquad \text{and} \\
\mathbf{T}^{\text{mag}} &= \underbrace{-\mu_0 \left[\frac{m_s^2}{\chi} \ln \left(\cosh \left(\frac{\chi}{m_s} \sqrt{\mathbf{h} \cdot \mathbf{h}} \right) \right) \right] \mathbf{1} + \mu_0 \left[\frac{m_s}{\sqrt{\mathbf{h} \cdot \mathbf{h}}} \tanh \left(\frac{\chi}{m_s} \sqrt{\mathbf{h} \cdot \mathbf{h}} \right) (\mathbf{h} \otimes \mathbf{h}) \right]}_{\text{magnetization stress}}.
\end{aligned} \tag{32.4.36}$$

The total Piola stress is then given by

$$\mathbf{T}_{\text{R}} = J \mathbf{T} \mathbf{F}^{-\top}. \tag{32.4.37}$$

32.4.5 **Magnetic flux density and magnetization**

Using the free energy (32.4.31) in the magnetic flux density relation (32.3.8) yields that the referential magnetic flux density is given by

$$\mathbf{b}_{\mathrm{R}} = J\mathbf{F}^{-1}\mathbf{b}, \qquad \text{with} \qquad \mathbf{b} = \mathbf{b}^{\text{maxw}} + \mathbf{b}^{\text{mag}}, \tag{32.4.38}$$

where

$$\begin{aligned} \mathbf{b}^{\text{maxw}} &= \underbrace{\mu_0\,\mathbf{h}}_{\text{field response}}, \qquad \text{and} \\ \mathbf{b}^{\text{mag}} &= \underbrace{\mu_0 \frac{m_s}{\sqrt{\mathbf{h}\cdot\mathbf{h}}} \tanh\left(\frac{\chi}{m_s}\sqrt{\mathbf{h}\cdot\mathbf{h}}\right)\mathbf{h}}_{\text{magnetization response}}. \end{aligned} \tag{32.4.39}$$

Using (32.2.33), (32.4.38), and (32.4.39), the spatial magnetization $\mathbf{m}$ is given by

$$\mathbf{m} = m_s \tanh\left(\chi\frac{|\mathbf{h}|}{m_s}\right)\frac{\mathbf{h}}{|\mathbf{h}|}. \tag{32.4.40}$$

32.4.6 **The evolution of $\mathbf{F}^{v(\alpha)}$**

The evolution equation for $\mathbf{F}^{v(\alpha)}$ in each of the non-equilibrium branches is

$$\dot{\mathbf{F}}^{v(\alpha)} = \mathbf{D}^{v(\alpha)}\,\mathbf{F}^{v(\alpha)}, \tag{32.4.41}$$

with the viscous stretching given by

$$\mathbf{D}^{v(\alpha)} = \dot{\bar{\gamma}}^{v(\alpha)}\left(\frac{\mathbf{M}_0^{e(\alpha)}}{2\,\bar{\tau}^{(\alpha)}}\right), \tag{32.4.42}$$

where

$$\bar{\tau}^{(\alpha)} \stackrel{\text{def}}{=} \frac{1}{\sqrt{2}}|\mathbf{M}_0^{e(\alpha)}|, \tag{32.4.43}$$

defines an *equivalent shear stress* for viscous flow, and with an *equivalent viscous shear strain rate* given by,

$$\dot{\bar{\gamma}}^{v(\alpha)} = f(\bar{\tau}^{(\alpha)}) \geq 0. \tag{32.4.44}$$

We take the equivalent viscous shear strain rate in (32.4.44) to be given by a simple linear relation

$$\dot{\bar{\gamma}}^{v(\alpha)} = \frac{\bar{\tau}^{(\alpha)}}{\eta^{(\alpha)}}, \tag{32.4.45}$$

with $\eta^{(\alpha)} > 0$ a constant *viscosity* parameter for each α. With this choice of $\dot{\bar{\gamma}}^{v(\alpha)}$ the dissipation inequality is

$$\mathcal{D} = \sum_{\alpha} \eta^{(\alpha)} (\dot{\bar{\gamma}}^{v(\alpha)})^2 \geq 0. \tag{32.4.46}$$

which is always satisfied since we have chosen positive viscosities, $\eta^{(\alpha)} > 0$.

Note that using (32.4.34) and (32.4.45) we may rewrite the viscous stretching relation (32.4.42) alternatively as

$$\mathbf{D}^{v(\alpha)} = \frac{\bar{\mathbf{C}}_0^{e(\alpha)}}{2\,\tau_r^{(\alpha)}} \qquad \text{with} \qquad \tau_r^{(\alpha)} \stackrel{\text{def}}{=} \frac{\eta^{(\alpha)}}{G_{\text{neq}}^{(\alpha)}}, \tag{32.4.47}$$

where $\tau_r^{(\alpha)}$ represents a *relaxation time* for the αth non-equilibrium branch.

32.4.7 Constitutive equation for the non-equilibrium stresses $\mathrm{T}^{(\alpha)}$ re-expressed

Consider the term $\mathbf{T}^{\alpha}$ for the non-equilibrium Cauchy stresses in eq. $(32.4.36)_2$. Fixing attention on one of these branches and suppressing the index α, we can write:

$$\mathbf{T} = J^{-1} G_{\text{neq}} \bar{\mathbf{B}}_0^e = J^{-5/3} G_{\text{neq}} \left(\mathbf{B}^e - \frac{1}{3} (\operatorname{tr} \mathbf{B}^e) \mathbf{1} \right). \tag{32.4.48}$$

Thus, noting that $\mathbf{B}^e = \mathbf{F}\mathbf{C}^{v-1}\mathbf{F}^{\top}$ and $\operatorname{tr} \mathbf{B}^e = \mathbf{C} : \mathbf{C}^{v-1}$, we obtain

$$\mathbf{T} = J^{-5/3} G_{\text{neq}} \left(\mathbf{F}\mathbf{C}^{v-1}\mathbf{F}^{\top} - \frac{1}{3} (\mathbf{C} : \mathbf{C}^{v-1}) \mathbf{1} \right), \tag{32.4.49}$$

which shows that the non-equilibrium Cauchy stress $\mathbf{T}$ may be calculated from a knowledge of $\mathbf{F}$ and $\mathbf{C}^v$. This observation is important in developing a time-integration procedure for viscoelasticity, in which evolution of viscoelastic flow phrased in terms of an evolution equation for $\mathbf{C}^v$, rather than an evolution equation for $\mathbf{F}^v$. We consider this below.

From (32.4.41) and (32.4.47), for a single branch, the evolution of $\mathbf{F}^v$ is

$$\dot{\mathbf{F}}^v = \mathbf{D}^v \, \mathbf{F}^v, \qquad \mathbf{D}^v = \frac{\bar{\mathbf{C}}_0^e}{2\tau_r}, \qquad \tau_r \stackrel{\text{def}}{=} \frac{\eta}{G_{\text{neq}}}. \tag{32.4.50}$$

Next, from $\mathbf{C}^v = \mathbf{F}^{v\top}\mathbf{F}^v$ we obtain that $\dot{\mathbf{C}}^v = \mathbf{F}^{v\top} (2\mathbf{D}^v) \, \mathbf{F}^v$. Hence, using $(32.4.50)_2$,

$$\dot{\mathbf{C}}^v = \mathbf{F}^{v\top} \left(\frac{\bar{\mathbf{C}}_0^e}{\tau_r} \right) \mathbf{F}^v. \tag{32.4.51}$$

Next, since $\bar{\mathbf{C}}^e = J^{-2/3}\left(\mathbf{F}^{v-\top}\mathbf{C}\mathbf{F}^{v-1}\right)$ and $\operatorname{tr}\bar{\mathbf{C}}^e = J^{-2/3}\left(\mathbf{C}\colon\mathbf{C}^{v-1}\right)$, we can write

$$\bar{\mathbf{C}}^e_0 = J^{-2/3}\left[\left(\mathbf{F}^{v-\top}\mathbf{C}\mathbf{F}^{v-1}\right) - \frac{1}{3}\left(\mathbf{C}\colon\mathbf{C}^{v-1}\right)\mathbf{1}\right], \tag{32.4.52}$$

and use of this in (32.4.51) gives the following evolution equation

$$\dot{\bar{\mathbf{C}}}^v = \frac{J^{-2/3}}{\tau_r}\left(\mathbf{C} - \frac{1}{3}\left(\mathbf{C}\colon\mathbf{C}^{v-1}\right)\mathbf{C}^v\right). \tag{32.4.53}$$

Finally, using (32.4.48) and reinstating the index α, eq. $(32.4.36)_2$ for the non-equilibrium Cauchy stresses may be replaced by

$$\mathbf{T}^{(\alpha)} = J^{-1}G^{(\alpha)}_{\text{neq}}\left(\bar{\mathbf{F}}(\mathbf{C}^{v(\alpha)})^{-1}\bar{\mathbf{F}}^\top - \frac{1}{3}(\bar{\mathbf{C}}\colon(\mathbf{C}^{v(\alpha)})^{-1})\mathbf{1}\right), \tag{32.4.54}$$

and using (32.4.53) the evolution equations (32.4.41) for $\mathbf{F}^{v(\alpha)}$ may be replaced by the following evolution equations for $\mathbf{C}^{v(\alpha)}$

$$\dot{\bar{\mathbf{C}}}^{v(\alpha)} = \frac{1}{\tau_r^{(\alpha)}}\left(\bar{\mathbf{C}} - \frac{1}{3}\left(\bar{\mathbf{C}}\colon(\mathbf{C}^{v(\alpha)})^{-1}\right)\mathbf{C}^{v(\alpha)}\right). \tag{32.4.55}$$

Note that in (32.4.54) and (32.4.55) the dependence on the internal variables $\mathbf{F}^{v(\alpha)}$ enters only through the *symmetric tensors* $\mathbf{C}^{v(\alpha)} = (\mathbf{F}^{v(\alpha)})^\top\mathbf{F}^{v(\alpha)}$ (cf., e.g., Wijaya et al., 2023; Lefévre et al., 2024).

32.4.8 **Summary of the governing partial differential equations and evolution equations for the specialized theory**

Since Ampere's law is automatically satisfied, the partial differential equations and evolution equations which govern the magneto-mechanical behavior of s-MREs are the following.

1. **The equation of motion:**
   This is given by

$$\operatorname{Div}\mathbf{T}_{\text{R}} + \mathbf{f}_{\text{R}} = \mathbf{0}, \tag{32.4.56}$$

with

$$\mathbf{f}_{\text{R}} = \mathbf{f}_{0\text{R}} - \rho_{\text{R}}\ddot{\mathbf{u}}, \tag{32.4.57}$$

and the Piola stress given by

$$\mathbf{T}_{\text{R}} = J\,\mathbf{T}\mathbf{F}^{-\top}, \qquad \text{with} \qquad \mathbf{T} = \mathbf{T}^{\text{eq}} + \sum_\alpha \mathbf{T}^{(\alpha)} + \mathbf{T}^{\text{maxw}} + \mathbf{T}^{\text{mag}}, \tag{32.4.58}$$

where

$$
\begin{aligned}
\mathbf{T}^{\text{eq}} &= J^{-1} G_{\text{eq}} \bar{\mathbf{B}}_0 - p\,\mathbf{1} \qquad \text{with} \qquad p = -K\,(J-1), \\
\mathbf{T}^{(\alpha)} &= J^{-1} G_{\text{neq}}^{(\alpha)} \left(\bar{\mathbf{F}} (\mathbf{C}^{v(\alpha)})^{-1} \bar{\mathbf{F}}^{\top} - \frac{1}{3} (\bar{\mathbf{C}} : (\mathbf{C}^{v(\alpha)})^{-1}) \mathbf{1} \right), \\
\mathbf{T}^{\text{maxw}} &= \mu_0 \left[\mathbf{h} \otimes \mathbf{h} - \frac{1}{2} (\mathbf{h} \cdot \mathbf{h}) \mathbf{1} \right], \qquad \text{and} \\
\mathbf{T}^{\text{mag}} &= -\mu_0 \left[\frac{m_s^2}{\chi} \ln \left(\cosh \left(\frac{\chi}{m_s} \sqrt{\mathbf{h} \cdot \mathbf{h}} \right) \right) \right] \mathbf{1} \\
&\quad + \mu_0 \left[\frac{m_s}{\sqrt{\mathbf{h} \cdot \mathbf{h}}} \tanh \left(\frac{\chi}{m_s} \sqrt{\mathbf{h} \cdot \mathbf{h}} \right) (\mathbf{h} \otimes \mathbf{h}) \right].
\end{aligned}
\tag{32.4.59}
$$

2. **Evolution of $\mathbf{C}^{v(\alpha)}$:**
The viscous right Cauchy–Green stretch tensors $\mathbf{C}^{v(\alpha)}$ evolve according to

$$
\dot{\mathbf{C}}^{v(\alpha)} = \frac{1}{\tau_r^{(\alpha)}} \left(\bar{\mathbf{C}} - \frac{1}{3} \left(\bar{\mathbf{C}} : (\mathbf{C}^{v(\alpha)})^{-1} \right) \mathbf{C}^{v(\alpha)} \right). \tag{32.4.60}
$$

3. **Gauss's law for magnetism:**
This is given by

$$
\operatorname{Div} \mathbf{b}_{\text{R}} = 0, \tag{32.4.61}
$$

with the referential magnetic flux density given by

$$
\mathbf{b}_{\text{R}} = J \mathbf{F}^{-1} \mathbf{b}, \qquad \text{with} \qquad \mathbf{b} = \mathbf{b}^{\text{maxw}} + \mathbf{b}^{\text{mag}}, \tag{32.4.62}
$$

where

$$
\begin{aligned}
\mathbf{b}^{\text{maxw}} &= \mu_0\, \mathbf{h}, \qquad \text{and} \\
\mathbf{b}^{\text{mag}} &= \mu_0 \frac{m_s}{\sqrt{\mathbf{h} \cdot \mathbf{h}}} \tanh \left(\frac{\chi}{m_s} \sqrt{\mathbf{h} \cdot \mathbf{h}} \right) \mathbf{h}.
\end{aligned}
\tag{32.4.63}
$$

Suitable boundary and initial conditions for these equations have been discussed previously in Section 32.3.3. Finally,

- The material parameters in the theory are $(G_{\text{eq}}, K, G_{\text{neq}}^{(\alpha)}, \tau_r^{(\alpha)}, \chi, m_s)$. These need to be determined from experiments.

32.5 A boundary-value problem formulation which accounts for the magnetic field in the air-space surrounding a magnetizable body

In this section we discuss a formulation of the boundary-value problem for MREs which accounts for the air-space surrounding a magnetizable body. Note that the air surrounding an MRE has nearly zero mechanical stiffness, whereas the magnetic field **h** in air is finite and of central importance in affecting the magnetization of the MRE. Our approach for modeling the effects of a magnetic field in the air-space surrounding a magnetizable body is guided by the work of Rambausek et al. (2022) and Rambausek and Schöbrel (2023).

32.5.1 Modeling approach for the combined air–solid domain

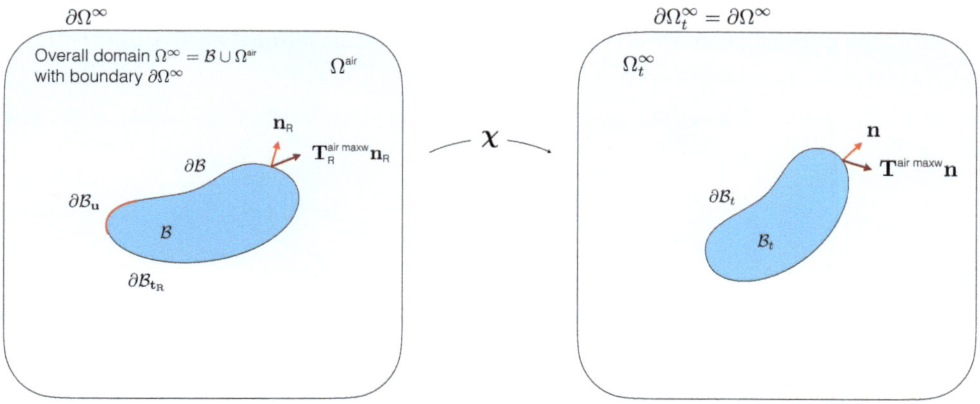

Fig. 32.2 Setting of the magneto-elastic boundary-value problem: A deformable and magnetizable body $\mathcal{B}$ supported at $\partial\mathcal{B}_{\mathrm{u}}$ and surrounded by a magnetically permeable air-like domain Ω^{air}. The overall domain is $\Omega^{\infty} = \mathcal{B} \cup \Omega^{\mathrm{air}}$. The boundary of the overall domain $\partial\Omega^{\infty}$ is sufficiently remote from $\partial\mathcal{B}$ such that the magnetic field perturbations caused by magnetization of $\mathcal{B}$ essentially vanish at $\partial\Omega^{\infty}$.

With respect to Fig. 32.2, we consider a deformable and magnetizable body $\mathcal{B}$ supported at $\partial\mathcal{B}_{\mathrm{u}}$ and surrounded by a magnetically permeable air-like domain Ω^{air}. The overall domain is the union of these two, $\Omega^{\infty} = \mathcal{B} \cup \Omega^{\mathrm{air}}$, and its boundary $\partial\Omega^{\infty}$ is presumed to be sufficiently remote from $\partial\mathcal{B}$ such that the magnetic field perturbations caused by magnetization of $\mathcal{B}$ essentially vanish at Ω^{∞}. In what follows:

- We denote the positions of points in the domain Ω^{∞} by $\mathbf{X}$, and extend the definition of our motion function $\mathbf{x} = \boldsymbol{\chi}(\mathbf{X}, t)$ to apply not only for points $\mathbf{X} \in \mathcal{B}$ but also to the extended region Ω^{∞}, so that the air-like region surrounding $\mathcal{B}$ is also taken to be *deformable*. Thus, we regard the initial configuration Ω^{∞} with boundary $\partial\Omega^{\infty}$ to be the reference configuration for the combined solid body and the air-like region surrounding it.
- The deformed domain is denoted by $\Omega_t^{\infty} = \boldsymbol{\chi}(\Omega^{\infty}, t)$.

- For definiteness, we assume that the motion function is suitably constrained so that the far-field boundary of the total domain remains fixed in shape, that is $\partial\Omega_t^\infty = \partial\Omega^\infty$.
- The initial boundary-value problem is stated in terms of the displacement field $\mathbf{u}(\mathbf{X},t)$ and the magnetic potential $\phi(\mathbf{X},t)$ which are defined for all $(\mathbf{X},t) \in \Omega^\infty \times [0,T]$. From $\mathbf{u}(\mathbf{X},t)$ and $\phi(\mathbf{X},t)$ one directly obtains

$$\mathbf{F}(\mathbf{X},t) = \mathbf{1} + \nabla\mathbf{u}(\mathbf{X},t) \qquad \text{and} \qquad \mathbf{h}_{\text{R}} = -\nabla\phi(\mathbf{X},t).$$

We emphasize that the deformation of the air domain is introduced here for the numerical convenience of applying far-field boundary conditions for the magnetic potential ϕ, and importantly also to account for the change in magnetic field immediately outside the boundary of the MRE due to the magnetization and deformation of the MRE.

Accordingly, in modeling the effects of air surrounding the body we assume that:

(i) The domain Ω^{air} is not magnetizable.

(ii) The non-magnetic air domain has a negligibly small stiffness relative to the stiffness of the magnetizable solid.

(iii) The spatial Maxwell stress in the domain Ω^{air} is given by the vacuum Maxwell stress,

$$\mathbf{T}^{\text{air maxw}} \stackrel{\text{def}}{=} \mu_0 \left(\mathbf{h} \otimes \mathbf{h} - \frac{1}{2}(\mathbf{h} \cdot \mathbf{h})\mathbf{1} \right), \tag{32.5.1}$$

and that this stress *does not* cause any deformation of the air domain (Rambausek and Schöbrel, 2023).

(iv) All effects of the Maxwell stress in the deformable air on the magnetizable solid are reflected by a referential magnetic traction defined by

$$\mathbf{T}_{\text{R}}^{\text{air maxw}}\mathbf{n}_{\text{R}} \stackrel{\text{def}}{=} \underbrace{\left[J\mu_0 \left(\mathbf{h} \otimes \mathbf{h} - \frac{1}{2}(\mathbf{h} \cdot \mathbf{h})\mathbf{1} \right) \mathbf{F}^{-\top} \right]}_{\mathbf{T}_{\text{R}}^{\text{air maxw}}} \mathbf{n}_{\text{R}}, \tag{32.5.2}$$

which acts on the boundary $\partial\mathcal{B}_{t_{\text{R}}}$ of the magnetizable solid $\mathcal{B}$. In the traction relation (32.5.2), the deformation gradient $\mathbf{F}$, with $J = \det\mathbf{F} > 0$, represents the value of deformation gradient of the solid on its boundary $\partial\mathcal{B}_{t_{\text{R}}}$, $\mathbf{h}$ is the spatial magnetic field in the air, and $\mathbf{n}_{\text{R}}$ is the outward unit normal to $\partial\mathcal{B}_{t_{\text{R}}}$.

32.5.2 Summary of the governing partial differential equations and boundary conditions for the combined air–solid domain

With these underlying requirements, the governing partial differential equations and boundary conditions for the boundary-value problem for MREs which accounts for the air-space surrounding a magnetizable body are taken as detailed below:

1. The governing partial differential equation for the mechanical problem is

$$\operatorname{Div}\mathbf{T}_{\mathrm{R}}+\mathbf{f}_{\mathrm{R}}=\mathbf{0}\qquad\forall\mathbf{X}\in\Omega^{\infty},\tag{32.5.3}$$

where,

(i) The Piola stress in the magnetizable body $\mathcal{B}$ is $\mathbf{T}_{\mathrm{R}}=J\mathbf{T}\mathbf{F}^{-\top}$ with the Cauchy stress $\mathbf{T}$ given by (32.4.35).

(ii) The Piola stress $\mathbf{T}_{\mathrm{R}}^{\mathrm{air}}$ in the mechanically deformable but non-magnetizable region Ω^{air} is given by

$$\begin{aligned}\mathbf{T}_{\mathrm{R}}^{\mathrm{air}}&\stackrel{\text{def}}{=}J\mathbf{T}^{\mathrm{air}}\mathbf{F}^{-\top}\qquad\text{with}\\ \mathbf{T}^{\mathrm{air}}&=J^{-1}G^{\mathrm{air}}\bar{\mathbf{B}}_0-p^{\mathrm{air}}\mathbf{1},\\ p^{\mathrm{air}}&\stackrel{\text{def}}{=}-K^{\mathrm{air}}(J-1)\end{aligned}\tag{32.5.4}$$

for all $\mathbf{X}\in\Omega^{\mathrm{air}}$. Here $(G^{\mathrm{air}},K^{\mathrm{air}})$ denote positive-valued shear and bulk moduli for the air surrounding the magnetizable body. Equation (32.5.4) reflects our assumption that the presence of a magnetic field in the air domain does not affect the mechanics of the non-magnetic air domain.

(iii) Also, the viscous right Cauchy–Green stretch tensors $\mathbf{C}^{v(\alpha)}$ in the magnetizable body $\mathcal{B}$ evolve according to (32.4.60), while the effects of viscoelasticity are neglected in the air-domain Ω^{air}.

(iv) The vector field $\mathbf{f}_{\mathrm{R}}$ is a generalized referential body force,

$$\mathbf{f}_{\mathrm{R}}=\mathbf{f}_{0\mathrm{R}}-\rho_{\mathrm{R}}\ddot{\mathbf{u}},\tag{32.5.5}$$

where $\mathbf{f}_{0\mathrm{R}}$ represents the conventional body force per unit volume of the reference body, and $(-\rho_{\mathrm{R}}\ddot{\mathbf{u}})$ represents the inertial body force, with ρ_{R} the mass density of the referential body and $\ddot{\mathbf{u}}$ the acceleration. The effects of inertia are neglected in the air domain Ω^{air}.

Next, with $\partial\mathcal{B}_{\mathrm{u}}$ and $\partial\mathcal{B}_{\mathrm{t_R}}$ denoting complementary subsurfaces of the boundary $\partial\mathcal{B}$ of the body $\mathcal{B}$, we consider a pair of boundary conditions:

$$\begin{aligned}\mathbf{u}&=\bar{\mathbf{u}}&&\text{on }\partial\mathcal{B}_{\mathrm{u}},\\ \mathbf{T}_{\mathrm{R}}^{\mathrm{solid}}\mathbf{n}_{\mathrm{R}}&=\bar{\mathbf{t}}_{\mathrm{R}}+\mathbf{T}_{\mathrm{R}}^{\mathrm{air\ maxw}}\mathbf{n}_{\mathrm{R}}+\mathbf{T}_{\mathrm{R}}^{\mathrm{air}}\mathbf{n}_{\mathrm{R}}&&\text{on }\partial\mathcal{B}_{\mathrm{t_R}},\end{aligned}\tag{32.5.6}$$

where the superscript $(\bullet)^{\mathrm{solid}}$ denotes a quantity defined within the body $\mathcal{B}$. In the boundary conditions above $\bar{\mathbf{u}}$ and $(\bar{\mathbf{t}}_{\mathrm{R}}+\mathbf{T}_{\mathrm{R}}^{\mathrm{air\ maxw}}\mathbf{n}_{\mathrm{R}}+\mathbf{T}_{\mathrm{R}}^{\mathrm{air}}\mathbf{n}_{\mathrm{R}})$ are prescribed displacement and tractions, both functions of $(\mathbf{X},t)$. Regarding the prescribed boundary tractions:

- The term $\bar{\mathbf{t}}_{\mathrm{R}}$ is a usual applied mechanical traction considered in boundary-value problems for solid media.

- The term $\mathbf{T}_{\mathrm{R}}^{\mathrm{air\ maxw}}\mathbf{n}_{\mathrm{R}}$ is a traction acting on the boundary of the magnetizable solid due the Maxwell stress in the air, given by (32.5.2).
- The term $\mathbf{T}_{\mathrm{R}}^{\mathrm{air}}\mathbf{n}_{\mathrm{R}}$ is a traction which arises due to the mechanical deformation of the non-magnetizable air—this contribution to the boundary traction will be small if the values of the shear and bulk moduli $(G^{\mathrm{air}}, K^{\mathrm{air}})$ of the air are chosen to be sufficiently small in magnitude relative to the corresponding moduli of the magnetizable solid.

Introducing the notation,[6]

$$[\![\mathbf{T}_{\mathrm{R}}]\!] \stackrel{\text{def}}{=} \mathbf{T}_{\mathrm{R}}^{\mathrm{air}} - \mathbf{T}_{\mathrm{R}}^{\mathrm{solid}}, \tag{32.5.8}$$

the traction boundary condition $(32.5.6)_2$ may be written as

$$[\![\mathbf{T}_{\mathrm{R}}]\!]\mathbf{n}_{\mathrm{R}} + \bar{\mathbf{t}}_{\mathrm{R}} + \mathbf{T}_{\mathrm{R}}^{\mathrm{air\ maxw}}\mathbf{n}_{\mathrm{R}} = \mathbf{0}. \tag{32.5.9}$$

We will show in the next chapter that when we consider a suitable weak form for the mechanical problem, this jump condition is automatically satisfied (in a weak sense).

Standard initial conditions involve the specifications:

$$\mathbf{u}(\mathbf{X},0) = \mathbf{u}_0(\mathbf{X}), \qquad \dot{\mathbf{u}}(\mathbf{X},0) = \dot{\mathbf{u}}_0(\mathbf{X}), \qquad \text{and} \qquad \mathbf{C}^{v(\alpha)}(\mathbf{X},0) = \mathbf{1}, \tag{32.5.10}$$

with $\mathbf{u}_0$ and $\dot{\mathbf{u}}_0$ prescribed functions on Ω^∞.

2. The governing partial differential equation for the magnetic problem is

$$\mathrm{Div}\,\mathbf{b}_{\mathrm{R}} = 0 \qquad \forall\,\mathbf{X} \in \Omega^\infty, \tag{32.5.11}$$

where

(i) the magnetic flux density in the body $\mathbf{b}_{\mathrm{R}} \in \mathcal{B}$ is given by (32.4.38), and

(ii) the magnetic flux density in the deformable air-like region Ω^{air} is given by

$$\mathbf{b}_{\mathrm{R}}^{\mathrm{air}} = J\mathbf{F}^{-1}\mathbf{b}^{\mathrm{air}} \qquad \text{with} \qquad \mathbf{b}^{\mathrm{air}} = \mu_0\mathbf{h}. \tag{32.5.12}$$

For simplicity, we only consider boundary conditions for ϕ on the outer boundary of the total domain of the form

$$\phi = \bar{\phi} \quad \text{on } \partial\Omega^\infty. \tag{32.5.13}$$

[6] As in (32.5.8), we use the notation

$$[\![\varphi]\!] \stackrel{\text{def}}{=} \varphi^{\mathrm{air}} - \varphi^{\mathrm{solid}}. \tag{32.5.7}$$

for the jump in a quantity φ in Ω^∞ across $\partial\mathcal{B}$, with $\mathbf{n}_{\mathrm{R}}$ pointing from "solid" to "air" across $\partial\mathcal{B}$. Whenever convenient, we suppress the superscript $(\bullet)^{\mathrm{solid}}$ for quantities defined within $\mathcal{B}$.

3. In addition, the magnetic fields $\mathbf{h}_R$ and $\mathbf{b}_R$ are required to satisfy the standard jump conditions (Dorfmann and Ogden, 2014)

$$\mathbf{n}_R \times [\![\mathbf{h}_R]\!] = \mathbf{0}, \qquad \mathbf{n}_R \cdot [\![\mathbf{b}_R]\!] = 0 \qquad \text{across } \partial\mathcal{B}, \tag{32.5.14}$$

which ensure that the tangential component of the magnetic field $\mathbf{h}_R$ and the normal component of the magnetic flux $\mathbf{b}_R$ remain continuous across the boundary $\partial\mathcal{B}$ of $\mathcal{B}$ with outward unit normal $\mathbf{n}_R$. We shall require that the scalar magnetic potential ϕ is a class C^0-continuous field, so that it is continuous across $\partial\mathcal{B}$; then $(32.5.14)_1$ is automatically satisfied.[7] It remains to consider the boundary condition $(32.5.14)_2$. We will show in the next chapter that when we consider a suitable weak form for the magnetic problem, this jump condition is also automatically satisfied (in a weak sense).

32.A Appendix: Detailed calculations for the stress and magnetic flux expressions for s-MREs

32.A.1 Detailed calculations for the stress and magnetic flux for the "ideal" theory of s-MREs

Calculation of the stress

The mechanical and magnetic contributions to the Piola stress are:

$$\mathbf{T}_R^{\text{mech}} \stackrel{\text{def}}{=} \mathbf{F}\left(2\frac{\partial \Omega_R^{\text{mech}}}{\partial \mathbf{C}}\right), \quad \mathbf{T}_R^{\text{totmag}} \stackrel{\text{def}}{=} \mathbf{F}\left(2\frac{\partial \Omega_R^{\text{totmag}}}{\partial \mathbf{C}}\right). \tag{32.A.1}$$

- First, straightforward calculations give (cf., Gurtin et al., 2010, Chapter 55):

$$2\frac{\partial \Omega_R^{\text{mech}}}{\partial \mathbf{C}} = \frac{2}{J^{2/3}}\left[\frac{\partial \Omega_R^{\text{mech}}(\bar{I}_1, J)}{\partial \bar{\mathbf{C}}} - \frac{1}{3}\left(\bar{\mathbf{C}} : \frac{\partial \Omega_R^{\text{mech}}(\bar{I}_1, J)}{\partial \bar{\mathbf{C}}}\right)\bar{\mathbf{C}}^{-1}\right] + J\frac{\partial \Omega_R^{\text{mech}}(\bar{I}_1, J)}{\partial J}\mathbf{C}^{-1}, \tag{32.A.2}$$

$$= \frac{2}{J^{2/3}}\frac{\partial \Omega_R^{\text{mech}}(\bar{I}_1, J)}{\partial \bar{I}_1}\left[\mathbf{1} - \frac{1}{3}(\operatorname{tr}\bar{\mathbf{C}})\bar{\mathbf{C}}^{-1}\right] + J\,K(J-1)\mathbf{C}^{-1}, \tag{32.A.3}$$

$$= J^{-2/3}G\left[\mathbf{1} - \frac{1}{3}(\operatorname{tr}\bar{\mathbf{C}})\bar{\mathbf{C}}^{-1}\right] + J\,K(J-1)\mathbf{C}^{-1}. \tag{32.A.4}$$

Hence the mechanical part of the Piola stress is

$$\mathbf{T}_R^{\text{mech}} = J^{-2/3}G\left[\mathbf{F} - \frac{1}{3}(\operatorname{tr}\mathbf{C})\mathbf{F}^{-\top}\right] + J\,K(J-1)\mathbf{F}^{-\top}. \tag{32.A.5}$$

- Next we calculate the magnetic Piola stress:

[7] A function Φ is of class C^0 if Φ is continuous and differentiable, but $\nabla\Phi$ is only piecewise continuous on $\mathcal{B}$.

$$\mathbf{T}_{\mathrm{R}}^{\mathrm{totmag}} = \mathbf{F}\left(2\,\frac{\partial}{\partial \mathbf{C}}\left(J\left(-\mu\,\frac{1}{2}\mathbf{h}_{\mathrm{R}}\cdot\mathbf{C}^{-1}\mathbf{h}_{\mathrm{R}}\right)\right)\right)$$
$$= -\mu\,\mathbf{F}\frac{\partial(J\,\mathbf{h}_{\mathrm{R}}\cdot\mathbf{C}^{-1}\mathbf{h}_{\mathrm{R}})}{\partial \mathbf{C}}.$$

Taking the derivative of the term

$$J\,\mathbf{h}_{\mathrm{R}}\cdot\mathbf{C}^{-1}\mathbf{h}_{\mathrm{R}},$$

with respect to $\mathbf{C}$ we find that

$$\frac{\partial}{\partial \mathbf{C}}\left(J\,\mathbf{h}_{\mathrm{R}}\cdot\mathbf{C}^{-1}\mathbf{h}_{\mathrm{R}}\right) = \frac{\partial J}{\partial \mathbf{C}}(\mathbf{h}_{\mathrm{R}}\cdot\mathbf{C}^{-1}\mathbf{h}_{\mathrm{R}}) + J\,\frac{\partial(\mathbf{h}_{\mathrm{R}}\cdot\mathbf{C}^{-1}\mathbf{h}_{\mathrm{R}})}{\partial \mathbf{C}}.$$

Recall that

$$\frac{\partial J}{\partial \mathbf{C}} = \frac{1}{2}J\,\mathbf{C}^{-1}.$$

Also, for an arbitrary invertible tensor $\mathbf{A}$, we recall the standard result that

$$\frac{\partial A^{-1}_{ij}}{\partial A_{kl}} = -A^{-1}_{ik}A^{-1}_{lj}.$$

Hence, suppressing the subscript "R" on $\mathbf{h}_{\mathrm{R}}$, consider the derivative

$$\begin{aligned}\frac{\partial(h_i C^{-1}_{ij} h_j)}{\partial C_{kl}} &= h_i\,h_j\frac{\partial C^{-1}_{ij}}{\partial C_{kl}}\\ &= -h_i\,h_j C^{-1}_{ik}C^{-1}_{lj}\\ &= -h_i\,h_j C^{-1}_{ki}C^{-1}_{lj} \qquad \text{(using the symmetry of } \mathbf{C}\text{)},\\ &= -(C^{-1}_{ki}h_i)(C^{-1}_{lj}h_j).\end{aligned}$$

Thus,

$$\frac{\partial(\mathbf{h}_{\mathrm{R}}\cdot\mathbf{C}^{-1}\mathbf{h}_{\mathrm{R}})}{\partial \mathbf{C}} = -(\mathbf{C}^{-1}\mathbf{h}_{\mathrm{R}})\otimes(\mathbf{C}^{-1}\mathbf{h}_{\mathrm{R}}), \tag{32.A.6}$$

and hence

$$\begin{aligned}\frac{\partial}{\partial \mathbf{C}}\left(J\,\mathbf{h}_{\mathrm{R}}\cdot\mathbf{C}^{-1}\mathbf{h}_{\mathrm{R}}\right) &= \frac{1}{2}J\,\mathbf{C}^{-1}(\mathbf{h}_{\mathrm{R}}\cdot\mathbf{C}^{-1}\mathbf{h}_{\mathrm{R}}) - J\,(\mathbf{C}^{-1}\mathbf{h}_{\mathrm{R}})\otimes(\mathbf{C}^{-1}\mathbf{h}_{\mathrm{R}}),\\ &= -J\left[(\mathbf{C}^{-1}\mathbf{h}_{\mathrm{R}})\otimes(\mathbf{C}^{-1}\mathbf{h}_{\mathrm{R}}) - \frac{1}{2}(\mathbf{h}_{\mathrm{R}}\cdot\mathbf{C}^{-1}\mathbf{h}_{\mathrm{R}})\mathbf{C}^{-1}\right]\end{aligned}$$

Finally since

$$\mathbf{T}_{\text{R}}^{\text{totmag}} = -\mu\,\mathbf{F}\frac{\partial(J\,\mathbf{h}_{\text{R}}\cdot\mathbf{C}^{-1}\mathbf{h}_{\text{R}})}{\partial\mathbf{C}},$$

we obtain

$$\mathbf{T}_{\text{R}}^{\text{totmag}} = J\,\mu\,\mathbf{F}\left[(\mathbf{C}^{-1}\mathbf{h}_{\text{R}})\otimes(\mathbf{C}^{-1}\mathbf{h}_{\text{R}}) - \frac{1}{2}\,(\mathbf{h}_{\text{R}}\cdot\mathbf{C}^{-1}\mathbf{h}_{\text{R}})\mathbf{C}^{-1}\right]. \tag{32.A.7}$$

- Thus, from (32.A.5) and (32.A.7), the Piola stress is given by,

$$\begin{aligned}\mathbf{T}_{\text{R}} &= J^{-2/3}G\left(\mathbf{F} - \frac{1}{3}(\operatorname{tr}\mathbf{C})\mathbf{F}^{-\top}\right) + K\,J\,(J-1)\mathbf{F}^{-\top}\\ &\quad + \mu\,J\mathbf{F}\left[(\mathbf{C}^{-1}\mathbf{h}_{\text{R}})\otimes(\mathbf{C}^{-1}\mathbf{h}_{\text{R}}) - \frac{1}{2}\,(\mathbf{h}_{\text{R}}\cdot\mathbf{C}^{-1}\mathbf{h}_{\text{R}})\mathbf{C}^{-1}\right].\end{aligned} \tag{32.A.8}$$

Using $\mathbf{T} = J^{-1}\mathbf{T}_{\text{R}}\mathbf{F}^{\top}$ and $\mathbf{h} = \mathbf{F}^{-\top}\mathbf{h}_{\text{R}}$, the Cauchy stress is given by

$$\mathbf{T} = J^{-1}G\bar{\mathbf{B}}_0 + K(J-1)\mathbf{1} + \mu\left[\mathbf{h}\otimes\mathbf{h} - \frac{1}{2}\,(\mathbf{h}\cdot\mathbf{h})\mathbf{1}\right], \tag{32.A.9}$$

where $\bar{\mathbf{B}}_0$ is the deviatoric part of the distortional left Cauchy–Green stretch tensor $\bar{\mathbf{B}} = J^{-2/3}\mathbf{F}\mathbf{F}^{\top}$. The second term in (32.A.9) is the usual form of the **Maxwell stress** in the material.

Calculation of the magnetic flux density

Next, the magnetic flux density is calculated using

$$\mathbf{b}_{\text{R}} = -\frac{\partial\Omega_{\text{R}}^{\text{totmag}}}{\partial\mathbf{h}_{\text{R}}} \qquad \text{with} \qquad \Omega_{\text{R}}^{\text{totmag}} = J\left(-\frac{1}{2}\,\mu\,\mathbf{h}_{\text{R}}\cdot\mathbf{C}^{-1}\mathbf{h}_{\text{R}}\right) \tag{32.A.10}$$

which gives

$$\begin{aligned}\mathbf{b}_{\text{R}} &= J\,\frac{1}{2}\,\mu\,\frac{\partial(\mathbf{h}_{\text{R}}\cdot\mathbf{C}^{-1}\mathbf{h}_{\text{R}})}{\partial\mathbf{h}_{\text{R}}},\\ &= J\,\frac{1}{2}\,\mu\,2\mathbf{C}^{-1}\mathbf{h}_{\text{R}},\\ &= J\,\mu\,\mathbf{C}^{-1}\mathbf{h}_{\text{R}}.\end{aligned} \tag{32.A.11}$$

Thus the referential magnetic flux density is given by

$$\mathbf{b}_{\text{R}} = J\,\mu\,\mathbf{C}^{-1}\mathbf{h}_{\text{R}}. \tag{32.A.12}$$

Using

$$\mathbf{h}_{\mathrm{R}} = \mathbf{F}^{\top}\mathbf{h} \qquad \text{and} \qquad \mathbf{b}_{\mathrm{R}} = J\mathbf{F}^{-1}\mathbf{b}, \tag{32.A.13}$$

the spatial form of (32.A.12) is

$$\mathbf{b} = \mu\mathbf{h}. \tag{32.A.14}$$

32.A.2 **Detailed calculations for the stress and magnetic flux for the general theory of s-MREs**

Calculation of the stress

REMARK

Let Ψ be defined on the space of symmetric tensors, and consider the function $\Lambda(\mathbf{F})$, defined on the set of all tensors through the requirement that

$$\Lambda(\mathbf{F}) = \Psi(\mathbf{C}), \qquad \mathbf{C} = \mathbf{F}^{\top}\mathbf{F}. \tag{32.A.15}$$

Then a standard result from tensor calculus is (Gurtin et al., 2010, eq. (3.30))

$$\frac{\partial\Lambda(\mathbf{F})}{\partial\mathbf{F}} = 2\mathbf{F}\frac{\partial\Psi(\mathbf{C})}{\partial\mathbf{C}}. \tag{32.A.16}$$

We shall use this result repeatedly in the calculations below. For example,

$$\frac{\partial(J\,\mathbf{h}_{\mathrm{R}}\cdot\mathbf{C}^{-1}\mathbf{h}_{\mathrm{R}})}{\partial\mathbf{F}} = 2\mathbf{F}\frac{\partial(J\,\mathbf{h}_{\mathrm{R}}\cdot\mathbf{C}^{-1}\mathbf{h}_{\mathrm{R}})}{\partial\mathbf{C}}.$$

From (32.2.67) and (32.4.31) it follows that the Piola stress is given by,

$$\mathbf{T}_{\mathrm{R}} = \mathbf{T}_{\mathrm{R}}^{\mathrm{eq}} + \mathbf{T}_{\mathrm{R}}^{\mathrm{neq}} + \mathbf{T}_{\mathrm{R}}^{\mathrm{maxw}} + \mathbf{T}_{\mathrm{R}}^{\mathrm{mag}}, \tag{32.A.17}$$

where

$$\begin{gathered}\mathbf{T}_{\mathrm{R}}^{\mathrm{eq}} \stackrel{\mathrm{def}}{=} \mathbf{F}\left(2\,\frac{\partial\Omega_{\mathrm{R}}^{\mathrm{eq}}}{\partial\mathbf{C}}\right), \quad \mathbf{T}_{\mathrm{R}}^{\mathrm{neq}(\alpha)} \stackrel{\mathrm{def}}{=} \mathbf{F}^{e(\alpha)}\left(2\,\frac{\partial\Omega_{\mathrm{R}}^{\mathrm{neq}(\alpha)}}{\partial\mathbf{C}^{e(\alpha)}}\right),\\ \mathbf{T}_{\mathrm{R}}^{\mathrm{maxw}} \stackrel{\mathrm{def}}{=} \mathbf{F}\left(2\frac{\partial\Omega_{\mathrm{R}}^{\mathrm{maxw}}}{\partial\mathbf{C}}\right), \quad \mathbf{T}_{\mathrm{R}}^{\mathrm{mag}} \stackrel{\mathrm{def}}{=} \mathbf{F}\left(2\frac{\partial\Omega_{\mathrm{R}}^{\mathrm{mag}}}{\partial\mathbf{C}}\right).\end{gathered} \tag{32.A.18}$$

- First, straightforward calculations give (cf., Gurtin et al., 2010, Chapter 55):

$$2\frac{\partial\Omega_{\mathrm{R}}^{\mathrm{eq}}}{\partial\mathbf{C}} = \frac{2}{J^{2/3}}\left[\frac{\partial\Omega_{\mathrm{R}}^{\mathrm{eq}}(\bar{I}_1,J)}{\partial\bar{\mathbf{C}}} - \frac{1}{3}\left(\bar{\mathbf{C}}:\frac{\partial\Omega_{\mathrm{R}}^{\mathrm{eq}}(\bar{I}_1,J)}{\partial\bar{\mathbf{C}}}\right)\bar{\mathbf{C}}^{-1}\right] + J\frac{\partial\Omega_{\mathrm{R}}^{\mathrm{eq}}(\bar{I}_1,J)}{\partial J}\mathbf{C}^{-1}, \tag{32.A.19}$$

$$= \frac{2}{J^{2/3}}\frac{\partial\Omega_{\mathrm{R}}^{\mathrm{eq}}(\bar{I}_1,J)}{\partial\bar{I}_1}\left[\mathbf{1} - \frac{1}{3}(\operatorname{tr}\bar{\mathbf{C}})\bar{\mathbf{C}}^{-1}\right] + J\,K(J-1)\mathbf{C}^{-1}, \tag{32.A.20}$$

$$= J^{-2/3}G_{\mathrm{eq}}\left[\mathbf{1} - \frac{1}{3}(\operatorname{tr}\bar{\mathbf{C}})\bar{\mathbf{C}}^{-1}\right] + J\,K(J-1)\mathbf{C}^{-1}. \tag{32.A.21}$$

Hence the mechanical equilibrium part of the Piola stress is

$$\mathbf{T}_{\mathrm{R}}^{\mathrm{eq}} = J^{-2/3}G_{\mathrm{eq}}\left[\mathbf{F} - \frac{1}{3}(\operatorname{tr}\mathbf{C})\mathbf{F}^{-\top}\right] + J\,K(J-1)\mathbf{F}^{-\top}. \tag{32.A.22}$$

- Similarly, for the non-equilibrium branch

$$2\frac{\partial\Omega_{\mathrm{R}}^{\mathrm{neq}(\alpha)}}{\partial\mathbf{C}^{e(\alpha)}} = \frac{2}{(J^{e(\alpha)})^{2/3}}\left[\frac{\partial\Omega_{\mathrm{R}}^{\mathrm{neq}(\alpha)}(\bar{I}_1^{e(\alpha)},J^{e(\alpha)})}{\partial\bar{\mathbf{C}}^{e(\alpha)}} - \frac{1}{3}\left(\bar{\mathbf{C}}^{e(\alpha)}:\frac{\partial\Omega_{\mathrm{R}}^{\mathrm{neq}(\alpha)}(\bar{I}_1^{e(\alpha)},J^{e(\alpha)})}{\partial\bar{\mathbf{C}}^{e(\alpha)}}\right)(\bar{\mathbf{C}}^{e(\alpha)})^{-1}\right], \tag{32.A.23}$$

$$= \frac{2}{(J^{e(\alpha)})^{2/3}}\frac{\partial\Omega_{\mathrm{R}}^{\mathrm{neq}(\alpha)}(\bar{I}_1^{e(\alpha)},J^{e(\alpha)})}{\partial\bar{I}_1^{e(\alpha)}}\left[\mathbf{1} - \frac{1}{3}(\operatorname{tr}\bar{\mathbf{C}}^{e(\alpha)})(\bar{\mathbf{C}}^{e(\alpha)})^{-1}\right], \tag{32.A.24}$$

$$= (J^{e(\alpha)})^{-2/3}G_{\mathrm{neq}}^{(\alpha)}\left[\mathbf{1} - \frac{1}{3}(\operatorname{tr}\bar{\mathbf{C}}^{e(\alpha)})(\bar{\mathbf{C}}^{e(\alpha)})^{-1}\right]. \tag{32.A.25}$$

Hence the mechanical non-equilibrium part of the Piola stress in each branch is

$$\mathbf{T}_{\mathrm{R}}^{\mathrm{neq}(\alpha)} = (J^{e(\alpha)})^{-2/3}G_{\mathrm{neq}}^{(\alpha)}\left[\mathbf{F}^{e(\alpha)} - \frac{1}{3}(\operatorname{tr}\mathbf{C}^{e(\alpha)})(\mathbf{F}^{e(\alpha)})^{-\top}\right] \tag{32.A.26}$$

- Next, we calculate the Piola stress contribution from the Maxwell energy:

$$\mathbf{T}_{\mathrm{R}}^{\mathrm{maxw}} = \mathbf{F}\left(2\frac{\partial}{\partial\mathbf{C}}\left(-(1/2)\,\mu_0\,J\,\mathbf{h}_{\mathrm{R}}\cdot\mathbf{C}^{-1}\mathbf{h}_{\mathrm{R}}\right)\right)$$
$$= -\mu_0\mathbf{F}\frac{\partial(J\,\mathbf{h}_{\mathrm{R}}\cdot\mathbf{C}^{-1}\mathbf{h}_{\mathrm{R}})}{\partial\mathbf{C}},$$

which yields, following a similar derivation to that in Section 32.A.1,

$$\mathbf{T}_{\mathrm{R}}^{\mathrm{maxw}} = J\,\mu_0\,\mathbf{F}\left[(\mathbf{C}^{-1}\mathbf{h}_{\mathrm{R}})\otimes(\mathbf{C}^{-1}\mathbf{h}_{\mathrm{R}}) - \frac{1}{2}(\mathbf{h}_{\mathrm{R}}\cdot\mathbf{C}^{-1}\mathbf{h}_{\mathrm{R}})\mathbf{C}^{-1}\right]. \tag{32.A.27}$$

- Next, consider the contribution

$$
\begin{aligned}
\mathbf{T}_{\mathrm{R}}^{\mathrm{mag}} &= \mathbf{F}\left(2\frac{\partial \Omega_{\mathrm{R}}^{\mathrm{mag}}(\mathbf{C},\mathbf{h}_{\mathrm{R}})}{\partial \mathbf{C}}\right) \\
&= 2\mathbf{F}\left[\frac{\partial}{\partial \mathbf{C}}\left(-J\,\mu_0\left[\frac{m_s^2}{\chi}\ln\left(\cosh\left(\frac{\chi}{m_s}\sqrt{\mathbf{h}_{\mathrm{R}}\cdot\mathbf{C}^{-1}\mathbf{h}_{\mathrm{R}}}\right)\right)\right]\right)\right] \\
&= 2\mathbf{F}\left[-\frac{\partial J}{\partial \mathbf{C}}\,\mu_0\left[\frac{m_s^2}{\chi}\ln\left(\cosh\left(\frac{\chi}{m_s}\sqrt{\mathbf{h}_{\mathrm{R}}\cdot\mathbf{C}^{-1}\mathbf{h}_{\mathrm{R}}}\right)\right)\right]\right. \\
&\qquad \left. -\,J\mu_0\left[\frac{m_s}{2\sqrt{\mathbf{h}_{\mathrm{R}}\cdot\mathbf{C}^{-1}\mathbf{h}_{\mathrm{R}}}}\tanh\left(\frac{\chi}{m_s}\sqrt{\mathbf{h}_{\mathrm{R}}\cdot\mathbf{C}^{-1}\mathbf{h}_{\mathrm{R}}}\right)\frac{\partial}{\partial \mathbf{C}}\left(\mathbf{h}_{\mathrm{R}}\cdot\mathbf{C}^{-1}\mathbf{h}_{\mathrm{R}}\right)\right]\right].
\end{aligned}
$$

Using

$$
\frac{\partial J}{\partial \mathbf{C}} = \frac{1}{2}J\,\mathbf{C}^{-1},
$$

and

$$
\frac{\partial(\mathbf{h}_{\mathrm{R}}\cdot\mathbf{C}^{-1}\mathbf{h}_{\mathrm{R}})}{\partial \mathbf{C}} = -(\mathbf{C}^{-1}\mathbf{h}_{\mathrm{R}})\otimes(\mathbf{C}^{-1}\mathbf{h}_{\mathrm{R}}),
$$

we obtain

$$
\begin{aligned}
\mathbf{T}_{\mathrm{R}}^{\mathrm{mag}} &= -J\,\mu_0\left[\frac{m_s^2}{\chi}\ln\left(\cosh\left(\frac{\chi}{m_s}\sqrt{\mathbf{h}_{\mathrm{R}}\cdot\mathbf{C}^{-1}\mathbf{h}_{\mathrm{R}}}\right)\right)\right]\mathbf{F}^{-\top} \\
&\quad + J\mu_0\left[\frac{m_s}{\sqrt{\mathbf{h}_{\mathrm{R}}\cdot\mathbf{C}^{-1}\mathbf{h}_{\mathrm{R}}}}\tanh\left(\frac{\chi}{m_s}\sqrt{\mathbf{h}_{\mathrm{R}}\cdot\mathbf{C}^{-1}\mathbf{h}_{\mathrm{R}}}\right)\mathbf{F}\left((\mathbf{C}^{-1}\mathbf{h}_{\mathrm{R}})\otimes(\mathbf{C}^{-1}\mathbf{h}_{\mathrm{R}})\right)\right].
\end{aligned}
\tag{32.A.28}
$$

Thus, the total Piola stress is given by

$$
\begin{aligned}
\mathbf{T}_{\mathrm{R}} &= J^{-2/3}G_{\mathrm{eq}}\left(\mathbf{F}-\frac{1}{3}(\mathrm{tr}\,\mathbf{C})\mathbf{F}^{-\top}\right)+J\,K(J-1)\mathbf{F}^{-\top} \\
&\quad + \sum_{\alpha=1}^{m}(J^{e(\alpha)})^{-2/3}G_{\mathrm{neq}}^{(\alpha)}\left[\mathbf{F}^{e(\alpha)}-\frac{1}{3}(\mathrm{tr}\,\mathbf{C}^{e(\alpha)})(\mathbf{F}^{e(\alpha)})^{-\top}\right] \\
&\quad + J\,\mu_0\,\mathbf{F}\left[(\mathbf{C}^{-1}\mathbf{h}_{\mathrm{R}})\otimes(\mathbf{C}^{-1}\mathbf{h}_{\mathrm{R}})-\frac{1}{2}(\mathbf{h}_{\mathrm{R}}\cdot\mathbf{C}^{-1}\mathbf{h}_{\mathrm{R}})\mathbf{C}^{-1}\right] \\
&\quad - J\,\mu_0\left[\frac{m_s^2}{\chi}\ln\left(\cosh\left(\frac{\chi}{m_s}\sqrt{\mathbf{h}_{\mathrm{R}}\cdot\mathbf{C}^{-1}\mathbf{h}_{\mathrm{R}}}\right)\right)\right]\mathbf{F}^{-\top} \\
&\quad + J\mu_0\left[\frac{m_s}{\sqrt{\mathbf{h}_{\mathrm{R}}\cdot\mathbf{C}^{-1}\mathbf{h}_{\mathrm{R}}}}\tanh\left(\frac{\chi}{m_s}\sqrt{\mathbf{h}_{\mathrm{R}}\cdot\mathbf{C}^{-1}\mathbf{h}_{\mathrm{R}}}\right)\mathbf{F}\left((\mathbf{C}^{-1}\mathbf{h}_{\mathrm{R}})\otimes(\mathbf{C}^{-1}\mathbf{h}_{\mathrm{R}})\right)\right],
\end{aligned}
\tag{32.A.29}
$$

and using $\mathbf{T} = J^{-1}\mathbf{T}_R\mathbf{F}^\top$ and $\mathbf{h} = \mathbf{F}^{-\top}\mathbf{h}_R$ the Cauchy stress is given by

$$\begin{aligned}\mathbf{T} &= J^{-1} G_{eq}\, \bar{\mathbf{B}}_0 + K(J-1)\mathbf{1} + \sum_{\alpha=1}^{m} (J^{e(\alpha)})^{-1} G_{neq}^{(\alpha)}\, \bar{\mathbf{B}}_0^{e(\alpha)} \\ &+ \mu_0 \left[\mathbf{h}\otimes\mathbf{h} - \frac{1}{2}(\mathbf{h}\cdot\mathbf{h})\mathbf{1}\right] \\ &- \mu_0 \left[\frac{m_s^2}{\chi}\ln\left(\cosh\left(\frac{\chi}{m_s}\sqrt{\mathbf{h}\cdot\mathbf{h}}\right)\right)\right]\mathbf{1} + \mu_0\left[\frac{m_s}{\sqrt{\mathbf{h}\cdot\mathbf{h}}}\tanh\left(\frac{\chi}{m_s}\sqrt{\mathbf{h}\cdot\mathbf{h}}\right)(\mathbf{h}\otimes\mathbf{h})\right],\end{aligned} \tag{32.A.30}$$

where $\bar{\mathbf{B}}_0$ is the deviatoric part of the distortional left Cauchy–Green stretch tensor $\bar{\mathbf{B}} = J^{-2/3}\mathbf{F}\mathbf{F}^\top$.

REMARK

Note that in the limit of small $|\mathbf{h}|$ we obtain,

$$\begin{aligned}\ln\left(\cosh\left(\frac{\chi}{m_s}\sqrt{\mathbf{h}\cdot\mathbf{h}}\right)\right) &\approx \frac{1}{2}\frac{\chi^2}{m_s^2}(\mathbf{h}\cdot\mathbf{h}),\\ \tanh\left(\frac{\chi}{m_s}\sqrt{\mathbf{h}\cdot\mathbf{h}}\right) &\approx \frac{\chi}{m_s}\sqrt{\mathbf{h}\cdot\mathbf{h}},\end{aligned}$$

In this limit the magnetic and Maxwell contributions to the Cauchy stress (32.A.30) reduce to

$$\begin{aligned}\mathbf{T}^{mag} + \mathbf{T}^{maxw} &= \mu_0\left[\mathbf{h}\otimes\mathbf{h} - \frac{1}{2}(\mathbf{h}\cdot\mathbf{h})\mathbf{1}\right] \\ &+ \mu_0\chi(\mathbf{h}\otimes\mathbf{h}) - \mu_0\,\chi\,\frac{1}{2}(\mathbf{h}\cdot\mathbf{h})\,\mathbf{1},\end{aligned} \tag{32.A.31}$$

or

$$\mathbf{T}^{mag} + \mathbf{T}^{maxw} = \mu\left[\mathbf{h}\otimes\mathbf{h} - \frac{1}{2}(\mathbf{h}\cdot\mathbf{h})\mathbf{1}\right], \tag{32.A.32}$$

where

$$\mu \stackrel{\text{def}}{=} \mu_0(1+\chi) \tag{32.A.33}$$

is the **magnetic permeability** of the material.

Calculation of the magnetic flux density

The magnetic flux density is calculated using

$$\mathbf{b}_{\mathrm{R}} = -\frac{\partial \breve{\Omega}_{\mathrm{R}}}{\partial \mathbf{h}_{\mathrm{R}}}. \tag{32.A.34}$$

First, the Maxwell energy

$$\Omega_{\mathrm{R}}^{\mathrm{maxw}} = J\left(-\frac{1}{2}\,\mu_0\,\mathbf{h}_{\mathrm{R}}\cdot\mathbf{C}^{-1}\mathbf{h}_{\mathrm{R}}\right) \tag{32.A.35}$$

gives a contribution, by similar calculations as in (32.A.11), of

$$\mathbf{b}_{\mathrm{R}}^{\mathrm{maxw}} = J\,\mu_0\,\mathbf{C}^{-1}\mathbf{h}_{\mathrm{R}}. \tag{32.A.36}$$

Next, the magnetic energy

$$\Omega_{\mathrm{R}}^{\mathrm{mag}} = -J\,\mu_0\left[\frac{m_s^2}{\chi}\ln\left(\cosh\left(\frac{\chi}{m_s}\sqrt{\mathbf{h}_{\mathrm{R}}\cdot\mathbf{C}^{-1}\mathbf{h}_{\mathrm{R}}}\right)\right)\right], \tag{32.A.37}$$

gives a contribution

$$\begin{aligned}
\mathbf{b}_{\mathrm{R}}^{\mathrm{mag}} &= J\,\mu_0\,m_s\tanh\left(\frac{\chi}{m_s}\sqrt{\mathbf{h}_{\mathrm{R}}\cdot\mathbf{C}^{-1}\mathbf{h}_{\mathrm{R}}}\right)\frac{\partial\sqrt{\mathbf{h}_{\mathrm{R}}\cdot\mathbf{C}^{-1}\mathbf{h}_{\mathrm{R}}}}{\partial\mathbf{h}_{\mathrm{R}}} \\
&= J\,\mu_0\,m_s\tanh\left(\frac{\chi}{m_s}\sqrt{\mathbf{h}_{\mathrm{R}}\cdot\mathbf{C}^{-1}\mathbf{h}_{\mathrm{R}}}\right)\frac{1}{2\sqrt{\mathbf{h}_{\mathrm{R}}\cdot\mathbf{C}^{-1}\mathbf{h}_{\mathrm{R}}}}2\,\mathbf{C}^{-1}\mathbf{h}_{\mathrm{R}} \\
&= J\,\frac{\mu_0\,m_s}{\sqrt{\mathbf{h}_{\mathrm{R}}\cdot\mathbf{C}^{-1}\mathbf{h}_{\mathrm{R}}}}\tanh\left(\frac{\chi}{m_s}\sqrt{\mathbf{h}_{\mathrm{R}}\cdot\mathbf{C}^{-1}\mathbf{h}_{\mathrm{R}}}\right)\mathbf{C}^{-1}\mathbf{h}_{\mathrm{R}}.
\end{aligned} \tag{32.A.38}$$

Hence

$$\mathbf{b}_{\mathrm{R}} = J\,\mu_0\left(1+\frac{m_s}{\sqrt{\mathbf{h}_{\mathrm{R}}\cdot\mathbf{C}^{-1}\mathbf{h}_{\mathrm{R}}}}\tanh\left(\frac{\chi}{m_s}\sqrt{\mathbf{h}_{\mathrm{R}}\cdot\mathbf{C}^{-1}\mathbf{h}_{\mathrm{R}}}\right)\right)\mathbf{C}^{-1}\mathbf{h}_{\mathrm{R}}. \tag{32.A.39}$$

Using

$$\mathbf{h}_{\mathrm{R}} = \mathbf{F}^{\top}\mathbf{h} \qquad \text{and} \qquad \mathbf{b}_{\mathrm{R}} = J\mathbf{F}^{-1}\mathbf{b}, \tag{32.A.40}$$

the spatial form of (32.A.39) is

$$\mathbf{b} = \mu_0\left(1+\frac{m_s}{\sqrt{\mathbf{h}\cdot\mathbf{h}}}\tanh\left(\frac{\chi}{m_s}\sqrt{\mathbf{h}\cdot\mathbf{h}}\right)\right)\mathbf{h}. \tag{32.A.41}$$

REMARK

Note that in the limit of small $|\mathbf{h}|$ and small stretches we obtain,

$$\tanh\left(\frac{\chi}{m_s}\sqrt{\mathbf{h}\cdot\mathbf{h}}\right) \approx \frac{\chi}{m_s}\sqrt{\mathbf{h}\cdot\mathbf{h}} \qquad \text{and} \qquad (\bar{I}_1 - 3) \approx 0.$$

In this limit the magnetic flux density (32.A.41) reduces to

$$\mathbf{b} = \mu_0\,(1+\chi)\,\mathbf{h}, \tag{32.A.42}$$

or

$$\mathbf{b} = \mu\mathbf{h}, \tag{32.A.43}$$

where $\mu \stackrel{\text{def}}{=} \mu_0(1+\chi)$ is the **magnetic permeability** of the material.

33 Numerical implementation of the theory for soft-MREs

33.1 Weak forms of the governing partial differential equations

1. **Mechanical problem**:
The balance of force is

$$\operatorname{Div}\mathbf{T}_{\mathrm{R}} + \mathbf{f}_{\mathrm{R}} = \mathbf{0}, \tag{33.1.1}$$

and the traction condition is

$$[\![\mathbf{T}_{\mathrm{R}}]\!]\mathbf{n}_{\mathrm{R}} + \bar{\mathbf{t}}_{\mathrm{R}} + \mathbf{T}_{\mathrm{R}}^{\text{air maxw}}\mathbf{n}_{\mathrm{R}} = \mathbf{0} \qquad \text{on } \partial\mathcal{B}_{\mathbf{t}_{\mathrm{R}}}. \tag{33.1.2}$$

Consider an admissible *test field* $\mathbf{u}_{\text{test}}(\mathbf{X})$, which is at least of class C^0 in Ω^∞ and which satisfies

$$\mathbf{u}_{\text{test}}(\mathbf{X}) = \mathbf{0} \qquad \text{on } \partial\mathcal{B}_{\mathbf{u}} \text{ and also on } \partial\Omega^\infty.$$

Then if the force balance (33.1.1) and the traction condition (33.1.2) are satisfied, then clearly

$$\int\limits_{\Omega^\infty} \operatorname{Div}\mathbf{T}_{\mathrm{R}}\cdot\mathbf{u}_{\text{test}}\,dv_{\mathrm{R}} + \int\limits_{\Omega^\infty} \mathbf{f}_{\mathrm{R}}\cdot\mathbf{u}_{\text{test}}\,dv_{\mathrm{R}} + \int\limits_{\partial\mathcal{B}_{\mathbf{t}_{\mathrm{R}}}} \left([\![\mathbf{T}_{\mathrm{R}}]\!]\mathbf{n}_{\mathrm{R}} + \bar{\mathbf{t}}_{\mathrm{R}} + \mathbf{T}_{\mathrm{R}}^{\text{air maxw}}\mathbf{n}_{\mathrm{R}}\right)\cdot\mathbf{u}_{\text{test}}\,da_{\mathrm{R}} = 0$$

holds for all admissible $\mathbf{u}_{\text{test}}$. Thus,

$$\begin{aligned}
\int\limits_{\partial\mathcal{B}_{\mathbf{t}_{\mathrm{R}}}} \left(\bar{\mathbf{t}}_{\mathrm{R}} + \mathbf{T}_{\mathrm{R}}^{\text{air maxw}}\mathbf{n}_{\mathrm{R}}\right)\cdot\mathbf{u}_{\text{test}}\,da_{\mathrm{R}} &= -\int\limits_{\Omega^\infty} \operatorname{Div}\mathbf{T}_{\mathrm{R}}\cdot\mathbf{u}_{\text{test}}\,dv_{\mathrm{R}} - \int\limits_{\Omega^\infty} \mathbf{f}_{\mathrm{R}}\cdot\mathbf{u}_{\text{test}}\,dv_{\mathrm{R}} \\
&\quad - \int\limits_{\partial\mathcal{B}_{\mathbf{t}_{\mathrm{R}}}} [\![\mathbf{T}_{\mathrm{R}}]\!]\mathbf{n}_{\mathrm{R}}\cdot\mathbf{u}_{\text{test}}\,da_{\mathrm{R}}, \\
&= -\int\limits_{\Omega^\infty} \operatorname{Div}\mathbf{T}_{\mathrm{R}}\cdot\mathbf{u}_{\text{test}}\,dv_{\mathrm{R}} - \int\limits_{\Omega^\infty} \mathbf{f}_{\mathrm{R}}\cdot\mathbf{u}_{\text{test}}\,dv_{\mathrm{R}} \\
&\quad - \int\limits_{\partial\mathcal{B}_{\mathbf{t}_{\mathrm{R}}}} [\![\mathbf{T}_{\mathrm{R}}^{\top}\mathbf{u}_{\text{test}}]\!]\cdot\mathbf{n}_{\mathrm{R}}da_{\mathrm{R}}.
\end{aligned} \tag{33.1.3}$$

Introduction to Coupled Theories in Solid Mechanics. Lallit Anand, Eric M. Stewart, Shawn A. Chester, Oxford University Press.
 DOI: 10.1093/oso/9780198986218.003.0034

Consider a vector field g on Ω^∞. By the generalized divergence theorem for a domain containing a discontinuity surface $\partial\mathcal{B}_{t_R}$ (Gurtin et al., 2010, eq. (32.27)$_1$),[1]

$$\int_{\partial\Omega^\infty} \mathbf{g}\cdot\mathbf{n}_R\,da_R = \int_{\Omega^\infty} \mathrm{Div}\,\mathbf{g}\,dv_R + \int_{\partial\mathcal{B}_{t_R}} [\![\mathbf{g}]\!]\cdot\mathbf{n}_R\,da_R. \tag{33.1.5}$$

Applying this generalized divergence theorem to $\mathbf{g} \equiv \mathbf{T}_R^\top\mathbf{u}_{\text{test}}$ gives

$$\begin{aligned}\int_{\partial\Omega^\infty} (\mathbf{T}_R^\top\mathbf{u}_{\text{test}})\cdot\mathbf{n}_R\,da_R &= \int_{\Omega^\infty} \mathrm{Div}\,\mathbf{T}_R\cdot\mathbf{u}_{\text{test}}\,dv_R + \int_{\Omega^\infty} \mathbf{T}_R : \nabla\mathbf{u}_{\text{test}}\,dv_R \\ &\quad + \int_{\partial\mathcal{B}_{t_R}} [\![\mathbf{T}_R^\top\mathbf{u}_{\text{test}}]\!]\cdot\mathbf{n}_R\,da_R.\end{aligned} \tag{33.1.6}$$

Thus, since $\mathbf{u}_{\text{test}}(\mathbf{X}) = \mathbf{0}$ on $\partial\Omega^\infty$, eq. (33.1.6) gives

$$0 = \int_{\Omega^\infty} \mathrm{Div}\,\mathbf{T}_R\cdot\mathbf{u}_{\text{test}}\,dv_R + \int_{\Omega^\infty} \mathbf{T}_R : \nabla\mathbf{u}_{\text{test}}\,dv_R + \int_{\partial\mathcal{B}_{t_R}} [\![\mathbf{T}_R^\top\mathbf{u}_{\text{test}}]\!]\cdot\mathbf{n}_R\,da_R, \tag{33.1.7}$$

use of which in (33.1.3) gives

$$\int_{\partial\mathcal{B}_{t_R}} (\bar{\mathbf{t}}_R + \mathbf{T}_R^{\text{air maxw}}\mathbf{n}_R)\cdot\mathbf{u}_{\text{test}}\,da_R = \int_{\Omega^\infty} \mathbf{T}_R : \nabla\mathbf{u}_{\text{test}}\,dv_R - \int_{\Omega^\infty} \mathbf{f}_R\cdot\mathbf{u}_{\text{test}}\,dv_R. \tag{33.1.8}$$

Hence, if the force balance $\mathrm{Div}\,\mathbf{T}_R + \mathbf{f}_R = \mathbf{0}$ on Ω^∞ and the traction condition $[\![\mathbf{T}_R]\!]\mathbf{n}_R + \bar{\mathbf{t}}_R + \mathbf{T}_R^{\text{air maxw}}\mathbf{n}_R = \mathbf{0}$ on $\partial\mathcal{B}_{t_R}$ are satisfied, then

$$\begin{aligned}\int_{\Omega^\infty} \mathbf{T}_R : \nabla\mathbf{u}_{\text{test}}\,dv_R - \int_{\partial\mathcal{B}_{t_R}} (\bar{\mathbf{t}}_R + \mathbf{T}_R^{\text{air maxw}}\mathbf{n}_R)\cdot\mathbf{u}_{\text{test}}\,da_R + \int_{\Omega^\infty} \rho_R\ddot{\mathbf{u}}\cdot\mathbf{u}_{\text{test}}\,dv_R \\ - \int_{\Omega^\infty} \mathbf{f}_{0R}\cdot\mathbf{u}_{\text{test}}\,dv_R = 0\end{aligned} \tag{33.1.9}$$

holds for all admissible test fields $\mathbf{u}_{\text{test}}$. Note that in writing (33.1.9) we have used eq. (32.5.5), that is $\mathbf{f}_R = \mathbf{f}_{0R} - \rho_R\ddot{\mathbf{u}}$.

[1] It is important to recall here that the notation for the jump $[\![\varphi]\!]$ in a quantity φ in $\mathcal{B}$ across $\partial\mathcal{B}$ that is adopted here is

$$[\![\varphi]\!] \stackrel{\text{def}}{=} \varphi^{\text{air}} - \varphi^{\text{solid}}. \tag{33.1.4}$$

with $\mathbf{n}_R$ pointing from "solid" to "air" across $\partial\mathcal{B}$.

2. **Magnetic problem**:
Consider an admissible scalar *test field* $\phi_{\text{test}}(\mathbf{X})$ in Ω^∞ which is at least of class C^0 in Ω^∞ and which satisfies

$$\phi_{\text{test}}(\mathbf{X}) = \mathbf{0} \quad \text{on } \partial\Omega^\infty.$$

If the balance equation $\operatorname{Div}\mathbf{b}_{\text{R}} = 0$ in Ω^∞ and the jump condition $[\![\mathbf{b}_{\text{R}}]\!]\cdot\mathbf{n}_{\text{R}} = 0$ on $\partial\mathcal{B}$ are satisfied then clearly

$$\int\limits_{\Omega^\infty} (\operatorname{Div}\mathbf{b}_{\text{R}})\,\phi_{\text{test}}\,dv_{\text{R}} + \int\limits_{\partial\mathcal{B}} ([\![\mathbf{b}_{\text{R}}]\!]\cdot\mathbf{n}_{\text{R}})\,\phi_{\text{test}}\,da_{\text{R}} = 0 \tag{33.1.10}$$

holds for all admissible ϕ_{test}. Thus, we have

$$\int\limits_{\Omega^\infty} (\operatorname{Div}\mathbf{b}_{\text{R}})\,\phi_{\text{test}}\,dv_{\text{R}} + \int\limits_{\partial\mathcal{B}} [\![\phi_{\text{test}}\,\mathbf{b}_{\text{R}}]\!]\cdot\mathbf{n}_{\text{R}}\,da_{\text{R}} = 0. \tag{33.1.11}$$

Applying the generalized divergence theorem (33.1.5) with $\mathbf{g} \equiv \phi_{\text{test}}\mathbf{b}_{\text{R}}$ gives

$$\int\limits_{\partial\Omega^\infty} (\phi_{\text{test}}\mathbf{b}_{\text{R}})\cdot\mathbf{n}_{\text{R}}\,da_{\text{R}} = \int\limits_{\Omega^\infty} (\operatorname{Div}\mathbf{b}_{\text{R}})\phi_{\text{test}}\,dv_{\text{R}} + \int\limits_{\Omega^\infty} \mathbf{b}_{\text{R}}\cdot\nabla\phi_{\text{test}}\,dv_{\text{R}} + \int\limits_{\partial\mathcal{B}} [\![\phi_{\text{test}}\mathbf{b}_{\text{R}}]\!]\cdot\mathbf{n}_{\text{R}}\,da_{\text{R}}. \tag{33.1.12}$$

Thus, since $\phi_{\text{test}}(\mathbf{X}) = \mathbf{0}$ on $\partial\Omega^\infty$, eq. (33.1.12) gives

$$0 = \int\limits_{\Omega^\infty} (\operatorname{Div}\mathbf{b}_{\text{R}})\phi_{\text{test}}\,dv_{\text{R}} + \int\limits_{\Omega^\infty} \mathbf{b}_{\text{R}}\cdot\nabla\phi_{\text{test}}\,dv_{\text{R}} + \int\limits_{\partial\mathcal{B}} [\![\phi_{\text{test}}\mathbf{b}_{\text{R}}]\!]\cdot\mathbf{n}_{\text{R}}\,da_{\text{R}}, \tag{33.1.13}$$

use of which in (33.1.11) gives

$$\int\limits_{\Omega^\infty} \mathbf{b}_{\text{R}}\cdot\nabla\phi_{\text{test}}\,dv_{\text{R}} = 0. \tag{33.1.14}$$

Thus, the coupled *weak forms* of the mechanical and magnetic problems are

$$\begin{aligned} &\int\limits_{\Omega^\infty} \mathbf{T}_{\text{R}} : \nabla\mathbf{u}_{\text{test}}\,dv_{\text{R}} - \int\limits_{\partial\mathcal{B}_{\text{t}_\text{R}}} (\bar{\mathbf{t}}_{\text{R}} + \mathbf{T}_{\text{R}}^{\text{air maxw}}\mathbf{n}_{\text{R}})\cdot\mathbf{u}_{\text{test}}\,da_{\text{R}} \\ &\qquad + \int\limits_{\Omega^\infty} \rho_{\text{R}}\ddot{\mathbf{u}}\cdot\mathbf{u}_{\text{test}}\,dv_{\text{R}} - \int\limits_{\Omega^\infty} \mathbf{f}_{0\text{R}}\cdot\mathbf{u}_{\text{test}}\,dv_{\text{R}} = 0, \\ &\qquad \int\limits_{\Omega^\infty} \mathbf{b}_{\text{R}}\cdot\nabla\phi_{\text{test}}\,dv_{\text{R}} = 0. \end{aligned} \tag{33.1.15}$$

33.2 Treatment of near-incompressibility of the s-MRE

In our numerical finite element implementation of the theory, rather than just using the displacement $\mathbf{u}$ and the magnetostatic potential ϕ as degrees of freedom, in order to model the near-incompressibility of the soft-elastomeric composite material as well as that of the air domain we will introduce another pressure-like field p which will be required to satisfy

$$\frac{p}{K} + (J-1) = 0, \tag{33.2.1}$$

with K taking on the value of the bulk modulus of the material within $\mathcal{B}$ and the value of the bulk modulus of the air domain K^{air} within the air domain. In our numerical implementation we will satisfy this pressure equation in a weak sense by introducing an arbitrary test field $p_{\text{test}}(\mathbf{X})$ in Ω^{∞}, and take the weak form of the pressure equation (33.2.1) as

$$\int\limits_{\Omega^{\infty}} \left[\frac{p}{K} + (J-1)\right] p_{\text{test}}\, dv_{\text{R}} = 0. \tag{33.2.2}$$

Summarizing, the coupled weak forms of the mechanical and magnetic problems are:

$$\left.\begin{aligned}
\int\limits_{\Omega^{\infty}} \mathbf{T}_{\text{R}} : \nabla \mathbf{u}_{\text{test}}\, dv_{\text{R}} - \int\limits_{\partial\mathcal{B}_{\mathbf{t}_{\text{R}}}} (\bar{\mathbf{t}}_{\text{R}} + \mathbf{T}_{\text{R}}^{\text{air maxw}} \mathbf{n}_{\text{R}}) \cdot \mathbf{u}_{\text{test}}\, da_{\text{R}} & \\
+ \int\limits_{\Omega^{\infty}} \rho_{\text{R}} \ddot{\mathbf{u}} \cdot \mathbf{u}_{\text{test}}\, dv_{\text{R}} - \int\limits_{\Omega^{\infty}} \mathbf{f}_{0\text{R}} \cdot \mathbf{u}_{\text{test}}\, dv_{\text{R}} &= 0, \\
\int\limits_{\Omega^{\infty}} \left[\frac{p}{K} + (J-1)\right] p_{\text{test}}\, dv_{\text{R}} &= 0, \\
\int\limits_{\Omega^{\infty}} \mathbf{b}_{\text{R}} \cdot \nabla \phi_{\text{test}}\, dv_{\text{R}} &= 0.
\end{aligned}\right\} \tag{33.2.3}$$

33.3 Time-integration procedure

The time domain $t \in [0, T]$ is discretized in a series of N_t time steps,

$$t_{\text{n}+1} = t_{\text{n}} + \Delta t, \qquad n = 0, 1, \ldots, N_t, \tag{33.3.1}$$

such that $t_{N_t} = T$. Then, in order to solve the residuals based on the weak forms (33.2.3) in a $(\mathbf{u}, p, \phi)$-based finite element solution procedure, given

$$(\mathbf{u}_{\text{n}}, \dot{\mathbf{u}}_{\text{n}}, \ddot{\mathbf{u}}_{\text{n}}, \mathbf{C}_{\text{n}}^{v(\alpha)}, \mathbf{T}_{\text{Rn}}, \mathbf{b}_{\text{Rn}}, \phi_{\text{n}})$$

at time $t_{\rm n}$, we need to calculate

$$(\dot{\mathbf{u}}_{\rm n+1}, \ddot{\mathbf{u}}_{\rm n+1}, \mathbf{C}^{v(\alpha)}_{\rm n+1}, \mathbf{T}_{\rm Rn+1}, \mathbf{b}_{\rm Rn+1})$$

in terms of $(\mathbf{u}_{\rm n+1}, p_{\rm n+1}, \phi_{\rm n+1})$ at time $t_{\rm n+1} = t_{\rm n} + \Delta t$.

Calculation of $\mathbf{C}^{v(\alpha)}_{\rm n+1}$, $\mathbf{T}_{\rm Rn+1}$ and $\mathbf{b}_{\rm Rn+1}$

Recall the evolution equation (4.3.58) for the internal variable $\mathbf{C}^{v(\alpha)}$, viz.

$$\dot{\mathbf{C}}^{v(\alpha)} = \frac{J^{-2/3}}{\tau_r^{(\alpha)}}\left(\mathbf{C} - \frac{1}{3}\left(\mathbf{C}:(\mathbf{C}^{v(\alpha)})^{-1}\right)\mathbf{C}^{v(\alpha)}\right). \tag{33.3.2}$$

A fully-implicit time integration of this evolution equation (33.3.2) using an Euler-backward rule gives,

$$\begin{aligned}\mathbf{C}^{v(\alpha)}_{\rm n+1} &= \mathbf{C}^{v(\alpha)}_{\rm n} + \Delta t\, \dot{\mathbf{C}}^{v(\alpha)}_{\rm n+1},\\ &= \mathbf{C}^{v(\alpha)}_{\rm n} + \Delta t\, \frac{J_{\rm n+1}^{-2/3}}{\tau_r^{(\alpha)}}\left(\mathbf{C}_{\rm n+1} - \frac{1}{3}\left(\mathbf{C}_{\rm n+1}:(\mathbf{C}^{v(\alpha)}_{\rm n+1})^{-1}\right)\mathbf{C}^{v(\alpha)}_{\rm n+1}\right),\end{aligned} \tag{33.3.3}$$

which may be rearranged as

$$\left[1 + \frac{\Delta t\, J_{\rm n+1}^{-2/3}}{3\tau_r^{(\alpha)}}\left(\mathbf{C}_{\rm n+1}:(\mathbf{C}^{v(\alpha)}_{\rm n+1})^{-1}\right)\right]\mathbf{C}^{v(\alpha)}_{\rm n+1} = \mathbf{C}^{v(\alpha)}_{\rm n} + \frac{\Delta t\, J_{\rm n+1}^{-2/3}}{\tau_r^{(\alpha)}}\mathbf{C}_{\rm n+1}, \tag{33.3.4}$$

and then written as

$$\begin{aligned}\mathbf{C}^{v(\alpha)}_{\rm n+1} &= \beta\left(\mathbf{C}^{v(\alpha)}_{\rm n} + \frac{\Delta t\, J_{\rm n+1}^{-2/3}}{\tau_r^{(\alpha)}}\mathbf{C}_{\rm n+1}\right)\\ \text{with}\quad \beta &\overset{\rm def}{=} \left(1 + \frac{\Delta t\, J_{\rm n+1}^{-2/3}}{3\tau_r^{(\alpha)}}\left(\mathbf{C}_{\rm n+1}:(\mathbf{C}^{v(\alpha)}_{\rm n+1})^{-1}\right)\right)^{-1}.\end{aligned} \tag{33.3.5}$$

Now, consider a non-singular tensor $\mathbf{A}$ and define its distortional part $\bar{\mathbf{A}}$ by

$$\bar{\mathbf{A}} \overset{\rm def}{=} (\det\mathbf{A})^{-1/3}\mathbf{A} \qquad \text{so that} \qquad \det(\bar{\mathbf{A}}) = 1. \tag{33.3.6}$$

Then for a real scalar a, the tensor $a\mathbf{A}$ has the property

$$\overline{a\mathbf{A}} = \bar{\mathbf{A}} \qquad (\text{since } \det(a\mathbf{A}) = a^3\det(\mathbf{A})). \tag{33.3.7}$$

Now, we desire that $\mathbf{C}^{v(\alpha)}_{\mathrm{n+1}}$ satisfy the volume-preserving constraint $\det(\mathbf{C}^{v(\alpha)}_{\mathrm{n+1}}) = 1$, that is

$$\mathbf{C}^{v(\alpha)}_{\mathrm{n+1}} = \overline{\mathbf{C}^{v(\alpha)}_{\mathrm{n+1}}}.$$

Thus, taking the distortional part of the right-hand side of eq. (33.3.5) and using the property (33.3.7) we obtain

$$\mathbf{C}^{v(\alpha)}_{\mathrm{n+1}} = \overline{\beta\left(\mathbf{C}^{v(\alpha)}_{\mathrm{n}} + \frac{\Delta t\, J^{-2/3}_{\mathrm{n+1}}}{\tau_r^{(\alpha)}}\mathbf{C}_{\mathrm{n+1}}\right)} = \overline{\left(\mathbf{C}^{v(\alpha)}_{\mathrm{n}} + \frac{\Delta t\, J^{-2/3}_{\mathrm{n+1}}}{\tau_r^{(\alpha)}}\mathbf{C}_{\mathrm{n+1}}\right)}, \tag{33.3.8}$$

which gives the update formula

$$\mathbf{C}^{v(\alpha)}_{\mathrm{n+1}} = \left(\det\left[\mathbf{C}^{v(\alpha)}_{\mathrm{n}} + \frac{\Delta t\, J^{-2/3}_{\mathrm{n+1}}}{\tau_r^{(\alpha)}}\mathbf{C}_{\mathrm{n+1}}\right]\right)^{-1/3}\left(\mathbf{C}^{v(\alpha)}_{\mathrm{n}} + \frac{\Delta t\, J^{-2/3}_{\mathrm{n+1}}}{\tau_r^{(\alpha)}}\mathbf{C}_{\mathrm{n+1}}\right). \tag{33.3.9}$$

This elegant update formula for $\mathbf{C}^{v(\alpha)}$—which is *fully implicit, volume-preserving, and iteration-free*—is due to Shutov et al. (2013).

Finally, recalling (32.4.58) and (32.4.59), the total Piola stress at the end of the step is given by

$$\begin{aligned}
\mathbf{T}_{\mathrm{R}\,\mathrm{n+1}} &= J_{\mathrm{n+1}}\,\mathbf{T}_{\mathrm{n+1}}\mathbf{F}^{-\top}_{\mathrm{n+1}}, \qquad \text{with}\\
\mathbf{T}_{\mathrm{n+1}} &= \mathbf{T}^{\mathrm{eq}}_{\mathrm{n+1}} + \sum_{\alpha}\mathbf{T}^{(\alpha)}_{\mathrm{n+1}} + \mathbf{T}^{\mathrm{maxw}}_{\mathrm{n+1}} + \mathbf{T}^{\mathrm{mag}}_{\mathrm{n+1}}, \qquad \text{where}\\
\mathbf{T}^{\mathrm{eq}}_{\mathrm{n+1}} &= J^{-1}_{\mathrm{n+1}}\,G_{\mathrm{eq}}\,\bar{\mathbf{B}}_{0\,\mathrm{n+1}} - p_{\mathrm{n+1}}\,\mathbf{1},\\
\mathbf{T}^{(\alpha)}_{\mathrm{n+1}} &= J^{-1}_{\mathrm{n+1}}\,G^{(\alpha)}_{\mathrm{neq}}\left(\bar{\mathbf{F}}_{\mathrm{n+1}}(\mathbf{C}^{v(\alpha)}_{\mathrm{n+1}})^{-1}\bar{\mathbf{F}}^{\top}_{\mathrm{n+1}} - \frac{1}{3}(\bar{\mathbf{C}}_{\mathrm{n+1}} : (\mathbf{C}^{v(\alpha)}_{\mathrm{n+1}})^{-1})\mathbf{1}\right),\\
\mathbf{T}^{\mathrm{maxw}}_{\mathrm{n+1}} &= \mu_0\left[\mathbf{h}_{\mathrm{n+1}}\otimes\mathbf{h}_{\mathrm{n+1}} - \frac{1}{2}(\mathbf{h}_{\mathrm{n+1}}\cdot\mathbf{h}_{\mathrm{n+1}})\mathbf{1}\right], \qquad \text{and}\\
\mathbf{T}^{\mathrm{mag}}_{\mathrm{n+1}} &= -\mu_0\left[\frac{m_s^2}{\chi}\ln\left(\cosh\left(\frac{\chi}{m_s}\sqrt{\mathbf{h}_{\mathrm{n+1}}\cdot\mathbf{h}_{\mathrm{n+1}}}\right)\right)\right]\mathbf{1}\\
&\quad + \mu_0\left[\frac{m_s}{\sqrt{\mathbf{h}_{\mathrm{n+1}}\cdot\mathbf{h}_{\mathrm{n+1}}}}\tanh\left(\frac{\chi}{m_s}\sqrt{\mathbf{h}_{\mathrm{n+1}}\cdot\mathbf{h}_{\mathrm{n+1}}}\right)(\mathbf{h}_{\mathrm{n+1}}\otimes\mathbf{h}_{\mathrm{n+1}})\right].
\end{aligned} \tag{33.3.10}$$

The calculation of $\mathbf{b}_{\mathrm{R}\,\mathrm{n+1}}$ is straightforward and given later in equation (33.5.8).

Calculation of $\ddot{\mathbf{u}}_{\mathrm{n+1}}$ and $\dot{\mathbf{u}}_{\mathrm{n+1}}$

Next, to solve the weak form $(33.2.3)_1$, a kinematic relationship is required between $(\ddot{\mathbf{u}}_{\mathrm{n+1}}, \dot{\mathbf{u}}_{\mathrm{n+1}})$ and $\mathbf{u}_{\mathrm{n+1}}$ so that the inertia term in the residual $(33.2.3)_1$ becomes a function solely of $\mathbf{u}_{\mathrm{n+1}}$.

We compute these terms using the trapezoidal Newmark time-integration scheme (Newmark, 1959):

$$\begin{aligned} \ddot{\mathbf{u}}_{\mathrm{n}+1} &= \frac{\mathbf{u}_{\mathrm{n}+1} - \mathbf{u}_{\mathrm{n}} - \Delta t\, \dot{\mathbf{u}}_{\mathrm{n}}}{(1/4)\Delta t^2} - \ddot{\mathbf{u}}_{\mathrm{n}}, \\ \dot{\mathbf{u}}_{\mathrm{n}+1} &= \dot{\mathbf{u}}_{\mathrm{n}} + \frac{1}{2}\Delta t\, (\ddot{\mathbf{u}}_{\mathrm{n}} + \ddot{\mathbf{u}}_{\mathrm{n}+1})\,. \end{aligned} \tag{33.3.11}$$

33.4 Solution procedure in the FEniCS code

For ease of notation we will write

$$(\bullet) \equiv (\bullet)_{\mathrm{n}+1} \qquad \text{and} \qquad (\bullet)_{\mathrm{old}} \equiv (\bullet)_{\mathrm{n}},$$

for all time-varying quantities $(\bullet)$ evaluated at times $\mathrm{n}+1$ and n, respectively.

In FEniCS we will use the following degrees of freedom

$$\{\mathbf{u},\, p,\, \phi\}, \tag{33.4.1}$$

as well as their known values from the previous step

$$\{\mathbf{u}_{\mathrm{old}},\, p_{\mathrm{old}},\, \phi_{\mathrm{old}}\}, \tag{33.4.2}$$

to calculate the *residuals* resulting from the weak forms of the governing equations, viz.

$$\begin{aligned} R_{\mathrm{u}} &= \int\limits_{\Omega^\infty} \mathbf{T}_{\mathrm{R}} \colon \nabla \mathbf{u}_{\mathrm{test}}\, dv_{\mathrm{R}} - \int\limits_{\partial \mathcal{B}_{\mathrm{t_R}}} (\bar{\mathbf{t}}_{\mathrm{R}} + \mathbf{T}_{\mathrm{R}}^{\mathrm{air\ maxw}} \mathbf{n}_{\mathrm{R}}) \cdot \mathbf{u}_{\mathrm{test}}\, da_{\mathrm{R}} \\ &\quad + \int\limits_{\Omega^\infty} \rho_{\mathrm{R}} \ddot{\mathbf{u}} \cdot \mathbf{u}_{\mathrm{test}}\, dv_{\mathrm{R}} - \int\limits_{\Omega^\infty} \mathbf{f}_{0\mathrm{R}} \cdot \mathbf{u}_{\mathrm{test}}\, dv_{\mathrm{R}}, \\ R_p &= \int\limits_{\Omega^\infty} \left[\frac{p}{K} + (J-1) \right] p_{\mathrm{test}}\, dv_{\mathrm{R}}, \\ R_\phi &= \int\limits_{\Omega^\infty} \mathbf{b}_{\mathrm{R}} \cdot \nabla \phi_{\mathrm{test}}\, dv_{\mathrm{R}}. \end{aligned} \tag{33.4.3}$$

FEniCS then automatically differentiates these residuals with respect to the separate degrees of freedom to obtain *tangents*. At each step in time, FEniCS uses the residuals and tangents in a global Newton–Raphson solver to determine current values of the degrees of freedom $\{\mathbf{u},\, p,\, \phi\}$ for which the residuals are zero (to within some small numerical tolerance) and which therefore satisfy the governing equations as well as the initial and boundary conditions (in a weak sense).

REMARK

A mixed finite element method is utilized to solve for the displacement $\mathbf{u}$ the pressure p, and the magnetic potential ϕ, concurrently. We use second-order Lagrangian interpolation functions for the displacement $\mathbf{u}$ and discontinuous Galerkin (DG) interpolation functions for the pressure p, which suffers a jump discontinuity across the s-MRE boundary $\partial\mathcal{B}$. We also use second-order Lagrangian interpolation functions for the magnetic potential ϕ, and use FEniCS quadrature-representation elements for the internal variables $\mathbf{C}^{v(\alpha)}$.

33.5 **Algorithm for calculation of the residuals**

The algorithm we follow in our code to calculate the residuals in (33.4.3) is summarized below.

1. Begin the step with known values of

$$\left\{\mathbf{u}_{\mathrm{old}},\ \dot{\mathbf{u}}_{\mathrm{old}},\ \ddot{\mathbf{u}}_{\mathrm{old}},\ p_{\mathrm{old}},\ \mathbf{C}^{v(\alpha)}_{\mathrm{old}},\phi_{\mathrm{old}}\right\}, \tag{33.5.1}$$

   where for the first step these quantities are given by the initial conditions for $\mathbf{u}$, $\dot{\mathbf{u}}$, $\ddot{\mathbf{u}}$, p, $\mathbf{C}^{v(\alpha)}$, ϕ. We also have temporary values of the degrees of freedom at the end of the step,

$$\left\{\mathbf{u},\ p,\ \phi\right\}. \tag{33.5.2}$$

2. Calculate the acceleration and velocity using the Newmark scheme:

$$\begin{aligned}\ddot{\mathbf{u}} &= \frac{\mathbf{u}-\mathbf{u}_{\mathrm{old}}-\dot{\mathbf{u}}_{\mathrm{old}}\,\Delta t}{(1/4)\,\Delta t^2}-\ddot{\mathbf{u}}_{\mathrm{old}},\\ \dot{\mathbf{u}} &= \dot{\mathbf{u}}_{\mathrm{old}}+\frac{1}{2}\Delta t\,\left(\ddot{\mathbf{u}}_{\mathrm{old}}+\ddot{\mathbf{u}}\right).\end{aligned} \tag{33.5.3}$$

3. Calculate necessary kinematical and magnetic field quantities for the current step:

$$\begin{aligned}&\mathbf{F}=\mathbf{1}+\nabla\mathbf{u},\quad J=\det\mathbf{F},\quad \mathbf{C}=\mathbf{F}^{\top}\mathbf{F},\\ &\bar{\mathbf{F}}=J^{-1/3}\mathbf{F},\quad \bar{\mathbf{C}}=\bar{\mathbf{F}}^{\top}\bar{\mathbf{F}},\quad \bar{\mathbf{B}}=\bar{\mathbf{F}}\bar{\mathbf{F}}^{\top},\\ &\mathbf{h}_{\mathrm{R}}=-\nabla\phi,\quad \mathbf{h}=\mathbf{F}^{-\top}\mathbf{h}_{\mathrm{R}}.\end{aligned} \tag{33.5.4}$$

4. Update the internal variables $\mathbf{C}^{v(\alpha)}$:

$$\mathbf{C}^{v(\alpha)}=\left(\det\left[\mathbf{C}^{v(\alpha)}_{\mathrm{old}}+\frac{\Delta t}{\tau_r^{(\alpha)}}\bar{\mathbf{C}}\right]\right)^{-1/3}\left(\mathbf{C}^{v(\alpha)}_{\mathrm{old}}+\frac{\Delta t}{\tau_r^{(\alpha)}}\bar{\mathbf{C}}\right). \tag{33.5.5}$$

5. Calculate the Piola stress,

$$\mathbf{T}_{\mathrm{R}} = J\,\mathbf{T}\mathbf{F}^{-\top}, \qquad \text{with}$$

$$\mathbf{T} = \begin{cases} \mathbf{T}^{\mathrm{eq}} + \sum_\alpha \mathbf{T}^{(\alpha)} + \mathbf{T}^{\mathrm{maxw}} + \mathbf{T}^{\mathrm{mag}}, & \mathbf{X} \in \mathcal{B} \\ \mathbf{T}^{\mathrm{eq}}, & \mathbf{X} \in \Omega^{\mathrm{air}} \end{cases} \tag{33.5.6}$$

where

$$\begin{aligned} \mathbf{T}^{\mathrm{eq}} &= J^{-1} G_{\mathrm{eq}}\,\bar{\mathbf{B}}_0 - p\,\mathbf{1}, \\ \mathbf{T}^{(\alpha)} &= J^{-1}\,G^{(\alpha)}_{\mathrm{neq}} \left(\bar{\mathbf{F}}(\mathbf{C}^{v(\alpha)})^{-1}\bar{\mathbf{F}}^{\top} - \frac{1}{3}(\bar{\mathbf{C}} : (\mathbf{C}^{v(\alpha)})^{-1})\mathbf{1} \right), \\ \mathbf{T}^{\mathrm{maxw}} &= \mu_0 \left[\mathbf{h} \otimes \mathbf{h} - \frac{1}{2}(\mathbf{h}\cdot\mathbf{h})\mathbf{1} \right], \qquad \text{and} \\ \mathbf{T}^{\mathrm{mag}} &= -\mu_0 \left[\frac{m_s^2}{\chi} \ln\left(\cosh\left(\frac{\chi}{m_s}\sqrt{\mathbf{h}\cdot\mathbf{h}} \right) \right) \right] \mathbf{1} \\ &\quad + \mu_0 \left[\frac{m_s}{\sqrt{\mathbf{h}\cdot\mathbf{h}}} \tanh\left(\frac{\chi}{m_s}\sqrt{\mathbf{h}\cdot\mathbf{h}} \right) (\mathbf{h}\otimes\mathbf{h}) \right], \end{aligned} \tag{33.5.7}$$

with G_{eq} taking on the value of the shear modulus of the material within $\mathcal{B}$ and the value of the shear modulus of the air-domain G^{air} within the air domain.

6. Calculate the magnetic flux density,

$$\mathbf{b}_{\mathrm{R}} = J\mathbf{F}^{-1}\mathbf{b}, \qquad \text{with}$$

$$\mathbf{b} = \begin{cases} \mathbf{b}^{\mathrm{maxw}} + \mathbf{b}^{\mathrm{mag}}, & \mathbf{X} \in \mathcal{B} \\ \mathbf{b}^{\mathrm{maxw}}, & \mathbf{X} \in \Omega^{\mathrm{air}} \end{cases} \tag{33.5.8}$$

where

$$\begin{aligned} \mathbf{b}^{\mathrm{maxw}} &= \mu_0\,\mathbf{h}, \qquad \text{and} \\ \mathbf{b}^{\mathrm{mag}} &= \mu_0 \frac{m_s}{\sqrt{\mathbf{h}\cdot\mathbf{h}}} \tanh\left(\frac{\chi}{m_s}\sqrt{\mathbf{h}\cdot\mathbf{h}} \right) \mathbf{h}. \end{aligned} \tag{33.5.9}$$

7. Compute the residuals:

$$R_{\mathbf{u}} = \int_{\Omega^{\infty}} \mathbf{T}_{\mathrm{R}} : \nabla \mathbf{u}_{\mathrm{test}} \, dv_{\mathrm{R}} - \int_{\partial \mathcal{B}_{\mathbf{t}_{\mathrm{R}}}} (\bar{\mathbf{t}}_{\mathrm{R}} + \mathbf{T}_{\mathrm{R}}^{\mathrm{air\ maxw}} \mathbf{n}_{\mathrm{R}}) \cdot \mathbf{u}_{\mathrm{test}} \, da_{\mathrm{R}}$$
$$+ \int_{\Omega^{\infty}} \rho_{\mathrm{R}} \ddot{\mathbf{u}} \cdot \mathbf{u}_{\mathrm{test}} \, dv_{\mathrm{R}} - \int_{\Omega^{\infty}} \mathbf{f}_{0\mathrm{R}} \cdot \mathbf{u}_{\mathrm{test}} \, dv_{\mathrm{R}},$$
$$R_p = \int_{\Omega^{\infty}} \left[\frac{p}{K} + (J-1) \right] p_{\mathrm{test}} \, dv_{\mathrm{R}}, \tag{33.5.10}$$
$$R_{\phi} = \int_{\Omega^{\infty}} \mathbf{b}_{\mathrm{R}} \cdot \nabla \phi_{\mathrm{test}} \, dv_{\mathrm{R}}.$$

with K taking on the value of the bulk modulus of the material within $\mathcal{B}$ and the value of the bulk modulus of the air-domain K^{air} within the air domain.

8. FEniCS solves for the current values of $\{\mathbf{u}, p, \phi\}$.

9. Set the current values as the "old" values for the next step in time,

$$\{\mathbf{u}, \dot{\mathbf{u}}, \ddot{\mathbf{u}}, p, \mathbf{C}^{v(\alpha)}, \phi\} \implies \{\mathbf{u}_{\mathrm{old}}, \dot{\mathbf{u}}_{\mathrm{old}}, \ddot{\mathbf{u}}_{\mathrm{old}}, p_{\mathrm{old}}, \mathbf{C}^{v(\alpha)}_{\mathrm{old}}, \phi_{\mathrm{old}}\}. \tag{33.5.11}$$

10. Repeat the steps above for desired number of steps in time.

34 Representative simulations of soft-MREs

In the applications of the theory shown in this chapter we will use the unit system shown in Table 34.1 and the material parameters shown in Table 34.2.

Table 34.1 Units.

Quantity	SI (m)	SI (mm)	Conversion factor C_f
Length	m	mm	$1\,\mathrm{m} = 10^3\,\mathrm{mm}$
Mass	kg	kg	1
Time	s	s	1
Electric charge	C	kC	$1\,\mathrm{C} = 10^{-3}\,\mathrm{kC}$
Force	N	mN	$1\,\mathrm{N} = 10^3\,\mathrm{mN}$
Energy	J	mJ	$1\,\mathrm{J} = 10^3\,\mathrm{mJ}$
Mass density	$\mathrm{kg/m^3}$	$\mathrm{kg/mm^3}$	$1\,\mathrm{kg/m^3} = 10^{-9}\,\mathrm{kg/mm^3}$
Stress	Pa	kPa ($\mathrm{mN/mm^2}$)	$1\,\mathrm{Pa} = 10^{-3}\,\mathrm{kPa}$
Current	A (C/s)	kA (kC/s)	$1\,\mathrm{A} = 10^{-3}\,\mathrm{kA}$
Magnetic field strength	A/m	kA/mm	$1\,\mathrm{A/m} = 10^{-6}\,\mathrm{kA/mm}$
Magnetic flux density	T (N s/C m)	mT (mN s/kC mm)	$1\,\mathrm{T} = 10^3\,\mathrm{mT}$

Introduction to Coupled Theories in Solid Mechanics. Lallit Anand, Eric M. Stewart, Shawn A. Chester, Oxford University Press.
 DOI: 10.1093/oso/9780198986218.003.0035

Table 34.2 Representative values of material parameters for an s-MRE, based on the s-MRE material studied by Wang et al. (2023, 2024). We use three Maxwell branches to model the viscoelasticity of the material.

Quantity	Value SI (m)	Value SI (mm)
ρ_R, mass density	2833 kg/m^3	2.833×10^{-6} kg/mm^3
G_{eq}, equilibrium shear modulus	411×10^3 Pa	411 kPa
K, bulk modulus	411×10^6 Pa	411×10^3 kPa
$G_{neq}^{(1)}$, non-equilibrium shear modulus	150×10^3 Pa	150 kPa
$\tau_r^{(1)}$, relaxation time	0.0159 s	0.0159 s
$G_{neq}^{(2)}$, non-equilibrium shear modulus	100×10^3 Pa	100 kPa
$\tau_r^{(2)}$, relaxation time	0.159 s	0.159 s
$G_{neq}^{(3)}$, non-equilibrium shear modulus	100×10^3 Pa	100 kPa
$\tau_r^{(3)}$, relaxation time	1.59 s	1.59 s
χ, magnetic susceptibility	1.75	1.75
m_s, magnetic saturation strength	0.495×10^6 A/m	0.495 kA/mm
μ_0, permeability of free space	$4\,\pi \times 10^{-7}$ T / (A/m)	$4\,\pi \times 10^{2}$ mT / (kA/mm)

In some classical experiments, Diguet (2010)—for his PhD thesis—fabricated cm-sized specimens of spherical and cylindrical shapes from isotropic s-MREs made from a soft silicone rubber filled with spherical iron particles $\approx 5\ \mu$m in diameter, at several different volume fractions ranging from 0.05 to 0.35. He conducted several experiments aimed at investigating the magnetostriction response of these specimens by subjecting them to a (roughly) uniaxial magnetic field of increasing magnitude, and monitoring their deformed geometry as a function of the applied field strength. The six differently shaped cylindrical s-MRE specimens studied by Diguet (2010) are shown in Fig 34.1(a), a schematic of the s-MRE microstructure in Fig 34.1(b), and example photographs of a magnetostriction experiment by Diguet (2010) in Fig 34.1(c). In his thesis and in an associated paper Diguet and his co-workers reported that:

- The *spherical*-shaped specimens showed a positive diametrical strain and became ellipsoidal in shape upon application of the magnetic field.
- However, for their *cylindrical*-shaped specimens they found that the magnetostriction was strongly *shape-dependent*. The measured average tensile strain of their cylindrical-shaped specimens displayed a strong dependence on the ratio of the length ℓ to the diameter d of their specimens, with a low aspect ratio ℓ/d resulting in a larger strain. They measured a maximum average strain on the order of 10% upon application of a 1 MA/m magnetic field, and then a marked hysteresis of the strain when the field was decreased back to zero; see Fig. 34.1(c).

This strong dependence of the overall magnetostriction on the specimen geometry of the cylindrical specimens—the "shape effect"—is physically non-intuitive, and it is important to

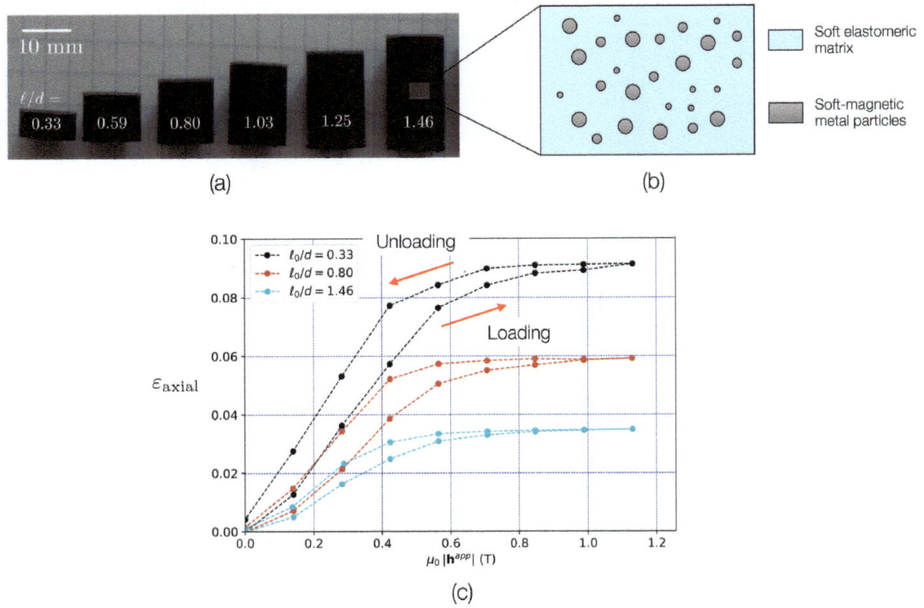

Fig. 34.1 (a) The six differently shaped cylindrical s-MRE specimens of Diguet (2010). (b) Schematic of the s-MRE microstructure. (c) Axial strain versus axial magnetic field strength for the different ℓ/d values studied by Diguet (2010). Arrows indicate loading and unloading for the case $\ell/d = 0.33$.

adequately model this phenomenon so as to be able to rationally design actuators based on s-MREs.

In this section, we first demonstrate the basic features of the theory by modeling the magnetostriction of a spherical s-MRE. Then, we model the shape-dependent magnetostriction behavior of cylindrical s-MREs. Then, as an interesting practical application of our theory we study the actuation response of a trilayer beam consisting of an s-MRE layer sandwiched between two non-magnetic elastomeric layers.

34.1 Magnetostriction of a spherical s-MRE

In order to uncover the basic mechanism of magnetostriction of s-MREs, in this section we will model the magnetostriction of a *spherical* domain of the material embedded in air and subjected to an applied magnetic field $\mathbf{h}^{\text{app}}$. This section is organized as follows:

- In Section 34.1.1 we describe our approach to modeling the application of a magnetic field to an s-MRE sphere embedded in an air domain. An explanation for the choice of the material parameters that we employ is also provided.
- In Section 34.1.2 we describe the magnetostriction behavior of a sphere which is composed of an *ideal magnetoelastic* s-MRE material as described in Section 32.4.1.

- In Section 34.1.3 we describe the magnetostriction behavior of a sphere composed of a non-idealized *nonlinear paramagnetic* s-MRE material as described in Section 32.4.2.
- Finally, in Section 34.1.4 we will examine the magnetic fields generated in the air domain and the solid domain. This will help provide an explanation of the roles that fields induced by the magnetization of the material play in the magnetostriction mechanism of s-MREs.

34.1.1 Modeling approach and material parameters

A spherical magnetorheological body which is embedded in air is numerically modeled using an axisymmetric idealization, as shown schematically in Figs. 34.2(a) and (b). The solid sphere has diameter $d_0 = 10\,\text{mm}$, and is surrounded by an air domain with radius $s = 50\,\text{mm}$ and height $2s = 100\,\text{mm}$. The radius s of the air domain is selected as ten times the radius of the sphere, so that the magnetostatic potentials applied at the edges of the domain are not (significantly) affected by the magnetization field generated in the sphere. The computational domain with the finite element mesh used in the simulation is shown revolved 180 degrees in Fig. 34.2(c). During magnetostriction the deformed diameter of the sphere d is measured using the displacements of points indicated by black dots in Figs. 34.2(d) and (e), and the associated diametrical stretch is defined as $\lambda \stackrel{\text{def}}{=} d/d_0$.

Boundary conditions: Referring to Figs. 34.2(a) and (b) the imposed boundary conditions are as follows:

- Mechanical boundary conditions: The horizontal centerline of the sphere indicated in Fig. 34.2(a) by a red dashed line is held mechanically fixed in the $\mathbf{e}_2$-direction to prevent translation of the sphere, and the top, bottom, and right edges of the air domain in Fig. 34.2 (b) are held mechanically fixed in the $\mathbf{e}_1$- and $\mathbf{e}_2$-directions.
- Magnetostatic potential boundary conditions: The magnetostatic potential on the top boundary of the air domain is kept fixed at $\phi = 0$, while the potential on the bottom boundary of the air domain is ramped up to a large positive value $\phi = 2s|\mathbf{h}^{\text{app}}|$; cf. Fig. 34.2(b). Recalling the definition of the magnetic field $\mathbf{h}_{\text{R}} = -\nabla\phi$, the difference in these two potentials over a distance of $2s$ in the $\mathbf{e}_2$-direction corresponds to a far-field applied magnetic field $\mathbf{h}^{\text{app}}$ oriented in the $\mathbf{e}_2$-direction.

We use the material parameters listed in Table 34.2, which are representative of the specific s-MRE material studied by Wang et al. (2023, 2024) which has a mass ratio of carbonyl iron particles, PDMS, and curing agent of 75:25:1. Our rationale for the choice of the values of material parameters is as follows:

- The value for the shear modulus G_{eq} and the mass density ρ_{R} are those measured by Wang et al. (2023, 2024) for their specific formulation of s-MRE material.
- The bulk modulus K for the composite material is taken to be much larger than the shear modulus, $K = 1000\,G_{\text{eq}}$, which is representative of slightly compressible elastomers.
- The magnetic susceptibility χ and magnetic saturation value m_s are fitted to an $\mathbf{h}$–$\mathbf{m}$ magnetization curve which was experimentally measured for the specific formulation of s-MRE studied by Wang et al. (2024), cf. Fig. 34.3.

- The non-equilibrium shear moduli $G_{\text{neq}}^{(\alpha)}$ and relaxation times $\tau_r^{(\alpha)}$ in our viscoelasticity formulation were fitted to the hysteresis curve data from cyclic simple shear experiments conducted by Wang et al. (2024) on samples of their s-MRE at frequencies of $f = 0.1$ Hz, 1.0 Hz, and 10 Hz and a shear strain amplitude of 15%. We found that three Maxwell branches in our model were sufficient to yield a reasonable match with their experimental curves, cf. Fig. 34.4.

In our theory we have used a simple linear form of the viscosity function, cf. eq. (32.4.45), and this is the reason behind the discrepancy in the numerically calculated and

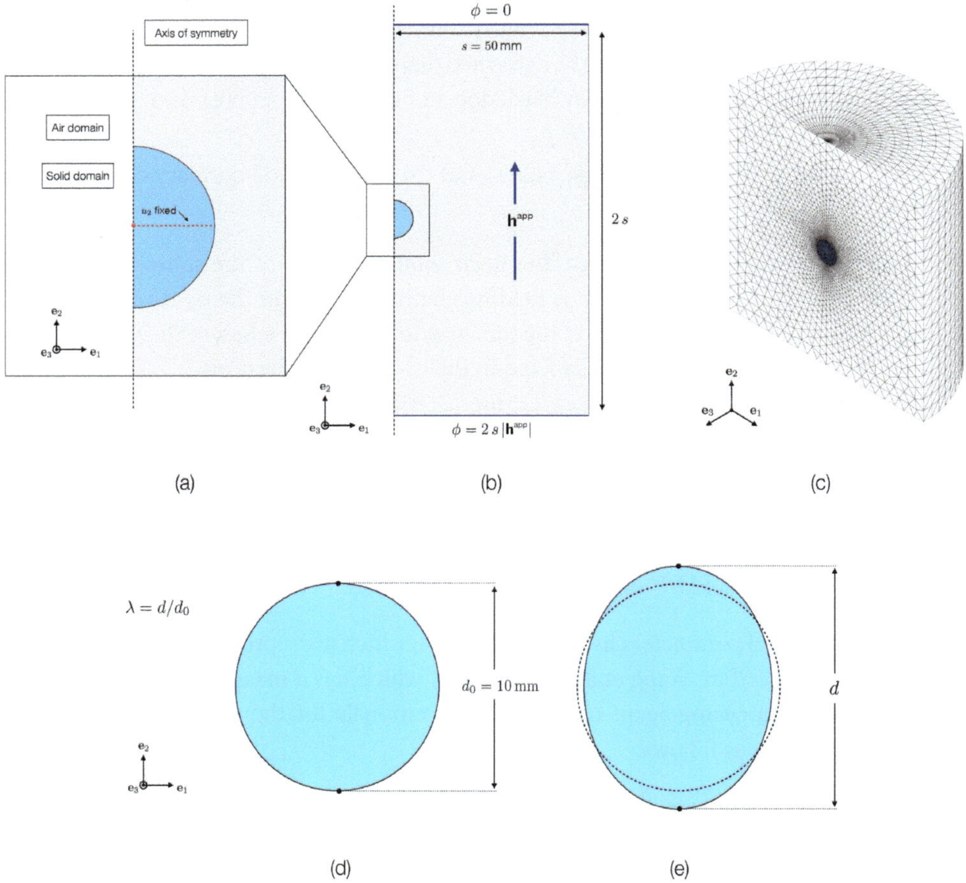

Fig. 34.2 (a) A schematic of the spherical s-MRE embedded in an air domain. (b) The combined axisymmetric air–solid domain, showing the applied magnetic field $\mathbf{h}^{\text{app}}$ in the $\mathbf{e}_2$-direction and the corresponding boundary conditions for the magnetostatic potential ϕ. (c) The computational domain revolved 180 degrees about the axis of symmetry, with mesh edges shown. (d) and (e) show the reference and deformed configurations of the spherical s-MRE; these figures also show the points across which the diametrical stretch $\lambda = d/d_0$ is measured.

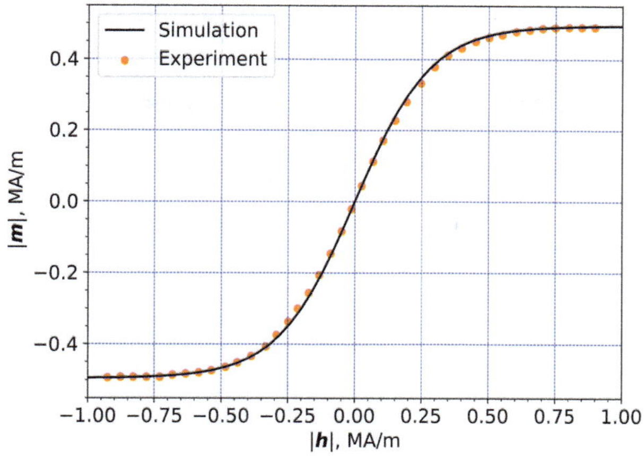

Fig. 34.3 The **h**–**m** magnetization curve predicted by our theory with $\chi = 1.75$ and $m_s = 0.495$ MA/m overlaid with the experimentally measured magnetization data of Wang et al. (2024).

the experimentally measured curve at $f = 10$ Hz. A better fit may be obtained by using more elaborate nonlinear viscosity functions, but we do not pursue such a refinement here.

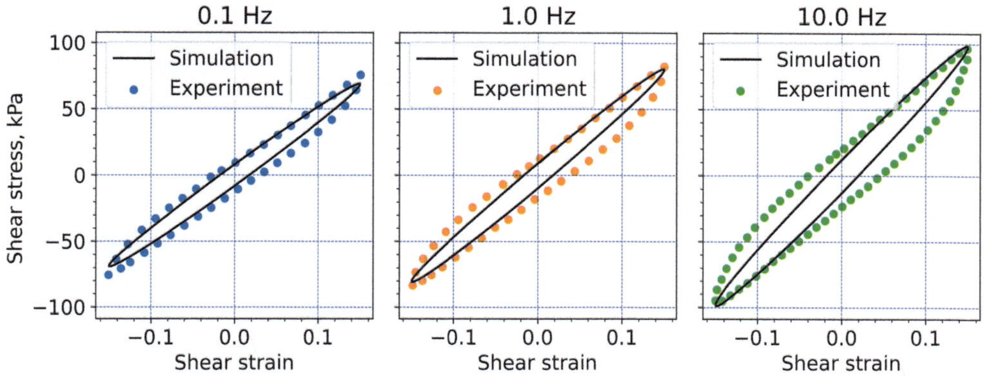

Fig. 34.4 Simple shear cyclic hysteresis curves numerically calculated using our theory for frequencies of $f = 0.1$ Hz, 1.0 Hz, and 10 Hz at a shear strain amplitude of 15% compared with the corresponding experimentally measured curves from Wang et al. (2024).

- Also, the shear modulus of air G^{air} is chosen to have a value of 1 Pa which is significantly smaller than that of the solid, so that the stiffness of the air does not significantly interfere with the deformation of the solid. The bulk modulus of air is taken as its actual value $K^{\text{air}} = 101\,\text{kPa}$.

34.1.2 Actuation of a sphere made of an ideal magnetoelastic material

First we consider a sphere whose response is modeled as that of an *ideal magnetoelastic material* using the constitutive relations discussed in Section 32.4.1. In this simulation we linearly ramp up the applied field $\mathbf{h}^{\text{app}}$ from zero to a strength of 0.8 MA/m, or equivalently $\mu_0|\mathbf{h}^{\text{app}}| = 1.0\,\text{T}$, in the $\mathbf{e}_2$-direction as shown in Fig. 34.5(a), and calculate the resulting diametrical engineering strain $(\lambda - 1)$ of the sphere in the $\mathbf{e}_2$-direction.

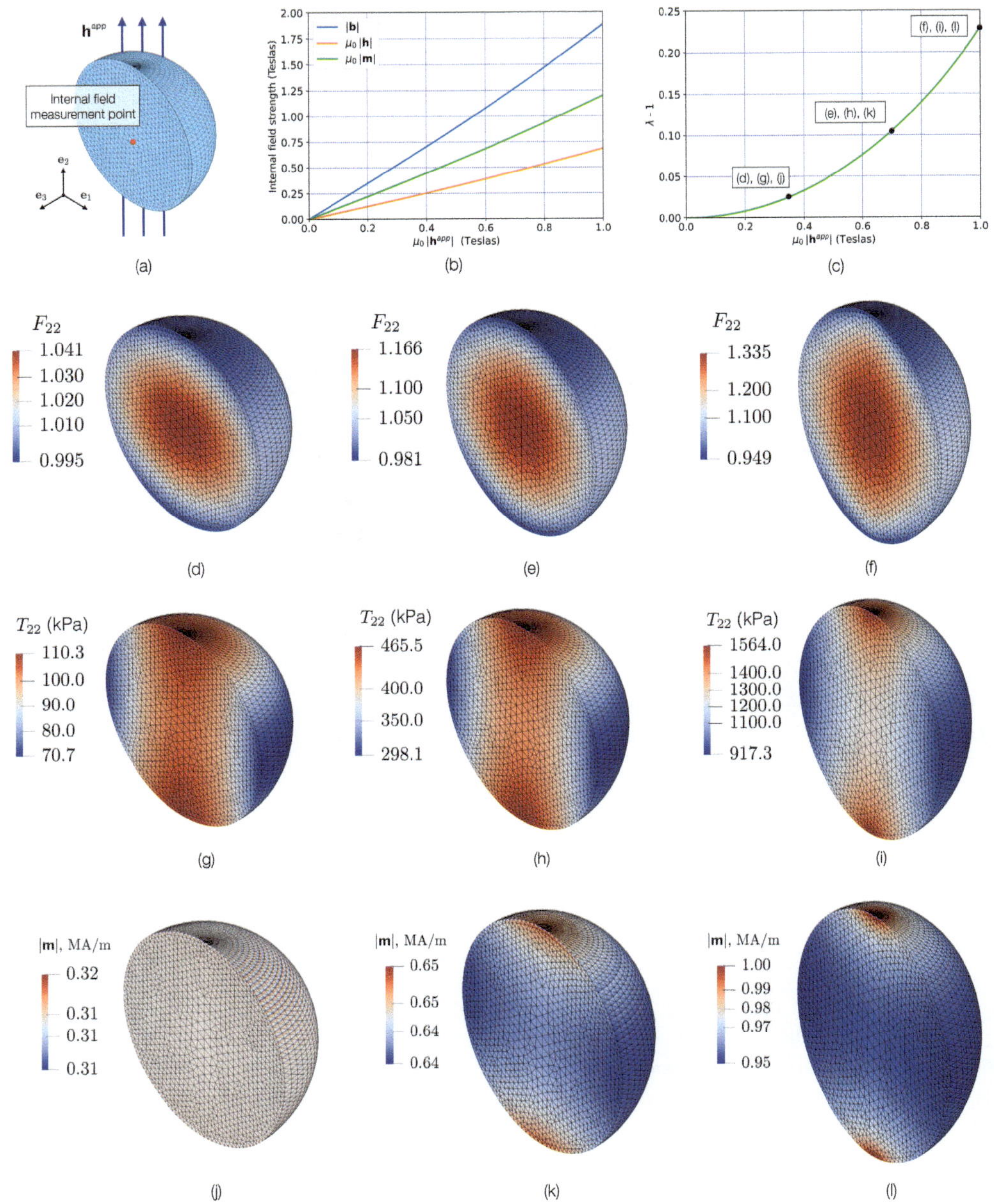

Fig. 34.5 (a) The undeformed sphere, together with the orientation of the applied magnetic field $\mathbf{h}^{\text{app}}$ and the point indicated by a red dot at which the internal fields are measured. (b) The internal fields $\mathbf{h}$, $\mathbf{m}$, and $\mathbf{b}$ measured at the center of the sphere during actuation as a function of $\mathbf{h}^{\text{app}}$. We have multiplied the $\mathbf{h}$ and $\mathbf{m}$ fields by μ_0 to emphasize the relation $\mathbf{b} = \mu_0(\mathbf{h} + \mathbf{m})$. (c) The diametrical strain $(\lambda - 1)$ of the sphere increases quadratically as $\mathbf{h}^{\text{app}}$ increases. (d) through (l) show contours of F_{22}, T_{22}, and $|\mathbf{m}|$ plotted on the deformed body for applied field strengths of $\mu_0|\mathbf{h}^{\text{app}}| = 0.35,\ 0.70,$ and 1.0 T.

The magnetization $\mathbf{m}$ and associated $\mathbf{h}$ and $\mathbf{b}$ fields which develop in the center of the sphere, the point indicated by a red dot in Fig. 34.5(a), are shown in Fig. 34.5(b) as a function of the applied field $\mu_0|\mathbf{h}^{\text{app}}|$. In plotting this figure we have multiplied the $\mathbf{h}$ and $\mathbf{m}$ fields by the vacuum permeability μ_0 in order to emphasize the relation $\mathbf{b} = \mu_0(\mathbf{h} + \mathbf{m})$. From this figure we note that:

- As $\mu_0|\mathbf{h}^{\text{app}}|$ is ramped up, the strength of the magnetization of the sphere $\mu_0|\mathbf{m}|$ (green line) is equal to 1.75 times the strength of the magnetic field $\mu_0|\mathbf{h}|$ (orange line) at all times; this is because the magnetic susceptibility has a value of $\chi = 1.75$. Also, the magnetic flux density $|\mathbf{b}|$ (blue line) obeys the relation $|\mathbf{b}| = \mu_0|(\mathbf{h} + \mathbf{m})|$, as required.
- Then, since the component of the $\mathbf{b}$-field in the $\mathbf{e}_2$-direction along the axis of the sphere must be continuous across the air–solid boundary, and the air is not magnetizable so that in that region $\mathbf{m} = \mathbf{0}$, the relation $\mathbf{b} = \mu_0(\mathbf{h} + \mathbf{m})$ implies that:
  - The magnetic field $\mathbf{h}$ in the sphere is *reduced* relative to the magnetic field $\mathbf{h}$ in the air.
  - It is the mis-match between the $\mathbf{h}$ field in the sphere versus the $\mathbf{h}$ field in the air around the sphere, which is *central* to the actuation behavior of the s-MRE. We will discuss this in greater detail in Section 34.1.4.

The diametrical actuation strain $(\lambda - 1)$ of the sphere is plotted as a function of $\mu_0|\mathbf{h}^{\text{app}}|$ in Fig. 34.5(c). The actuation strain $(\lambda - 1)$ increases quadratically with the applied field $\mu_0|\mathbf{h}^{\text{app}}|$, up to a value of around 23% for $\mu_0|\mathbf{h}^{\text{app}}| = 1.0$ T.[1] This quadratic-dependence of the actuation strain follows directly from the quadratic-dependence of the Maxwell stress part of the Cauchy stress on the $\mathbf{h}$ field, cf. eq. (32.4.14),

$$\mathbf{T}^{\text{maxw}} = \mu\left(\mathbf{h} \otimes \mathbf{h} - \frac{1}{2}(\mathbf{h} \cdot \mathbf{h})\mathbf{1}\right). \tag{34.1.1}$$

Next:

- Figures 34.5(d), (e), and (f) show contours of the deformation gradient component F_{22} in the actuation direction plotted on the deformed body for applied field strengths of $\mu_0|\mathbf{h}^{\text{app}}| = 0.35,\ 0.70,$ and 1.0 T, respectively. From these figures it can be observed that the sphere becomes ellipsoidal during actuation.

  Note that F_{22}—which is a measure of stretch in the actuation direction—is highly non-uniform in the sphere. F_{22} has values greater than one in the core of the sphere indicating extension in the core region, but it has values less than one in a small region near the poles of the sphere indicating compression near the poles.
- Figures 34.5(g), (h), and (i) show contours of the T_{22} component of the Cauchy stress in the actuation direction in the sphere. The contours of T_{22} are highly non-uniform in the sphere, and their magnitude increases nonlinearly as $\mu_0|\mathbf{h}^{\text{app}}|$ increases.

[1] A maximum applied field strength $\mu_0|\mathbf{h}^{\text{app}}| = 1.0$ T was chosen due to the large actuation strains observed for the ideal magnetoelastic sphere. At field strengths greater than this value the actuation strain exceeds 23%, and numerical convergence issues occur.

- Finally, Figs. 34.5(j), (k), and (l) show contours of the magnetization strength $|\mathbf{m}|$ in the sphere. Figure 34.5(j) shows that initially the magnetization strength is essentially uniform. However, Figs. 34.5(k) and (l) show that as the initially spherical geometry becomes ellipsoidal in shape, $|\mathbf{m}|$ attains locally higher values near the poles.

Summarizing, the overall character of the magnetostriction behavior of a sphere of soft-magnetic soft magnetorheological material which exhibits an *ideal magnetoelastic* response is that:

- The sphere stretches inhomogenously in the direction of the applied field $\mathbf{h}^{\text{app}}$, and becomes ellipsoidal in shape.
- The actuation strain is proportional to the square of the applied field $|\mathbf{h}^{\text{app}}|$. This quadratic-dependence of the actuation strain follows directly from the quadratic-dependence of the Maxwell stress on $\mathbf{h}$.

34.1.3 **Actuation of a sphere made of a nonlinear paramagnetic material**

Next we consider actuation of a sphere composed of a nonlinear paramagnetic material as described by the constitutive relations in Section 32.4.2, which not only account for nonlinear paramagnetism, but also for viscoelasticity of the elastomeric matrix.

At first, we ramp up the applied field $|\mathbf{h}^{\text{app}}|$ over a long time period of 1000 s so that the effects of viscoelasticity on the magnetostriction response of the s-MRE are negligible. Following that, we will study the effects of viscoelasticity on the magnetostriction response of a sphere.

Magnetostriction with viscoelastic effects suppressed

The field $\mathbf{h}^{\text{app}}$ is ramped up from zero to a strength of 1.19 MA/m, or equivalently $\mu_0|\mathbf{h}^{\text{app}}| = 1.50$ T, in the $\mathbf{e}_2$-direction over a time period of 1000 s so that viscoelastic effects on the magnetostriction response are negligible. The magnetization $\mathbf{m}$ and associated $\mathbf{h}$ and $\mathbf{b}$ fields which develop in the center of the sphere, the point indicated by a red dot in Fig. 34.6(a), are shown in Fig. 34.6(b) as a function of the applied field $\mu_0|\mathbf{h}^{\text{app}}|$. In plotting this figure we have multiplied the $\mathbf{h}$ and $\mathbf{m}$ fields by the vacuum permeability μ_0 in order to emphasize the relation $\mathbf{b} = \mu_0(\mathbf{h} + \mathbf{m})$. From this figure we note that:

- As $\mu_0|\mathbf{h}^{\text{app}}|$ is ramped up, the strength of the magnetization of the sphere $\mu_0|\mathbf{m}|$ (green line) is initially equal to 1.75 times the strength of the magnetic field $\mu_0|\mathbf{h}|$ (orange line) since the magnetic susceptibility has a value of $\chi = 1.75$. Then, as $\mu_0|\mathbf{h}^{\text{app}}|$ increases further, the magnetization $\mu_0|\mathbf{m}|$ shows a clear *saturation* behavior towards a saturation level of $\mu_0 m_s = 0.622$ T.
- As in the case of the linear paramagnetic material in Section 34.1.2, the effect of the magnetization $\mathbf{m}$ of the sphere is to *decrease* the $\mathbf{h}$ field within the sphere relative to the $\mathbf{h}$ field outside the sphere, which—as we will argue later in Section 34.1.4—leads to the observed magnetostriction response of the sphere.
- As the magnetization saturates, the rate of increase of the magnetic field $\mu_0|\mathbf{h}|$ in the sphere with the applied field $\mu_0|\mathbf{h}^{\text{app}}|$ increases, eventually becoming one-to-one.

Figure 34.6 (c) shows the diametrical actuation strain $(\lambda - 1)$ of the sphere plotted as a function of $\mu_0|\mathbf{h}^{\text{app}}|$. Initially, when $\mu_0|\mathbf{h}^{\text{app}}| \ll \mu_0 m_s$ the actuation strain $(\lambda - 1)$ increases quadratically with the applied field $\mu_0|\mathbf{h}^{\text{app}}|$. However, as $\mu_0|\mathbf{h}^{\text{app}}|$ further increases and the magnetization saturates, the axial actuation strain also moves through an inflection point before saturating at a value of 6.3%.

Next:

- Figures 34.6(d), (e), and (f) show contours of the deformation gradient component F_{22} in the actuation direction plotted on the deformed body for applied field strengths of $\mu_0|\mathbf{h}^{\text{app}}| = 0.5,\ 1.0,$ and 1.5 T, respectively.

  As in the case of the ideal magnetoelastic sphere, the contours show extension $(F_{22} > 1)$ in the core of the sphere, but compression $(F_{22} < 1)$ in a small region near the poles of the sphere. Again, the sphere becomes ellipsoidal during actuation.

- Figures 34.6(g), (h), and (i) show contours of the normal component T_{22} of the Cauchy stress for the same applied field strengths of $\mu_0|\mathbf{h}^{\text{app}}| = 0.5,\ 1.0,$ and 1.5 T, respectively.

  Here, interestingly, although increasing $\mu_0|\mathbf{h}^{\text{app}}|$ beyond magnetic saturation from 1.0 T to 1.5 T does not appreciably change the axial actuation strain $(\lambda - 1)$ it does increase the maximum Cauchy stress component T_{22} significantly, from 806.5 kPa to 1485.3 kPa.

  To understand this behavior, we argue in the Remark below that for $|\mathbf{h}^{\text{app}}| \gg m_s$ the magnetization stress $\mathbf{T}^{\text{mag}}$ is much smaller than the Maxwell stress $\mathbf{T}^{\text{maxw}}$. As a result, as the magnetic field $\mathbf{h}$ increases beyond magnetic saturation the Maxwell stress in the air continues to increase at roughly the same rate as that in the s-MRE, and therefore although the internal stress increases, *no overall motion of the air–solid boundary occurs.*

REMARK

Consider a magnetic field

$$\mathbf{h} = h\,\mathbf{e}_1,$$

for an orthonormal triad $\{\mathbf{e}_1, \mathbf{e}_2, \mathbf{e}_3\}$. The resulting non-zero components of $\mathbf{T}^{\text{mag}}$ are

$$\begin{aligned} T^{\text{mag}}_{11} &= -\mu_0 \frac{m_s^2}{\chi} \ln\left(\cosh\left(\frac{\chi}{m_s} h\right)\right) + \mu_0 \frac{m_s}{h} \tanh\left(\frac{\chi}{m_s} h\right) h^2, \\ T^{\text{mag}}_{22} &= T^{\text{mag}}_{33} = -\mu_0 \frac{m_s^2}{\chi} \ln\left(\cosh\left(\frac{\chi}{m_s} h\right)\right). \end{aligned} \tag{34.1.2}$$

In the limit of large magnetic fields, that is as $h \to \infty$, we can show that

$$\lim_{h\to\infty} T^{\text{mag}}_{11} = \lim_{h\to\infty} \left[\mu_0\, m_s\, h \tanh\left(\frac{\chi}{m_s} h\right) - \mu_0 \frac{m_s^2}{\chi} \ln\left(\cosh\left(\frac{\chi}{m_s} h\right)\right)\right] = \mu_0 \frac{m_s^2}{\chi} \ln 2,$$

and similar limit analyses show that

$$\lim_{h\to\infty} T^{\text{mag}}_{22} = \lim_{h\to\infty} T^{\text{mag}}_{33} = -\mu_0\, m_s\, h.$$

Summarizing, the components of $\mathbf{T}^{\text{mag}}$ as $h \to \infty$ are

$$T_{11}^{\text{mag}} = \mu_0 \frac{m_s^2}{\chi} \ln 2, \qquad T_{22}^{\text{mag}} = T_{33}^{\text{mag}} = -\mu_0\, m_s\, h, \tag{34.1.3}$$

which are at most linear in h. Meanwhile, the components of $\mathbf{T}^{\text{maxw}}$ are

$$T_{11}^{\text{maxw}} = \frac{1}{2}\mu_0 h^2, \qquad T_{22}^{\text{maxw}} = T_{33}^{\text{maxw}} = -\frac{1}{2}\mu_0 h^2, \tag{34.1.4}$$

which are all quadratic in h. Therefore, clearly

$$\lim_{h\to\infty} (\mathbf{T}^{\text{maxw}} + \mathbf{T}^{\text{mag}}) = \mathbf{T}^{\text{maxw}}, \tag{34.1.5}$$

that is, *the Maxwell stress dominates the magnetization stress when* $|\mathbf{h}| \gg m_s$.

- Finally, Figs. 34.6(j), (k), and (l) show contours of the magnetization strength $|\mathbf{m}|$ in the sphere. In contrast to the response of the idealized magnetoelastic sphere, here $|\mathbf{m}|$ remains nearly uniform as the applied field is ramped up. This is because although the sphere has become ellipsoidal, as shown in in Figs. 34.6(k) and (l), at the same time the magnetization has increased close to its saturation value of $m_s = 0.495$ MA/m. This acts as an upper limit on the magnetization strength, and suppresses the spatial variation of $|\mathbf{m}|$.

Summarizing, the magnetostriction behavior of a *nonlinear* paramagnetic sphere is more complicated than that of the idealized *linear* paramagnetic sphere. The key features are:

- The sphere stretches inhomogenously in the direction of the applied field $\mathbf{h}^{\text{app}}$, and becomes ellipsoidal in shape.
- As the applied field strength $\mu_0|\mathbf{h}^{\text{app}}|$ increases, the increase in the diametrical stretch $(\lambda - 1)$ of the sphere
  - is proportional to the square of the applied field when $\mu_0|\mathbf{h}^{\text{app}}| \ll \mu_0 m_s$,
  - is roughly proportional to the applied field when $\mu_0|\mathbf{h}^{\text{app}}| \approx \mu_0 m_s$, and it then
  - saturates towards a constant value when $\mu_0|\mathbf{h}^{\text{app}}| \gg \mu_0 m_s$.
- Although the axial stretch saturates when $\mu_0|\mathbf{h}^{\text{app}}| \gg \mu_0 m_s$, the axial stress in the sphere continues to grow with increasing $\mu_0\mathbf{h}^{\text{app}}$.

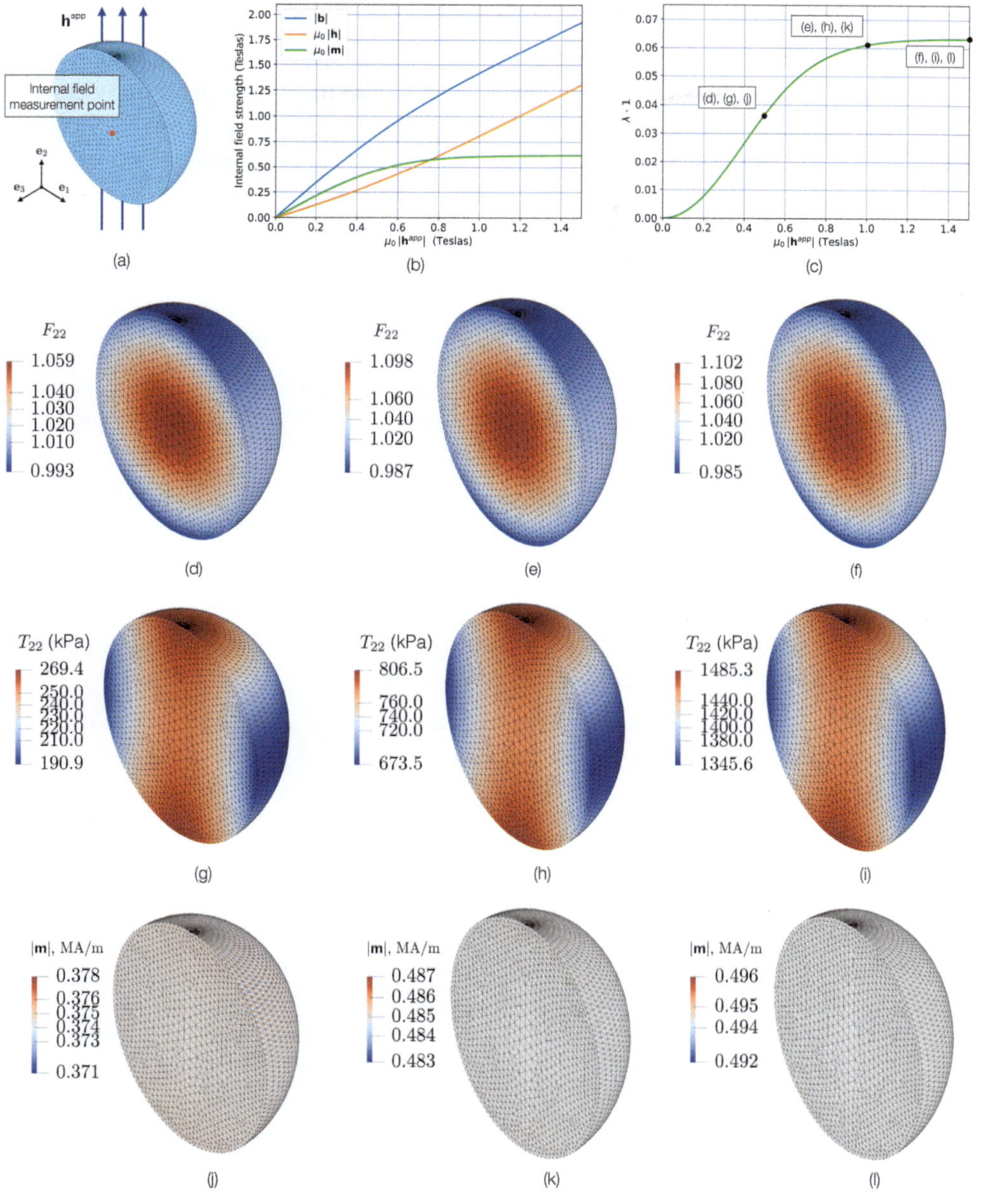

Fig. 34.6 (a) The undeformed sphere, showing the orientation of the applied magnetic field $\mathbf{h}^{app}$ and the point indicated by a red dot at which internal fields are measured. (b) The internal $\mathbf{h}$, $\mathbf{m}$, and $\mathbf{b}$ fields measured at the center of the sphere during actuation as a function of $\mathbf{h}^{app}$. We have multiplied the $\mathbf{h}$ and $\mathbf{m}$ fields by μ_0 to emphasize the relation $\mathbf{b} = \mu_0(\mathbf{h} + \mathbf{m})$. (c) The diametrical strain $(\lambda - 1)$ of the sphere initially increases, but then saturates for large $\mathbf{h}^{app}$. (d) through (l) show contours of F_{22}, T_{22}, and $|\mathbf{m}|$ plotted on the deformed body for applied field strengths of $\mu_0|\mathbf{h}^{app}| = 0.5,\ 1.0,\ \text{and}\ 1.5$ T. Note that the displacements in these figures have been *multiplied by a factor of three* for ease of visualization.

The effects of viscoelasticity on magnetostriction

The magnetostriction simulations presented thus far have been conducted at quasi-static rates so that viscoelastic effects are negligible. To illustrate the effects of viscoelasticity on magnetostriction of a sphere, Fig. 34.7 reports the diametrical strain versus applied field strength $\mu_0|\mathbf{h}^{\text{app}}|$ for a simulation in which the applied field is instead ramped up from zero to $\mu_0|\mathbf{h}^{\text{app}}| = 1.5$ T at a high rate of 0.15 T/s, and then reduced back to zero.

Since the applied magnetic field $\mathbf{h}^{\text{app}}$ induces a stress in the material, the effects of viscoelasticity of the material on the magnetostriction response of the material may be understood by analogy to the effects of stress on a viscoelastic material in a stress-controlled experiment. Specifically:

- The diametrical strain $(\lambda - 1)$ during loading at a high rate (solid green curve) is lower than the quasi-static response (dashed black curve), because of the rate-dependent stiffening of the material.
- The diametrical strain during unloading at a high rate (solid green curve) is higher than the quasi-static response (dashed black curve), due to the creep of the material under the applied field.
- The sphere retains a residual diametrical strain of 0.3% when the magnetic field is fully unloaded to $\mu_0|\mathbf{h}^{\text{app}}| = 0$ T at the higher rate (solid green curve)—as would be observed in a viscoelastic material under an applied stress cycle.

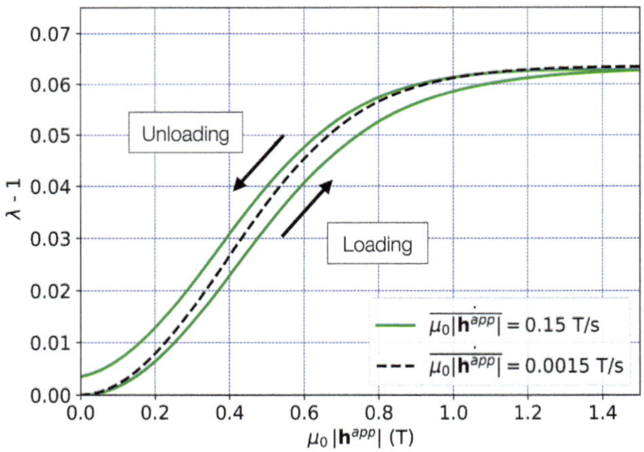

Fig. 34.7 Loading–unloading curves of the diametrical strain $(\lambda - 1)$ versus applied field strength $\mu_0|\mathbf{h}^{\text{app}}|$ for simulations in which the applied field is ramped up from zero to 1.5 T and then ramped back down to zero at two different rates of 0.15 T/s and 0.0015 T/s.

34.1.4 **Magnetic fields in the combined air–solid domain and the mechanism of magnetostriction**

Up to this point, we have not examined the magnetic fields in the air surrounding the s-MRE sphere; however, these fields are central to the magnetostriction mechanism of s-MREs. First, we will examine the perturbations to the magnetic field in and around the s-MRE which arise due to the magnetization $\mathbf{m}$ of the s-MRE.

The s-MRE material develops a magnetization $\mathbf{m}$ in response to the applied field $\mathbf{h}^{\text{app}}$, cf. Fig. 34.8(a). This $\mathbf{m}$ field perturbs the magnetic field in the specimen and the surrounding air to generate the total magnetic field $\mathbf{h}$ in the combined air–solid domain, as shown in Figs. 34.8(b) and (c) at a point in time when $|\mathbf{h}^{\text{app}}| = 1.19\,\text{MA/m}$. Let

$$\mathbf{h}^{\text{per}} \stackrel{\text{def}}{=} \mathbf{h} - \mathbf{h}^{\text{app}}$$

denote the perturbation to the applied magnetic field. Figure 34.8(d) shows contours of the magnitude $|\mathbf{h}^{\text{per}}|$ in and around the sphere. From this figure we observe that the perturbation field $\mathbf{h}^{\text{per}}$ in the sphere acts in a direction *opposite* to the applied magnetic field $\mathbf{h}^{\text{app}}$—leading to the well-known "demagnetization" effect.

Next, let the perturbed magnetic field in the solid near the s-MRE/air boundary be denoted by $\mathbf{h}^{\text{solid}}$, and the perturbed magnetic field in the air near this boundary be denoted by $\mathbf{h}^{\text{air}}$. From Figs. 34.8(b) and (c) we see that the effect of the perturbation field $\mathbf{h}^{\text{per}}$ is to *decrease* $\mathbf{h}^{\text{solid}}$, and at the same time to *increase* $\mathbf{h}^{\text{air}}$, which therefore introduces a mis-match in the fields $\mathbf{h}^{\text{solid}}$ and $\mathbf{h}^{\text{air}}$ across the s-MRE/air interface.

The difference between $\mathbf{h}^{\text{solid}}$ and $\mathbf{h}^{\text{air}}$ across the s-MRE/air boundary is at the root of the overall magnetostrictive response of s-MREs, as we explain next.

The mechanism of magnetostriction

The traction jump condition (32.5.9) across the s-MRE/air boundary, when expressed spatially (with no applied mechanical tractions and neglecting any contribution from $\mathbf{T}^{\text{air}}\mathbf{n}$), requires that

$$\mathbf{T}^{\text{solid}}\mathbf{n} = \mathbf{T}^{\text{air maxw}}\mathbf{n} \qquad \text{on } \partial\mathcal{B}_t. \tag{34.1.6}$$

Using $(32.4.58)_2$ we can rewrite the jump condition (34.1.6) as

$$\Big(\underbrace{\mathbf{T}^{\text{eq}} + \mathbf{T}^{(\alpha)}}_{\mathbf{T}^{\text{mech}}} + \mathbf{T}^{\text{maxw}} + \mathbf{T}^{\text{mag}}\Big)\,\mathbf{n} = \mathbf{T}^{\text{air maxw}}\mathbf{n} \qquad \text{on } \partial\mathcal{B}_t, \tag{34.1.7}$$

from which it is clear that any mis-match between the magnetic stresses $\mathbf{T}^{\text{air maxw}}$ and $(\mathbf{T}^{\text{maxw}} + \mathbf{T}^{\text{mag}})$ must be accommodated by a stress $\mathbf{T}^{\text{mech}}$ in the solid, which in turn causes a deformation of the solid. Thus:

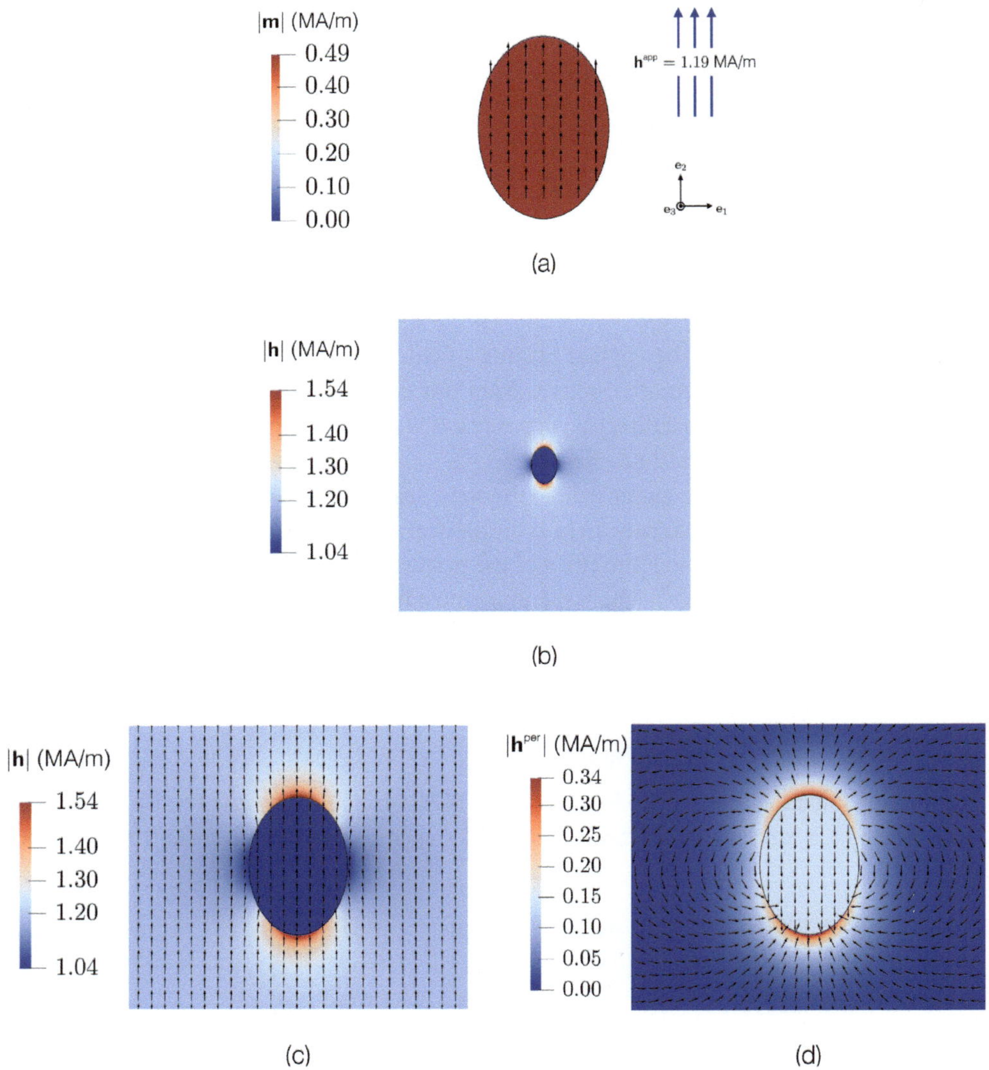

Fig. 34.8 (a) The magnetization **m** of the sphere when $|\mathbf{h}^{\text{app}}| = 1.19$ MA/m, at which point the sphere is magnetically saturated and $|\mathbf{m}| = m_s \equiv 0.495$ MA/m. (b) Contours of the magnetic field strength $|\mathbf{h}|$ in the complete combined air–solid domain. (c) Detailed view of the total magnetic field **h** near the s-MRE sphere. (d) Detailed view of the perturbation to the magnetic field $\mathbf{h}^{\text{per}}$ near the s-MRE sphere. Overlaid glyphs in (c) and (d) show field orientations. Displacements have been magnified by a factor of three for ease of visualization.

- *It is the difference between the Maxwell stress in air $\mathbf{T}^{\text{air maxw}}$ and the magnetic stress $(\mathbf{T}^{\text{maxw}} + \mathbf{T}^{\text{mag}})$ in the solid at the s-MRE/air interface which is the fundamental driving force for the deformation of the s-MRE—that is the "magnetostriction" of the s-MRE.*

Such a mismatch between $\mathbf{T}^{\text{air maxw}}$ and $(\mathbf{T}^{\text{maxw}} + \mathbf{T}^{\text{mag}})$ arises due to two factors: (i) a difference between the magnetic fields $\mathbf{h}^{\text{air}}$ and $\mathbf{h}^{\text{solid}}$ at the s-MRE/air interface, and (ii) a difference between the magnetic permeability of the air μ_0 and the magnetic permeability of the solid μ. This assertion is easily verified by examining the following transparent forms for the expressions for $\mathbf{T}^{\text{air maxw}}$ and $(\mathbf{T}^{\text{maxw}} + \mathbf{T}^{\text{mag}})$ for small values of $|\mathbf{h}|$:[2]

$$\begin{aligned}\mathbf{T}^{\text{air maxw}} &= \mu_0 \left(\mathbf{h}^{\text{air}} \otimes \mathbf{h}^{\text{air}} - \frac{1}{2} (\mathbf{h}^{\text{air}} \cdot \mathbf{h}^{\text{air}}) \mathbf{1} \right), \\ \mathbf{T}^{\text{maxw}} + \mathbf{T}^{\text{mag}} &= \underbrace{\mu_0 (1 + \chi)}_{=\mu} \left(\mathbf{h}^{\text{solid}} \otimes \mathbf{h}^{\text{solid}} - \frac{1}{2} (\mathbf{h}^{\text{solid}} \cdot \mathbf{h}^{\text{solid}}) \mathbf{1} \right).\end{aligned} \tag{34.1.8}$$

34.2 Magnetostriction of cylindrical s-MREs and the shape effect

34.2.1 Modeling approach and material parameters

A cylindrical body embedded in air is numerically modeled using an axisymmetric idealization, as shown schematically in Figs. 34.9(a) and (b). The solid cylinder has a diameter $d = 11.8\,\text{mm}$ to match the specimens studied in the magnetostriction experiments of Diguet et al. (2010), and is surrounded by an air domain with radius $s = 50\,\text{mm}$ and height $2s = 100\,\text{mm}$. As in the experiments of Diguet et al. (2010), the undeformed axial length of the cylinder ℓ_0 can be varied across simulations to achieve different length-to-diameter ratios ℓ_0/d.[3] The specific cylinder shown in Fig. 34.9(a) has a length $\ell_0 = 9.44\,\text{mm}$ and therefore $\ell_0/d = 0.80$. The computational domain with the finite element mesh used in the simulation is shown revolved 180 degrees in Fig. 34.9(c). The deformed axial length ℓ of the cylinder during magnetostriction is measured using the points indicated by black dots in Figs. 34.9(d) and (e), and the associated axial stretch is $\lambda \overset{\text{def}}{=} \ell/\ell_0$.

Boundary conditions: Referring to Fig. 34.9(a) and (b) the imposed boundary conditions are as follows:

- Mechanical boundary conditions: The horizontal centerline of the cylinder indicated in Fig. 34.9(a) by a red dashed line is held mechanically fixed in the $\mathbf{e}_2$-direction to prevent translation of the cylinder, and the top, bottom, and right edges of the air domain in Fig. 34.9(b) are held mechanically fixed in the $\mathbf{e}_1$- and $\mathbf{e}_2$-directions.

[2] These relations can be obtained from the more general relations by using the approximations $\ln(\cosh(x)) \approx x^2/2$ and $\tanh(x) \approx x$ for small x.

[3] Rounded corners with a small radius of 0.2 mm were used to prevent concentrations of the magnetic field $\mathbf{h}$ and associated difficulties in the numerical simulations.

- Magnetostatic potential boundary conditions: The magnetostatic potential on the top boundary of the air domain is kept fixed at $\phi = 0$, while the potential on the bottom boundary of the air domain is ramped up to a large positive value $\phi = 2s|\mathbf{h}^{\text{app}}|$; cf. Fig. 34.9(b). Recalling the definition of the magnetic field $\mathbf{h}_{\text{R}} = -\nabla\phi$, the difference in these two potentials over a distance of $2s$ in the $\mathbf{e}_2$-direction corresponds to a far-field applied magnetic field $\mathbf{h}^{\text{app}}$ oriented in the $\mathbf{e}_2$-direction.

Material parameters: The material parameters which we use to model the response of the s-MRE cylinders are the same as those listed in Table 34.2, with two modifications which serve to model the specific s-MRE material of Diguet et al. (2010):

- The value for the equilibrium shear modulus G_{eq} of the s-MRE material (30% volume fraction of $5\,\mu$m diameter iron particles in a silicone elastomer) is estimated by using the Young's modulus measured by Diguet et al. (2010) ($\approx$ 700 kPa) and assuming incompressibility so that $G_{\text{eq}} = E/3 \approx 230$ kPa.
- We use a magnetic susceptibility value $\chi = 2.5$ and magnetic saturation value $m_s = 0.40$ MA/m which are calibrated to match experimental data from Diguet et al. (2010) for the axial stretch λ as a function of the applied magnetic field $\mathbf{h}^{\text{app}}$ for an s-MRE cylinder with $\ell_0/d = 0.80$.

REMARK

The experimental data reported by Diguet et al. (2010) is insufficient for us to determine the complete set of material parameters appearing in our theory. In this section, we therefore use the mass density ρ_{R} and the viscoelasticity parameters in Table 34.2 from experimental data from a similar s-MRE published by Wang et al. (2023, 2024). The two s-MRE materials used by these authors are compositionally similar, both consisting of approximately 30% volume fraction of iron micro-particles with average diameter 5–7 μm embedded in a silicone-based elastomeric matrix—we therefore expect their mechanical properties to be similar.

34.2.2 Modeling the magnetostriction of a cylinder with $\ell_0/d = 0.80$

Here we simulate the magnetostrictive response of a cylindrical s-MRE specimen with $\ell_0/d = 0.80$, and compare the results from the simulation against the corresponding experimentally measured axial stretch data from Diguet et al. (2010).

The applied field $\mathbf{h}^{\text{app}}$ is linearly ramped up from zero to a strength of 1.0 MA/m, or equivalently $\mu_0|\mathbf{h}^{\text{app}}| = 1.257\,$T, in the $\mathbf{e}_2$-direction as shown in Fig. 34.10(a) and then linearly decreased back to zero. We ramp up the applied field $\mathbf{h}^{\text{app}}$ at a rate of 0.1 (MA/m)/s and then reduce it back to zero at the same rate, in order to observe the hysteretic magntostrictive response of the cylinder.[4]

[4] Diguet et al. (2010) do not report the rate at which the magnetic field strength was changed in their magnetostriction experiments. Here, we selected a rate which resulted in good agreement between the simulated axial stretch and the experimental data of Diguet et al. (2010) for a cylinder with $\ell_0/d = 0.80$.

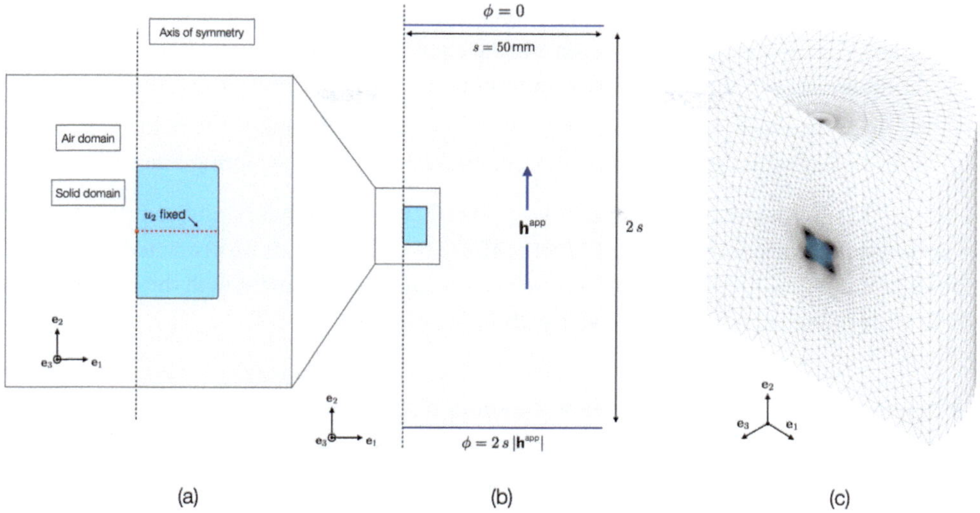

Fig. 34.9 (a) A schematic of an s-MRE cylinder embedded in an air domain. (b) The combined axisymmetric air–solid domain, showing the applied magnetic field $\mathbf{h}^{app}$ in the $\mathbf{e}_2$-direction and the corresponding boundary conditions for the magnetostatic potential ϕ. (c) The computational domain revolved 180 degrees about the axis of symmetry, with mesh edges shown. The (d) reference and (e) deformed configurations of the cylindrical s-MRE, indicating the points where the axial stretch λ is measured.

The magnetization $\mathbf{m}$ and associated $\mathbf{h}$ and $\mathbf{b}$ fields which develop in the center of the cylinder, the point indicated by a red dot in Fig. 34.10(a), are shown in Fig. 34.10(b) as a function of the applied field $\mu_0|\mathbf{h}^{app}|$. In plotting this figure we have multiplied the $\mathbf{h}$ and $\mathbf{m}$ fields by the vacuum permeability μ_0 in order to emphasize the relation $\mathbf{b} = \mu_0(\mathbf{h} + \mathbf{m})$. From this figure we note that:

- As $\mu_0|\mathbf{h}^{app}|$ is ramped up, the strength of the magnetization of the cylinder $\mu_0|\mathbf{m}|$ (green line) is initially equal to 2.5 times the strength of the magnetic field $\mu_0|\mathbf{h}|$ (orange line) since the magnetic susceptibility has a value of $\chi = 2.5$. Then, as $\mu_0|\mathbf{h}^{app}|$ increases further, the magnetization $\mu_0|\mathbf{m}|$ shows a clear *saturation* behavior towards a saturation level of $\mu_0 m_s = 0.503$ T.

- The effect of the magnetization $\mathbf{m}$ of the cylinder is to *decrease* the $\mathbf{h}$ field within the cylinder relative to the $\mathbf{h}^{\text{app}}$ field outside the cylinder (the *demagnetization effect*), which drives the observed magnetostriction response.
- As the magnetization saturates, the rate of increase of the magnetic field $\mu_0|\mathbf{h}|$ in the cylinder with the applied field $\mu_0|\mathbf{h}^{\text{app}}|$ increases, eventually becoming one-to-one.

Next, Fig. 34.10(c) shows the axial strain $(\lambda - 1)$ predicted by our theory, and comparison with the corresponding experiment from Diguet et al. (2010). The axial strain predicted by our theory and its numerical implementation is in good *quantitative* agreement with the experimental data of Diguet et al. (2010) for a cylinder with $\ell_0/d = 0.80$.

Then:

- Figurs 34.10(d), (e), and (f) show contours of the deformation gradient component F_{22} in the actuation direction plotted on the deformed body for field strengths of $\mu_0|\mathbf{h}^{\text{app}}| = 0.3,\ 0.6,$ and $1.2\,\text{T}$, respectively, during loading.

  The contours show a *non-homogeneous* extension ($F_{22} > 1$) in the bulk of the cylinder, but also some compression ($F_{22} < 1$) in small regions near the corners of the cylinder, where large gradients of F_{22} occur.
- Figures 34.10(g), (h), and (i) show contours of the normal component T_{22} of the Cauchy stress for the same applied field strengths of $\mu_0|\mathbf{h}^{\text{app}}| = 0.3,\ 0.6,$ and $1.2\,\text{T}$, respectively, during loading.

  Just as for the spherical s-MRE, although increasing $\mu_0|\mathbf{h}^{\text{app}}|$ beyond magnetic saturation from $0.6\,\text{T}$ to $1.2\,\text{T}$ does not appreciably change the axial actuation strain $(\lambda - 1)$, it does significantly increase the maximum value of the Cauchy stress component T_{22}, from $337.5\,\text{kPa}$ to $957.9\,\text{kPa}$.
- Finally, Figs. 34.10(j), (k), and (l) show contours of the magnetization strength $|\mathbf{m}|$ in the cylinder which are *non-homogeneous* in the specimen until magnetic saturation occurs.

  The details of the spatial variation of the magnetic field within the cylinder depend primarily upon the shape of the cylinder as measured by the ratio ℓ_0/d, which is intimately related to the shape effect observed in the experiments of Diguet et al. (2010). We will discuss this phenomenon further in the next section.

Summarizing, the key features of the magnetostriction response are:

- The cylinder stretches in the direction of the applied field $\mathbf{h}^{\text{app}}$, cf. Figs. 34.10(d)–(f), but does not simply elongate along the axis of the cylinder—instead it undergoes a complicated shape change which is most pronounced at the corners of the cylinder.
- As the applied field strength $\mu_0|\mathbf{h}^{\text{app}}|$ increases, the increase in the axial strain $(\lambda - 1)$ of the cylinder
  - is proportional to the square of the applied field when $\mu_0|\mathbf{h}^{\text{app}}| \ll \mu_0 m_s$,
  - is roughly proportional to the applied field when $\mu_0|\mathbf{h}^{\text{app}}| \approx \mu_0 m_s$, and it then
  - saturates towards a constant value when $\mu_0|\mathbf{h}^{\text{app}}| \gg \mu_0 m_s$.
- Although the axial stretch saturates during loading when $\mu_0|\mathbf{h}^{\text{app}}| \gg \mu_0 m_s$, the axial stress in the cylinder continues to grow with increasing $\mu_0\mathbf{h}^{\text{app}}$, cf. Figs. 34.10(g)–(i).

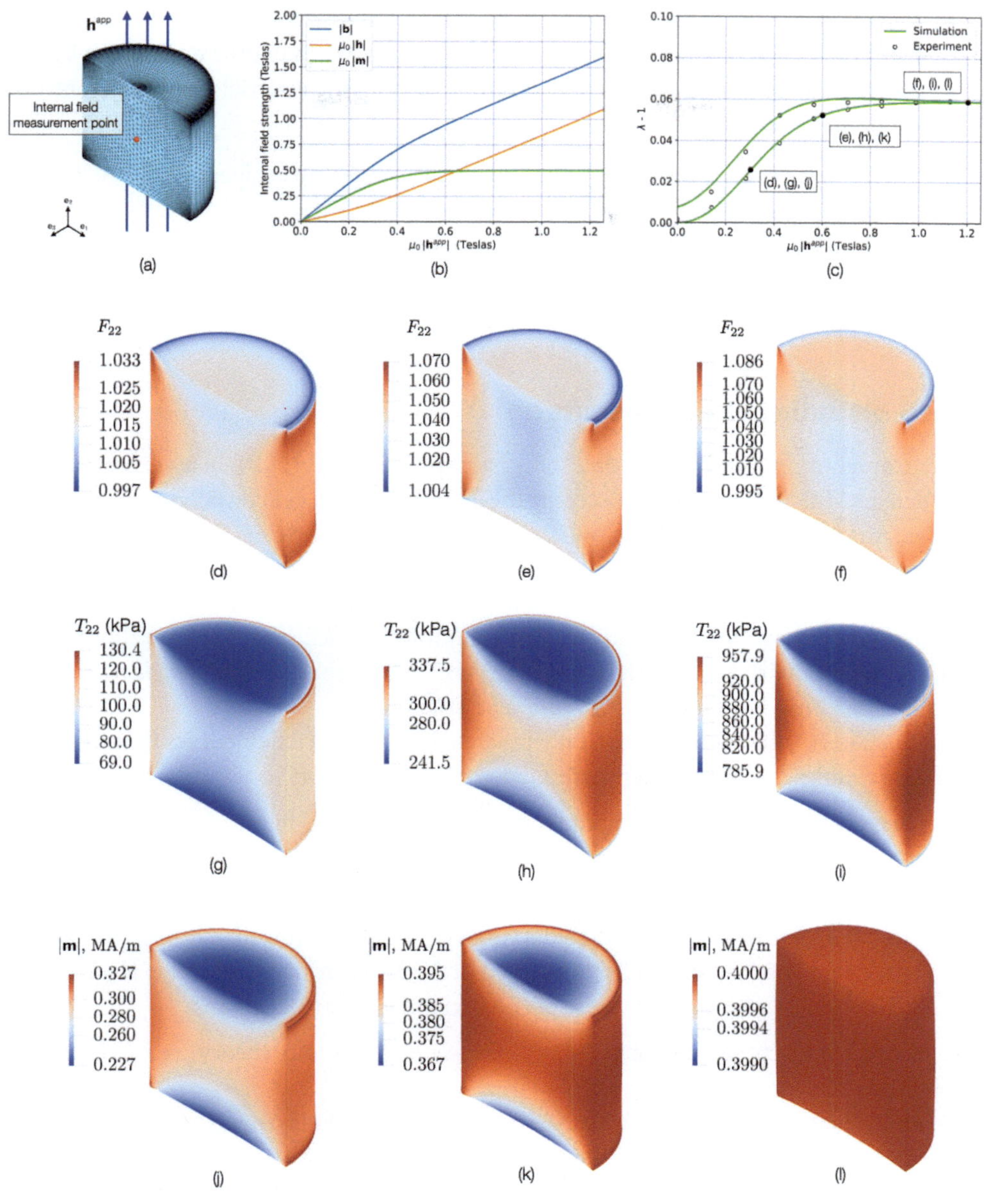

Fig. 34.10 (a) The undeformed cylinder, showing the orientation of the applied magnetic field $\mathbf{h}^{\text{app}}$ and the point, indicated by a red dot, for which the internal fields are reported. (b) The internal **h**, **m**, and **b** fields measured at the center of the cylinder during actuation as a function of $\mathbf{h}^{\text{app}}$. We have multiplied the **h** and **m** fields by μ_0 to emphasize the relation $\mathbf{b} = \mu_0(\mathbf{h} + \mathbf{m})$. (c) The axial strain $(\lambda - 1)$ of the cylinder overlaid with experimental data from Diguet et al. (2010). (d)–(l) show contours of the deformation gradient component F_{22}, Cauchy stress component T_{22}, and magnitude of magnetization $|\mathbf{m}|$ plotted on the deformed body for applied field strengths of $\mu_0|\mathbf{h}^{\text{app}}| = 0.3,\ 0.6,\ \text{and}\ 1.2\,\text{T}$. The displacements in these figures have been magnified by a factor of three for ease of visualization.

- Regarding the effects of viscoelasticity, the applied magnetic field $\mathbf{h}^{\text{app}}$ induces a stress in the material and therefore the viscoelastic effects in s-MREs may be understood by analogy to the effects of stress on a non-magnetic viscoelastic material in a stress-controlled experiment. Specifically:
  - The axial strain during unloading is higher than that during loading, due to the creep of the material under the applied field.
  - The cylinder retains a residual axial strain of 0.8% when the magnetic field is fully unloaded to $\mu_0|\mathbf{h}^{\text{app}}| = 0\,\text{T}$, as would be observed in a viscoelastic material under an applied stress cycle.

34.2.3 **Modeling the shape effect**

Next, in order to study the "shape effect" observed by Diguet et al. (2010) in his experiments, we conducted additional simulations. Keeping the diameter d fixed, we considered two additional axial lengths ℓ_0 for specimen aspect-ratios of $\ell_0/d = 0.33$ and $\ell_0/d = 1.46$, corresponding to the extreme values of aspect-ratio ℓ_0/d studied by Diguet et al. (2010) in his experiments.

The experimental data of Diguet et al. (2010) for the axial strain versus applied magnetic field measured in various differently shaped cylindrical s-MRE specimens are shown in Fig. 34.11(a), where the shape effect is clear—*cylinders with lower* ℓ_0/d *exhibit higher axial strain* $(\lambda - 1)$ *at the same level of the applied magnetic field.*

Figure 34.11(b) shows the numerically predicted magnetostriction curves from our corresponding simulations for cylinders with $\ell_0/d \in \{0.33, 0.80, 1.46\}$:

- The individual magnetostriction curves in this figure show the same qualitative magnetic-saturation and viscoelastic-hysteretic behavior as observed in the experimental curves of Diguet et al. (2010).
- Also, the trend in the value of the axial strain $(\lambda - 1)$ versus the ℓ_0/d ratio is in qualitative agreement with the shape effect observed in the experiments of Diguet et al. (2010)—in that cylinders with smaller values of ℓ_0/d exhibit higher axial strains.

The essential aspects of the "shape effect" are captured by our theory and its numerical implementation.

We note that while the results from the simulations show similar trends as those observed in the experiments, the magnitude of the changes to axial strain $(\lambda - 1)$ in the simulations with changes to ℓ_0/d is smaller than that observed in the experiments. This discrepancy may be due either (a) to physical phenomena which are not included in our theory, or (b) to imprecision and uncertainty in the experiments of Diguet et al. (2010) (which is a distinct possibility). More refined experimental and theoretical work needs to be done to bring theory and experiment in closer accord with each other.

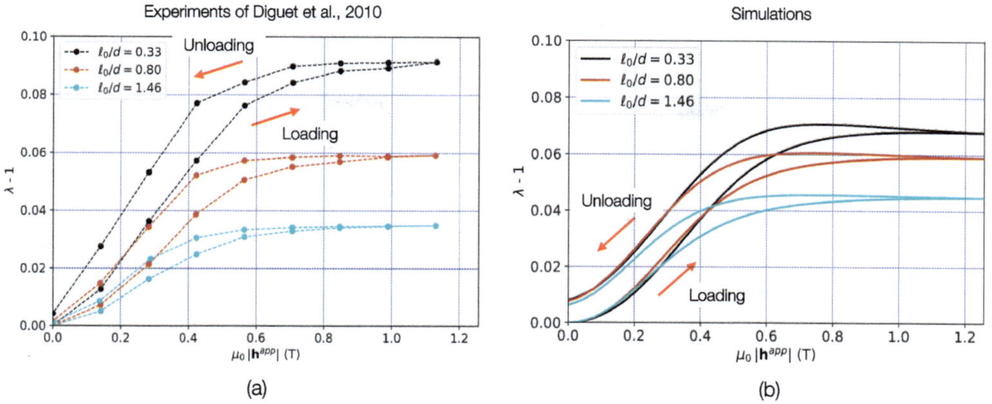

Fig. 34.11 (a) Axial strain $(\lambda - 1)$ versus applied magnetic field $\mu_0|\mathbf{h}^{\text{app}}|$ curves for s-MRE cylinders with different aspect-ratios ℓ_0/d; results from the experiments by Diguet et al. (2010). (b) The corresponding results for cylinders with $\ell_0/d \in \{0.33, 0.80, 1.46\}$ predicted by our theory and its numerical implementation.

Understanding the shape effect

We now use the capabilities of our numerical implementation to examine the magnetic fields in the combined air–solid domain predicted by our theory in order to understand the physical origins of the shape effect.

The s-MRE material develops a magnetization $\mathbf{m}$ in response to the applied field $\mathbf{h}^{\text{app}}$, cf. Figs. 34.12(a) and (b). This $\mathbf{m}$ field perturbs the magnetic field in the specimen and the surrounding air to generate the total magnetic field $\mathbf{h}$ in the combined air–solid domain, as shown in Figs. 34.12(c) and (d) for the cylinders with aspect-ratios of $\ell_0/d = 0.33$ and $\ell_0/d = 1.46$, respectively, at a point in time when $|\mathbf{h}^{\text{app}}| = 1.0$ MA/m.

Next, recall from Section 34.1.4 that

- it is the difference between the magnetic fields $\mathbf{h}^{\text{air}}$ and $\mathbf{h}^{\text{solid}}$ at the s-MRE/air interface which drives the overall magnetostriction behavior of s-MREs.

From Figs. 34.12(c) and (d) we see that the mis-match between $\mathbf{h}^{\text{solid}}$ and $\mathbf{h}^{\text{air}}$ is much larger in the cylinder with the smaller aspect-ratio, than in the cylinder with the larger aspect-ratio, due to "magnetic fringing" at the lateral boundaries of the larger aspect-ratio cylinder. This shape-dependent difference between $\mathbf{h}^{\text{solid}}$ and $\mathbf{h}^{\text{air}}$ across the s-MRE/air boundary is at the root of the overall shape-dependent magnetostrictive response of s-MREs.

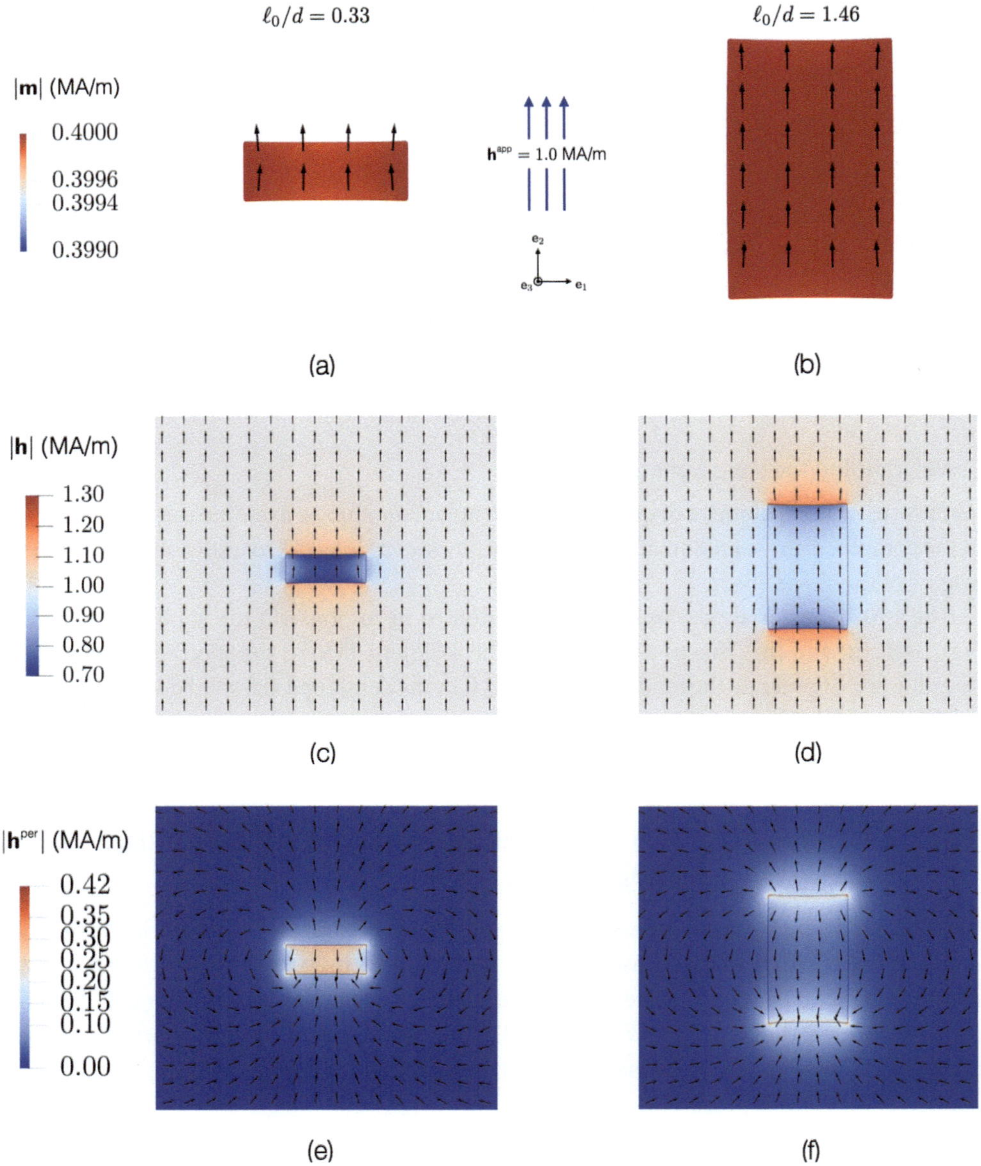

Fig. 34.12 (a) and (b) show contours of the magnetization $\mathbf{m}$, (c) and (d) show contours of the total magnetic field $|\mathbf{h}|$, and (e) and (f) show contours of the perturbation field $|\mathbf{h}^{\text{per}}| = |\mathbf{h} - \mathbf{h}^{\text{app}}|$ for two differently shaped s-MRE cylinders with aspect-ratios $\ell_0/d = 0.33$ and $\ell_0/d = 1.46$, all at a point in time when $|\mathbf{h}^{\text{app}}| = 1.0$ MA/m. Glyphs indicate the local field orientation. The perturbation field $\mathbf{h}^{\text{per}}$ field has significantly larger magnitude inside the shorter cylinder with $\ell_0/d = 0.33$, relative to the longer cylinder with $\ell_0/d = 1.46$.

34.3 Modeling magnetic actuation of a trilayer s-MRE beam

In this section, as an interesting practical application of our theory, we study the actuation response of a trilayer s-MRE composite beam. The trilayer beam is composed of three perfectly bonded layers, with the middle layer made from an s-MRE material and the upper and lower layers made from a magnetically inert elastomeric material which is similar in composition to the s-MRE layer but has no embedded carbonyl iron particles.

The simulation set-up is shown schematically in Fig 34.13. The trilayer beam is taken to be 20 mm long. The middle s-MRE layer is 0.5 mm thick, while the upper and lower non-magnetic support layers are each 0.25 mm thick. We fix the strength of the applied magnetic field at $|\mathbf{h}^{\mathrm{app}}| = 40$ kA/m and examine the actuation behavior of the beam under various angles θ of the applied field relative to the line perpendicular to the length of the beam, cf. Fig. 34.13(a).[5] In our simulations we found that the non-magnetic support layers were necessary to suppress geometric instabilities under an applied magnetic field; the support layers ensure controllable bending behavior.

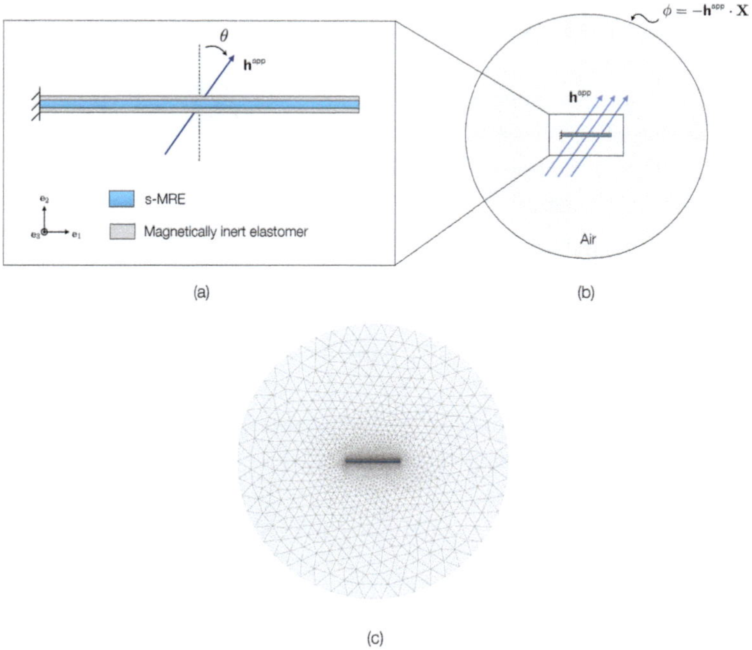

Fig. 34.13 (a) A schematic of the trilayer s-MRE composite beam and the applied magnetic field $\mathbf{h}^{\mathrm{app}}$ oriented at an angle θ. (b) The combined air–solid domain, indicating the magnetostatic potential prescribed on the circular boundary to produce a far-field applied field of $\mathbf{h}^{\mathrm{app}}$. (c) The mesh used in the simulations.

[5] We use an applied magnetic field $\mathbf{h}^{\mathrm{app}}$ with a moderate strength of 40 kA/m in order to keep the deflection of the beam in the range of several mm. Larger applied magnetic field strengths result in larger displacements and significant distortion of the air mesh.

For the s-MRE material we use the material parameters listed in Table 34.2. To model the magnetically inert support layers we use material properties for a non-magnetic elastomer with a mass ratio of PDMS to curing agent of 10:1 reported by Wang et al. (2024). Specifically,

- we take the shear modulus of the non-magnetic elastomer to be $G_{eq} = 375\,\text{kPa}$ and the referential mass density to be $\rho_R = 1045\,\text{kg/m}^3$ as measured by Wang et al. (2024), and
- we neglect any magnetization response of the material and associated contributions to the Cauchy stress or magnetic **b**-field;

all other material properties are taken to be the same as in Table 34.2.

A trilayer beam which is embedded in air is numerically modeled using a plane-strain idealization, as shown schematically in Fig. 34.13(b). The radius of the air domain is selected as $50\,\text{mm}$, which is ten times the half-length of the beam, so that the magnetostatic potentials applied at the edges of the domain are not (significantly) affected by the magnetization field generated in the beam. The computational domain with the finite element mesh is shown in Fig. 34.13(c).

Boundary conditions: Referring to Figs. 34.13(a) and (b) the imposed boundary conditions are as follows:

- Mechanical boundary conditions: The left edge of the beam indicated in Fig. 34.13(a) is held mechanically fixed, and the circular edge of the air domain in Fig. 34.13(b) is also held mechanically fixed.

  Additionally, since the support layers are magnetically inert, their magnetic response is similar to that of the air. Accordingly, we enforce the jump condition for the force balance (32.5.9) over the boundary of the s-MRE layer alone rather than the boundary of the entire trilayer beam.
- Magnetostatic potential boundary conditions: The applied magnetic field $\mathbf{h}^{\text{app}}$ is oriented at an angle θ according to

$$\mathbf{h}^{\text{app}} = |\mathbf{h}^{\text{app}}| \left(\sin\theta\, \mathbf{e}_1 + \cos\theta\, \mathbf{e}_2\right). \tag{34.3.1}$$

  The spatially varying magnetostatic potential corresponding to this magnetic field is applied on the circular boundary of the air domain,

$$\phi(\mathbf{X}) = -\mathbf{h}^{\text{app}} \cdot \mathbf{X}, \tag{34.3.2}$$

  cf. Fig. 34.13(b). Recalling the definition of the magnetic field $\mathbf{h}_R = -\nabla\phi$, this spatially varying potential $\phi(\mathbf{X})$ corresponds to a uniform far-field applied magnetic field of $\mathbf{h}^{\text{app}}$.

34.3.1 **Quasi-static actuation**

To understand the basic bending actuation behavior of the trilayer beam we first study its response to a magnetic field $\mathbf{h}^{\text{app}}$ whose angle of incidence θ is varied, cf. Fig. 34.14(a). For a given angle θ, we ramp up the strength of the magnetic field $|\mathbf{h}^{\text{app}}|$ from zero to 40 kA/m slowly, over a period of 1000 s, so that dynamic and viscoelastic effects are negligible and the resulting tip displacement of the beam at the end of the simulation is its "steady-state" value.

Contours of the magnetization strength $|\mathbf{m}|$ are plotted on the steady-state deformed configuration of the trilayer beam under a magnetic field of 40 kA/m oriented at an angle of $\theta = 35°$ in Fig. 34.14(b), where the tip displacement of the beam measures 5.3 mm in the $\mathbf{e}_2$-direction. Note that the magnetization in the magnetically inert layers is zero.

Figure 34.14(c) then plots the variation of the steady-state tip displacement as a function of the angle θ for angles in the range -90 degrees to 90 degrees. This figure reveals several important characteristics for the actuation response of the beam:

- The tip displacement response is anti-symmetric about $\theta = 0°$, with positive θ leading to positive tip displacement and negative θ leading to negative tip displacement.
- The cases $\theta = 0°$ and $\theta = \pm 90°$ do not induce any displacement of the beam.[6] The tip displacement is more sensitive to imperfections in the incidence angle θ near $\theta = 0°$ than it is near $\theta = \pm 90°$.
- The incidence angles of $\theta = \pm 35°$ correspond to the minimum and maximum attainable vertical tip displacements of ± 5.3 mm.

The steady-state deformed configurations of the beam upon application of a 40 kA/m magnetic field with various angles $\theta \in \{-35°, -10°, 0°, 10°, 35°\}$ are shown in Fig. 34.14(d).

From these simulations, we see that the actuation behavior of a trilayer s-MRE composite beam is highly controllable in terms of both direction and magnitude through the angle at which an external magnetic field is applied — a degree of control which is potentially advantageous for actuators made from s-MREs.

[6] Note the lack of a bending response of the s-MRE beam in the case $\theta = 0°$. This is in stark contrast with the actuation behavior of beams made from an h-MRE material, in that an h-MRE beam with remanent magnetization $\mathbf{m}^{\text{rem}}$ pointing along its axis will deform dramatically in response to an externally applied magnetic field oriented perpendicular to the beam (cf., e.g., Zhao et al., 2019). The reason behind this different response is that the actuation mechanisms at work in the two cases are quite different: (i) in an h-MRE beam the bending is driven by magnetic body-torques associated with mis-alignment between $\mathbf{m}^{\text{rem}}$ and $\mathbf{h}$, whereas (ii) in an s-MRE beam the magnetization $\mathbf{m}$ of the s-MRE is always aligned with $\mathbf{h}$, cf. eq. (4.21), and thus no magnetic body-torques can develop, hence there is no bending in the case of $\theta = 0°$.

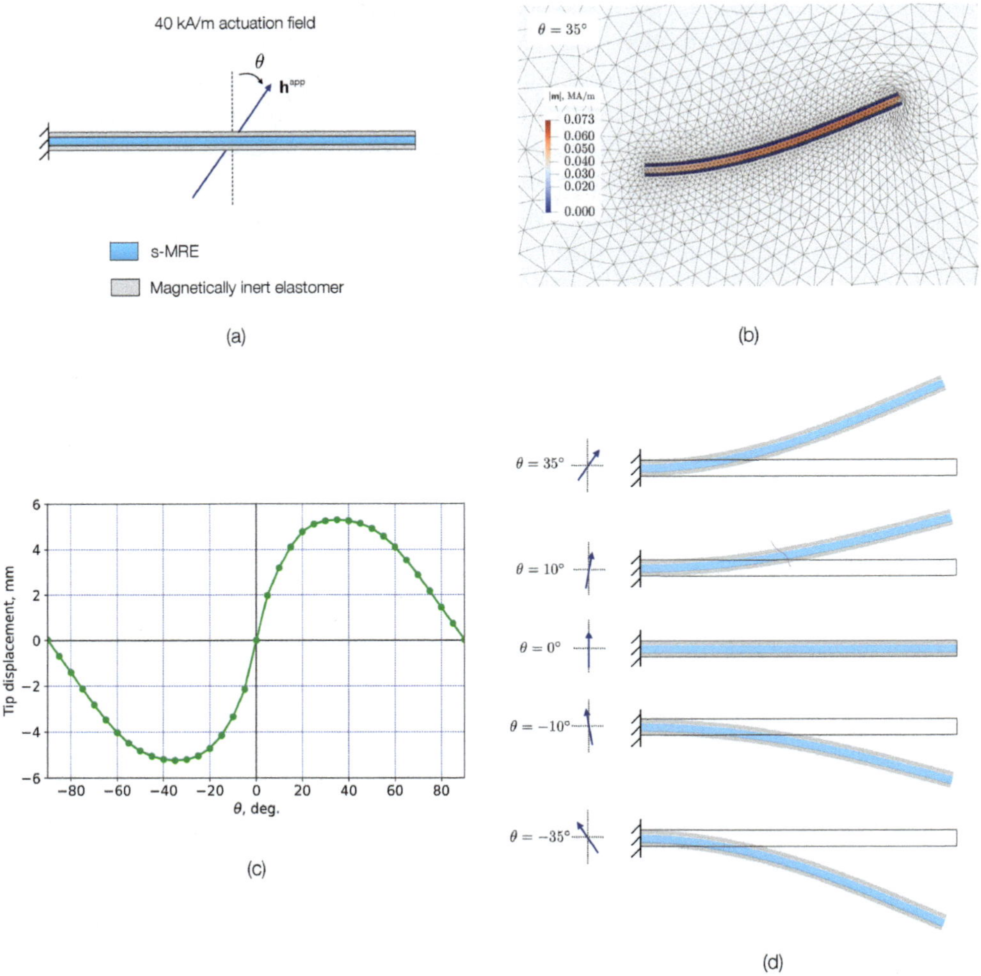

Fig. 34.14 (a) A schematic of the magnetic field $\mathbf{h}^{\text{app}}$ applied to the trilayer s-MRE composite with strength of 40 kA/m and angle of incidence θ. (b) Contours of the magnetization of the trilayer beam in the deformed configuration for $\theta = 35°$. (c) The steady-state tip displacement of the trilayer s-MRE composite beam as a function of the angle θ at which a 40 kA/m magnetic field is applied. (d) The deformed shape of the trilayer beam under application of 40 kA/m magnetic field with various angles $\theta \in \{-35°, -10°, 0°, 10°, 35°\}$.

34.3.2 **Dynamic actuation including viscoelastic and gravitational effects**

We next show the results of a simulation of the actuation response of a trilayer s-MRE beam in which we include the effects of inertia, gravitational body force, as well as viscoelasticity—all of which greatly affect the performance of these materials when used as actuators. The gravitational body force is taken as

$$\mathbf{f}_{0,\mathrm{R}}^{\mathrm{grav}} = \rho_{\mathrm{R}}\,\mathbf{a}^{\mathrm{grav}}, \tag{34.3.3}$$

where $\mathbf{a}^{\mathrm{grav}}$ is the acceleration due to gravity at the surface of the Earth, taken to be $9.81\ \mathrm{m/s}^2$ in the negative $\mathbf{e}_2$-direction as shown in Fig. 34.15(a).

In this simulation we apply a magnetic field at an incidence angle of $\theta = 35°$, cf. Fig. 34.15(a), which is selected since it leads to the largest positive steady-state tip displacement (cf. Section 34.3.1). The magnitude $|\mathbf{h}^{\mathrm{app}}|$ of the applied field is taken to vary over time according to a linear interpolation of the tabulated values shown in Table 34.3.

Figure 34.15(b) shows curves of the applied $|\mathbf{h}^{\mathrm{app}}|$ field as a function of time, and the corresponding displacement of the tip of the beam in the $\mathbf{e}_2$-direction in response to this signal:

- From time $t = 0\,\mathrm{s}$ to $0.2\,\mathrm{s}$ there is no applied magnetic field, $|\mathbf{h}^{\mathrm{app}}| = 0$, but the beam deflects downwards under its own weight, exhibiting overshoot and damped oscillations around a downward tip displacement of $\sim 2\,\mathrm{mm}$.
- Then, from $t = 0.2\,\mathrm{s}$ to $0.25\,\mathrm{s}$ $|\mathbf{h}^{\mathrm{app}}|$ is ramped up from zero to $40\,\mathrm{kA/m}$, and the magnetically induced stresses in the beam overcome the gravitational force, causing the tip of the beam to shoot upwards.
- Next, $|\mathbf{b}^{\mathrm{app}}|$ is held at 40 kA/m from $t = 0.25$ s to 0.45 s, during which time the tip motion displays damped oscillations about a positive displacement of ~ 2.5 mm.
- Finally, from $t = 0.45$ s to 0.50 s $|\mathbf{h}^{\mathrm{app}}|$ is ramped down from $40\,\mathrm{kA/m}$ back to zero and subsequently held constant at zero until a final time of $t = 1.0\,\mathrm{s}$. In response to this removal of the $|\mathbf{h}^{\mathrm{app}}|$ field, the tip of the beam drops downwards under the self-weight of the beam, and again displays overshoot and damped oscillations around a downward tip displacement of $\sim 2\,\mathrm{mm}$. Subsequently, the amplitude of these oscillations reduces significantly due to dissipative effects of viscoelasticity of the beam.

Table 34.3 History of applied magnetic flux density $|\mathbf{h}^{\mathrm{app}}|$ at an incidence angle of $\theta = 35°$.

Time, s	0	0.20	0.25	0.45	0.5	1.0
$\lvert\mathbf{h}^{\mathrm{app}}\rvert$, kA/m	0	0	40	40	0	0

This simulation demonstrates some important practical aspects of the dynamic actuation response of s-MRE beams.[7] Specifically, with reference to Fig. 34.15(b):

- Since s-MREs have a low shear modulus, the deformations induced by gravity under the material's own weight can be quite large—2.5 mm in this simulation—and similar in magnitude to the deformation due to applied magnetic actuation fields with strengths of tens of kA/m.
- Since the elastic shear wave speed in the material is low, again because the shear modulus is low, the response time of the actuator increases. In this simulation the response time is on the order of tens of milliseconds.
- Since the elastomeric matrix of s-MREs displays marked viscoelastic behavior, significant overshoot and long settling times are expected during actuation. In this simulation the overshoot is as large as 1.5 mm, and the settling time is on the order of one second.

It is clear from this simulation that inertia, gravity, and viscoelasticity all greatly affect the performance of s-MRE actuators.

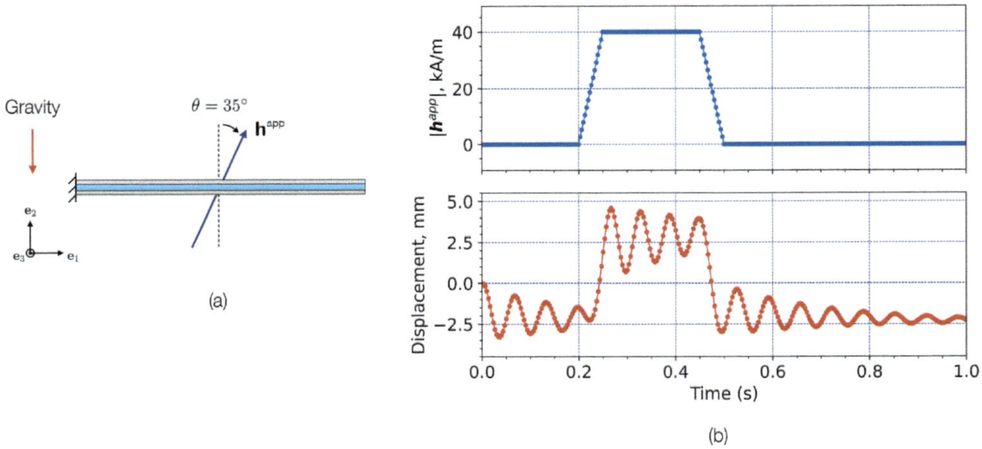

Fig. 34.15 (a) A schematic of the magnetic field $\mathbf{h}^{app}$ applied to the trilayer s-MRE beam with angle of incidence $\theta = 35°$, and a red arrow indicating the direction of acceleration due to gravity. (b) Plots of the applied magnetic field strength $|\mathbf{h}^{app}|$ and resulting tip displacement of the beam in the $\mathbf{e}_2$-direction as a function of time.

[7] Which are not made clear by their quasi-static actuation behavior discussed in Section 34.3.1.

Part XII

Appendices

Appendix A A brief on electromagnetics in matter

In this appendix we present a brief summary of the specific forms of Maxwell's equations and the electromagnetic energy transport expression in the (i) electrostatic and (ii) magnetostatic limits which are pertinent to the theories and applications discussed in this textbook.

A.1 Basic fields. Maxwell's equations

A discussion of electromagnetics in matter involves the following **fields**:[1]

(i) $\mathbf{e}$, electric field [V/m];

(ii) $\mathbf{d}$, electric displacement [$\mathrm{C/m^2}$];

(iii) $\mathbf{h}$, magnetic field [A/m];

(iv) $\mathbf{b}$, magnetic flux density [Tesla];

(v) q, *free* charge per unit volume [$\mathrm{C/m^3}$];

(vi) $\mathbf{j}$, *free* charge current density [$\mathrm{C/(m^2 s)}$] .

The equations that govern the behavior of electromagnetic fields in a region of space containing a material body are the following Maxwell's equations:

- **Gauss's law**:

$$\operatorname{div}\mathbf{d} = q. \tag{A.1.1}$$

- **Faraday's law**:

$$\operatorname{curl}\mathbf{e} + \frac{\partial \mathbf{b}}{\partial t} = \mathbf{0}. \tag{A.1.2}$$

- **Gauss's law for magnetism**:

$$\operatorname{div}\mathbf{b} = 0. \tag{A.1.3}$$

- **Maxwell–Ampère's law**:

$$\operatorname{curl}\mathbf{h} - \frac{\partial \mathbf{d}}{\partial t} = \mathbf{j}. \tag{A.1.4}$$

[1] See Appendix B for a listing of units for electromagnetic quantities in the SI (MKS) system.

REMARK

Note that in materials that are subject to *electric polarization and induced magnetization*, it is conventional to split the total charge and the total current into "bound" and "free" parts. The "bound" parts are the *fields generated in the material in response to external electromagnetic fields*. One has no direct control over the "bound" charges and currents, so we have written Maxwell's equations in a form which makes explicit reference only to "free" charges and currents.

A.2 Galilean invariant forms of Maxwell's equations

Of importance in classical physics is a change in frame $\{\mathbf{x}, t\} \mapsto \{\acute{\mathbf{x}}, t\}$ which has the form

$$\acute{\mathbf{x}} = \mathbf{x} - \mathbf{u}\,t, \qquad \mathbf{u} = \text{constant}, \tag{A.2.1}$$

and is known as a **Galilean change of frame**. We suppose that under such a frame change the motion $\mathbf{x} = \boldsymbol{\chi}(\mathbf{X}, t)$ transforms as

$$\acute{\boldsymbol{\chi}}(\mathbf{X}, t) = \boldsymbol{\chi}(\mathbf{X}, t) - \mathbf{u}t, \tag{A.2.2}$$

so that the velocities of material points transform as

$$\acute{\mathbf{v}}(\acute{\mathbf{x}}, t) = \mathbf{v}(\mathbf{x}, t) - \mathbf{u}. \tag{A.2.3}$$

Thus, dropping arguments, the Galilean transformation law for material velocity $\mathbf{v}$ is

$$\acute{\mathbf{v}} = \mathbf{v} - \mathbf{u}. \tag{A.2.4}$$

It may be shown that the fields $\{\mathbf{d}, \mathbf{b}, q\}$ are *invariant* under a Galilean change in frame, that is,

$$\acute{\mathbf{d}} = \mathbf{d}, \qquad \acute{\mathbf{b}} = \mathbf{b}, \qquad \acute{q} = q; \tag{A.2.5}$$

however, the electric field $\mathbf{e}$, the magnetic field $\mathbf{h}$, and the current $\mathbf{j}$ are not invariant, but the fields

$$\begin{aligned} \mathcal{E} &\stackrel{\text{def}}{=} \mathbf{e} + \mathbf{v} \times \mathbf{b}, \\ \mathcal{H} &\stackrel{\text{def}}{=} \mathbf{h} - \mathbf{v} \times \mathbf{d}, \\ \mathcal{J} &\stackrel{\text{def}}{=} \mathbf{j} - q\mathbf{v}, \end{aligned} \tag{A.2.6}$$

with $\mathbf{v}$ the velocity of a material particle, are invariant under a Galilean frame change. That is,

$$\acute{\mathcal{E}} = \mathcal{E}, \qquad \acute{\mathcal{H}} = \mathcal{H}, \qquad \text{and} \qquad \acute{\mathcal{J}} = \mathcal{J}. \tag{A.2.7}$$

- The field $\mathcal{E}$ is known as the **electromotive intensity**,
- the field $\mathcal{H}$ is known as the **magnetomotive intensity**, while
- the field $\mathcal{J}$ is known as the **conduction current density**.

Using these relations, Maxwell's differential equations for a deformable body may be expressed as (cf., e.g., Kovetz, 2000):

$$\begin{aligned} \operatorname{div}\mathbf{d} &= q, \\ \operatorname{curl}\mathcal{E} + \overset{*}{\mathbf{b}} &= \mathbf{0}, \\ \operatorname{curl}\mathcal{H} - \overset{*}{\mathbf{d}} &= \mathcal{J}, \\ \operatorname{div}\mathbf{b} &= 0, \end{aligned} \tag{A.2.8}$$

where for any vector $\mathbf{a}$,

$$\overset{*}{\mathbf{a}} = \dot{\mathbf{a}} + (\operatorname{div}\mathbf{v})\mathbf{a} - \mathbf{L}\mathbf{a} \tag{A.2.9}$$

denotes its *flux derivative*, in which $\dot{\mathbf{a}}$ is the *material derivative*, $\mathbf{v}$ the velocity, and $\mathbf{L} = \operatorname{grad}\mathbf{v}$ the velocity gradient. Maxwell's equations (A.2.8) are written in a form which involves only Galilean invariant quantities. In the next two sections we discuss the (i) electrostatic and (ii) magnetostatic limits of Maxwell's equations which are pertinent to the subject matter of this book.

A.3 Electrostatic limit of Maxwell's equations

The definition of the *electrostatic limit* of Maxwell's equations that we employ here is one in which any effects of coupling of the magnetic fields with electric fields are neglected by setting

$$\mathbf{b} = \mathbf{0} \qquad \text{and} \qquad \overset{*}{\mathbf{b}} = 0\,. \tag{A.3.1}$$

Under these assumptions Faraday's law $(A.2.8)_2$ becomes,

$$\operatorname{curl}\mathcal{E} = \mathbf{0}. \tag{A.3.2}$$

Since the curl of any gradient field is zero ($\operatorname{curl}\operatorname{grad} f \to e_{ijk} f_{,kj} = 0$), in the quasi-static limit we assure that (A.3.2) will always be satisfied by assuming that $\mathcal{E}$ is the gradient of some scalar field. By convention we write

$$\mathcal{E} = -\operatorname{grad}\phi \tag{A.3.3}$$

where ϕ is known as the *electric potential* (or voltage) field. The negative sign in (A.3.3) follows the convention that the electric field points from a region of high potential to low potential, whereas the vector $\operatorname{grad}\phi$ by definition points in the direction of increasing ϕ.

Also, Gauss's law $(A.2.8)_1$ states that

$$\operatorname{div}\mathbf{d} = q. \tag{A.3.4}$$

Equations (A.3.2) and (A.3.4) constitute the governing equations of *electrostatics* in $\mathcal{B}_t$.

A.3.1 Energy transport in the electrostatic limit

The flux of *electromagnetic energy* is described by the *Poynting vector* (see, e.g., Kovetz (2000), Section 45 and Section 54),

$$\boldsymbol{\mathcal{S}} \stackrel{\text{def}}{=} \boldsymbol{\mathcal{E}} \times \boldsymbol{\mathcal{H}}. \tag{A.3.5}$$

This vector describes the flux of electromagnetic energy (energy per unit area per unit time). Accordingly, the flux of electromagnetic energy *into* part $\mathcal{P}_t$ of the deformed body $\mathcal{B}_t$ is given by,

$$\mathcal{E}(\mathcal{P}_t) = -\int_{\partial\mathcal{P}_t} \boldsymbol{\mathcal{S}}\cdot\mathbf{n}\,da\,, \tag{A.3.6}$$

where $\mathbf{n}$ is the outward unit normal to its boundary $\partial\mathcal{P}_t$. Using the divergence theorem,

$$-\int_{\partial\mathcal{P}_t} \boldsymbol{\mathcal{S}}\cdot\mathbf{n}\,da = -\int_{\mathcal{P}_t} \operatorname{div}\boldsymbol{\mathcal{S}}\,dv = -\int_{\mathcal{P}_t} \operatorname{div}(\boldsymbol{\mathcal{E}}\times\boldsymbol{\mathcal{H}})\,dv\,. \tag{A.3.7}$$

Next, use of the vector identity

$$\operatorname{div}(\boldsymbol{\mathcal{E}}\times\boldsymbol{\mathcal{H}}) = \boldsymbol{\mathcal{H}}\cdot(\operatorname{curl}\boldsymbol{\mathcal{E}}) - \boldsymbol{\mathcal{E}}\cdot(\operatorname{curl}\boldsymbol{\mathcal{H}}), \tag{A.3.8}$$

allows us to write (A.3.7) as

$$-\int_{\partial\mathcal{P}_t} \boldsymbol{\mathcal{S}}\cdot\mathbf{n}\,da = -\int_{\mathcal{P}_t} \Big(\boldsymbol{\mathcal{H}}\cdot(\operatorname{curl}\boldsymbol{\mathcal{E}}) - \boldsymbol{\mathcal{E}}\cdot(\operatorname{curl}\boldsymbol{\mathcal{H}})\Big)\,dv\,. \tag{A.3.9}$$

In the electrostatic limit $\operatorname{curl}\boldsymbol{\mathcal{E}} = 0$, and since $\operatorname{curl}\boldsymbol{\mathcal{H}} = \overset{*}{\mathbf{d}} + \boldsymbol{\mathcal{J}}$ (cf. eq. $(A.2.8)_3$), the *inward* electromagnetic energy flux is then given by,

$$-\int_{\partial\mathcal{P}_t} \boldsymbol{\mathcal{S}}\cdot\mathbf{n}\,da = \int_{\mathcal{P}_t} \boldsymbol{\mathcal{E}}\cdot(\overset{*}{\mathbf{d}} + \boldsymbol{\mathcal{J}})\,dv \tag{A.3.10}$$

$$= -\int_{\mathcal{P}_t} \operatorname{grad}\phi\cdot(\overset{*}{\mathbf{d}} + \boldsymbol{\mathcal{J}})\,dv \tag{A.3.11}$$

$$= -\left[\int_{\mathcal{P}_t} \operatorname{div}(\phi(\overset{*}{\mathbf{d}} + \boldsymbol{\mathcal{J}}))\,dv - \int_{\mathcal{P}_t} \phi\operatorname{div}(\overset{*}{\mathbf{d}} + \boldsymbol{\mathcal{J}})\,dv\right] \tag{A.3.12}$$

$$= -\left[\int_{\mathcal{P}_t} \operatorname{div}(\phi(\overset{*}{\mathbf{d}} + \boldsymbol{\mathcal{J}}))\,dv - \int_{\mathcal{P}_t} \phi\underbrace{\operatorname{div}(\operatorname{curl}\boldsymbol{\mathcal{H}})}_{\equiv 0}\,dv\right], \tag{A.3.13}$$

or

$$\mathcal{E}(\mathcal{P}_t) = -\int_{\partial\mathcal{P}_t} \phi\,(\overset{*}{\mathbf{d}} + \mathcal{J}) \cdot \mathbf{n}\, da\,. \tag{A.3.14}$$

A.3.2 Connection between spatial and referential formulations of the basic fields for electrostatics

Recall the transformation laws

$$\mathbf{dx} = \mathbf{FdX}, \qquad dv = J dv_{\mathrm{R}}, \qquad \mathbf{n}da = J\mathbf{F}^{-\top}\mathbf{n}_{\mathrm{R}}\, da_{\mathrm{R}}, \tag{A.3.15}$$

between elemental lengths, volumes, and oriented area elements in the deformed and the reference bodies.

Next, Faraday's law (A.2.8) in the electrostatic limit (A.3.1) is,

$$\mathrm{curl}\,\boldsymbol{\mathcal{E}} = \mathbf{0} \qquad \text{with} \qquad \boldsymbol{\mathcal{E}} = \mathbf{e} + \mathbf{v} \times \mathbf{b},$$

and since we also neglect any effects of the magnetic flux $\mathbf{b}$ in the electrostatic limit, this law reduces to

$$\mathrm{curl}\,\mathbf{e} = \mathbf{0}. \tag{A.3.16}$$

Consider the integral form of Faraday's law (A.3.16),

$$0 = \int_{\mathcal{S}_t} \mathrm{curl}\,\mathbf{e} \cdot \mathbf{n}da = \int_{\mathcal{C}_t} \mathbf{e} \cdot \mathbf{dx}, \tag{A.3.17}$$

where $\mathcal{S}_t$ is a positively oriented surface with boundary $\mathcal{C}_t$ which is a closed curve. Equation (A.3.17) may be written in referential form as follows

$$0 = \int_{\mathcal{C}_t} \mathbf{e} \cdot \mathbf{dx} = \int_{\mathcal{C}} (\mathbf{F}^{\top}\mathbf{e}) \cdot \mathbf{dX} = \int_{\mathcal{S}} \mathrm{Curl}\,(\mathbf{F}^{\top}\mathbf{e}) \cdot \mathbf{n}_{\mathrm{R}} da_{\mathrm{R}}.$$

Defining

$$\mathbf{e}_{\mathrm{R}} \overset{\mathrm{def}}{=} \mathbf{F}^{\top}\mathbf{e}, \tag{A.3.18}$$

we obtain the following local referential form of Faraday's law,

$$\mathrm{Curl}\,\mathbf{e}_{\mathrm{R}} = \mathbf{0}. \tag{A.3.19}$$

Consider next the integral form of Gauss's law $(\text{A.2.8})_1$ over a part $\mathcal{P}_t$,

$$\int_{\mathcal{P}_t} q\, dv = \int_{\mathcal{P}_t} \mathrm{div}\,\mathbf{d}\, dv = \int_{\partial\mathcal{P}_t} \mathbf{d} \cdot \mathbf{n}\, da. \tag{A.3.20}$$

Equations (A.3.20) may be written in referential form as follows,

$$\begin{aligned}\int_{\mathcal{P}_t} q\,dv = \int_{\mathcal{P}_t} \operatorname{div}\mathbf{d}\,dv &= \int_{\partial\mathcal{P}_t} \mathbf{d}\cdot\mathbf{n}\,da,\\ \int_{\mathcal{P}} J\,q\,dv_{\mathrm{R}} &= \int_{\partial\mathcal{P}} \mathbf{d}\cdot(J\mathbf{F}^{-\top})\mathbf{n}_{\mathrm{R}}\,da_{\mathrm{R}},\\ \int_{\mathcal{P}} J\,q\,dv_{\mathrm{R}} &= \int_{\partial\mathcal{P}} (J\mathbf{F}^{-1}\mathbf{d})\cdot\mathbf{n}_{\mathrm{R}}\,da_{\mathrm{R}}.\end{aligned}$$

Defining

$$\mathbf{d}_{\mathrm{R}} \stackrel{\text{def}}{=} J\mathbf{F}^{-1}\mathbf{d} \qquad \text{and} \qquad q_{\mathrm{R}} \stackrel{\text{def}}{=} J\,q, \tag{A.3.21}$$

we obtain the local referential form of Gauss's law,

$$\operatorname{Div}\mathbf{d}_{\mathrm{R}} = q_{\mathrm{R}}. \tag{A.3.22}$$

Summarizing, the electric field $\mathbf{e}(\mathbf{x},t)$, electric displacement $\mathbf{d}(\mathbf{x},t)$, and volumetric free charge density $q(\mathbf{x},t)$, have their referential counterparts defined by,

$$\begin{aligned}\mathbf{e}_{\mathrm{R}} &\stackrel{\text{def}}{=} \mathbf{F}^{\top}\mathbf{e},\\ \mathbf{d}_{\mathrm{R}} &\stackrel{\text{def}}{=} J\mathbf{F}^{-1}\mathbf{d},\\ q_{\mathrm{R}} &\stackrel{\text{def}}{=} J\,q.\end{aligned} \tag{A.3.23}$$

Finally, recall the expression (A.3.14) for the influx of electromagnetic energy into part $\mathcal{P}_t$ though its boundary, viz.

$$\mathcal{E}(\mathcal{P}_t) = -\int_{\partial\mathcal{P}_t} \phi\,(\overset{*}{\mathbf{d}} + \mathcal{J})\cdot\mathbf{n}\,da\,. \tag{A.3.24}$$

We may express this referentially as

$$\begin{aligned}-\int_{\partial\mathcal{P}_t} \phi\,(\overset{*}{\mathbf{d}} + \mathcal{J})\cdot\mathbf{n}\,da &= -\int_{\partial\mathcal{P}} \phi\,(\overset{*}{\mathbf{d}} + \mathcal{J})\cdot(J\mathbf{F}^{-\top})\mathbf{n}_{\mathrm{R}}\,da_{\mathrm{R}},\\ &= -\int_{\partial\mathcal{P}} \phi\,(J\mathbf{F}^{-1}\overset{*}{\mathbf{d}})\cdot\mathbf{n}_{\mathrm{R}}\,da_{\mathrm{R}} - \int_{\partial\mathcal{P}} \phi(J\mathbf{F}^{-1}\mathcal{J})\cdot\mathbf{n}_{\mathrm{R}}\,da_{\mathrm{R}}.\end{aligned} \tag{A.3.25}$$

Using the identity

$$\dot{\mathbf{d}}_{\mathrm{R}} = J\mathbf{F}^{-1}\overset{*}{\mathbf{d}}, \tag{A.3.26}$$

and introducing the definition

$$\mathcal{J}_{\mathrm{R}} \overset{\text{def}}{=} J\mathbf{F}^{-1}\mathcal{J}, \tag{A.3.27}$$

we obtain that the referential form of the influx of energy into a part of the body is given by

$$\mathcal{E}(\mathcal{P}) = -\int_{\partial\mathcal{P}} \phi\,\dot{\mathbf{d}}_{\mathrm{R}} \cdot \mathbf{n}_{\mathrm{R}}\, da_{\mathrm{R}} - \int_{\partial\mathcal{P}} \phi\,\mathcal{J}_{\mathrm{R}} \cdot \mathbf{n}_{\mathrm{R}}\, da_{\mathrm{R}}. \tag{A.3.28}$$

A.3.3 **Dielectrics**

Dielectric materials are non-conducting and do not possess free charges. Thus for a dielectric

$$q = 0, \qquad \mathcal{J} = \mathbf{0}. \tag{A.3.29}$$

Hence, for a dielectric material the referential free charge and the conduction current also vanish,

$$q_{\mathrm{R}} = 0, \qquad \mathcal{J}_{\mathrm{R}} = \mathbf{0}. \tag{A.3.30}$$

In this case (A.3.28) reduces to

$$\mathcal{E}(\mathcal{P}) = -\int_{\partial\mathcal{P}} \phi\,\dot{\mathbf{d}}_{\mathrm{R}} \cdot \mathbf{n}_{\mathrm{R}}\, da_{\mathrm{R}}. \tag{A.3.31}$$

A.3.4 **Summary of the referential forms of the basic laws of electrostatics for dielectric materials**

The referential counterpart of Faraday's law (A.3.2) is

$$\mathrm{Curl}\,\mathbf{e}_{\mathrm{R}} = \mathbf{0}. \tag{A.3.32}$$

Further, with $\phi(\mathbf{X},t) \equiv \phi(\boldsymbol{\chi}(\mathbf{X},t),t)$ the referential description of the electric potential,[2]

$$\mathbf{e}_{\mathrm{R}}(\mathbf{X},t) \overset{\text{def}}{=} -\nabla\phi(\mathbf{X},t). \tag{A.3.33}$$

The referential counterpart of Gauss' law (A.3.22) is

$$\mathrm{Div}\,\mathbf{d}_{\mathrm{R}} = 0. \tag{A.3.34}$$

Further, the influx of electromagnetic energy into a part $\mathcal{P}$ of the reference body $\mathcal{B}$ is,

$$\mathcal{E}(\mathcal{P}) = -\int_{\partial\mathcal{P}} \phi\,\dot{\mathbf{d}}_{\mathrm{R}} \cdot \mathbf{n}_{\mathrm{R}}\, da_{\mathrm{R}}. \tag{A.3.35}$$

[2] For economy of notation, we do not employ a different symbol for the referential description of the potential.

A.4 Magnetostatic limit of Maxwell's equations

The definition of the *magnetostatic limit* of Maxwell's equations that we employ here is one in which any coupling of the electric fields with magnetic fields are neglected by setting

$$\mathbf{d} = 0, \qquad \text{and} \qquad \overset{*}{\mathbf{d}} = 0, \tag{A.4.1}$$

and assuming that the conduction current $\mathcal{J}$ is *steady*. Under these assumptions Ampere's law $(A.2.8)_3$ becomes,

$$\operatorname{curl}\boldsymbol{\mathcal{H}} = \boldsymbol{\mathcal{J}}. \tag{A.4.2}$$

In the magnetically coupled theory and applications shown in this textbook, we only consider situations in which there is *no conduction current*, such that $\mathcal{J} = \mathbf{0}$ and Ampere's law (A.4.2) further simplifies to

$$\operatorname{curl}\boldsymbol{\mathcal{H}} = \mathbf{0}. \tag{A.4.3}$$

Since the curl of any gradient field is zero ($\operatorname{curl}\operatorname{grad} f \to e_{ijk} f_{,kj} = 0$), in the quasi-static limit we assure that (A.4.3) will always be satisfied by assuming that $\boldsymbol{\mathcal{H}}$ is the gradient of some scalar field. By convention we write

$$\boldsymbol{\mathcal{H}} = -\operatorname{grad}\phi \tag{A.4.4}$$

where ϕ is known as the *magnetic potential.* The negative sign in (A.4.4) follows the convention that the magnetic field points from a region of high potential to low potential, whereas the vector $\operatorname{grad}\phi$ by definition points in the direction of increasing ϕ.

Also, Gauss's law for magnetism $(A.2.8)_4$ states that

$$\operatorname{div}\mathbf{b} = 0. \tag{A.4.5}$$

Equations (A.4.3) and (A.4.5) constitute the governing equations of magnetostatics in $\mathcal{B}_t$.

A.4.1 Energy transport in the magnetostatic limit

Recall that the flux of *electromagnetic energy* (energy per unit area per unit time) is described by the *Poynting vector* (cf., eq. (A.3.5))

$$\boldsymbol{\mathcal{S}} \overset{\text{def}}{=} \boldsymbol{\mathcal{E}} \times \boldsymbol{\mathcal{H}} \tag{A.4.6}$$

such that the flux of electromagnetic energy *into* part $\mathcal{P}_t$ of the deformed body $\mathcal{B}_t$ is given by,

$$\mathcal{E}(\mathcal{P}_t) = -\int_{\partial\mathcal{P}_t} \boldsymbol{\mathcal{S}}\cdot\mathbf{n}\,da, \tag{A.4.7}$$

where $\mathbf{n}$ is the outward unit normal to its boundary $\partial\mathcal{P}_t$. Using the divergence theorem,

$$-\int_{\partial\mathcal{P}_t} \boldsymbol{\mathcal{S}}\cdot\mathbf{n}\,da = -\int_{\mathcal{P}_t} \operatorname{div}\boldsymbol{\mathcal{S}}\,dv_{\mathrm{R}} = -\int_{\mathcal{P}_t} \operatorname{div}(\boldsymbol{\mathcal{E}}\times\boldsymbol{\mathcal{H}})\,dv_{\mathrm{R}}\,. \tag{A.4.8}$$

Next, use of the vector identity

$$\operatorname{div}(\boldsymbol{\mathcal{E}} \times \boldsymbol{\mathcal{H}}) = \boldsymbol{\mathcal{H}} \cdot (\operatorname{curl} \boldsymbol{\mathcal{E}}) - \boldsymbol{\mathcal{E}} \cdot (\operatorname{curl} \boldsymbol{\mathcal{H}}), \tag{A.4.9}$$

allows us to write (A.4.8) as

$$-\int_{\partial \mathcal{P}_t} \boldsymbol{\mathcal{S}} \cdot \mathbf{n}\, da = -\int_{\mathcal{P}_t} \Big(\boldsymbol{\mathcal{H}} \cdot (\operatorname{curl} \boldsymbol{\mathcal{E}}) - \boldsymbol{\mathcal{E}} \cdot (\operatorname{curl} \boldsymbol{\mathcal{H}})\Big)\, dv_{\mathrm{R}}. \tag{A.4.10}$$

In the magnetostatic limit $\operatorname{curl} \boldsymbol{\mathcal{H}} = \mathbf{0}$, and since $\operatorname{curl} \boldsymbol{\mathcal{E}} = -\overset{*}{\mathbf{b}}$ (cf. eq. (A.2.8)$_2$), the *inward* electromagnetic energy flux is then given by,

$$-\int_{\partial \mathcal{P}_t} \boldsymbol{\mathcal{S}} \cdot \mathbf{n}\, da = \int_{\mathcal{P}_t} \boldsymbol{\mathcal{H}} \cdot \overset{*}{\mathbf{b}}\, dv_{\mathrm{R}} \tag{A.4.11}$$

$$= -\int_{\mathcal{P}_t} \operatorname{grad} \phi \cdot \overset{*}{\mathbf{b}}\, dv_{\mathrm{R}} \tag{A.4.12}$$

$$= -\left(\int_{\mathcal{P}_t} \operatorname{div}(\phi\, \overset{*}{\mathbf{b}})\, dv_{\mathrm{R}} - \int_{\mathcal{P}_t} \phi\, \underbrace{\operatorname{div} \overset{*}{\mathbf{b}}}_{\equiv 0}\, dv_{\mathrm{R}}\right) \tag{A.4.13}$$

$$= -\int_{\mathcal{P}_t} \operatorname{div}(\phi \overset{*}{\mathbf{b}})\, dv_{\mathrm{R}}, \tag{A.4.14}$$

or,

$$\mathcal{E}(\mathcal{P}_t) = -\int_{\partial \mathcal{P}_t} (\phi\, \overset{*}{\mathbf{b}}) \cdot \mathbf{n}\, da. \tag{A.4.15}$$

A.4.2 Connection between spatial and referential formulations of the basic fields for magnetostatics

Recall the transformation laws

$$\mathbf{dx} = \mathbf{F} \mathbf{dX}, \qquad dv_{\mathrm{R}} = J dv_{\mathrm{R}}, \qquad \mathbf{n} da = J \mathbf{F}^{-\top} \mathbf{n}_{\mathrm{R}}\, da_{\mathrm{R}}, \tag{A.4.16}$$

between elemental lengths, volumes, and oriented area elements in the deformed and the reference bodies.

Next, Ampere's law (A.2.8)$_3$ in the magnetostatic limit (A.4.1) and absence of conduction currents $\boldsymbol{\mathcal{J}}$, viz.

$$\operatorname{curl} \boldsymbol{\mathcal{H}} = \mathbf{0} \qquad \text{with} \qquad \boldsymbol{\mathcal{H}} = \mathbf{h} - \mathbf{v} \times \mathbf{d},$$

and since we also neglect any effects of the electric displacement $\mathbf{d}$ in the magnetostatic limit, this law reduces to

$$\operatorname{curl} \mathbf{h} = \mathbf{0}. \tag{A.4.17}$$

The integral form of Ampere's law (A.4.17) is,

$$0 = \int_{\mathcal{S}_t} \operatorname{curl}\mathbf{h} \cdot \mathbf{n}da = \int_{\mathcal{C}_t} \mathbf{h} \cdot d\mathbf{x}, \tag{A.4.18}$$

where $\mathcal{S}_t$ is a positively oriented surface with boundary $\mathcal{C}_t$, which is a closed curve. Equation (A.4.18) may be written in referential form as follows

$$0 = \int_{\mathcal{C}_t} \mathbf{h} \cdot d\mathbf{x} = \int_{\mathcal{C}} (\mathbf{F}^\top\mathbf{h}) \cdot d\mathbf{X} = \int_{\mathcal{S}} \operatorname{Curl}(\mathbf{F}^\top\mathbf{h}) \cdot \mathbf{n}_{\mathrm{R}} da_{\mathrm{R}}.$$

Defining

$$\mathbf{h}_{\mathrm{R}} \stackrel{\text{def}}{=} \mathbf{F}^\top\mathbf{h}, \tag{A.4.19}$$

we obtain the following local referential form of Ampere's law,

$$\operatorname{Curl}\mathbf{h}_{\mathrm{R}} = \mathbf{0}. \tag{A.4.20}$$

Consider next the integral form of Gauss's law $(\text{A.2.8})_4$ over a part $\mathcal{P}_t$,

$$0 = \int_{\mathcal{P}_t} \operatorname{div}\mathbf{b}\, dv_{\mathrm{R}} = \int_{\partial\mathcal{P}_t} \mathbf{b} \cdot \mathbf{n}\, da. \tag{A.4.21}$$

Equations (A.4.21) may be written in referential form as follows,

$$
\begin{aligned}
0 &= \int_{\partial\mathcal{P}_t} \mathbf{b} \cdot \mathbf{n}\, da, \\
0 &= \int_{\partial\mathcal{P}} \mathbf{b} \cdot (J\mathbf{F}^{-\top})\mathbf{n}_{\mathrm{R}}\, da_{\mathrm{R}}, \\
0 &= \int_{\partial\mathcal{P}} (J\mathbf{F}^{-1}\mathbf{b}) \cdot \mathbf{n}_{\mathrm{R}}\, da_{\mathrm{R}}.
\end{aligned}
$$

Defining

$$\mathbf{b}_{\mathrm{R}} \stackrel{\text{def}}{=} J\mathbf{F}^{-1}\mathbf{b}, \tag{A.4.22}$$

we obtain the local referential form of Gauss's law,

$$\operatorname{Div}\mathbf{b}_{\mathrm{R}} = 0. \tag{A.4.23}$$

Summarizing, the magnetic field $\mathbf{h}(\mathbf{x}, t)$ and magnetic flux $\mathbf{b}(\mathbf{x}, t)$ have their referential counterparts defined by,

$$\begin{aligned} \mathbf{h}_{\mathrm{R}} &\stackrel{\text{def}}{=} \mathbf{F}^{\top}\mathbf{h}, \\ \mathbf{b}_{\mathrm{R}} &\stackrel{\text{def}}{=} J\mathbf{F}^{-1}\mathbf{b}. \end{aligned} \tag{A.4.24}$$

Finally, recall the expression (A.4.15) for the influx of electromagnetic energy into part $\mathcal{P}_t$ though its boundary, viz.

$$\mathcal{E}(\mathcal{P}_t) = -\int_{\partial\mathcal{P}_t} (\phi\,\overset{*}{\mathbf{b}}) \cdot \mathbf{n}\, da\,. \tag{A.4.25}$$

We may express this referentially as

$$\begin{aligned} -\int_{\partial\mathcal{P}_t} (\phi\,\overset{*}{\mathbf{b}}) \cdot \mathbf{n}\, da &= -\int_{\partial\mathcal{P}} (\phi\,\overset{*}{\mathbf{b}}) \cdot (J\mathbf{F}^{-\top})\mathbf{n}_{\mathrm{R}}\, da_{\mathrm{R}}, \\ &= -\int_{\partial\mathcal{P}} \phi\,(J\mathbf{F}^{-1}\overset{*}{\mathbf{b}}) \cdot \mathbf{n}_{\mathrm{R}}\, da_{\mathrm{R}}\,. \end{aligned} \tag{A.4.26}$$

Using the identity

$$\dot{\mathbf{b}}_{\mathrm{R}} = J\mathbf{F}^{-1}\overset{*}{\mathbf{b}}, \tag{A.4.27}$$

we obtain that the referential form of the influx of energy into a part of the body is given by

$$\mathcal{E}(\mathcal{P}) = -\int_{\partial\mathcal{P}} (\phi\,\dot{\mathbf{b}}_{\mathrm{R}}) \cdot \mathbf{n}_{\mathrm{R}}\, da_{\mathrm{R}}. \tag{A.4.28}$$

A.4.3 Summary of the referential form of the basic laws of magnetostatics

The referential counterpart of Ampere's law is

$$\mathrm{Curl}\,\mathbf{h}_{\mathrm{R}} = \mathbf{0}. \tag{A.4.29}$$

Further, with $\phi(\mathbf{X}, t) \equiv \phi(\boldsymbol{\chi}(X, t), t)$ the referential description of the magnetic potential is[3]

$$\mathbf{h}_{\mathrm{R}}(\mathbf{X}, t) \stackrel{\text{def}}{=} -\nabla\phi(\mathbf{X}, t). \tag{A.4.30}$$

[3] For economy of notation, we do not employ a different symbol for the referential description of the potential.

The referential counterpart of Gauss's law is

$$\operatorname{Div}\mathbf{b}_{\mathrm{R}} = 0. \tag{A.4.31}$$

Further, the influx of electromagnetic energy into a part $\mathcal{P}$ is,

$$\mathcal{E}(\mathcal{P}) = -\int_{\partial\mathcal{P}} \phi\,\dot{\mathbf{b}}_{\mathrm{R}}\cdot\mathbf{n}_{\mathrm{R}}\,da_{\mathrm{R}}. \tag{A.4.32}$$

A.4.4 Jump conditions in magnetostatics

Since we will account for the magnetic field in the air-space surrounding an s-MRE body, we must also satisfy the jump conditions which are associated with the governing equations of magnetostatics across the boundary between the solid s-MRE domain and the air domain.

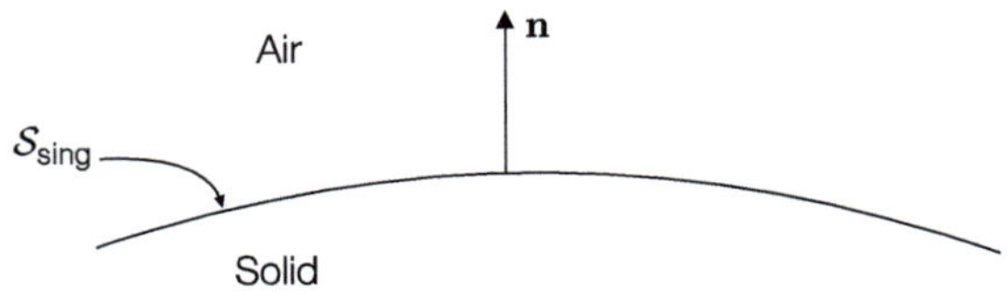

Fig. A.1 $\mathcal{S}_{\text{sing}}$ denotes a singular surface with unit normal $\mathbf{n}$. The two sides of the singular surface are denoted by "solid" and "air", with $\mathbf{n}$ pointing from "solid" to "air."

With reference to Fig. A.1, let $\mathcal{S}_{\text{sing}}$ denote a singular surface with unit normal $\mathbf{n}$. Denote the two sides of the singular surface by "solid" and "air," with $\mathbf{n}$ pointing from "solid" to "air." Consider a field φ which is continuous on either side of $\mathcal{S}_{\text{sing}}$, but suffers a jump discontinuity across $\mathcal{S}_{\text{sing}}$. We define the jump in φ across $\mathcal{S}_{\text{sing}}$ as the function

$$[\![\varphi]\!] = \varphi^{\text{air}} - \varphi^{\text{solid}}, \tag{A.4.33}$$

where φ^{air} and φ^{solid} denote the limits of φ as $\mathcal{S}_{\text{sing}}$ is approached from its two sides. Thus, for example, $\mathbf{b}$ has the discontinuity

$$[\![\mathbf{b}]\!] \stackrel{\text{def}}{=} \mathbf{b}^{\text{air}} - \mathbf{b}^{\text{solid}},$$

and similarly for other quantities.

The discontinuity conditions associated with the basic laws of magnetostatics,

$$\operatorname{curl}\mathbf{h} = \mathbf{0} \quad \text{and} \quad \operatorname{div}\mathbf{b} = 0, \tag{A.4.34}$$

are given by

$$\mathbf{n}\times[\![\mathbf{h}]\!] = \mathbf{0} \qquad \text{and} \qquad \mathbf{n}\cdot[\![\mathbf{b}]\!] = 0. \tag{A.4.35}$$

We prove the jump condition results (A.4.35) below.

The jump condition on the tangential component of the magnetic field

Consider the Maxwell equation $(\text{A.4.34})_1$ integrated over an open surface $\mathcal{S}$ with bounding curve $\mathcal{C}$,

$$\int_{\mathcal{S}} \mathbf{n} \cdot (\operatorname{curl} \mathbf{h})\, da = 0. \tag{A.4.36}$$

Recall Stokes' theorem for a vector field $\mathbf{v}$, viz.

$$\int_{\mathcal{C}} \mathbf{v} \cdot \mathbf{dx} = \int_{\mathcal{S}} \mathbf{n} \cdot \operatorname{curl} \mathbf{v}\, da.$$

Applying Stokes' theorem to the surface integral in (A.4.36) we obtain

$$\int_{\mathcal{C}} \mathbf{h} \cdot \mathbf{dx} = 0. \tag{A.4.37}$$

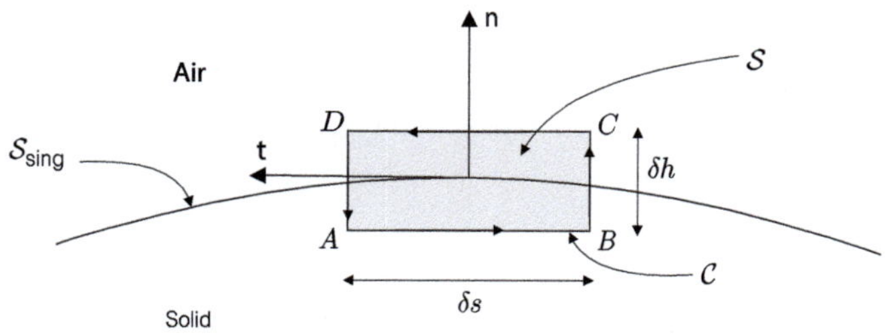

Fig. A.2 $\mathcal{S}_{\text{sing}}$ denotes the singular surface with unit normal n and a unit tangent vector t; the two sides of the singular surface by "solid" and "air," with n pointing from "solid" to "air." $\mathcal{S}$ is an infinitesimal rectangular surface with bounding curve $\mathcal{C}$ identified by the contour $ABCD$ lying in the plane of the unit normal n to $\mathcal{S}_{\text{sing}}$.

Let $\mathcal{S}_{\text{sing}}$ denote a singular surface with unit normal $\mathbf{n}$ and a unit tangent vector $\mathbf{t}$, as shown in Fig. A.2. Denote the two sides of the singular surface by "solid" and "air," with $\mathbf{n}$ pointing from "solid" to "air." Next, let $\mathcal{S}$ be an infinitesimal rectangular surface with oriented bounding curve $\mathcal{C}$ identified by the contour $ABCD$ lying in the plane of the unit normal $\mathbf{n}$ to $\mathcal{S}_{\text{sing}}$. The sides AB and CD of $\mathcal{C}$ are parallel to $\mathbf{t}$ and have lengths δs. The sides BC and DA are parallel to $\mathbf{n}$ and have lengths δh. The magnetic fields in the solid and air domains are denoted $\mathbf{h}^{\text{solid}}$ and $\mathbf{h}^{\text{air}}$, respectively. Then, application of (A.4.37) to $\mathcal{C}$ of Fig. A.2 yields the result,

$$\underbrace{\mathbf{h}^{\text{solid}} \cdot (-\mathbf{t}\,\delta s)}_{\text{AB}} + \int_{BC} \mathbf{h} \cdot \mathbf{n}\,\delta h + \underbrace{\mathbf{h}^{\text{air}} \cdot (\mathbf{t}\,\delta s)}_{\text{CD}} + \int_{DA} \mathbf{h} \cdot (-\mathbf{n})\,\delta h = 0. \tag{A.4.38}$$

Taking the limit of $\delta h \to 0$ and then dividing by δs, we obtain

$$\mathbf{h}^{\text{air}} \cdot \mathbf{t} - \mathbf{h}^{\text{solid}} \cdot \mathbf{t} = 0 \qquad \Rightarrow \qquad \mathbf{t} \cdot [\![\mathbf{h}]\!] = 0,$$

which holds for arbitrary unit vectors $\mathbf{t}$ which are normal to $\mathbf{n}$. Recall the vector identity

$$\mathbf{a} \times (\mathbf{b} \times \mathbf{c}) = \mathbf{b}\,(\mathbf{a} \cdot \mathbf{c}) - \mathbf{c}\,(\mathbf{a} \cdot \mathbf{b}).$$

Apply the identity with $\mathbf{a} \equiv \mathbf{t}$, $\mathbf{b} \equiv \mathbf{n}$, and $\mathbf{c} \equiv [\![\mathbf{h}]\!]$:

$$\mathbf{t} \times (\mathbf{n} \times [\![\mathbf{h}]\!]) = \mathbf{n} \underbrace{(\mathbf{t} \cdot [\![\mathbf{h}]\!])}_{=0} - [\![\mathbf{h}]\!] \underbrace{(\mathbf{t} \cdot \mathbf{n})}_{=0} = \mathbf{0}.$$

Thus, since $\mathbf{t} \neq \mathbf{0}$ we must have

$$\mathbf{n} \times [\![\mathbf{h}]\!] = \mathbf{0}, \tag{A.4.38}$$

which is $(\text{A.4.35})_1$. This jump condition encodes the requirement that the tangential component of the magnetic field $\mathbf{h} \cdot \mathbf{t}$ must be continuous across the singular surface $\mathcal{S}_{\text{sing}}$.

The jump condition on the normal component of the magnetic flux density

Now consider a cylindrical pill-box $\mathcal{P}_t$ of infinitesimal height δh and cross-sectional area $\mathbf{n}\delta a$ straddling the singular surface $\mathcal{S}_{\text{sing}}$, as depicted in Fig. A.3. Equation $(\text{A.4.34})_2$ integrated over the volume of the cylinder gives,

$$\int_{\mathcal{P}_t} \operatorname{div}\mathbf{b}\, dv_{\text{R}} = 0,$$

and applying the divergence theorem to the left-hand side gives

$$\int_{\partial\mathcal{P}_t} \mathbf{b} \cdot \mathbf{n}\, da = 0, \tag{A.4.39}$$

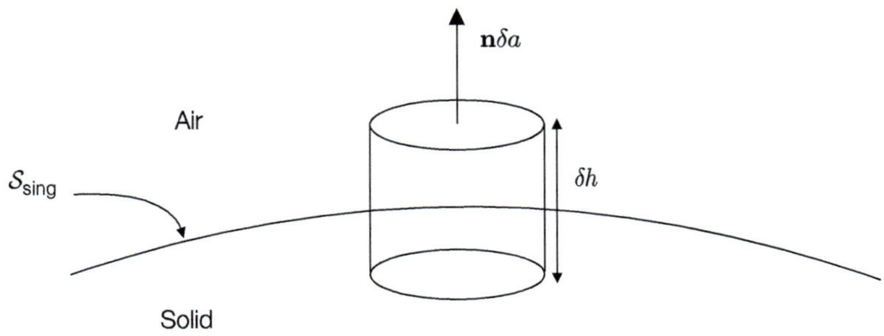

Fig. A.3 A pill-box of height δh and cross-sectional area δa straddling the singular surface $\mathcal{S}_{\text{sing}}$, with $\mathbf{n}$ pointing from "solid" to "air."

where $\partial\mathcal{P}_t$ is the bounding surface of the cylinder. Since δh is infinitesimal and the flux of $\mathbf{b}$ across the lateral surface of the cylinder is negligible as $\delta h \to 0$, the only contribution to the surface integral comes from the top and bottom faces of the cylinder. In the limit as $\delta h \to 0$, equation (A.4.39) therefore evaluates as

$$\mathbf{b}^{\text{air}} \cdot (\mathbf{n}\,\delta a) - \mathbf{b}^{\text{solid}} \cdot (\mathbf{n}\,\delta a) = 0,$$

or, dividing by δa,

$$\mathbf{n} \cdot [\![\mathbf{b}]\!] = 0, \tag{A.4.39}$$

which is $(\text{A.4.35})_2$. This jump condition encodes the requirement that the normal component of the magnetic flux density $\mathbf{b} \cdot \mathbf{n}$ must be continuous across the singular surface $\mathcal{S}_{\text{sing}}$.

Referential forms of the jump conditions

Similar arguments, when constructed in the reference configuration, show that the referential magnetic fields $\mathbf{h}_{\text{R}}$ and $\mathbf{b}_{\text{R}}$ are required to satisfy the referential jump conditions (cf., e.g. Dorfmann and Ogden, 2014)

$$\mathbf{n}_{\text{R}} \times [\![\mathbf{h}_{\text{R}}]\!] = \mathbf{0}, \qquad \mathbf{n}_{\text{R}} \cdot [\![\mathbf{b}_{\text{R}}]\!] = 0 \qquad \text{across } \mathcal{S}_{\text{R sing}}, \tag{A.4.40}$$

which require that the tangential component of the magnetic field $\mathbf{h}_{\text{R}}$ and the normal component of the magnetic flux $\mathbf{b}_{\text{R}}$ remain continuous across the referential singular surface $\mathcal{S}_{\text{R sing}}$ with outward unit normal $\mathbf{n}_{\text{R}}$.

Appendix B Units for mechanical, electrical, and magnetic quantities

B.1 Units in SI system

In Table B.1 we summarize the basic physical units necessary for the description of mechanical, electrical, and magnetic quantities using the SI system of units[1] which are based on the meter–kilogram–second (MKS) system.[2]

The fundamental electric quantities $\{\mathbf{e}, \mathbf{p}, \mathbf{d}\}$ obey the relation

$$\mathbf{d} = \epsilon_0\, \mathbf{e} + \mathbf{p} \qquad \text{with} \qquad \epsilon_0 = 8.854187817 \times 10^{-12}\, (\mathrm{C/m^2})/(\mathrm{V/m}), \tag{B.1.1}$$

where ϵ_0 is the electric permittivity of vacuum, and the fundamental magnetic quantities $\{\mathbf{b}, \mathbf{m}, \mathbf{h}\}$ obey

$$\mathbf{h} = \mu_0^{-1}\mathbf{b} - \mathbf{m} \qquad \text{with} \qquad \mu_0 = 4\pi \times 10^{-7}\, \mathrm{T/(A/m)}, \tag{B.1.2}$$

where μ_0 is the magnetic permeability of vacuum. The units of ϵ_0 and μ_0 may also be expressed more compactly as

$$\epsilon_0 = 8.854187817 \times 10^{-12}\, \mathrm{F/m} \qquad \text{and} \qquad \mu_0 = 4\pi \times 10^{-7}\, \mathrm{H/m}. \tag{B.1.3}$$

The constants ϵ_0 and μ_0 are related to the speed of light c by

$$c = \frac{1}{\sqrt{\epsilon_0 \mu_0}}. \tag{B.1.4}$$

[1] French: Le Systéme International d'unités

[2] Another, often used framework in electrodynamics is the Gaussian system of units based on the centimeter–gram–second (CGS) system. These days SI units are used by most engineers, and also by many physicists. The trend since the latter half the twentieth century has been in favor of a universal adoption of the SI units. Further discussion of electromagnetic field variables and their units may be found in, e.g., Rosensweig (1989) and Kovetz (2000).

Table B.1 Physical quantities in SI units (MKS system).

Symbol	Name	SI unit	SI base units
l	length	meter	m
m	mass	kilogram	kg
t	time	second	s
$\mathbf{f}$	force	newton	$\mathrm{N = kg\,m\,s^{-2}}$
$\mathcal{E}$	energy	joule	$\mathrm{J = Ws = CV = kg\,m^2\,s^{-2}}$
$\mathcal{P}$	power	watt	$\mathrm{W = J\,s^{-1} = kg\,m^2\,s^{-3}}$
q	charge	coulomb	C
$\tilde{\mathbf{j}}$	current	ampere	$\mathrm{A = C\,s^{-1}}$
$\mathbf{j}$	current density	$\frac{\text{ampere}}{\text{meter}^2}$	$\mathrm{A\,m^{-2} = C\,m^{-2}\,s^{-1}}$
ϕ	electric potential	volt	$\mathrm{V = J\,C^{-1} = kg\,m^2\,s^{-2}\,C^{-1}}$
$\mathbf{e}$	electric field	$\frac{\text{volt}}{\text{meter}}$	$\mathrm{V\,m^{-1} = kg\,m\,s^{-2}\,C^{-1}}$
$\mathbf{d}$	electric displacement	$\frac{\text{coulomb}}{\text{meter}^2}$	$\mathrm{C\,m^{-2}}$
$\mathbf{p}$	polarization	$\frac{\text{coulomb}}{\text{meter}^2}$	$\mathrm{C\,m^{-2}}$
C_p	electric capacitance	farad	$\mathrm{F = C\,V^{-1} = kg^{-1}\,m^{-2}\,s^2\,C^2}$
$\mathbf{b}$	magnetic flux density	tesla	$\mathrm{T = N\,A^{-1}\,m^{-1} = kg\,s^{-1}\,C^{-1}}$
$\mathbf{h}$	magnetic field	$\frac{\text{ampere}}{\text{meter}}$	$\mathrm{A\,m^{-1} = m^{-1}\,s^{-1}\,C}$
$\mathbf{m}$	magnetization	$\frac{\text{ampere}}{\text{meter}}$	$\mathrm{A\,m^{-1} = m^{-1}\,s^{-1}\,C}$
L_i	magnetic inductance	henry	$\mathrm{H = T\,m^2\,A^{-1} = kg\,m^2\,C^{-2}}$

Image Credits

Chapter 28:

- Fig. 28.1(a)–(b), and Fig. 28.5 Photographs of a cantilever ionic polymer–metal composite (IPMC) beam under a 3 V step voltage, at $t = 0\,\mathrm{s}$ and $t = 20\,\mathrm{s}$ (Nguyen et al., 2007). Reprinted with permission from Elsevier.

Chapter 31:

- Fig. 31.3(b) Photograph of the deformation of a 3D-printed magnetically responsive shape-morphing structure under application of a 200 mT magnetic **b**-field (Kim et al., 2018). Reprinted with permission from Shawn A. Chester.
- Fig. 31.7(c) Photograph of a bistable hard-magnetic soft-elastomeric arch actuator (Tan et al., 2022). Reprinted with permission from Elsevier.

Chapter 34:

- Fig. 34.1(a) Photograph of cylindrical s-MRE specimens with varied aspect ratios (Diguet et al., 2010). This work is in the public domain and available from `https://theses.hal.science/tel-00488910/`.

Bibliography

Abaqus. *Abaqus Standard, V 6.17*, 2017.

L. Anand. Constitutive equations for the rate-dependent deformation of metals at elevated temperatures. *ASME Journal of Engineering Materials and Technology*, 104:12–17, 1982.

L. Anand. Constitutive equations for hot-working of metals. *International Journal of Plasticity*, 1(3):213–231, 1985.

L. Anand. A constitutive model for compressible elastomeric solids. *Computational Mechanics*, 18(5):339–355, 1996.

L. Anand. A Cahn–Hilliard-type theory for species diffusion coupled with large elastic-plastic deformations. *Journal of the Mechanics and Physics of Solids*, 60:1983–2002, 2012.

L. Anand and S. Govindjee. *Continuum Mechanics of Solids*. Oxford University Press, 2020.

L. Anand, K. Kamrin, and S. Govindjee. *Introduction to Mechanics of Solid Materials*. Oxford University Press, 2022.

I. Ang, N. Bouklas, and B. Li. Stabilized formulation for phase-field fracture in nearly incompressible hyperelasticity. *International Journal for Numerical Methods in Engineering*, 123(19):4655–4673, 2022.

M. J. Ansari, R. R. Rajendran, S. Mohanto, U. Agarwal, K. Panda, K. Dhotre, R. Manne, A. Deepak, A. Zafar, M. Yasir, et al. Poly (N-isopropylacrylamide)-based hydrogels for biomedical applications: A review of the state-of-the-art. *Gels*, 8(7):454, 2022.

J. Argyris and J. Doltsinis. On the natural formulation of large deformation couple thermomechanical problems. *Computer Methods in Applied Mechanics and Engineering*, 25:195–253, 1981.

A. M. Arnold, J. Su, and E. M. Sabolsky. Influence of environmental conditions and voltage application on the electromechanical performance of Nafion-Pt IPMC actuators. *Smart Materials and Structures*, 31:115031, 2022.

E. M. Arruda and M. C. Boyce. A three-dimensional constitutive model for the large stretch behavior of rubber elastic materials. *Journal of the Mechanics and Physics of Solids*, 41(2):389–412, 1993.

S. Baek and A. Srinivasa. Diffusion through an elastic solid undergoing large deformation. *Journal of Non-Linear Mechanics*, 39:201–218, 2004.

A. J. Bard, L. R. Faulkner, and H. S. White. *Electrochemical methods: fundamentals and applications*. John Wiley & Sons, 2022.

K.-J. Bathe. *Finite element procedures*. Prentice Hall, 1996.

M. Z. Bazant. Theory of chemical kinetics and charge transfer based on nonequilibrium thermodynamics. *Accounts of Chemical Research*, 46:1144–1160, 2013.

T. Beda. An approach for hyperelastic model-building and parameters estimation a review of constitutive models. *European Polymer Journal*, 50:97–108, 2014.

J. Bergstrom and M. Boyce. Constitutive modeling of the large strain time-dependent behavior of elastomers. *Journal of the Mechanics and Physics of Solids*, 46:931–954, 1998.

R. Bhavsar, N. Y. Vaidya, P. Ganguly, A. Humphreys, A. Robisson, H. Tu, N. Wicks, G. H. McKinley, and F. Pauchet. Intelligence in novel materials. *Oilfield Review*, 20(1):32–41, 2008.

M. Biot. Thermoelasticity and irreversible thermodynamics. *Journal of Applied Physics*, 27:240–253, 1956.

J. E. Bischoff, E. M. Arruda, and K. Grosh. A new constitutive model for the compressibility of elastomers at finite deformations. *Rubber Chemistry and Technology*, 74:541–559, 2001.

J. Bleyer. Numerical tours of Computational Mechanics with FEniCSx, Jan. 2024. `https://doi.org/10.5281/zenodo.10470942`.

B. A. Boley and J. H. Weiner. *Theory of Thermal Stresses*. Wiley, 1960.

J. Bonet and R. D. Wood. *Nonlinear Continuum Mechanics for Finite Element Analysis*. Cambridge University Press, 1997.

J. Bonet, A. Gil, and R. Wood. *Nonlinear Solid Mechanics for Finite Element Analysis: Dynamics*. Cambridge University Press, 2021.

M. C. Boyce and E. M. Arruda. Constitutive models of rubber elasticity: a review. *Rubber Chemistry and Technology*, 73:504–523, 2000.

S. B. Brown, K. H. Kim, and L. Anand. An internal variable constitutive model for hot working of metals. *International Journal of Plasticity*, 5:95–130, 1989.

W. F. Brown. *Magnetoelastic Interactions*, volume 9. Springer, 1966.

J. Cahn. On spinodal decomposition. *Acta Metallurgica*, 9:795–801, 1961.

J. Cahn and J. Hilliard. Free energy of a nonuniform system-I: Interfacial free energy. *The Journal of Chemical Physics*, 28:258–267, 1958.

D. Carlson. *Handbuch der Physik Bd. VI/2a*, chapter Linear thermoelasticity, pages 297–346. Springer, Berlin, 1972.

Y. Cha and M. Porfiri. Mechanics and electrochemistry of ionic polymer metal composites. *Journal of the Mechanics and Physics of Solids*, 71:156–178, 2014.

P. Chadwick. Thermodynamics of rubberlike materials. *Philosophical Transactions of the Royal Society, London*, A276:371–403, 1974.

P. Chadwick and C. Creasy. Modified entropic elasticity of rubberlike materials. *Journal of the Mechanics and Physics of Solids*, 32:337–357, 1984.

Y.-C. Chen and L. Wheeler. Derivatives of the stretch and rotation tensors. *Journal of Elasticity*, 32:175–182, 1993.

S. A. Chester and L. Anand. A coupled theory of fluid permeation and large deformations for elastomeric materials. *Journal of the Mechanics and Physics of Solids*, 58:1879–1906, 2010.

S. A. Chester and L. Anand. A thermo-mechanically coupled theory for fluid permeation in elastomeric materials: Application to thermally responsive gels. *Journal of the Mechanics and Physics of Solids*, 59:1978–2006, 2011.

J. M. D. Coey. *Magnetism and Magnetic Materials*. Cambridge University Press, 2010.

A. Cohen. A Pade approximant to the inverse Langevin function. *Rheologica Acta*, 30:270–273, 1991.

B. D. Coleman and W. Noll. The thermodynamics of elastic materials with heat conduction and viscosity. *Archive for Rational Mechanics and Analysis*, 13:167–178, 1963.

K. Danas. Effective response of classical, auxetic and chiral magnetoelastic materials by use of a new variational principle. *Journal of the Mechanics and Physics of Solids*, 105:25–53, 2017.

K. Danas. A Unified Theoretical Modeling Framework for Soft and Hard Magnetorheological Elastomers. In K. Danas and O. Lopez-Pamies, editors, *Electro- and Magneto-Mechanics of Soft Solids:*

Constitutive Modelling, Numerical Implementations, and Instabilities, pages 59–139. Springer Nature, Switzerland, 2024.

K. Danas, S. Kankanala, and N. Triantafyllidis. Experiments and modeling of iron-particle-filled magnetorheological elastomers. *Journal of the Mechanics and Physics of Solids*, 60:120–138, 2012.

R. DeHoff. *Thermodynamics in Materials Science*. CRC Press, Boca Raton, 2006.

V. Deshpande and R. McMeeking. Models for the interplay of mechanics, electrochemistry, thermodynamics, and kinetics in lithium-ion batteries. *Applied Mechanics Reviews*, 75:010801, 2023.

C. Di Leo, E. Rejovitzky, and L. Anand. A Cahn–Hilliard-type phase-field theory for species diffusion coupled with large elastic deformations: application to phase-separating Li-ion electrode materials. *Journal of the Mechanics and Physics of Solids*, 70:1–29, 2014.

E. J. Dickinson and A. J. Wain. The Butler–Volmer equation in electrochemical theory: Origins, value, and practical application. *Journal of Electroanalytical Chemistry*, 872:114145, 2020.

G. Diguet. *Huge Magnetostriction of Magneto-rheological composite*. PhD thesis, Université Joseph-Fourier - Grenoble I, Jan. 2010.

G. Diguet, E. Beaugnon, and J. Y. Cavaillé. Shape effect in the magnetostriction of ferromagnetic composite. *Journal of Magnetism and Magnetic Materials*, 322(21):3337–3341, Nov. 2010.

M. Doi. *Introduction to Polymer Physics*. Clarendon Press, Oxford, 1996.

M. Doi. Gel dynamics. *Journal of the Physical Society of Japan*, 78:052001, 2009.

J. Dolbow, E. Fried, and H. Ji. A numerical strategy for investigating the kinetic response of stimulus-responsive hydrogels. *Computer Methods in Applied Mechanics and Engineering*, 194:4447–4480, 2005.

A. Dorfmann and R. Ogden. Magnetoelastic modelling of elastomers. *European Journal of Mechanics – A/Solids*, 22:497–507, 2003.

A. Dorfmann and R. Ogden. Nonlinear magnetoelastic deformations of elastomers. *Acta Mechanica*, 167:13–28, 2004.

L. Dorfmann and R. Ogden. Nonlinear electroelasticity. *Acta Mechanica*, 174:167–183, 2005.

L. Dorfmann and R. Ogden. *Nonlinear Theory of Electroelastic and Magnetoelastic Interactions*. Springer, New York, 2014.

L. Dorfmann and R. Ogden. Nonlinear electroelasticity: material properties, continuum theory and applications. *Proceedings of the Royal Society A*, 473:20170311, 2017.

J. Dorn. Some fundamental experiments on high temperature creep. *Journal of the Mechanics and Physics of Solids*, 3:85–116, 1955.

M. Doyle, T. Fuller, and J. Newman. Modeling of galvanostatic charge and discharge of the lithium/polymer/insertion cell. *Journal of the Electrochemical Society*, 6:1526–1533, 1993.

F. P. Duda, A. C. Souza, and E. Fried. A theory for species migration in finitely strained solid with application to polymer network swelling. *Journal of the Mechanics and Physics of Solids*, 58:515–529, 2010.

M. Duduta, E. Hajiesmaili, H. Zhao, R. J. Wood, and D. R. Clarke. Realizing the potential of dielectric elastomer artificial muscles. *Proceedings of the National Academy of Sciences*, 116(7):2476–2481, 2019.

C. Durning and K. Morman. Nonlinear swelling of polymer gels. *Journal of Chemical Physics*, 98:4275–4293, 1993.

A. Eringen. On the foundations of electroelastostatics. *International Journal of Engineering Science*, 1:127–153, 1963.

A. Eringen and G. Maugin. *Electrodynamics of Continua I Foundations and Solid Media & II Fluids and Complex Media*. Springer, New York, 1990.

S. Erlicher, L. Bonaventura, and O. S. Bursi. The analysis of the generalized-α method for non-linear dynamic problems. *Computational Mechanics*, 28:83–104, 2002.

P. Flory. Thermodynamics of high polymer solutions. *Journal of Chemical Physics*, 10:51–61, 1942.

P. Flory. *Principles of Polymer Chemistry*. Cornell University Press, Ithaca, 1953.

P. Flory. Thermodynamic relations for high elastic materials. *Transactions of the Faraday Society*, 57:829–838, 1961.

C. D. Fuerst and E. G. Brewer. High-remanence rapidly solidified Nd-Fe-B: Die-upset magnets. *Journal of Applied Physics*, 73(10):5751–5756, May 1993.

S. Galante, A. Lucantonio, and P. Nardinocchi. The multiplicative decomposition of the deformation gradient in the multiphysics modeling of ionic polymers. *International Journal of Non-Linear Mechanics*, 51:112–120, 2013.

D. Garcia-Gonzalez and M. Hossain. A microstructural-based approach to model magneto-viscoelastic materials at finite strains. *International Journal of Solids and Structures*, 208–209:119–132, 2021.

F. Garofalo. An empirical relation defining the stress dependence of minimum creep rate in metals. *Trans Metall Soc AIME*, 227:351–355, 1963.

A. Gent. A new constitutive relation for rubber. *Rubber Chemistry and Technology*, 69:59–61, 1996.

P. Germain. The method of virtual power in continuum mechanics. Part 2: microstructure. *SIAM Journal of Applied Mathematics*, 25:556–575, 1973.

J. Gough. A description of a property of caoutchouc or Indian rubber; with some reflections on the case of the elasticity of this substance. *Memoirs of the Literary and Philosophical Society of Manchester*, 1:288–295, 1805.

A. Green and J. Adkins. *Large Elastic Deformations and Nonlinear Continuum Mechanics*. Clarendon Press, Oxford, 1960.

A. Green and R. Rivlin. The mechanics of non-linear materials with memory – part I. *Archive for Rational Mechanics and Analysis*, 1:1–21, 1957.

D. Guan-Suo. Determination of the stretch and rotation in the polar decomposition. *Journal of Elasticity*, 50:197–208, 1998.

M. Gurtin. On the thermodynamics of elastic materials. *Mathematical Analysis and Applications*, 18:39–44, 1967.

M. Gurtin and L. Anand. The decomposition $\mathrm{F} = \mathrm{F_e F_p}$, material symmetry, and plastic irrotationality for solids that are isotropic-viscoplastic or amorphous. *International Journal of Plasticity*, 21:1686–1719, 2005.

M. E. Gurtin. A gradient theory of single-crystal viscoplasticity that accounts for geometrically necessary dislocations. *Journal of the Mechanics and Physics of Solids*, 50(1):5–32, 2002.

M. E. Gurtin, E. Fried, and L. Anand. *The Mechanics and Thermodynamics of Continua*. Cambridge University Press, Cambridge, 2010.

B. Häggblad and J. A. Sundberg. Large strain solutions of rubber components. *Computers & Structures*, 17:835–843, 1983.

E. Hajiesmaili and D. Clarke. Dielectric elastomer actuators. *Journal of Applied Physics*, 129:151102, 2021.

K. Haldar. Constitutive modeling of magneto-viscoelastic polymers, demagnetization correction, and field-induced poynting effect. *International Journal of Engineering Science*, 165:103488, 2021.

K. Haldar, B. Kiefer, and A. Menzel. Finite element simulation of rate-dependent magneto-active polymer response. *Smart Materials and Structures*, 25:104003, 2016.

M. Harmon, M. Tang, and C. Frank. A microfluidic actuator based on thermoresponsive hydrogels. *Polymer*, 44:4547–4556, 2003.

D. Henann, S. Chester, and K. Beroldi. Modeling of dielectric elastomers: design of actuators and energy harvesting devices. *Journal of the Mechanics and Physics of Solids*, 61:2047–2066, 2013.

H. M. Hilber, T. J. Hughes, and R. L. Taylor. Improved numerical dissipation for time integration algorithms in structural dynamics. *Earthquake Engineering & Structural Dynamics*, 5:283–292, 1977.

A. Hoger and D. E. Carlson. Determination of the stretch and rotation in the polar decomposition of the deformation gradient. *Quarterly of Applied Mathematics*, 42:113–117, 1984.

G. Holzapfel and J. Simo. Entropy elasticity of isotropic rubber-like solids at finite strains. *Computer Methods in Applied Mechanics and Engineering*, 132:17–44, 1996.

W. Hong, X. Zhao, J. Zhou, and Z. Suo. A theory of coupled diffusion and large deformation in polymeric gel. *Journal of the Mechanics and Physics of Solids*, 56:1779–1793, 2008.

M. Hossain, D. Vu, and P. Steinmann. Experimental study and numerical modelling of VHB4910 polymer. *Computational Materials Science*, 59:65–74, 2012.

R. C. Huang and L. Anand. Non-linear mechanical behavior of the elastomer polydimethylsiloxane (PDMS) used in the manufacture of microfluidic devices. *Innovation in Manufacturing Systems and Technology (IMST)*, 2005. `https://dspace.mit.edu/handle/1721.1/7456`.

M. Huggins. Some properties of solutions of long-chain compounds. *Journal of Physical Chemistry*, 46:151–158, 1942.

K. Hutter and A. van de Ven. *Field Matter Interactions in Thermoelastic Solids*. Springer, Berlin, 1978.

J. D. Jackson. *Classical Electrodynamics*. John Wiley & Sons, 1999.

C. Jo, D. Pugal, I.-K. Oh, K. J. Kim, and K. Asaka. Recent advances in ionic polymer–metal composite actuators and their modeling and applications. *Progress in Polymer Science*, 38(7):1037–1066, 2013.

J. P. Joule. On some thermodynamic properties of solids. *Philosophical Transactions of the Royal Society of London A*, 149:91–131, 1859.

C. Kadapa and M. Hossain. A robust and computationally efficient finite element framework for coupled electromechanics. *Computer Methods in Applied Mechanics and Engineering*, 372:113443, 2020.

C. Kadapa and M. Hossain. A linearized consistent mixed displacement–pressure formulation for hyperelasticity. *Mechanics of Advanced Materials and Structures*, 29(2):267–284, 2022.

S. Kankanala and N. Triantafyllidis. On finitely strained magnetorheological elastomers. *Journal of the Mechanics and Physics of Solids*, 52:2869–2908, 2004.

K. Kim and M. Shahinpoor. Ionic polymer–metal composites: II. Manufacturing techniques. *Smart Materials and Structures*, 12:65–79, 2003.

Y. Kim, H. Yuk, R. Zhao, S. Chester, and X. Zhao. Printing ferromagnetic domains for untethered fast-transforming soft materials. *Nature*, 558:274–279, 2018.

A. Kovetz. *The Principles of Electromagnetic Theory*. Cambridge University Press, Cambridge, 1990.

A. Kovetz. *Electromagnetic Theory*. Oxford University Press, 2000.

E. Kröner. Allgemeine kontinuumstheorie der versetzungen und eigenspannungen. *Archive for Rational Mechanics and Analysis*, 4:273–334, 1960.

H. Langtangen and A. Logg. *Solving PDEs in Python. The FEniCS Tutorial Volume I.* Springer, 2017.

E. H. Lee. Elastic-plastic deformation at finite strains. *Journal of Applied Mechanics*, 36(1):1–6, 1969.

H. Lee, J. Zhang, H. Jiang, and N. X. Fang. Prescribed pattern transformation in swelling gel tubes by elastic instability. *Physical Review Letters*, 108(21):214304, 2012.

V. Lefèvre, K. Danas, and O. Lopez-Pamies. A general result for the magnetoelastic response of isotropic suspensions of iron and ferrofluid particles in rubber, with applications to spherical and cylindrical specimens. *Journal of the Mechanics and Physics of Solids*, 107:343–364, 2017.

V. Lefévre, F. Sozio, and O. Lopez-Pamies. Abaqus implementation of a large family of finite viscoelasticity models. *Finite Elements in Analysis and Design*, 232:104114, 2024.

A. Leronni and L. Bardella. Modeling actuation and sensing in ionic polymer metal composites by electrochemo-poromechanics. *Journal of the Mechanics and Physics of Solids*, 148:104292, 2021.

C. Linder, M. Tkachuk, and C. Miehe. A micromechanically motivated diffusion-based transient network model and its incorporation into finite rubber viscoelasticity. *Journal of the Mechanics and Physics of Solids*, 59:2134–2156, 2011.

Z. Liu, A. McBride, A. Ghosh, L. Heltai, W. Huang, T. Yu, P. Steinmann, and P. Saxena. Computational bifurcation analysis of hyperelastic thin shells. *arxiv.org/pdf/2210.15854 [math.NA]*, 2022.

J. Lubliner. A model of rubber viscoelasticity. *Mechanics Research Communications*, 12:93–99, 1985.

Y. Mao, S. Lin, X. Zhao, and L. Anand. A large deformation viscoelastic model for double-network hydrogels. *Journal of the Mechanics and Physics of Solids*, 100:103–130, Mar. 2017.

G. Marckmann and E. Verron. Comparison of hyperelastic models for rubber-like materials. *Rubber Chemistry and Technology*, 79:835–858, 2006.

G. Maugin. *Continuum Mechanics of Electromagnetic Solids*. North-Holland, Amsterdam, 1988.

R. McMeeking and C. Landis. Electrostatic forces and stored energy in deformable dielectric materials. *Journal of Applied Mechanics*, 72:581–590, 2005.

R. M. McMeeking, C. M. Landis, and S. M. A. Jimenez. A principle of virtual work for combined electrostatic and mechanical loading of materials. *International Journal of Non-Linear Mechanics*, 42:831–838, 2007.

M. Mehnert, M. Hossain, and P. Steinmann. Numerical modeling of thermo-electro-viscoelasticity with field-dependent material parameters. *International Journal of Non-Linear Mechanics*, 106:13–24, 2018.

M. Mehnert, M. Hossain, and P. Steinmann. Experimental and numerical investigations of the electro-viscoelastic behavior of VHB 4905TM. *European Journal of Mechanics – A/Solids*, 77:103797, 2019.

C. Miehe. Entropic thermoelasticity at finite strains. Aspects of the formulation and numerical implementation. *Computer Methods in Applied Mechanics and Engineering*, 120:243–269, 1995.

C. Miehe and S. Goktepe. A micro–macro approach to rubber-like materials. part II: The micro-sphere model of finite rubber viscoelasticity. *Journal of the Mechanics and Physics of Solids*, 53:2231–2258, 2005.

N. Minaian, Z. J. Olsen, and K. J. Kim. Ionic polymer–metal composite (IPMC) artificial muscles in underwater environments: Review of actuation, sensing, controls, and applications to soft robotics.

In D. A. Paly and N. M. Werely, editors, *Bioinspired Sensing, Actuation, and Control in Underwater Soft Robotic Systems*, pages 117–139. Springer, 2021.

A. Miranville. The Cahn–Hilliard equation and some of its variants. *AIMS Mathematics*, 2(3):479–544, 2017.

G. Montella, S. Govindjee, and P. Neff. The exponentiated Hencky strain energy in modeling tire derived material for moderately large deformations. *ASME Journal of Engineering Materials and Technology*, 138:031008–(1–12), 2016.

M. A. Moreno, J. Gonzalez-Rico, M. L. Lopez-Donaire, A. Arias, and D. Garcia-Gonzalez. New experimental insights into magneto-mechanical rate dependences of magnetorheological elastomers. *Composites Part B: Engineering*, 224:109148, Nov. 2021.

D. Mukherjee, M. Rambausek, and K. Danas. An explicit dissipative model for isotropic hard magnetorheological elastomers. *Journal of the Mechanics and Physics of Solids*, 151:104361, 2021.

S. Narayan and L. Anand. A large deformation elastic–viscoplastic model for lithium. *Extreme Mechanics Letters*, 24:21–29, 2018.

S. Narayan, E. M. Stewart, and L. Anand. Coupled electro-chemo-elasticity: Application to modeling the actuation response of ionic polymer–metal composites. *Journal of the Mechanics and Physics of Solids*, 152:104394, 2021.

P. Nardinocchi and M. Pezzulla. Curled actuated shapes of ionic polymer metal composites strips. *Journal of Applied Physics*, 113(22):224906, 2013.

P. Nardinocchi, M. Pezzulla, and L. Placidi. Thermodynamically based multiphysic modeling of ionic polymer metal composites. *Journal of Intelligent Material Systems and Structures*, 22(16):1887–1897, 2011.

P. Nardinocchi, M. Pezzulla, B. J. Akle, M. Guenther, and T. Wallmersperger. Actuation and buckling effects in IPMCs. In Y. Bar-Cohen, editor, *Electroactive Polymer Actuators and Devices (EAPAD) 2014*, volume 9056, pages 98–108. International Society for Optics and Photonics, SPIE, 2014.

S. Nemat-Nasser. Micromechanics of actuation of ionic polymer–metal composites. *Journal of Applied Physics*, 92(5):2899–2915, 2002.

S. Nemat-Nasser. Electrochemomechanics of ionic polymer–metal composites. In W. N. Sharpe, editor, *Springer Handbook of Experimental Solid Mechanics*, pages 187–201. Springer, 2008.

S. Nemat-Nasser and J. Y. Li. Electromechanical response of ionic polymer–metal composites. *Journal of Applied Physics*, 87(7):3321–3331, 2000.

S. Nemat-Nasser and C. W. Thomas. Ionomeric polymer–metal composites. In Y. Bar-Cohen, editor, *Electroactive Polymer (EAP) Actuators as Artificial Muscles: Reality, Potential, and Challenges*, volume PM 136, pages 171–230. SPIE Press, 2004.

J. Newman and K. Thomas-Alyea. *Electrochemical Systems*. Wiley, 2012.

N. Newmark. A method of computation for structural dynamics. *J. Engrg. Mech. Div. ASCE*, 85:67–94, 1959.

V. K. Nguyen, J. W. Lee, and Y. Yoo. Characteristics and performance of ionic polymer–metal composite actuators based on Nafion/layered silicate and Nafion/silica nanocomposites. *Sensors and Actuators B: Chemical*, 120(2):529–537, 2007.

A. Norris. Invariants of $\mathbf{C}^{1/2}$ in terms of invariants of $\mathbf{C}$. *Journal of Mechanics of Materials and Structures*, 2:1805–182, 2007.

F. H. Norton. *The Creep of Steel at High Temperatures*. McGraw-Hill, 1929.

W. Nowacki. *Thermoelasticity*. Addison-Wesley, Reading, Mass., 1962.

J. Oden. Finite element analysis of nonlinear problems in the dynamical theory of coupled thermoelasticity. *Nuclear Engineering and Design*, 10:465–475, 1969.

R. W. Ogden. Large deformation isotropic elasticity: On the correlation of theory and experiment for compressible rubberlike solids. *Proceedings of the Royal Society London*, A328:567–583, 1972.

K. Oguro, Y. Kawami, and H. Takenaka. An actuator element of polyelectrolyte gel membrane–electrode composite. *Osaka Kogyo Gijutsu Shikensho Kiho*, 43(1):21–24, 1992.

Y. Pao. Electromagnetic forces in deformable continua. In S. Nemat-Nasser, editor, *Mechanics Today*, 4:209–305, 1978.

R. Pelrine, R. Kornbluh, Q. Pei, and J. Joseph. High-speed electrically actuated elastomers with strain greater than 100%. *Science*, 28:836–839, 2000.

J.-P. Pelteret, B. Walter, and P. Steinmann. Application of metaheuristic algorithms to the identification of nonlinear magneto-viscoelastic constitutive parameters. *Journal of Magnetism and Magnetic Materials*, 464:116–131, Oct. 2018.

A. Pipkin and T. Rogers. A non-linear integral representation for viscoelastic behaviour. *Journal of the Mechanics and Physics of Solids*, 16:59–72, 1968.

A. Plummer, C. Adkins, J.-F. Louf, A. Košmrlj, and S. S. Datta. Obstructed swelling and fracture of hydrogels. *Soft Matter*, 20:1425–1437, 2024.

M. Porfiri, A. Leronni, and L. Bardella. An alternative explanation of back-relaxation in ionic polymer metal composites. *Extreme Mechanics Letters*, 13:78–83, 2017.

J. H. Poynting. On pressure perpendicular to the shear planes in finite pure shears, and on the lengthening of loaded wires when twisted. *Proc. R. Soc. Lond. A*, 82:546–559, 1909.

Y. Qiu, E. Zhang, R. Plamthottam, and Q. Pei. Dielectric elastomer artificial muscle: materials innovations and device explorations. *Accounts of Chemical Research*, 52(2):316–325, 2019.

M. Rambausek and J. Schöbrel. Curing spurious magneto-mechanical coupling in soft non-magnetic materials. *Numerical Methods in Engineering*, 124:2261–2291, 2023.

M. Rambausek, D. Mukherjee, and K. Danas. A computational framework for magnetically hard and soft viscoelastic magnetorheological elastomers. *Computer Methods in Applied Mechanics and Engineering*, 391:114500, 2022.

S. Reese and S. Govindjee. A theory of finite viscoelasticity and numerical aspects. *International Journal of Solids and Structures*, 35:3455–3482, 1998.

R. S. Rivlin. Large elastic deformations of isotropic materials. IV. Further developments of the general theory. *Philosophical Transactions of the Royal Society of London. Series A*, 241:379–397, 1948.

R. S. Rivlin and D. W. Saunders. Large elastic deformations of isotropic materials. VII. Experiments on the deformation of rubber. *Philosophical Transactions of the Royal Society of London. Series A*, 243:251–288, 1951.

R. E. Rosensweig. Thermodynamics of Electromagnetism. In G. Astarita, editor, *Thermodynamics: An Advanced Textbook for Chemical Engineers*, pages 365–438. Springer US, 1989.

M. Rossi and T. Wallmersperger. Thermodynamically consistent three-dimensional electrochemical model for polymeric membranes. *Electrochimica Acta*, 283:1323–1338, 2018.

M. Rossi, T. Wallmersperger, J. A. Ramirez, and P. Nardinocchi. Thermodynamically consistent electro-chemo-mechanical model for polymer membranes. In Y. Bar-Cohen, editor, *Electroactive*

Polymer Actuators and Devices (EAPAD) XX, volume 10594, pages 76–86. International Society for Optics and Photonics, SPIE, 2018.

A. Salvadori, R. McMeeking, D. Grazioli, and M. Magri. A coupled model of transport-reaction-mechanics with trapping. part I – small strain analysis. *Journal of the Mechanics and Physics of Solids*, 114:1–30, 2018.

K. Sawyers. Comments on the paper determination of the stretch and rotation in the polar decomposition of the deformation gradient by A. Hoger and D. E. Carlson. *Quarterly of Applied Mathematics*, 44:309–311, 1986.

H. Schild. Poly (N-isopropylacrylamide): experiment, theory, and application. *Progress in Polymer Science*, 17:163–249, 1992.

J. Schröder, T. Wick, S. Reese, P. Wriggers, R. Müller, S. Kollmannsberger, M. Kästner, A. Schwarz, M. Igelbüscher, N. Viebahn, et al. A selection of benchmark problems in solid mechanics and applied mathematics. *Archives of Computational Methods in Engineering*, 28:713–751, 2021.

J. Schröder, N. Viebahn, P. Wriggers, F. Auricchio, and K. Steeger. On the stability analysis of hyperelastic boundary value problems using three- and two-field mixed finite element formulations. *Computational Mechanics*, 60(3):479–492, 2017.

N. Scott. $\mathbf{U} = \mathbf{C}^{1/2}$ and its invariants in terms of invariants of $\mathbf{C}$ and its invariants. *Journal of Elasticity*, 141:363–379, 2020.

M. Shahinpoor and K. J. Kim. Ionic polymer–metal composites: I. Fundamentals. *Smart Materials and Structures*, 10(4):819–833, 2001.

M. Shahinpoor and K. J. Kim. Ionic polymer–metal composites: III. Modeling and simulation as biomimetic sensors, actuators, transducers, and artificial muscles. *Smart Materials and Structures*, 13(6):1362–1388, 2004.

Y. Shi, E. Askounis, R. Plamthottam, T. Libby, Z. Peng, K. Youssef, J. Pu, R. Pelrine, and Q. Pei. A processable, high-performance dielectric elastomer and multilayering process. *Science*, 377(6602):228–232, 2022.

A. V. Shutov, R. Landgraf, and J. Ihlemann. An explicit solution for implicit time stepping in multiplicative finite strain viscoelasticity. *Computer Methods in Applied Mechanics and Engineering*, 265:213–225, 2013.

F. Sidoroff. Un modele viscoelastique non lineaire avec configuration intermmiaire. *Journal de Mecanique*, 13:679–713, 1974.

J. C. Simo and R. L. Taylor. Quasi-incompressible finite elasticity in principal stretches. Continuum basis and numerical algorithms. *Computer Methods in Applied Mechanics and Engineering*, 85:273–310, 1991.

E. M. Stewart and L. Anand. Magneto-viscoelasticity of hard-magnetic soft-elastomers: Application to modeling the dynamic snap-through behavior of a bistable arch. *Journal of the Mechanics and Physics of Solids*, 179:105366, 2023.

E. M. Stewart and L. Anand. Magnetostriction of soft-magnetorheological elastomers. *Journal of the Mechanics and Physics of Solids*, 194:105934, Jan. 2025.

E. M. Stewart, S. Narayan, and L. Anand. On modeling the infiltration of water in a peg-da hydrogel and the resulting swelling under unconstrained and mechanically constrained conditions. *Extreme Mechanics Letters*, 54:101775, 2022.

E. M. Stewart, S. A. Chester, and L. Anand. Example codes for coupled theories in solid mechanics, Jan. 2025. `https://doi.org/10.5281/zenodo.14680876`.

Z. Suo. Theory of dielectric elastomers. *Acta Mechanica Solida Sinica*, 23:449–578, 2010.

Z. Suo, X. Zhao, and W. Greene. A nonlinear field theory of deformable dielectrics. *Journal of the Mechanics and Physics of Solids*, 56:581–590, 2008.

T. Sussman and K.-J. Bathe. A finite element formulation for nonlinear incompressible elastic and inelastic analysis. *Computers & Structures*, 26:357–409, 1987.

K. Tan, L. Chen, S. Yang, and Q. Deng. Dynamic snap-through instability and damped oscillation of a flat arch of hard magneto-active elastomers. *International Journal of Mechanical Sciences*, 230:107523, Sept. 2022.

T. Tanaka and D. Fillmore. Kinetics of swelling of gels. *Journal of Chemical Physics*, 70:1214–1218, 1979.

L. Tang, L. Wang, X. Yang, Y. Feng, Y. Li, and W. Feng. Poly(N-isopropylacrylamide)-based smart hydrogels: Design, properties and applications. *Progress in Materials Science*, 115:100702, 2021.

H. Tiersten. *A Development of the Equations of Electromagnetism in Material Continua*. Springer, New York, 1990.

H. F. Tiersten. On the nonlinear equations of thermoelectroelasticity. *International Journal of Engineering Science*, 9:587–604, 1971.

R. Toupin. The elastic dielectric. *J. Ration. Mech. Anal.*, 5:849–915, 1956.

L. R. G. Treloar. *The Physics of Rubber Elasticity*. Clarendon Press, Oxford, 1975.

C. Truesdell and R. Toupin. *The Classical Field Theories*. Cambridge University Press, 1960.

F. Vogel, S. Goktepe, P. Steinmann, and E. Kuhl. Modeling and simulation of viscous electroactive polymers. *European Journal of Mechanics-A/Solids*, 48:112–128, 2014.

B. Wang, R. Bustamante, L. Kari, H. Pang, and X. Gong. Modelling the influence of magnetic fields to the viscoelastic behaviour of soft magnetorheological elastomers under finite strains. *International Journal of Plasticity*, 164:103578, May 2023.

B. Wang, L. Kari, H. Pang, and X. Gong. Modelling the dynamic magnetic actuation of isotropic soft magnetorheological elastomers. *International Journal of Mechanical Sciences*, 266:108908, 2024.

J. Wang, Z. Chen, M. Mauk, K. Hong, M. Li, S. Yang, and H. Bau. Self-actuated, thermo-responsive hydrogel valves for lab on a chip. *Biomedical Microdevices*, 7:313–322, 2005.

Q. Wang, M. Tahir, L. Zhang, and X. Zhao. Electro-creasing instability in deformed polymers: experiment and theory. *Soft Matter*, 7(14):6583–6589, 2011.

S. Wang, M. Decker, D. Henann, and S. Chester. Modeling of dielectric viscoelastomers with application to electromechanical instabilities. *Journal of the Mechanics and Physics of Solids*, 95:213–229, 2016.

G. Weber and L. Anand. Finite deformation constitutive equations and a time integration procedure for isotropic, hyperelastic-viscoplastic solids. *Computer Methods in Applied Mechanics and Engineering*, 79(2):173–202, 1990.

I. P. Wijaya, O. Lopez-Pamies, and A. Masud. A unified determinant-preserving formulation for compressible/incompressible finite viscoelasticity. *Journal of the Mechanics and Physics of Solids*, 177:105312, 2023.

P. Wriggers. Discretization methods for solids undergoing finite deformations. In *Advanced Finite Element Technologies*, pages 17–51. Springer, New York, 2016.

D. Yan, B. F. Aymon, and P. M. Reis. A reduced-order, rotation-based model for thin hard-magnetic plates. *Journal of the Mechanics and Physics of Solids*, 170:105095, 2023.

C. Yoon, R. Xiao, J. Park, J. Cha, T. D. Nguyen, and D. H. Gracias. Functional stimuli responsive hydrogel devices by self-folding. *Smart Materials and Structures*, 23(9):094008, 2014.

R. Zhao, Y. Kim, S. A. Chester, P. Sharma, and X. Zhao. Mechanics of hard-magnetic soft materials. *Journal of the Mechanics and Physics of Solids*, 124:244–263, 2019.

X. Zhao and Q. Wang. Harnessing large deformation and instabilities of soft dielectrics: Theory, experiment, and application. *Applied Physics Reviews*, 1:021304, 2014.

X. Zhao, W. Hong, and Z. Suo. Electromechanical hysteresis and coexistent states in dielectric elastomers. *Phys. Rev. B*, 76:134113, 2007.

Index